Municipal Storm Water Management

Municipal Storm Water Management

Thomas N. Debo
Andrew J. Reese

LEWIS PUBLISHERS
Boca Raton   Ann Arbor   London   Tokyo

Library of Congress Cataloging-in-Publication Data

Debo, Thomas N. (Thomas Neil), 1941-
  Municipal storm water management / by Thomas N, Debo, Andrew J.
Reese.
        p.   cm.
  Includes bibliographical references and index.
  ISBN 0-87371-981-6
    1. Urban runoff—management. I. Reese, Andrew J.   II. Title
  TD657.D43   1995
  628′.21′068—dc20

                                                                                94-37141
                                                                                    CIP

© 1995 by CRC Press, Inc.
Lewis Publishers is an imprint of CRC Press

No claim to original U.S. Government works
International Standard Book Number 0-87371-981-6
Library of Congress Card Number 94-37141
Printed in the United States of America 1 2 3 4 5 6 7 8 9 0
Printed on acid-free paper

Preface

Just what the world needs, another text on storm water management!

The field of municipal storm water management has expanded to the extent that it would take many volumes to exhaustively cover the subject. But, in the day-to-day design and management of the storm water system, the engineer, planner or administrator should not need to resort to a large number of documents to find what he or she needs. It should be readily available in one spot, carefully explained where necessary, but not overly cluttered with detailed theory and derivation. That is the approach and major aim of this book.

From a technical design standpoint the authors have combed the literature to find and present clear treatments of each of the key design situations the practitioner will encounter. A wealth of knowledge has been compiled from Federal, state and local design publications including extensive use of the American Association of State Highway and Transportation Officials (AASHTO) Model Drainage Manual. In each case the authors have attempted to provide a stand-alone discussion including all relevant charts, figures and tables. In addition is much original information gathered from experience in many cities and counties.

To properly understand and manage storm water requires more than a knowledge of design procedures and hydrologic or hydraulic practice. Actually "making it happen" in a real world municipality, filled with the demands of regulations, financial pressures, stakeholder groups and politics is also an important part of storm water management. While harder to quantify, the tools and understanding necessary for institutional, as well as technical, success are also included in this book. They are derived from literally hundreds of experiences and projects in cities and counties across the United States. A careful reading will provide both a wealth of information and principles and a sense of what works and what does not.

Storm water management is a fast evolving field. Any treatment of it is only a snapshot of a changing picture. The authors have included the most up to date information available in each of the key areas of the practice. Also gathered together and included are little known but valuable tables, charts and procedures covering both common designs and more specialized situations. The evolution of storm water management, as any science or practice, builds on basic understandings and concepts. In each case the authors have attempted to provide, first, sufficient discussion to lay a foundation and framework upon and within which any new developments can be built.

The book is primarily designed for use by engineers, designers, and planners involved in municipal storm water management. Information contained should cover most of the design applications within urban areas and most of the institutional aspects of storm water management the planner or administrator faces daily. The book has also been designed to be applicable to civil and environmental undergraduate and graduate courses from both a planning or administration and engineering perspective.

Thomas N. Debo
Andrew J. Reese

Dr. Thomas N. Debo

Andrew J. Reese

The Authors

Dr. Thomas N. Debo is Associate Dean and Professor in the City Planning Program in the College of Architecture at the Georgia Institute of Technology and President of Debo & Associates, Inc., a consulting firm specializing in the urban storm water management area. He received his B.S. in Civil Engineering from Michigan Technological University in 1963 and his Master of City Planning and Ph.D. in Civil Engineering from Georgia Institute of Technology in 1972 and 1975 respectively. He has over two decades of experience dealing with storm water management including development of hydrologic/hydraulic computer programs, engineering design handbooks, local ordinances and policies, and technical and engineering studies and designs. He has worked with numerous municipalities throughout the United States, state and Federal agencies, and large consulting and legal firms, and has extensive experience as an expert witness. Dr. Debo has published several books and over forty articles in professional journals and other publications. He has been very active in conducting storm water related workshops throughout the country and has been invited to give presentations and speeches extensively throughout the United States and several European countries. He is a member of the American Society of Civil Engineers, American Water Resources Association and a registered engineer in the State of Georgia. Dr. Debo is married and the father of four daughters and presently resides in Atlanta, GA.

Andrew J. Reese has over 20 years experience in a wide variety of water resource, hydraulic and hydrologic engineering and management roles. His responsibilities have included management of large and complex municipal storm water program development, storm water quality permitting, performing research in many aspects of hydraulic engineering, teaching at Vanderbilt University, conducting short courses nationally on storm water management and for the Corps of Engineers, developing storm water utilities, writing municipal ordinances and technical manuals, developing and executing public awareness programs, software development, and hydraulic and hydrologic planning, computer modelling and design. His clients have included several Federal agencies, states, municipalities and the private sector. Mr. Reese has published and delivered over thirty articles in the areas of hydraulic design and urban storm water management in several countries and has been invited to speak at numerous seminars and conferences. He holds a bachelor of science degree from Cornell University, and masters degrees in hydraulic engineering and business from Colorado State University and Boston University respectively. He is a registered professional engineer and professional hydrologist. Mr. Reese is currently a principal engineer and technical director for water resources for Ogden Environmental and Energy Services, Inc. He is married and the father of four children and presently resides in Nashville, TN.

The authors would like to dedicate this book to their families for having fun without us for a year .... and accepting us back afterwards.

Table Of Contents

Chapter 1 Introduction To Municipal Storm Water Management

1.1 An Historical Perspective

A hundred years ago the primary concern with surface water was draining it from wetlands and floodplains to make way for agricultural development. Later, as flooding of developed areas and demands for cheaper power increased, the concern broadened to include flood control and hydropower development. This was the hey-day of the Corps of Engineers' and Bureau of Reclamation's river projects. Suburban growth in the 1950's through 1970's coupled with higher expectations for quality of development led to development of efficient curb and gutter drainage systems to quickly convey rainwater off streets and yards and into streams. Protection of roads and property was of paramount importance. Throughout the later part of this period it became apparent that localized flooding was often caused by this type of development and the importance of on-site and regional storm water detention basins became recognized. Detention became a standard site development requirement in many of the more progressive urban areas. The development of personal computers enhanced the ability to actually plan drainage systems for existing and future development. Modern storm water <u>quantity</u> management was given birth.

Later, on the strength of an emerging body of information from Section 208 studies, the experience of various cities, and other data gathering efforts such as the National Urban Runoff Program (NURP), the importance of urban areas as sources of non-point pollution grew. This culminated in the early 1990's with the publication of Federal regulations mitigating and controlling pollution runoff in urban storm water. Comprehensive storm water <u>quality</u> management began.

Each of these changes in the way we have looked at surface water runoff was a by-product of urban growth. Each engineering response to growth pressures, in turn, created other unforeseen problems. Drainage of more marginal lands for agriculture removed the natural buffering action of floodplains and wetlands, thus increasing flooding and non-point pollution. Large river projects modified, sometimes adversely, stream regimes and environments. Efficient drainage of urban areas often led to nuisance flooding, the gutting of small urban streams and the destruction of in-stream ecosystems. Multiple privately owned detention basins have often proven unmanageable and unmaintainable while becoming ineffective eyesores, often after only three to five years. Pollution control programs and structural controls may suffer a similar fate if appropriate institutional support mechanisms (legal, financial, organizational and technical) are not in place.

In spite of the many problems in municipal storm water management, success stories abound. This book is about successful municipal storm water management. It looks at both the institutional and technical sides of storm water management showing the intimate relationship between the two. For to be successful, municipal storm water management must be based on sound technical information and procedures supported by a stable institutional and administrative foundation.

1.2 Understanding Storm Water Management Problems and Solutions

It is vitally important to understand the linkage between physical problems and deeper institutional root causes of those problems. Many municipalities have not understood this linkage and, as a result, wrestle continuously with the same problems never arriving at permanent solutions. Figure 1-1 illustrates the dynamics of this technical-institutional relationship using a "five whys" methodology which comes out of a Total Quality Management (TQM) consideration of storm water management.

Typically, a storm water administrator, public works engineer or political leader gets a drainage complaint call: "I have a flooding problem and I want you to fix it." This is Level 1: the complaint. The complaint could just as easily have been an erosion or pollution complaint. Following the same methodology would eventually lead to the same conclusions.

When the question is asked: "Why is there a flooding problem?" (Level 2), there is usually one of (or a combination of) four reasons: (1) obstructed or damaged structures; (2) high risk residence location; (3) undersized structures; or (4) increased flow due to the impacts of urban development. The typical solution is to go out and fix the problem or, more commonly, to tell the resident that the problem is not the municipality's responsibility.

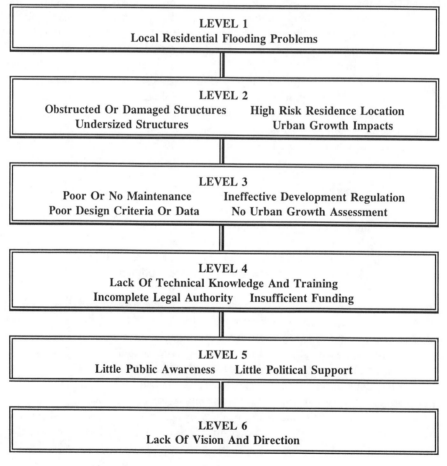

Figure 1-1 Multi-level Storm Water Problems

However, if the Level 3 question is asked: "Why did this flooding problem (and ones like it) occur in the first place?" a matching set of four, more foundational, reasons is uncovered.

1. Structures are obstructed or damaged because they are not inspected and maintained. Municipalities typically maintain very little of the drainage system. Most maintenance that is done is in response to complaints and performed within the street right-of-way only. Much of it is done only to protect streets or public property. The fifty to eighty percent of the drainage system closest to private houses and other structures is rarely or never inspected or maintained. This is true despite the fact that much of the water carried in these drainage systems is derived from public streets and is, in some sense, therefore, "public water". Over years of neglect pipes and channels inevitably begin to fill with debris and sediment, structures begin to weather or are damaged, and erosion eats away at culvert headwalls and tail sections. Drainage systems work flawlessly when it is not raining. Finally the system is tested and over-whelmed by a storm whose intensity is often less than the system was designed to carry. Homes are flooded, roads are overtopped, damage is incurred, complaints flood in.

2. Homes are located too close to streams because no one properly regulates the location. Because of the demands of the Federal Flood Insurance Program, most municipalities control the location and elevation of new construction within regulated floodplains. However, the vast majority of complaints are received from residents remote from regulated floodplains in locations where there is no such control. Municipalities which would not consider allowing development within the 100-year floodplain below the mandatory flood protection elevation daily approve plans for developments where a number of homes, unwittingly located within unregulated floodplains, would be inundated by smaller more localized 100-year floods.

3. Structures are undersized because of poor or inappropriate methodology and incomplete data. Most municipalities allow drainage structures to be sized using the Rational Method. While this is not wrong, per se, the limits of this method are rarely understood by designers and plans reviewers. In cases where backwater effects predominate or under other special circumstances such methods may give non-conservative results. Additionally, many munici-palities have little data on actual rainfall values, inlet capacities, tidal influences or actual expected future maintenance-related condition of structures. Without this information design-ers may produce inferior designs unknowingly.

4. Upstream development floods downstream development because it is not accounted for in the design of the downstream structures and/or it is not accounted for far enough downstream in the drainage system. Few municipalities require designers to account for either their own flow related impacts or for the flow increases from expected development located upstream from themselves. Higher and faster peaks, greater flow volumes, increased velocities, warmer and dirtier water and lower base flow are all the result of urban development. Those municipalities that do account for impacts with a detention ordinance or policy rarely assess the impacts beyond the site boundaries. Therefore the mitigative effect of the detention basin is not analyzed very far downstream; and the accumulative consequences of development, even with detention, result in growing systemic flooding problems.

If again the "why" question (Level 4) is asked, three basic causative factors emerge.

1. Cities do not require appropriate levels of technical analysis because they are not sure what to require and how to implement these requirements. In spite of the wealth of computer software for drainage system analysis and the ability to remotely collect rainfall and runoff information, most municipalities have not had the time or the knowledge to investigate and invest in such solutions. Programs are often staffed on a day-to-day basis by junior or mid-level engineers without the authority or experience to make such changes. Their superiors have multiple other pressing duties and responsibilities and, without prompting and education,

do not see storm water as having the same importance or the same clear solutions as roads or solid waste. Thus, overall master planning for storm water management is seldom done.

2.   Municipalities do not impose certain flood mitigation measures, development controls or maintain off the public right-of-way because there is no legal authority to do so, and there is little impetus to establish such authority. To extend control of development beyond Federal mandates or to extend maintenance beyond the bare minimums requires gaining the support of political leaders, key staff members and "stakeholders". It is often difficult to stimulate such desires when so few of these individuals have anything clear to gain by doing so. Environmental Protection Agency (EPA) mandates, local citizen groups and/or a big flood event are often the necessary catalysts to action.

3.   Even if these last two factors were solved the bottom line is that there is no stable, adequate and equitable funding source for storm water management. Storm water usually cannot compete effectively with such things as solid waste and street repair for general tax based funds. Therefore a shift toward dedicated storm water funding is occurring throughout the country. This can take the form of such things as sales taxes, earmarked tax revenues and user fee systems (storm water utilities).

Each of these three foundational issues (master planning, legal authority and financing) will be covered in later chapters.

The more basic factors emerge with the next "why" question (Level 5). Even when key storm water staff understand the problems they must ask: who else is aware that flooding, erosion or pollution problems exist? Who supports a growth in storm water management? Who must support it for a successful program to be established?

The public is usually little aware of flooding problems and municipal staff have little long-term political support to solve such problems. If actions are not taken and decisions made within a month or two after a flood event support quickly dries up and memories fade. Other pressing demands thrust aside flooding, erosion and surface water pollution problems. However the problem remains, largely invisible, until the next time a large storm moves through the area. Building and maintaining consensus and support for the storm water program is necessary for its establishment and survival. The subject of public and political awareness is addressed in chapter 3 in this book.

The basic reason for lack of success in storm water management is the same as for lack of success in anything else: lack of vision or direction (Level 6). In almost every case where a successful program has been developed, one or several individuals had or developed a vision for what their streams and creeks should and could be. A southwestern city wanted its citizens to be safe from major floods and developed a vision for attaining that goal. A northwestern suburb wanted its streams to remain pristine spawning grounds for salmon and places for recreation. A midwestern city wanted its miles of underground pipes to function as they were designed and to remove the thousands of traffic barricades, often set in concrete, throughout the city. An eastern city wanted its streams to become urban greenway trails, attractive to its citizens. In each case a dedicated few charted the course and set the pace for the many. They built consensus, informed the public, developed effective technology, established policies and procedures and implemented well thought-out plans.

Had the original complaint been an erosion or sedimentation problem an analogous set of four factors causing erosion and sedimentation can be given:

•   the banks and/or bed have been attacked by the flow so as to reduce their ability to resist erosive forces or are not maintained;

•   more sediment from upstream reaches overwhelms the stream's ability to carry it away and sediment deposition occurs; or less sediment is present in the inflowing stream, than the natural situation, which may cause the water to attempt to pick up additional sediment (up to its carrying capacity) from the bed and banks;

- the basic slope or channel configuration has been constructed or modified in such a way as to place the channel in an unnatural configuration and without appropriate bank protection (termed "out of regime") natural forces tend to bring the channel back in regime; or
- more water creates more and longer shear stresses on the sides and bed of the channel causing more erosion.

Had the original complaint been pollution, a set of four other factors can be derived:

- pollution control devices are no longer operating properly;
- illegal dumping or illicit connections are present;
- pollution control devices were not placed or designed properly; or
- more uncontrolled development-related pollution is entering the system.

In each of these other two cases the five whys could be asked. The tracks will merge at Level 3 with the same correspondence.

1.3 Institutional And Technical Aspects

Notice that the first levels of assessment contain primarily physical and technical problems for which structural technical solutions are appropriate. Water is impacted by some physical means. However, when the later levels are considered, the solutions are institutional, programmatic and non-structural in nature. People are impacted by administrative means. These foundational problems allow or generate the more visible physical problems. If the root institutional problems are not eventually solved there will be a continual need to respond to an overwhelming number of flooding, erosion and pollution complaints. The basic municipal storm water management philosophy will continue to be that damage must be suffered before corrective action can be taken rather than taking preemptive action to avoid problems. Successful municipal storm water management programs account for and deal with both the technical and institutional aspects.

Storm water management programs which have been developed to solve these problems are usually developed and implemented by technically based persons. This results from the dominance of the technical aspects which surround most storm water management problems and the engineering and technical analysis and design procedures used to solve these problems. However, to implement a successful and effective storm water management program, the technical side of storm water must be complemented by the institutional side. Non-technical bureaucrats and community leaders will never, and should not be expected to, become fully technically adept. Therefore, technical persons must become involved in and adept at the institutional side of storm water management, and must become proficient at explaining technical issues in clear non-technical terms.

The storm water management structure must bring together the institutional goals, objectives, and administration and the technical solutions using models and master plans by means of regulations, policies and ordinances. When properly conceived, legal authority spans the gap between the two by pairing institutional goals or concerns with technical solutions through the use of performance oriented criteria.

1.4 The Organization Of The Rest Of The Book

The remainder of the book is organized into three sections. The first section deals with the institutional aspects of storm water management: the foundations. The second section of the book deals with technical and design aspects. The third section deals with the implementation aspects of storm water management facilities and programs.

It is difficult to combine these three sections into one text without doing injustice to one or the other or without developing a book that is too lengthy for publication. Every effort has been made to ensure that this is a stand-alone text. Where detailed treatment is beyond the scope of this text, appropriate references are given to lead the interested reader deeper and wider. The authors would appreciate any suggestions for improvements to this text, which can be sent to us through our publisher.

Chapter 2 Storm Water Management Programs

2.1 The Storm Water Management Program

The term "storm water management" can take on as many different meanings as there are storm water management programs. Some are narrow, single purpose programs while others are broad, basin or political entity wide. There are also a number of general and specific ways to group the various functions and components of storm water management for consideration. A storm water management program may have many or few of these components and each of these components may be well or poorly developed. We will look at several ways to understand and specify storm water management.

The basic underlying purpose of storm water management is to <u>keep people from the water, to keep the water from people, and to protect or enhance the environment</u> while doing so. Keeping people from the water involves a wide ranging number of non-structural damage mitigation measures including zoning, floodplain management, home acquisition, flood proofing, etc. Keeping the water from people involves direct interference with the water through the use of flow amount or flow elevation reduction measures such as retention or detention storage facilities or channel enlargement.

Maintenance or enhancement of water quality is taking on added importance through the mandates of the 1987 Water Quality Act and its implementing regulations. Aesthetics and multi-objective planning of water resources are also growing rapidly as expressed through stream restoration, greenway construction and integration of flood control with parks or multi-use spaces.

2.2 Functional Perspective Of Storm Water Management

One way of looking at storm water management is to divide it into specific general categories which are either defined duties or physical products. These components are: (1) the legal, technical, organizational and financial underpinnings required to implement the storm water management program; (2) long range planning guidelines and goals; and (3) day-to-day management tools, responsibilities and procedures. Day-to-day management carries out the long range goals and objectives using the tools and capabilities afforded by the underpinnings. Following is a discussion of the legal, technical, organizational and financial underpinnings of municipal storm water management programs.

Legal

For a storm water program to be effective it must be enforceable. Several storm water or floodplain management programs have failed legal tests for a variety of reasons. However, most programs will stand these tests if they are: not in violation of state legislation or municipal charters, equitable, fairly enforced, can be shown to be in the interest of the general public, and are well documented. State legislation may be necessary to establish local regulatory authority, to levy taxes or fees to finance such a program, or to allow creation of a special utility or control district.

There are specific programs, policies, and laws which establish the authority of a municipality to implement a storm water management program. Some of these are:

- Federal Flood Insurance program,
- Federal wetlands regulation,
- NPDES permitting,
- state authority and legislation,
- local municipal charter,
- local ordinances,
- local regulations,
- local policies (written and unwritten),
- checklists and informal guidance.

Local control may take the form of detailed ordinances requiring the vote of municipal governing bodies for amendment or, preferably, a general ordinance which incorporates an easily updated regulation or guidance document by reference. Municipal storm water management ordinances are sometimes developed as a series of special ordinances such as: detention ordinance, floodplain ordinance, erosion control ordinance, floodway ordinance, etc. though a comprehensive single ordinance reduces inconsistencies and confusion and helps ensure that gaps do not exist. Care must be taken to ensure there is a clear differentiation between actual legal requirements and mere statements of preferred policy. Many municipalities manage storm water on the basis of "understandings" of how things will be done. It is important to write down and define these understandings. Lines of authority and division of authority/responsibility should be clearly understood and well documented.

Measurable, performance-oriented design criteria must be given and goals should be stressed above technical methodology. For example, a requirement may be that a detention pond must reduce both the 2-year and 10-year peak flows to pre-existing conditions both at the site exit and down stream at a testing point where the newly controlled drainage area makes up ten percent of the total drainage area above that testing point. Or a pollution reduction best management practice (BMP) should reduce total suspended solids by 85 percent. However, within the confines of the performance criteria, innovation in design is allowed and encouraged.

Consistently applied, well documented and technically sound policies and procedures are of vital importance to cities and counties administering development. This will become even more important as storm water quality issues grow. Chapter 4 (Storm Water Ordinances And Regulations) gives details on storm water ordinances.

Technical

The technical support component usually consists of: a storm water management design manual or technical guide, a family of engineering software, some sort of master plan for new development and data collection or databases on such things as infrastructure, land use, soils, flow and rainfall.

Most municipalities have some form of a drainage or storm water management manual, or, at least, drainage information contained in subdivision regulations, standards and specifications, or similar documents. The sections in the manual should, as a minimum, make clear reference to all technical aspects of regulatory requirements. The ordinance tells what must be done. The manual shows how.

The technical manual may be comprehensive or may make reference to commonly available design documents or methods. It should be designed to make use of local data and environment, local procedures where appropriate, and local capabilities. The actual design procedures should be clear and not cluttered by superfluous discussion. Special sections on theory, etc. may be separate from the step-by-step instruction and design examples should abound.

No technical manual can cover all eventualities. There must always be room left for engineering innovation and the use of methods not found in the manual. However, if methods not found in the manual are used it should be made clear in the manual and/or ordinance that there may be additional review-time and submittal requirements. Several newer developments in technical guidance will make increasing impacts in the development of future drainage and storm water management manuals.

Information Mapping - Information Mapping is a term used describing the placement of guidance information on a page for maximum ease of use. It utilizes bold headings, underlining, grouping, key word blocks, tabs, etc. Technical manuals using this technique will be common in the future.

Paired Software - Paired Software refers to the development of a hand calculation method in the manual that mimics the methodology in a user friendly micro computer environment. Thus, a technical manual not only stands on its own but serves as a software users guide helping engineers make the transition from hand methods to computers. Software should be well documented, user-friendly, menu driven, interactive and public domain where available.

This concept can be extended beyond the manual to streamline master planning. "Hierarchical modeling" is a term used to extend the paired software concept into a three level structure where hand methods contained in the manual and supported by simple software at the review level are able to be input into complex basin models for assessment of overall impacts to a particular basin. Methods such as the U. S. Soil Conservation Service methodology for unit hydrograph generation or the Federal Highway Administration culvert procedure are conducive for this purpose.

Computer Models - The use of Computer Aided Drafting and Design (CADD) and Geographical Information Systems (GIS) is coming rapidly into common use for support of all aspects of storm water management. Such information layers as infrastructure inventories, soils, land use, outfalls, sampling sites, floodplains, complaints, etc. are in common use in many municipalities and integration of this digital and graphical information with such things as master planning, program management, maintenance management, complaint response, and sampling program tracking is growing.

While basic CADD systems (often with add-on packages) are very useful for a variety of applications, the use of GIS is rapidly growing in popularity. GIS is a term loosely (and often incorrectly) used to describe a variety of combinations of databases and graphical computer systems. Systems sometimes called GIS range from simple mapping software on the low end of complexity to sophisticated graphical and database systems at the high end. The user's task is to define exactly what types of information will be needed, what the future requirements will be and who the users will be. Depending on the actual day-to-day uses a raster based system or a vector based system may have advantages. A raster system essentially covers an area with a fine mesh grid. Each grid block can then be assigned any number of attributes. It works best with remotely acquired data. A vector based system uses lines, points, and polygons and is best suited for definition of boundaries, lines, and determining specific information about a polygon or line. It also tends to have a faster information retrieval time for many applications.

A true GIS can relate many layers of information to each other and perform any number of complex data manipulations and analyses from any combination of information and then display the information graphically to any scale. This level of sophistication may not be necessary or affordable for every municipality. However, this type of information is very useful for such things as hydrologic modeling, infrastructure maintenance scheduling, budgeting, taxation, and total master planning. Often the cost of GIS can be shared across several departments who will share the same databases. This will require a "champion" to sell and coordinate the activities and to ensure all graphical layers have been specified with sufficient exactness to ensure continued and wide usefulness. Municipalities can "trade up" to a GIS system over time without loss of data and information if they use wisdom in how CADD and database information is digitized and organized.

Organizational

In many municipalities, perhaps most, storm water management grew as an adjunct duty of an office which had primary responsibility in another area. Street maintenance forces often maintain culverts, pipes, catch basins and tail ditches in conjunction with street maintenance and repair. However, their primary concern is not the drainage system but the street. Planners often administer the floodplain program as an ancillary duty to zoning. Because the storm water function lacks focused leadership and is dispersed throughout the municipal organization it is often neglected or deferred. Badly needed storm water funds are diverted to more important needs. Those municipalities which have successful storm water programs have given the program focus through organizational change. In most the drainage system is seen as a "public system" in much the same way as the water supply and wastewater systems are viewed. This leads naturally into also seeing the storm water function as a "utility".

Administrative procedures should be clearly defined and efficient. Municipalities can operate in partnership with the development community if their policies and procedures are logical, consistent and efficient. Many municipalities have moved toward "one-stop shopping" for the development process wherein developers need only to submit plans to one office which then coordinates the review process and ensures a timely processing of plans. Other municipalities have a "sketch plan" or "concept plan" requirement for storm water where a potential site is discussed and flow and pollution control requirements placed on the development up front before a great investment in site layout or design has occurred.

Public identity and cultivation of public and political support is important to the survival and acceptability of storm water (and all public) programs. Storm water management must be seen as "customer" oriented and responsive. More discussion of public awareness is contained in the Public Awareness And Involvement chapter (Chapter 3).

Financial

What happens in many municipalities is that a large flood event will trigger a public outcry for something to be done. This then results in a study of the problem. The time required to make the study is long enough for the memory of the flood to have faded in the minds of the political decision makers and their constituents. Procrastination sets in. Funds are not allocated to make necessary improvements, the study is shelved, and the municipality awaits the next flood event. Someone has called this succession from flood to panic to planning to procrastination the "hydro-illogical cycle" since it seems to parallel the "hydrologic cycle" of rainfall to runoff to evaporation and back again. Experience has shown that the key issue to be solved in breaking out of this cycle is primarily one of financing. When adequate financing is available many of the other functions take care of themselves.

The use of a user fee based system (often termed a "storm water utility") of funding is growing in popularity. A true storm water utility is simply a funding entity or umbrella under which exists any number of funding and operational activities related to storm water management.

Just as most municipalities now consider both water and wastewater public systems, the drainage system should be considered a public system and a public responsibility. After all, public water makes up a large portion of the flow through the system, and systems quickly grow to such a size that a single property owner has not the technical nor financial wherewithal to manage the system. Even if they did have the resources, a piecemeal management of the system would eventually be as bad as none at all. And just like the fee charged to water and wastewater customers is based on their use of those systems, storm water charges are based on how much water an individual property owner puts through the system or how much of the system they use. Total impervious area or some combination of total impervious area and total land area is often used as a surrogate for this use. When a person paves over a field or woods they are putting a demand on the drainage system it was not "designed" to carry. Cost is incurred because of this demand, and they, the user, should bear that cost.   More information is given on storm water utilities in the Financing Storm Water Management Programs chapter (Chapter 5).

Long Range Aspects

Long range storm water management aspects can be thought of as divided into two areas: program wide and basin specific. Program wide storm water management defines just what it is that a municipality wants to accomplish through its storm water management program; what are its program priorities and objectives. A municipality must seek to establish answers to questions about its desire and authority to promote public safety and well being, protect the environment, encourage commerce and control development. These basic objectives find definition in general and specific policy statements which may address such things as drainage levels-of-service, service areas, standards, construction requirements, erosion control, water quality goals, floodplain control and zoning restrictions.

Typical basic policy statements are designed to accomplish the following seven foundational goals:

- protect life and health,
- minimize property losses,
- enhance floodplain use,
- ensure a functional drainage system,
- protect and enhance the environment,
- encourage aesthetics, and
- guide development.

A storm water ordinance is the usual document used to communicate these long range goals and policies and to translate them into specific activities and programs.

The primary vehicle by which long range, drainage basin specific, storm water management is accomplished is the storm water master plan. Master planning refers to a wide range of types of comprehensive plans which: mitigate known flooding, erosion and pollution problems, plan to avoid future problems and/or seek to take advantage of opportunities to enhance the environment, improve flood control capabilities and seek multi-objective use goals.

The storm water master planning process moves from identification of problems, needs, and opportunities through preliminary solutions and constraints, to final recommendations and actions. In the past, master planning has often been driven almost solely by existing acute flooding problems. Recently, communities have taken steps to provide both preventive measures (such as zoning or pre-planned regional detention facilities) and remedial measures (such as channel improvement). Storm water master planning is discussed in the Storm Water Master Planning chapter (Chapter 14).

Day-To-Day Storm Water Management

The day-to-day aspects of storm water management, supported by the legal, technical and financial underpinnings, carry the storm water management program to accomplishment of long term goals, defined by the storm water master plans and goal and policy statements. Day-to-day management tools and procedures can include a number of items, following are some examples.

- Calibrated hydrologic and hydraulic models are used to assess the impact of planned development or capital improvement structures on flood flows and water surface elevations for existing and ultimate development conditions. They are kept up to date as development occurs or periodically. These models may be tied to GIS or other graphic systems.

• Engineering software is used to allow the reviewer to quickly check proposed development plans, design of storm water management facilities, and to determine if the design is optimal or simply adequate. Often checklists for designs are filled in by the designer to facilitate such automated review. This software can be part of a hierarchical modeling structure.

• Storm water related databases can be anything from simple inventories of existing structures to a full blown general application GIS system using overlays for topography, land use, soils, roads, utilities, etc. Procedures for maintaining the databases and decision making are also important. A computerized infrastructure management system (IMS) may be used to serve this role.

• Data collection systems are invaluable for the development and refinement of future hydrologic and hydraulic modeling efforts. A system of recording rain and crest-stage gages are a necessity, at least in key basins, to improve the accuracy of results. Some municipalities have gone to remote collection of data and even remote control of flood gates, weirs and other structures.

• Flood studies have normally been done as part of the National Flood Insurance Program. When kept up to date, they serve as a valuable reference to control development. Often flood studies are extended beyond the one-square-mile limit set by the Federal program to afford consistent control of development wherever flooding is likely to occur.

• A flood warning system or other hazard mitigation system has been found to be useful in areas of flash flooding and in areas where structural solutions are impractical (such as railroad underpasses).

• The implementation of a system to collect funds is a day-to-day activity which is an integral part of most storm water management programs. This may include databases, mailings, fee determinations, etc.

• Procedures for review, inspection and enforcement of storm water management requirements should be clearly defined and streamlined. A single authority is desirable where feasible.

• A computerized complaints system has proved very valuable, not only for satisfying the needs of the public, but for identifying nuisance problems which may be leading edge indicators of a growing problem within a basin. An equitable system of prioritizing the complaints and dealing with them in a timely manner is also necessary. Customer service is also important as is the "customer" mentality it fosters.

• A coordinated maintenance management system ensures that the existing infrastructure is capable of functioning to its full capacity. Limits of maintenance (both physical and financial) should be clearly defined to avoid making decisions under pressure. Maintenance should be preventative as well as remedial.

• Appropriate supporting literature can assist the manager in answering common questions or explaining often used procedures simply and easily.

2.3 Functions In Storm Water Management

Table 2-1 lists the basic functions of a storm water management program from a management point of view. A typical municipality will have many of these functions dispersed throughout the municipality's governing structure.

Effective management of storm water usually requires that a single staff position be identified as the central figure in storm water control. This person should have primary control of all storm water management functions (e.g., drainage, floodplains, water quality, erosion and sediment control).

Table 2-1 MAJOR STORM WATER MANAGEMENT FUNCTIONAL COST CENTERS	
1. **Administration** General Administration Gen Prog Planning and Development 2. **Special Programs** Public Awareness and Involvement GIS and Database Management Spec Prog Planning and Dev 3. **Billing and Finance** Billing Operations Customer Service Financial Management Capital Outlay 4. **Indirect Cost Allocation** Overhead Costs Cost Control Support Services 5. **Storm Water Quality Mgmt** Quality Master Planning Retrofitting Program Comprehensive Monitoring Program Struc and Non-Struc BMP Progs Pest, Herb and Fertilizer Used Oil & Toxic Materials Street Maint Prog Spill Response and Clean Up	(continued) Prog for Pub Ed and Reporting Leakage and Cross Connections Industrial Program Gen Com and Residential Program Illicit Con and Illegal Dumping Landfills and Other Waste Facilities 6. **Engineering** Des Criteria, Stds and Guidance Field Data Collection Quantity Master Planning Design, Field and Ops Engineering Hazard Mitigation 7. **Operations** General Maintenance Management General Routine Maintenance Emergency Response Maintenance 8. **Regulation and Enforcement** Gen Code Dev and Enforcement General Permit Administration Gen Drainage Sys Insp & Reg Multi-Obj Floodplain Management Erosion Control Program 9. **Capital Improvements** Major Capital Improvements Minor Capital Improvements Land, Easement, and Right-of-Way

Both large centralized organizations having almost all the functions in-house and small decentralized organizational structures hiring many of the functions both within municipal staff and outside have proven to be successful. Storm water quality has been identified separately though, in practice, it should be integrated with all the quantity functions in each functional area.

2.4 Typical Storm Water Management Problem Areas

A series of studies and inventories was performed during 1986, 1987 and 1992 in some leading cities and associated counties. These inventories (combined with experience in a number of other municipalities across the country) uncovered common problem areas and isolated a set of character-istics of successful municipal storm water programs. Certain characteristic problems are, or had been, the experience of most surveyed. Those municipalities which developed successful programs have exhibited fewer of these problem traits. No municipality exhibited all of them, and all experienced different levels of success and failure in each area. These typical problems and success characteristics are given below under the different categories of storm water management: long term, legal/ financial/ organizational /technical support, and day-to-day management.

Long Term Problem Areas

- Specific goals or policies had never been articulated in writing. Therefore such a thing as enforcement of design standards, storm water quality programs, development plans approval, detention requirements or longer term planning was without concrete purpose or ability to assess achievement, but rather was based on a vague sense of doing some good.
- Planners had an inability to get a grasp on what the extent of potential flooding, erosion, or pollution problems is, or will be in the future. Capital improvement planning was incomplete and poorly coordinated with other planning functions.
- If storm water master plans or other types of storm water assessment were available, they were often difficult or cumbersome to interpret, out-of-date, not directly usable as analysis tools, or had used advanced engineering techniques or computer analyses incompatible with every day design application or beyond the capabilities of engineering staff to modify easily and reuse. They tended to be static snapshots of pieces of an ever emerging larger picture.

Legal/Financial/Organizational/Technical Underpinning Problem Areas

- Storm water management was often seen as a low priority item when compared to such things as roads and sanitation. This, in itself, is not a problem since it usually is of lower priority. However, the result was that storm water was often managed piecemeal by various persons as an additional duty, poorly financed and unplanned.
- There was little knowledge of the state of repair of the drainage infrastructure and what was actually out there. Therefore future maintenance could not be scheduled or budgeted. The impact of this is that failed parts of the system remained defective for long periods.
- Many local flooding problems existed which were beyond the financial and/or technical ability of homeowners to repair but were not under the jurisdiction of the municipality. Causes of this were numerous including: poor design, poor maintenance, unanticipated upstream development overtaxing the system, and the unavoidable deterioration or natural adjustment of any man-made drainage way.
- Comprehensive and usable planning, policy, or design documents backed by effective ordinances and enforcement powers rarely existed or were enforced.
- Many runoff problems were caused by flow from areas outside the jurisdiction of the municipality. Inter-jurisdictional cooperation was lacking.
- Local, on-site detention was in a poor state of repair. This is particularly true of residential sites. The municipality did not have the funds, manpower, and/or legal ability to make the necessary repairs.
- Funding was inadequate and poorly targeted to meet actual needs. Impact or other fees could only be used in designated areas and general funds were lacking for needed infrastructure maintenance. No analysis of needed funding had ever been done except on a year to year basis. Legal challenges to fees existed.
- Engineering methods used by designers were not uniform with regard to the type of method used, degree of accuracy, or presentation.
- Little data existed in directly usable form. Thus many design engineers used inferior methods rather than develop the necessary information. Few rain or flow gages existed even in basins of known future development.
- There was little knowledge of or concern for the environmental aspects of urban runoff in spite of the fact that Congress had mandated certain requirements in the future for virtually all municipalities.

Day-To-Day Problem Areas

- Planners had an inability to predict the impacts of potential developments or the impacts of required storm water controls such as detention. Therefore detention was required in a haphazard fashion or by using general rules-of-thumb.
- Day-to-day maintenance was driven by complaints and political pressures. Regular preventive maintenance was not done. Prioritization of maintenance was haphazard.
- Use and enforcement of proper erosion control measures was seen as unimportant or not possible.
- Those responsible for storm water management spent a perceived disproportionate amount of time in plans review, often recalculating laboriously various drainage calculations by hand. There was little or no time or ability to assess whether designs were adequate or optimal.
- The development process was not defined well enough to ensure compliance with technical requirements, timely inspection, and as-built certifications.
- Many development control policies were simply "understood" by local engineers but never written, thought through in detail, or coordinated.

Characteristics Of Successful Programs

Those municipalities that appeared to have the most successful programs exhibited some common characteristics. Understanding these characteristics will help any municipality become more successful. First of all, they had found solutions for most of the problems stated above. But also, they had gone beyond simply overcoming most of the problems enumerated above and had gone on to develop a coordinated program which took advantage of the unique political and physical environments in which they functioned. Several other specific comments which do not fall under the three categories above could be given.

Public Relations - Those successful programs had executed public relations programs to ensure the cooperation of the general public. One city had conducted public meetings in every neighborhood presenting problems that would be solved in that neighborhood. Many municipalities have established hotlines, logos, and published attractive brochures describing the programs. To provide greater identity some had special uniforms for the storm water maintenance crews. Other municipalities created multi-media presentations, built information displays, held public forums and provided informative news releases. Citizens had a sense of ownership of the surface water system. Many had citizen monitoring or adopt-a-stream type programs. Student education and involvement was important.

Key Champions - Most successful programs depended on key individuals or groups to champion the cause of storm water management. It is important to have key persons in politics, government staff and citizen groups. The staff person must be low enough within the staff structure to care about storm water management and high enough to make a difference. Many municipalities had citizen advisory committees. Most had developed open and effective communication and reporting channels with political leaders and other staff components.

Financing - Almost every successful program in a larger urban area had developed a storm water utility district. This funding mechanism has been found to be superior to any other in being equitable to all citizens, generating the needed funding in a relatively painless way, and withstanding legal challenges.

Basin-wide Control - To help overcome multi-jurisdictional problems, several municipalities established a regional entity which performed some storm water work. The Denver, Colorado metropolitan area established the Denver Urban Drainage and Flood Control District by special legislative act. The City of Albuquerque, New Mexico established the Albuquerque Metropolitan Arroyo Flood Control Authority. The Louisville, Kentucky metropolitan area administers storm water through the Metropolitan Sewer District.

Clear Procedures And Goals - Each had an efficient sense of purpose with well defined goals and objectives in each key program area, and the means to achieve these goals.

Focused Authority - In many successful programs the total storm water function was consolidated or, at least, controlled and coordinated under one authority. This included maintenance (or at least maintenance management), capital budgeting, floodplain management, plans review, inspection, enforcement, emergency management, planning and design. The most successful programs were not hidden as adjuncts or blended into other utilities such as wastewater or water supply. The storm water administrator was senior enough to speak for the program at the highest levels.

Strong Technical Guidance And Capabilities - The most successful programs had developed strong technical capabilities, resources and approaches both in-house and through the use of outside consultants. Guidance manuals were clear and complete. Master planning was done and used to guide development and capital improvements. Technical review was thorough and state-of-the-art designs were strongly encouraged or required. Water quality was considered equally with flood control and efficient rainwater removal. Aesthetics and multi-objective best management practices were encouraged and sometimes mandated through greenway plans and landscaping requirements.

Guided Development - Each successful program had well developed procedures to guide and control development, detailed submittal requirements, adequate inspection staffs, and streamlined enforcement capabilities. Development assumed responsibility in that it bore most or all of the cost of system improvement. Most administrators interviewed stated that there was some resistance and a need for re-education at first, but the program ran smoothly after that. And most stated that there were no adverse repercussions from charging developments for drainage improvements. Often master plans guided development.

Comprehensive Maintenance - Successful programs had made a shift away from primarily reactive maintenance toward proactive and preventive maintenance. The problem of maintenance responsibility for residential subdivisions was overcome in many of the successful programs. Most provided defined levels of maintenance service to the whole system without regard to right-of-way location. Most private detention facilities were not maintained to function as designed. Government operated maintenance of these facilities (including an abandonment program) proved to be most effective in several communities. Those municipalities with storm water utilities often figured charges such that all detention facilities could be maintained by the local government if private maintenance failed. Strong crediting mechanisms often were used to encourage appropriate design, construction and ongoing maintenance of drainage facilities.

2.5 Costs And Description Of A Storm Water Program

There are a large number of storm water programs in the United States in various stages of development. One measure of the maturity of a storm water program is the amount of money spent on it. Levels of funding for the basic programs (less any extraordinary items like large capital programs, large flood control maintenance programs, costs shared with counties, etc.) tend to be the same across the country. Table 2-2 indicates a typical range of program costs expressed in cost per acre (assuming a normal mix of urban development). Improvement or addition of the indicated capabilities are reflected in the cost ranges.

Table 2-2 Typical Storm Water Program Costs		
Program Type	Cost Ac/Yr	Typical Program Features
Incidental	15-35	Reactive incidental maintenance and regulation as part of other programs
Minimum	30-50	ADD: ROW maintenance, better regulation and inspection, more staff, erosion control
Moderate	40-70	ADD: additional maintenance programs and levels of service, better regulation and inspection, some planning, minor capital program, general upgrade of capabilities
Advanced	70-140	ADD: maintenance (of some sort) of the whole system, master planning, regional treatment, some water quality, data collection, multi-objective planning, strong control of development and other programs, utility funding
Exceptional	over 200	ADD: storm water quality, advanced flood control and levels of service for maintenance, aesthetics become more important, public programs

Most programs were originally drainage programs appended to either a street maintenance, water and sewer or general public works function. Other public services take precedence over the storm water infrastructure. In most municipalities knowledge of the location and condition of the drainage infrastructure, even that on public rights-of-way, is not known. As services or requirements were added (such as FEMA administration) they were added to various departments or divisions of government. Therefore, operations may be part of a street's division or water and sewer. FEMA and other planning is part of a planning department. Engineering lodges in public works. Regulation and enforcement may be found in many or all of the other departments depending on whether it is development related and requires a subdivision or rezoning process for development. In few cases is there an individual who has the authority, knowledge and desire to manage all aspects of the storm water program. These would be termed "incidental" storm water programs. All operations are complaint driven and normally limited to the right-of-way and a clear public interest. Planning is nil; regulation is limited to the minimum required for development.

Technical guidance is poor or copied from other nearby municipalities or other government publications.

Better maintenance, perhaps with dedicated personnel, a more defined regulatory process, technical support at a higher level and erosion control would be added to the incidental program to make it a minimum storm water management program. Further additions would increase the quality of the program as shown in Table 2-2.

Chapter 3 Public Awareness And Involvement

3.1 Introduction

Public awareness and education are the foundations of a successful urban storm water management program. From before the program's inception and throughout its growth and service life, interaction with, and education of, the public is not an adjunct to the program's purpose; they are (or should be) a main purpose of the program. This is becoming especially important as the demands of the NPDES permit and its source control philosophy comes more into play in major and minor municipalities across the country.

Public awareness and education are carried out in storm water management programs in two main ways: specific public awareness campaigns and ongoing "baseline" public information programs and activities. These two aspects differ in that a campaign has a beginning and an end while the ongoing program goes through transformations but does not envision an ending. But for either case there are specific factors that should be kept in mind and methodologies that will make a storm water management program successful.

3.2 Defining The "Public"

The first step in designing a public program is to ask the question: Who is the public?" Who am I trying to reach and with what message? Who will have which specific concerns? Who is influenced by this or that aspect of the program? Different sectors or segments of the "public" will participate or be interested in different issues and in different stages of the same issue:

- the development community will be vitally interested in regulatory and financial aspects;
- the environmental community will be vitally interested in water quality issues;
- specific neighborhoods will be interested in specific provisions for drainage controls, regional storm water treatment, safety, park integration, greenways, etc.;
- clubs or social organizations may be interested in participating in some programs;
- school children can be interested in the environment or clean creeks;
- social classes may be interested in utility fees or charges;
- tax exempt and governmental properties will be interested in new user fees;
- commercial and industrial concerns may have similar interests in fees and charges (and credits); and
- design professionals will have an interest in the technical criteria and regulatory requirements.

Figure 3-1 illustrates the basic public awareness and involvement understanding. There are different levels of involvement for different people based on such things as job duties, technical expertise, level of concern and willingness to invest time and effort. Different types of public education and involvement will be necessary to reach different groups.

The center core of the project is key staff; the decision makers. They are the ones tasked with implementing or carrying-on a successful storm water management program. They are the engine

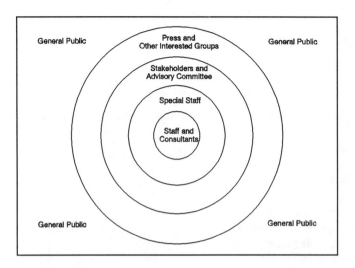

Figure 3-1 Public Awareness Strategy

that drives the program. They are often assisted by outside expertise in public awareness program planning, development and execution. They have full-time involvement. The next circle out contains the specialized staff and consultants necessary for part-time support (such as legal, GIS, or financial staff).

The next two levels of involvement come from stakeholders, political leaders or a specialized citizen group. People organize around common interests: political, financial, moral, environmental, cultural, social, locational, etc. Since some aspect of the storm water management program impacts them or something they value they will be "stakeholders" in the program's processes.

The difference in the two circles is the sense of "officialness" of the involvement or the intensity of the concern. Those most involved or those most affected by the storm water management programs of the community should and do have the greatest influence. From experience, a small vocal group of dedicated stakeholders often controls the direction of storm water management program growth and activities. Such a group can fill many roles from policy setting to simple advisory duties and from technical to strictly policy oriented activities. Stakeholders normally include the development community, representatives from any large institutions or industries, members representing socioeconomic groups and environmental groups. The press is also involved at the outer of these two levels.

The final level is the general public. Storm water is not normally vitally important to most of the local residents . . . unless there is flooding. It can be of interest to the general public if there is an effective public education and awareness program. For example, based on the results of a statistically derived phone survey conducted in an eastern city, clean water is important to the average citizen; so much so that they were willing to pay over $5 per month to help clean it up.

The closer to the center of the target one moves the greater is the level of effort required to maintain involvement and influence. Also, as one tries to influence those in the more outer rings, the more effort it takes to maintain that awareness.

3.3 Plan To Include The Public

For storm water success, especially in campaigns such as initiation of a storm water utility fee, it is important to have a "champion" in three areas: political leader, staff leader and respected community leader.   These champions must be well educated on the issues and articulate in explaining the goals of the program.   The other factor for success (or at least a factor to avert disaster) is to identify early the needs, interests, and level of concern of the various stakeholder groups and plan a way to address those concerns.   The greater the controversy the greater the public interest.

The step-wise and thoughtful development of a public awareness plan is the most vital tool in the success of a public awareness program.   The public should be involved at every stage of the process - but the stages must be carefully defined.   The plan describes in detail the objectives, activities, sequencing, timing, costs and responsibilities for every aspect of the public program. Development of the plan is important because it (EPA, 1990):

*   forces a careful analysis of how the public plan fits the storm water management program;
*   brings together in planning and (hopefully) agreement all entities or departments which will be involved in the plan and program; and
*   communicates to the public that they are of vital importance and that the public entity is "contracted" to involve them.

The plan is normally developed as a group participation activity, even involving certain stakeholders at times. There is an old and true adage in the public awareness business: "bring me in early I'm your partner; bring me in late, I'm your judge." It takes longer on the front end to do this, but ensures success on the tail end. Often potential stakeholders can be interviewed about their concerns as a prelude to inviting them to fuller involvement. Appropriate screening can be done to avoid troublemakers, turf warriors, and illogical or uncompromising radicals.

The process to develop the plan is as important as the plan itself. In the process stakeholders grow to respect positions held by others (if there is potential controversy involved) or gain a vision for the objectives of the program.   Often working groups are sub-divided from the full group to develop certain parts of the plan or at least collect thoughts on parts of the plan. Table 3-1 gives the elements of a typical public awareness plan.

Table 3-1 Elements Of A Public Awareness Plan

1. Describe the history leading up to this point and major past players or causative events.
2. Describe the objectives and goals of the program and potential major issues likely to emerge in a campaign or in a general public awareness program.
3. Estimate the level of public interest likely to be generated or hoped to be generated and develop ways to gage public reaction or program effectiveness.
4. Identify the major potential stakeholders and their level and type of concern.
5. List and describe stages in the campaign, activity, or program the public awareness program must support and the public involvement support stages.
6. Outline a sequential and interrelated plan of activities and products to support each stage of the public program and the storm water management program.
7. List key milestones when the public program will be reassessed and mid-course corrections made if appropriate.
8. Estimate costs and level of effort required for each phase of the program.

It must be remembered that the public program is to support and follow the storm water management program not lead and shape it. The program drives the public campaign not vice versa. There is often a tendency for the public program to take on a life of its own losing sight of the "real world" objectives of the storm water management program.

For example, assume a municipality has a project in mind that has the following three goals:
1. To obtain a federally mandated permit to discharge storm water from the municipal separate storm sewer system to creeks and streams.
2. To plan, organize and take steps to establish the foundation for a comprehensive storm water quality and quantity management program.
3. To develop and carry out a storm water utility to fund the storm water management program.

Based on these goals the following general objectives could be developed (Ogden, 1992):
1. To inform and educate the stakeholders and the public at large about the following:
   - Storm water management programs and needs
   - The Federal storm water quality regulations and permit requirements
   - Storm water management and program growth and direction
   - Storm water management program costs and financing alternatives with emphasis on the storm water utility
   - The Advisory Committee and Council actions and discussions
2. To seek input from the stakeholders and public at large on the following:
   - The existing storm water problems and future needs
   - Desired storm water management program direction, activities and structure
   - Willingness to pay for the program
3. To involve the stakeholders and public at large in program development through the following ways:
   - Through the use of a citizen task force
   - By meeting with stakeholders and other groups
   - Through the use of special events
   - Through the use of hotlines
   - Through the use of public hearings
4. To gain consensus for the storm water quality management plan, storm water management program and the creation of the utility.
5. To monitor the Public Education program in the following ways:
   - By measuring public acceptance of the total program
   - By monitoring all media, i.e., TV, newspaper, and radio to keep track of public opinion
   - To obtain feedback at all public events
   - To obtain feedback through a hotline

The following target audiences could then be defined:
1. Area neighborhoods
   - Business (owners)
   - Residential
     - Associations
     - Party precinct leaders
     - Churches
     - Council wards
2. Local media
   - Print
     - News departments
     - Feature writers
     - Editorial writers
   - Broadcast
     - News departments
     - Talk show hosts
     - Public service directors
3. Elected officials
   - Federal and State

- City council
- County commission
4. Government staff
   - City or County Managers
   - City or County Public Service and Information Director
   - Other staff
5. Developers
   - Residential
   - Commercial
6. Environmental Groups
   - Sierra Club
   - Other groups (local, regional, national)
7. General Public

With this basic information the public awareness plan could be formulated following the guidance in Table 3-1 and assessing the techniques in the next section.

3.4 Public Program Techniques

There are almost an infinite number of techniques, combinations of techniques and variations to inform and educate the public. Table 3-2 is an example list of some of the most common used in urban storm water management public programs. Each has different advantages and effectiveness and none of the techniques can be used by itself. Each must be used with others to create a synergy. Some are most effective to introduce general information to masses of people while others are more "rifle shots" targeted toward even a single individual. Some are participatory and some are quite passive.

The point is that each activity or product must be clearly understood within the context of the target audience, the objectives, the phase of the campaign or program, the nature of what is being advertised and the message communicated, etc.

Table 3-2 Example Techniques Used In A Public Involvement Plan

- media kit and white paper
- media tour
- news releases and op-ed pieces
- feature stories on surface water problems
- B-roll footage for local news stations
- paid advertisement
- developed pamphlets
- free government informational and educational pamphlets
- slide-based or other video
- bill stuffers
- talk shows
- public access programming
- speaker's bureau availability
- political leader meetings
- one-on-one key person meetings
- booths

- special neighborhood or stakeholder meetings (such as Chamber of Commerce, etc.)
- information workshops
- customer service representatives
- hotline
- identity creation - logo, letterhead, vehicle markings
- phone survey research

Written public information pieces must be factual and unbiased. There will certainly be a backlash if a piece about a subject where controversy can be expected is seen to be a sales job for the community officials. When this happens, the staff and the program lose credibility . . . their greatest asset.

Public Participation

Some of the techniques are specially conceived to invite participation of various target audiences. Table 3-3 lists the most common participatory techniques together with features, advantages and disadvantages.

Table 3-3 Participation Techniques Used In A Public Involvement Plan

Technique	Features	Advantages	Disadvantages
Advisory Groups/ Task Forces	Stakeholders are invited to participate. May be technical, advisory or directive in nature. Should be balanced and representative.	Provides oversight to the process or campaign. Promotes communications between parties and gets all sides exposed. Can provide a forum for reaching consensus. Obtains "ownership" of solutions, policies or programs.	Consensus can fall apart if members have no authority to make or enforce decisions. If advice is not taken is potential for controversy. Requires substantial commitment of staff time.
Focus Groups	Small discussion groups to give "typical" reactions of the general public. Normally conducted by a professional facilitator. May be several parallel groups or sessions.	Provides in-depth reaction and detailed input. Good for predicting emotional reactions.	May not be representative of the general public or a specific group. Might be perceived as manipulative.
Hotline	Widely advertised phone number to handle questions and provide information.	Gives a sense of people knowing who to call. Gives a customer service feel. Provides one-stop service when done right. Can handle two-way communication with the general public.	Requires effective and trained hotline operator on standby.

Table 3-3 Participation Techniques Used In A Public Involvement Plan

Technique	Features	Advantages	Disadvantages
Interviews	Face-to-face interviews with key persons or stakeholders.	Can be used to anticipate reactions or gain key individual support and provide targeted education.	Requires extensive staff time and an effective interviewer.
Hearings	Formal meetings where people present formal speeches and presentations.	May be used for introductory or "wrap-up" meetings. Useful for legal purposes or to handle general emotional public input safely.	Can exaggerate differences without opportunity for feedback or rebuttal. Does not permit dialogue. Requires time to organize and conduct.
Meetings	Less formal meetings of persons to present information, ask questions, etc.	Highly legitimate form for public to be heard on issues. May be structured to allow public to be heard on issues and small group interaction.	May permit only limited dialogue. May get exaggerated positions or grandstanding. May be dominated by forceful individuals.
Workshops	Smaller meeting designed to complete a task or communicate detailed or technical information.	Very useful to handle specific tasks or to communicate, in a hands-on way, technical information. Permits maximum use of dialogue and consensus building.	Inappropriate for large audiences. May require several different workshops due to size limitations. Requires much staff time in detailed preparations and many meetings.
Plebiscite	City-wide election to decide a particular issue.	Provides a definite and usually binding decision to end a stalemate or drawn out process.	Campaign can be expensive and time consuming. General public may be susceptible to uninformed or emotional arguments.
Volunteer Programs	Programs where citizens volunteer to perform some necessary function in support of the storm water management program.	Builds shared ownership and a sense of personal pride. Allows stakeholders to experience some of the realities of the program.	May not be high quality work. May be liability issues. Takes a good deal of staff supervision.
Polls	Carefully designed questions are asked of a statistically selected portion of the "public."	Provides a quantitative estimate of general public opinion.	Susceptible to specific wording of questions. Provides only a static snapshot of a changing public opinion. Costly.

After: EPA, 1990

Public Participatory Groups

There is much to be wary of in public participatory groups. Generally speaking for example, public hearings, at best, rarely provide useful information and are largely ignored by the general public. At worst they turn into a shouting match where misinformation rules the day with little or no chance for logical and reasoned rebuttal.

Advisory groups can be helpful but under controlled and well defined circumstances. There is a built-in dichotomy with such groups in that the members selected to represent stakeholders on such advisory groups are normally selected because they are decision makers who carry weight with their constituency. Such people are rarely satisfied with only providing advice or being involved in situations where their advice is not perceived as being heard. Alternately the persons sent to the advisory committee may have little or no authority within their constituency but are sent to be eyes and ears. If decisions are reached that the stakeholders do not like, the participation of such individuals will be somewhat meaningless as issues are revisited politically.

Workshops can serve a very valuable role in providing information of a factual or technical nature while attempting to skirt the more emotionally charged side of things. This allows a mutual understanding to be reached and a common baseline of information to be gained before tackling controversial issues.

If the public is invited to participate, the storm water manager should make sure their input is recognized and considered fully . . . and that they know it. This is accomplished through summary feedback of what was heard.

Elected officials, in particular, need to be kept informed of all aspects of the program or campaign. Elected officials should never be taken by surprise, should never find something out by reading it in the paper, and should never give a speech or talk on the subject without adequate preparation by the staff.

Remember that a key for any successful campaign is to ensure that the municipality has taken all appropriate steps to meet any expected concerns of the citizens. For example, before implementing a user fee or rate increase for a storm water utility it is prudent to define and document problems the fee will solve and to take a hard look at the program costs and activities and ensure (EPA, 1989):

- inefficiencies have been eliminated;
- labor forces have been trimmed and are lean; and
- the program is perceived as the minimum necessary to meet customer expectations or regulatory demands.

3.5 Volunteer Programs

One highly successful method of public involvement and education is through volunteer programs. This has been especially successful in the areas of recycling and citizen monitoring of stream quality. Once skeptical, states and local agencies now see citizen monitoring as a key in the total sampling program (Nichols, 1992). These programs have more than tripled since 1988. Other examples of the growth and importance of volunteer programs include the following.

- EPA has sponsored several conferences on the use of volunteers.
- A complete magazine, *The Volunteer Monitor*, is published in San Francisco.
- EPA publishes a *National Directory of Citizen Volunteer Environmental Monitoring Programs* that lists programs by state.
- The Izaak Walton League has used citizen monitors since 1927.

The Izaak Walton League offers a training course in simplified monitoring techniques including a simplified macroinvertebrate monitoring methodology through its Save Our Streams program.

Urban storm water management programs such as the Bellevue Washington "Stream Teams" (Bellevue, 1989), the Charlotte, North Carolina "Streamwalk Program", or the more recent Seattle "Water You Doing" program (Seattle, 1992) have proven that citizens want to be involved in their streams. Those municipalities that have fostered a sense of community ownership of the streams have seen their surface water quality and aesthetics improve measurably over those who still see the streams as abandoned and unsightly urban fields. Volunteers can reach people and locations that staff cannot, and they can build community support in an effective way when staff personnel cannot.

There has been much learned about how to conduct a successful citizen participation program over the years. Generally there are six basic principles which together form the foundation for volunteer success (University of Wisconsin, 1993). These are given in Table 3-4.

EPA has compiled a source book for citizen involvement and education materials for water quality. The publication contains information on hundreds of: pamphlets, books, bumper stickers, catalogs, citizen action guides, fact sheets, software, handbooks and manuals, newsletters and magazines, booklets, posters, slide shows, student activities, videos and other ideas (EPA, 1993a, EPA 1993b).

Table 3-4 Keys For Successful Volunteer Programs

- *Program and Staff Commitment* - The program and staff must, at all levels, continually acknowledge that the program counts on volunteers to be successful; that they have something important to contribute. Staffs must be trained and encouraged to cultivate this attitude. They should be trained on how to make best use of and train volunteers.
- *Recruitment Strategy* - Includes development of volunteer job descriptions and targeting ways and locations to look for volunteers.
- *Communicating Effectiveness* - Formal or informal assessments of achievement are important to let volunteers know they are important and to make corrections to problems.
- *Enjoyment* - This is achieved through both social interaction and a pleasant working environment. It is important to set a congenial working environment and find tactful ways to deal with problems.
- *Personal Development* - Gaining knowledge and skills is important for many volunteers. Project staff should provide opportunities for personal development in skills necessary for the volunteer program.
- *Shared Ownership* - The commitment of the volunteers rests on a sense that they own part of the program. Staff members need to acknowledge this sense of shared ownership of the program and be willing to share management and coordination duties. Citizen advisory committees have more of a sense of ownership if they are given responsibilities for setting agendas, schedules and facilitating meetings. Ownership is also communicated when volunteers are given opportunities to speak publicly about the program and represent it to other potential volunteers or citizen groups.

3.6 Dealing With The Media

Tables 3-5 through 3-7 are bullet lists of tips for developing a trusted and effective relationship with the media and for conducting interviews, television appearances and news conferences effectively (Harris, 1986, AWWA, 1989, Newsom and Scott, 1981, Bernstein, 1988, Wilcox and Nolte, 1990).

Table 3-5 Media Tips - Television Interviews

- Listen to questions and remember then in detail.
- Smile when talking.
- Speak in 30 second clips.
- Be cordial... you don't need that kind of publicity.
- Give education and information in an entertaining way.
- Deliver a positive message about the subject.
- Give the information asked.
- Dress conservatively.
- You know more than anyone on this subject but brush up.
- Talk to a tenth grade education - no technical jargon.
- All microphones are always "on".

Table 3-6 Media Tips - News Conferences

- Remember that there is competition for the news media's attention and they often have several conferences to choose from.
- Make sure a news conference is more appropriate than an interview or press release.
- Pick a good central location familiar and accessible to the media.
- Choose a day early in the week and one without competing conferences.
- Avoid the rush by holding it early in the afternoon - say 1:30.
- Give notification in the form of a "media advisory" six working days prior to the conference.
- Address information to the "news planning desk" and include: Event, Time, Place, Speakers and Background.
- List your event in the local news calendar (termed the "daybook") one day prior.
- Provide footage for TV media. Have visuals in any case.
- Hire a photographer to photograph the event on film . . . black and white is fine.
- Prepare the room - remember to allow for sound hookups and camera platforms where necessary. Find the pay phones. Put up directional signs.
- Brief a spokesperson and prepare (see Table 3-7).
- Make up simple press kits: news release summarizing the conference; lists of names and titles of speakers; bio and photo of primary spokesperson; fact sheet on municipal program or campaign and the municipal department in charge; additional articles, brochures, etc.
- Offer phone interviews to no shows and send along press kits.
- Follow-up the day before.

Table 3-7 Media Tips - Interviews

- Be interviewed in a familiar setting.
- Anticipate touchy questions and have answers ready.
- If you cannot answer say so. State that the information is not for public disclosure and politely decline to answer.
- Stay cordial. Do not argue.
- Talk from the viewpoint of the public's interest.
- Talk in personal terms whenever possible.
- Allow time for the interviewer to meet his or her deadline.
- Know your topic thoroughly.
- Have printed support material on hand to give the interviewer.
- Set ground rules for the interview if possible.
- Build personal rapport prior to the interview.
- Avoid off-the-record remarks.
- Help the reporter make only one stop for all necessary information and follow-up interviews.
- Give the reporter the story they are after.
- Offer to get answers to questions unanswered and then get them to the reporter quickly.
- Do not ask when the story will run or how big it will be.
- State the most important fact at the beginning.
- Do not repeat a question or words in a question if it is offensive, even to deny them . . . they could be printed.
- Direct questions deserve direct answers.
- Tell the truth and do not exaggerate.
- Keep in touch on a continuing basis.
- Allow for the fact that the reporter knows little about storm water.
- Offer to help in the follow-up if necessary.

The performance and objectives of any storm water management program or campaign become known to the public through the reporting of the print and broadcast media. The media can be the best friend and support or it can become the worst nightmare of the storm water staff person or political leader.

Staff and political leaders often complain that reporters publish stories that are inaccurate, sloppy, misleading, incomplete and biased against them. Editors and reporters state that municipal staff and leadership are often less than candid, defensive, provide partial information in an unwieldy or untimely fashion, hide key facts and limit access to key information persons. They also state that there is a difference between bias and simple unfavorable news which is not understood by the subjects of the reports.

There are some things that can be done to improve the relationship and make it a win-win situation. The reporter has a job with a tight deadline. He or she is looking for news: controversy, provocative stories, interesting stories, human interest, etc. If a subject is dry, highly technical, without pictures or news footage or a "hook" for the reader or listener it is not news.

Deadlines for general news in the daily morning paper are generally four p.m. the day before publication. For the evening paper it is anywhere before four p.m. the day of publication. Feature stories normally should be sent or developed several weeks prior to the expected publication time to ensure they are given ample opportunity for inclusion. Television news can be included up to about two hours prior to air time if there is no film involved. If film is involved there is normally a longer lead time required of about two additional hours. Radio general news releases should be done the day prior to broadcast but there is great flexibility in radio broadcast media. News magazines should be contacted individually for interest in stories and submittal deadlines and formats. Colored photos are always a must to ensure consideration.

News releases should be short, use short words and deal with the key elements in the first two paragraphs: who, what, when, where, why, how. Head it with Municipal name and department, address, phone number, date, contact person, and "NEWS RELEASE" (AWWA, 1989).

3.7 Risk Communications

Storm water management facilities and conveyance structures are dangerous when there is flooding. People are injured or killed and property is damaged by flooding, erosion and pollution. One important community education and awareness role for urban storm water and floodplain managers is that of risk communication. Risk communication is the exchange of information between the storm water manager and the general or particular (such as floodplain residents) public concerning a particular hazard and what can be done and is being done to manage the hazard and its consequences (EPA, 1990).

Examples of risk based communication needs in a community pertaining to urban storm water management include such things as communicating effectively to citizens: floodplain hazards; water quality hazards; spill hazards; stream crossing hazards; risks of flooding from a levee or flood wall project; dam break hazards; hazards from the construction of regional structural controls; etc.

The Federal Emergency Management Agency's (FEMA) Community Rating System (CRS) is a methodology by which communities can reduce the flood insurance rates its citizens pay through proper storm water and floodplain risk management (FEMA, 1993). One of the categories under which points can be gained is in the area of Public Information. Under this category the community can gain credit if it promotes programs and activities such as:

- issuance and maintenance of elevation certificates which indicate flood hazard elevations for new construction;
- risk communication of flood hazards and availability of federal insurance;
- real estate agent disclosure of flood hazards;
- maintenance of flood information at a public library; and
- technical advice on flood hazards and flood protection measures.

The Corps of Engineers (as well as other agencies) is turning to "risk-based analysis" for the sizing and evaluation of flood damage reduction projects. In this way they have an opportunity to effectively communicate to the public the actual risks and uncertainties involved in the project and the likelihood of project failure or floods that overwhelm project capabilities (COE, 1993).

Risk communication is successful if it raises the level of understanding of the relevant issues or actions and satisfies those involved that they are adequately informed within the limits of available knowledge. Risk communication is very difficult to do correctly due both to the technical nature of the risks being communicated and the potential for both over- or under-reaction to the risks involved.

There are typical steps in the development of a generic risk communication program as developed by EPA (EPA, 1990). These are given in Table 3-8. The Risk Message Checklist, example given in Table 3-9, is a convenient list of questions to be asked whenever a community is considering construction or initiation of a program in which there are both hazards and benefits.

Table 3-8 Risk Communication Steps

1. Identify risk communication objectives for each step of a public awareness campaign or for each phase or aspect of a particular program.
2. Determine the information exchange needed to take place for each part decided in step one. Use a risk message checklist like the one in Table 3-9.
3. Identify the interests with whom information must be exchanged.
4. Develop appropriate risk messages for each targeted group or audience.
5. Identify the appropriate channels, techniques or means for communicating risks to various segments of the public.

After: EPA, 1990

People perceive risk in different ways. Some risks are seen as more acceptable than others. There are certain characteristics of risk perception that the storm water manager should keep in mind when deciding how to communicate to the public at large concerning risk related topics (Hance, et al., 1987). Table 3-10 gives some basic principles when considering risk as citizens perceive it.

If a crisis happens, be accessible; be accurate; and take responsibility for what is yours (AWWA, 1989). Tell the bad news right away. Explain what is confidential and why and tell all the facts you know but no more. Update frequently and regularly and monitor news coverage and correct any mistakes.

Table 3-9 Risk Message Checklist

Information About the Nature of Risks
- What are the hazards of concern and their characteristics?
- What is the probability of exposure to hazards?
- What is the hazard's distribution or location?
- What is the probability of each type of harm from the hazard?
- What are the sensitivities of the population groups to hazards?
- What is the total population risk?

Information About the Nature of the Benefits of the Program or Proposed Project
- What are the benefits associated with the program or proposed project and their characteristics?
- Will the benefits actually occur?
- Who benefits and in what way?
- How many people benefit and for how long?
- Does any group get disproportionate benefits?
- What is the total benefit?

Information on Alternatives
- Are there any alternatives to the hazard?
- What is the effectiveness of each alternative?
- What are the risks and hazards of alternatives and of doing nothing?
- What are the costs and how are they distributed?

Uncertainties About Risks
- What are the weaknesses in the available data?
- What are the inherent assumptions?
- How sensitive are the estimates to changes or inaccuracies in assumptions?
- How sensitive are decisions made to changes in assumptions?

Information on Management
- Who is responsible for decisions?
- What issues have legal importance?
- What constraints and resources are on the decisions?

After: EPA, 1990

Table 3-10 Basic Risk Principles

- Voluntary risks are more acceptable than imposed risks.
- Risks under individual control are more acceptable than those under government control.
- Risks that seem fair are more acceptable than those that seem unfair.
- Risk information that comes from a reputable source is better than information from a less trustworthy source.
- Risks that are "dreaded" are less acceptable than those that carry less dread.
- Risks that are unseen or come without warning are less acceptable than those which are detectable or come with warning.
- Physical distance from a site influences acceptability of risk.
- Rumor, misinformation, dispute and sheer volume of information all may interact to give a misperception of risk.

After: EPA, 1990

People who are affected by the impacts of decisions made by storm water managers expect some kind of mitigation. Most impacts affect neighborhoods, and local residents demand some sort of equivalent benefit for any perceived damages. Mitigation is most effective if it can be in some way tied to the project. For example, siting a detention pond in a neighborhood can be mitigated through the creation of a small neighborhood park in the area, especially if the neighborhood has young children. However, safety issues normally cannot be mitigated by anything less than reducing the unsafe aspect of a project (EPA, 1990). It is also true that most citizens see the present situation as risk free while any changes made will be closely questioned. For example, a channel widening project may alert citizens to the safety issues of a neighborhood creek while there were few concerns prior to the project. Many mitigation issues concern procedure as much as result. The questions of who makes the decisions, how they are made, and who has input are vital.

3.8 Technical Communications

The problem of communicating technical information to a non-technical public is a very real one encountered every day by storm water managers. Table 3-11 gives a checklist of steps for the development of a technical communication campaign.

In one city a woman called the storm water manager and, upon learning she had to pay a portion of an assessment to all homes within a watershed for a flood control facility, stated, "I want to know how I got in this watershed and I want out"! In another instance an individual builder called the local municipality to inquire about the cost of obtaining a "hydrograph" which was required by local ordinance.

Table 3-11 Technical Communication Checklist

1. *Anticipate the issues that will emerge.*
    - Use interviews and polls.
    - Identify studies that will be necessary to resolve these issues.
    - Develop an interim strategy for dealing with questions about these issues prior to study completion.
    - Identify and evaluate alternative strategies to mitigate potential impacts.
2. *Get participation in developing the study plan.*
    - Use an outside unbiased consultant.
    - Use a "scoping" process to identify issues from every stakeholder.
    - Use workshops to identify issues.
    - Use both technical citizens for technical input and non-technical citizens to get a "sanity check".
3. *Validate methodological assumptions.*
    - Use Table 3-8 or rely on workshops and round-table discussions.
4. *Invite public involvement in consultant selection.*
    - Reduces the sense of staff advocacy for a project.
5. *Provide technical assistance to the public.*
    - Use citizen education workshops prior to study evaluation.
    - Hire outside consultants to serve as advocates for the citizens.
6. *Use an outside body to review a technical study.*
    - Can be a technical citizen body or university staff.
    - Can be hired outside objective consultant.
    - Remember that an objective review might find the study inadequate and require parts of it to be redone.
7. *Present technical information in understandable language.*
    - Have non-technical personnel such as public relations experts review language in handouts, etc. and translate technical information into common language.
    - Use advisory groups to "pretest" news releases, reports, etc.

Source: EPA, 1990.

Managers often become experts in dealing with complex issues in simple ways. There are many pitfalls in this. Over simplification can lead to false assumptions of knowledge. There is a predisposition for the public to mistrust technical information. Also, there is a history of misuse of such information for many subjects leaving all parties assumed guilty. Unsupported rhetoric is often believed more than state-of-the-art studies. Who speaks is often more important than what is said.

Two goals of any storm water manager are building the credibility of technical information in the eyes of the public and improving the relevance of technical studies to public concerns (EPA, 1990). The fundamental way to do this is to create visibility and participation on the front end of a study (such as a capital improvement plan or flood study) to reduce resistance to the study results. How a study is done is as important as what the findings are. Key stakeholders can oversee the plans for the study, oversee the methodology and assumptions, participate in hiring technical expertise, and review the results.

References:

American Water Works Association, "Public Information", Denver CO, 1989.

Bellevue, Washington, "Stream Team Guidebook", Storm and Surface Water Utility, 1989.

Bernstein, G., "Meet the Press", Public Relations Journal, March, 1988.

Environmental Protection Agency, "Building Support for Increasing User Fees", EPA/430-09-89-006, 1989.

Environmental Protection Agency, "Sites for Our Solid Waste", EPA/530-SW-90-019, 1990.

Environmental Protection Agency, Office of Wastewater Management and Region V, "Urban Runoff Management Information/ Education Products", OWEC (EN-336), 401 M St. SW Washington, DC, 20460, (202) 260-8328, 1993a.

Environmental Protection Agency, "Guide to Federal Water Quality Programs and Information", EPA-230-B-93-001, 1993.

Federal Emergency Management Agency, "National Flood Insurance Program Community Rating System Coordinator's Manual", With Supplement 1, FIA 15, July, 1993.

Hance, B. J., C. Chess and P. M. Sandman, "Improving Dialogue with Communities: Risk Communication Manual for Government". Prepared for the New Jersey Dept. of Environmental Protection, Rutgers University, New Brunswick, NJ, 1987.

Harris, J., "Get the Most Out of News Conferences", Public Relations Journal, September, 1986.

Newsom, D., and A. Scott, "This is PR", Wadsworth Publishing, 1981.

Nichols, A., "Citizens Monitor Water Quality", Water Environment & Technology, March, 1992.

Ogden Environmental and Energy Services, Inc., "Public Education Plan for the City of Charlotte, NC - Draft", 1992.

Seattle Engineering Department, "Water You Doing", Seattle Municipal Building, 600 4th Ave., Seattle, WA, 98104-1879.

US Army Corps of Engineers, "Risk-based Analysis for Sizing and Performance Evaluation of Flood Damage Reduction Projects", EC 1105-2-XXXX (DRAFT), November, 1993.

University of Wisconsin, Keeping Current, Environmental Resources Center, Rm. 216 Agricultural Hall, Madison, WI (Sara M. Steele and Cathaleen Finley), March/ April, 1993.

Wilcox, D. L. and L. W. Nolte, "Public Relations Writing and Media Techniques", Harper and Row, NY, 1990.

Chapter 4 Storm Water Ordinances And Regulations

4.1 Introduction

Storm water ordinances and regulations are a basic tool or method available to local governments for the control and/or management of urban drainage and flood control problems. Specifically, ordinances or regulations can be used for floodplain management, on-site detention and retention, erosion and sediment control, development regulation, and water quality control and enhancement. There is a difference between ordinances, resolutions, regulations, policies etc. However, they are considered together in this chapter since the implementation and application of each is often similar. There are a number of regulatory drivers and resulting ordinance types used in municipalities throughout the country.

Passage of the National Flood Insurance Act of 1968 and subsequent amendments have provided an incentive for local governments to adopt floodplain regulations and ordinances. If a community wishes its citizens to be eligible for subsidized flood insurance and certain post flood disaster assistance it must meet minimum requirements of the National Flood Insurance Program (NFIP) which includes the adoption of floodplain regulations for designated flood hazard areas. Most local governments with designated flood hazards are a part of the NFIP.

With regard to erosion and sediment control some states have passed legislation requiring local governments to enact erosion and sediment control regulations. Maryland was one of the first states to require such measures and continues to be a leader in this area. Another example is the State of Georgia which passed an Erosion and Sedimentation Act of 1975 that requires the governing authority of each county and municipality to adopt a comprehensive erosion and sedimentation ordinance establishing procedures governing land disturbing activities. The State of South Carolina has passed a comprehensive Stormwater Management and Sediment Regulation Act. This act mandates that storm water management and water quality controls be enacted by all municipalities within the state. This is one of the most comprehensive storm water management state laws.

Federally funded area-wide water quality planning activities in the late 1970's and early 1980's performed pursuant to Section 208, pl 92-500, The Federal Water Quality Control Act, resulted in recommending erosion and sediment control ordinances. Also, Federal Environmental Protection Administration (EPA) grants for sewage treatment and interceptor construction require municipalities to adopt erosion and sediment control ordinances as a grant condition. Recently, EPA has required municipalities to comply with water quality regulations related to the Clean Water Act.

The 1987 Water Quality Act and implementing regulations have led many municipalities into regulatory control of illicit discharges, illegal connections, industrial discharges and a variety of other non-point discharge related controls and ordinances.

All of these and other activities have resulted in the development and adoption of numerous municipal ordinances, regulations, and policies related to storm water management.

4.2 Legal Basis And Considerations

Ordinances, policies, regulations and even city codes and charters contain within their language applications of basic principles of law. These principles are based on both common law, legislation and case law. Therefore, understanding the legal basis of ordinances and regulations is important to an understanding of how to apply and amend them. Should a municipality be challenged legally a clear understanding of the legal basis for decisions will lend credibility and viability to staff policy development and implementation.

Legal consideration of storm water and flood water (part of "water law") has developed across the United States based on property rights law in slightly different ways and usually in response to a narrowly defined "crisis". Thus it is not to be considered a seamless whole but a patchwork of cases and judgments which together can comprise both individually contradictory findings and generally consistent applications.

In the course of defining individual property rights, the right of society as a whole and of various governmental bodies is also defined. This definition has followed along separate lines depending on the various locations and states of water: ground water law, watercourse law and diffused surface water law (Cox, 1982). While the distinction is sometimes blurred, different decisions for and against municipalities have been based on these separate considerations.

Watercourse Law

Watercourse law has developed differently in different groups of states owing to their different concerns and history of cases. This grouping has led to the development of two basic doctrines within watercourse law: the riparian doctrine generally in the more humid East and the doctrine of prior appropriation generally in the arid West.

Riparian Doctrine - The basic concept of the riparian doctrine is that private water rights exist as an incidence of the ownership of land bordered or crossed by watercourses. All owners of such riparian land (defined as a contiguous tract of land in contact with a stream and within the same watershed as the stream) have a constitutionally protected right to certain use of water in the stream while non-riparian land owners generally do not. That right cannot be "taken" without due process. A riparian right is reciprocal to other riparian rights on the same watercourse and all rights along a watercourse must be respected equally.

In terms of water supply the riparian doctrine does not fix amounts of water rights but allows use to vary with conditions. Unexercised rights cannot be lost by prescription and continue to be attached to property, developed or not.

Historically the riparian rights doctrine was exercised through the theory of natural flow. According to this theory each downstream land owner was entitled to the natural flow of a stream except as diminished by domestic use. As water supplies changed and were made public this theory of the application of the riparian doctrine gave way to the reasonable use theory of riparian rights.

In the reasonable use theory each land owner has the right to make use of any water, provided that his use is reasonable in relation to the use of other riparian land owners. The application of this reasonable use theory required considerable interpretation for individual cases and therefore much ambiguity and seeming (and real) contradiction in similar court cases (APWA, 1981). In application for flood control a land owner has the right to protect his land from flooding but only if this protection does not harm another riparian land owner. This then has implications in: the construction of flood walls, diking land which floods upstream property, the undersizing of conveyance structures at crossings, filling in floodplains to remove property from flooding, draining of wetlands, etc.

While engineers have looked at development in terms of protecting the upstream land owner from on site flooding, judges have increasingly taken an approach which considers the downstream property owner's rights not to have the watercourse altered to his detriment.

Prior Appropriation Doctrine - The prior appropriation doctrine evolves out of a scarcity of water resources related to common law application in mining claims and irrigation. Like the riparian doctrine the appropriative right is a property right. But it differs considerably in application. The basic premise of the appropriation doctrine is "first in time - first in right". Rights are not equal nor attached inseparably to the land but ranked in a hierarchy established by the date beneficial uses were first initiated and tied to place and type of use, not location. A long period of non-use presumes abandonment and the allowance of transfer of that water right.

In relation to storm water management for flood control and water quality the prior appropriation doctrine has more indirect applications. Flood control activities should not conflict with established water rights. The scarcity of water has implications on water use and reuse in the west... both of which could be covered under NPDES storm water and other permitting programs. As true water resources grow scarcer in the east there may be instances of application of the prior appropriation doctrine beyond its present agriculture association of today.

Diffused Surface Water Law

Diffused surface water is the term applied to runoff after it has hit the ground but before it enters a defined watercourse. With few exceptions the right to use diffused surface water belongs to the property on which it is found. Three doctrines have been developed which apply to diffused surface water. They have been in predominance in about a third of the states each. They are termed the common enemy doctrine, the civil law doctrine and the reasonable use doctrine. The reasonable use doctrine is gaining more popularity in recent years though it was the last to arise historically (APWA, 1981).

Common Enemy Doctrine - Under this property law-based rule a land owner can do anything he wants with the water on his land including diversion, concentration, diking, etc. It is a strict interpretation of property rights. Since water is a common enemy each property owner is required to protect himself as best he can... might makes right. A more recent modernization of this rule has limited the modifications of water flow to those which are incidental to ordinary use, protection, or development of the land. In any case the lower property is required to accept the flow from the upper property even with increased volume, velocity and peak.

Civil Law Doctrine - Under this property law-based rule a land owner should be liable for downstream injury caused by interference with the natural flow of water from his property. However, the upper property has an implied easement for the natural flow across the lower property and is termed the "dominant estate". The key word is "natural", and interpretation of that word has led to many court cases.

Reasonable Use Doctrine - This rule is a compromise rule which falls somewhere between the other two with regard to whether alteration of the natural drainage amount and pattern is lawful. In its application diversion and concentration of runoff is an expected part of development but is limited to some reasonable amount and proper engineering care and design is expected. Thus there is a case-by-case application balancing the utility with the harm of the development.

Liability And Damages

The application of the reasonable use rule is different from the other two in that it is based on tort law rather than property law. In tort law liability is based on negligence. There is no need to show negligence but simple injury or damage under property law.

Individual Liability - Defenses under property law include acts of God, third party activities which contributed to the injury, or that the plaintiff himself is at fault. Under the reasonable use rule, factors considered in judgments include such things as: the social value of the activity, its suitability for the location in question, the difficulty in preventing damage, the foreseeability of the harm, the motive with which the developer acted, and the impact on the activity if compensation is required for resulting injury (Cox, 1982, APWA, 1981). The extent of damages is based on the extent of the injury, the character of the injury, the social value of the injured activity, its own suitability to the location in question, and the difficulty for the injured party to avoid harm.

Municipal Liability - Municipalities are not ordinarily liable for failure to provide drainage systems or poor system planning. Providing drainage systems is a legislative power with which courts do not ordinarily interfere. Liability for construction as opposed to design is a much more common cause of municipal liability. This is also normally true of poor or negligent maintenance of constructed facilities. Citizens should have a reasonable expectation and reliance on municipal constructed systems to operate as designed and planned. Municipalities have been found liable for concentrating and funneling drainage on to private property and for trapping water by taking away natural outlets. They have also been challenged in court in cases where inadequate openings were supplied for stream crossings. Also, theories of nuisance are being applied to cities to avoid the need to prove negligence. Repeated flooding, sewer backups, and urban development which renders existing systems inadequate have all been used in nuisance cases with municipalities found liable.

It is clear that the "jury is not in" on a municipality's or individual's liability for storm water decisions. However, it is clear that the "reasonable use" doctrine is being widely interpreted in recent cases. Jury awards have been given for concentration, increase, or diversion of flows which may have passed the reasonable use test years ago. Cities and counties have been named as co-defendants in a number of areas. A changing expectation of storm water service levels is growing and spreading. Even developing and estimating a detailed estimate of the risk in a design may not protect from liability (Lewis, 1992, Kenworth, 1972). Some recent examples include the following.

- Assuming even one-time maintenance of a certain site may imply a continuing responsibility.
- Not maintaining sites where public water makes up a significant portion of the total flow may also open a municipality to some level of liability.
- Allowing accumulated flow increases which damage downstream properties may be seen as negligent even if a municipality follows a standard practice of development controls.
- Drainage structures which may back water into upstream property, even if properly designed according to common practice, and even for storms in excess of design may be seen as inappropriate by the courts.
- Municipalities which provide certain types of design assistance to developers (such as establishing flow elevations along creeks) may incur design liability.

"Takings" - The definition of a "taking" of private land by a municipality for storm water control or flood control purposes is often a murky issue. Cases have been decided which seem to contradict each other. Generally:

- ordinances and regulations adopted for a valid public purpose may substantially reduce land values without a taking;
- the impact of regulations on the entire land must be reviewed before a judgment of a taking is rendered;
- no property owner has a right to create a nuisance or threaten public safety; and
- regulations are, in general, a taking if they deny all use or all economic use of an entire property including investment backed expectations.

Municipalities can avoid takings in a number of ways (ASFPM, 1990):
- providing variances or special permit abilities in ordinances to deal with difficult cases;
- emphasizing health and safety, avoidance of pollution and prevention of nuisances in written findings of investigations or hearings in permit denials;
- tie regulations to state or federal programs where possible;
- apply large lot zoning to areas where intense development would be seen as detrimental (such as floodplains or sensitive areas);
- document with great care the need for regulations, the denial of permits in areas where land values are very high;
- encourage preapplication meetings where mutually agreeable designs and developments can be envisioned prior to design investment... be flexible;
- apply regulations consistently and fairly maximizing the opportunities for public hearings and minimizing decision making based on politics or non-technical issues;
- if a development moratorium is adopted do so for a clearly stated and legitimate reason and for a fixed period;
- coordinate regulatory, tax and public works policies to ensure that the fiscal burden on land owners for community services is consistent with permitted uses;
- apply, in extreme cases, transferrable development rights to help relieve burdens on land owners; and
- use acquisition rather than regulation to protect public interests and to avoid single land owners or groups of land owners bearing the cost for a general public good. Past examples of the use of acquisition include Tulsa (municipal purchase of homes when they come up for sale) and Birmingham (Federal home purchase and relocation). In other cases attempted residential condemnation proceedings were so politically unpopular that viable projects were eventually abandoned.

Water Quality And Water Law

Many states and localities have enacted water quality control legislation which is only partly based on the various doctrines and rules for watercourse law and diffuse surface water law. Florida's or Virginia's approaches to storm water control are good examples of such legislation. For example, the reasonable use concept of the riparian doctrine can serve as a limit on quality changes as well as quantity changes. However, the application of the various water law theories has limited use for water pollution. Also, most legal actions involving water pollution involve violations of state or federal laws and are not handled in court by individuals. Most individual suits in the water quality area are not based on property law but on tort law. It is not uncommon for such suits to oppose pollution on the basis that it constitutes a nuisance or trespass or that it results from negligent conduct.

4.3 Municipal Ordinances

Storm water management ordinances for municipalities normally have four major components and today normally address both storm water quantity and quality (EPA Region V, 1990).

Legal Authority And Context - Depending on the type of state government, municipalities may or may not have the derived authority to control their storm water management quantity and quality. Chapter 1 describes briefly basic derived legal authority for storm water management. Often storm water provisions are scattered throughout other codes. In these cases municipalities sometimes develop a compendium of all such related codes and publish, informally, such a collection for the aid of developers and engineers (not to mention regulators) in the municipality.

The context for whom and for what the ordinances is written is quite important. Grandfather and severability clauses are important in case any portion of the ordinance is invalidated.

Technical Basis - The ordinance should provide clear technically based and measurable performance standards rather than design criteria standards. For example, "control the 2- and 10-year storm such that the post-development peak is equal to or less than the pre-development 2- and 10-year peaks" is preferable to a set of plans and specifications on how detention ponds must look.

Administrative Apparatus - A clearly identified administrator of each ordinance should be identified in the ordinance. If procedures under the ordinance are required, checklists, forms and steps should be clearly laid out and provision made within the local government for efficient implementation of the requirements. These should be included in implementing regulations or policy documents to ease the revision process.

Enforcement Provisions - Enforcement must be fair, consistent, stepped, and it must be swift. Several steps of enforcement beginning with a warning and ending with a penalty large enough to stop and punish any would-be offender should be included. The mechanism and will-power to carry out such enforcement actions must be provided with the ordinance. It is better not to have such an ordinance than to have it and not be able to carry out its provisions. The ability to obtain and execute a stop work order is important in situations where development is harming the environment or causing a health or safety hazard. Rigorously training and then empowering field inspectors as the first line of defense is the best method for effective enforcement.

Special Water Quality Considerations - The drafting of an ordinance for storm water quality involves considerations beyond those for storm water quantity such as the following (EPA Region V, 1990, Ogden, 1992).
- Many municipalities have, but have not taken advantage of, provisions in charters or codes for public health protection which can include storm water quality provisions. These should be spelled out by the municipality or enabling legislation developed to clearly indicate the authority for storm water quality. The legal authority for storm water must be sufficient to prohibit illicit discharges and illegal dumping, allow for inspections and require compliance with any program or policy provisions.
- Technical criteria for storm water quality should be carefully thought out and spelled out. Capture of the first one or one-half inch of runoff can be shown to make sense in most climates and for most urban sized basins. However, other provisions (for example, prohibition of curb and gutter) must have appropriate technical backing, program, and political support before extension of pilot or literature results to the whole municipality is implemented.

- Technical analysis required, even tacitly, within an ordinance for general development should not be beyond the means of the general civil engineering public to perform. This can be accomplished through both education and sound design criteria manuals.
- Administration of a storm water quality ordinance must be based on easily measurable performance criteria and not be arbitrary or allow too much room for "judgment" when such judgment is nearly impossible to provide by the available staff (or by any staff).
- Violations of storm water quality criteria are more difficult to prove without exhaustive monitoring programs. The provision of self-monitoring for industry may be applicable in some cases. In most cases the fact that a given BMP is being maintained to physical standards should be sufficient. In other words, it is easier and more fair to hold land owners and developers to a physical design criteria than to a criteria which must be stated and enforced in terms of measured pollution reduction. Therefore it is incumbent on ordinance writers to attempt to link pollution reduction amounts to physical designs. These designs and their maintenance then become the measure of compliance, not the chemical condition of the receiving water. Adjustments can always be made in the criteria in the future if acceptable results are not obtained.
- Storm water quality ordinances must allow room for both innovation and regionalization or physical treatment. This may be difficult but sufficient safeguards can be built in terms of extra monitoring to ensure performance of unknown and untried BMPs or designs.

Detention - With regard to drainage and flood control, many local governments are requiring on-site detention facilities to reduce downstream runoff peaks because they feel it is good storm water management practice to do so. Residents downstream of new developments are increasingly questioning the wisdom of allowing upstream developments to cast more water downstream for them to deal with. Such concern puts pressure on local governments to require on-site detention in new developments. In addition, some states have passed laws related to use of detention facilities. The State of Pennsylvania passed a Storm Water Management Act in 1978. This act requires the development of storm water management plans for all watersheds within the State. Section 13 of this law entitled, Duty of Persons Engaged in the Development of Land, states:

Any landowner and any person engaged in the alteration or development of land which may affect storm water runoff characteristics shall implement such measures consistent with the provisions of the applicable watershed storm water plan as are reasonably necessary to prevent injury to health, safety or other property. Such measures shall include such actions as are required:

(1) to assure that the maximum rate of storm water runoff is no greater after development than prior to development activities; or

(2) to manage the quantity, velocity and direction of resulting storm water runoff in a manner which otherwise adequately protects health and property from possible injury.

Implementation of item (1) almost always involves the use of detention storage facilities. Such a requirement is the major focus of many municipal storm water ordinances that have been adopted. This has resulted in the proliferation of many small detention storage facilities in many urban areas.

As a result of the increased use of ordinances and regulations to control quantity and quality problems related to storm runoff, municipalities have gained experience with implementing these regulations and some areas are experiencing the positive and negative effects resulting from their implementation. From this experience, professionals involved with urban storm water management are posing several interesting questions. Is the adoption of ordinances which generally require the use of detention facilities on all developments the answer to flooding and drainage problems? Do sediment control regulations adequately protect water resources? What engineering and technical methods should be included in a community's storm water management program? What controls should be included in storm water quantity and quality control?

If a detention ordinance is implemented three additional factors should be present. First, the municipality should have a mechanism to encourage and implement regional storage facilities. Second, the detention pond should be designed to control downstream impacts, not just peak flows at the outlet. Third, a comprehensive inspection and maintenance program should be implemented.

4.4 Drafting Local Ordinances And Regulations

As stated earlier, storm water management ordinances can be used to address a variety of subjects. Separate ordinances can be written for each of the areas identified or a single comprehensive ordinance can be adopted to address all storm water related problems. For example, Austin, TX adopted a comprehensive type "watersheds" ordinance which covers site development and subdivision regulations, watershed protection measures, maintenance, and enforcement (Austin, 1986). For the purposes of this section, a storm water management ordinance will refer to a comprehensive ordinance that sets forth a local municipality's policies related to drainage, flood control, and water quality aspects as well as the legal framework for permitting actual implementation of the controls. A storm water management program will include the ordinance and supporting program elements (i.e., technical tools needed, financing strategy, engineering design manuals).

Before any municipality drafts and adopts a storm water management ordinance, some preliminary study and evaluation should be made. The municipality should identify the problem areas that the ordinance must consider and basic decisions should be made concerning the municipality's philosophy in dealing with storm water management problems. The general format of the storm water management ordinance should be developed to effectively reflect the goals and objectives of the municipality.

Table 4-1 and the discussion that follows describes a five-part process involved in drafting a storm water management ordinance.

Table 4-1 Process For Drafting Storm Water Management Ordinances

Part 1 - Identify Problems And Issues
Part 2 - Formulate Objectives
Part 3 - Develop Policies
Part 4 - Draft the Storm Water Management Ordinance
Part 5 - Develop Storm Water Management Design Manual

Identify Problems And Issues

Before any aspect of a storm water management program is considered, the problems and issues related to storm water quantity and quality must be determined. Identification of the problems will define in general terms the scope and extent of the storm water management program. Although each municipality should identify its own problems and issues, listed below are some examples that have been addressed in several different municipalities.
- control of local storm water management problems
- flood control
- water quality enhancement
- erosion and sediment control

- economic efficiency in storm water system design
- improving engineering design and analysis
- protection for present and future development
- equity in storm water design and implementation
- control of development within floodplain areas
- control of runoff from proposed developments
- maintenance of storm water management facilities
- funding sources for storm water management programs
- controlling nuisance problems within the drainage system
- enhancement of the local environment
- addressing citizen complaints and problems
- providing training and education related to storm water management
- efficiency in review and inspection procedures
- providing public participation in program development

The local municipality must determine which problems they need to include in their storm water management program and then proceed with the formulation of objectives.

Formulation Of Objectives

Many times in developing storm water management ordinances and programs, very little consideration is given to stating objectives to be accomplished by the program, keeping the objectives up-to-date as the program is developed, or evaluating the adequacy and making appropriate changes to the objectives as the program is developed and implemented. The results of an extensive study of urban floodplain programs in five metropolitan areas indicated the important role objectives can play (Debo, 1975). Many of the persons interviewed as part of this study gave what they considered to be the objectives of their storm water management program but also stated this represented their opinion and not an official position of the local government. In contrast to what was done, when the subject of objectives was brought up during interviews, both technical and nontechnical personnel felt the establishment of a definite set of objectives is not only important but should be one of the first steps in formulating a storm water management program. Furthermore, many of those interviewed felt that objectives should be used to give direction and feedback to the overall program. Experience in several municipalities since this study was completed confirms the results and further indicates the importance of establishing definite objectives for the storm water management program.

For example, Charlotte, NC developed a complete set of policy statements through a process of citizen, consultant and staff interaction over a period of many months. Louisville MSD established policy goals in each key area of storm water management in the development of its program. These policy goals serve to guide future development of both ordinances and program elements. Prince George's County, MD is a good example of overall goals for a storm water program (Prince Georges County, 1984):

"...to protect, maintain, and enhance the public health, safety and general welfare by establishing minimum requirements and procedures to control the adverse results of increased stormwater runoff associated with land development. Proper management of stormwater runoff will minimize damage to public and private property, reduce the effects of development on land and stream channel erosion, assist in the attainment and maintenance of water quality standards, reduce local flooding, and maintain nearly as possible, the pre-development runoff characteristics of the area."

It is also important that all personnel be informed of the storm water management objectives and when changes are made. Such information should be published and promulgated by the municipality. In the municipalities studied, personnel within the local governments as well as outside consultants complained they did not know about several aspects of the total program and were not kept informed of program changes (Debo, 1975). One means of keeping everyone up-to-date would be for the local government to publish a newsletter documenting changes and other aspects of their storm water management program.

The opinions of numerous elected officials were obtained as part of this study concerning different aspects of their local storm water management program. All those interviewed said that objectives were essential. They felt objectives help evaluate long term planning, help interpret regulations and ordinances, and give direction to the program. Some elected officials said objectives should be separately stated but contained in the ordinance while others felt they should be contained in another document such as a general plan or resolution. Others said objectives were inherent in any document and need not be written separately. The politicians did not offer any specific objectives but several stated that they should be specific and simply stated in contrast to some of the more general and elaborate ones contained in many planning documents. Most of the politicians saw no particular problems in arriving at stated objectives for their storm water management programs, although they could not give specific reasons why their present program existed without some stated objectives.

Without properly defined objectives, both known to the people involved in the storm water management program and actually representing the local municipality's view, the major effort of the personnel involved with the program is with the day-to-day problems and their immediate solution. Little if any time is given to any of the following important aspects of the program, which would have an impact on the implementation of the program.
1. Long-range implications of present actions.
2. The direction and scope of the storm water management program.
3. The effectiveness of implementing the different program elements.
4. Changes that should be made in the program to achieve desired objectives.
5. Determination of what is being accomplished by the present program and what the local community wants to accomplish.

It should be pointed out that stating objectives will not guarantee that the above five aspects will receive proper consideration, but it should at least bring some of these aspects to the attention of the policy makers and those implementing the storm water management program. Thus it is important to have stated objectives for the storm water management program rather than inherent objectives. Stated objectives should enhance evaluations of different aspects of the program and lessen the chance of misinterpretations and forgetting what the objectives are.

It is often the case that "specialty" ordinances are necessary. In this case there will be narrowly defined objectives to meet specific needs in support of a more comprehensive program. These special ordinances must fit into the framework of an overall storm water program. Some municipalities, for example, have instituted detention ordinances without commensurate programs for inspection, enforcement and maintenance. The result has been a general failure of detention due to clogging, structural failure and vandalism. There have been municipalities which have actually begun abandonment proceedings for detention to eliminate redundant or ineffective detention ponds. Examples of such specialty ordinances include such topics as lawn and fertilizer control (Shoreview, 1985, 1988), a wetlands overlay district (Wayzata, 1988), a hazardous materials ordinance for spill control (Louisville MSD, 1988) or the plethora of floodplain management regulations.

Developing Policies

The storm water management ordinance must be officially adopted by a governing body before it can be legally applied. As a result, the political process becomes the controlling force behind such a document. Therefore, it should be designed so that it can be understood by elected officials and aid them in the execution of their responsibility.

Elected officials should not be dealing with engineering criteria or design standards. Also most elected officials are not administrators and should not be concerned with the detailed administrative aspects of a storm water management program except as they relate to budget control. Elected officials should be more concerned with determining the overall policies within their jurisdiction. Thus to make the storm water management ordinance a viable part of the political process, it should serve as a policy statement for the municipality. Technical and engineering details related to the program should be included in other documents.

Some administrators prefer ordinances which are very detailed and attempt to cover all problems that may arise. In addition, they tend to include engineering criteria and design standards in order to encourage uniformity. The administrators who must deal with the public feel they can be more firm in enforcing the provisions of the storm water management program if the provisions are written into law. They contend if many members of the public, or a few influential ones, complain to elected officials, these officials may find it easier to use their political pressure to allow exceptions if a specific point of contention is not written into law. In contrast, other administrators feel that storm water management ordinances should be limited to stating specific policies. Engineering criteria and technical details should be contained in other documents in order to fully discuss these criteria and details and keep them up-to-date. Elected officials could then refer technical questions to the engineering or public works departments and spend more time and effort dealing with policy matters. In order to prevent the engineering or public works departments from becoming completely autonomous from the public and political influence, engineering documents should be subject to the approval of elected officials, without formally being enacted into law. Or they may be adopted, by reference, in the ordinance. This would encourage some uniformity and give administrators a ready reference for dealing with problems and combating political pressures to make exceptions. Bellevue, Washington used the "by reference" method to adopt a storm water master plan and system map (Bellevue, 1974).

Generally, the ordinance and supporting documents should state "why" and "what" should be accomplished. Technical manuals and other guidance should provide the "how" information. Ordinances should give performance-based criteria but leave the means and methods of accomplishing these criteria up to other more technical sources. For example, a detention ordinance may give the design criteria (such as: "control the 2-year and 10-year storms to the predeveloped peak flow values at the outlet and to a designated point downstream") while the design manual gives analysis and design specifics.

Some examples of policies that should be considered in a storm water management ordinance are the following.

- Which aspects of the storm water management ordinance should address frequent flood events (i.e., 10-year flood), and which should address rare events (i.e., 100-year flood), and which aspects should address both types of flood events?
- Should the ordinance provisions be applied only in large drainage areas or also in small ones, and if so how small?
- Should the intent of the ordinance be to pass the increased runoff from new developments downstream or control runoff on-site?
- Should the regulations and standards of the storm water management program vary from one part of the municipality to another or be uniform throughout the municipality?

- Is the local government going to maintain the entire drainage system, certain portions of the system, or require private maintenance?
- Should flood storage facilities be centralized or decentralized, publicly owned or privately owned?
- What restrictions should be placed on development related to flood control and control of drainage problems?
- What erosion control and water quality enhancement practices should be required by the ordinance?

These and other questions must be answered in order to develop policies that can be included in the storm water management ordinance. Sometimes this can be done by obtaining a consensus from the local politicians, personnel from the planning, engineering, or public works department. Other times technical studies may be needed to determine appropriate policies to accomplish specific tasks. Following is an example of the results of a study which were used to develop a policy concerning downstream limits for hydraulic analysis of detention storage facilities.

Detention Analysis Example

After years of requiring engineers to analyze peak flows from proposed developments at the exit to these developments, municipalities are starting to realize that limiting the hydraulic analysis of storm water management facilities to the exit of the proposed development or detention site is not sufficient to prevent increased downstream flooding and drainage problems. Thus there is a need for a policy to address this problem which could be incorporated into local ordinances and regulations. What is agreed is that hydraulic analysis should extend downstream from proposed developments but how far downstream the limits of this analysis should extend is difficult to determine. Requiring that all analysis extend a standard distance downstream (i.e., one mile) may not be a sufficient distance for a very large development or may be much too far for very small developments. It must be remembered that any required analysis will cost money that must eventually be paid in higher development costs. Thus some procedure needs to be developed to account for developments of different size, location, and land use density. The policy should also not require expensive engineering and technical studies that will be a burden to the development process.

A study completed for the City and County of Greenville, South Carolina and Raleigh, NC demonstrates how such a policy could be developed (Debo and Reese, 1992). This study used a hydrologic-hydraulic computer model to analyze the downstream effects of storm runoff from developments of different size, shape, physical characteristics, and location within larger drainage basins. This study also examined different size flood events and different types of downstream drainage systems. The results of numerous computer runs of watersheds located throughout the United States were analyzed and Figure 4-1 shows a typical summary of these results. It can be seen from this figure that the effects of the development process stabilizes at the point where the proposed development represents from 5 to 10 percent of the drainage area, depending on the size of the development and the amount of increased impervious area. This analysis was then used as the basis for developing the following policy statement that has been incorporated in the Greenville ordinance and several other municipal ordinances. This policy is included in the model ordinance given in Appendix A at the end of this chapter.

"In determining downstream effects from storm water management structures and the development, hydrologic-hydraulic engineering studies shall extend downstream to a point where the proposed development represents less than ten (10) percent of the total watershed draining to that point."

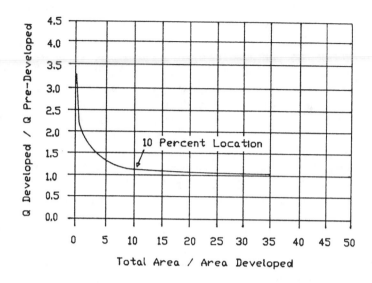

Figure 4-1 Determining Downstream Analysis Limits

Similar computer model analyses can be effectively used to develop other reasonable and adequate policies which can provide storm water management facilities designed to protect local residents without undue cost to the development process. Modeling the effects of different policies can provide valuable information in the policy development process.

Drafting The Ordinance

To be an effective part of the municipality's storm water management program, the initial draft of the storm water management ordinance should be a team effort. The team should consist of representatives from the engineering, public works, planning, environmental management, legal and any other affected departments. All of these departments will play a significant role in solving local flooding, drainage and water quality problems and excluding them from active participation in the drafting of the ordinance would be a major mistake. It should be remembered that the ordinance, if properly drafted and implemented, will play a major role in guiding the entire storm water management program for the municipality. If the final ordinance is not acceptable to any of these departments, its prospects for implementation are usually quite low. The team effort in drafting the ordinance seems quite logical but in many municipalities, ordinances have been drafted by only one or two departments without consulting the other departments.

Storm water management ordinances will eventually affect, either directly or indirectly, many different interests groups in the local community including citizens, developers, builders, realtors, investors, politicians, environmentalists, consulting engineers, landscape architects, surveyors, and local government employees. Some groups will be involved in the implementation and enforcement of the ordinance while others will have to comply with specific ordinance provisions. In order to get as much local support as possible, it will be necessary to involve representatives from all these groups in the formulation and drafting of the ordinance. Public participation should be integrated into the initial phases of the process (i.e., when problems and issues are established) and not used just to endorse the final draft of the ordinance.

The involvement of local "stakeholders" is usually important to obtain a "buy in" to the process, and to help avoid regulatory language which may prove to be unworkable or unreasonable. Grass roots consensus is normally preferable to top-down imposition of regulations. A balance of both development and business interests and environmental or homeowner groups allows an even consideration of all aspects of policies. The use of "Blue Ribbon Committees" has worked quite well in many municipalities. Under this arrangement, representatives from local interest groups are placed on a committee whose function is to review and comment on proposed provisions of the ordinance. The committee also provides a means of public input into the process of drafting the ordinance. The final ordinance then goes to the local governing body with the support of this committee.

These storm water ad hoc task forces or advisory committees can have varying levels of authority and responsibility depending on how decisions are made within any given community. In one mid-west city the advisory board retained great authority in the development of policies and presented an independent report to council recommending the final policy statements. In another eastern city a group provided only advice to the staff and had little say in the final ordinance.

Many times municipalities will hastily adopt a storm water management ordinance in order to conform with Federal or state mandates. To comply with these mandates municipalities may be tempted to obtain ordinances from nearby municipalities or even municipalities thousands of miles away. The local municipality then changes the names in the ordinance to fit local names and adopts the ordinance. It seems quite obvious that flood, drainage and water quality problems in Los Angeles, California would be quite different from those in Tampa, Florida or Atlanta, Georgia. Thus, an ordinance drafted for one area may not be appropriate to deal with the problems in another area. This distinction might not be quite so obvious with two municipalities that are geographically very close. However, when using an ordinance as a policy statement, one has to consider more than just the physical differences between municipalities, as each may have very different goals and objectives, local laws, or past flooding and water quality practices which might affect the specific provisions of an acceptable ordinance.

There are also numerous ordinances available in the storm water management area such as that developed by Maryland (1983) or northern Illinois (IDOT, undated) to assist its local counties and municipalities by providing minimums for their own program's development. Care and judgment should be exercised when using any model ordinance. Because model ordinances are written to cover many aspects of storm water management, they are general in nature and not aimed at specific local problems. The Maryland case is a good example in that it required the use of control measures including infiltration practices. Many of those communities which required infiltration practices did not have the technical or organizational wherewithal to carry out the provisions of their own ordinances resulting in a very high failure rate of the infiltration practices (Galli, 1992, Schueler, et al., 1992).

While model ordinances and ordinances from other areas should be evaluated, each community should draft its own ordinance tailored to its goals, objectives and problems. Municipalities would be wise to spend adequate time in formulating and drafting the ordinance before it is adopted. Otherwise, they may find that the ordinance must be continually revised or amended -- a process that would eventually erode the public's confidence in the municipality's ability to deal with storm water management problems.

If a municipality does not have an engineering and/or a planning department that can draft the ordinance, they should seek professional assistance from another level of government (regional, state, or Federal) or employ the services of an experienced consulting engineering or planning firm. Obtaining assistance from a professional outside of the municipal government is usually recommended in order to provide some unbiased input to the ordinance development process. It is important to obtain the proper professional assistance during the drafting of the ordinance in order to prevent as many problems as possible.

Logically, development of a local ordinance using a model ordinance should follow these steps:

- develop the means to provide input from politicians, local citizens, technical groups and others interested in being involved;
- identify and characterize the storm water management problems in the local area;
- establish goals and objectives for the ordinance;
- examine each provision of the model ordinance to determine its applicability to the local area;
- eliminate those provisions that do not apply and make revisions where necessary to adapt the model provisions to local needs (this may also include adding some new provisions);
- draft the local ordinance using the revised provisions;
- have a legal review made of the ordinance to ensure its compatibility with local, state, and Federal laws. In addition, the legal staff should be involved during development of the ordinance to avoid any legal problems.

Appendix A at the end of this chapter contains an example storm water management ordinance that is tailored after similar ordinances that have been used by several municipalities in the southeast section of the United States. This ordinance only provides an example. Many changes and additions may be needed before it would be an adequate ordinance for a particular municipality.

Storm Water Management Design Manual

If the ordinance is designed to function as a policy statement then the municipality should publish a manual which would contain the engineering criteria and design standards needed for implementation of the storm water management program. This manual would contain the technical information needed for both the flood and drainage aspects, and the water quality control aspects of the program.

If it is desired that the manual have the force of law then it must be either incorporated by reference in the original enactment (ordinance) or adopted at some later time by the governing body as a whole. This will necessitate that the municipality formally adopt any changes to the manual after it is adopted. In most cases the storm water management design manual would not be officially enacted into law by the municipality but could be subject to its approval and kept in a form such that it could be continually updated and changed to conform to the latest criteria and procedures. One of the earlier comprehensive manuals, developed by the Denver Urban Drainage District, uses this procedure and publishes the manual in loose-leaf notebooks so that they can be easily updated (Storm Drainage Criteria Manual, 1969). Following is a quotation from this manual which further emphasizes the need for updating such a manual.

"A compilation of engineering criteria such as the Urban Drainage Criteria Manual is a dynamic rather than static volume and needs to be reviewed and updated to keep it abreast of developments in the important and rapidly expanding field of urban storm drainage. It is the intent of those responsible for the conception and development of the Manual to periodically issue revisions to the Manual which incorporate new data, methods or criteria and such other information as may be deemed appropriate."

This manual has been updated through the years with several revisions.

The major thrust of a storm water management design manual is to present criteria and examples such that a design engineer can determine generally what is expected by the municipality in different flood, drainage, and water quality designs. The manual should not be a limiting document forcing the engineer to use standard designs and procedures, but instead should present performance standards. Numerous design examples should be given. The manual should have the connotation of a document that suggests and informs rather than limits or prescribes. The manual should not replace good engineering judgment.

The municipality's engineering department should be assigned the responsibility of keeping the storm water management design manual up-to-date. Periodic reviews of this document should be made by elected officials or a special committee selected for this purpose (i.e., every five years) or when major changes in the manual occur. This would enable the elected officials to be kept informed of engineering and technical changes which might affect stated policies and objectives, and also provide a check to ensure a sound basis for a given update. If needed, elected officials or committee members could obtain outside opinions on engineering and technical matters to augment those provided by the local engineering department.

Storm water management design manuals have recently been developed for several municipalities in North and South Carolina. As an example of the general content of these manuals, Table 4-2 gives the table of contents for the Greenville, South Carolina Storm Water Management Design Manual.

Table 4-2
Storm Water Management Design Manual Greenville, South Carolina

Table Of Contents

Chapter 1 - Introduction
Chapter 2 - Storm Water Management Ordinance
Chapter 3 - Data Availability
Chapter 4 - Documentation
Chapter 5 - Hydrology
Chapter 6 - Storm Drainage Systems
Chapter 7 - Design of Culverts
Chapter 8 - Open Channel Hydraulics
Chapter 9 - Storage Facilities
Chapter 10 - Energy Dissipation
Chapter 11 - Infiltration Facilities Design
Chapter 12 - Best Management Practices

4.5 Flexibility In A Storm Water Management Program

One of the initial steps in formulating a storm water management program is to evaluate thoroughly the general policies and objectives to be sought by the program, and whether a "flexible" or "rigid" ordinance should be used to implement the program. A "flexible ordinance" would be limited to stating the general policies of the storm water management program and allows judgment to be used to determine the best solution for given storm water management problems. "Rigid ordinances" are very specific both technically and administratively and require standard solutions for storm water management problems. As an example the following ordinance provisions might be included in flexible and rigid ordinances.

Flexible Ordinance Provision - If the required hydrologic and hydraulic studies reveal that the proposed development would cause increased flood stages, for the design storm, so as to increase

the flood damages to existing developments or property, or increase flood elevations beyond the vertical limits set for the floodplain district, then the development permit shall be denied unless one or more of the following flood mitigation measures are used to solve anticipated flooding and drainage problems:
- on-site storage,
- off-site storage,
- improve existing drainage system, or
- flood proofing.

Rigid Ordinance Provision - All developments over five acres must use on-site storage to limit the peak runoff rate from the site to that which would have been expected from predevelopment conditions. The design of the storage facility shall be based on the Rational Formula Method with a $C = 0.3$ for predeveloped conditions using a design storm of 10-year frequency.

A flexible storm water management program encourages the design engineer to evaluate each storm water management problem and try to devise an optimum solution for that problem. Also, the local government retains flexibility in administering and implementing its program. Thus it is possible that different restrictions and designs could be applied to different areas of the community. However, a flexible storm water management program will introduce several possible problems in administering the program.
1. The municipality will have to hire sufficient qualified personnel to evaluate each development on a case-by-case basis.
2. In addition to studying the drainage within a development, the drainage interrelationships between all the developments within a watershed and among watersheds will need evaluation.
3. The municipality will need to obtain the technical aids necessary to do the above evaluations.
4. The municipality should consider taking an active role in the design, operation, and maintenance of the drainage system.

In contrast, many municipalities have written rigid ordinances which legislate standard evaluations, designs, and solutions to storm water management problems. Personnel from these municipalities feel it is easier to administer a rigid program because a standard set of rules can be developed and applied to all developments. They also contend that the administration of a rigid program requires fewer, less qualified personnel than would be required for a flexible program.

The major disadvantage of a rigid program is that it may force too much conformity in engineering designs. Each storm water management problem involves certain aspects that are unique to a particular situation. These unique aspects are difficult to take into account in formulating general engineering rules and criteria. As an example, allowing an increase of 5 cfs or 10 cfs in a drainage design may be appropriate for a large drainage system but may be totally inappropriate for a small design where the capacity of a grate inlet would be exceeded and runoff would flow into an adjacent building. In addition, design methods and procedures change with time while many times provisions of ordinances and programs do not. Thus, a rigid program may not allow for the use of the most economical or cost effective solutions.

In many instances, the basis for including in the storm water management ordinance certain design and engineering criteria was that these criteria were adopted many years ago and had been accepted because of their continual use. Thus, the use and acceptance of engineering criteria can result from their inclusion in an ordinance or because of their continued use. Adopting a rigid program for administrative ease has advantages and disadvantages which should be thoroughly evaluated. It should be remembered that much of the ingenuity and imagination that could be used to solve storm water management problems can be stifled by rigid storm water management ordinances and programs.

In practice most storm water management programs incorporate ordinances which are somewhere between a flexible and rigid approach. A program with specific guidelines that also allows for deviation from the standard if it can be shown that such deviation is in the best interest of the municipality is usually the best approach. It is usually an advantage to have some uniformity in engineering designs but standard solutions should not preclude the use of good engineering judgment in the evaluation of storm water management problems. Each municipality will have to determine how much flexibility it will allow in its program depending on local conditions (e.g., available staff and resources, severity of local problems, political climate). The decision is not either a flexible or rigid ordinance but how much flexibility should be incorporated into the ordinance.

4.6 When To Adopt An Ordinance

When to adopt a storm water management ordinance can be almost as important as what is included in the ordinance. Storm water management problems have some unique characteristics which separate them from the usual problems involved in other land use controls. As an example, if some land use controls (or zoning ordinances) are not used to prohibit certain combinations of development from taking place, you may end up living next to a large factory or other non-compatible use. Once the factory is constructed, local residents must coexist with it twenty-four hours a day, seven days a week. Thus, the conflicting use is always present. Urban storm water management problems, especially flooding problems, are quite different. On a warm sunny day, the stream that meanders by a residential lot can be a very pleasant neighbor and can offer hours of enjoyment for adjacent residents. But, on some occasions, which may be very rare, this neighbor may expand, overtop its banks and create a ravaging flood which damages property and destroys land areas. Days later when the sun shines, this same neighbor is safely back in its banks.

Thus the types of problems created by urban flooding have several unique characteristics.

*   Urban flooding and water quality problems normally affect only a small portion of the total citizens in a community, those who happen to live near the ditches, streams and rivers (though all citizens are affected if streets are flooded and unsafe).
*   In most cases, those citizens who are damaged by flooding are the same ones who enjoy the benefits of the stream or river during non-flooding periods.
*   During or shortly after flooding events, interest and public sympathy are high for those damaged, but quickly diminishes with time.
*   Flood-prone areas offer many advantages for development in spite of the potential dangers (e.g., aesthetic qualities of streams and rivers, relatively flat easily developable land).
*   Correction of storm water management problems can range from simple inexpensive flood proofing to complex expensive flood control projects.
*   Major flooding events may occur very rarely with long periods where flood-prone areas are unaffected by flood waters.
*   Surface drainage problems cause many problems, particularly of an inconvenience nature, which are difficult to assign monetary values.

Figure 4-2 graphically depicts the interest cycle associated with flood related problems. The interest cycle is referred to as a citizen and government interest and involvement hydrograph. The basic point being depicted in the figure is that too many urban storm water management ordinances are being adopted either as a result of Federal or state mandates or as an emotional result of a major flooding event. These ordinances are adopted without a community commitment to their implementation and the costs and resource needs this commitment would entail. An ordinance is

Figure 4-2 Citizen And Government Interest And Involvement Hydrograph

only one element of a complete drainage program, is relatively inexpensive to write and adopt, and can represent only an outward commitment to correct the associated problems.

Adopting an ordinance early in the development of a storm water management program can have several disadvantages:

1. The ordinance should represent the policies of the municipality relative to storm water management problems. These policies should be thoughtfully developed after careful study.

2. Adopting an ordinance can be an end in itself. Once an ordinance is "on the books", the body politics of a community can feel their job is done and the problems solved. A typical response to future citizen complaints could be "we have adopted an ordinance which should solve your problems". Obtaining further commitments such as money and staff can be very difficult.

3. The ordinance should reflect the municipality's financial and technical ability to implement its provisions. Thus, the financial and technical requirements associated with a storm water management program should be known before the ordinance is adopted. The optimum time to adopt an ordinance will vary from one municipality to another, but an ordinance should not be adopted until the community has a definite commitment to its implementation and has developed a storm water management program to a point where the ordinance reflects a realistic approach to deal with the local problems. Thus, municipalities should stop adopting ordinances that are only intended to comply with Federal and state programs or result from emotional consequences, but, instead, adopt well thought out and locally supported ordinances. It might be better not to adopt any ordinance rather than adding one more meaningless regulation to what might be a long list of existing ones.

4.7 The Complete Storm Water Management Program

Drafting and adopting a storm water management ordinance, although an important part of any storm water management program, cannot be viewed as encompassing a complete program. Several other important aspects of the program must be considered and are discussed in the following sections.

Land Use Planning Aspects

A storm water management program has several land use planning aspects that deserve mention. One question that should be discussed is, "where should the ordinance and other aspects of the program be applied in the planning process?" Should the ordinance be applied during the initial planning stages or not until specific development proposals are reviewed? The answer is "the ordinance and other aspects of the drainage program should be applied at several stages in the planning process".

A storm water management ordinance usually states specific objectives related to the storm water management program. These objectives should be incorporated into an overall planning program. The ordinance also states specific land use policies (areas included in ordinance restrictions, uses allowed in specific areas, special permits, etc.) which should be included in all land use planning efforts. Provisions of the ordinance have specific consequences with regard to the zoning process (improvements required, nonconforming uses, hydrologic and hydraulic studies required, and provisions related to land uses). Those provisions which affect different land use classifications should be taken into account when preparing zoning districts and when deciding on zoning changes and variances.

Municipal Role In Encouraging Innovative Solutions

One of the major purposes of adopting a flexible storm water management program is to encourage innovative solutions to storm water management problems. In many municipalities, the private engineering sector has not been particularly innovative in finding new solutions to storm water management problems. The primary interest of the engineer, most likely spurred by pressure from developers, is to get development plans approved by the local government so proposed construction can proceed. Also, many engineers feel that rigid rules and regulations which must be followed in order to get plans approved discourage innovation. Engineers and developers know it is much easier to get approval for standard designs than for new or untried ones. As a result, if innovative and imaginative designs are going to be applied to flooding problems, it will be up to the local government to take the lead and provide the climate necessary to encourage such designs. To accomplish this, the local government will have to hire qualified personnel who can use their knowledge and expertise to propose, encourage, and efficiently review new and different approaches to storm water management problems. However, many times, those people who appropriate funds disagree with the need to hire sufficient qualified personnel and thus limit the effectiveness of a storm water management program.

In defense of the "not-so-innovative", it should be noted that people in public works often need to be conservative. Mistakes, poor and ineffective designs, are often lived with for a long time. Also, if a public works facility or concept fails the consequences can be serious and sometimes life threatening. Innovative solutions are often unproven and care must be exercised in their application. Thus, the benefits of an "innovative" versus a "tried and true" approach must be weighted against consequences and the results of possible failure.

Administrative Problems

It is one thing to adopt regulations requiring detailed engineering evaluations and another to administer and enforce such regulations. Appendix A at the end of this chapter presents a storm water management ordinance which requires detailed engineering studies and evaluations. Strict compliance with this ordinance could require costly engineering studies for many developments where common sense would indicate that such studies may not be necessary to make needed decisions. Evaluation of upstream and downstream effects of proposed development and associated storm water management facilities, evaluation of the effects of a proposed development on flood elevations, design and evaluation of storage facilities, and the interrelationships between several storage facilities and drainage structures all present complex hydrologic and hydraulic problems. Solutions are difficult if not impossible with traditional engineering methods, and most municipal engineering staffs and local consultants do not have the resources necessary to effectively implement such provisions. Thus, the administrative, planning, and engineering aspects of a storm water management program should be coordinated so that one aspect does not produce problems for another.

Technical Requirements

Adequate administration of a storm water management program requires that the municipality have the necessary technical aids to properly assess the effects of proposed developments. In some communities (e.g., Cobb County, Georgia; Greenville, South Carolina; Chapel Hill, North Carolina) hydrologic computer simulation models have been developed and used for this purpose. Such computer simulation models are extremely helpful in administering a comprehensive storm water management program. If municipalities adopt a program dealing with such areas as the combined effects of several developments, drainage interrelationships between different watersheds and sub-watersheds, floodplain map updates, and detailed evaluations of the drainage effects of proposed developments, then it should accept the responsibility of providing the technical tools and financial resources needed to administer the program. Development of these technical tools can be quite expensive and time consuming; thus a commitment to their development and use should be obtained before adopting the storm water management ordinance.

Staff And Financial Resources Needed

Developing and implementing a comprehensive storm water management program will necessitate employing qualified professionals with expertise in hydrology and water resources planning. Adequate funding of the program will also be critical to effectively implementing all of the elements involved. The staff and financial resources needed should be assessed early in the development stage of the program so appropriate budgets can be established and political commitments of support obtained.

Field Inspection

Although it is relatively easy and inexpensive for a local government to adopt regulations, it becomes much more difficult and expensive to implement them. Not only should the municipality anticipate the need to hire qualified office personnel to administer the storm water management program but sufficient qualified field personnel will be needed.

Adequate field inspection is a major problem associated with implementing storm water management programs. Both the quantity and quality of inspectors are often below what would be adequate to do a good job of implementing the programs. Field inspection is especially essential for implementing the erosion, sediment and water quality aspects of the program. Sediment ponds fill up, hay bale barriers get destroyed or moved, undisturbed buffer zones are often encroached by development, and berms and other erosion and sediment control facilities get damaged or for some reason do not operate as designed. Unless construction sites are adequately inspected, needed repairs and maintenance to erosion and sediment control facilities are usually not done and these facilities quickly become ineffective. Although a final inspection can usually determine if the proper drainage facilities have been installed, there are also problems with not having more frequent inspections of these facilities. It is often impossible to tell what is beyond the physical features of facilities - for example, was structural backfill placed correctly, was properly graded riprap used, are the foundations placed as designed, etc.

Inspection is conducted and paid for in a number of ways:

- In order to decrease the field inspection work, some municipalities have been successful in using a system of spot checking construction sites with large fines for violating storm water management regulations. Random sampling with fines could encourage self-maintenance by owners, require fewer inspections, and raise funds for the inspection program. Thus, in lieu of hiring a large staff of inspectors, some form of random sampling combined with fines, as provided by law, could prove effective in enforcing ordinance provisions.
- In one southern city inspectors are hired by developers from an approved list. While the developer pays for the inspector, the inspector reports to the city.
- Many municipalities have conducted studies to require the development community to pay the partial or complete cost of the plans review and inspection. The fee is levied on a variety of bases including: actual hourly accounting, number of lots, acreage or length of water, waste water and storm drain systems. For example, Hampton, VA uses $0.25 per linear foot of these conveyance systems (Whitley, 1993).

Personnel and resources available plus local political and economic conditions will determine, to a large extent, what administrative procedures are used in implementing the program. Each municipality should adopt procedures which prove to be most effective for their particular conditions and circumstances.

A relatively new type of ordinance which has storm water management implications is the "Infrastructure Permit" or the "Private Development Permit". It allows for ongoing inspection of infrastructure as it is being built with the inherent ability for stop-work orders being enacted prior to the infrastructure adoption into the public system proceedings (Whitley, 1993).

Before any storm water management program is adopted, the financial responsibility of providing adequate field inspection should be assessed and provided. Getting storm water management facilities properly designed and included in the construction plans is an important part of the program, but getting them constructed, operating, and maintained has proved to be much more difficult.

Enforcement

Enforcement of ordinance provisions is very important if the desired program objectives are to be reached. Basically, enforcement can be divided into two areas. First, all the required paperwork such as development plans, hydrologic and hydraulic studies, drainage plans, erosion and sediment control plans must be completed and approved by the municipality. Second, adequate inspection must be provided during and after the development phase to ensure that the proposed measures are installed and function as designed. In addition, adequate penalties must be provided to ensure compliance with the provisions of the ordinance.

Enforcement must be phased, swift and effective. Phased enforcement allows for warnings and advice prior to fines or work stoppages. Swift enforcement reduces reconstruction costs and irreversible environmental damage. Effective enforcement is consistent and fair. Each development is handled in the same way without undue reliance on an individual inspector's subjective judgment. In many municipalities the enforcement is uneven, unsupported by the political leadership and cumbersome to execute. Inspector time lost in paperwork and court appearances significantly reduces the effectiveness of any program. "Environmental courts" have expedited such legal proceedings in some municipalities. However, there is no substitute for having adequate authority, well documented and consistently and professionally executed and supported.

Legal Considerations

Before any municipality adopts a storm water management ordinance, the municipality's legal staff should complete a detailed review of the ordinance to ensure that provisions do not conflict with Federal, state or local laws, nor goes beyond the bounds established by state law and municipal charter.

4.8 Conclusion

Storm water management programs are becoming more and more complex both in their administrative and technical aspects. Drafting and adopting a storm water management ordinance is an important part of such a program but it is not the complete program. Determining community objectives and policies related to drainage and flood control, determining water quality standards, acquiring the needed technical personnel, providing adequate field inspection, providing the technical tools, and giving overall direction to the program are also important parts of any storm water management program. As a result of Federal and state mandates, many municipalities are drafting and adopting storm water management ordinances. It is now time to integrate these ordinances into a complete program in order to achieve the desired results. Without a complete program, these ordinances will prove ineffective and will only add to the frustration of those trying to deal with our complex urban storm water management problems.

References

Association of State Flood Plain Managers (ASFPM), The Insider, Article by Jan Kusler, pp. 5-6, 1990.

American Public Works Association, "Urban Storm Water Management" Chicago, Ill., Special Report No. 49, 1981.

Austin, TX, "Comprehensive Watershed Ordinance", Ordinance # 860508-V, 1986.

Bellevue Washington, "Ordinance 2003, an ordinance relating to storm and surface water...", February, 1974.

Cox, W. E., "Water Law Primer", Proc. Wat. Res. Planning and Mgmt. Div. ASCE, Vol. 108, No. WR1, March, 1982.

Galli, J., "Analysis of Urban BMP Performance and Longevity in Prince George's County, Maryland", Final Rpt., Metro Wash. Council of Govts., Aug., 1992.

Debo, T. N., "Survey and Analysis of Urban Drainage Ordinances and a Recommended Model Ordinance", thesis presented Georgia Institute of Technology, Feb. 1976.

Debo, T. and Reese, A.J., "Downstream Impacts of Detention", Proc. NOVATECH 92, Nov 3-5, 1992, Lyon, France.

Environmental Protection Agency (EPA), Region V, "Stormwater Management Ordinances for Local Governments", Chicago, Ill., Dec. 1990.

Illinois Department of Transportation (IDOT), "Model Flood Plain Ordinance".

Kenworth, W. E., "Urban Drainage: Aspects of Public and Private Liability", Dicta, Denver, CO., July-Aug, 1972.

Lewis, G.L., "Jury Verdict: Frequency Versus Risk-Based Culvert Design", ASCE J. of Water Res. Plan. & Mgmt., Vol. 118, No. 2, Mar/April, 1992, pp. 166-185.

Louisville MSD, "Hazardous Materials Ordinance", Nov. 1988.

Maryland, "Model Stormwater Management Ordinance", Dept. of the Environment, 1983.

Ogden Environmental and Energy Services, Inc., "Storm Water Quality Program Development - Charlotte, NC", Ogden - Nashville, TN, 1992.

Prince Georges County, MD, Storm Water Ordinance, Bill No. CB-52-1984, 1984.

Schueler, T. R., P. A. Kumble and M. A. Heraty, "A Current Assessment of Urban Best Management Practices", Metro Wash. Council of Govts., March, 1992.

Shoreview, MN, "An Ordinance ... Relating to Lawn Fertilizer Application", Ordinance No. 477, 1985.

Shoreview, MN, "An Ordinance ... Relating to Lawn Fertilizer Application and Pesticide Control", Ordinance No. 477, 1988.

Wayzata, MN, "An Ordinance...Adding Wetlands District Definitions", Ordinance No. 515, 1988.

Whitley, Fred, "Permitting Public Infrastructure Installation of Private Developments", in Public Works, August, 1993.

Appendix A - Example Storm Water Management Ordinance

Table Of Contents

ARTICLE I.   GENERAL PROVISIONS

SECTION A.   Title; purpose

1. The provisions of this ordinance shall constitute and be known as the "Storm Water Manage-
    ment Ordinance For The Municipality".

2.   The purpose of this Ordinance is to protect, maintain, and enhance the public health, safety, and general welfare by establishing minimum requirements and procedures to control the adverse effects of increased storm water runoff associated with both future land development and existing developed land within the Municipality. Proper management of storm water runoff will minimize damage to public and private property, ensure a functional drainage system, reduce the effects of development on land and stream channel erosion, assist in the attainment and maintenance of water quality standards, enhance the local environment associated with the drainage system, reduce local flooding, maintain as nearly as possible the pre-developed runoff characteristics of the area, and facilitate economic development while mitigating associated flooding and drainage impacts.

3.   The application of this Ordinance and the provisions expressed herein shall be the minimum storm water management requirements and shall not be deemed a limitation or repeal of any other powers granted by State statute. In addition, if site characteristics indicate that complying with these minimum requirements will not provide adequate designs or protection for local property or residents, it is the designer's responsibility to exceed the minimum requirements as necessary. The Municipal Engineer or designee shall be responsible for the coordination and enforcement of the provisions of this ordinance.

SECTION B.   Definitions

For the purpose of this Ordinance, the following terms, phrases and words, and their derivatives, shall have the meaning given herein:

1.   As-built plan shall mean a set of engineering or site drawings that delineate the specific permitted storm water management facility as actually constructed.

2.   Best management practices shall mean a wide range of management procedures, schedules of activities, prohibitions on practices and other management practices which have been demonstrated to effectively control the quality and/or quantity of storm water runoff and which are compatible with the planned land use.

3.   Cross-drain culvert shall mean a culvert located under a roadway.

4.   Design report shall mean the report that accompanies the storm water management plan and includes data used for engineering analysis, results of all analysis, design and analysis calculations (including results obtained from computer programs), and other engineering data that would assist the Municipal Engineer in evaluating proposed storm water management facilities.

5.   Designer shall mean a professional who is permitted to prepare plans and studies required by this ordinance.

6.   Detention structure shall mean a permanent storm water management structure whose primary purpose is to temporarily store storm water runoff and release the stored runoff at controlled rates.

7.   Development should generally mean any of the following actions undertaken by a public or private individual or entity:
   •   the division of a lot, tract or parcel of land into two (2) or more lots, plots, sites, tracts, parcels or other divisions by plat or deed, or
   •   any land change, including, without limitation, clearing, tree removal, grubbing, stripping, dredging, grading, excavating, transporting and filling of land.

8.   Develop land shall mean to change the runoff characteristics of a parcel of land in conjunction with residential, commercial, industrial, or institutional construction or alternation.

9. Developed land use conditions shall mean the land use conditions according to the current Municipality Land Use Map or proposed development plan.

10. Easement shall mean a grant or reservation by the owner of land for the use of such land by others for a specific purpose or purposes, and which must be included in the conveyance of land affected by such easement.

11. Erosion shall mean the wearing away of land surface by the action of wind, water, gravity, ice, or any combination of those forces.

12. Erosion and sediment control shall mean the control of solid material, both mineral and organic, during a land disturbing activity to prevent its transport out of the disturbed area by means of air, water, gravity, or ice.

13. Existing land use conditions shall mean the land use conditions existing at the time of the most recent official aerial photography available from the Municipality.

14. Grading shall mean excavating, filling (including hydraulic fill) or stockpiling of earth material, or any combination thereof, including the land in its excavated or filled condition.

15. Impervious shall mean the condition of being impenetrable by water.

16. Imperviousness shall mean the degree to which a site is impervious.

17. Infiltration shall mean the passage or movement of water through the soil profile.

18. Interior culvert shall mean a culvert that is not located under a roadway.

19. Land disturbing activity shall mean any use of the land by any person that results in a change in the natural cover or topography that may cause erosion and contribute to sediment and alter the quality and/or quantity of storm water runoff.

20. Maintenance shall mean any action necessary to preserve storm water management facilities in proper working condition, in order to serve the intended purposes set forth in Article I of this Ordinance and to prevent structural failure of such facilities. Maintenance shall not include actions taken solely for the purpose of enhancing the aesthetics aspects associated with storm water management facilities.

21. Municipal Engineer shall mean the duly designated Department Head of the municipal engineering department or department of public works, or his duly authorized agent. The Municipal Engineer shall be licensed and registered to perform the duties of engineer, as herein specified.

22. Municipal Engineering Department shall mean the department responsible for all storm water management activities and implementation of the provisions of this ordinance.

23. Municipality shall mean a city or county that is adopting this ordinance.

24. Natural waterways shall mean waterways that are part of the natural topography. They usually maintain a continuous or seasonal flow during the year and are characterized as being irregular in cross-section with a meandering course. Construction channels such as drainage ditches shall not be considered natural waterways.

25. Nonerodible shall mean a material, e.g., natural rock, riprap, concrete, plastic, etc., that will not experience surface wear due to natural forces of wind, water, ice, gravity or a combination of those forces.

26. Nonpoint source pollution shall mean pollution contained in storm water runoff from ill-defined, diffuse sources.

27. One hundred year frequency storm shall mean a storm that is capable of producing rainfall expected to be equaled or exceeded on the average of once in 100 years. It also may be expressed as an exceedance probability with a 1 percent chance of being equaled or exceeded in any given year.

28. On-site storm water management shall mean the design and construction of a facility necessary to control storm water runoff within and for a single development.

29. Person responsible for the land disturbing activity shall mean:

    a.   the person who has or represents having financial or operational control over the land disturbing activity; and/or

    b.   the landowner or person in possession or control of the land who directly or indirectly allowed the land disturbing activity or has benefitted from it or who has failed to comply with any provision of this ordinance.

30. Post-development conditions shall mean the conditions which exist following the completion of the land disturbing activity in terms of topography, vegetation, land use and rate, volume or direction of storm water runoff.

31. Pre-developed conditions shall mean those land use conditions that existed prior to the initiation of the land disturbing activity in terms of topography, vegetation, land use and rate, volume or direction of storm water runoff.

32. Preliminary plat shall mean the preliminary plat of a residential subdivision submitted pursuant to the Municipality's Subdivision Regulations.

33. Record survey shall mean a final field survey which locates the visible surface features of a constructed storm water facility on the ground, but without locating non-visible or sub-surface features such as the actual route and elevation of buried pipe.

34. Regional storm water management shall mean the design and construction of a facility necessary to control storm water runoff within or outside a development and for one or more developments.

35. Registered Civil Engineer shall mean a civil engineer properly registered and licensed to conduct work within the Municipality.

36. Registered Land Surveyor shall mean a land surveyor properly registered and licensed to conduct work within the Municipality.

37. Registered Landscape Architect shall mean a landscape architect properly registered and licensed to conduct work within the Municipality.

38. Responsible personnel shall mean any foreman, superintendent, or similar individual who is the on-site person in charge of land disturbing activities.

39. Retention structure shall mean a permanent structure whose primary purpose is to permanently store a given volume of storm water runoff. Release of the given volume is by infiltration and/or evaporation.

40. Sediment shall mean solid particulate matter, both mineral and organic, that has been or is being transported by water, air, ice, or gravity from its site of origin.

41. Stabilization shall mean the installation of vegetative or structural measures to establish a soil cover to reduce soil erosion by storm water runoff, wind, ice and gravity.

42. Stage work or stage construction shall mean a plan for the staged construction of storm water facilities where portions of the facilities will be constructed as different stages of the proposed development are started or completed.

43. Storm water concept plan shall mean the overall proposal for a storm drainage system, including storm water management structures, and supporting documentation as specified in the Storm Water Management Design Manual, for each proposed private or public development to the extent permitted by law. Also included are the supporting engineering calculations and results of any computer analysis, if necessary.

44. Storm water management shall mean the collection, conveyance, storage, treatment and disposal of storm water runoff in a manner to minimize accelerated channel erosion, increased flood damage, and/or degradation of water quality and in a manner to enhance and ensure the public health, safety, and general welfare, which shall include a system of vegetative or structural measures, or both, that control the increased volume and rate of storm water runoff caused by manmade changes to the land.

45. Storm water management design manual shall mean the manual of design, performance, and review criteria for storm water management practices, prepared under the direction of the Municipal Engineer. Copies of this manual can be obtained from the Municipal Engineering Department.

46. Storm water management facilities shall mean those structures and facilities that are designed for the collection, conveyance, storage, treatment and disposal of storm water runoff into and through the drainage system.

47. Storm water management plan shall mean the set of drawings and other documents that comprise all of the information and specifications for the drainage systems, structures, concepts and techniques that will be used to control storm water as required by this ordinance and the Storm Water Management Design Manual. Also included are the supporting engineering calculations and results of any computer analysis.

48. Storm water management qualitative control shall mean a system of vegetative, structural, or other measures that reduce or eliminate pollutants that might otherwise be carried by storm water runoff.

49. Storm water runoff shall mean the direct response of a watershed to precipitation and includes the surface and subsurface runoff that enters a ditch, stream, storm drain or other concentrated flow during and following the precipitation.

50. Subdivision shall mean (1) The creation of one or more new streets, alleys or other public ways; or, the changing of any rights-of-way of any existing streets, alleys or other public ways. (2) Any division or redivision of lot, tract, or parcel or land, regardless of its prospective use. Such subdivision may be accomplished by platting or by description of metes and bounds or otherwise into two (2) or more lots or other divisions for sale or improvement. The following are not defined as subdivisions:
    a. The combination or recombination of portions of previously platted lots where the total number of lots is not increased and the resultant lots are in accordance with the rules and regulations contained in the Municipality's Subdivision Regulations and with the Municipality's Zoning Ordinance.
    b. Division or sale of land by judicial decree which shall not be deemed a division for purposes of this ordinance.
    c. The acquisition of land for the purpose of widening or opening of streets when the acquisition and work is done by the Municipality, State or other governmental agency.
    d. The division of land into parcels greater than five (5) acres where no street right-of-way dedication is involved.

51. Swale shall mean a structural measure with a lining of grass, riprap or other materials which can function as a detention structure and convey storm water runoff without causing erosion.

52. Ten-year frequency storm shall mean a storm that is capable of producing rainfall expected to be equaled or exceeded on the average of once in 10 years. It may also be expressed as an exceedance probability with a 10 percent chance of being equaled or exceeded in any given year.

53. Twenty-five year frequency storm shall mean a storm that is capable of producing rainfall expected to be equaled or exceeded on the average of once in 25 years. It may also be expressed as an exceedance probability with a 4 percent chance of being equaled or exceeded in any given year.

54. Two-year frequency storm shall mean a storm that is capable of producing rainfall expected to be equaled or exceeded on the average of once in 2 years. It may also be expressed as an exceedance probability with a 50 percent chance of being equaled or exceeded in any given year.

55. Variance shall mean the modification of the minimum storm water management requirements for specific circumstances where strict adherence of the requirements would result in unnecessary hardship and not fulfill the intent of this ordinance.

56. Waiver shall mean the relinquishment from storm water management requirements by the Municipal Engineer for a specific land disturbing activity on a case-by-case review basis.

57. Water quality shall mean those characteristics of storm water runoff from a land disturbing activity that relate to the physical, chemical, biological, or radiological integrity of water.

58. Water quantity shall mean those characteristics of storm water runoff that relate to the rate and volume of the storm water runoff to downstream areas resulting from land disturbing activities.

59. Watershed shall mean the drainage area contributing storm water runoff to a single point.

SECTION C.   Scope of ordinance

No person shall develop any land without having provided for appropriate storm water management measures that control or manage runoff, in compliance with this Ordinance, unless exempted in Article I, Section D below.

SECTION D.   Exemptions from requirements

The following development activities are exempt from the provisions of the Ordinance and the requirements of providing storm water management measures.

1. Land disturbing activities on agricultural land for production of plants and animals useful to man, including but not limited to: forages and sod crops, grains and feed crops, tobacco, cotton, and peanuts; dairy animals and dairy products; poultry and poultry products; livestock, including beef cattle, sheep, swine, horses, ponies, mules, or goats, including the breeding and grazing of these animals; bees; fur animals and aquaculture, except that the construction of an agricultural structure of one or more acres, such as broiler houses, machine sheds, repair shops and other major buildings and which require the issuance of a building permit shall require the submittal and approval of a storm water management plan prior to the start of the land disturbing activity.

2. Land disturbing activities undertaken on forest land for the production and harvesting of timber and timber products.

3. Construction or improvement of single family residences or their accessory buildings which are separately built and not part of multiple construction of a subdivision development.

SECTION E.   Storm water management design manual

To assist in the design and evaluation of storm water management facilities in the Municipality, a Storm Water Management Design Manual has been developed. Recommended design procedures and criteria are presented for conducting hydrologic and hydraulic evaluations. Although the intention of the manual is to establish uniform design practices, it neither replaces the need for engineering judgment nor precludes the use of information not presented. Other accepted engineering procedures may be used to conduct hydrologic and hydraulic studies if approved by the Municipal Engineer.

ARTICLE II.   STORM WATER CONCEPT AND PRELIMINARY DEVELOPMENT PLANS

SECTION A.   Scope of development plans

1. a. In developing plans for residential subdivisions, individual lots in a residential subdivision development shall not be considered to be separate land disturbing activities and shall not require individual permits. Instead the residential subdivision development, as a whole, shall be considered to be a single land disturbing activity. Hydrologic parameters that reflect the ultimate subdivision development shall be used in all engineering calculations.

    b. If individual lots or sections in a residential subdivision are being developed by different property owners, all land disturbing activities related to the residential subdivision shall be covered by the approved storm water management plan for the residential subdivision. Individual lot owners or developers shall sign a certificate of compliance that all activities on that lot will be carried out in accordance with the approved storm water management plan for the residential subdivision.

    c. Residential subdivisions which were approved prior to the effective date of these regulations are exempt from these requirements. Development of new phases of existing subdivisions which were not previously approved shall comply with the provisions of these regulations.

2. For land disturbing activities involving two (2) acres or less of actual land disturbance which are not part of a larger common plan of development or sale, the person responsible for the land disturbing activity shall submit a simplified storm water management control plan meeting the requirements listed below. This plan does not require approval by the Municipal Engineering Department and does not require preparation or certification by the designers specified in Section K of Article II.

    a. A narrative description of the storm water management facilities to be used.

    b. A general description of topographic and soil conditions of the development site.

    c. A general description of adjacent property and a description of existing structures, buildings, and other fixed improvements located on surrounding properties.

    d. A sketch plan to accompany the narrative which shall contain:

    • a site location drawing of the proposed project, indicating the location of the proposed project in relation to roadways, jurisdictional boundaries, streams and rivers;

    • the boundary lines of the site on which the work is to be performed;

    • all areas within the site which will be included in the land disturbing activities shall be identified and the total disturbed area calculated;

    • a topographic map of the site;

    • anticipated starting and completion dates of the various stages of land disturbing activities and the expected date the final stabilization will be completed.

    • the location of temporary and permanent vegetative and structural storm water management control measures.

    e. Storm water management plans shall contain certification by the persons responsible for the land disturbing activity that the land disturbing activity will be accomplished pursuant to the plan.

    f. Storm water management plans shall contain certification by the person responsible for the land disturbing activity of the right of the Municipal Engineer to conduct on-site inspections.

3. For land disturbing activities disturbing more than two acres, the requirements of Article II, Sections B - K shall apply.

SECTION B.   Storm water concept and storm water management plans

1. A storm water concept plan for each development shall be submitted for review by the Municipal Engineer prior to submission of the storm water management plan and construction plans for the entire development, or any portion thereof.
2. All preliminary plats of the development shall be consistent with the storm water concept plan required in Paragraph 1 above.
3. Upon approval of the concept plan, the applicant shall submit a final storm water management plan (as part of the construction plans) to the Municipal Engineer for review and approval; provided that the Municipal Engineer may accept and submit into the review process a storm water concept plan if it identifies the location and type of facilities to be constructed in sufficient detail to accurately estimate construction costs and the Municipal Engineer determines that a storm water management plan is not needed. If accepted under this provision, the storm water concept plan then becomes the storm water management plan for this development.
4. Should any storm water management plan involve any storm water management facilities or land to be dedicated to public use, the same information shall also be submitted for review and approval to the department having jurisdiction over the land or other appropriate departments or agencies identified by the Municipal Engineer for review and approval. This storm water management plan shall serve as the basis for all subsequent construction.
5. The storm water concept plan may be reviewed, if needed, with the designer, after Municipal review, where it will either be approved, approved with changes, or rejected. If rejected, changes, additional analysis, or other information needed to approve the next submittal of the concept plan shall be identified. The Municipal review of the storm water concept plan will be completed within five (5) working days from and after the receipt of the plan.
6. Within ten (10) working days from and after the receipt of the storm water management plan, the Municipal Engineer shall issue a decision approving, rejecting or conditionally approving the plan with modification.

SECTION C.   Permit requirements

1. No final occupancy permit shall be issued without the following:
   a. Recorded easements for storm water management facilities.
   b. Receipt of an as-built plan which includes a certification of the storm drainage system.
2. No site grading permit shall be issued or modified without the following:
   a. Right of entry for emergency maintenance if necessary.
   b. Right of entry for inspections.
   c. Any off-site easements needed.
   d. An approved storm water concept plan or storm water management plan, as appropriate.
3. The approved storm water management plan shall contain certification by the applicant that all land clearing, construction, development and drainage will be done according to the storm water management plan or previously approved revisions. Any and all site grading permits may be revoked at any time if the construction of storm water management facilities is not in strict accordance with approved plans.
4. In addition to the plans and permits required from the Municipality, applicants shall obtain all state and federal permits required for the proposed development.

SECTION D.   Fees

A list of fees for plan review and other fees associated with this ordinance can be obtained from the Municipal Engineering Department.

SECTION E.   Permit suspension and revocation

1. A site grading permit may be suspended or revoked if one or more of the following violations have been committed:
    a. violation(s) of the conditions of the storm water management plan approval;
    b. construction not in accordance with the intent of the approved plans;
    c. non-compliance with correction notice(s) or stop work orders(s); or
    d. the existence of an immediate danger in a downstream area in the judgment of the Municipal Engineer.
    If one or more of these conditions is found, a written notice of violations shall be served upon the owner or authorized representative and an immediate stop-work order may be issued. The notice shall set forth the measures necessary to achieve compliance with the plan. Correction of these violations must be started immediately or the owner shall be deemed in violation of this ordinance.

SECTION F.   Minimum runoff control requirements

1. The minimum storm water control requirements shall provide management measures necessary to accomplish the following:
    a. Install storm water management facilities to limit the 2-year and 10-year developed peak discharge rates to pre-developed peak discharge rates. The design of these facilities shall be based on procedures contained in the Storm Water Management Design Manual or approved by the Municipal Engineer.
    b. The requirements, or portions thereof, of item (a.) may be waived by the Municipal Engineer if it can be shown by detailed engineering calculations and analysis which are acceptable to the Municipal Engineer that one of the following exists:
        1. the installation of storm water management facilities would have insignificant effects on reducing downstream flood peaks; or
        2. storm water management facilities are not needed to protect downstream developments and the downstream drainage system has sufficient capacity to receive any increase in runoff for the design storm; or
        3. it is not necessary to install storm water management facilities to control developed peak discharge rates at the exit to a proposed development and installing such facilities would increase flood peaks at some downstream locations; or
        4. the Municipal Engineer determines that storm water management facilities are not needed to control developed peak discharge rates and installing such facilities would not be in the best interest of the Municipality.
    c. The requirements, or portions thereof, of item (a.) may not be waived if the Municipal Engineer determines that not controlling downstream flood peaks would increase known flooding problems, or exceed the capacity of the downstream drainage system.
    d. A waiver shall only be granted after a written request is submitted by the applicant containing descriptions, drawings, and any other information that is necessary to evaluate the proposed land disturbing activity. A separate written waiver request shall be required if there are subsequent additions, extensions, or modifications which would alter the

approved storm water runoff characteristics to a land disturbing activity receiving a waiver. The Municipal Engineer will conduct a review of the request for a waiver within ten (10) working days. Failure of the Municipal Engineer to act by the end of the tenth working day will result in the automatic approval of the waiver.

   e. Discharge velocities shall be reduced to provide a nonerosive velocity flow from a structure, channel, or other control measure or the velocity of the 10-year design storm runoff in the receiving waterway prior to the land disturbing activity, whichever is greater.

2. For all storm water management facilities, a hydrologic-hydraulic study shall be done showing how the drainage system will function with and without the proposed facilities. For such studies the following land use conditions shall be used. Existing land use data shall be taken from the most recent aerial photograph and field checked and updated.

   a. For the design of the facility outlet structure, use developed land use conditions for the area within the proposed development and existing land use conditions for upstream areas draining to the facility.

   b. For any analysis of flood flows downstream from the proposed facility, use existing land use conditions for all downstream areas.

   c. All storm water management facilities emergency spillways shall be checked using the 100-year storm and routing flows through the facility and emergency spillways. For this analysis, developed land use conditions shall be used for all areas within the analysis.

   d. If accepted for municipal maintenance, the effects of existing upstream detention facilities can be considered in the hydrologic-hydraulic study.

SECTION G.   Storm water management facilities

1. Storm water management facilities may include both structural and nonstructural elements. Natural swales and other natural runoff conduits shall be retained where practicable.

2. Where additional storm water management facilities are required to satisfy the minimum control requirements, the following measures are examples of what may be used:

   a. storm water detention structures (dry basins);

   b. storm water retention structures (wet ponds);

   c. facilities designed to encourage overland flow, slow velocities of flow, and flow through buffer zones;

   d. infiltration practices.

3. Where detention and retention structures are used, designs which consolidate these facilities into a limited number of large structures will be preferred over designs which utilize a large number of small structures.

4. Storm water management plans can be rejected by the Municipal Engineer if they incorporate structures and facilities that will demand considerable maintenance, will be difficult to maintain, or utilize numerous small structures if other alternatives are physically possible.

5. The drainage system and all storm water management structures within the Municipality (including both public and private portions) will be designed to the same engineering and technical criteria and standards. The Municipal Engineering Department's review will be the same whether the portion of the drainage system will be under public or private control or ownership.

6. All storm water management measures shall be designed in accordance with the design criteria contained in the Storm Water Management Design Manual using procedures contained in this manual or procedures approved by the Municipal Engineer.

SECTION H.   Plan requirements

Storm water management plans shall include as a minimum the following.

1. A vicinity map indicating a north arrow, scale, boundary lines of the site, and other information necessary to locate the development site.
2. The existing and proposed topography of the development site except for individual lot grading plans in single family subdivisions.
3. Physical improvements on the site, including present development and proposed development.
4. Location, dimensions, elevations, and characteristics of all storm water management facilities.
5. All areas within the site which will be included in the land disturbing activities shall be identified and the total disturbed area calculated.
6. The location of temporary and permanent vegetative and structural storm water management control measures.
7. An anticipated starting and completion date of the various stages of land disturbing activities and the expected date the final stabilization will be completed.
8. A determination that no occupied first floor elevation of any structure is below the 100-year plus one foot flood elevation.
9. At the discretion of the Municipal Engineer, for all portions of the drainage system which are expected to carry between 50 and 150 cfs for the 100-year storm, the 100-year plus one foot flood elevation analysis shall be required. To require the 100-year plus one foot flood elevation analysis, the Municipal Engineer should determine that one of the following conditions may exist.
    a. The estimated runoff would create a hazard for adjacent property or residents.
    b. The flood limits would be of such magnitude that adjacent residents should be informed of these limits.
10. For all portions of the drainage system which are expected to carry 150 cfs or more for the 100-year storm, the 100-year plus one foot flood elevation analysis shall be done and flood limits shall be shown on the storm water management plans.
11. Storm water management plans shall include designation of all easements needed for inspection and maintenance of the drainage system and storm water management facilities. As a minimum, easements shall have the following characteristics.
    a. Provide adequate access to all portions of the drainage system and structures.
    b. Provide sufficient land area for maintenance equipment and personnel to adequately and efficiently maintain the system with a minimum of ten (10) feet along both sides of all drainage ways, streams, channels, etc., and around the perimeter of all detention and retention facilities, or sufficient land area for equipment access for maintenance of all storm water management facilities. This distance shall be measured from the top of the bank or toe of the dam whichever is applicable.
    c. Restriction on easements shall include prohibiting all fences and structures which would interfere with access to the easement areas and/or the maintenance function of the drainage system.
12. To improve the aesthetic aspects of the drainage system, a landscape plan for all portions of the drainage system shall be part of the storm water management plan. This landscape plan shall address the following.
    a. Tree saving and planting plan.
    b. Types of vegetation that will be used for stream bank stabilization, erosion control, sediment control, aesthetics and water quality improvement.
    c. Any special requirements related to the landscaping of the drainage system and efforts necessary to preserve the natural aspects of the drainage system.

13. To improve the water quality aspects of the drainage system, the storm water management plan shall include best management practices to control the water quality of the runoff during the land disturbing activities and during the life of the development. This requirement is in addition to the requirements contained in the Municipality Sediment Control Ordinance.

14. The storm water management plan shall include all engineering calculations needed to design the system and associated structures including pre- and post-development velocities, peak rates of discharge, and inflow and outflow hydrographs of storm water runoff at all existing and proposed points of discharge from the site.

15. Description of site conditions around points of all surface water discharge including vegetation and method of flow conveyance from the land disturbing activity.

16. Construction and design details for structural controls.

17. The expected timing of flood peaks through the downstream drainage system shall be assessed when planning the use of detention facilities.

18. In determining downstream effects from storm water management structures and the development, hydrologic-hydraulic engineering studies shall extend downstream to a point where the proposed development represents less than ten (10) percent of the total watershed.

19. All storm water management facilities and all major portions of the conveyance system through the proposed development (i.e., channels, culverts) shall be analyzed, using the design and 100-year storms, for design conditions and operating conditions which can reasonably be expected during the life of the facility. The results of the analysis shall be included in the hydrologic-hydraulic study.

20. If the storm water management plan and/or design report indicates that there may be a drainage or flooding problem at the exit to the proposed development or at any location between the exit point and the 10 percent downstream point, the Municipal Engineer may require:
    a. water surface profiles plotted for the conditions of pre- and post-development for the 10-year design storm;
    b. water surface profiles plotted for the conditions of pre- and post-development for the 100-year design storm;
    c. elevations of all structures potentially damaged by 10- and/or 100-year flows.

21. All storm water management plans submitted for approval shall contain certification by the person responsible for the land disturbing activity that the land disturbing activity will be accomplished pursuant to the approved plan and that responsible personnel will be assigned to the project.

22. All storm water management plans shall contain certification by the person responsible for the land disturbing activity, of the right of the Municipal Engineer to conduct on-site inspections.

23. The storm water management plan shall not be considered approved without the inclusion of an approval stamp with a signature and date on the plans by the Municipal Engineering Department. The stamp of approval on the plans is solely an acknowledgement of satisfactory compliance with the requirements of these regulations. The approval stamp does not constitute a representation or warranty to the applicant or any other person concerning the safety, appropriateness or effectiveness of any provision, or omission from the storm water management plan.

24. Approved storm water management plans remain valid for five (5) years from the date of an approval. Extensions or renewals of the plan approvals will be granted by the Municipal Engineer upon written request by the person responsible for the land disturbing activity.

SECTION I.   Plan hydrologic criteria

The hydrologic criteria to be used for the storm water concept and storm water management plans shall be as follows:
1. 25-year design storm for all cross-drain culverts and drainage designs.
2. 10-year design storm for all interior culverts and drainage designs.
3. 2- and 10-year design storms for all detention and retention basins using procedures contained in the Storm Water Management Design Manual or approved by the Municipal Engineer.
4. All drainage designs shall be checked using the 100-year storm for analysis of local flooding, and possible flood hazards to adjacent structures and/or property.
5. All hydrologic analysis will be based on land use conditions as specified in Section II F 2.
6. For the design of storage facilities, a secondary outlet device or emergency spillway shall be provided to discharge the excess runoff in such a way that no danger of loss of life or facility failure is created. The size of the outlet device or emergency spillway shall be designed to pass the 100-year storm as a minimum requirement.

SECTION J.   Plan water quality criteria

Following are the criteria related to using storm water management facilities for water quality purposes.

Ponds, lakes and reservoirs

1. When the land disturbing activity consists of the construction of a pond, lake or reservoir which is singly built and not part of a permitted land disturbing activity, the following procedures will apply:
    a. A storm water management plan will not be required if the pond, lake or reservoir is permitted under the State Dams and Reservoirs Safety Act or has received a Certificate of Exemption from the State Dams and Reservoirs Safety Act. Best management practices should be used to minimize the impact of erosion and sediment.
    b. A storm water management plan will be required for the construction of all ponds, lakes or reservoirs not meeting the conditions in (a) above that otherwise meet the size requirements for storm water management plan approval.
2. When ponds are used for water quality protection, the ponds shall be designed as both quantity and quality control structures. Sediment storage volume shall be calculated considering the clean out and maintenance schedules specified by the designer during the land disturbing activity. Sediment storage volumes may be predicted by the Universal Soil Loss Equation or methods acceptable to the Municipal Engineer.
3. Storm water runoff and drainage to a single outlet from land disturbing activities which disturb ten (10) acres or more shall be controlled during the land disturbing activity by a sediment basin where sufficient space and other factors allow these controls to be used until the final inspection. The sediment basin shall be designed and constructed to accommodate the anticipated sediment loading from the land-disturbing activity and meet a removal efficiency of 80 percent suspended solids or 0.5 ML/L peak settable solids concentration, whichever is less. The outfall device or system design shall take into account the total drainage area flowing through the disturbed area draining to the basin.
4. Other practices may be acceptable to the Municipal Engineer if they achieve an equivalent removal efficiency of 80 percent for suspended solids or 0.5 ML/L peak settable solids

concentration, which ever is less. The efficiency shall be calculated for disturbed conditions for the 10-year 24-hour design storm event.

5. Permanent water quality ponds having a permanent pool shall be designed to store and release the first 1/2 inch of runoff from the site over a 24 hour period. The storage volume shall be designed to accommodate, at least, 1/2 inch of runoff from the entire site.
6. Permanent water quality ponds, not having a permanent pool, shall be designed to release the first inch of runoff from the site over a 24-hour period.
7. The use of measures other than ponds to achieve water quality improvement are recommended on sites containing less than ten (10) disturbed acres.

Infiltration practices

1. Permanent infiltration practices, when used, shall be designed to accept, at a minimum, the first inch of runoff from all impervious areas.
2. Areas draining to infiltration practices must be established and vegetative filters established prior to runoff entering the system. Infiltration practices shall not be used if a suspended solids filter system does not accompany the practice. If vegetation is the intended filter, there shall be at least a 20 foot width of vegetative filter prior to storm water runoff entering the infiltration practice.
3. The bottom of the infiltration practice shall be at least 2.0 feet above the seasonal high water table, whether perched or regional, determined by direct piezometer measurements which can be demonstrated to be representative of the maximum height of the water table on an annual basis during years of normal precipitation, or by the depth in the soil at which mottling first occurs.
4. The infiltration practice shall be designed to completely drain of water within 72 hours.
5. Soils must have adequate permeability to allow water to infiltrate. Infiltration practices are limited to soils having an infiltration rate of at least 0.30 inches per hour. Initial consideration will be based on a review of the appropriate soil survey, and the survey may serve as a basis for rejection. On-site soil borings and textural classifications must be accomplished to verify the actual site and seasonal high water table conditions when infiltration is to be utilized.
6. Infiltration practices greater than three feet deep shall be located at least 10 feet from basement walls.
7. Infiltration practices designed to handle runoff from impervious parking areas shall be a minimum of 150 feet from any public or private water supply well.
8. The design of infiltration practice shall provide an overflow system with measures to provide a non-erosive velocity of flow along its length and at the outfall.
9. The slope of the bottom of the infiltration practice shall not exceed five percent. Also, the practice shall not be installed in fill material as piping along the fill/natural ground interface may cause slope failure.
10. An infiltration practice shall not be installed on or atop a slope whose natural angle of incline exceeds 20 percent.
11. Clean outs will be provided, at a minimum, every 100 feet along the infiltration practice to allow for access and maintenance.

SECTION K.   Professional registration requirements

Storm water concept and storm water management plans and design reports that are incidental to the overall or ongoing site design shall be prepared, certified, and stamped/sealed by a qualified registered Professional Engineer, Land Surveyor or Landscape Architect, using acceptable engineering standards and practices. All other Storm water concept and storm water management plans and design reports shall be prepared, certified, and stamped/sealed by a qualified registered Professional Engineer, using acceptable engineering standards and practices.

The engineer, surveyor, or landscape architect shall perform services only in areas of his/her competence, and shall undertake to perform engineering or land surveying assignments only when qualified by education and/or experience in the specific technical field. In addition, the engineer, surveyor, or landscape architect must verify that the plans have been designed in accordance with this ordinance and the standards and criteria stated or referred to in this ordinance.

ARTICLE III.   OWNERSHIP AND MUNICIPALITY PARTICIPATION

SECTION A.   Ownership of storm water management facilities

1. All storm water management facilities shall be privately owned and maintained unless the Municipality accepts the facility for Municipality ownership and maintenance. The owner of all private facilities shall grant to the Municipality, a perpetual, non-exclusive easement which allows for public inspection and emergency repair.
2. All storm water management measures relying on designated vegetated areas or special site features shall be privately owned and maintained as defined on the storm water management plan.
3. Most regional storm water management facilities will be publicly owned and/or maintained.

SECTION B.   Municipality participation

When the Municipal Engineer determines that additional storage capacity beyond that required by the applicant for on-site storm water management is necessary in order to enhance or provide for the public health, safety and general welfare, to correct unacceptable or undesirable existing conditions or to provide protection in a more desirable fashion for future development, the Municipal Engineer may:
   a. require that the applicant grant any necessary easements over, through or under the applicant's property to provide access to or drainage for such a facility;
   b. require that the applicant attempt to obtain from the owners of property over, through or under where the storm water management facility is to be located, any easements necessary for the construction and maintenance of same (and failing the obtaining of such easement the Municipality may, at its option, assist in such matter by purchase, condemnation, dedication or otherwise, and subject to (c) below, with any cost incurred thereby to be paid by the Municipality); and/or
   c. participate financially in the construction of such facility to the extent that such facility exceeds the required on-site storm water management as determined by the Municipal Engineer.
To implement this provision both the municipality and developer must be in agreement with the proposed facility that includes the additional storage capacity and jointly develop a cost sharing plan which is agreeable to all parties.

ARTICLE IV.   MAINTENANCE, CONSTRUCTION AND INSPECTION

Section A.   <u>Maintenance</u>

1. Any storm water discharge control facility which services a single lot or commercial and industrial developments shall be privately owned and maintained; provided, however, the owner thereof shall grant to the Municipality, a perpetual, non-exclusive easement which allows for public inspection and emergency repair, in accordance with the terms of the maintenance agreement set forth in Article IV, Section B, below.
2. All regional storm water discharge control facilities, identified on municipal storm water discharge control masterplans, shall be publicly owned and/or maintained.
3. All other storm water discharge control facilities shall be publicly owned and/or maintained only if accepted for maintenance by the Municipality.
4. Private maintenance requirements shall be a part of the deed to the affected property.

Section B.   <u>Maintenance agreement (privately owned facilities only)</u>

1. A proposed inspection and maintenance agreement shall be submitted to the Municipal Engineer for all private on-site storm water discharge control facilities prior to the approval of the storm water management plan. Such agreement shall be in form and content acceptable to the Municipal Engineer and shall be the responsibility of the private owner. Such agreement shall provide for access to the facility by virtue of a non-exclusive perpetual easement in favor of the Municipality at reasonable times for regular inspection by the Municipal Engineer. The agreement will identify who will have the maintenance responsibility. Possible arrangements for this maintenance responsibility might include the following:
    *   use of homeowner associations,
    *   arrangements to pay the Municipality for maintenance,
    *   private maintenance by development owner(s), or
    *   contracts with private maintenance companies.
    All maintenance agreements shall contain without limitation the following provisions:
    a. A description of the property on which the storm water management facility is located and all easements from the site to the facility;
    b. Size and configuration of the facility;
    c. A statement that properties which will be served by the facility are granted rights to construct, use, reconstruct, repair, maintain, access to the facility;
    d. A statement that each lot served by the facility is responsible for repairs and maintenance of the facility and any unpaid ad valorem taxes, public assessments for improvements and unsafe building and public nuisance abatement liens charged against the facility, including all interest charges together with attorney fees, cost and expenses of collection. If an association is delegated these responsibilities, then membership into the association shall be mandatory for each parcel served by the facility and any successive buyer, the association shall have the power to levy assessments for these obligations, and that all unpaid assessments levied by the association shall become a lien on the individual parcel; and
    e. A statement that no amendments to the agreement will become effective unless approved by the municipality.
2. The agreement shall provide that preventive maintenance inspections of storm water management facilities may be made by the Municipal Engineer, at his option. Without limiting the generality of the foregoing, the Municipal Engineer's inspection schedule may include an

inspection during the first year of operation and once every year thereafter, and after major storm events (i.e., 5- or 10-year floods).

3. Inspection reports shall be maintained by the Municipal Engineer.

4. The agreement shall provide that if, after an inspection, the condition of a facility presents an immediate danger to the public health, safety or general welfare because of unsafe conditions or improper maintenance, the Municipality shall have the right, but not the duty, to take such action as may be necessary to protect the public and make the facility safe. Any cost incurred by the Municipality shall be paid by the owner.

5. The agreement shall be recorded by the owner in the Register of Deeds prior to the final inspection and approval.

6. The agreement shall provide that the Municipal Engineer shall notify the owner(s) of the facility of any violation, deficiency or failure to comply with this Ordinance. The agreement shall also provide that upon a failure to correct violations requiring maintenance work, within ten (10) days after notice thereof, the Municipal Engineer may provide for all necessary work to place the facility in proper working condition. The owner(s) of the facility shall be assessed the costs of the work performed by the Municipal Engineer pursuant to this subsection and subsection 4 above and there shall be a lien on all property of the owner which property utilizes or will utilize such facility in achieving discharge control, which lien, when filed in the Register of Deeds, shall have the same status and priority as liens for ad valorem taxes. Should such a lien be filed, portions of the affected property may be released by the Municipality following the payments by the owner of such owner's pro-rata share of the lien amount based upon the acreage to be released with such release amount to be determined by the Municipal Engineer, in his reasonable discretion.

7. The Municipal Engineer, at his sole discretion, may accept the certification of a registered engineer in lieu of any inspection required by this Ordinance.

Section C.   Construction and inspection

1. Prior to the approval of the storm water management plan, the applicant shall submit a proposed staged construction and inspection control schedule. This plan shall indicate a phase line for approval; otherwise the construction and inspection control schedule will be for the entire drainage system.

2. No stage work, related to the construction of storm water management facilities, shall proceed until the next preceding stage of work, according to the sequence specified in the approved staged construction and inspection control schedule, is inspected and approved.

3. Any portion of the work which does not comply with the storm water management plan shall be promptly corrected by the permittee.

4. The permittee shall notify the Municipal Engineer before commencing any work to implement the storm water management plan and upon completion of the work.

5. The permittee shall provide an "as-built" plan certified by a registered professional (as outlined in Article II, Section K) to be submitted upon completing of the storm water management facilities included in the storm water management plan. The registered professional shall certify that:
   a. the facilities have been constructed as shown on the "as-built" plan, and
   b. the facilities meet the approved storm water management plan and specifications or achieves the function for which they were designed.

6. A final inspection shall be conducted by the Municipal Engineer upon completion of the work included in the approved storm water management plan to determine if the completed work is constructed in accordance with the plan.

7. The Municipal Engineer shall maintain a file of inspection reports and provide copies of all inspection reports to the permittee that include the following.
    a. The date and location of the site inspection.
    b. Whether the approved plan has been properly implemented.
    c. Any approved plan deficiencies and any actions taken.
8. The Municipal Engineer will notify the person responsible for the land disturbing activity in writing when violations are observed describing the following.
    a. Nature of the violation.
    b. Required corrective actions.
    c. The time period for violation correction.

ARTICLE V.   MISCELLANEOUS PROVISIONS

SECTION A.   Variances from requirements

1. The Municipal Engineer may grant a variance from the requirements of this Ordinance if there are exceptional circumstances applicable to the site such that strict adherence to the provisions of the Ordinance will result in unnecessary hardship and not fulfill the intent of the Ordinance.
2. A written request for a variance shall be required and shall state the specific variance sought and the reasons, with supporting data, for their granting. The request shall include descriptions, drawings, calculations and any other information that is necessary to evaluate the proposed variance.
3. Any substantial variance from the storm water management plan or concept plan shall be referred to all agencies which reviewed the original plan.
4. The Municipal Engineer will conduct a review of the request for a variance within ten (10) working days. Failure of the Municipal Engineer to act by the end of the tenth working day will result in the automatic approval of the variance.

SECTION B.   Appeals

Any person aggrieved by a decision of the Municipal Engineer (including any decision with reference to the granting or denial of a variance from the terms of this Ordinance) may appeal same by filing a written notice of appeal with the Municipal Engineer within thirty (30) calendar days of the issuance of said decision by the Municipal Engineer. The Municipal Engineer can then reverse his/her decision or send this notice to the Appeals Board with comments. A notice of appeal shall state the specific reasons why the decision of the Municipal Engineer is alleged to be in error and the Municipal Engineer shall prepare and send to the Appeals Board and Appellant, within fifteen (15) days of receipt of the notice of appeal, a written response to said notice of appeal.

All such appeals shall be heard by the Appeals Board which is hereby granted specific authority to hear and determine such appeals in a quasi-judicial capacity. Said appeal shall be heard by the Appeals Board at its next regularly scheduled meeting date, not to exceed thirty (30) days after receipt of the notice of appeal, or at such other time as may be mutually agreed upon in writing by the Appellant and the Chairperson of the Appeals Board. The Appeals Board will then render a decision within fifteen (15) days after the appeal has been heard.

Each party to the appeal shall be entitled to a hearing before the Appeals Board under judicial forms of procedure, at which hearing each party shall have the right to present evidence and sworn testimony of witnesses, to cross-examine witnesses, and to cause a transcription of the proceedings to be prepared.

Should either party be dissatisfied with the decision of the Appeals Board, any appeal of said decision may be appealed to the Superior Court by writ of certiorari.

SECTION C.   Penalties

1. Upon determination that a violation of this ordinance has occurred the owner shall be given a written notice of the violations and the time in which to correct the deficiencies.
2. If construction violations of the approved plan are occurring, an immediate stop-work order may be issued by the Municipal Engineer. If the Municipality issues a stop work order, the Municipality must show cause within forty eight (48) hours.
3. Any person violating this ordinance or any part thereof, including failing to stop work upon order, shall upon conviction thereof, be fined not more than two hundred dollars or imprisoned not more than thirty (30) days for each offense. Each separate interval of 24 hours, or every day, such violations shall be continued, committed or existing, shall constitute a new and separate offense and be punished, as aforesaid, for each separate period of violation.
4. The Municipal Attorney may institute injunctive, mandamus or other appropriate action or proceedings at law or equity for the enforcement of this Ordinance or to correct violations of this Ordinance, and any court of competent jurisdiction shall have the right to issue restraining orders, temporary or permanent injunctions, mandamus or other appropriate forms of remedy or relief.

SECTION D.   Grandfather clause

Any applicant or owner of a parcel of land within the jurisdiction of the Municipality who has constructed the required storm water management facility or who is in the process of meeting the storm water management requirements of the law at the time of the effective date of this Ordinance may elect to apply to the Municipal Engineer for reconsideration under the provisions of this ordinance.

SECTION E.   Conflict with other laws

Whenever the provisions of this ordinance impose more restrictive standards than are required in or under any other ordinance, the regulations herein contained shall prevail. Whenever the provisions of any other ordinance require more restrictive standards than are required herein, the requirements of such shall prevail.

SECTION F.   Severability

If any term, requirement or provision of this Ordinance or the application thereof to any person or circumstance shall, to any extent, be invalid or unenforceable, the remainder of this Ordinance or the application of such terms, requirements and provisions to persons or circumstances other than those to which it is held invalid or unenforceable, shall not be affected thereby and each term, requirement or provision of this Ordinance shall be valid and be enforced to the fullest extent permitted by law.

SECTION G.   Amendments

This ordinance may be amended in the manner as prescribed by law for its original adoption. Before the Municipality governing body amends this Ordinance, it must seek the advice of the Municipal Engineer who will make a recommendation for each amendment within thirty (30) days of this request.

SECTION H.   Liability

Neither the approval of a plan under the provisions of this ordinance nor the compliance with the provisions of this ordinance shall relieve any person from the responsibility for damage to any person or property otherwise imposed by law nor shall it impose any liability upon the Municipality for damage to any person or property.

SECTION I.   Effective date

The Ordinance shall be effective within sixty (60) days after adoption of this Ordinance by the Municipality and by reference to the Storm Water Management Design Manual.

Chapter 5 Financing Storm Water Management Programs

5.1 Financing Needs Of Storm Water Management Programs

Storm water management is rarely and then only temporarily the highest priority program within most municipalities. Unless a home or property is affected by flooding or drainage problems, other services provided by the municipality are much more important and will obtain more support from local politicians. If flood awareness is high, psychological factors could increase a temporary willingness to pay (Thunberg and Shabman, 1991). But this is not generally true for the whole of the municipal population.

Improving the drainage system, installing regional detention facilities, and upgrading culverts under local roadways are not seen as projects that will gain votes or support from a large segment of the voting population. Single purpose drainage bond issues have had major problems in gaining enough support for passage (Debo and Williams, 1979). Building roads, providing recreation areas, upgrading schools, and providing funds for police and programs dealing with drugs are much more likely to obtain funding before storm water management projects.

Until recently, storm water management in many municipalities has either been ignored or received attention only when major problems resulted during major storm events. Unless developers could be required to provide needed storm water management facilities as part of new developments these facilities were not provided. Also, except for facilities under public roadways, maintenance of the drainage system was delegated to private adjacent land owners. As a result little if any maintenance was done and drainage systems within urban areas across the country are inadequate to provide the protection needed for adjacent property and structures. In most municipalities, storm water management is seen as an added function of the public works and engineering department and sufficient funds to implement this added function are not provided.

With growing public expectation for service, the passage of the EPA water quality regulations, and some state mandates to provide water quantity and quality management, municipalities are being required to develop and implement comprehensive storm water management programs. Many of these programs will require spending several times current drainage budgets to provide the needed services. As a result, municipalities are now struggling with the problem of how to fund these programs without adversely impacting other municipal services. Different approaches are being taken by different municipalities depending on local politics, funding sources available, and citizen support for different funding programs. A recent survey found that citizens of one city were willing to pay over five dollars per month for clean water in the streams while there was very little willingness to pay for flood control (Charlotte, 1991).

The major new source of funding for storm water management is in the form of a user fee system under the auspices of a storm water utility (Cyre and Reese, 1992). This form of funding has several advantages over other competing forms of finance including its equitability, stability and adequacy.

5.2 Definition Of A Storm Water Utility

Municipalities employ a variety of funding methods, including service charges, several types of taxes, franchises and other fees, fines, and penalties. The various funding methods have distinctive characteristics which separate them legally, technically, and in terms of public perceptions. Four major categories of municipal revenue generation methods are taxes, service charges, exactions, and assessments.

- Taxes are intended primarily as revenue generators, without any particular association with the activities or improvements that they fund.
- Service charges are not established simply to generate revenue, but are tied to the objectives of a specific program to which they are associated. For example, water and sewer service charges are structured to cover the cost of those programs, not to simply generate revenue which is used for other purposes as well.
- Exactions, including dedications of right-of-way and payments related to the granting of franchises, are related to the extension of an approval or privilege to use.
- Assessments are geographically or otherwise limited taxes levied for improvements or activities of direct and special benefit to those who are being charged.

Municipal storm water management programs have been funded using a number of mechanisms as the primary generator of funds including: property taxes, sales taxes, state revolving funds, road funding, user fees, bonding, and surcharges on other utility fees. By far the most common current funding method is tax based. However, the concept of a storm water utility based funding method is fast growing. In the early 1970's there were only one or two true storm water utilities in existence. In the early 1990's there are over 200. This number is expected to more than triple in the next decade as the financial impacts of storm water quality legislation reach the many small municipalities.

A storm water utility falls primarily under the second of these funding categories: a service charge. It is based on the premise that the urban drainage system is a public system, similar to a waste water or water supply system. When a demand is placed on either of these two later systems the user pays. In the same way when a forested or grassy area is paved a greater flow of water is placed on the drainage system. This is the demand. The greater the demand (i.e. the more the parcel of land is paved), the greater the user fee should be. A storm water utility differs from the other two water-related utilities in several key ways. First of all, there is no way to remove or discontinue services for non-payment. Secondly the service is provided to all citizens without choice. Third, the demand placed on the system can only roughly be measured or approximated. Also, the actual service rendered to a particular property is often difficult to quantify. Despite these drawbacks, the utility concept for storm water financing is a viable and growing funding method.

Few if any storm water utilities have failed court challenges if: (1) they are fair and reasonable, (2) the costs are related to the services rendered, (3) they are legal by charter or legislation, and (4) the proper procedures were followed in setting up the utility.

The distinctions of the four revenue categories are very important. One of the critical issues which typically must be resolved if a utility service charge of any type is legally challenged is whether the service charge is clearly related to and incidental to the activities and improvements of the utility, or is in fact merely a means of creating revenue for all governmental purposes generally (a tax), or is a special assessment (which is supposed to reflect a direct and special benefit). Thus a storm water utility must be based on a storm water program and not simply a perceived financial need or willingness to pay.

A storm water utility must be seen as an umbrella under which individual communities address their own specific needs in a manner consistent with local problems, priorities and practices. It is understood in three ways as summarized in the box below. Development of adequate funding is the bottom-line solution which allows other, more technical programmatic solutions to flourish and be successful. With the expected needs for organizational changes within most municipalities to manage storm water and the demands placed on municipalities by the EPA, NPDES permit for storm water discharges, the stability and adequacy of a utility is a great advantage.

A storm water utility provides a vehicle for:

- consolidating or coordinating responsibilities that were previously dispersed among several departments and divisions;
- generating funding that is adequate, stable, equitable and dedicated solely to the storm water function; and
- developing programs that are comprehensive, cohesive and consistent year-to-year.

A storm water utility is equitable

WHAT IS A STORM WATER UTILITY?

- **A FUNDING METHOD**

  A method or mix of methods for providing adequate, stable and equitable funding for the comprehensive storm water program.

- **A PROGRAM CONCEPT**

  A comprehensive storm water quantity and quality program with an effective balance of: capital, operational, regulatory, engineering, planning and administrative activities.

- **AN ORGANIZATIONAL ENTITY**

  A legal entity with the authority to control storm water, operate systems and assess fees and charges.

---

because the cost is borne by the user on the basis of demand placed on the drainage system. It is stable because it is not as dependent on the vagaries of the annual budgetary process as taxes. And it is adequate because a typical storm water program can be financed with payments below the normal customer willingness to pay (EPA, 1989).

No two utilities are identical just as no two communities are just alike (Lindsey, 1988). Therefore it is not prudent to follow a pre-fabricated "one size fits all" approach, but seek to carefully understand the make-up of the community, its problems and its goals. There must be a clear understanding of the community's storm water systems, capabilities and issues. A questionnaire can be very helpful in gaining understanding of a community. An example questionnaire is provided at Appendix A at the end of this chapter.

Some communities have simply attempted to clone a storm water program concept or rate methodology adopted by another city or county. Any city or county should carefully guard against that temptation if they want to be successful. A storm water program or rate structure cloned from somewhere else cannot sustain intense scrutiny if a staff, advisory committee, elected officials, or interest groups really take a hard look at it. The program then often fails politically. Conversely, a well-thought out program and funding strategy can survive the scrutiny that would accompany controversy if questions should develop during or after the program development process.

The real danger of the cloning approach is that it inevitably falls short of efficiently and effectively solving the local storm water problems because it is not founded on addressing them. The local problems, needs, and circumstances must drive the form, priorities, and pace of transition of the program. The success of programs in Charlotte, North Carolina; Louisville, Kentucky;

Tulsa, Oklahoma; Cincinnati, Ohio; Mecklenburg County, North Carolina; Bellevue, Washington and elsewhere is not based on offering the same solutions everywhere, but rather tailoring the program and financing strategies to the local needs. Solving real short-term and long-term storm water problems in a municipality must be the focus of the program and utility development.

5.3 Utility Funding Methods

There are a number of funding methods, under the utility's umbrella, available for financing both storm water capital improvement projects and financing storm water infrastructure maintenance programs. In most cases, the method selected in any given municipality is based on a few factors:
- the authority to use different funding methods available through state statutes or home rule charters,
- the scope of program or projects to be funded,
- related local funding policies and practices,
- the general financial health of the local governments and their individual revenue funds, and
- the political atmosphere in which elected officials must make funding decisions.

A base funding method (or methods) must include the following if the storm water management program is to be successful:
- have sufficient revenue capacity to meet short-term program and capital project needs;
- provide a relatively stable revenue stream;
- be appropriate for most if not all of the different types of projects and activities; and
- be flexible enough to adjust as the program evolves during the next decade and beyond.

The financing method package developed for a particular utility is divided into three modules: (1) the basic rate methodology; (2) modification factors which can be applied to any of the rate concepts to enhance equity, reduce costs, and meet other objectives; and (3) the secondary funding methods that can be adopted in concert with the service charges. Typical modification factors include: flat rates for single-family residences, fixed costs per account, a crediting mechanism, location charges or credits, etc. Secondary funding methods are evaluated primarily to ensure that they would be consistent with the basic rate methodology.

The development of a preferred rate methodology involves the following steps.
1. Selection of a menu of appropriate funding methods for the program.
2. Definition of a primary funding method.
3. Selection of a basic rate concept from a menu of options.
4. Selection of appropriate rate modifiers from a menu of options.
5. Selection of appropriate "secondary" funding methods from the remaining funding methods.

Normally a rate structure analysis looks at basic rate methodologies. Each method has advantages and disadvantages. Some typical basic methods for calculating demand on the system consist of consideration of:
- impervious area;
- both impervious area and gross area;
- impervious area and impervious percentage;
- gross area and an intensity of development factor; or
- gross area only, with extensive use of modifying factors.

The basic rate methodology is selected on the basis of (Rosholt & Pigott, 1991):
- workability, ease of administration;
- fairness and equity;
- consistency with local policy;
- ability to meet revenue requirements;

- objectivity versus discretional determination of individual costs;
- applicability to entire service area; and
- effectiveness of desired incentives.

After the basic rate methodology is selected modification factors are examined. The modification factors are used to enhance equity or improve ease of utility implementation and management without unduly sacrificing equity. Typical modification factors include:

- a flat rate single-family residential charge;
- a base rate for certain costs which are fixed per account;
- basin-specific surcharges for major capital improvements;
- a surcharge for properties located in floodplains;
- credits against the monthly service charge for properties which have on-site detention or retention systems;
- a water quality impact factor;
- a development and land use factor; and
- a level of service factor.

Evaluation criteria are employed in the funding study to assess each method of primary or secondary funding:

- financial impact on citizens and businesses;
- equity and public acceptance;
- revenue sufficiency;
- timeliness, and the process required for implementation;
- the cost of development, implementation, and upkeep; and
- consistency with capital project and program needs.

While storm water is fundamentally different from wastewater or water supply many of the funding methods available in these utilities have analogous methods for storm water. Bonding, short-term financing and credit enhancers are all available to the water and wastewater utility (Raftelis, 1989, AWWA, 1986) and the storm water utility.

Capital planning for storm water utilities has traditionally not been very sophisticated. Water and Wastewater utilities have used various capital-budgeting techniques (such as return on investment, internal rate of return or net present-value) for a number of years. Unlike other utilities Federal or state financing is normally not available. There may be opportunities for capitalization grants under a state revolving fund (SRF) program though this is not a certainty from state to state nor is the program expected to continue indefinitely (Houmis, 1992, Davis, et al., 1989). Because storm water does not normally have treatment plants and does not own or operate much of its system such methods are difficult to apply. However, as infrastructure rehabilitation and replacement needs increase and master planning for major flood control projects takes place such accounting methods may take on more importance.

The more popular financing methods available to most municipalities include the following (Raftelis, 1989, EPA, 1988, Cyre, 1987, EPA, 1990):

1. Storm Water Drainage Services Charges - This method of financing fits well with storm water and flood control program needs in most municipalities. It is normally the primary financing mechanism for the utility and is often called "the utility". The functions and costs associated with storm water management are similar to those for a water or sanitary sewer utility. Thus, this "storm water utility" would be set up and function in a manner similar to other existing utilities. The principal difference between water and sanitary sewer services charges and storm water service charges is the basis of measurement employed for setting rates. The typical basis for storm water charges is the runoff production potential of a particular property. Usually, to avoid excessive utility administration and start-up costs, only industrial/commercial/multi-

family sites are considered on an individual basis while single family residences are covered under blanket charges.

2. Revenue Bonding - This method of municipal funding is most commonly used for facility construction, the purchase of equipment, or other major capital outlays. Unlike general obligation bonds, which are backed by the full faith and credit of the issuing governmental agency, revenue bonds are backed by revenues from specifically defined sources. Thus, it is intended to work in conjunction with some other ongoing funding system. "Anticipation notes" are somewhat similar with the anticipated revenues from a future utility covering the notes issued to generate funds to set up the utility.

3. System Development Charges - System development charges are also best used in conjunction with other funding methods as a mechanism for balancing financial participation in the cost of services and facilities. They provide a funding mechanism through which owners of properties which develop in the future share in the cost of projects built in anticipation of their needs. They can properly be described as a deferral mechanism through which a property owner's financial participation in a project is delayed until development occurs which uses the additional capacity originally built into the system.

4. Special Assessments And Improvement Districts - A number of different methods of levying special assessments on benefitted properties have been used throughout the United States for storm water and flood control improvements. Projects funded through special assessments must have a special benefit to the properties included in the assessment area and charges for each parcel must be consistent with the relative benefit to each property. Special assessment mechanisms are most often used for small local projects since they localize the cost of projects which serve a limited area and a limited constituency.

5. Plan Review And Inspection Fees - This method of funding is typically used by public works departments and utilities which operate storm water drainage systems and must maintain and regulate their use. The fee is used to cover the cost of plan review and inspection of project sites during the construction of the proposed development or to fund ongoing inspection programs of private structural facilities such as detention ponds or water quality best management practices. Many municipalities charge only a token fee while others try to determine the actual costs involved and set a fee to cover these costs.

6. In Lieu Of Construction Fees - Many municipalities require the construction of a storm water detention facility for all non-residential properties unless it can be demonstrated that construction of such a facility would adversely affect storm water peak discharge rates downstream from the site. The In Lieu of Construction Fee provides a funding mechanism whereby construction of numerous individual on-site detention basins or other storm water conveyance structures can be waived with developers contributing a fee to a fund to build a regional facility or make improvements in storm water conveyance at locations remote from the development site. One drawback of this funding method is that the facility must be built ahead of the development and, therefore, an alternate funding source must be available. Rarely do the funds for construction become available prior to the construction start.

7. General Facilities Charges - Utility service charge rate structures sometimes incorporate surcharges for facilities which are necessary to provide adequate service to all rate payers. These charges are known by various names but the term "General Facilities Charges" describes their purpose. General facilities charges for storm water and flood control improvements would be most appropriate for regional detention facilities or drainage systems serving only roads, parks or other public properties used by the citizens in a municipality. Such charges could also be used for the development of a data base or mapping information which serves all properties generally rather than just a certain sector or area.

8. Impact Fees - Impact fees are also designed to be used in conjunction with other funding mechanisms. They are most appropriate as a mechanism to balance financial participation with the cost of service, especially the short term cost resulting from the development of private property. Impact fees must pass the "rational nexus" test and should only be used where impacts to be mitigated are specific rather than general; a consistent methodology is available for quantifying the impact; a separate accounting for each impact fee project can be maintained; and a mechanism exists for returning the unused portion of an impact fee after a mitigation period has passed. Impact fees have lately come under close legal scrutiny (Schlette, 1989). In some cases, the cost of providing enough information to develop an appropriate impact fee method might make this funding method unattractive to the local municipality.

9. Developer Extension/Late-Comer Fees - Storm water management facilities built as part of private developments often should be oversized to accommodate service beyond the immediate confines of a single project. This results in additional construction costs to serve the needs of these external areas. A financing method can be employed which allows an original developer to be compensated for these front-end expenses by developers subsequently building in the same area and served by the oversized facilities. These fees are particularly useful where an overall development master plan is available which defines future storm water requirements.

Other secondary funding methods include sales taxes, surcharges, grants and loads, penalties, and many more.

5.4 Storm Water Utility Policy Issues

The analysis of storm water utility funding has numerous policy implications (Hardten, et al., 1990, Priede, 1990, Scholl, 1991a, 1991b, Zielke, 1990). Many issues involve deciding how service charges should be implemented and applied to specific properties in a consistent and fair manner. Timing is also important. Some issues will be resolved early in the process, some will wait until the utility is functioning.

Although policy-making in the highest sense is reserved to the Mayor and Council, day-to-day policy decisions are in fact often made at several levels. The Municipal Council formally adopts many of the major policy decisions which guide the municipality.

The Mayor also makes policy decisions often based on Municipal Council positions. Other policy decisions are made by Municipal Manager and staff administrators pursuant to the general directives spelled out by the Mayor and Council. Recognizing this dispersed policy-making environment, a simple hierarchy is recommended for the level of review of the important issues:

*   key staff and consultants,
*   other involved staff,
*   advisory committee,
*   manager's office, and
*   municipal council.

An initial screening of possible issues must consider the following (Cyre, 1986):

*   impacts of policy decision alternatives on costs and manpower;
*   the appropriate level(s) of municipal government at which the issue should be addressed and resolved;
*   the relationship of each specific issue to other policy issues; and
*   the priority and timing associated with the issue given the municipality's objective of implementing alternative funding for storm water management.

Policy issues in the development of a storm water utility can be divided into those dealing primarily with program issues, those dealing primarily with funding issues, and those dealing primarily with billing technical issues. Following is a list of typical policy issues in the three categories:

Program Related Policy Issues
- Program Mission
- Program Service Description
- Extent of Service
- Storm Water Quality Strategy
- Privatization
- Relationship With Other Programs
- Public Input or Advisory Groups

- Major Program Priorities
- Service Area
- Levels of Service
- Organization and Staffing
- Interlocal Agreements and Responsibilities
- Public Relations

Funding Related Policy Issues
- Types of Storm Water Services Funded
- Basis for Cost Distribution
- Prior Investment
- Future Use of Storm Water Systems
- Accounting Method
- Rate Methodology
    - Basic Funding Methodology
    - Secondary Funding Methods
    - Modification Factors
- Overall Funding Strategy
- Credits
- Equivalent Residential Unit (ERU) Base
    - Public Streets and Property
    - State and Federal Property

Billing Related Policy Issues
- Billing and Collection Methods
    - New Stand Alone System
    - Independent Database System Tie-in
    - Modification of Existing Billing System
- Appeals and Adjustments
- Billing Period
- Collections and Delinquencies
    - Water Bill Tie-in
    - Property Liens
    - Enforcement Procedural Issues
- Management Reporting
- Master Account File Development Process and Accuracy
    - Use of Other Databases
    - Resolution Procedures for Discrepancies
    - Number and Type of Data Fields Required
- Impervious Area Methodology
    - Rounding and Ranges
    - Use of Street Centerline Data
    - Impervious or Total Area Measurement Accuracy
- Master Account File Database Maintenance and Updating Processes
- Billing Cost Allocations
- Customer Service Procedures
- Billing Owners or Tenants

- Case Exceptions Including
  - Multiple Owners
  - Multi-Story Condominiums
  - Consolidated Billing
- Use of GIS, Mapping or CADD
- Information to Put On Bill

- Undivided Interest, Common Areas
- Storm Water Only Accounts

5.5 Typical Costs And Rates

The cost of managing storm water can be quantified in terms of cost per developed acre per year. Based on experience across the country Table 5-1 has been developed.

In a typical municipality which uses a storm water utility form of financing a charge of one dollar per residential unit per month (plus equivalent charges for non-residential properties based on impervious area) will generate between about $25 and $45 dollars per acre per year. Based on these two figures it can be seen that a typical advanced program with costs per acre per year of $100.00 would require a user fee of about $2.00 - $4.00 per month.

Typical fees are hard to compare because they are based on differently configured programs. Some municipalities have included a charge for their own streets in the rate to be paid by the municipality itself. This would tend to reduce the monthly rate about 12 to 20 percent and shift this cost to tax based funding. Other municipalities have elected to fund portions of the program through other methods such as state street funds or bonding. The definition of what is included under the storm water utility is also differently interpreted. This is especially true when storm water quality related programs are funded fully or partially by a utility charge. Table 5-2 provides approximate information on monthly charges for a number of cities across the country based on surveys (APWA, 1991, Black and Vetch, 1992, Hartigan, 1989, HDR, 1990) and other sources.

5.6 Steps In A Typical Financing Study

Each municipality will need to develop a financing program that takes into account the local politics, existing municipal programs, available resources, state programs, etc. Thus each program will be tailored for that specific municipality but there are some common steps that should be followed by all municipalities. Rate making for storm water is similar to that for water and wastewater and similar procedures and opportunities for automation and use of GIS apply (Raftelis, 1898, Hargett and Eggeman, 1992, Tomaselli, 1989). While there are many ways to approach utility development (Hodges, 1991, Hartigan, 1989, Rosholt, 1990, Cyre, 1990) utility development generally follows along three parallel tracks each playing off the other. The three tracks (not including administration and project management) are:

- Program Track - what will the program accomplish and when will it be accomplished, how will it be organized, what are its priorities?
- Utility Track - who makes up the rate base, what factors will be used, what secondary funding methods will be used, how will a large number of policy issues be answered?
- Billing Track - how will the bill be sent out, how will the master account file be developed, what databases now exist, etc.?

The efforts involved in the development of the utility can be costly and time consuming. However, the total setup costs amount to only a few months revenue. Often these funds are obtained by establishing the legal entity of the storm water enterprise fund and obtaining a loan

Table 5-1 Typical Costs Of Storm Water Management Programs

Program Type	Cost/Ac/Yr	Typical Program Features
Incidental	15-25	Reactive incidental maintenance and regulation as part of other programs
Minimum	30-50	ADD:right-of-way maintenance, better regulation and inspection, more staff, erosion control
Moderate	40-70	ADD:additional maintenance programs and levels of service, better regulation and inspection, some planning, minor capital program, general upgrade of capabilities
Advanced	70-140	ADD: maintenance (of some sort) of the whole system, master planning, regional treatment, some water quality, data collection, multi-objective planning, strong control of development and other programs, utility funding
Exceptional	over 200	ADD: storm water quality, advanced flood control and levels of service for maintenance, aesthetics become more important, public programs

Table 5-2 Monthly Storm Water Management Fees

CITY	CHRG/MO ($)	CITY	CHRG/MO ($)
Louisville MSD, KY	2.75	Tallahassee, FL	5.00
Cincinnati, OH	1.28	St. Petersburg, FL	4.30
Bellevue, WA	7.98	Tulsa, OK	2.00
Sacramento, CA	9.06	Norfolk, VA	4.00
Durham, NC	3.41	Greensboro, NC	2.68
Austin, TX	3.32	Modesto, CA	3.08
Sarasota, FL	3.00	Miami, FL	2.50
Corvallis, OR	2.15	Daytona Beach, FL	1.75
Winter Park, FL	4.00	Gainesville, FL	3.30
Ann Arbor, MI	3.63	Seattle, WA	2.49
Salt Lake City, UT	3.00	Loveland, CO	2.50
Charlotte, NC	3.05	Tacoma, WA	4.27

from another fund such as the water or wastewater reserve fund. Anticipation notes are another way to fund the establishment of the utility (Raftelis, 1989).

A typical scope of services for the development of a storm water utility is outlined below:

Project Administration, Management And Support

Task A-1   Project Administration And Management
Task includes all project management, general meetings, technical track coordination, municipal information briefs and project updates, and coordination with sub-contractors.

Task A-2   Public Involvement Program
The Team will develop a plan for a public information and involvement program. The plan will address briefly the involvement of staff, political leaders, stakeholders, the general public and the news media and provide a menu of activities which can be used to support the storm water program and utility development. The Consultant will assist the staff on an as-needed basis to advise on public information and awareness activities and to execute various portions of the plan.

Storm Water Program Development

Task S-1   Storm Water Program And Financing Strategy
The Team will define a three to five-year storm water program strategy and identify the type of funding needed to support the program concept.
- Program priorities will be established. The mix of operational and capital investment program elements and the types of costs to be incurred will be projected, and the critical steps in achieving the strategy determined.
- The results of funding policy decisions will be incorporated into a strategy for funding development. The strategy will spell out the specifics of what must be done to develop and implement the funding and rate methodology(ies) selected by the municipality.

The deliverable product will be a program and financing strategy report.

Task S-2   Organization And Staffing
The Team will assess the impacts of storm water program decisions on the current organizational, management, and staffing. Future organization and staffing will be projected consistent with storm water program and utility funding policy decisions (Tasks S-1 and U-1). The deliverable product will be an organization and staffing report.

Task S-3   Cost Of Service Analysis
The Team will develop an estimate of the total cost of storm water services and facilities (revenue requirements) for the period to be funded by the initial service charge. The costs will be based on policy, program and staffing decisions made by the municipality. An examination will be made of which types of costs will be funded from the utility user fee and which can be funded from other sources. The deliverable product will be a cost of service report.

Task S-4   Program Implementation Assistance
There will invariably be numerous policy and implementation decisions which must be handled quickly "on the fly". This task allows the Team to work together on an "on-call" basis in implementing the various policy decisions that have been made and developing new policies for situations encountered after the first bill goes out.

Storm Water Utility Development

Task U-1   Storm Water Funding Policy Issues
A number of storm water funding policy issues must be identified and decisions reached in a timely manner. The policy issues will include those related to the costs and levels of service, interlocal agreement terms, rate structure modifications, secondary funding methods billing, collections, and other matters. The policy issues will be developed and examined, alternative policy positions for the municipality described, and recommendations presented in the form of short policy statement papers. Results of these policy decisions will guide Tasks S-1 and U-2.

Task U-2   Rate Structure Analysis
A range of funding options will be quickly explored and preliminary decisions made prior to subjecting them to more detailed analysis during the rate study. Municipal staff and the Team will participate in funding policy discussions. Basic user fee rate methodology (basis for charges), modification factors and secondary funding methods will be explored and recommendations made. The deliverable will be a report describing the analysis and recommendations.

Task U-3   Budget And Cash Flow Analysis and Detailed Rate Study
The Cost of Service Analysis is primarily used for rate-making decisions, whereas a Budget and Cash Flow Analysis would be used for financial management in the context of the municipality's accounting systems and practices. This task facilitates the translation of the cost of service information to the municipality's budget format and provides a projection of the cash flow that will result from the implementation of a storm water service charge.

The Detailed Rate Study will apply the cost of service information to the rate methodology for various classes of rate payer similar to water rates (AWWA, 1983). A rate algorithm will be developed to calculate the service charges for all properties subject to the initial storm water service charge and the basic user fee will be calculated. The Rate Study will also consider how other funding methods might be used both within and outside the umbrella of a storm water rate ordinance, and what their impact would be on service charge rates and rate design. The deliverable product will be a rate study report.

Task U-4   The Rate Ordinance(s)
The Team will develop a draft rate ordinance consistent with the technical findings of the Cost of Service Analysis and Rate Study. The Team will investigate other legal changes necessary for the establishment of a utility. The final ordinance will be prepared by the municipal attorney's office.

Billing And Database Development

Task D-1   Data, Materials And Information
The Team will:
* identify and evaluate materials (i.e., parcel maps, site maps) and information (i.e., detention maintenance agreements, lists of impervious areas) needed during the development and implementation of storm water service charges;
* recommend how and when the materials and information should be assembled;
* recommend scheduling of associated analytical, system development, and implementation work;
* identify gaps and deficiencies in the information, systems, and procedures, and determine a remedy;

- propose data evaluation criteria consistent with the types of information to be used in various aspects of the funding implementation work; and
- develop a detailed methodology and schedule for development of the master account file.

The deliverable product will be a material and information evaluation and implementation report and a Methods of Procedure manual for the master account file.

Task D-2   Base Master Account File

This task provides for gathering together the working materials, data, information on support systems, and other resources that will be used in the preparation of a master account file. Because of uncertainties at this time, it is possible that data and/or system deficiencies may necessitate creation of new information or approaches. The Team will prepare the actual master account file to be used for the storm water service charge billing. The municipality will specify the format for the file for use in billing purposes. The deliverable product will be an account file designed to the municipality's specifications.

Task D-3   Data For Initial Billing

This task provides for the Team to assemble and/or generate the actual data to be used in calculating storm water service charges for individual properties. Maintenance of the data after the initial billing will be a municipal responsibility. The work should be accomplished largely by data processing rather than by examining the site conditions on each property. Deliverable product will be rate charges and basis for all billable properties within the municipal boundaries.

The Team will identify and correct errors (to the extent possible) which occur in the generation and assembly of data prior to the initial storm water service charge billing. Should it be jointly determined that a test run of the delivery system is required to check for billing errors all mailing costs will be borne by the municipality.

Task D-4   Billing System Development

The Team will develop a billing system for delivery of the storm water bill. This task can include anything from development of a turn-key stand-alone billing system to simply adding storm water accounts to an existing and adequate billing system.

Task D-5   Inquiry And Complaint Response Measures

The Team will develop inquiry and complaint response measures to handle an expected initial public reaction to the billing. The Team will develop response strategies and response sequences for a number of types of inquiries including billing questions and basic technical answers.

5.7 Coordinating Financing And Program Needs

Municipalities starting to develop their storm water management programs must consider the cash flow problems that might arise. As an example if the municipality decides to use service charges as the basis for their funding, the municipality will need to have the funds required to fund the study to set rates, prepare master account file, modify billing, etc. Such a study can be quite expensive especially if the municipality has not developed any aspects of their storm water management program such as master plans, ordinances, etc. Also as soon as the municipality announces that they have a storm water management program and especially if local citizens are required to pay service charges, the municipality must be in a position to deliver adequate services. Citizens will expect the municipality to be able to respond to flooding and drainage problems, provide some maintenance to the local drainage system, require needed storm water management facilities in new developments, etc. (Debo and Williams, 1979).

Coordinating financing and program needs is very important especially during the first few years of the storm water management program development and implementation. Thus, municipalities might want to consider several financing options such as passing a major bond issue to finance the development and initial implementation of the program and then use service charges and other financing methods to pay back the bonds. Also priorities should be established to determine which needs should be met with initial funding and which should be scheduled for later implementation. Implementation should also be distributed throughout the municipality so that all local residents will benefit from the program, and help build support and assist in obtaining future funding.

Finally, whatever funding method or methods are used by a municipality, the amount of funds raised should be sufficient to implement a program at a service level acceptable to the local citizens. To provide inadequate funding will limit implementation of different elements of the program and will lead to erosion of support for the entire storm water management program. The funding sources should also be protected from political pressures that might redirect funding to other services during times when flooding and drainage problems are not evident. Thus the funding source should be dedicated to the storm water management program and provide an adequate level of consistent funding over the long term.

References

American Public Works Association, Financing Stormwater Facilities: A Utility Approach", APWA, 1991.

American Water Works Assoc., "Water Rates", AWWA Manual M1, AWWA, Denver, CO.,1986.

American Water Works Assoc., "Water Rates and Related Charges", AWWA Manual M26, AWWA, Denver, CO.,1986.

American Water Works Assoc., "Water Utility Capital Financing", AWWA Manual M29, AWWA, Denver, CO.,1988.

Charlotte, NC., Results of Citizen Survey for Storm Water Management, 1991.

Cyre, H. J., "Key Feasibility Issues in Developing and Implementing a Stormwater Management Utility", Int. Pub. Works Congress and Eq. Show, New Orleans, Sept. 23, 1986.

Cyre, H. J., "Refinements in Stormwater Utility Rate Structures", ASCE 14th Annual Wat. Res. Plan. & Mgmt. Conf., Kansas City, MO, March 16-18, 1987.

Cyre, H. J., "Financing Urban Storm Water Management Case Studies of Change in Progress", Proc. WPCF Annual Conf., Oct. 7-11, Washington, DC., 1990.

Cyre, H.J. and Reese, A. J., "Storm Water Utilities in the United States of America", Proc. NOVATECH 92, Nov 3-5, 1992, Lyon, France.

Davis, P. E., A. E. Major and J. N. Dunlap, "Making the State Revolving Fund Work: The Tennessee Experience", Water Envir. and Tech., Nov., 1989.

Debo, T. N. and J. T. Williams, "Voter Reaction To Multiple-Use Drainage Projects", ASCE J. of WR Planning and Mgmt., Vol. 105, No. WR2, Sept., 1979.

Environmental Protection Agency, "Financing Marine and Estuarine Programs: A Guide to Resources", USEPA Ofc. of Water Rpt. 503/8-88/001, 1988.

Environmental Protection Agency, "Building Support for Increasing User Fees", USEPA, Ofc. of Water, Rpt. No. 430/09-89-006, 1989.

Environmental Protection Agency, "Financing Mechanisms for BMP's", USEPA Region V, Chi., Ill., December, 1990.

HDR Engineering, Inc., "Storm Water" News Bulletin, February, 1990.

Hardten, R. D., R. B. Bensen, and K. D. Thomson, "How Much to Charge and How to Collect it: Stormwater Rate Setting and the Billing System", Proc. WPCF Annual Conf., Oct. 7-11, Washington, DC., 1990.

Hartigan, J. P., "Use of Stormwater Utility to Meet New Water Quality Requirements", Presented to Virginia Sec. ASCE, Oct. 20, 1989.

Hargett, C. W., and G. L. Eggeman, "Database Utility Rate Model", Public Works, August, 1992.

Hodges, R. H., "How to Create a Stormwater Utility", Public Works, Oct., 1991.

Houmis, N., "Infrastructure Financing", Am. Consulting Engr., Winter, Vol. 2, No. 1, 1992.

Lindsey, G., "Financing Stormwater Management: The Utility Approach", Sed. & Stormwater Administration, State of Maryland, 1988.

Lindsey, G., "Update to Survey of Stormwater Utilities", Proc. WPCF Annual Conf., Oct. 7-11, Washington, DC., 1990.

Pardiwala, S. D., and J. R. Leserman, "Gaining Acceptance for Utility Rate Increases", Public Works, June, 1992.

Priede, N., "Managing Stormwater Utilities", Proc. WPCF Annual Conf., Oct. 7-11, Washington, DC., 1990.

Raftelis, George, "Water and Wastewater Finance and Pricing", Lewis Publishers, 1989.

Rosholt, J. E., "Program for Successful Implementation of a Storm Water Utility", Proc. WPCF Annual Conf., Oct. 7-11, Washington, DC., 1990.

Rosholt, J. E. and S. P. Pigott, "Financing and Service Charge Alternatives for Storm & Surface Water Management", URS Consultants, Inc., Personnel Communication, 1991.

Schlette, T. C., "Funding Wastewater Projects with Impact Fees", Water Environment and Technology, Nov., 1989.

Scholl, J. E., "Rate Structure Development for Stormwater Management Utility Billing", CH2M Hill, Gainesville, FL, Personnel Communication, 1991a.

Scholl, J. E., "Stormwater Management Utility Billing Rate Structure", Water Environment and Technology, Jan., 1991b.

Thunberg, E., and L. Shabman, "Determinants of Landowner's Willingness to Pay for Flood Hazard Reduction", AWRA Water Res. Bull., Vol 27., No. 4, August, pp. 657-665, 1991.

Tomaselli, L. K., "A Geographic Information Systems Approach to Fiscal Impact Analysis", PhD Dissertation, U of Minn., Minneapolis, MN, 1989.

Zielke, L. J., "Legalizing Stormwater Service Fees and Avoiding Issues of Taxation", Proc. WPCF Annual Conf., Oct. 7-11, Washington, DC., 1990.

Appendix A - Financing Storm Water Management Programs

QUESTIONNAIRE - MUNICIPALITY OF ANYPLACE
STORM WATER MANAGEMENT

Distribution Of Questionnaire

The following functional or interested personnel should each receive a copy of the questionnaire:

planning	design	engineering	maintenance	operations
enforcement	inspection	zoning	regulation	plans review
financial	floodplain mgmt	codes and stds	emergency svc	police
fire	legal	parks and rec	health & safety	complaints
water & sewer	utility	councilmen	mayor	other staff

and: ■ key community technical individuals such as state, Federal or university offices;
■ the development community; and
■ concerned citizens group representatives.

Purpose Of Questionnaire

The Municipality of Anyplace has contracted with the Consultant Team to conduct an analysis of storm water management in Municipality of Anyplace. One purpose of this study is to look at the policies regulating storm water management in Anyplace and to make recommendations for improvements and modifications as well as developing an ordinance to regulate storm water management activities.

The study is not budgeted to allow for in depth interviews with all individuals involved in some aspect of storm water management. This questionnaire is designed to serve as a basis for and to supplement the interviews that will be conducted.

Directions To Respondents

Respondents should use a separate piece of paper to answer the questions and may wish to dictate answers for secretarial transcription. There may be some areas where the individual feels he or she is not qualified to respond. They should not hesitate to leave these questions unanswered or simply insert the name and office of the individual who they feel is best qualified to address the question. A broad distribution of this questionnaire should ensure that all questions are answered. Respondents should answer those questions they can in sufficient detail so that their responses can stand alone and complete. Do not hesitate to include reports, documents, etc. with the response.

The Consultant wishes to thank each respondent for their participation in this information gathering process.

PLEASE PROVIDE YOUR ANSWERS ON A SEPARATE SHEET KEYED TO QUESTION NUMBERS HERE

A. Physical Drainage Problems

A1.   Complaints. About how many drainage complaints are on file? How are complaints handled? What is the response procedure? What type of flooding complaints predominate?

A2.   Flooding. How would you characterize the flooding problems in the Municipality of Anyplace? Are they primarily related to uncontrolled overflow of creeks, overland flow, undersized systems, major or minor systems, nuisance or life-threatening, etc.? Are they in open or closed conveyances? Are any of the problems related to aging of the system, clogged systems, damaged systems, etc.?

A3.   Erosion. How would you characterize erosion problems? Are they minor backyard problems, major creek erosion, only in spots along creeks, minor feeder streams, only downstream from urban development sites, etc.? What are the predominant causes?

A4.   Water Quality. Describe any Municipality of Anyplace activities in urban water quality? What type of information is available? Sampling program? Are there any known areas of water quality impairment?

A5.   Locations. What is the geographical extent of the various types of problems? Are they concentrated in a few areas along the major watercourses, widely scattered minor problems? Name the areas or reaches/streets. Do they vary from one part of the municipality to another? How? Are they seasonal, year-around or limited to special weather events?

A6.   Impact. How directly do the problems impact people? What are some specific examples? Is there significant exposure to personal hazard or risk or significant property damage due to flooding?

A7.   Safety Concerns. Have drainage conditions ever posed a problem for fire, police or emergency aid vehicles whether by flooding major transportation routes or contributing to poor driving conditions? Does potential exist even though no known problems have yet been encountered? Where? Have there been any other safety concerns related to flooding or conveyance structures?

A8.   Sanitary System. Have there been any water quality related health or safety problems due to leaking sanitary systems, septic system malfunction, etc?

A9.   Toxics. Are there any known areas of hazardous or toxic materials storage which may pose a problem during flooding conditions? Describe the emergency response capability of the municipality.

A10.   Responses. Describe the municipal responses to flooding, erosion, water quality and/or safety concerns. What types of solutions are typical? Are there special designs which seem to work best?

A11.   Root Causes. Can you characterize any deeper problems you feel might be deeper root causes of drainage problems?

A12.   Limits. What are the major constraints on responding to complaints or correcting flooding, quality and/or erosion problems?

B. Administration, Finance And Program Development

B1.   Organization and Responsibility. Use a functional table to identify who, if anyone, has primary and secondary responsibility in each area now performed in the Municipality of

Anyplace. Provide an organization chart if available for one or more agencies. Describe: principal storm water related duties, annual budgets related to storm water management and legal authority.

B2. Financial Overview. What are the basic financing methods used to support storm water management in all its various aspects within the municipality? What other financing methods are presently used for other municipal programs that are well accepted by the community? Which methods are least popular? What level of acceptance do the municipal's utility service charges have generally? What is the level of acceptance of property, sales and other taxes which generate local revenue?

B3. Financial Priorities. Is there a perceived balance in funding various municipality programs or is there a perception that one or more programs receive strong support while others lack public or political support?

B4. Utility Charge Structure. Does the Municipality of Anyplace presently charge schools, public hospitals, municipal and other government agencies, churches, other non-profit charitable organizations, state or federal buildings or properties for water and sanitary sewer services?

B5. Utility Condition. What is the financial condition of the municipal's utilities? Have they issued revenue bonds for capital improvements or other purposes? What was the bond rating?

B6. Economic Condition. What is the condition of the local economy? What is the level of unemployment? Does the Municipality of Anyplace have large concentrations of low income and/or elderly citizens? Are delinquencies on water and sewer utility charges increasing or decreasing recently? How many residential mortgage foreclosures have there been recently?

B7. Bonding Health. What is the municipal's general obligation bond rating? How much uncommitted general obligation bonding capacity exists?

B8. Willingness to Pay. How much do you think residents would be willing to pay monthly for drainage control if it were funded through a service charge?

B9. Billing Area. What is the present area within the municipal limits? Is annexation being aggressively pursued? Does the planning function go beyond municipal limits?

B10. Public Involvement. What types of special interest groups or other governmental authorities have some impact on municipal storm water management? Is there a constituency supporting storm water in the general public or neighborhood groups?

B11. Political Awareness and Support. Describe the political concern and involvement in storm water management? Are they concerned and knowledgeable?

B12. Current Storm Water Issues. What issues or circumstances have led to the current interest in storm water management? What are the key issues in the minds of staff, political leaders and the public?

B13. Current Other Issues. Are there any other issues now ongoing which may impact the ability of the municipality to respond aggressively to storm water needs?

B14. Program Planning. What plans are now underway to change, add to, or in any way impact any existing storm water programs or activities?

B15. Limits. In your opinion, what now limits the storm water management program from an organizational, financial and administrative point-of-view?

C. Planning, Design And Engineering

C1.   Design Criteria. What types of design guidance or design criteria documents or software packages exist that are used by engineers in Anyplace? What policies (written or unwritten) exist in the Municipality of Anyplace for design criteria?

C2.   Technical Studies. What types of technical studies exist which describe the storm water system and address either quality, quantity or erosion problems? Has there been any storm water master planning? Include soil surveys and flood program documents.

C3.   Design Methods. What are the typical design methodologies used for common drainage designs whether or not they are included in municipal documents or manuals?

C4.   Water Supply. What is the municipal water supply? If it is surface waters how are they protected?

C5.   Hazard Mitigation. Is there any type of flood hazard mitigation program in place such as flood warning, emergency evacuation, floodplain zoning, flood proofing, public education, etc.?

C6.   Limits. In your opinion, what now limits the storm water management program from a planning, engineering and design point-of-view?

D. Operations And Maintenance

D1.   Basic Scope and Operation. What portions of the storm water conveyance system are maintained (e.g., catch basins, minor streams, crossings, etc.)? Who performs this maintenance? What is done typically (schedule, activities, etc.)?

D2.   Priority. Is the maintenance program for each part of the conveyance system periodic or driven mainly by complaints? How is maintenance prioritized? What is the emergency response capability?

D3.   Budget and Manpower. What is the level of manpower, organization, equipment and budget for each entity that performs maintenance? Is any maintenance contracted out (e.g., remedial, minor reconstruction)?

D4.   Erosion. How are erosion and sediment control addressed in the program?

D5.   Limits. In your opinion, what now are the major constraints on expanding the maintenance system and operation (legal, financial, etc.)?

E. Regulation And Enforcement

E1.   Development Process. Describe the development process as it relates to grading and drainage. What permits are required, and when? What inspections are performed, when and by whom?

E2.   Written Guidance. Describe the documents, policies, checklists, etc. which govern all aspects of storm water management (For example: ordinances, regulations, written policies, etc.).

E3.   Authority. Who establishes and enforces storm water related regulations? What is the involvement of various agencies? What enforcement penalties are available?

E4.   FEMA. Describe the municipal's implementation of the FEMA requirements.

E5.   Floodplain Management. Describe the Municipality of Anyplace efforts in greenway planning and general floodplain management. Has there been any other park and recreation planning in conjunction with storm water management?

E6.   Water Quality. Which department within the municipality will have responsibility for any water quality regulations (including pre-treatment, toxicity and water quality)? How do other departments support this department?   Who would run a storm water quality program?

E7.   <u>Other Government Levels.</u> Describe the role of state and Federal government in storm water management in the Municipality of Anyplace in flood control, erosion and environmental quality.

F. Capital Improvements And Expenditures

F1.   Describe the municipal's capital improvement program for storm water management. Have there been any recent projects? How were they financed?

F2.   What is the expected level of need for major or minor capital improvements?

F3.   Describe the municipal's land acquisition or easement program.

Chapter 6 Data Availability And Collection

6.1 Overview

Introduction

The need for accurate and applicable data and information to assist the storm water manager has never been greater. With the onset of both complex water quantity control and the need to assess and seek to control storm water quality the usefulness of pertinent data cannot be overstated. Knowing what data are needed, what may be available, where to find it, what to do if data are missing or partial, how to assess data, and how to plan and organize a data collection effort are part of the day-to-day responsibilities of the modern storm water manager. Yet it is an area where many storm water managers have little training or experience.

The purpose of this chapter is to outline the types of data that are normally required for storm water management analysis and design, possible sources, and details of common types of data collection. Identification of storm water management data needs should be a part of the early planning phase of any project or program. Several categories of data may be relevant to a particular storm water management project, including published data on precipitation, soils, land use, topography, streamflow, instream water quality, point and nonpoint discharges and flood history. Field investigations are generally necessary to determine drainage areas, identify pertinent features, obtain high water information, and survey channel sections and bridge and culvert crossings. Because of its complexity and the relative difference in character, environmental sampling is considered separately at the end of this chapter.

Data Collection Effort

The principal use of storm water management data is to establish the hydrologic, hydraulic and water quality characteristics of a watershed or conveyance system in order to evaluate storm water runoff, flooding conditions and water quality values. Depending on the type of study or design being done both existing and future watershed conditions may be considered. Data should be collected before calculations are initiated, using the following general guidelines:

1. Identify data needs, sources, accuracy required, and uses.

2. Collect published data, based on sources identified in Step 1 and information presented in the following sections of this chapter.

3. Compile and document the results of Step 2, and compare data needs and uses with published data availability. Identify any additional field data needs.

4. Collect field data based on needs identified in Step 1 and Step 3.

5. Compile and document the results of Step 4.

The effort necessary for data collection and compilation should be tailored to the importance of the project. Not all of the data discussed in this chapter will be needed for every project. Some data will be necessary for submission to a municipal engineering or public works department as part of engineering studies and analysis. Other data will be needed to make engineering and design decisions, apply hydrologic and hydraulic procedures, etc. Finally, data may be needed for administration or judicial review. Examples might include historic flood elevations, recent flood elevations and associated damages, recent land use and drainage facility changes, etc.

A well planned data collection program leads to a more orderly and effective analysis and design that is commensurate with:
*   project or program scope,
*   project or program cost,
*   the complexity of site hydraulics, and
*   regulatory requirements.
    Data collection for a specific project must be tailored to:
*   site conditions,
*   scope of the engineering analysis,
*   social, economic and environmental requirements,
*   unique project requirements, and
*   regulatory requirements.

6.2 Sources And Types Of Data

There are many potential sources of data typically required for storm water management projects. Identifying these sources can be difficult, and making the subsequent necessary contacts can be time consuming. In addition to the data sources discussed in this chapter, existing watershed master plans, flood studies or other local studies provide the best source of data for the watersheds studied and provide a starting point for watersheds not studied. Appendix A at the end of this chapter gives a list of data sources that will prove useful for storm water design and analysis projects.

Often the types and accuracy of existing data determine the types and detail of studies that need to be done. For example, historical rainfall-runoff data for a particular watershed will naturally lead to individual storm calibration of hydrologic models. If such data are not available synthetic calibration or data transfer is necessary.

Data Categories

There are many ways to categorize the types of data needed for various studies. Some data are continuous while others are discreet or binary (yes or no) data. Some are man made while others are natural. Table 6-1 provides a listing of commonly encountered primary urban storm water management data acquisition categories. Development of estimates for these data needs can be done using either field or office methods or a combination of both. For more specialized studies many more categories and types of data could be listed. Specific parameters required for each category depend on the methods chosen for making calculated estimates.

Table 6-1 Categories Of Primary Data Needs For Urban Storm Water Management

Precipitation
- rainfall intensity-duration-frequency
- rainfall temporal distribution
- rainfall areal extent
- storm movement
- annual and seasonal precipitation
- past studies

Losses
- interception
- depression storage
- evaporation parameters
- infiltration/ soil parameters
- rainfall runoff gaging information
- past studies

Open Channel Flow
- channel system layout
- channel shape and geometry parameters
- channel slope
- flow resistance
- natural controls
- channel structures locations, types and routing information
- channel conveyance structures
- historical data on peak flows
- stage-discharge rating curves
- velocity and discharge measurements
- flow frequency information from stream gaging
- debris and ice flow information
- adjacent or regional information or parameters
- past studies

Basin Characteristics
- size
- basin slope
- basin shape - length, width
- drainage density
- land use - agricultural and urban
- surface and sub-surface geology
- karst or depression storage information
- past studies

Flood Damages
- stage-frequency curves
- property appraisals and types

- business types and susceptibility to flooding
- generalized stage-damage curves
- historical damage information
- future development and population estimates
- existing flood control structural or non-structural programs
- transportation information

Pipe Systems
- inlet size, type, location
- outlet types and protection
- backwater and submergence information
- locations sizes, slopes and shapes of pipes
- ground cover
- flow bypass information
- connectivity
- maintenance and structural condition and blockages
- ages of system
- maintenance records
- material types
- special structures
- past studies

Water Quality
- buildup-wash off parameters
- land use loadings
- Event Mean Concentration (EMC) values - regional, site specific, national
- receiving water quality information - storm and base flow
- designated uses of receiving waters
- specific modeling parameters
- Best Management Practice (BMP) design information and efficiency values
- rainfall and runoff volumes
- chemical sampling information
- biological or fish tissue testing information
- known use impairment
- atmospheric deposition
- point sources and types
- non-structural programs and effectiveness
- existing structural BMPs

Groundwater
- surface and sub-surface geology
- soils information and parameters
- observation well information
- ground water level measurements
- well, geologic, or stratigraphic log information
- aquifer test information
- infiltration test information
- electrical, seismic or gravimetric information
- laboratory analysis of soils

- groundwater quality analysis
- downhole groundwater quality data

Sediment Transport and Geomorphic Studies
- channel sinuosity and planform statistics
- channel type: meandering, braided, straight
- sediment transport measurements: suspended load, bed load and wash load
- flow duration information or curves
- bed and bank shear stress
- turbulence and dispersion
- bed and bank material and sizes and roughness
- vegetation
- bedforms and bar locations
- local or general scour information
- historical information on channel size, form or alignment changes
- surface and sub-surface geology
- backwater information from receiving stream
- existing bank or bed protection locations and types
- dredging records, potomology studies, reservoir sedimentation records
- past studies

6.3 Major Data Topics

Following is a brief discussion of some of the more common major data topics that relate to storm water management facility analysis and design.

Watershed Characteristics

The three most important general physical watershed characteristics for municipal storm water management are contributing size, slope and land use.

Contributing Size - The size of the contributing drainage area expressed in acres or square miles is determined from some or all of the following.
- Direct field surveys with conventional surveying instruments.
- Use of topographic maps together with field checks to determine any changes in the contributing drainage area such as may be caused by:
  - terraces,
  - urban storm drain systems,
  - lakes, sinks,
  - debris or mud flow barriers,
  - reclamation/flood control structures, and
  - irrigation diversions.
  Note: U.S. Geological Survey (U.S.G.S.) topographic maps are available for many areas of the United States. Topographic maps can also be obtained from municipal and county entities and local developers.
- Use of State Highway Planning Survey Maps.

- Aerial maps or aerial photographs, these may be available from local municipalities, state agencies, and real estate companies.

In determining the size of the contributing drainage area, any subterranean flow or any areas outside the physical boundaries of the drainage area that have runoff diverted into the drainage area being analyzed should be included in the total contributing drainage area. In addition, it must be determined if flood waters can be diverted out of the basin before reaching the site.

Slopes - The slope of the stream, the average slope of the watershed (basin slope), and other characteristics of the terrain important to the study or design should be determined. Hydrologic and hydraulic procedures in other chapters of this book are dependent on watershed slopes and these other physical characteristics.

The average watershed slope together with the length and flow retardance are the major factors affecting the runoff rate through the watershed (SCS, 1968). The determination of average watershed slope can be standardized in such a way as to produce consistent and fairly representative results. The determination of average watershed slope consists of (SCS, 1977):

1. Selecting four to eight representative slope sections that typify the slopes found in the watershed and drawing lines from the top of these slopes to the bottom.
2. Calculating the slope for each line as the elevation difference divided by the length of the line.
3. Adjust each slope by multiplying it by the estimated percentage of the watershed it represents. A weighted product is then computed to obtain the average slope for the entire watershed.

Watershed Land Use - Watershed land use information is particularly important, especially how it relates to the amount of impervious area for existing and future land use and how this impervious area is distributed throughout the watershed. Directly connected impervious areas in smaller watersheds exhibit runoff characteristics different from areas where runoff flows over grassy areas first. If the demarcation line between pervious and impervious areas is clear, two separate hydrologic models are sometimes developed for the directly connected areas and the non-directly connected areas. If the impervious areas are located in concentrated places in the watershed, the watershed is often broken into sub-areas to reflect consistent land use within each sub-area. Defining land use categories to break along zoning category lines is often convenient to permit wider application of results to land use decisions.

It may not be necessary to collect specific land use categories if the impervious areas are known from aerial photographs, digital mapping, multi-spectral photography and raster processing, or some other means. In terms of flow the only concern is whether the area is directly connected and small, or simply impervious. However, for water quality concerns it may be necessary to determine the actual land use categories, specific industrial category, and age of the neighborhood, because different categories can exhibit different event mean concentrations of certain pollutants. Also, some land uses are known for higher incidence of illicit connections, point sources, or illegal dumping. For watershed land use the following should be done.

- Present and expected future land use, particularly the location, degree of anticipated urbanization, and data source should be defined and documented.
- Information on existing use and future trends may be obtained from:
  - aerial photographs (conventional and infrared) and satellite images,
  - zoning maps and Master plans,
  - U.S.G.S. and other maps, and
  - municipal planning or public works agencies.
- Specific information about particular tracts of land can often be obtained from the court house, filed plans in a public works or engineering office, owners, developers, realtors, and local residents. Care should be exercised in using data from some of these sources since their

reliability may be questionable and these sources may not be aware of future development within the watershed which might affect specific land uses.
- Existing land use data for small watersheds can best be determined or verified from a field survey. Field surveys should also be used to update information on maps and aerial photographs, especially in basins that have experienced changes in development since the maps or photos were prepared. Infrared aerial photographs may be particularly useful in identifying types of urbanization and vegetation at a point in time. Aerial photography to track crop use may be available from the Department of Agriculture or SCS State representatives.

Streams, Rivers, Ponds, Lakes And Wetlands

At all streams, rivers, ponds, lakes, and wetlands that will affect or may be affected by the proposed structure or construction, the following data should be secured. These data are essential in determining the expected hydrology and may be needed for regulatory permits.
- Outline the boundary (perimeter) of the water body for the ordinary highwater.
- Determine the elevation of normal as well as high water for various frequencies.
- Obtain a detailed description of any natural or manmade controls, spillway, outlet works, or emergency spillway works including cross sections, dimensions, elevations, and operational characteristics.
- Obtain a description of adjustable gates, soil and water control devices.
- Determine the use of the water resource (stock water, fish, recreation, power, irrigation, municipal or industrial water supply, etc.).
- Note the existing conditions of the stream, river, pond, lake or wetlands as to turbidity and silt.
- Determine riparian ownership(s) as well as any water rights if appropriate.

Roughness Coefficients

Roughness coefficients, ordinarily in the form of Manning's "n" values should be estimated for the entire flood limits of the stream. A tabulation of Manning's "n" values with descriptions of their applications can be found in the Open Channel chapter. Some municipalities have developed a Manning's n book for their area which show pictures of local stream channels and floodplain areas with appropriate Manning's n values given (Greenville, South Carolina, 1992)

Stream Profile

Stream bed profile data should be obtained and these data should extend sufficiently upstream and downstream to determine the average slope and to encompass any proposed construction or aberrations. Identification of "headcuts" which could migrate to the site under consideration is particularly important. Profile data on live streams should be obtained from the water surface, where there is a stream gage relatively close; the discharge, date and hour of the reading should be obtained.

Stream Cross-Sections

Stream cross-section data should be obtained that will represent the typical conditions at a structure site as well as other locations where stage-discharge and related calculations will be necessary. The actual locations may depend on the type of modeling being done. Cross sections should be "typical" for the reach in question and may, in fact, not be a particular section at all but a typical composite. Cross sections should be taken wherever channel curvature makes the parallel streamline assumption invalid, where roughness changes significantly, and where conveyance changes by an abnormal amount.

Existing Structures

The location, size, description, condition, observed flood stages, and channel section relative to existing structures on the stream reach and near the site should be secured in order to determine their capacity and effect on the stream flow. Any structures, downstream or upstream, which may cause backwater or retard stream flow should be investigated. Also, the manner in which existing structures have been functioning with regard to such things as scour, overtopping, debris and ice passage, fish passage, etc. should be noted. With bridges these data should include span lengths, type of piers, and substructure orientation which usually can be obtained from existing structure plans. The necessary culvert data includes other things such as size, inlet and outlet geometry, slope, end treatment, culvert material, and flow line profile. Photographs and high water profiles or marks of flood events at the structure and past flood scour data can be valuable in assessing the hydraulic performance of the existing facility.

Improvements, property use, and other developments adjacent to the proposed site both upstream and downstream may determine acceptable flood levels. Incipient inundation elevations of these improvements or fixtures should be noted. In the absence of upstream development acceptable flood levels may be based on freeboard requirements. In these instances, the presence of downstream development becomes particularly important as it relates to potential overflow points along the road grade or drainage system.

Flood History

The history of past floods and their effect on existing structures are of exceptional value in making flood hazard evaluation studies, as well as needed information for sizing structures. Information may be obtained from newspaper accounts, local residents, flood marks or other positive evidence of the height of historical floods. Changes in channel and watershed conditions since the occurrence of the flood should be evaluated in relating historical floods to present conditions.

Debris And Ice

The quantity and size of debris and ice carried or available for transport by a stream during flood events should be investigated and such data obtained for use in the design of structures. In addition, the times of occurrence of debris and ice in relation to the occurrence of flood peaks should be determined; and the effect of backwater from debris and ice jams on recorded flood heights should be considered in using stream flow records. Data related to debris and ice considerations can be obtained from municipal records and state agencies.

Scour Potential

Scour potential is an important consideration relative to the stability of the structure over time. Scour potential will be determined by a combination of the stability of the natural materials at the facility site, tractive shear force exerted by the stream and sediment transport characteristics of the stream. Data on natural materials can be obtained from agencies or by tests at the site.

Bed and bank material samples sufficient for classifying channel type, stability, and gradations, as well as a geotechnical study to determine the substrata if scour studies are needed, will be required. The various alluvial river computer model data needs will help clarify what data are needed. Also, these data are needed to determine the presence of bed forms so a reliable Manning's n as well as bed form scour can be estimated.

Controls Affecting Design Criteria

Many controls will affect the criteria applied to the final design of storm water management structures including allowable headwater level, allowable flood level, allowable velocities, and resulting scour, and other site specific considerations. Data and information related to such controls can be obtained from Federal, State and local regulatory agencies and site investigations to determine what natural or man-made controls should be considered in the design. In addition there may be downstream and upstream controls which should be documented.

Downstream Control - Includes any ponds or reservoirs, along with their spillway elevations and design levels of operation, should be noted as their effect on backwater and/or stream bed aggradation may directly influence the proposed structure. Also, any downstream confluence of two or more streams should be studied to determine the effects of backwater or streambed change resulting from that confluence.

Upstream Control - Upstream control of runoff in the watershed should be noted. Conservation and/or flood control reservoirs in the watershed may effectively reduce peak discharges at the site and may also retain some of the watershed runoff. Capacities and operation designs for these features should be obtained. The Soil Conservation Service, Corps of Engineers, Bureau of Reclamation, consulting engineers, and other reservoir sponsors often have complete reports concerning the operation and design of proposed or existing conservation and/or flood control reservoirs.

The redirection of flood waters can significantly affect the hydraulic performance of a site. Some actions that redirect flows are irrigation facilities, debris jams, mud flows, and highways or railroads.

6.4 Data Acquisition, Survey Information And Field Reviews

Remote Data Acquisition

It is clear that efficient remote data acquisition, interpretation, analysis and use is a fast changing and emerging science/art. Aerial photography is a common method for obtaining both physical representations of the terrain and digital base maps. Aerial photography involves decisions on scale, accuracy, important information, ground control, processing equipment, final format of information, coverages to be mapped, etc. There is the possibility to use multispectral photography to capture information which is not available in the visible light range such as vegetation type differences, moisture, heat radiation, impervious areas, etc. The wide array of mapping products can be confusing. For example, there is a great deal of difference in terms of both accuracy and potential applications between mapping and aerial photography. And within photography a great difference between aerial photography, scaled photographs, rectified photographs, and digital orthophotography. The type and accuracy of use drives the technological decisions. It is good advice to seek several opinions before investing in new mapping or photography.

Satellite data acquisition has evolved over the last several years to become a very valuable tool even for urban storm water managers.   The most common source of both remotely sensed information and data in the United States is the U.S.G.S. EROS Data Center located north of Souix Falls, SD (customer service is at 605-594-6507). It was established in 1971 to support the NASA LANDSAT program. It is responsible for collecting, processing, archiving and distributing a wide range of remotely sensed land data and has an inventory of over 3 million satellite and 7 million aerial photographs of the Earth's surface.   It is accessible through CD-ROM disk readers or through an on-line system called the Global Land Information System (GLIS).

Global positioning systems (GPS) use satellites to accurately determine position in three-space through signals received from such satellites.   Such use of satellite information allows for very accurate and fast surveys and positioning even in remote areas.   These systems are often tied to Geographical Information Systems (GIS) for automated mapping and even real time information processing.

A loose consortium of Federal, state and local agencies and industry is developing a protocol and format for the compatible exchange of spatial data and information called the National Spatial Data Infrastructure (NSDI).   NSDI will make possible the interchange of all sorts of spatial data by making their format compatible. Industry, for its part, is building standardized products which use or are compatible with these formats.   The main coordinating entity for this is the U.S. Geological Survey at its Reston, VA National Mapping Division, Branch of Geographic Data Coordination (703-648-5725).

Geographical Information Systems

Geographic information systems (GIS) are an emerging tool somewhat akin to the emergence of CADD over ten years ago. GIS can provide the following types of information related to storm water applications (after Berry, 1994a).

- Mapping of land and underground features, land uses, soils, rainfall amounts, watershed boundaries, slopes, land cover, etc., to any scale and in any combination is easy once the information is in the system.
- Database and map linking provides management type answers to where things are and gives ready information about them. For example, information about storm water pipes can be placed in a database tied to the pipe.
- Changes over time can be easily portrayed in GIS format or in charts and graphs generated in GIS systems. For example, temporal water quality changes can be graphically portrayed as color changes moving across the map in a demonstration or as a series of maps showing the color changes.
- Inter-relationships between mapped objects can be determined by GIS. This works very well for such things as determining storm water runoff with the intersection of soils, land use and watershed basin boundaries. Distances and slopes can be automatically determined for time of concentration calculations. Three-dimensional plots can be developed to get an idea of the appearance of a site for multi-objective use.
- Suitability type analysis is easy with GIS through the use of various rating schemes. For example, suitability for park reserves can be determined by intersecting soils, land use, vegetation, ownership, utilities, and species information. The best areas (those with the highest rating) can be colored to "jump off the map" for even the most obtuse citizen.
- Cause and effect type questions about natural or man made systems can be asked and answered, such as autocorrelation between rainfall gage sites or the relationship between industrial SIC coded locations and in-stream pollution monitoring results.

- What if analysis can easily be accomplished with GIS. Land use decisions and their impacts on pollution or flooding can be tried in endless combinations to determine the best solutions or to determine the outcome of current decisions.
- Convincing graphics is the final major category of GIS use. Overhead GIS projections or maps have been used to educate or convince citizens and political leadership concerning a course of action or a project's viability. Proper implementation of GIS applications for storm water management involves planning for both storm water only applications and to integrate these applications with other potential users within the municipality. For example, the street centerline network serves as a good "skeleton" on which to hang other coverages and features. But the ultimate accuracy with which it is established depends on its ultimate use and users.

Four basic issues must be fully considered before and during implementation (Berry, 1994b): hardware/software, databases, applications, and human impact. Hardware and software issues are relatively simple to solve once the basic uses for the system are identified. However some municipalities have found that the "devil is in the details", and that some systems delivered at lower cost do not perform quite as expected when it comes to database development or transfer. Then the cost can be enormous. The database development can cost many times the hardware and software development cost. Some databases are available and should be used if at all possible.

Telemetry

The first telemetered data was transmitted via phone lines in the 1930's. Since that time there have been thousands of applications using hundreds of different types of telemetry systems including phone lines, radio, satellite and micro wave. These systems are used in handling of data such as water level, temperature, flow rate, gate movement, precipitation, humidity and other data types. There are two general ways to transmit the data: analog and digital. The digital method is now by far the most common. Digital output is either serial or parallel (one bit of information at a time or all bits at a time).

The ability for two-way communication with remote sites is also valuable for the urban storm water manager (termed Supervisory Control and Data Acquisition or SCADA). There is the possibility to remotely control flood or pollution control gates, dams, weirs, and all types of hydraulic devices from a central control room. For example, for flood control this capability has been demonstrated in Bellevue, WA. This is bringing about the rapid convergence of automated mapping and facilities management (AM/FM) and SCADA. For example, interior drainage pumping stations can be remotely controlled to catch first flush of pollutants and pump back to sanitary treatment sites, but also be opened for necessary interior drainage flood control. A series of detention ponds can be controlled for maximum flood reduction through a series of rain gages, a SCADA system and a knowledge-rule based hydrologic model. Combined sewer capacities can be maximized using the same principles.

Survey Information

Complete and accurate survey information is necessary to develop a design that will best serve the requirements of a site. The individual in charge of the drainage survey should have a general knowledge of storm water management design and as such should coordinate the data collection with the designer. The amount of survey data gathered should be commensurate with the importance and cost of the proposed structure and the expected flood hazard.

At many sites photogrammetry is an excellent method of securing the topographical components of drainage surveys. Planimetric and topographic data covering a wide area are easily and cost effectively obtained in many geographic areas. A supplemental field survey is required to provide data in areas obscured on the aerial photos (underwater, under trees, etc.).

Data collection should be as complete as possible during the initial survey in order to avoid repeat visits. Thus, data needs must be identified and tailored to satisfy the requirements of the specific location and size of the project early in the project design phase. Coordination with all departments requiring drainage related survey data before the initial field work is begun will help ensure the acquisition of sufficient, but not excessive survey data. An example form and check list for collecting field data are provided in Appendix B at the end of this chapter.

Field Reviews

Watershed, stream reach and site characteristic data, as well as data on other physical characteristics, can be obtained from a field reconnaissance of the site. Some level of field or aerial drainage survey of the site and its contributing watershed should always be undertaken as part of the analysis and design. Survey requirements for small storm water management facilities such as 36 inch culverts are much less extensive than those for major facilities such as bridges. However, the purpose of each survey is to provide an accurate picture of the conditions within the zone of hydraulic influence of the facility. Data that can be obtained or verified includes the following.
- Contributing drainage area characteristics.
- Stream reach data - cross sections and thalweg profile.
- Existing structures.
- Location and survey for development, existing structures, etc., that may affect the determination of allowable flood levels, capacity of proposed storm water management facilities, or acceptable outlet velocities.
- Drift/debris characteristics.
- General ecological information about the drainage area and adjacent lands.
- High water elevations including the date of occurrence.

Much of these data must be obtained from an on-site inspection. It is often much easier to interpret published sources of data after an on-site inspection. Only after a study of the area and a complete collection of required information should the designer proceed with the design of the storm water management facility. All pertinent data and facts gathered through the survey should be documented as explained in the Documentation chapter.

There are several criteria that should be established before making the field visit. Does the magnitude of the project warrant an inspection, or can the same information be obtained from maps, aerial photos, or by telephone calls? What kind of equipment should be taken, and most important, what exactly are the critical items at this site? Photographs should be taken. As a minimum, photos should be taken looking upstream and downstream from the site as well as along any contemplated structure centerline in both directions. Details of the stream bed and banks should also be photographed along with structures in the vicinity both upstream and downstream. Close up photographs complete with a scale or grid should be taken to facilitate estimates of the stream bed gradation.

6.5 Data Evaluation

Once the needed data have been collected, the next step is to compile it into a usable format. The designer must ascertain whether the data contains inconsistencies or other unexplained anomalies which might lead to erroneous calculations or results. The main reason for analyzing the data is to draw all of the various pieces of collected information together, and to fit them into a comprehensive and accurate representation of the hydrologic and hydraulic characteristics of a particular site.

Experience, knowledge, and judgment are important parts of data evaluation. It is in this phase that reliable data should be separated from that which is less reliable and historical data combined with that obtained from measurements. Historical data should be reviewed to determine whether significant changes have occurred in the watershed and whether these data can be used. Data acquired from the publications of established sources such as the U.S.G.S. can usually be considered as valid and accurate. Maps, aerial photographs, Landsat images, and land use studies should be compared with one another and with the results of the field survey and any inconsistencies resolved.

Data Accuracy

Errors can creep into any measuring or sampling program, and may take the forms of: bias, random measuring errors, and gross mistakes and errors. Bias is a consistent over- or under-estimation of the true values of a measurement. Errors may enter measurements through such things as improperly zeroing an instrument or not taking into account tape sag or vertical angles in distance measurements. Bias errors can only be reduced through carefully planning and constantly checking the measurement activities.

Random errors are inherent in reading of any instrument or measuring device or in performing any measurement activity. These errors can be reduced through repetition of the measurements and attention to quality control procedures. Gross mistakes and errors can be caught through careful and deliberate measurement processes, repetition, quality assurance programs, and common sense checks. While bias was previously defined, two other terms are important to data collection: precision and accuracy. Precision is a measure of the closeness of agreement among individual measurements; the scatter. Accuracy is a measure of how closely the measurements reflect the true value. Little or no bias and high precision lead to high accuracy.

It is important to establish data accuracy needs or availability before data are collected. For example, the publication "Accuracy of Computed Water Surface Profiles," U.S. Army Corps of Engineers, Dec. 1986, focuses on determining relationships between:
- survey technology and accuracy employed for determining stream cross sectional geometry,
- degree of confidence in selecting Manning's roughness coefficients, and
- the resulting accuracy of hydraulic computations.

The report also presents methods for determining the upstream and downstream limits of data collection for a hydraulic study requiring a specified degree of accuracy.

It is also important to realize that data collected for one purpose is often then used for a different purpose by the same or a different agency. The accuracy of the data should be sufficient to meet defined needs and its limitations stated.

Sensitivity Studies

Often sensitivity studies can be used to evaluate data and the importance of specific data items to the final design. Sensitivity studies consist of conducting a design or study with a range of values for specific data items. Following are the steps in such a study.

Step 1   Evaluate the model(s) which will be used, and determine which independent parameters are key for sensitivity testing. Determine which parameters are correlated with each other as well as the dependent variables. Attempt to isolate only basic independent parameters.

Step 2   Determine, if possible, the relationship of the dependent variable with the independent
          input parameters mathematically. Differentiating the equation with respect to the parame-
          ter will then provide an exact measure of the sensitivity.

Step 3   For other parameters determine a "standard" set of variables which will serve as the
          starting point for the sensitivity study. This set should center on the expected values over
          the range of the study conditions.

Step 4   Vary each key parameter in turn holding all other parameters at the standard value. The
          range of variation should cover any expected range for the study site conditions. Be aware
          of parameters which may have secondary amplifying effects on each other at specific high
          or low ranges.

Step 5   Develop a dimensionless plot of the varying output variable (such as peak runoff, pollutant
          load, etc.) and varying parameter divided by the standard values. This allows for
          percentage comparisons. Check the plot for the reasonableness of the variation. Further
          investigate any apparently anomalous relationships.

Step 6   Note those parameters which cause greater variation in the dependent variable for the
          conditions expected to be tested. Establish standards for measurement accuracy attainable
          and determine the effects of measurement or estimation error on the final results.

          The effect on the final design can then be established. This is useful in determining what
specific data items have major effects on the final design and the importance of possible data
errors. Time and effort should then be spent on the more sensitive data items making sure these
data are as accurate as possible. This does not mean that inaccurate data are accepted for less
sensitive data items, but it allows prioritization of the data collection process given a limited budget
and time allocation.

Statistical Analysis

          Statistical analysis in urban storm water management usually falls into one of four topics: time-
series analysis, extreme event analysis, geo-statistical analysis or general statistical analysis
(containing hypothesis testing, probability function fitting, extrapolation, etc.).

          For hydrologic data or other types of data that have inherent randomness statistically developed
sampling may be important. While there is a physical reason for all variability, much of it cannot
be accounted for within the model we are using to attempt to mimic nature. Three reasons account
for much of the difficulty in predicting hydrologic or environmental phenomena (Hirsch, et. al.,
1993): inherent randomness, sampling error and incorrect understanding of the processes involved.
For example, in frequency analysis there is rarely sufficient data to accurately estimate the risk
of exceedance of a certain flow, pollution loading, rainfall intensity, etc. Assumptions are made
about the probability distributions which the maxima follow. However, these distributions are only
approximations of the real distribution of maxima. When a short period of record is extended,
large errors can result in predicting, for example the flood event for spillway design.

          Time Series Analysis - This analysis is often called trend sampling, systematic sampling or
sampling along a line. Many hydrologic and environmental variables lend themselves to this type
of sampling including rainfall, peak flows, volumes, pollution peaks or concentrations, low flows,
etc. Urban runoff time series analysis is often used in continuous modeling applications or to
establish statistical properties of a particular parameter for other analysis. Thus a time series is
nothing more than a sequence of values arrayed in chronological order which can be characterized
by statistical properties (Chow, 1964). It is a subset of overall statistical applications to measured
parameters termed: stochastic hydrology (Yevjevich, 1972a).

To develop a time series a continuous variable (such as discharge) is sampled over a discreet time interval. The series of data is then subjected to statistical tests to determine means, variances, correlations, etc. A time series is said to be "autocorrelated" if the measurements of the variable from time-step to time-step depend somewhat on each other. This can happen, for example, when flows depend on slow storage release in a watershed and thus are related to each other (Salas, 1993). Series are said to be "cross correlated" if two different measurement locations show a relationship to each other. This can happen when two or more rain gages are in the same region and thus influenced by the same storm. Trends may become apparent when both a random and non-random element of a variable are separated through statistical analysis.

Using time series analysis rain gages can be corrected, data can be transferred, statistics can be derived and used to generate years of simulated data, seasonal variation can be quantified, relationships can be tested and quantified, etc. The reader is referred to the given references for further information.

Extreme Event Analysis - This analysis is really a subset of general statistical fitting of data to probability distributions. However, its use is so common in hydrology and environmental work that a separate brief discussion is warranted. It has been shown that distributions of the largest events over N periods of time with M events occurring in each period approach a limit as M increases. Depending on the distribution of the M variables the final probability distribution (called an extremal distribution) will take one of three basic forms (Chow, 1964). A generalized extremal distribution has been developed which encompasses all three of the forms (Hosking, et. al., 1985). Different distributions are used to attempt to model different phenomena through the use of plotting positions and goodness-of-fit testing (Stedinger et. al., 1993):

- The Normal distribution is used for average annual stream flow or average annual pollution loadings where the central limit theorem dictates a normal distribution of the averages of many observations.
- Lognormal transformations are used for variables which are positively skewed such as event mean concentrations for urban storm water runoff (USEPA, 1983).
- A Gumbell distribution is used for maxima of certain events such as rainfall (Hershfield, 1961).
- A Weibull distribution characterizes distributions of minima bounded by zero such as low flows.
- A Log-Pearson Type III distribution is recommended by the Water Resource Council for flood distributions (Interagency Committee, 1982).

Geostatistical Analysis - This is the analysis of spatially distributed information and the drawing of inferences from that data. For example, it is used in urban storm water management to estimate rainfall distributions over an area from a number of rain gages or to estimate groundwater height or transmissivity properties based on well data (Kitanidis, 1993). In storm water quality it is used to estimate the distribution of pollutants in a waterbody from samples or to estimate regional contamination based on a grid of measurements (Gilbert, 1986).

Geostatistical analysis can estimate unknown quantities or plot a contour of quantity estimates, evaluate the reliability and risk involved in the estimates, and predict the effect on the results of adding another sampling or measuring location. The two most commonly used methods in geostatistics are weighted moving average methods and trend surface analysis. Weighted moving averages is the most flexible of the methods and lends itself well to computer applications. Kriging is a method for applying a weighting to the values from contiguous locations based on location on a laid out grid (Clark, 1979).

Other Statistical Methods - Other methods include a wide variety of statistics used in various aspects of hydrology and environmental science including (Yevjevich, 1972b, Chow, 1964, Bowker and Lieberman, 1972):

- probability functions and frequency distributions to fit certain phenomena;
- combinatorial analysis;
- risk and uncertainty;
- statistical analysis of empirical distributions;
- descriptive statistics;
- estimation statistics;
- regression and correlation analysis;
- sampling theory;
- hypotheses testing;
- goodness of fit; and
- multi-variate analysis.

A note of caution is in order with statistical analysis. Spurious correlation of parameters, false regression fitting, biased data and interpretation, unknown mathematical errors, etc. all can come into play (Gumbel, 1926, Benson, 1965). Correlations of dimensionless groupings of numbers will look better than they are if the same variable is used in each of the dimensionless groups. The use of logarithmic plotting can cause an otherwise poor fit to look better when large value differences are hidden in the log compression. Percentage comparisons of large numbers where small variations make a difference will always look good while they will always look poor for smaller numbers. Means, modes and medians are all termed "averages" at times. Even fitting a straight line to data can take on complications depending on the purpose for that fitting (Hirsch and Gilroy, 1984). Graphical scale selection or distortion can inadvertently magnify variations or hide them. Sampling or monitoring can miss the events below the lower detection limit biasing the data. It is easy to "lie with statistics" even when one is not trying to (Huff, 1952).

6.6 Precipitation Data Collection

Overview

Depending on the data needed and the physical conditions in the field there are many different techniques and variations on these techniques for collecting data and information. New technology is making the acquisition of data and information easier. For example:

- it is possible through multi-spectral digital analysis and aerial photography or satellite imagery to estimate the percentage imperviousness of an area without direct measurements or mapping or to classify land uses;
- new survey equipment and techniques allow for the location of points in space through a "point-and-shoot" technique without recourse to a reflector or back sighting;
- hand held data loggers allow for the logically complex acquisition of inventory information quickly and easily and the automated transfer of such information directly into CADD or GIS equipment;
- NEXRAD radar precipitation information will be available by 1996 for most of the United States allowing for future convenient GIS and spatially distributed model applications in hydrology while software using NEXRAD is coming to market (Smith, 1993);
- topographic maps can be obtained for areas throughout the United States on CD-ROM; and
- large on-line databases of hydrologic data are available nation-wide.

This acquisition methodology matches the need for more and better data by the newer computer models designed to take advantage of faster micro-computers and better graphical interfaces.

However, there will always be the need for the accurate measurement of such data as flow, rainfall, channel dimensions, etc. The next sections discuss such measurements for the benefit of the storm water manager who needs to develop estimates of such values for design, plans review and modelling of non-complex systems. The collection of precipitation data and flow data are the two most important surface water quantity data collection needs.

Data Collection

The collection and proper interpretation of precipitation data for the development of design storms or for continuous hydrologic modeling is one of the more complex field operations in urban storm water management. Even on relatively small urban catchments lack of knowledge about the spatial distribution of rainfall is the greatest source of error in runoff simulation (Shilling, 1984). There are a number of assumptions inherent in precipitation estimation which may not be fully true in any field operation, such as (Arnell, et al., 1984):

- measurements reflect the true values of precipitation - there is no gage, radar or human error in measurement;
- data are consistent - and there are no internal changes to the system during the data collection period (double mass analysis is used to check this assumption);
- data are homogeneous - there are no external changes to the system during data collection (double mass analysis is again used to check this assumption);
- data are stationary - there is no cyclic climatic change, trends or periodicity happening (statistical testing is used to check this);
- data are independent - there is no correlation between events (statistical testing is used);
- data series are sufficiently long to reflect the population;
- precipitation extremes follow a prescribed probability distribution; and
- estimated parameters fully describe the presumed probability distribution.

For a complete discussion of the topic see Arnell, et al., (1984).

For example, gage data collected at an often distant airport or National Weather Service site are often used to try to calibrate hydrologic models to known storm runoff. The gage data are then assumed uniform and applied uniformly over the watershed. The correlation between measured runoff hydrograph and measured rainfall hyetograph is often poor. This poor correlation can lead to harmful calibrations when the modeler sets other parameters unrealistically to "fit" rainfall and runoff. The obvious solution is to increase the number of gages and number of storms for calibration and to distribute the rainfall more realistically among gages.

Factors Influencing Data Collection

There are a number of factors which make the measurement and application of precipitation data difficult. Depending on the catchment characteristics (size, slope, land use, orientation, etc.), the need for accuracy, and the available time and financial budget, there are different approaches and considerations which may be important. It should be remembered that for small catchments the peak flow estimate is dependent on the estimate of the peak intensity of the rain while for large catchments the peak outflow is more dependent on accurate estimates of the volume of the storm.

Storm Types - Precipitation in the United States generally falls into one of two patterns. *Stratiform* systems are the frontal or precipitation line systems which move out of the Pacific and

Gulf carrying large amounts of relatively steady rainfall over large areas. These are often called extra tropical cyclones (Gilman, 1964, Smith, 1993). *Convective* systems are made up of local thunderstorms which form quickly, often repeatedly, releasing short intense "cloud bursts" over limited areas. These midlatitude thunderstorms are very dependent on the land use over which they form. It has been shown on several continents that urban areas generate a greater frequency of high intensity thunderstorms due to the heating effects of the paved surfaces (Huff, 1975, Vogel and Huff, 1975, and Niemczynowicz, 1987). This "shadow effect" is felt both in the urban area and downwind, and especially during hot summer afternoons.

In smaller urban catchments the convective type storms normally cause flooding of the minor feeder streams, ditches, neighborhoods and streets. But for the large creeks and streams it is often the frontal systems whose long and intense storms cause the major system flooding, though there are many exceptions.

Storm Movement - Studies in both Lund, Sweden (Niemczynowicz, 1987) and St. Louis, MO (Vogel, 1984) show that storm movement has an important effect on hydrograph generation, especially when the storm is moving downstream along the watershed axis. Many hydrologic models have the ability to switch rainfall on and off for calibration purposes. For long basins oriented along the axis of normal storm movement (such as might be the case when orographic effects predominate or where prevailing winds dominate) this may be appropriate for design purposes.

Storm movement information may be approximated through the use of high altitude wind, long term wind, and wind rose data available from the National Weather Service (Niemczynowicz, 1988). In fact Niemczynowicz (1988) has shown that the use of storm movement in model calibration allows for the use of fewer rain gages (3 instead of 12 over a 25 square kilometer area) to achieve the same level of accuracy. This is because modeling storm movement automatically takes into account an areal reduction and a peak duration increase. The peak intensity moves as a longer duration streak across the watershed rather than falling all at once simultaneously throughout the watershed.

Storm Decay - Storm cells build and decay over time. The scale of such a cycle varies considerably but can be neglected for most smaller urban catchments, perhaps on the order of 10 square miles for most of the United States. For example, average path length for cells was about 5 miles in St. Louis, MO (Huff, 1975).

Precipitation Data Acquisition

The National Oceanic and Atmospheric Administration (NOAA) National Climatic Data Center, (NCDC) Climate Services Branch, is responsible for the collection of precipitation data. Data on hourly, daily and monthly precipitation at a large number of stations across the United States is available to the public on diskette, microfiche or hard copy. Orders can be placed by calling (704) 259-0682 or by writing the center at Federal Building, Asheville, NC, 28071-2733.

The National Weather Service (NWS) of NOAA can provide historic, present and predicted weather to aid in sampling operations. Weather radio is available from 380 NWS stations via weather band radio and provides continuous broadcasts of the most current weather information (EPA, 1992).

There are basically two ways to physically obtain rainfall data: rain gages and radar or satellite imagery (USGS, 1984).

Rain Gages - Gages are of two types: recording and non-recording. All gages have a collector which leads to a funnel and thence to some sort of holding container with a measuring device. Non-recording gages are two standard sizes: 4 or 8 inches with either a graduated container or a dip-stick. Snow boards or snow stakes or a modified rain gage are used for snowfall measurement. Recording gages are of three types all with a clock driven drum and continuous pen trace.

* Weighing Type gages record continuously the weight of the empty container and rainfall. This type of gage has the disadvantage in that it cannot empty itself. But it is effective for solid precipitation since melting is not necessary prior to recording amounts and thus is good in colder climates.
* Tipping Bucket Type gages work using a balanced pair of small "buckets". When sufficient rainfall enters the collecting bucket the gage is unbalanced. It tips, placing the second bucket in position to collect rain while the first empties into a holding container. Each tip of the bucket records a specific amount of rainfall: normally not greater than 2 mm.
* Float Type gages drain rainfall into a container with a float mounted pen trace. Several gages have automatic emptying devices or siphoning mechanisms. Heaters are necessary in freezing weather and water must be retained in the bottom of the float chamber at all times. This may require filling during long dry periods.

The accuracy of rain gages is not a given. It has been shown that wind induced turbulence can reduce the catch of the gage by as much as fifty percent (Court, 1960). A wind speed of 10 mph causes about a 15 percent deficiency while a speed of 25 mph causes a 40 percent deficiency (Wilson, 1954). Since turbulence varies with height above ground the lower the gage the better. Shields are available which improve both rain and snow catch (Gray, 1970). Also the gage should be in a non-exposed area but open to rainfall.

Radar Precipitation Data Acquisition - This data acquisition method has advanced greatly since its inception in the 1940's (Vogel, 1984). Radar can provide temporal resolutions to 5 minutes and spatial resolutions to 1 km. The Next Generation Weather Radar System (NEXRAD) will, by 1996, be a network of 120 radar sites able to provide spatial resolution of about a 4 by 4 km square and one hour, though finer temporal and spatial information can be produced by these WSR-88D systems.

Radar operates by sending and receiving echoes of radar waves in the length range of from 1 to 10 cm. The backscatter is then interpreted based on both the radar equipment and storm target characteristics and reflectivity versus rainfall equations are developed (Smith, 1993). Automated methods are being developed which will provide incredible power to the precipitation modeler in terms of accuracy and spatial detail. This type of raster or grid based data fits well with GIS hydrologic modelling applications for both model calibration and real-time forecasting and flood warning. Radar estimates are subject to a number of types of error. It is normally prudent to use field gage observations to calibrate the radar and provide "ground truth".

A system in Chicago pioneered this type of forecasting using both radar information and rain gage information (to calibrate the radar) (Vogel, 1984). By using a combination system such as this many fewer rain gages can be employed while spatial and temporal accuracy is maintained. Satellite imagery is also used for rainfall measurement though it is not as accurate, being based on infrared determination of relationships between cloud height and brightness and actual rainfall.

Rain Gaging Networks

The design of a gaging network depends on the analytic purpose, economics, climatic stress and access. The shorter the duration, the smaller the area of a storm, or the more variability in geography, the more gages are necessary to obtain absolute estimates. The effect of increasing gage density is to increase the average rainfall value over the watershed. The reason is that with more gages the chances of one of the gages recording the peak rainfall track is higher.

To measure convective storms with some degree of accuracy, a dense rain gage grid is required. For short term rainfall with a one to ten minute storm increment (averaging time), with rain cells reported to vary between 1 to 5 square kilometers, a gage spacing of about 1.8 kilometers will explain about 90 percent of the variation (Huff, 1979). Simanton and Osborn (1980) have shown that the correlation between values of contiguous gages falls off dramatically when they are separated more than 5 to 10 kilometers for air-mass thunderstorms. McGuinness (1963) has developed an equation for watersheds in Ohio to estimate the error when a lesser gage density is used for thunderstorm type rainfall as:

$$E = 0.03 \ P^{0.54} \ G^{0.24} \tag{6.1}$$

Where: E = absolute error, in.
         P = "true" value of rainfall, in.
         G = gage network density, sq mi/gage

Alternately Hershfield (1965) using data collected from 15 watersheds across the United States provides information by which the distance between gages to obtain a correlation coefficient of 0.9 can be obtained from the 2-year 1-hour and 24-hour storms. Figure 6-1 gives the appropriate information.

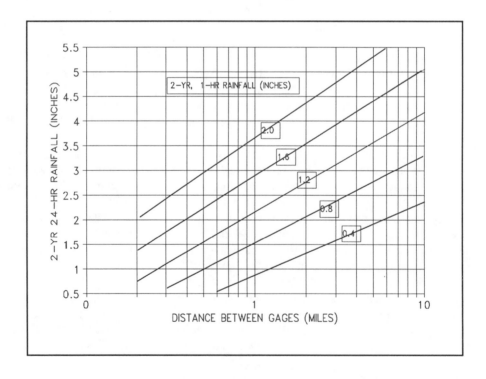

Figure 6-1 Estimating Gage Spacing Distance As Function Of The 2-year, 1-hour And 24-hour Storms

It may not be cost effective to attempt such a gage density. For smaller catchments a single rain gage is often placed near the centroid of the watershed to monitor a few storms. As mentioned above, the use of storm movement can enhance even sparse data.

A uniform network is normally found to be best when developing gage spacing though the spacing should reflect orographic or other changes. Special networks in these areas work well and can be integrated into an overall grid. For larger areas, such as over a whole urban area, the concept of "unit source areas" can save time and money (Gray, 1970). A unit source area is a representative gaging network which represents a much larger area. These smaller areas are gaged intensively and the results are then extrapolated to other areas. Care must be taken when performing modeling of the whole area that concentrations of rainfall from the unit source area are not extrapolated throughout the area as if a thunderstorm of limited extent was covering the whole of the city simultaneously.

Often a more dense grid system can be greatly reduced over time as certain gages are found to be representative of greater areas (some gages are redundant) and knowledge is gained of storm cell size and spacing.

Rainfall Average Depth Estimation

We are seldom interested in precipitation at a point but in its areal distribution so that runoff can be predicted. If only one gage is available the areal reduction equation provided in the Urban Hydrology chapter can be used. The application of areal reduction factors to small areas has been proven in Swedish studies (Niemczynowicz, 1982). It has even theoretically been shown that the areal reduction factor could be greater than 1 (Nguyen, et al., 1981).

The average depth of rainfall over an area can be estimated from the totals from several gages in the area using several methods. All of the methods applied mathematically or graphically have the same form:

$$P = x_1 p_1 + x_2 p_2 + \dots x_n p_n \tag{6.2}$$

Where P is the mean areal precipitation at a watershed or site of interest, and p_i and x_i are the n station rainfalls and weighting factors respectively. The sum of all the x's is 1. The only difference in the methods is the method of weighting the point precipitation values. Both Kelway (1974) and Gottschalk and Jutman (1983) have investigated the various methods without reaching strong conclusions on which is most accurate. Since precipitation is non-normally distributed none of the methods reduces the uncertainty in estimate since the true value between observation points is never known (Arnell, et al., 1984). An arithmetic mean can be used for flat areas with similar climatic conditions where each of the x's are equal to 1/n.

The Thiessen polygon method, a graphical application (Thiessen, 1911), assumes that the amount of rainfall at any gage site can be applied to the distance half way to the nearest station in any direction. The applicable areas are found by drawing lines between each station. The perpendicular bisector of each line is then joined where they cross, enclosing an area around each gage site.

The isohyetal method consists of plotting rainfall totals, for the storm or time increment, on a map and drawing rainfall isohyets (rainfall contours) through the points. A modification of this method, the percentage-of-mean-annual method (Gilman, 1964), is used for storm averages in areas with orographic effects. The following steps are employed.
- The ratio of the storm amount to the mean or seasonal average amount is calculated for each gage site and plotted on a map.
- Isolines of the ratios are estimated and plotted.

- The means of the ratios are determined for each incremental area between isolines.
- Each mean ratio is then multiplied by the appropriate mean or seasonal amount to obtain rainfall depth averages.
- Multiplying each incremental area by the depth for that area, adding and dividing by the total area, then gives the area weighted average for the total area of concern.

The reciprocal-distance method has been used to estimate point thunderstorm induced rainfall amounts with accuracy in both the South (Dean and Snyder, 1977) and South West (Simantona and Osborn, 1980). In this method the rainfall at any point is an average of the gage readings from nearby gages weighted by the distance to those gages to some power. If the point rainfall calculated can be related to an incremental area, such as a watershed sub-basin at its centroid, then average rainfall over an area can be calculated. The equation for n gages is of the form:

$$P_p = [P_1/d_1^b + P_2/d_2^b + \dots P_n/d_n^b]/[1/d_1^b + 1/d_2^b + \dots 1/d_n^b] \qquad (6.3)$$

Where: P_p = weighted average point rainfall, in.
        P_i = rainfall at gage i, in.
        d_i = distance from point to gage, mi
        b = distance exponent, (1 in Southwest, 2 in South Piedmont area).

Missing gage information can be estimated from three surrounding gages by weighting it using the normal annual precipitation or some other seasonal or annual precipitation measure.

$$P_p = 1/3 [(N_p/N_1)P_1 + (N_p/N_2)P_2 + (N_p/N_3)P_3] \qquad (6.4)$$

Where: N = normal annual or seasonal precipitation and other variables are as indicated in the previous equation.

This method only works well when there is a high correlation between the gages used. If there is not, a better approach would be to sketch an isohyetal map and estimate the missing data from that or correlate using fewer gages.

As mentioned earlier a "kriging technique" has been applied to spatial precipitation data as well as fitting of polynomials and Fourier series (Thorpe, et al., 1979). This method does not provide better estimation of spatial distribution but does allow a better estimate of the confidence limits on the estimate. The most promising future development is the application of better data acquisition through the use of radar and GIS applications.

6.7 Flow Data Collection - Overview

Introduction

The collection of flow data in urban applications is normally limited to discharge and velocity measurement in open systems (ditches, feeder streams, channels and streams and in partially full conduits) and pressure flow in closed conduits. The measurement of flow includes the use of devices (structures, meters, etc.) and of techniques which may or may not involve complex devices. Basic references for further information and details on various techniques include: Kirkpatrick and Shelley, 1975, USGS, 1983, Brater & King, 1976, Grant, 1989, Leupold & Stevens, 1991, Rantz, 1982, and Bureau of Reclamation, 1981.

The measurement of urban runoff flow is complicated by several factors including:
- the poor quality of the flow and the amount of sediment and debris which are washed into the flow;

- the likelihood of vandalism of gaging sites and devices;
- the flashy and intermittent nature of the flows; and
- high velocities and presence of many sharp bends leading to nonuniform or abnormal velocity distributions throughout much of the flow area.

Therefore the selection of flow measuring devices and the locations of measurement are of prime importance. Flow measurement can be targeted toward continuous measurement, peak flow only, low flow or miscellaneous measurements in a relatively continuous flow. One-time measurements can also be made to calibrate a continuous device.

Basic Equipment And Techniques Overview

Flow measurement belongs to two broad categories: quantity measurement and rate measurement. In quantity measurement the primary measurement device weighs volumes of water or measures certain masses of water. For rate measurement the primary device interacts with the fluid in a certain way dependent on one or more properties of the fluid. This interaction with a continuous stream of fluid is then counted or measured in such a way that dependence on fluid properties, physical laws or empirical equations convert the interaction to a flow velocity and rate. Under these two basic categories there are many basic ways to measure flow and a number of instruments or techniques under each of the methods. Table 6-2 is a summary of the most common ways to measure flow in open channel and closed systems.

Common types of measuring devices for pressure flow applications include: venturi meters, ultrasonic meters (both doppler and transit time), flow nozzles, orifice meters, magnetic flow meters, rotameter, elbow meters and pitot tube meters. The most common methods for flow measurement in open channels and partially full pipes include flumes, weirs, pitot tubes, flow meters, orifices, nozzles and slope and velocity measurements with application of theoretical or empirical formulae.

Shelley and Kirkpatrick (1975) give some basic criteria in assessing which type of equipment might be applicable for any situation including:

1. Range - covering a 100 to 1 range is preferable.
2. Accuracy - an accuracy of $\pm$ 10 percent of the reading is necessary; $\pm$ 5 percent is highly desirable and repeatability of $\pm$ 2 percent is desired in all instances.
3. Flow effects - the unit should be able to retain its accuracy over a range of conditions found in urban flow.
4. Gravity and Pressure Flow - it is desirable the unit have the ability to handle both types of flow in a closed conduit situation.
5. Submergence or Backwater Sensitivity - backwater or even reversed flow should be sensed by the unit even if it cannot continue to function accurately under such conditions.
6. Solids - sediments and floatables should move through the unit without clogging or blockage.
7. Flow Obstruction - the unit should function without undue obstruction of the flow.
8. Head Loss - the unit should induce as little head loss as possible.
9. Manhole Operation - the unit would be able to be installed and operate within a manhole.
10. Power Requirements - the unit should require little or no power and be able to operate remotely.
11. Site Requirements - unit should not require extraordinary site support facilities.
12. Installation Restrictions - the unit should be flexible enough to operate in a number of installation types without restrictions or limitations.
13. Simplicity and Reliability - the unit should be simple to operate and calibrate, with a minimum of moving parts and be reliable over a long period and many different flow and site types.
14. Unattended Operations - the unit should function reliably while unattended.

Table 6-2 Categories Of Primary Flow Measurement Devices

Classification	Type	Sub-type
Quantity Devices		
gravimetric	weigher, tilting trap, weigh dump	
volumetric	metered tank or bucket	
	displacement piston	
	nutating disk	
	sliding or rotating vane	
Rate Devices		
differential	venturi	
pressure	flow nozzle	
	round edge orifice	
	square edge orifice	concentric, eccentric segmented, gate or variable area
	centrifugal	elbow, long radius bend, guide vane speed ring
	impact tube	pitot-static, pitot venturi
	linear resistance	pipe section, capillary tube, porous plug
variable area	gate	
	cone & float	
head-area	weir	broad crested, sharp crested and combination
	flume	venturi, parshall, palmer-bowlus, cutthroat, san dimas, trapezoidal, SCS types HS, H and HL
	open flow nozzle	
flow velocity	float	simple, integrating
	tracer	
	vortex	vortex and eddy shedding
	rotating element (meter)	horizontal, vertical
force-	vane	
displacement	jet deflection	
thermal	hot & cold tip	
other	electromagnetic, acoustic, doppler, dilution, electrostatic, nuclear resonance	

Source: Shelley and Kirkpatrick, 1975, Leupold & Stevens, 1991

15. Maintenance Requirements - routine maintenance should be minimal and troubleshooting should be able to be handled in the field with ease.
16. Adverse Ambient Effects - the unit should be free from being effected by adverse climatic or chemical environments.
17. Ruggedness - the unit should be able to withstand rough handling and be as vandal and theft proof as possible.
18. Self Contained - the unit should be self contained as much as possible.
19. Precalibration - the unit should not require field calibration at each site.
20. Ease of Calibration - any necessary calibration should be accomplished with ease in the field.
21. Maintenance of Calibration - the unit should operate accurately for extended periods of time.
22. Cost - the unit should be affordable in terms of installation, maintenance and operation.
23. Portability - the unit should be able to be transported easily to one or many measurement points using standard equipment.

   In addition, the units should also have the following characteristics.
*   Adaptability - the unit should be able to measure a wide variety of flow types and in different modes and interface with many types of secondary communication and recording equipment.
*   Submersion Proof - the unit should be able to be submerged without damage.

   Not every type of equipment will need to have all these characteristics for every application. But, given the applications, different characteristics will be crucial. For example, gravimetric and volumetric methods work well for low flow and illicit connection measurement when a simple bucket or tank can collect all the flow for a certain time. Flow meter sampling works well for specific one time applications or measurements of flow in open channels but does not work to supply a continuous record of flow. For a continuous record a fixed structure with a known stage-discharge or pressure-discharge relationship works well and thus orifice plates and venturi tubes work well for pressure flow while weirs, flumes and orifices work well for channel and partially full pipe flow.

   With this set of criteria (numbers 1 through 23) Shelley and Kirkpatrick evaluated a number of types of flow measurement equipment. Table 6-3 gives a summary for the most common types of urban storm water runoff measurement equipment. The numbers across the top refer to the numbered criteria on the previous pages. Much of the work was subjective.

Common Errors

   There are both avoidable and unavoidable errors which will be encountered in the installation and operation of any flow measurement site. Mort (1955), Thomas (1957), USBR (1974), Leupold and Stevens (1991) and Grant (1989) provide descriptions of these errors and ways to minimize or avoid them. The most commonly mentioned errors are:
*   faulty fabrication or construction;
*   improper gage or head measuring location or head measurement;
*   incorrect zero setting;
*   use of device outside its range of accuracy;
*   improper maintenance or checking against settling, shifting, etc.;
*   turbulence in approach channels; and
*   unnoticed clogging and debris.

6.8 Flow Data Collection - Devices And Structures

Gravimetric And Volumetric

The most common type of gravimetric device is the tipping bucket rain gage. Other types are used for calibration of other flow meters where a tank is placed on a scale and flow is timed and weighed. If the fluid density can be assumed constant then a volumetric method is also appropriate and weighing can be avoided if the contained volume is known. The simple bucket and stop watch or metering tank are forms of this method. This method works well if the ditch or pipe outfall is elevated and the flow is low. The diverting of flow should happen rapidly at the beginning and end to increase accuracy and height measured after the oscillations have dampened (Rouse, 1949). Accuracies of $\pm$ 1 percent of the reading can be obtained over a range of 10:1 or better.

Existing equipment or structures can be used for volumetric measurements such as pump wet wells or small well defined detention ponds. There are a number of volumetric meters for such applications such as water meters with various types of disks or impellers. The vane meter called the Dethridge Wheel has been used in Australia and New Zealand for irrigation water flows, but not in the United States.

Venturi Meter Tubes

A venturi meter is a pipe segment consisting of a converging section, a throat and a diverging section. It is used for flow measurement under pressure conditions in pipes. Figure 6-2 illustrates a typical venturi meter layout. The venturi operates on the Bernoulli principle where fluid passing through a reduced area increases in velocity under decreased pressure. This change in pressure can be measured and related to the flow rate. A venturi meter is self cleansing, operates with a large concentration of solids, has high pressure loss recovery, and can operate reliably over a long period of time. The major drawback is initial cost of installation.

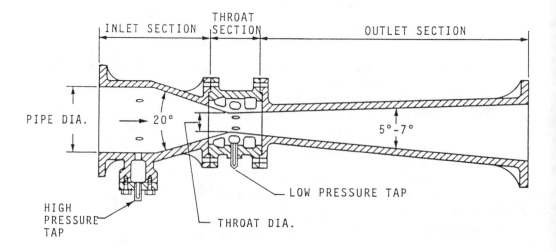

Figure 6-2 Typical Venturi Meter Tube (after Shelley and Kirkpatrick, 1975)

Table 6-3 Evaluation Of Primary Flow Measurement Devices

Equipment Type	Flow Equipment Evaluation Criteria																						
	1	2	3	4	5	6	7	8	9	10	11	12	13	14	15	16	17	18	19	20	21	22	23
Gravimetric - all	G	G	H	Y	L	H	H	H	P	M	H	H	P	Y	H	M	F	Y	Y	G	F	H	N
Volumetric - all	P	G	H	Y	L	H	H	M	P	L	H	H	F	Y	H	M	F	Y	Y	G	F	H	N
Venturi Tube	P	G	S	N	L	S	S	L	P	L	H	H	G	Y	M	M	G	Y	Y	G	G	H	N
Flow Nozzle	P	F	S	N	L	H	H	H	P	L	H	S	G	Y	H	M	F	Y	Y	G	P	L	Y
Elbow Meter	P	F	S	N	L	S	S	L	P	L	H	S	G	Y	L	M	G	Y	N	F	G	L	N
Sharp Crested Weir	F	F	M	N	M	H	H	H	F	L	M	M	G	Y	H	M	G	Y	Y	G	P	L	Y
Broad Crested Weir	F	F	S	N	H	H	M	M	G	L	M	M	G	Y	L	M	G	Y	N	F	F	L	N
Parshall Flume	G	F	S	N	M	S	S	L	F	L	M	M	G	Y	L	M	G	Y	Y	G	G	M	Y
Palmer-Bowles*	F	F	S	N	M	S	S	L	G	L	S	S	G	Y	L	M	G	Y	Y	G	G	L	Y
Cutthroat Flume	G	F	S	N	L	S	S	L	P	L	S	S	G	Y	L	M	G	Y	Y	G	G	L	N
Trapezoidal Flume	G	F	S	N	L	S	S	L	F	L	M	M	G	Y	L	M	G	Y	Y	G	G	L	N
SCS Flumes	G	F	S	N	H	M	S	H	G	L	M	M	G	Y	M	M	G	Y	Y	G	F	L	Y
Open Flow Nozzle	G	F	S	N	H	M	S	H	G	L	M	M	G	Y	M	M	G	Y	Y	G	F	L	Y

Equipment Type	1	2	3	4	5	6	7	8	9	10	11	12	13	14	15	16	17	18	19	20	21	22	23
Float Velocity	G	P	H	N	L	L	S	S	L	G	L	S	S	G	N	L	H	G	N	-	-	L	Y
Tracer Velocity	F	F	M	Y	L	S	S	L	G	M	S	S	F	Y	M	S	F	N	N	G	G	H	Y
Rotating Meter	F	F	S	Y	L	H	M	L	F	L	S	S	G	N	H	H	G	N	Y	G	G	L	Y
Force-Momentum	P	G	S	N	L	M	M	L	P	H	H	H	P	Y	H	S	P	Y	Y	G	G	H	N
Doppler Meter	P	G	S	Y	L	H	S	L	F	M	M	M	F	Y	M	S	F	Y	Y	G	G	H	N
Optical Meter	F	P	S	N	L	S	S	L	F	L	S	S	G	N	L	H	G	N	Y	G	G	L	Y
Dilution	G	G	M	Y	L	M	S	L	G	M	S	S	F	Y	M	S	F	N	N	G	G	H	Y
Acoustic Meter	G	G	S	Y	L	M	S	L	F	M	M	M	F	Y	M	S	F	F	Y	G	G	H	N
Electromagnetic	F	G	S	Y	L	S	S	L	P	H	M	M	F	Y	M	S	F	Y	Y	G	G	H	N
Hot-Tip Meter	F	P	S	Y	L	H	M	L	F	M	M	M	F	Y	H	M	F	Y	Y	G	F	H	N

Flow Equipment Evaluation Criteria

NOTES: F-Fair, G-Good, H-High, L-Low, M-Medium or Moderate, N-No, P-Poor, S-Slight, Y-Yes
* includes the Leopold-Lagco Flume

Source: Shelley and Kirkpatrick, 1975

To specify a venturi the flow range and pipe size should be known. There are different ratios of throat diameter to pipe inside diameter (termed the *beta ratio*) for each size of venturi. The venturi should be installed with at least five pipe diameters upstream and two pipe diameters downstream from any pipe fittings. The manufacturer's rating curve should be consulted to obtain flow ranges, but generally they are about 5:1 with an accuracy of $\pm$ 1-2 percent of readings. They can be attached to any number of readout devices for continuous or periodic readings. The equation of flow for a venturi is:

$$Q = \frac{C}{\sqrt{1 - \left[\dfrac{D_2}{D_1}\right]^4}} \; A\sqrt{2gH}$$

$$(6.5)$$

Where: Q =   discharge, cfs
        g =   acceleration due to gravity, 32.2 ft/s^2
        D_1 =   pipe inside diameter, ft
        D_2 =   throat diameter, ft
        H =   difference in piezometric head, ft
        C =   coefficient of discharge
        A =   flow area, ft^2

The coefficient C varies with the Reynolds number, with the diameter and velocity for the Reynolds number determined by the throat diameter. C varies in a smooth curve from about 0.94 for a Reynolds number of 5×10^3 to a constant value of 0.985 for at and above a Reynolds number of 5×10^5. The Reynolds number is expressed as:

$$\mathbf{R = V_2 D_2 / \nu}$$

$$(6.6)$$

Where: R =   Reynolds number
        V_2 =   velocity in venturi throat, ft/x
        ν =   kinematic viscosity of water, ft^2/s
For more information see manufacturers' literature and Shelley and Kirkpatrick, 1975.

Orifice Meters and Nozzles

Any constriction in the flow in a pressure conduit or pipe will cause a velocity increase and a pressure reduction. A number of methods to measure flow falling under the general headings of orifice meter or flow nozzle are available on the market, the thin plate orifice being the most commonly used orifice meter. Flow nozzles range from a venturi-like insert at one end to a thin nozzle approaching an orifice plate at the other.

Flow nozzles are more expensive than orifice plates and are sensitive to upstream conditions. Twenty or more diameters upstream to the nearest bend or fitting are recommended. Accuracies of flow nozzles can approach that of a venturi when calibrated in place. Some nozzles are designed to fit on the end of a freely discharging pipe flowing full or partially full. Figure 6-3 shows a typical flow nozzle.

The orifice is the least expensive of the pressure reducing flow measurement devices for pipe full-flow. It is normally a thin plate with a hole cut in it clamped at a pipe flange or joint. It also produces the greatest head loss across the orifice and is the most sensitive to upstream disturbances. Flows heavy in debris or sediment will often reduce the accuracy, although eccentric orifices with the opening at the pipe invert reduce this somewhat. Range is limited to about 5:1 and accuracies of ±1 percent are achievable. Orifices are also used commonly for detention and retention pond outlet works design.

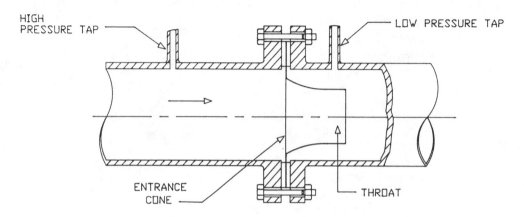

Figure 6-3 Typical Flow Nozzle (after Shelley and Kirkpatrick, 1975)

The basic discharge equation for the orifice is:

$$Q = C_o A (2gH)^{0.5} \qquad\qquad (6.7)$$

Where: Q = discharge, cfs
        A = cross-section area of pipe, ft^2
        g = acceleration due to gravity, 32.2 ft/s^2
        C_o = orifice discharge coefficient
        H = head from center of pipe to the water surface, ft

The discharge coefficient varies with orifice geometry and edge conditions and its value is quite sensitive when not a simple square edge orifice. The discharge coefficient is the product of the velocity coefficient and the contraction coefficient. The velocity coefficient is obtained experimentally by dividing the actual velocity at the *vena contracta* (the smallest area of the contracting jet emerging from the orifice) by the theoretical velocity. The value is on the order of 0.94 to 0.99 (Nelson, 1983). The contraction coefficient is the ratio of the flow area at the *vena contracta* to the area of the orifice. Typical values are in the range of 0.61 to 0.67. However any number of factors may reduce or suppress the full contraction giving a significantly higher discharge coefficient including: rounding the edges, providing a short tube downstream or placing the orifice such that the orifice edge coincides with the wall, top or bottom of the entrance area.

For sharp square edge entrance conditions the coefficient, C_o, can be taken to be 0.6 for a wide range of sizes for both circular and rectangular orifices. For orifices with a sharp edge, but having a length of pipe discharging to the atmosphere downstream, the discharge coefficient is taken as 0.8 (Bauer, et al., 1969). If part of the boundary is suppressed due to improving of the inlet through beveling, rounding or placement at the boundary, an empirical correction can be applied to the coefficient as (USBR, 1974):

$$C_s = 1 + 0.15r \tag{6.8}$$

Where: C_s = correction coefficient multiple of C_o
       r = ratio of the suppressed portion of the perimeter of the orifice to the total orifice perimeter.

See Bauer, et al., (1969) for a discussion of combinations of conditions.

If the approach velocity is significant the head in the discharge equation must be corrected by adding to it the velocity head of the approaching flow ($V^2/2g$). Pipes smaller than 12 inches may be analyzed as a submerged orifice if H/D is greater than about 1.5. For other conditions the opening operates as a weir up to a certain point and culvert analysis with inlet conditions is appropriate. When in doubt construct two stage-discharge curves, one for each condition, and use the one yielding the higher stage for a given flow (Barfield, Warner & Haan, 1981). For rectangular orifices with low head a corrected equation taking head into account is:

$$Q = 2/3 \ C_o \ W \ (2g)^{0.5} \ (H_2^{3/2} - H_1^{3/2}) \tag{6.9}$$

Where: W = width of the rectangular orifice, ft
      H_1 = distance from top of orifice to water surface, ft
      H_2 = distance from bottom of orifice to water surface, ft

Variable gates have also been used as orifices. For a discussion of various types of variable gates and discharge coefficients see COE (1987), Bauer, et al., (1969), Rantz (1982) and USBR (1974).

Weirs - Overview

A weir is a type of a head-area flow control and measuring device used in open channels and pipes flowing partially full. It is essentially a dam or overflow structure built across the channel or within a pipe. The edge or surface over which the water flows is called the crest. The overflowing sheet of water is termed the nappe. When the weir is thin and beveled on the downstream side such that a sharp upstream edge exists the nappe springs clear of the weir (except for very low heads). This is called a sharp-crested weir. If not it is called a broad-crested weir (or sometimes "weirs not sharp-crested"). Under the two categories of sharp-crested and broad-crested weirs there are many subcategories based on weir shape. Sharp crested weirs are used for flow measurement while broad crested weirs, due to greater durability, are often used in conjunction with flow structures where flow measurement is a secondary concern. Sharp-crested weirs are quite susceptible to floating debris and sediment deposition in the slack water area upstream from the weir. This deposition will eventually modify the approach conditions requiring an adjustment for approach velocity or, better, cleaning out of the approach channel or pool.

For a weir of a proper size and shape, with steady state flow conditions, and weir-to-pool height relationship, critical flow will exist near the overflow point. Thus, only one depth of flow can exist for any discharge, largely independent of channel roughness or other uncontrollable circumstances. Discharges are calculated by measuring the depth of flow from the crest of the weir vertically upward to the water surface elevation at a point upstream from the weir where the drawdown effects of the accelerated flow are negligible - about three times the head on the weir. This measurement is done using any number of "secondary" measuring devices including staff gages and float gages. For detailed information on the construction and maintenance of float wells and other secondary measuring facilities see Rantz (1982) and Leupold and Stevens (1991).

Sharp-Crested Weirs

Commonly used shapes for sharp-crested weirs are rectangular, trapezoidal (or Cipolletti), and triangular (or V-notched). The sharp-crested weirs consist of a thin plate, 1/8 to 1/4 inch thick, with a bevel on the downstream side and the upstream side sharpened. The crest plate should be vertical and the upstream side smooth. The crest should be exactly level to ensure accurate measurement. The connection to the channel should be watertight. The lower nappe of the weir should be vented with a pipe through the side wall if the weir is suppressed. To lower the approach velocity factor, the height of the weir from the channel bottom should be at least twice the maximum expected head and the upstream flow area eight times the nappe cross sectional area for a distance 15 to 20 times the head on the crest (Grant, 1989). The weir height should be over 1 foot. The approach should be straight at least 20 times the maximum expected head and have as little slope as possible. Submerged flow conditions should be investigated and the weir crest elevation set to avoid submergence. The secondary flow measurement device to measure stage should be located at least three times the head upstream, and must be zeroed exactly. The crest should be kept clean through periodic inspection and the upstream side cleaned of sediment and other debris. Often a low flow door for flushing can be built into the weir. The weir size should be set only after considerable study of the range of potential flow rates.

Sharp crested weirs in the laboratory can achieve an accuracy of $\pm 1\%$ of measurement but $\pm 5\%$ is expected in the field. The typical range of performance is about 20:1. It must be remembered that the head lost will be at least equal to the head measured. Details on weir construction can be obtained from USBR (1974, 1978).

The standard generic weir flow equation is:

$$Q = CLH^n \qquad\qquad (6.10)$$

Where: Q = discharge, cfs
   H = head above weir crest excluding velocity head, ft
   L = horizontal "effective" weir length, ft
   n = 1.5 for non-variable flow width weirs such as rectangular and other values for other weirs

Submerged Weir Correction - If the water surface downstream of the weir rises to the point that there is not free ventilation of the underside of the nappe, with free fall distance less than about twice the head on the weir, the discharge may increase due to the formation of a partial vacuum under the nappe. If the tailwater rises to a point equal to the crest elevation, the weir is considered submerged and an incomplete vertical contraction occurs. Dependable measurements cannot be guaranteed in this area though correction factors exist. When the tailwater height (H_t), measured relative to the crest elevation, above the crest rises to about 2/3 that of the head on the crest (H), the discharge becomes appreciably affected. A correction for submergence for any of the shapes of weirs discussed below can be given by the Villemonte equation (Villemonte, 1947):

$$Q_S = Q_f(1 - (H_2/H_1)^n)^{0.385} \qquad\qquad (6.11)$$

Where: Q_s = submergence flow, cfs
   Q_f = free flow, cfs
   H_1 = upstream head above crest, ft
   H_2 = downstream head above crest, ft
   n = 1.5 for non-variable flow width weirs and other values for other weirs

Approach Velocity Correction - If the conditions for neglecting approach velocity are violated it may be prudent to include an estimate of the approach velocity as a correction to the calculated discharge. The approach velocity, V, is simply the discharge divided by the cross sectional area of the approach channel. The approach velocity head is then:

$$h_v = V^2/2g \tag{6.12}$$

Where: V = approach velocity, ft/s
g = acceleration of gravity, 32.2 ft/s^2
h_v = approach velocity head, ft

The corrected head above the weir is then:

$$H_v = [(H+h_v)^{3/2} - h^{3/2}]^{2/3} \tag{6.13}$$

Where: H_v = corrected head on weir, ft
h_v = approach velocity head, ft
H = head above weir crest excluding velocity head, ft

Sharp-Crested Rectangular Weirs - Sharp-crested rectangular weirs with no end contractions and with two end contractions are illustrated in Figures 6-4 and 6-5. The discharge equation coefficient in equation 6.10 for a vertical sharp-crested rectangular weir configuration is given by Rehbock as quoted in Rouse (1949):

$$C = 3.237 + 0.428(H/H_c) + 0.0175/H \tag{6.14}$$

Where: H_c = height of weir crest above channel bottom, ft
H = head above weir crest excluding velocity head, ft

An average value of this equation, 3.33, is often used in practice and in flow measurement manuals. This equation is accurate for H/H_c ratios less than about 10. For H/H_c ratios greater than about 15 the weir acts as a sill and the flow equation becomes (Chow, 1959):

$$C = 5.68 (1 + H_c/H)^{1.5} \tag{6.15}$$

When the ends of the weir do not coincide with the sides of the approach pool or channel, the flow from the sides of the channel approaches the weir opening at an angle. The momentum of this flow carries it past the opening edge and causes a contraction of the flow as it corners through the opening and turns downstream. If the weir ends do coincide with the pool or channel sides, this contraction is suppressed, and more flow can pass through the weir opening. Provision must be made for ventilation of the underside of the weir. The correction for contracted sides for rectangular weirs is given in equation 6.16. Corrections are built into the discharge coefficients of other weir equations. For the contraction to be complete the distance from the walls to the weir ends must be greater than twice the head on the weir.

$$L = L' - 0.1NH \tag{6.16}$$

Where: L' = measured horizontal weir length, ft
N = number of contracted sides, $N=2$ for both sides contracted, $N=1$ for one side contracted and $N=0$ for no contraction (suppressed weir)
H = head above weir crest excluding velocity head, ft

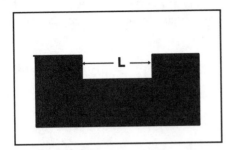

Figure 6-4 Sharp-Crested Weir
(Two End Contractions)

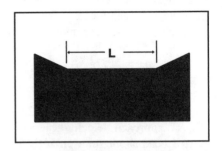

Figure 6-5 Sharp-Crested Weir
(No End Contractions)

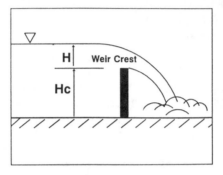

Figure 6-6 Sharp-Crested Weir And Head

Rantz, 1982, provides an alternate way to obtain both the discharge coefficient and the contraction correction coefficient. Figure 6-7 gives the variation of the discharge coefficient with sloped approach conditions and for various values of H/H_c. Figure 6-8 gives corrections (k_c) for various ratios of contraction ratios (b/B) and H/H_c ratios, where b is the measured weir length and B is the channel width. If the rectangular weir is set between two rounded vertical abutments instead of sharp edges, a further correction is made to increase the discharge (Rantz, 1982). Assume the radius of curvature of the abutment is "r". The following rule applies:

for: $r/b = 0$ correction factor $= k_c$
for: $0 \leq r/b \leq 0.12$ correction factor $=$ interpolated   (6.17)
for: $r/b > 0.12$ correction factor $= (1+k_c)/2$

The height of the weir crest above the bottom of the approach channel and the distance from the sides of the notches to the sides of the channel should each be more than twice the head on the crest. The weir should not be used for heads less than 0.2 feet nor for heads greater than 1/2 the crest length (USBR, 1974). Discharge coefficients lose accuracy for crest lengths less than one foot and triangular weirs should be used instead (Grant, 1989).

Trapezoidal and Cipolletti Weirs - Trapezoidal weirs of all dimensions have been fabricated. In all cases but the special case of the Cipolletti weirs the discharge is assumed to be a combination of a rectangular weir and a triangular weir. However, the discharge coefficients have to be experimentally determined in the field using flow meter measurements at various stages. For the special case of the Cipolletti weir the side slopes of the trapezoidal section are 1V:4H. This slightly greater area, for the same weir length, in the bottom of the trapezoid, b, is designed to counteract the contraction effects and make the Cipolletti weir act as if it were a fully suppressed rectangular weir of length of b. The equation of discharge for the Cipolletti weir is:

$$Q = 3.367 \; L \; H^{1.5} \tag{6.18}$$

Considering the velocity of approach the formula becomes (USBR, 1974):

$$Q = 3.367 \; L \; (H + 1.5h_v)^{1.5} \tag{6.19}$$

Where: h_v = approach velocity, ft/s

The height of the weir crest above the bottom of the approach channel and the distance from the sides of the notches to the sides of the channel should each be more than twice the head on the crest. The weir should not be used for heads less than 0.2 feet nor for heads greater than 1/3 the crest length (USBR, 1974).

V-Notch Weirs - The V-notch weir is particularly suited for small flows as it concentrates the low flow in the center allowing the nappe to spring free from the crest. The minimum distance from the sides of the weir to the channel banks should be twice the head on the weir. Also the minimum distance from the point of the notch to the channel bottom should be more than twice the head on the weir (USBR, 1974). The 90 degree V-notch weir can handle flows up to about 10 cfs easily.

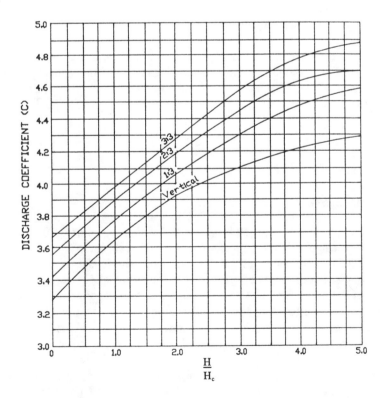

Figure 6-7 Discharge Coefficients For Sharp-Edged Suppressed Rectangular Weirs

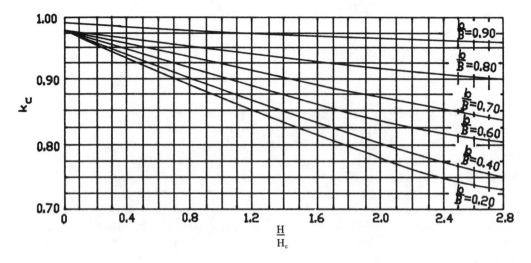

Figure 6-8 Adjustment Factor k_c For Contracted Rectangular Sharp-Edged Weirs

The discharge through a V-notch weir can be calculated from the following equation (Brater and King, 1976):

$$Q = 2.5 \tan(\theta/2)H^{2.5} \tag{6.20}$$

Where: Q =   discharge, cfs
         θ =   angle of v-notch, degrees
         H =   head on apex of notch, ft

Much more data are available on standard 90 degree and 120 degree notches than other size openings, the wider angle accommodating a much wider range of flows than the narrower angle (Leupold & Stevens, 1991). For these the standard equations can more accurately be taken respectively as (Rantz, 1982):

$$Q = 2.47\ H^{2.50} \quad (90^\circ) \tag{6.21}$$

and

$$Q = 4.35\ H^{2.50} \quad (120^\circ) \tag{6.22}$$

Broad-Crested Weirs

The equation generally used for the broad-crested weir is equation 6.10. The typical form the weir takes is a rectangular block in the bottom of the open channel. Rounding of the upstream edge of the rectangular block improves the flow efficiency and renders the weir less sensitive to sediment deposition and weir height effects. If R is the radius of curvature and P the height of the weir, ratios of R/P less than 0.094 cause no significant reduction in the C factor. Rounding of the upstream edge in the R/P range from 0.094 up to 0.25 causes a gradual increase in the C factor for all values of H/P (H is the head on the crest). Rounding above the R/P ratio of 0.25 has no further effect on the C factor (Ramamurthy, et al., 1988). The effects of an inclined upstream face can be found in Clemmens, et al., (1984) and Replogle (1978). Information on C values as a function of sharp-edged (upstream edge) rectangular weir crest breadth and head is given in Table 6-4.

Table 6-4 Broad-Crested Weir Coefficient C Values As A Function Of Weir Crest Breadth And Head

Measured Head, H[1] (ft)	Breadth Of The Crest Of Weir (ft)										
	0.50	0.75	1.00	1.50	2.00	2.50	3.00	4.00	5.00	10.00	15.00
0.2	2.80	2.75	2.69	2.62	2.54	2.48	2.44	2.38	2.34	2.49	2.68
0.4	2.92	2.80	2.72	2.64	2.61	2.60	2.58	2.54	2.50	2.56	2.70
0.6	3.08	2.89	2.75	2.64	2.61	2.60	2.68	2.69	2.70	2.70	2.70
0.8	3.30	3.04	2.85	2.68	2.60	2.60	2.67	2.68	2.68	2.69	2.64
1.0	3.32	3.14	2.98	2.75	2.66	2.64	2.65	2.67	2.68	2.68	2.63
1.2	3.32	3.20	3.08	2.86	2.70	2.65	2.64	2.67	2.66	2.69	2.64
1.4	3.32	3.26	3.20	2.92	2.77	2.68	2.64	2.65	2.65	2.67	2.64
1.6	3.32	3.29	3.28	3.07	2.89	2.75	2.68	2.66	2.65	2.64	2.63
1.8	3.32	3.32	3.31	3.07	2.88	2.74	2.68	2.66	2.65	2.64	2.63
2.0	3.32	3.31	3.30	3.03	2.85	2.76	2.27	2.68	2.65	2.64	2.63
2.5	3.32	3.32	3.31	3.28	3.07	2.89	2.81	2.72	2.67	2.64	2.63
3.0	3.32	3.32	3.32	3.32	3.20	3.05	2.92	2.73	2.66	2.64	2.63
3.5	3.32	3.32	3.32	3.32	3.32	3.19	2.97	2.76	2.68	2.64	2.63
4.0	3.32	3.32	3.32	3.32	3.32	3.32	3.07	2.79	2.70	2.64	2.63
4.5	3.32	3.32	3.32	3.32	3.32	3.32	3.32	2.88	2.74	2.64	2.63
5.0	3.32	3.32	3.32	3.32	3.32	3.32	3.32	3.07	2.79	2.64	2.63
5.5	3.32	3.32	3.32	3.32	3.32	3.32	3.32	3.32	2.88	2.64	2.63

[1]Measured at least 2.5H upstream of the weir. Reference: Brater and King (1976).

Submergence Effects - Submergence effects cause a reduction in the discharge coefficient. Table 6-5 gives approximate values of reduction in flow for a range of R/P values. H_2 and H_1 are the downstream and upstream heads with reference to the crest elevation respectively, and Q_s is the submerged flow while Q is the unsubmerged flow.

Table 6-5 Broad-Crested Weir Submergence Discharge Reduction As A Function Of Head Ratio

H_2/H_1	R/P=0.0 Q_s/Q	R/P = 0.125 Q_s/Q	R/P = 1.00 Q_s/Q
0.80	0.95	0.98	1.00
0.85	0.87	0.90	0.96
0.90	0.77	0.80	0.88
0.95	0.48	0.60	0.70

Reference: Ramamurthy, R. S., U. S. Tim and M. V. J. Rao, "Characteristics of Square-Edged and Round-Nosed Broad-Crested Weirs", ASCE J. of Irrig. Drain. Engrg., Vol. 114, No. 1, 1988. Reproduced by permissions of ASCE.

Triangular Channel Broad-crested Weirs - Broad crested weirs with upstream ends suitably rounded can be used in triangular channels to measure flow. Gill (1985) collected data from many authors and slightly modified equations for discharge coefficients. The discharge coefficient is independent of vertex angle over the range tested. His equations are:

$$Q = 0.64 \ C_d \ (0.4 \ g)^{1/2} \ H^{2.5} \ \tan (\theta/2) \tag{6.23}$$

and:

$$C_d = [(H/L) - 0.0718]^{0.032} \tag{6.24}$$

Where: Q = discharge, cfs
   C_d = discharge coefficient
   g = acceleration of gravity, 32.2 ft/s^2
   H = static head on crest, ft
   θ = bottom vertex angle of channel
   L = stream-wise length of weir crest, ft

Ogee Shapes - The shape of the nappe over a flat broad crested weir is often unstable leading to a varying flow rate for the same head. In an attempt to correct for this deficiency the ogee shaped spillway was developed. The purpose of the ogee crest is to structurally mimic the falling stream of flow over a sharp crested weir and support the nappe at suitable pressures over a range of flows. Extensive model testing has been accomplished for this ogee shape. A complete description of the design procedure for this type of crest can be found in Reese and Maynord, 1987, COE, 1988 and USBR, 1977. The latest procedure is to fit the upstream quadrant with an equation which is in the form of an ellipse, and the downstream quadrant with another equation of the falling free flowing nappe.

Special Weirs

A variety of specially shaped sharp edged weirs have been developed for different purposes. Most of them have attempted to provide proportional or linear stage-discharge curves for irrigation, sediment control and flood control purposes, as well as flow measurement (Rao and Chandrasekaran, 1977, Murthy and Pillai, 1977, 1978a, 1978b, Murthy and Giridhar, 1989, Swamee, et al., 1991 and Sandvic, 1985). For erosion control or settling design the velocity can be controlled, as any function of depth, to meet any relationship. Different shapes can be superimposed on each other to achieve almost any physically possible desired effect.

One type of weir is the inverted triangle weir whose discharge relationship is given in equation 6.25. The limits of the equation are $0.22D_t \le H \le 0.94D_t$. Flows less than that are termed base flows and cannot be approximated by this linear relationship.

$$Q = 0.4481 \ C_d/g \ W_t \ D_t^{1/2} \ (H_t\text{-}0.0817D_t) \tag{6.25}$$

Where: Q = discharge, cfs
   C_d = discharge coefficient taken as 0.615
   g = acceleration of gravity, 32.2 ft/s^2
   W_t = bottom width of triangle, ft
   D_t = height of triangle, ft
   H_t = depth of flow in triangular weir, ft

The proportional weir (or "Sutro" weir, though the Sutro shape is only one of many similar shapes) is distinguished from other control devices by having a linear head-discharge relationship achieved by allowing the discharge area to vary nonlinearly with head. Although more complex to design and construct, a proportional weir may significantly reduce the required storage volume for a given site for detention design. Design equations for proportional weirs are (Sandvik, 1985, Murthy and Pillai, 1978, Rouse, 1949, Pratt, 1914):

$$Q = 4.97 \ a^{0.5} \ b(H - a/3) \tag{6.26}$$
$$x/b = 1 - (1/3.17) \ (\arctan \ (y/a)^{0.5}) \tag{6.27}$$

Where: Q = discharge, cfs
         Dimensions a, b, h, x, and y are shown in Figure 6-9.
Dimension "a" is assumed and, for design discharges, "b" is calculated until reasonable dimensions are determined. Using these values and equation 6.27 the coordinates of the weir are found.

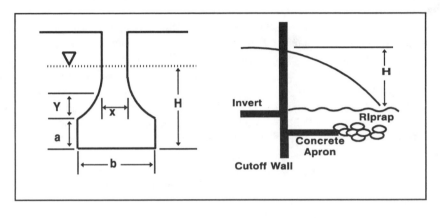

Elevation   Section

Figure 6-9 Proportional Weir Dimensions

<u>Flumes - General</u>

A flume is basically a specially designed constricted section within an open channel whose purpose is to measure flow through measuring the depth reduction caused by an increase in velocity. Flumes can operate in subcritical, critical and supercritical (standing wave) modes and are basically the open channel equivalent of the venturi meter (Rouse, 1949). In the last two modes only one depth measurement need be made upstream from the flume. In subcritical, or submerged flow, a measurement must also be made in a section downstream from the flume. Critical flow flumes give more sensitive measurement than supercritical flow flumes but are more susceptible to errors if there is deposition of sediment in the approach section.

While there are many types of flumes in use for open channels, the basic requirement of any kind of non-subcritical flume is that its constriction produce critical depth in the channel in a measurable and predictable way with the least head loss possible. All flumes have a discharge equation which takes the form of:

$$Q = C_f \ W_f \ H^{1.5} \tag{6.28}$$

Where: Q = discharge, cfs
        C_f = discharge coefficient
        W_f = width of flume throat, ft
        H = head on flume, ft

Calibration is necessary to determine the value of the discharge coefficient, and careful construction is necessary to maintain accuracy. Kilpatrick (1965) gives six approaches in the design of flumes and should be consulted for basic flume theory and design (also found in Shelley and Kirkpatrick, 1975 and Grant, 1989). Parshall flumes will be described in detail to give an example of flume selection and sizing. All other types of flumes can be selected and sized using the same general procedures.

Flume Versus Weir - The use of weirs is preferred when low flow measurement is of concern, cost is a great factor, or great variability of flows is to be expected. However, a flume is more advantageous than a weir in the following situations:

- Heavy sediment load of small to moderate size particles may clog a weir while it will generally pass through a flume. Therefore flumes have a great advantage in pipe flow measurements.
- Flumes are less sensitive to approach velocity. Flumes are more suitable in steeper channels where the Froude number can get above 0.75 while weirs lose accuracy due to approach velocity considerations.
- Flumes operate with less head loss than weirs (about one-fourth) and therefore it is advantageous to use a flume if the channel has low banks, though their lesser range make a controlled flow (such as in a diversion or channel with high flow bypass) or more narrow range of interest situation most applicable (Rantz, 1982).
- A flume can operate at a higher flow rate than a comparably sized weir (Grant, 1989).
- A flume is not subject to wear like a sharp-crested weir and has no moving parts like other types of meters.

General Flume Requirements - The most common type of flumes are of the critical or supercritical type. They pass the flow through critical depth and operate accurately for a specific head range, provided the flume is accurately constructed and properly installed. There are general requirements to ensure accurate flume measurements. The flume should be located in a straight section without bends immediately upstream. The approach flow should be evenly distributed across the flume entrance and relatively free from turbulence and waves. Velocity of approach is not of great concern if there are no great boils or current concentrations. Backwater submergence should be carefully considered and accounted for. Curved wingwalls should be used upstream from the flume to correct approach flow conditions rather than to try to make a mathematical correction in the discharge coefficient (USBR,1974).

Flumes - Parshall Flumes

The Parshall (after the late Dr. Ralph L. Parshall) flume was developed by the Soil Conservation Service in 1922 by taking the normal subcritical flume and dropping the floor to produce supercritical flow. Figures 6-10 and 6-11 give the layout and tabular dimensions of a typical Parshall flume. The constricted throat of the flume produces a discharge that is related to the width of the flume and the head on the flume. All other dimensions are related to the width. The level converging section followed by the downward sloping throat gives the flume a relatively high tolerance for submergence conditions. The flume is not proprietary in design or construction. Accuracies of $\pm 5\%$ are obtainable with flow ranges as expressed in Figure 6-11.

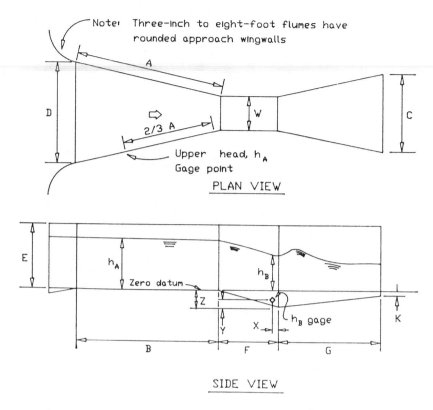

Figure 6-10 Parshall Flume Dimensions - Layout (Rantz, 1982)

Flow Equations - Where free flow conditions exist over the entire range of flow the flow is dependent only on the upstream head and the entire downstream diverging section can be omitted. This simplification has been used in the design of small portable Parshall flumes often called Montana flumes. Although derived experimentally all the free flow relations can be expressed by the following equations for throat widths 1 to 8 ft and for throat widths 10 to 50 ft respectively:

$$Q = 4\ W\ H^N \qquad (6.29)$$
$$N = 1.522\ W^{0.026} \qquad (6.30)$$
$$Q = (3.69\ W + 2.5)\ H^{1.6} \qquad (6.31)$$

Where: Q = discharge, cfs
           W = throat width, ft
           H = head measured relative to throat floor, ft

Sizes smaller than one foot fit the general equation $Q = KH^n$ where values of K and n can be found in Table 6-6.

Very careful construction and zeroing of the staff gages is important. For example, an error in 0.01 foot in setting the flume elevation and a similar error in the staff gage can combine to cause an eight percent error in the flow reading (USBR, 1974). It is important to ensure the flume walls extend well into the bank to prevent flanking and that freeboard is provided to prevent bank overtopping. As a general rule the most economical flume is one which has a throat width that is from 1/3 to 1/2 the channel width.

| | Widths | | | Axial Lengths | | | Wall Depth in Converging Section | Vertical Distance below crest | | Converging wall length | h_A dist. upstream from crest | Gage Points | | Free Flow Capacities | |
| | | | | | | | | | | | | | h_B | | |
Size; Throat width W	Upstream end D	Down-stream end C	Con-verging Section B	Throat Section F	Diverging Section G		Dip at Throat Z	Lower end of flume K	A		X	Y	Min.	Max.
inches	feet	feet	feet	feet	feet	feet	feet	feet	feet	feet	feet	feet	ft^3/s	ft^3/s
1	0.549	0.305	1.17	0.250	0.67	0.5-0.75	0.094	0.062	1.19	0.79	0.026	0.042	0.005	0.15
2	.700	.443	1.33	.375	.83	0.50-0.83	.141	.073	1.36	.91	.052	.083	.01	.30
3	.849	.583	1.50	.500	1.00	1.00-2.00	.188	.083	1.53	1.02	.083	.125	.03	1.90
6	1.30	1.29	2.00	1.00	2.00	2.0	.375	.25	2.36	1.36	.167	.25	.05	3.90
9	1.88	1.25	2.83	1.00	1.50	2.5	.375	.25	2.88	1.93	.167	.25	.09	8.90
feet														
1.0	2.77	2.00	4.41	2.0	3.0	3.0	.75	.25	4.50	3.00	.167	.25	.11	16.1
1.5	3.36	2.50	4.66	2.0	3.0	3.0	.75	.25	4.75	3.17	.167	.25	.15	24.6
2.0	3.96	3.00	4.91	2.0	3.0	3.0	.75	.25	5.00	3.33	.167	.25	.42	33.1
3.0	5.16	4.00	5.40	2.0	3.0	3.0	.75	.25	5.50	3.67	.167	.25	.61	50.4
4.0	6.35	5.00	5.88	2.0	3.0	3.0	.75	.25	6.00	4.00	.167	.25	1.30	67.9
5.0	7.55	6.00	6.38	2.0	3.0	3.0	.75	.25	6.50	4.33	.167	.25	1.60	85.6
6.0	8.75	7.00	6.86	2.0	3.0	3.0	.75	.25	7.0	4.67	.167	.25	2.60	103.5
7.0	9.95	8.00	7.35	2.0	3.0	3.0	.75	.25	7.5	5.00	.167	.25	3.00	121.4
8.0	11.15	9.00	7.84	2.0	3.0	3.0	.75	.25	8.0	5.33	.167	.25	3.50	139.5
10	15.60	12.00	14.0	3.0	6.0	4.0	1.12	.50	9.0	6.00			6	300
12	18.40	14.67	16.0	3.0	8.0	5.0	1.12	.50	10.0	6.67			8	520
15	25.0	18.33	25.0	4.0	10.0	6.0	1.50	.75	11.5	7.67			8	900
20	30.0	24.00	25.0	6.0	12.0	7.0	2.25	1.00	14.0	9.33			10	1340
25	35.0	29.33	25.0	6.0	13.0	7.0	2.25	1.00	16.5	11.00			15	1660
30	40.4	34.67	26.0	6.0	14.0	7.0	2.25	1.00	19.0	12.67			15	1990
40	50.8	45.33	27.0	6.0	16.0	7.0	2.25	1.00	24.0	16.00			20	2640
50	60.8	56.67	27.0	6.0	20.0	7.0	2.25	1.00	29.0	19.33			25	3280

Figure 6-11 Tabular Dimensions Parshall Flume (Rantz, 1982)

Table 6-6 Parshall Flume Discharge Parameters - Widths < 1'

Width (inches)	K	n
1"	0.3380	1.550
2"	0.6760	1.550
3"	0.9920	1.547
6"	2.060	1.580
9"	3.070	1.530

Source: Grant, 1989

Submergence - If H_b is the head on the downstream side and H_a is the head on the upstream side relative to the level floor in the converging section, then submergence can be expressed as H_b/H_a. Discharge is not reduced until the submergence ratios given in Table 6-7 are reached. H_a is measured at a distance of 2/3 the length of the converging section upstream from the throat section's leading edge for smaller flumes and a distance 2/3 (W/2 + 4) upstream from the same point for large flumes. W is the throat width. Submergence based on these limits is corrected for using equation 6.32. The factors k_s and Q_c can be read from Figures 6-12 and 6-13. Figure 6-14 gives head loss values for Parshall flumes one to eight feet wide. Enter the graph with a known percent submergence and read up to the discharge. From the discharge read horizontally to determine throat width and down to head loss.

Table 6-7 Parshall Flume Submergence Limits

Width Range	Submergence Ratio Limit
1 to 3 inches	0.50
6 to 9 inches	0.60
1 to 8 feet	0.70
8 to 50 feet	0.80

Source: USBR, 1974.

$$Q_s = Q - k_s Q_c \qquad (6.32)$$

Where: Q_s = submerged flow, cfs
$\quad\quad\ Q$ = discharge for free flow from equation 6.29 or 6.31, cfs
$\quad\quad\ k_s$ = discharge correction coefficient
$\quad\quad\ Q_c$ = discharge correction, cfs

Example Design - Selecting the size of a Parshall flume is a trial and error procedure illustrated by the following example. Assume a flume is to be sized for a channel 10 feet wide with moderate grade, and a 2.5 foot flow depth at a maximum flume design flow of 20 cfs. From Figure 6-12 flumes in the size range from 2 foot to 8 feet would do the job. From equations 6.29 and 6.30 (rearranged to solve for H) the following information is calculated:

Flume Size (ft)	H (ft)	Flume Size (ft)	H (ft)
2	1.80	6	0.89
3	1.38	7	0.81
4	1.15	8	0.75
5	1.00		

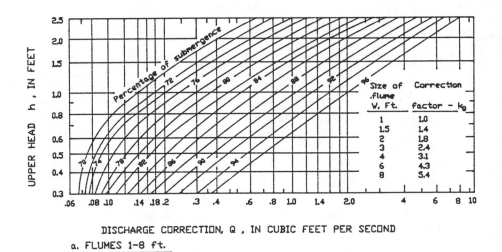

Figure 6-12 Submergence Correction Flumes 1 Ft To 8 Ft (Rantz, 1982)

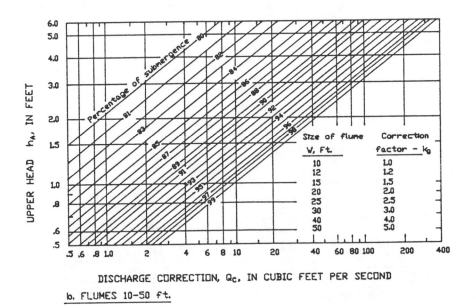

Figure 6-13 Submergence Correction Flumes 10 Ft To 50 Ft (Rantz, 1982)

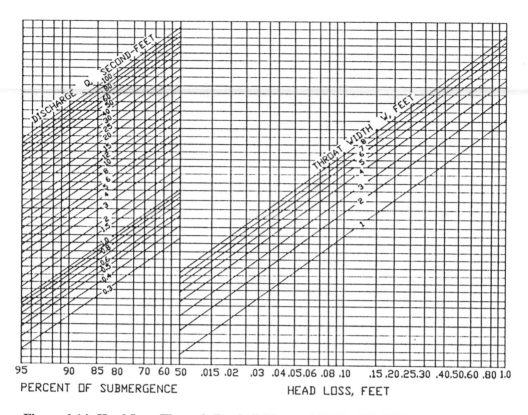

Figure 6-14 Head Loss Through Parshall Flumes 1 Ft To 8 Ft Wide (USBR, 1974)

From Table 6-7 the submergence ratio is 0.70 for flumes in this range. This is equal to H_b/H_a. The total depth is 2.5 feet downstream from the flume. Therefore the crest height plus H_b must be equal to 2.5. This crest height above the channel bottom is equal to K in Figure 6-10 if the lower end of the flume is flush with the channel bottom. However, there are many cases where the flume diverging section ends above the channel bottom and a scour protected ramp or step down is constructed. The rest of the analysis will assume the value is "K" for simplicity.

For free flow conditions using the four foot flume size as a trial:

$H_b/H_a = H_b/1.1 = 0.70$

$H_b = 0.81$ ft

$H_b + K = 2.5$ ft

$K = 2.5 - 0.81 = 1.69$ ft = crest height above bottom.

The flume will obstruct the channel and cause a rise in water surface upstream, the backwater effect. The difference in water surface elevation for the flow upstream of the flume with and without the flume is the headloss caused by the flume. The head loss through the flume can be found from Figure 6-14. Follow the 70 percent submergence vertically up to the 20 cfs line. From there extend horizontally to the right to the 4 foot throat width and down to the head loss. From this figure the head loss is determined to be 0.43 ft. Then the upstream water surface elevation is 2.5 + 0.43 or 2.93 feet. Then determine the rest of the site dimensions and see if there is sufficient freeboard on the channel to handle this situation.

Using the same analysis for the 3 and 2 foot flumes:

3 foot flume: H_b = 0.97 ft   K = 1.53 ft   Loss = 0.54 ft   WSEL = 3.04 ft

2 foot flume: H_b = 1.27 ft   K = 1.23 ft   Loss = 0.71 ft   WSEL = 3.21 ft

Cost comparisons for wingwall construction to cut off flanking flow and freeboard comparisons are then made.

It is normally desirable to design the flume for free flow. But if it is necessary to lower the upstream water surface as much as possible then the flume could be designed for submerged flow at 20 cfs. This would lower the entire structure in the channel and provide more channel bank freeboard (USBR, 1974). Using 95 percent submergence as the maximum:

H_b/H_a = 0.95 = $H_b/1.15$

H_b = 1.09 ft

H_b + K = 2.5 ft

K = 2.5 - 1.09 = 1.41 ft = crest height above bottom.

LOSS = 0.077 ft   WSEL = 2.58 ft

This lowers the upstream water surface elevation 2.93 - 2.58 = 0.35 ft.

Larger flumes are generally designed in the same way but much more care and analysis is done to properly set the crest elevation, determine freeboard, and ensure no interference with flood flow levels through harmful backwater effects.

Other Open Channel Flumes

There are many other types of open channel flumes used for various purposes and having various advantages and disadvantages. Several of the most popular will be mentioned here.

Cutthroat Flumes - The cutthroat flume was developed for use on flatter gradient streams where both free flow and submerged flow conditions would exist. In plan view the Cutthroat flume looks like a Parshall flume with the throat section missing (ergo "cutthroat") (See Figure 6-15). It is a flat bottomed flume which has both placement (it can be placed directly on the channel bed) and fabrication advantages over the Parshall flume. Flumes of this type were extensively tested by Skogerboe, et al., (1967,1974), who found that a depth measurement in the diverging section and near the entrance in the converging section produced a more stable measurement than in the throat section. Rectangular cutthroat flume sizes of 1, 2, 3, 4, and 6 feet have been studied. Submergence ratios vary smoothly from 79% to 88% over this range. For a complete list of flow equations see Skogerboe, et al., (1974).

Referring to Figure 6-15 the dimensions are:

$$B = W + 2 L_1/3 = W + L_2/3 \tag{6.33}$$
$$L_a = 2L/9 \tag{6.34}$$
$$L_b = 5L/9 \tag{6.35}$$

Palmer-Bowlus Flumes - The name Palmer-Bowlus is given to a class of closed conduit flume, which is generally a constriction in the flow characterized by a uniform cross section and a length approximately one pipe diameter. The flume is effective for open channel type flow in a pipe. Commercially available flumes of this type are sold in a variety of sizes for easy installation in circular conduits. The trapezoidal shape is the most common among commercial manufacturers. See Figure 6-16 for the various cross-sectional shapes of Palmer-Bowlus flumes.

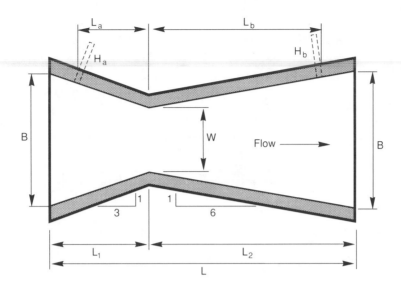

Figure 6-15 Rectangular Cutthroat Flume (after Grant, 1989)

Palmer-Bowlus flumes are designated by the pipe size rather than the flume throat size and are available for pipes from four inches to 42 inches. The flumes come in three types: permanent in-pipe installation, temporary in-pipe installation, and installation at the end of pipe outlet sections. The flows to be measured as well as the pipe size may be of concern for low flows in a large pipe. A smaller than pipe diameter sized flume can be used if low flows are of concern though some construction of an approach and bulkheads will be necessary (Grant, 1989).

Measurements are taken one-half pipe diameter upstream from the flume not in the drawdown transition zone prior to the throat. The elevation of the water above the throat is the standard head measurement. Manufacturers provide rating curves for their flumes. Critical flow must be maintained through the throat of the flume without submergence. Submergence generally occurs at a ratio of about 85 percent for most trapezoidal flumes. The slope of the pipe must be such that subcritical flow occurs in the approach section and that the flow is uniform and not overly turbulent. This condition can be assured if there are no bends upstream for a distance of about 25 pipe diameters. The flume is installed with a near horizontal throat section. See manufacturers' literature, Wells and Gotaas (1958) or Ludwig and Ludwig (1951) for more information.

Long-throated Flume Broad-crested Weirs - Some types of broad-crested weirs essentially act as long throated flumes but without the normal flume's (such as the Parshall flume) demands for exacting dimensions and construction if suitable accuracy is to be maintained. Figure 6-17 gives a typical layout for this type of broad-crested weir measuring device. These "bottom-contracted" devices have advantages in that they: are simple and inexpensive to construct; have reasonable construction tolerances; are not easily clogged with sediment; have low absolute head loss; can measure flow within ±3 percent of the true discharge measurement; have a wide variety of shapes to fit many situations; and have a very predictable head-discharge relationship due to nearly parallel flow lines and hydrostatic pressures if tailwater effects can be neglected. For detailed information see Clemmens, et al., 1984a.

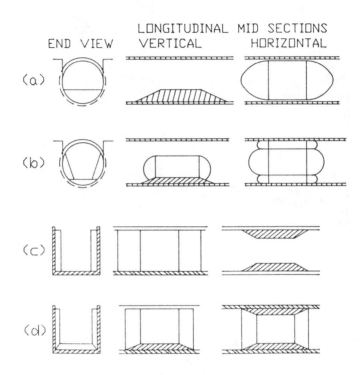

Figure 6-16 Cross-Sectional Shapes Of Palmer-Bowlus Flumes

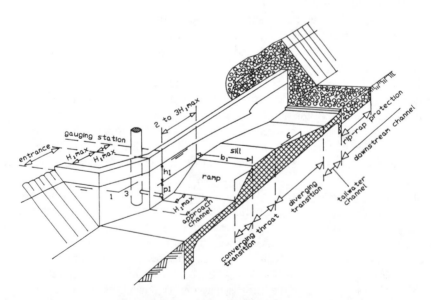

Figure 6-17 Typical Long-throated Broad-crested Weir Layout
Source: Clemmens, A. J., J. A. Replogle and M. G. Bos, "Rectangular Measuring Flumes for Lined and Earthen Channels", ASCE J. of Irrig. Drain. Engrg., Vol. 110, No. 2, 1984a. Reproduced by permission of ASCE.

Leopold-Lagco Flumes - This type of flume is a proprietary flume manufactured by the F. B. Leopold Company. It is actually a Palmer-Bowlus type flume with a rectangular cross sectional throat designed specifically for installation in circular conduits. This type of flume is noted for high accuracy with low head loss. Accuracies as great as ±3 percent can be obtained with depths up to 90 percent of full pipe and submergences up to 85 percent. A minimum throat length of 60 percent of the pipe diameter is necessary and the measuring point should be near one-half the pipe diameter upstream from the beginning of the transition section to the throat section (Grant, 1989). Standard Leopold-Lagco sizes fit pipes from 4 to 72 inches. The discharge equation for free flow in a Leopold-Lagco flume is:

$$Q = 2.407 \ D^{0.953} \ H^{1.547} \tag{6.36}$$

Where: Q = discharge, cfs
          D = pipe diameter, ft
          H = head measured at $D/2$ upstream from the flume entrance with the floor of the flume
              at zero elevation, ft

Flume size requirements based on either maximum head or maximum flow can be estimated from the following equations:

$$S = 17.18205 \ H_m^{1.00525} \tag{6.37}$$
$$S = 10.52642 \ Q_m^{0.401289} \tag{6.38}$$

Where: S = flume size, inches
         H_m = maximum head, feet
         Q_m = maximum discharge, cfs

Trapezoidal Flumes - Trapezoidal flumes form a family of many sizes and types from super-critical-flow flumes, with sloping throat floors, to flat floor throated flumes to flumes without floors at all but a triangular base and zero throat width. Trapezoidal flumes are used to obtain a wider range of flows than is possible in a Parshall flume since the side walls expand with depth. Trapezoidal flumes have been constructed for flows ranging from less than 0.1 cfs to 26,000 cfs. The trapezoidal shape conforms well to most ditch shapes and has the advantage of having a flat bottom conforming to the channel slope. Figure 6-18 shows a typical trapezoidal flume and dimensions for a 1 foot and 2 foot flume (Robinson, 1968 and Robinson and Chamberlain, 1960).

The 1 foot flume has a maximum flow rate of 7 cfs and head range of from 0.20 to 1.30 feet. The 2 foot flume has a maximum flow rate of 50 cfs and a head range from 0.30 to 2.60 feet. Submergence corrections found from the original references should be made for submergences greater than 80 percent. Equations for free flow respectively are:

$$Q = 3.23 \ H^{2.5} + 0.63 \ H^{1.5} + 0.05 \tag{6.39}$$
$$Q = 4.27 \ H^{2.5} + 1.67 \ H^{1.5} + 0.19 \tag{6.40}$$

Where: H = head above the throat section, ft
         Q = discharge, cfs

Trapezoidal flumes are generally installed level although conformance to channel slope is permissible if flow is subcritical and correction on head readings is made. The flume should be placed higher than the ditch bottom for natural ditches or in all cases where submergence is a concern. The supercritical trapezoidal flume with a sloping bottom is used commonly by the US Geological Survey. Typical sizes and configurations are given in Rantz (1992).

HS, H, and HL Flumes - The HS, H and HL flumes have had proven performance for the measurement of low flows, flows from small urban areas and flows in pollution abatement work by the Soil Conservation Service (SCS) since the 1930's. Figures 6-19 and 6-20 show the general shape and give the dimensions for the three related kinds of flumes. The wide span shape and narrow opening make these flume types ideal for urban storm water measurements where both high and low flows are important. They are easy to fabricate and can be portable and temporarily installed. These flumes actually act much like open channel flow nozzles and attempt to combine the accuracy of a sharp crested weir with the self cleansing properties of a flume. Their flow is contracted through the opening which increases flow velocity and flushes sediment through. The opening is trapezoidal in shape giving sensitivity to low flows and capacity for high flows.

These flumes are designated according to the maximum head, and the maximum depth in the entrance section, of the flume. Thus a one foot flume has a maximum depth in the entrance of one foot and the same maximum head. All flume dimensions given in Figure 6-20 are proportional to this maximum depth. HS flumes were designed for small flows in the range of 0.08 to 0.82 cfs. H flumes were designed for medium flows with maximum flows in the range of 0.35 to 31.0 cfs. And HL flumes were designed for larger flows whose maximum value is in the range of 20.7 to 117.0 cfs. Table 6-8 gives flow limits for various sizes.

The approach channel should have a length 3 to 5 times the head on the flume. The preferred shape is rectangular and the flume generally has a free spill off the end. Submergence starts to effect the flow when the ratio of the downstream head above the measuring section to the head in the measuring section is about 0.3 (less than 1 percent effect) and grows to about a 3 percent effect when the ratio is 0.5 (Grant, 1978). Often it is necessary to enlarge the channel downstream from the flume to maintain a non-submerged condition. Details on the design and flow equations are given by Gwinn and Parson, 1976 and ARS Handbook 224, 1962.

Table 6-8 H-Type Flumes Flow Limits

Maximum Capacity, CFS

Depth, ft	HS Flume	H Flume	HL Flume
0.4	0.085	-	-
0.5	0.14	0.347	-
0.6	0.23	-	-
0.75	-	0.957	-
0.8	0.47	-	-
1.0	0.82	1.97	-
1.5	-	5.42	-
2.0	-	11.1	20.7
2.5	-	19.3	-
3.0	-	30.7	57.0
3.5	-	-	83.9
4.0	-	-	117.0
4.5	-	84.5	-

Source: Grant, 1989

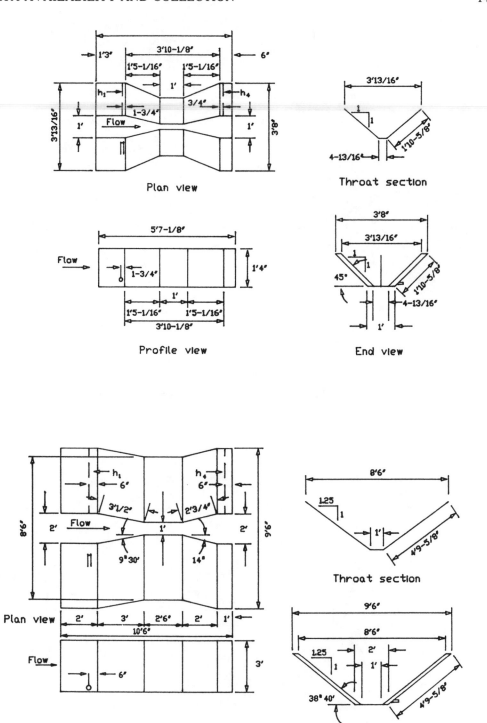

Figure 6-18 Trapezoidal Flume (after: Robinson, 1968)

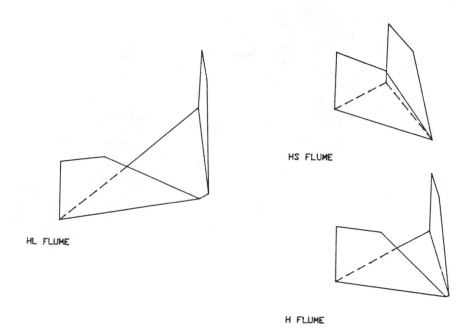

Figure 6-19 Isometric View Of HS, H And HL Flumes
(after: Shelley and Kirkpatrick, 1975)

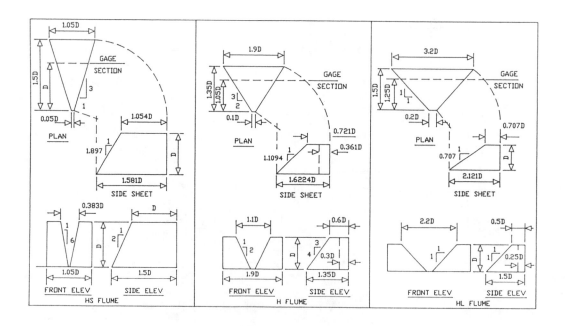

Figure 6-20 HS, H And HL Type Flumes
(after Shelley and Kirkpatrick, 1975)

Open Flow Nozzles

For continuous and accurate measurement of flow at the end of a pipe the open flow nozzle is inexpensive and practical. The nozzle must discharge to a free fall and is used for pipes flowing partially full. Figure 6-21 shows two types of open flow nozzles in common use: the Kennison and the parabolic nozzles.

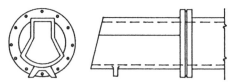

A. Kennison nozzle (Q proportional to H)

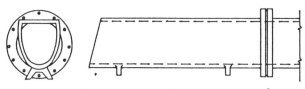

B. Parabolic nozzle (Q proportional to H²)

Figure 6-21 Kennison And Parabolic Open Flow Nozzles (after Grant, 1989)

This device works like all head-area devices in that a predetermined relationship exists between the head in the nozzle and the flow through it. Flow nozzles are factory calibrated and offer reasonable readings ± 5 percent of reading, even under difficult conditions. Continuous readings can be taken through the use of a float or air bubbler within the nozzle. This type of nozzle is capable of a 20:1 range ratio which is better than either a Parshall flume or a weir. Parabolic nozzles are roughly four times the pipe diameter in length while the Kennison nozzle is about twice the diameter in length. For the Kennison nozzle, Q is proportional to H while for the parabolic nozzle, Q is proportional to H^2. Table 6-9 gives dimensions and approximate capacities for commercially available flow nozzles. Factory calibration will give flow ratings.

Trajectory Flow Measurement

The estimation of flow freely discharging from a horizontal pipe can be estimated through measurement of the trajectory of flow and application of a series of empirical relationships. The most common of these methods is called the California pipe method. This method is for rough estimates of flow because accuracies of only ±10 percent are common. The method was developed by Vanleer (1922). This method is best for measurement of comparatively small flows in pipes such as dry weather flows or small storm flows.

The method was derived for pipes from 3 to 10 inches in diameter though it has been success-fully tested in pipes up to three feet in diameter. A horizontal end section of the pipe should be at least six diameters and can be added to the end of a sloping outfall pipe. The empirically derived rating formula is then:

$$Q = 8.69 \ (1\text{-}a/d)^{1.88} \ d^{2.48} \tag{6.41}$$

Where: Q = flow in pipe, cfs
        a = distance from the top of the inside surface of the pipe to the liquid surface, mea-
            sured in the plane of the end of the pipe, ft
        d = diameter of pipe, ft

If the method is to be used with a depth measurement at the end of the pipe instead of "a" the
equation can be rewritten as:

$$Q = 8.69 \ d^{0.60} \ D^{1.88}$$ (6.42)

Where: D = depth of flow at the pipe outlet, ft

Table 6-9 Open Flow Nozzles
Dimensions And Approximate Capacities

Nozzle Diameter (inches)	Parabolic Type Nozzle		Kennison Type Nozzle	
	Nozzle Length inches	Maximum Flow Capacity (cfs)	Nozzle Length inches	Maximum Flow Capacity (cfs)
6	28	0.42	12	0.42
8	35	0.88	16	0.70
10	43	1.50	20	1.31
12	50	2.32	24	1.94
16	66	4.52	32	4.19
20	81	7.60	40	6.97
24	96	11.6	48	11.5
30	119	19.4	60	17.9
36	142	30.1	72	30.1

Source: Grant, 1978

Measurement Of Stage

All of the open channel and closed conduit devices described above require a measurement
of stage or pressure head to calculate flow. Closed conduit measurement is normally taken care
of through manometers or pressure transducers. Open channel stage measurement can take place
in many different ways.

Stilling Wells - Stilling wells are basically a vertical riser connected to the measuring device
or structure by a small diameter pipe or tube at the point of head measurement. The small diameter
tube or pipe damps out the pressure and wave surges inherent in the open system, giving a stable
and accurate measurement of the head on the primary device. The size of the vertical stilling well
depends on the type of secondary stage measuring device to be used. Often galvanized-iron
culverts are used, sealed on the lower end, and provided with perforated openings throughout the
height. To effectively dampen surges, openings in stilling wells should be about 1/1000 of the
inside of the stilling well, unless the well is some distance from the channel (USBR, 1974). Table
6-10 provides float well opening information.

Table 6-10 Port And Pipe Sizes For Stilling Well Inlets

Size of Float Well	Inlet Port Diameter (inches)	Inlet Conduit Diameter (inches)
12 inches	1/2	1/2
16 inches	1/2	3/4
20 inches	5/8	3/4
24 inches	3/4	1
30 inches	1	1 1/2
36 inches	1 1/4	2
3 X 3 feet square	1 1/4	2
3 X 4 feet rectangle	1 1/2	3
4 X 5 feet rectangle	2	4

Source: USBR, 1974

Various methods are used to keep stilling wells free from ice and silt, including submersible heaters, light bulbs, or an oil layer for ice and a purging pump for permanent installations or hand cleaning by using a snake for sediment.

Stage Measuring Devices - Floats have been the most common way to measure stage in a stilling well. Such a device is little more than a float attached to a cable over a pulley with a counterweight at the other end and some sort of recording device, often a chart drum with timer driven motor or a solid state digital recorder. Other more common devices include (Rantz, 1982, Grant, 1989, Shelley and Kirkpatrick, 1975):

- A bubble gage system consists of a source of pressurized gas, a sensitive pressure control system, a tube extending into the water at a fixed and known elevation, and a manometer. The gas is fed into the tube and bubbles freely into the water. The pressure at the bubble outlet is essentially the pressure head of the water. This head is read on the manometer.
- Pressure transducers directly sense the pressure of the head of water.
- Surface detector methods rely on the fact that water is a good conductor of electricity. A conducting bob is lowered to the water surface and kept in contact with the surface through the use of control circuitry.
- Ultrasonic devices are a non-contact device which senses the depth of flow by measuring the time for a sonic pulse to bounce off the water surface and back to the same or a paired sensor. Velocity can also be measured ultrasonically by measuring the difference in time for a pulse of sound to travel upstream versus downstream.
- Electromagnetic devices take advantage of Faraday's law which states that if a conductor (the water) is passed through a magnetic field, a voltage will be induced.
- A staff gage is simply a graduated rod placed at a known elevation in the water at the head measuring site. However, this method is non-recording.

6.9 Flow Data Collection - Natural Controls

The development of a gaging site for the installation of a permanent stage-discharge station is similar to the development of artificial controls such as a weir or flume with the exception that the stage-discharge relationship is developed, not through known relationships between head and discharge, but through the measurement of point velocities in the flow area and the integration of these point velocities (and the small cross sectional areas they represent) to produce an estimate of discharge for a given stage. When this is done for a number of stages a "permanent" relationship between stage and discharge is developed.

Selection Of Gaging-Station Sites

The selecting of a gaging site is guided by several considerations. First of all, the purpose for which the data are needed will often dictate the location or the reach of stream. For example, if the purpose is to measure discharge in conjunction with a proposed project the site will be near the project. The other criterion is that the site be an adequate control.

The physical element that governs the stage-discharge relationship is called the control. The control may be a point where the flow passes through critical depth such as a water fall or break in the grade. It may be a constriction in a channel which controls the flow depth upstream. Rarely will a single control dictate depth over the whole range of flows. Often the control will shift from a slope break point for lower flows to channel constrictions for larger flows. There are three attributes of a satisfactory control: stability, permanence and sensitivity.

Stability refers to the one-to-one correspondence between stage and discharge over all expected conditions. If the cross section is too close to a downstream tributary which may cause backwater or a dam with variable gate control there may be times when a certain stage reflects a different flow than the one measured during stream gaging. Changing vegetation may also be reflected in different stages for given flows in different seasons. Vegetation that suddenly leans over at a given velocity will cause gage readings to be inaccurate as will low and dense canopy cover levels which interfere with high flow levels during the summer season.

Permanence means the control will remain constant over a long period of time. Rock outcroppings have this characteristic while sand bars and pools do not. Streams that are undergoing degradation or sedimentation due to urban development will have to be remeasured periodically. Sand bed streams scour during high flows and redeposit sediment on the backside of the flow hydrograph. The depth of this scour will need to be measured or estimated to gain accuracy in high flow measurements.

Sensitivity means that, at low flow, a small change in flow will be reflected in a relatively large change in stage. Vee shaped channels or those with small low flow channels are often best for this. Channels with small pools at the site during low flow will ensure a measurement can be made for even the lowest flows.

The ideal gage site has the following criteria (Rantz, 1982).

- The general course of the stream is straight up and downstream for some distance (200 to 300 feet for medium sized urban streams).
- The total flow is confined to one channel for all stages without bypass flow.
- The stream bed is not subject to scour and fill and is free from excessive aquatic growth.
- Banks are permanent, high enough to contain the flow and free from heavy brush.
- Unchanging natural controls are present in the form of a bedrock outcrop or riffle at low flow and channel constriction at higher flows, or a falls that remains unsubmerged.
- A pool is present upstream from the control at low stages.
- The site is far enough upstream from tidal influence or the influence from confluence with a tributary.
- A satisfactory reach for measuring discharge is available within reasonable proximity of the gage site.
- The site is accessible for installation and maintenance of the gaging station.

Rarely will any site meet all these criteria. The storm water manager should weight pluses and minuses to maximize accuracy and mitigate any factors which will cause error. See Rantz (1982) for ways to minimize poor locational aspects of a site.

Velocity Measurements

Velocity measurements are taken in a series of verticals across a section in the vicinity of the gaging station location. The site for taking velocity measurements should preferably have the following characteristics.
- Reach is straight and streamlines are parallel to each other.
- Velocities are greater than 0.5 ft/s and depths are greater than 0.3 ft.
- Streambed is relatively uniform and free from boulders, aquatic growth, excessive turbulence, and dead flow areas.
- The section is close enough to the gaging station site to avoid intervening tributary inflow or the need to calculate storage for rapidly changing flow situations.

After the site has been selected the width perpendicular to the flow is measured and vertical sections laid out. No vertical strip subsection should contain more than 10 percent of the flow. For medium to large urban streams this may equate to 25 to 30 subsections though no sections should be less than one foot apart. A tagline stretched across the stream is used to mark the locations.

Velocity is measured in the vertical using one of two common methods: the 0.6 depth method and the 0.2-0.8 depth method. For depths less than 2.5 feet use the method where the velocity is measured once per vertical subsection at a point 0.6 of the depth from the surface. This is the theoretical mean velocity in the vertical. If the depth is greater use both the 0.2 and 0.8 points and take the arithmetic mean.

This measurement procedure assumes that the stream can be waded. For details on other means of performing measurements see Rantz (1982). A rule of thumb for wading safety which has been proven in experiments on mountain streams in Colorado is: if the sum of the velocity in feet per second and depth in feet is greater than your height in feet it is unsafe to wade in the stream. For example unless you are over six feet do not wade a stream which is two feet deep flowing at four feet per second. Use a float such as a ball, sticks or leaves and a stopwatch to estimate the velocity for safety purposes. When wading the hydrographer should stand at least 1.5 feet downstream from the flow meter holding the rod to which the meter is attached vertical.

Figure 6-22 shows the layout for flow calculations. Distances are measured from one bank and always referenced to that bank. Proper use of the meters can produce accuracies of ±5 percent.

Equation 6.43 gives the method for calculation of discharge in a subsection. The nomenclature is given in Figure 6-22.

$$q_x = v_x \left[\frac{b_{(x+1)} - b_{(x-1)}}{2} \right] d_x \qquad (6.43)$$

Where: q_x = discharge through subsection x, cfs
v_x = mean velocity at vertical x, ft/s
b_{x-1} = distance from initial point to preceding vertical, ft
b_{x+1} = distance from initial point to next vertical, ft
d_x = depth of water at vertical x, ft

For example, the discharge at subsection 4 (heavily outlined in Figure 6-22) is:

$$q_4 = v_4 \left[\frac{b_5 - b_3}{2} \right] d_4$$

For the first and last sections the equations are respectively:

$$q_1 = v_1 \left[\frac{b_2 - b_1}{2} \right] d_1$$ (6.44)

$$q_n = v_n \left[\frac{b_n - b_{(n-1)}}{2} \right] d_n$$ (6.45)

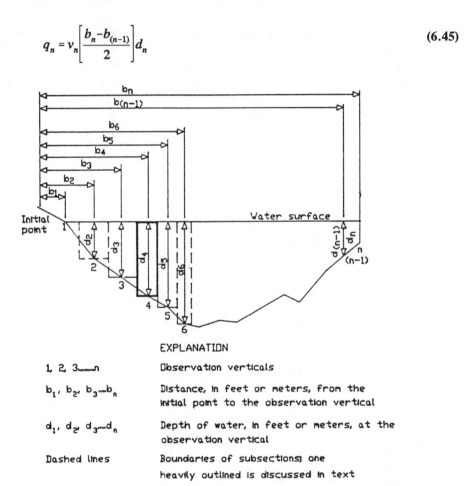

EXPLANATION

1, 2, 3......n   Observation verticals

b_1, b_2, b_3...b_n Distance, in feet or meters, from the
                              initial point to the observation vertical

d_1, d_2, d_3...d_n Depth of water, in feet or meters, at the
                              observation vertical

Dashed lines   Boundaries of subsections; one
                              heavily outlined is discussed in text

Figure 6-22 Flow Measurement Layout (Source: Rantz, 1982)

Fulford and Sauer (1986) have shown that considerable time can be saved if the cross section is fairly uniform and well defined by simply estimating the velocities in many of the verticals rather than measuring them. This is especially true for velocities in sections which are nearly uniform. They found that a linear interpolation scheme was as accurate as any complex scheme and that only five vertical measurements were necessary for fairly uniform channel sections. Mean absolute errors of under 5 percent were measured. This method is especially effective for determining flows for other stages after the initial cross sectional survey and detailed measurements are done for one stage. In this case the velocity is measured in five representative verticals. The mean velocity in a subsection between two measured verticals (verticals 2 and 3) is:

$$v_i = (a/L) v_3 + [(L-a)/L] v_2 \qquad (6.46)$$

Where: v_i = interpolated velocity in an intermediate subsection, ft/s

a = distance between verticals, v_i is calculated and v_2 is measured, ft

L = distance between the two measured velocity verticals, ft

v_2, v_3 = measured velocities at section 2 and 3 respectively, ft/s

Bank velocity is approximated as zero giving the estimated velocity between the bank and the first measured velocity vertical as:

$$v_i = (a/L) v_1 \qquad (6.47)$$

Velocity Meters

Current meters come in two types: vertical-axis with cups or vanes and horizontal-axis (propeller) with vanes. The two give similar results for similar situations. The Price AA vertical-axis cup meter is a U.S.G.S. standard for most situations where depths exceed 1.5 feet. A Price pygmy meter is used for depths down to 0.3 feet. For lesser depths use a portable weir or flume. Other popular meters include the Ott, Neyrpic, Haskell, Hoff and Braystoke. Details on these meters and their use can be found in Rantz (1982) and from the various manufacturers.

6.10 Flow Data Collection - Tracer Methods

Discharge can be estimated using any of a number of related tracer methods in which a known quantity, rate or concentration of a substance is injected into the flow at some upstream location. It is them observed, measured or weighed at some downstream location and the total flow determined.

Salt-Velocity Method - In this method a slug of salt solution is forced into the stream and measured downstream at two separate locations through the use of electrodes. Since salt water conducts electricity better than fresh water the electrodes will register a "hydrograph" of electrical conductance with a recognizable center of mass. The time taken between the passage of the center of mass at the two electrode sites can be used along with the distance to calculate the mean velocity.

Salt Or Color-Dilution Methods - In this method salt solution of known concentration, C_1, is introduced at a constant rate q_s. After the solution has flowed far enough to mix completely it produces a uniform concentration, C_2. The natural salt concentration of the flow is C_0. Then the discharge can be calculated directly from:

$$Q = q_s \left[\frac{C_1 - C_2}{C_2 - C_0} \right] \qquad (6.48)$$

This method will also work with any number of dyes and measurements taken using florescence or color analysis. Dyes used include: flourescein or potassium permanganate. This method also works with radioisotopes and the use of a Geiger counter. See Rantz (1982) for complete details.

6.11 Environmental Data Considerations And Collection

The need for environmental data in engineering analysis and design stems from the need to investigate and mitigate possible impacts of urban runoff due to specific development or other ongoing activities in a watershed. Nowhere in urban storm water management is the complexity of the situation mismatched by the dearth of good data, information and analysis techniques. Neither the actual physiological effects of various storm water pollution constituents nor the concentrations required to cause these effects are generally known, even with the most complete measurements. The great variation in flow volumes, concentrations, bioavailability, chemical interactions, durations of high pollution loading, etc., make it difficult to do more than roughly estimate actual impacts of pollution in many situations. For example, the toxicity of metals found in urban storm water runoff depends on such things as: pH, organic and inorganic ligands, hardness, other chemicals and temperature (Davies, 1986). Herricks (1986) has gone so far as to say that there are no simple cause and effect relationships in the natural environment. A complete understanding of the typical storm water quality situation is not possible under current scientific knowledge.

However, great improvements in urban storm water quality can be made without a complete understanding of cause and effect. Rules of thumb about design criteria, chemical concentrations and capture and treatment volumes can be useful in reducing pollution loadings and actual impacts, with sufficient accuracy. Measurements may be significant in fact without being statistically significant due to too few data points to form a reliable confidence interval (Loftis, et al., 1991). This suggests that "measuring" be balanced with "doing". While it may never be known, for example, just how much reduction in lead loading will prevent damage to a certain organism in a local stream, it can usually be said with some certainty that abnormally high lead levels should be reduced and logical engineered steps can be taken to do just that... and to monitor the results. This "do something" approach is recommended by Sonnen (1983, 1986) and tacitly adopted by USEPA in its "Maximum Extent Practicable" standard set in the storm water quality regulations (See Storm Water Quality Management Plans, Chapter 15).

Data Needs

Environmental data needs may be summarized as follows.
* Information necessary to define the environmental sensitivity of the facility's site relative to impacted surface waters, e.g., water use, water quality and standards, aquatic and riparian wildlife biology, and wetlands information. Some of this information is available in the water quality standards and criteria published by State Water Control Boards, studies from various Federal agencies (USEPA, USGS, etc.) and targeted literature searches.
* Physical, chemical and biological data for many streams are also available from State and Federal water pollution control agencies, the U.S.G.S., municipalities, water districts and industries which use surface waters as a source of water supply. In unique instances a data collection program possibly lasting several years and tailored to the site may be required.
* Information necessary to determine the most environmentally compatible design, e.g., circulation patterns and sediment transport data. Data on circulation, tides, water velocity, water quality and wetlands are available from the U.S. Coastal and Geodetic Survey, U.S. Army Corps of Engineers, universities, marine institutes, State, Federal, and local agencies and

organizations. Information on sediment transport is vital in defining the suitability of a stream for most beneficial uses including fish habitat, recreation and water supply. Such information may be essential for projects in critical water use areas such as near municipal or industrial water supply intakes.

- Information necessary to define the need for and design of mitigation measures should be obtained, e.g., fish characteristics (type, size, migratory habits), fish habitat (depth, cover, pool-riffle relationship), sediment analysis and water use and quality standards. Fish and fish habitat information is available from State and Federal Fish and Game Agencies.

Environmental Sampling Objectives

The development of a monitoring program depends on the study objectives, physiography of the site, duration of the data collection effort, expertise of the staff, and funding for the study (Terstriep, 1986). It is therefore very important to first establish clear objectives for the sampling program. Sonnen (1983) and Reinelt, et al., (1988) suggest a list of typical types of, or objectives for, environmental sampling:
- monitoring of a physical process for research purposes;
- problem identification or reconnaissance monitoring;
- alternative solution monitoring;
- design performance monitoring;
- regulatory compliance or enforcement monitoring; and
- operational performance, trend or ambient monitoring.

While research monitoring to establish physical principles requires a great amount of data over a long time period and over many monitoring situations, problem identification monitoring may require only five or six storms at several sites and in receiving streams. Constituents in the latter type of study could include only indicator species such as BOD, TSS, fecal coliform and nitrogen or phosphorus. An approach for problem identification monitoring which quickly isolates locations and approaches was developed by Reinelt, et al., 1988. Their approach is to define objectives; choose levels of detail; analyze the watershed to find critical monitoring locations; develop the monitoring program and statistical support for the results; and prioritize the monitoring tasks based on program objectives.

Alternative solution monitoring must be sufficient for calibration of a model used to test different alternatives. In design monitoring the only data collected are the minimum necessary to obtain input to the design criteria model. In environmental design the parameters are often estimated based on typical values of such things as event mean concentrations.

Compliance monitoring normally consists of measuring one or several constituents over a period of time (or forever) and comparing results to a regulatory standard. If no standard exists then the monitoring is expected to report on trends or the effectiveness of a BMP program. Statistical testing may be important in this type of monitoring though most of the tests were derived from an assumption of a normal distribution of the variable. This is not always the case with environmental data as asymmetrical and skewed data are the norm (Gilbert, 1987). In these cases other methods or transformation of the data can be accomplished. Many urban storm water variables (such as event mean concentration or flood flows) appear to follow a log-normal distribution (Driscoll, 1986). When monitoring operations, two different objectives can be met: checking the operation against the design expectation and building a record of long-term performance. One must always be aware that the samples taken to estimate long term performance may be tainted by auto- or serial-correlation (lack of independence of events). This correlation is dependent on both the temporal and spatial scale and procedures used in the measurements (Loftis, et al., 1991). For example, measurement of dissolved oxygen in the same place at 8:00 am each morning would not provide trend analysis for any other time of day, nor in another spot in the stream. There is no "right" way to do it except to use sound judgment and to fully document assumptions and procedures.

It is also important to consider which types of and concentrations of constituents are likely to be encountered or are of importance and what tests are necessary for detection. For example, for a designated use of swimming some measure of the presence of pathogenic organisms should be of importance. For irrigation the presence of dissolved solids is important (USEPA, 1983). In storm water quality considerations thresholds often exist and should be identified if possible. Considering dissolved oxygen, for example, any value above the threshold for survival of a certain species may be acceptable. Any value below a level where the species cannot survive will be unacceptable. Making improvements in the dissolved oxygen value which do not raise it above the survival threshold will not achieve any beneficial result with regard to that species even though funds and efforts are expended. In the same way the mere presence of a toxic metal in the environment does not automatically mean it is available biologically for uptake. It may be physically bound to sediment particles or chemically bound in some other way and thus unavailable.

The impacts of storm water on quality are both physical and chemical. Therefore a simple analysis of chemical constituents will not suffice for assessment of the quality of a particular stream. Erosion and sedimentation in many cases does far more harm to a benthic community than does chemical contamination.

Typical Sampling Program Steps

A typical sampling program uses the following steps (Gilbert, 1987, USEPA, 1992, Valiela and Whitfield, 1989, Loftis, et al., 1991, USGS, 1984, Ogden 1992):

Step 1   Clearly define the study and sampling objectives.
Step 2   Define the time-space population of interest developing as recognizable boundaries on the population as possible.
Step 3   Perform a literature search sufficient to determine best methods, ranges of variables to be encountered, possible problems, data trends, etc.
Step 4   Define the hypothesis to be tested and estimate the sampling size required.
Step 5   Define the types of tests and sample analysis to be made on any samples or the types of statistical or quality assurance analyses to be made of measurements.
Step 6   Collect field information necessary to define a sampling plan.
Step 7   Define the types of samples to be collected (e.g., composite, grab, continuous) or measurements to be made.
Step 8   Define field procedures and field equipment including risks and safety procedures.
Step 9   Define quality assurance procedures for all phases of the program from collection through analysis or measurement through quality checks.
Step 10 Conduct the study according to plans.
Step 11 Summarize, plot and analyze results as appropriate.
Step 12 Assess possible error and uncertainty.
Step 13 Assess results compared to objectives and report any changes in procedure for future studies.

Key Constituents

An understanding of the typical pollutants found in urban storm water runoff and typical concentrations will help direct any sampling program. Storm water runoff quality has been characterized through the results of numerous studies. The National Urban Runoff Program

(NURP) was an EPA sponsored program from 1978 to 1983. It included 28 projects across the United States (USEPA, 1983). The runoff related findings of the NURP study are instructive when considering a storm water quality sampling and monitoring program (USEPA, 1983).

- Heavy metals are the most prevalent priority pollutant in urban runoff. Freshwater acute criteria violations and chronic exceedances were common for copper (47 and 82 percent) and lead (23 and 94 percent). Zinc and cadmium had chronic exceedance frequencies of 77 and 48 percent respectively.
- Organic pollutants were detected less frequently than heavy metals and at lower concentrations. Sixty three of a possible 106 organics were detected. The most common were a plasticizer, four pesticides, three phenols, four polycyclic aromatics and a single halogenated aliphatic. Organics tend to be very site or municipality specific rather than a general national problem.
- Coliform bacteria are present at high levels and can be expected to exceed EPA water quality criteria during and immediately after storm events.
- Nutrients are present in moderate quantities (an order of magnitude less than most POTWs) in almost all samples.
- Oxygen demanding substances are present in concentrations equivalent to secondarily treated effluent from POTWs, although no NURP site specifically attributed low DO to urban runoff.
- Total suspended solids in runoff is high compared to treated effluent (an order of magnitude). While Total Suspended Solids (TSS) has no EPA water quality standard, TSS is a carrier of adsorbed toxics and other pollutants and may cause sedimentation or other physical stream problems.

The NURP program sampled for 120 priority pollutants. Of these 77 were detected in urban storm water runoff from commercial, residential and light industrial sites. Those detected in ten percent of the samples are listed in Table 6-11. A significant number of these pollutants exceeded various freshwater criteria.

Different constituents have different impacts and sources. Some are sampled in different ways or have specialized handling and preservation considerations. Reference should be made to basic sources such as USEPA (1992 a,b), Standard Methods (1989) and USEPA (1983) and to the literature to determine the various forms of each constituent and instream chemical interaction.

Metals - Trace metals are often toxic to both man and animals. Concentrations of zinc, cadmium, lead and copper were found in frequencies and concentrations high enough to warrant concern, though actual impairment due to urban storm water runoff is not easy to prove. Other metals (arsenic, beryllium, chromium, cyanide, mercury, nickel, selenium, and thallium) were also detected in a few samples and in minute quantities. Large fractions of the metals adsorbed to sediments and are not immediately available for bioaccumulation (though they can be resuspended during high flows). The dilution of urban runoff when it enters receiving streams is also a factor in the consideration of metals impacts on receiving waters. Copper and zinc are the most soluble of the metals and therefore of the most concern.

Nutrients - Nitrogen (N) and phosphorous (P) (in all the various compounds) are found in urban runoff and contribute to undesirable algal blooms in lakes and ponds. Control of the flow of the limiting nutrient is essential to slow the eutrophication process of water bodies. Available nitrogen is generally obtained from the nitrates plus ammonia nitrogen compounds. The available phosphorous fraction is generally the soluble orthophosphorus plus a fraction (20 to 30 percent) of the particulate. The theoretical uptake rate of these nutrients in algae is 7.5N to 1 P. If the ratio of N:P is greater than 10 the limiting nutrient is most likely phosphorous. If the ratio is less than 5 it is nitrogen. In between it may vary (Wanielista, et al., 1981). Physicochemical and biological processes determine the eventual fate of the nutrients in that they may: remain in solution or suspension, settle out, chemically interact and precipitate or adsorb to sediments, be taken up biologically, or be released back to solution through chemical changes, and plant growth and death cycles.

Table 6-11 Priority Pollutants
Detected In At Least 10 Percent Of NURP Samples

Metals and Inorganic	Frequency of Detection
Antimony	13
Arsenic	52
Beryllium	12
Cadmium	48
Chromium	58
Copper	91
Cyanads	23
Lead	94
Nickel	43
Selenium	11
Zinc	94
Pesticides	-
Alpha-hexachlorocyclohex-ane	20
Alpha-endosulfan	19
Chlordane	17
Lindane	15
Halogenated Aliphatics	-
Methane, dichloro-	11
Phenols and Cresols	-
Phenol	14
Phenol, pentachloro-	19
Phenol, 4-nitro-	10
Phthalate Esters	-
Phthalate, bis(2-ethylhexyl)	22
Polycyclic Aromatic Hy-drocarbons	-
Chysene	10
Fluoranthene	16
Phenanthrene	12
Pyrene	15

Source: USEPA, 1983

Nitrogen exists, in order of decreasing oxidation state, as: nitrate (NO_3^--N), nitrite (NO_2^--N), ammonia (NH_3-N) and organic nitrogen. Organic nitrogen and ammonia are often measured together and are termed "kjeldahl" nitrogen after the test used to determine their concentration. Phosphorous exists primarily in the form of phosphate, either soluble or suspended.

Human Pathogens - The most significant pathogens present in untreated wastewater can be classified into four groups: bacteria, protozoa, helminths, and viruses (USEPA, 1992). For a human to be infected with one of these organisms a number of steps must occur, beginning with the presence of an organism in sufficient numbers and a long enough exposure to elicit infection. Because it is technically difficult and expensive to measure all potential pathogens found in stormwater it has been common practice (as with wastewater) to use a surrogate measure of fecal contamination. The most common indicator of the presence of human pathogens has been the presence of either total coliforms or fecal coliform bacteria. More recently only E. coli and enterococci have been used in recreational waters. Several features of these coliforms make them useful as surrogates for pathogenic contamination: they occur naturally in the feces of warm-blooded animals in higher concentrations than the pathogenic organisms; they are easily and definitively detected; their presence is usually positively correlated with fecal contamination; and they usually respond similarly to environmental conditions and treatment processes as do the pathogenic bacteria (Ogden, 1994).

While the presence of such coliforms indicate some fecal contamination they are not considered a fool-proof indicator of the presence of pathogenic organisms (Field, 1993). Many bather illnesses are not related to enteric bacteria but to other organisms that cause infections of the skin, eye, ear, nose and upper respiratory systems. There is a need for epidemiological studies which relate illness to specific pathogens and attempt to trace the source of such pathogens. One such study was conducted by Ogden (1994) to determine the actual impacts of combined sewer overflows for Sacramento County, California. A retrospective public health risk assessment of sickness reports among certain outdoors occupations was done and matched to probable overflow dates. Such data can conclude if there is a correlation from cases reported or found in the database. However, it is more difficult to determine the "no impact" hypothesis since the data are not very exact.

Oxygen Demand - Dissolved oxygen is a necessary component of aquatic life. Decomposition of organic matter by micro organisms depletes the available supply of oxygen, particularly in slower moving waters (COG, 1987). The direct test for oxygen is the dissolved oxygen test (DO). Indirect methods include the biochemical oxygen demand (BOD) and chemical oxygen demand (COD) tests. Neither of these tests have been proven to be very effective in correlating or predicting DO problems and fish kills. The BOD test measures the oxygen utilized, during a specified time period - normally 5 days, for the biochemical degradation of organic material and the oxygen used to oxidize inorganic material. The COD test measures all the oxidizable matter present in the fluid through the use of a strong chemical reaction. It oxidizes even some materials which would not normally provide natural oxygen demands. Correlations with BOD can be established in wastewaters or fairly uniform waters from a specific source. Variability in storm water runoff often makes this correlation difficult and inaccurate.

Sediment - Sediments may be the most destructive pollutant of all. Urban development can increase the amount of sediment entering streams manyfold. Because, in some cases, the stream cannot transport the increased sediment load it settles out, filling ponds, wetlands and reservoirs, smothering benthic communities. It inhibits visual feeders from seeing prey, clogs fish gills, reduces spawning and juvenile survival rates, carries toxics and trace metals and makes water supplies more difficult to treat. The scouring of sediment from bank and beds destroys habitat and adjacent vegetation. Sediments are measured in terms of total suspended solids, settleable solids, turbidity, and dissolved solids. Volatile solids are those that burn off at 550 degrees centigrade and are organic in nature (Standard Methods, 1989).

Oil And Grease - Scheuler (COG, 1987) and Stenstrom, et al., (1984) provide excellent summaries and analysis of the oil and grease problem in urban runoff. The actual impacts are not easily measurable but are very visually apparent in both the oily sheen on urban runoff and the oily muck found in catch basins and at outfalls. Oil and grease compounds can be toxic at concentrations found in urban storm water. In testing "oil and grease" is any substance which is soluble in trichlorotriflouroethane and not a specific compound or chemical family. Thus it is defined by the test done to identify it.

Other pollutants are also of importance in urban runoff sampling including a list of the priority pollutants, chlorides from road salting, thermal impacts, trash and floatables.

Pollution Sources

Urban pollution accumulation and washoff into streams and ponds is the summation of many, seemingly unimportant, individual activities and physical processes... thus it is called non-point. Each typical pollutant has one or several principal sources. These sources introduce the pollutant to the environment. The pollutant is then transported, transformed, deposited, resuspended, biologically taken up, and so on.

Erosion And Sedimentation - Sediment loads from construction sites have been variously reported in the literature to be as high as 50 tons/acre/year. Post construction values are several orders of magnitude less than this. Even during post construction the dust and dirt from erosion is the main source of street particulant matter and a main carrier of adsorbed contaminants such as ammonium ions, toxic materials and phosphates. The most common sources of sediment include: natural weathering, construction, urban bare soil erosion, agricultural erosion, litter and dust, highway erosion, stream bank and bed erosion. Erosion and sediment yield is estimated a number of ways including (Novotny and Chesters, 1981): stream sampling, reservoir sediment surveys, sediment delivery equations, transport functions, and watershed modelling (such as the Opus Model, Smith, 1992). Of these the well known Universal Soil Loss Equation for upland erosion is the most common.

Atmospheric Fallout - Dust from both dryfall and wetfall is the most important source of pollutants on impervious surfaces. Total average annual dustfall values range from about three to five tons per square mile per month in the Tennessee and Ohio valleys, to 15 to 25 in California and the upper Mississippi, to 30 to 50 in the lower Mississippi and Texas Southwest areas, to over 140 in the Upper Colorado River areas. Table 6-12 gives specific pollution deposition rates for the Washington DC area.

It is estimated that about 90 percent of all the fallout on impervious surfaces eventually makes its way into the drainage system. The sources of this pollution include emissions from industry and vehicles, smoke, wind erosion, volcanoes, coal burning plants, pesticide drift, naturally occurring forest emissions, and domestic heating. Rainfall in urban areas can contain levels of COD in the 15 to 70 mg/l range, Ammonia-N in the 0.3 to 0.7 mg/l, and lead in the 0.03 mg/l range. Acid rain, due to anthropogenic emissions of sulphur and nitrogen oxides, is a serious problem the world over caused by rainfall pH in the 4 range. For more information see Novotny and Chesters, 1981, or NAPAP, 1993.

Vehicles - Vehicles cause pollution by abrasion of road surfaces and the subsequent dust, resuspension of other deposition and transfer to areas adjacent to roads and curb sections, and inorganic and organic pollutants such as lead, asbestos, oil and grease, copper, chromium, nickel, phosphorus, zinc and rubber.

**Table 6-12 Average Annual Atmospheric
Deposition Rates For The Washington D. C. Area**
(lbs/ac/yr)

Pollutant	Rural	Suburban	Urban
Total Solids	99	155	245
COD	199	133	210
Total Nitrogen	19.9	12.8	17.0
Nitrate-N	9.4	5.6	6.8
Ammonia-N	5.5	1.1	1.0
TKN	10.5	7.2	10.2
Total P	0.71	0.50	0.84
Ortho-P	0.28	0.26	0.35
Cadmium	-	0.09	0.003
Copper	-	0.21	0.61
Lead	0.06	0.44	0.53
Iron	-	1.57	5.60
Zinc	0.67	1.35	0.65

Source: MWCOG, 1983

Other Human Activities - Other human activities contribute to the pollution problem through normal industrial activities, illegal dumping, illicit connections, household hazardous waste disposal practices, fertilizer and pesticide applications, pet excretions, leaf and litter, and many more sources.

Most of these pollutants, those which are not directly introduced into the drainage system, accumulate on the road surface and within a short distance (less than two feet) from the curb. Therefore, many urban pollution studies report pollution loadings and removals in terms of curb length instead of loadings per unit area. There is a strong correlation between urban density and curb miles which can be expressed as (APWA, 1969):

$$CL = 234.5 - 200 * 0.631^{PD} \tag{6.49}$$

or

$$CL = -1.049 + 1.4098\ I + 0.0149\ I^2 \tag{6.50}$$

Where: CL = curb length density, ft/acre
PD = population density, persons/acre
I = impervious percentage (half impervious equals 50)

Refuse accumulation along curbs is relatively rapid. Sartor, et al., (1974) found accumulations in eight American cities from 48 grams per curb mile per day for residential and 66 for multi-family to 69 and 127 for commercial and industrial respectively.

While the typical street dirt accumulation has more than 80 percent of its volume of sizes greater than 100 μm, the pollution particulate sizes range much smaller. Pollution associated with the dust and dirt fraction of general curb dirt and litter can be significant. Sartor, et al., (1974) found 77% of COD, 57% of BOD_5, 59% of TKN and 51% of metals having particulate sizes less than 100 μm. Table 6-13 gives typical values in milligram per gram of total solids collected from street curb sections for various land uses.

Table 6-13 Pollutants Associated With Street Dirt And Litter

Constituent	Residential (mg/l)	Industrial (mg/l)	Commercial (mg/l)	Transportation (mg/l)[1]
BOD$_5$[2]	9.16	7.50	8.33	2.3
COD[2]	20.82	35.71	19.44	54.
Volatile[2] Solids[2]	71.67	53.57	77.00	51.
TKN[2]	1.666	1.392	1.111	0.156
PO$_4$-P[2]	0.916	1.214	0.833	0.61
NO$_3$-N[2]	0.050	0.064	0.500	0.079
Pb[3]	1.468	1.339	3.924	12.
Cr[3]	0.186	0.208	0.241	0.08
Cu[3]	0.095	0.055	0.126	0.12
Ni[3]	0.022	0.059	0.059	0.19
Zn[3]	0.397	0.283	0.506	1.5
Total Coli forms (No./g)[2]	160,000	82,000	110,000	-
Fecal Coli-forms (No./g)[2]	16,000	4,000	5,900	925

[1] Shaheen, 1975
[2] Sartor, et al., 1974
[3] Amy, et al., 1975

Source: cited in Novotny and Chesters, 1981

Impacts On Receiving Waters

The impacts of urban storm water on receiving waters is not as well understood as that of point sources. Urban runoff differs from point source discharges in that it is intermittent and of short duration, highly variable from storm to storm, can be highly but variably diluted as it enters receiving waters, and carries a relatively high load of total suspended solids. Therefore comparison of concentration limits and other numeric standards, low flow criteria, lethal dose testing and other point source methods of analysis do not apply readily to storm water analysis. Wet weather criteria have been studied by EPA and proposed by several authors though a standard has never been adopted. The impacts on receiving waters can be divided into three categories (Mancini and Plummer, 1986):
•   short term changes in water quality during and immediately after a storm event;

- long term impacts associated with contaminants adsorbed to settled sediment or nutrients in lakes and ponds with long detention times; and
- scour and deposition of bed and banks.

The NURP study adopted a three level definition of "problems" to assist in evaluating urban storm water impacts (USEPA, 1983):

- impairment or denial of beneficial use;
- water quality criterion violation; and
- local public perception.

Based on these criteria NURP concluded that the impacts of urban runoff are highly site specific. It is clear from various studies that if the ambient water quality of a stream, lake, pond or estuary is already near a threshold level the boost given by a slug of urban storm water runoff could push it beyond that threshold (Bastian, 1986). Frequent exceedances of EPA's heavy metal criteria are common though actual impairment of aquatic life from the criteria violations were not observed. Of the heavy metals copper, lead and zinc are the most prevalent. Copper was concluded to be a significant threat to aquatic life in the South and Southeastern United States. Generally speaking organic pollutants are not a typical urban runoff hazard, though freshwater intakes in the vicinity of urban runoff locations should be tested for such pollutants. This is not to say that in specific cases where there are spills, dumping or misuse they cannot cause problems.

The physical aspects of urban runoff cause as much or more habitat destruction and species diversity limitation as do the chemical aspects. Pollution adsorption to sediments and possible resuspension is a suspected, though unmeasured, problem. Coliform bacteria are present in great numbers in all urban runoff. Nutrient and bacterial loadings can severely impact urban lakes and ponds in terms of trophic state and recreational use impairment. A survey of fisheries experts indicated that turbidity, temperature and nutrient loadings are the three most important urban storm water impacts on fisheries habitat (Heaney, 1988).

It is clear that urban storm water managers are interested in the mitigation of real, discernible impacts of urban storm water runoff: aquatic organisms, aesthetics, human health and safety. Measurement of constituents such as BOD, COD and TOC and application of numeric criteria are a step away from these real impacts toward a more easily measured standard. A further step away is the development of simplistic design criteria (e.g. capture of the first half-inch). These steps away from assessing real impacts have been necessary because of the combined need to do something immediate to stop the rapid degradation of urban waters and the general lack of understanding of the details of cause and effect (Huber, 1988). However, there will be (and is in some states) an eventual shift in consideration of urban storm water impacts from an emphasis on numerical criteria and simple design criteria toward site-specific, risk-based and biologically-based criteria and standards (Huber, 1988, Heaney, 1988).

Land Use Baseline Data

Actual sampling information can be compared to baseline information both to test for inconsistencies and to, perhaps, avoid needless sampling if initially sampled values compare with average values. The NURP, and other, data provide such a basis. Because of the variability of measurements within storms, among different storms at one site and among sites, it was desirable to use a measure which tended to reduce this variability somewhat. The measure of the magnitude of urban runoff pollution chosen is termed the event mean concentration (EMC). It is defined as the total constituent mass discharge divided by the total runoff volume for a given storm event. It is defined as the event load divided by the event flow volume. It as calculated as:

$$EMC = \frac{\sum\limits_{i=1}^{n} V_i C_i}{\sum\limits_{i=1}^{n} V_i} \qquad\qquad (6.51)$$

Where: V_i = flow volume per time increment i, liters

$\quad\quad\ \ C_i$ = average concentration for time period i, mg/L

$V_i C_i$ is the mass of pollution for a given time increment i. The summation of all the $V_i C_i$ values gives the total event loading. The summation of all the V_i values gives the total event flow volume. The division of the two gives the event mean concentration (EMC).

Table 6-14 demonstrates this calculation for a runoff event sampling TSS. Assume that the sampling increment is 10 minutes. Then dividing the total mass of pollutant by the total volume of discharge gives: EMC = 14,445,600mg/150,600L = 95.9 mg/L

Table 6-14 Example EMC Calculation For TSS

Average Discharge Rate (L/sec)	Measured Concentration (C_i) (mg/L)	Incremental Runoff Volume (V_i) (L)	Incremental Mass of Pollutant (mg)
8	25	4,800.00	120,000.00
16	145	9,600.00	1,392,000.00
45	180	27,000.00	4,860,000.00
60	120	36,000.00	4,320,000.00
48	80	28,800.00	2,304,000.00
35	45	21,000.00	945,000.00
22	20	13,200.00	264,000.00
12	23	7,200.00	165,600.00
5	25	3,000.00	75,000.00
TOTALS		150,600.00	14,445,600.00

With few exceptions EMC's were found to not vary significantly for similar land uses from site to site for the same constituent and were found to be distributed lognormally. Therefore measures of central tendency (median and mean) and scatter (standard deviation, coefficient of variation) could be calculated as well as expected values at any frequency of occurrence by using the natural logarithmic transformation of the raw data. Standard statistical tests and sampling theory can also be used on the lognormally distributed data.

EMC's also did not vary with storm volume. That is, over experienced storm depths or duration the mean concentration of the pollutant over the storm event did not vary significantly with runoff volume. What this implies, and what has been shown by various authors, is that the total volume of pollution (as opposed to the concentration) will vary directly with the volume of runoff. And since the volume of runoff varies directly with the amount of impervious area, the total pollution volume varies directly with impervious area.

Based on these findings certain calculations of pollutant loading become simplified. After taking the natural logarithm of each of "N" observations the statistics can be calculated as:

mean	$M = \Sigma(x_i)/N$	(6.52)
standard deviation	$S = [\Sigma(x_i - M)^2/(N - 1)]^{1/2}$	(6.53)
coefficient of variation	$C = S/M$	(6.54)
median	$m = M/(1 + C^2)^{1/2}$	(6.55)
expected value	$X = \exp(M + ZS)$	(6.56)

Where: M = natural logarithm of the mean value of the EMC observations
 x = natural logarithm of an individual EMC observation
 N = number of observations
 S = standard deviation of the logarithms of the observations
 C = coefficient of variation of the logarithm of the observations
 m = median of the natural logarithm of the observations
 Z = the standard normal probability from probability table for normal distribution found in any statistics textbook
 X = expected value (non-logarithm) of the logarithms of the observations

It should be noted that for standard deviation calculations the following identity is helpful:

$$\Sigma(x_i - M)^2 = \Sigma x_i^2 - (\Sigma x_i)^2/N \qquad (6.57)$$

While the median value is of concern for more acute pollution situations, the mean value is of value when considering long term chronic or cumulative pollution. Conversion between the two can be obtained from equation 6.55. Table 6-15 gives median event mean concentrations for various land uses for a number of common constituents from several sources. The measured coefficient of variation from the NURP data is also presented to facilitate calculations of mean values. While the analysis of the NURP information did not prove a statistical relationship between land use and EMC (except for open areas) the differences in values for different constituents, for different land uses, are instructional and have been distinguished in use (SQTF, 1993) .

An example conversion between the median and the mean can be done for the Commercial/Industrial value for BOD5. The median value is 9.3 mg/l and the coefficient of variation is 0.31. To compute annual loadings the mean is desired. Recall that these are the log transformed values in the equations. Therefore from equation 6.55:

$m = M/(1 + C^2)^{1/2}$ $\ln(9.3) = M/(1 + 0.31^2)^{1/2}$
$M = 2.3347$ $e^{0.988} = 10.33$ mg/l $= EMC$

The total annual loading of pollution can then easily be approximated by multiplying the mean concentration in the runoff by the total annual runoff as (SMFT, 1993):

$$L = 0.2266 * EMC * [0.15 + 0.75 \ I] * P * A \qquad (6.58)$$

Where: L = pollution loading, lbs/year
 EMC = mean event mean concentration, mg/l

Table 6-15 Median Event Mean Concentrations
And Coefficients Of Variation For Various Land Uses

Pollutant	Residential		Mixed/NURP Urban Site Median[4]		Commercial/Industrial		Highways[3]	Open/Non-urban	
	Median	CV	Median	CV/90%[5]	Median	CV	Mean	Median	CV
BOD (mg/l)	10.0	0.41	7.8/9[4]	0.52/15[5]	9.3	0.31	9.7/24[6]	8.0[1]	-
COD (mg/l)	73	0.55	65/65[4]	0.58/140[5]	57	0.39	130-360/14.76[6]	40/51[1]	0.78
TSS (mg/l)	101	0.96	67/100[4]	1.14/300[5]	69	0.85	150-450	70/216[1]	2.92
Total Lead (µg/l)	144	0.75	114/144[4]	1.35/350[5]	104	0.68	500-1300/960[6]	30/0.0[1]	1.52
Total Copper (µg/l)	33	0.99	27/34[4]	1.32/93[5]	29	0.81	70-180/103[6]	0.0[1]	-
Total Zinc (µg/l)	135	0.84	154/160[4]	0.78/500[5]	226	1.07	300-750/410[6]	195/0.0[1]	0.66
Cadmium							0-30		
TKN (mg/l)	1.9	0.73	1.29/1.5[4]	0.50/3.30[5]	1.18	0.43	1.78/2.996[6]	0.965 0.610[2] 1.36[1]	1.00
NO$_2$ + NO$_3$ (mg/l)	0.736	0.83	0.558/0.68[4]	0.67/1.75[5]	0.572	0.48	0.830/1.14[6]	0.543 0.730[1]	0.91
Total P (mg/l)	0.383	0.69	0.263/0.33[4]	0.75/0.70[5]	0.201	0.67	0.440/0.79[6]	0.121 0.150[2] 0.230[1]	1.66
Soluble P (mg/l)	0.143	0.46	0.056/0.12[4]	0.75/0.21[5]	0.080	0.71	0.170	0.026 0.040[2] 60[1]	2.11

[1] from SMTF, 1993 (no coefficient of variation)
[2] from MWCOG, 1987, Virginia hardwood forest (no coefficient of variation)
[3] Highway data from SMTF, 1993 and FHWA, 1990.
[4] NURP Median EMC values for combined urban sites (USEPA, 1983 Table 6-14)
[5] NURP 90th percentile values for combined urban sites (USEPA, 1983 Table 6-14)
[6] Smith and Lord (1990) results from 6 street sites.
General Source: USEPA, 1983

I = fraction of impervious area, acres

P = annual rainfall, in.

A = watershed area, acres

For the example above the annual loading of BOD5 for an area of 10 acres, 35 percent impervious and an annual rainfall of 45 inches is:

$$L = 0.2266 * 10.33 * [0.15 + 0.75 (0.35)] * 45 * 10 = 434 \text{ lbs}$$

A refinement of this method, termed the Simple Method, is given in the chapter on BMPs.

Urban Runoff Quality Data Collection

The importance of proper sampling is paramount since the decisions which will be based on samples can be little more accurate than the sample itself (Wullschleger, et al., 1976). Monitoring takes place either within the storm water collection system, at its outfalls or in the receiving waters. It is either storm related or storm event independent. This section will provide some basic data and information on urban storm water sampling.

Site Selection - Within-system sampling must take place at a defined "point source" which 40 CFR 122.2 defines for urban purposes as: any pipe, ditch, channel, tunnel, conduit, well, or discrete fissure. The sites should be easily accessible in adverse weather conditions, not inviting to vandals (private backyards work well for this), not subject to flow interruptions or backwater conditions, and physically close enough to the normal location of the sampling technician to meet start-of-sampling criteria, if first flush grab samples are necessary. The site must also represent only that land use or other phenomena which is to be measured. Commingling flows from several land sources will lead to misleading results. However, when other options are not feasible known single source concentrations can be used and tested against a site with several land uses and weighting used.

Receiving stream sites must clearly represent a particular watershed and reach of stream. The site should reflect the runoff from a particular area or region of interest and be sited to fairly represent this site. The limits of the stream reach for which the gage serves as a representative sample should be determined approximately by assessing land use homogeneity in the tributaries above and below the gage site and the distance upstream for which the reach flow is consistent.

Safety considerations should be paramount including use of the appropriate equipment, safety gear, the provision for two persons at all sites with any safety risk at all and ready or constant communications at other sites. Sampling in manholes normally requires training in confined space entry.

Storm Event Sampling - EPA has established guidelines for storm event discharge characterization sampling under the NPDES permit application. The storm depth must be greater than 0.1 inch accumulation to ensure adequate flow is generated. It must be preceded by a minimum 72 hour dry period to allow for pollution buildup before the storm. And the rainfall depth and duration should be representative (defined as not varying from the mean by more than 50 percent in either category).

It can not be known for sure if the event being sampled is characteristic until after all the criteria have been met. However, use of National Weather Service or private real time radar information and quick mobilization should allow enough time to setup and sample prior to the rainfall-runoff event beginning. Automated samplers can be used to avoid constant mobilization preparedness. They can be programmed to trigger based on either flow or rainfall. Automated samplers can not be used in all cases though because some constituents have very short holding times or the automated sampling tubing can be a source of contamination from sample to sample. Grab samples must be used in these cases.

Sample Types - There are many types of samples common to urban storm water quality programs including: sediment sampling, fish tissue sampling, water quality sampling, biological assessment, etc. Water quality sampling includes measurement of both the physical (temperature, pH, turbidity, etc.) and chemical (toxics, nitrogen, etc.) condition of the water. Water quality sampling consists of two basic types of samples: grab samples and composite samples (EPA, 1992). Either of these sample types can be collected manually or by automated equipment. Grab samples are discrete "snapshots" of the water quality at an "instant" in time (usually less than 5-15 minutes). Composite samples are a mixture of samples taken in discrete intervals over the storm event.

Details on sampling technique, preservation, quality control and handling can be found in EPA (1993, 1992a and 1992b, 1976) as well as standard laboratory analysis manuals such as Standard Methods (1989). The development of complex sampling programs for receiving waters is beyond the scope of this text. However, the principles provided previously concerning sampling objectives should be employed in such programs.

References

Agricultural Research Service, "Field Manual for Research in Agricultural Hydrology", Handbook No. 224, USDA, Washington, D. C., 1962.

American Association Of State Highway And Transportation Officials, Model Drainage Manual, Washington, D. C., 1991.

American Public Works Assoc., "Water Pollution Aspects of Urban Runoff", FWPCA (EPA), WP-20-15, 1969.

Amy, G., R. Pitt, R. Singh, W. L. Bradford and M. B. LaGraff, "Water Quality Management Planning for Urban Runoff", USEPA No. 44019-75-004, Washington, D.C., 1975.

Arnell, V., P. Harremoes, M. Jensen, N.B. Johansen and J. Niemczynowicz, "Review of Rainfall Data Application for Design and Analysis", Wat. Sci. Tech., Vol. 116, Copenhagen, 1984.

Barfield, B. J., R. C. Warner, and C. T. Haan, "Applied Hydrology and Sedimentology for Disturbed Areas", Oklahoma Technical Press, 1981.

Bastian, R. K., "Potential Impacts on Receiving Waters", Urban Runoff Quality, Proc. Engr. Foundation Conf., Henniker, NH, edited by B. Urbonas and L. A. Roesner, ASCE, 1986.

Bauer, W. J., D. S. Louie and W. L. Voordin, "Basic Hydraulics", Section 2 in Handbook of Applied Hydraulics, C. V. Davis and K. E. Sorenson editors, McGraw-Hill, 1969.

Benson, M. A., "Spurious Correlation in Hydraulics and Hydrology", ASCE J. of Hydraulics, Vol. 91, No. HY4, July, 1965.

Bowker, A.H. and G.J. Lieberman, "Engineering Statistics", Prentice Hall, 1972.

Brater and King, Handbook of Hydraulics for the Solution of Hydraulic Engineering Problems, 6th Edition, McGraw-Hill, Inc., New York, New York, 1976.

Bureau of Reclamation, "Water Measurement Manual", Second Ed., US Dept. of Interior, Denver, CO, 1981.

Burnham M. and D. Davis, Accuracy of Computed Water Surface Profiles, U. S. Army Corps of Engineers, Hydrologic Engineering Center, Davis, California, December 1986.

Cheremisinoff, Nicholas, "Flow Measurement and Instrumentation", National Environmental Journal, Nov./Dec., 1991.

Chow, V.T., "Handbook of Applied Hydrology", McGraw Hill, 1964.

Clark, I., "Practical Geostatistics", Applied Science Publishers, London, 1979.

Clemmens, A. J., J. A. Replogle and M. G. Bos, "Rectangular Measuring Flumes for Lined and Earthen Channels", ASCE J. of Irrig. Drain. Engrg., Vol. 110, No. 2, 1984a.

Clemmens, A. J., M. G. Bos, and J. A. Replogle "RBC Broad-crested Weirs for Circular Sewers and Pipes", J. of Hydrology, Vol. 68, pp. 349-368, 1984b.

Clesceri, L. S., A. E. Greenberg and R. R. Trussell, eds., "Standard Methods for the Examination of Water and Wastewater", APHA-AWWA-WPCF, 17th ed., 1989.

Court, A., "Reliability of Hourly Precipitation Data", J. Geophys. Res., Wash. D.C., Dec., 1960.

Davies, P.H., "Toxicology and Chemistry of Metals in Urban Runoff", in Urban Runoff Quality, Proc. of Engr. Found. Conf., June 23-27, Henniker, NH, Urbonas and Roesner, eds., 1986.

Dean, J.D., and W.M. Snyder, "Temporally and Areally Distributed Rainfall", ASCE J. of Irrigation and Drainage, Vol. 103, No. IR2, June, 1977.

Driscoll, E.D., "Lognormality of Point and Non-point Source Pollution Concentrations", Urban Runoff Quality, Proc. of Engr. Found. Conf., June 23-27, Henniker, NH, Urbonas and Roesner, eds., 1986.

Driscoll, E.D., P.E. Shelley and E.W. Strecker, "Pollutant Loadings and Impacts from Highway Stormwater Runoff", Vol. I, FHWA-RD-88-006, April, 1990.

Field, R., "Use of Coliform as an Indicator of Pathogens in Storm-generated Flows", Proc. 6th Int. Conf. on Urban Storm Drainage, Niagara, Ontario, Sept., 1993.

Fulford, J. M. and V. B. Sauer, "Comparison of Velocity Interpolation Methods for Computing Open Channel Discharge", Selected Papers in Hydrologic Science, USGS WSP 2290, January, 1986.

Gilbert, R.O., "Statistical Methods for Environmental Pollution Monitoring", Van Nostrand Reinhold, 1987.

Gill, M. A., "Flow Measurement by Triangular Broad-crested Weir", Water Power and Dam Construction, Vol. 37, No. 8, 1985.

Gilman, C.S., "Rainfall", in Handbook of Applied Hydrology, Ch. 9, V.T. Chow, ed., McGraw-Hill, 1964.

Gottschalk, L. and T. Jutman, "Calculation of Areal Means of Meteorological Variables for Watersheds", Proc. of Den 7ende Nordiske Hydrologiske Konferense, Forde, Norway, 1983.

Grant, D.M., "ISCO Open Channel Flow Measurement Handbook", Third Ed., 1989.

Gray, D.M., "Handbook on the Principles of Hydrology", National Research Council of Canada, ISBN 0-912394-07-2, 1970.

Greenville, South Carolina, Manning's n Manual, City/County of Greenville, 1992.

Gwinn, E. R. and D. A. Parsons, "Discharge Equations for HS, H and HL Flumes", ASCE J. Hydrau. Div., Vol. 102, No. HY1, January, 1976.

Gumbel, E. J., "Spurious Correlation and its Significance to Physiology", J. of Am. Statistical Assoc., June, 1926.

Heaney, J. P., "Cost Effectiveness and Urban Storm-Water Quality Criteria", Urban Runoff Quality Controls, Proc. Engr. Foundation Conf., Potosi, MO, edited by L. A. Roesner, B. Urbonas and M. B. Sonnen, ASCE, 1988.

Herricks, E.E., "Disciplinary Integration: The Solution", in Urban Runoff Quality, Proc. of Engr. Found. Conf., June 23-27, Henniker, NH, Urbonas and Roesner, eds., 1986.

Hershfield, D.M., "Rainfall Frequency Atlas of the United States", TP 40, U.S. Dept. of Commerce, Weather Bureau, 1961.

Hershfield, D.M., "On the Spacing of Rain Gages", IASH & WMO Symp., Design of Hydrometeorological Systems, Quebec City, 1965.

Hirsch, R. M., D.R. Helsel, T.A. Cohn and E.J. Gilroy, "Statistical Analysis of Hydrologic Data", Ch. 17 in Handbook of Hydrology, D.R. Maidment, ed., McGraw Hill, 1993.

Hirsch, R. M. and Gilroy, E. J., "Methods of Fitting a Straight Line to Data: Examples in Water Resources", Wat. Res. Bull., Vol. 20, No. 5, October, 1984.

Hosking, J.R.M., J.R. Wallis and E.F. Wood, "Estimation of the Generalized Extreme Value Distribution by the Method of Probability Weighted Moments", Technometrics Vol. 27, No. 3, 1985.

Huber, W. C., "Technological, Hydrological and BMP Basis for Water Quality Criteria and Goals", Urban Runoff Quality Controls, Proc. Engr. Foundation Conf., Potosi, MO, edited by Roesner, Urbonas and Sonnen, ASCE, 1988.

Huff, D. "How to Lie With Statistics", W. W. Norton & Co., New York, 1954.

Huff, F.A., "Radar Analysis of Urban Effects on Rainfall", 17th Conf. on Radar Meteorology, Am. Met. Soc., Boston Mass., 1975.

Huff, F.A., "Spatial and Temporal Correlation of Precipitation in Illinois", Ill. State Water Survey, Cir. 141, Urbana, Ill., 1979.

Interagency Advisory Committee on Water Data, "Guidelines for Determining Flood Flow Frequency", Bull. 17B U.S. Dept. of Int., USGS, Reston, VA, 1982.

Kelway, P.S., "A Scheme for Assessing the Reliability of Interpolated Rainfall Estimates", J. of Hydrology, Vol. 21, 1974.

Kilpatrick, F. A., "Use of Flumes in Measuring Discharges at Gaging Stations", Surface Water Techniques, Book 1 Ch. 16, USGS, Wash. DC, 1965.

Kitanidis, P.K. "Geostatistics", Ch. 19 in Handbook of Hydrology, D.R. Maidment, ed., McGraw Hill, 1993.

Leupold & Stevens, Inc., "Steven Water Resources Data Book", 5th Ed., Leupold & Stevens, Beaverton, OR, 1991.

Loftis, J.C., G.B. McBride, and J.C. Ellis, Considerations of Scale in Water Quality Monitoring and Data Analysis, Water Res. Bull., Vol. 27, No. 2. April 1991.

Ludwig, J. H. and R. G. Ludwig, "Design of Palmer-Bowlus Flumes", Sewage and Industrial Wastes, Vol. 23, No. 9, Sept., 1951.

Mancini, J. L. and A. H. Plummer, "Urban Runoff and Water Quality Criteria", Urban Runoff Quality, Proc. Engr. Foundation Conf., Henniker, NH, edited by B. Urbonas and L. A. Roesner, ASCE, 1986.

McGuiness, J.L., "Accuracy of Estimating Watershed Mean Rainfall", J. of Geophys. Research, Vol. 68, 1963.

McTrans Center, University of Florida, 512 Weil Hall, Gainesville, Florida 32611, (904) 392-0378.

Metropolitan Washington Council of Governments, "Urban Runoff in the Metropolitan Washington Area - Final Report", Prepared for EPA and WRPB, 777 North Capital Street N.W., Suite 300, Washington, D.C. 20002-4201, Telephone 202-962-3265, 1983.

Murthy, K. K., and K. G. Pillai, "Some Aspects of Quadratic Weirs", ASCE J. of Hydraulics, Vol 103:9, Sept., 1977.

Murthy, K. K., and K. G. Pillai, "Design of Constant Accuracy Linar Proportional Weir", ASCE J. of Hydraulics, Vol 104:4, April, 1978a.

Murthy, K. K., and K. G. Pillai, "Modified Proportional V-Notch Weirs", ASCE J. of Hydraulics, Vol 104:5, May, 1978b.

Murthy, K. K. and D. P. Giridhar, "Inverted V-Notch: Practical Proportional Weir", ASCE J. of Irr. & Drainage Engrg., Vol 115:6, December, 1989.

National Acid Precipitation Assessment Program, 1992 Report to Congress, 722 Jackson Place, N.W., Wshington, D.C. 20503, June, 1993.

Nelson, S. B., Water Engineering, Chapter 21 in Standard Handbook for Civil Engineers, F. S. Merritt, editor, McGraw-Hill, 1983.

Niemczynowicz, J, "Storm Tracking Using Raingage Data", J. Hydrology, 93: pp. 135-152, 1987.

Niemczynowicz, J, "Moving Storms as an Areal Input to Runoff Simulation Models", Int. Symp. on Urban hydrology and Municipal Engineering, Markham, June 15, 1988.

Niemczynowicz, J., "Areal Intensity-Duration Frequency Curves for Short Term Rainfall Events", Nordic Hydrology, Vol. 13, No. 4, 1982.

Nguyen, V.T.V., J. Rouselle, and M.B. McPherson, "Evaluation of Areal Versus Point Rainfall With Sparse Data", Canadian J. of Civ. Engineering, Vol. 8, No. 2, 1981.

Novotny, V. and G. Chesters, "Handbook of Nonpoint Pollution", Van Nostrand Reinhold, 1981.

Ogden Environmental and Energy Services, Inc., Retrospective Study Report Public Health Assessment for Outflows from the Combined Sewer System, 221 Main St., Suite 1400, San Francisco, CA, 94105, 1994.

Ogden Environmental and Energy Services, Inc., "Storm Water Sampling Guide", Fairfax, VA., 1992.

Pratt, E. A., "Another Proportional Flow- Sutro Weir", Engineering News-Record, Vol. 72:9, 1914.

Ramamurthy, R. S., U. S. Tim and M. V. J. Rao, "Characteristics of Square-Edged and Round-Nosed Broad-Crested Weirs", ASCE J. of Irrig. Drain. Engrg., Vol. 114, No. 1, 1988.

Rantz, S.E., "Measurement and Computation of Streamflow" 2 Vols., US Geological Survey Water Supply Paper 2175, 1982.

Rao N. S. and D. Chandrasekaran, "Proportional Weirs as Velocity Controlling Devices", ASCE J. of Hydraulics, Vol 103:6, June, 1977.

Reese, A. J. and S. T. Maynord, "Design of Spillway Crests", ASCE J. of Hy. Engr., Vol. 113, No. 4, 1987.

Reinelt, L. E., R. R. Horner and B. W. Mar, "Nonpoint Source Pollution Monitoring Program Design", J. Water Res. Plan. & Dev., ASCE, Vol. 114, No. 3, 1988.

Replogle, J. A., "Flumes and Broad Crested Weirs: Mathematical Modelling and Laboratory Ratings", Flow Measurement of Fluids, North-Holland Publishing Co., Amsterdam, pp 321-328, 1978.

Robinson, A. R. and A. R. Chamberlain, "Trapezoidal Flumes for Open Channel Flow Measurement", Trans. ASAE, Vol. 3, No. 2, 1960.

Robinson, A. R., "Trapezoidal Flumes for Measuring Flow in Irrigation Channels", ARS 41-140, Agricultural Research Service, USDA, 1968.

Rouse, Hunter, "Engineering Hydraulics", John Wiley & Sons, 1949.

Salas, J.D. "Analysis and Modelling of Hydrologic Time Series", Ch. 19 in Handbook of Hydrology, D.R. Maidment, ed., McGraw Hill, 1993.

Sartor, J. D., G. B. Boyd, and F. J. Agardy, "Water Pollution Aspects of Street Surface Contaminants", J. WPCF, Vol. 46, 1974.

Shaheen, D. G., "Contribution of Urban Roadway Usage to Water Pollutions", USEPA No. 600/2-75-004,

Shelley, P.E. and Kirkpatric, G.A., "Sewer Flow Measurement - a State-of-the-art Assessment", EPA 600/2-75-027, 1975. Wash. D. C., 1975.

Shilling, W., "A Quantitative Assessment of Uncertainties in Stormwater Modeling", Proc. Int. Conf. Urban Storm Drainage, Gothenburg, Sweden, 1984.

Simanton, J.R. and H.B. Osborn, "Reciprocal-distance Estimate of Point Rainfall", ASCE J. of Hydraulics, Vol. 106, No. HY7, July, 1980.

Skogerboe, G. V., M. L. Hyatt, R. K. Anderson and K. O. Eggleston, "Design and Calibration of Submerged Open Channel Flow Measurement Structures - Part 3, Cutthroat Flumes", Utah Water Research Laboratory, College of Engineering, Utah State University, Logan, UT, Rpt. WG 31-4, April, 1967.

Skogerboe, G. V., R. S. Bennett, and W. R. Walker, "Generalized Discharge Relations for Cutthroat Flumes", ASCE J. of Irr. and Drain. Div., Vol. 98, No. IR4, Dec. 1974.

Smith, D. L. and B. N. Lord, "Highway Water Quality Control - Summary of 15 Years of Research", Trans. Research Record, No. 1279, 1990.

Smith, J.A., "Precipitation" in Handbook of Hydrology, Ch. 3, D.R. Maidment, ed., McGraw-Hill, 1993.

Smith, R. E., "Opus: An Integrated Simulation Model for Transport of Nonpoint-Source Pollutants at the Field Scale", ARS-98, Agricultural Research Service, USDA, July, 1992.

Soil Conservation Service, " A Method for Estimating Volume and Rate of Runoff in Small Watersheds", SCS TP 149, January, 1968.

Soil Conservation Service, "Guide for the Use of Technical Release No. 55 - Urban Hydrology", SCS Albany, NY, 1977.

Sonnen, M.B., "Guidelines for the Monitoring of Urban Runoff Quality", EPA-600/2-83-124, Nov., 1983.

Sonnen, M.B., "Review of Data Needs and Collection Technology", in Urban Runoff Quality, Proc. of Engr. Found. Conf., June 23-27, Henniker, NH, Urbonas and Roesner, eds., 1986.

Stedinger, J.R., R.M. Vogel and E. Foufoula-Georgiou, "Frequency Analysis of Extreme Events", Ch. 18 in Handbook of Hydrology, D.R. Maidment, ed., McGraw Hill, 1993.

Stormwater Quality Task Force, State of California, "California Stormwater Best Management Practice Handbooks - Municipal", March, 1993.

Swamee, P. K., S. K. Pathak, M. Agarwal and A. S. Ansari, "Alternate Linear Weir Design", ASCE J. of Irr. & Drainage Engrg., Vol 117:3, May/June, 1991.

Terstriep, M. L., "Design of Data Collection Systems", Urban Runoff Pollution, NATO ASI Series, Torno, Marsalek and Desbordes editors, 1986.

Thiessen, A.H., "Precipitation for Large Areas", Monthly Weather Review, Vol. 39, pp. 1082-1084, July, 1911.

Thorpe, W.R., C.W. Rose and R.W. Simpson, "Areal Interpolation of Rainfall with a Double Fourier Series", J. of Hydrology, Vol. 42, 1979.

U. S. Bureau of Reclamation, "Water Measurement Manual", 2nd ed., 1974.

U. S. Dept. of the Interior, Bureau of Reclamation, "Design of Small Canal Structures", 1978.

U. S. Dept. of the Interior, Bureau of Reclamation, "Design of Small Dams", 1977.

U. S. Environmental Protection Agency (USEPA), in association with the US Agency for International Development, Manual: Guidelines for Water Reuse, EPA/625/R-92/004, Sept., 1992.

U. S. Environmental Protection Agency, "Methodology for the Study of Urban Storm Generated Pollution and Control", EPA-600/2-76-145, 1976.

U. S. Environmental Protection Agency, "Results of the Nationwide Urban Runoff Program", Vol 1 - Final Report, EPA PB 84-18552, Dec. 1983.

U. S. Environmental Protection Agency, Office of Water, "NPDES Storm Water Sampling Guidance", EPA 833-B-92-001, 1992a.

U. S. Environmental Protection Agency, "Guidance Manual for the Preparation of Part 2 of the NPDES Permit Applications for Discharges From Municipal Separate Storm Sewer Systems", EPA 833-B-92-002, 1992b.

U. S. Environmental Protection Agency, "Investigations of Inappropriate Pollutant Entries into Storm Drainage Systems", EPA/600/R-92/238, Jan. 1993.

U. S. Geological Survey, "National Handbook of Recommended Methods for Water-Data Acquisition", Office of Water Data Coordination, Reston, VA, 1984.

Valiela, D. and P.H. Whitfield. Monitoring Strategies to Determine Compliance with Water Quality Objectives. Water Res. Bull., Vol. 25, No. 1. February 1989.

Vanleer, B. R., "The California Pipe Method of Water Measurement", Engineering News Record, August 3, 1922 and August 21, 1924.

Villemonte, J. R., "Submerged Weir Discharge Studies", Engineering News Record, Dec. 25, 1947.

Vogel, J.L. and F.A. Huff, "Mesoscale Analysis of Urban Related Storms", Proc. of Nat. Symp. for Hydrologic Modeling, Am. Geo. Union, Davis, CA., 1975.

Vogel, J.L., "Potential Urban Rainfall Prediction Measurement System", Wat. Sci. Tech., Vol. 16, Copenhagen, 1984.

Wanielista, M. P., Y. A. Yousef, B. L. Golding and C. L. Cassagnol, "Stormwater Management Manual", University of Central Florida, Orlando, 1981.

Wells, E. A. and H. B. Gotaas, "Design of Venturi Flumes in Circular Conduits", Trans. ASCE, Vol. 123, 1958.

Wilson, W.T., "Discussion of Precipitation at Barrow, Alaska", Trans. Am. Geophys. Union, Vol. 35, pp. 206-207, 1954.

Wullschleger, R. E., A E. Zanoni and C. A. Hansen, "Methodology for the Study of Urban Generated Pollution and Control", EPA-600/2-76-145, August, 1976.

Yevjevich, V. "Stochastic Processes in Hydrology", Water Resources Publications, Ft. Collins, CO., 1972a.

Yevjevich, V. "Probability and Statistics in Hydrology", Water Resources Publications, Ft. Collins, CO., 1972b.

Appendix A - Sources Of Data

Following is a list of data sources that might prove useful for storm water management analysis and design.

Principle Hydrology Data Sources

- Meteorological Data
  National Oceanography and Atmospheric Agency (NOAA)
  Climatic Data Center
  Asheville, North Carolina 28801
- Regional and local flood studies
- U.S. Geological Survey regional and any site studies
- Surveyed high water marks and site visits by local agencies
- Hydrology data from others (see below)

Principle Watershed Data Sources

- U.S. Geological Survey maps ("Quad" sheets)
  U.S. Geological Survey
  Rocky Mountain Mapping Center
  Stop 504
  Denver Federal Center
  Denver, Colorado 80225
  (303) 236-5829
- EROS aerial photographs
  U.S. Geological Survey
  EROS Data Center
  Sioux Falls, South Dakota 57198
  (605) 594-6151
- U.S. Geological Survey local offices
- State and local maps and aerial photos
- State geological maps
- Soil Conservation Service and BLM Soils Maps
- County Soils Maps
- Site visits
- Watershed data from others

Principle Site Data Sources

- Local agency files of aerial drainage surveys
- Local agency files for existing facilities
- Site visits by local agency
- Field or aerial surveys from others (see below)

Principle Regulatory Data Sources

- Federal Flood Plain delineations and studies
  Federal Emergency Management Agency
  Flood Map Distribution Center
  6930 (A-F) San Tomas Road
  Baltimore, Maryland 21227-6227
  Watts 71-800-638-6620
- State floodplain delineations and studies
- FHWA design criteria and practices
  Federal Highway Administration
  U.S. Department of Transportation
  400 Seventh Street SW
  Washington, D.C. 20590
- State laws
- Local ordinances and master plans
- Local agency policy statements
- Corps of Engineers Section 404 permit program (see Environmental below)
- U.S. Coast Guard
- U.S. Environmental Protection Agency (EPA) (see Environmental below)
- State EPA's (see Environmental below)
- Federal Registers
  Superintendent of Documents
  U.S. Printing Office
  Washington, D.C. 20402
  (202) 783-3238

Principle Environmental Data Sources

- U.S. Environmental Protection Agency data and studies
- Corps of Engineers data and studies
- U.S. Geological Survey water quality data
- State water quality data
- Environmental statements prepared by other Federal, State, and local agencies as well as private parties
- Environmental data from others (see below)

Principle Demographic, Economic and Political Data Sources

- Local agency files for existing facilities
- Local agency plans for proposed facilities
- Local agency field or aerial surveys
- Site visits by local agency
- Local agency planning, budgeting and scope documents
- Internal local agency reports, memorandums, minutes, and verbal communications

Other Data Sources

- U.S. Bureau of Reclamation (USBR)
  U.S. Bureau of Reclamation Center
  Denver, Colorado 80225
  (303) 236-8098
- Regional and State U.S. Bureau of Land Management (BLM)
- Regional U.S. Environmental Protection Agency (EPA)
- Regional U.S. Federal Emergency Management Agency (FEMA)
- Regional and State U.S. Fish and Wildlife Service (USFWS)
- Regional and State U.S. Forest Service (USFS)
- Regional and State U.S. Soil Conservation Service (SCS)
- Regional and State U.S. Corps of Engineers (COE)
- Regional U.S. Coast Guard (USCG)
- Regional and State U.S. Geological Survey (USGS)
- Regional and State Federal Highway Administration (FHWA)
- National Weather Service (NWS)
- National Marine Fisheries Service (NMFS)
- National Oceanic and Atmospheric Administration (NOAA)
- Any State counterparts to the above Federal agencies
- Any local counterparts to the above Federal agencies
- State or local irrigation, drainage, flood control, and watershed districts
- Any indian councils
- Municipal governments
- Any "planning" districts
- Any regional water quality control boards
- Any river basin compacts, commissions, committees, and authorities
- Tennessee Valley Authority
- Private citizens
- Private industry

Appendix B - Field Investigation Form And Check List

Form 1
Field Visit Investigation Form

Date:_____ Project:_____   By:_____

Structure Type_____   Pier Type_____

Size or Span_____   Skew _____

# of Barrels or Spans_____   Inlet_____

Clear Ht_____   Outlet_____

Abut Types_____   % Grade of Road_____

Inlet Type_____   % Grade of Stream_____

Existing Wtwy Cover_____

Overflow Begins @ El._____   Length of Overflow_____

                                    Check for Debris_____

Max AHW (ft)_____   Check for Ice_____

Reason:_____   Side Slopes_____

_____   Height of Banks_____

Up or Downstream Restriction:

Outlet Channel, Base_____

Manning's n Value:

Type of Material in Stream_____

Ponding_____

Check Bridges Upstream and Downstream

Check Land Use Upstream and Downstream

Survey Required?   Yes____   No____

Remarks:

Form 2
Hydraulic Survey Field Inspection Check List

I. General Project Data

1. Project Number:_____ 2. County:_____
3. Road Name:_____
4. Site Name:_____, Station _____ M.P._____
5. Site Description: ( ) Cross Drain, ( ) Irrigation, ( ) Storm Drain, ( ) Long. Encroach, ( ) Channel,( ) Other _____
6. Survey Source: ( ) Field, ( ) Aerial, ( ) Other_____
7. Date Survey Received:_____, From_____

8. Site Inspected by_____ on _____

II. Office Preparation For Inspection

1. Reviewed:
   Aerial Photos - ( ) Yes, Photo #'s _____, ( ) None Available
   Mapping/Maps - ( ) Yes, Map #'s _____, ( ) None Available
   Reports - ( ) Yes, ( ) No, ( ) None Available at this time
   Municipal Permanent File - ( ) Yes, ( ) No, ( ) No file data found

2. Special Requirements and Problems Identified for Field Checking:
   ( )   Hydrologic Boundary - obtain hydrologic channel geometry
   ( )   Adverse Flood History - obtain HW Marks/dates/eye witness
   ( )   Irrigation Ditch - obtain several Water Right depths
   ( )   Permits Req'd - ( ) COE ( ) Dam, ( ) Coast Guard, ( ) State/Local
   ( )   Other_____
   ( )   Adverse Channel Stability and Alignment History - Check for headcutt-
          ing, bank caving, braiding, increased meander activity
   ( )   Structure Scour - check flow alignment, scour at culvert outlet or evi-
          dence of bridge scour
   ( )   Obtain bed/bank material samples at _____

III. Field Inspection
   (The following details obtained at the site are annotated on the Drainage
   Survey)

1. Survey appears correct: - ( ) Yes, ( ) Apparent errors are: _____
   _____
   which were resolved by: _____

2. Flooding Apparent? - ( ) No, ( ) Yes, HW marks obtained, ( ) Yes but HW
   marks not obtained because _____

3. Do all Floods Reach Site? - ( ) Yes, ( ) No and details obtained, ( ) No but
   details not obtained because _____

4. Do Floodwaters Enter Irrigation Ditch ? - ( ) N/A, ( ) No, ( ) Yes and details
   obtained, ( ) Yes but details not obtained because _____
   _____

5. Hydrologic Ch. Geom. obtained? - ( ) Yes, ( ) No because _____
   _____

6. Channel Unstable? - ( ) No , ( ) Yes because of ( ) headcutting observed and
   ( ) amount/location obtained, ( ) bank caving, ( ) braiding, ( ) increased meander
   activity, ( ) Other _____

7. Structure Scour in Evidence? - ( ) No, ( ) Minor, ( ) Yes and ( ) obtained
bed/bank samples and ( ) noted any flow alignment problems, ( )Yes and ( )
bed/bank material samples not obtained and ( ) flow alignment not noted because

_____

8. Irrigation facility? - ( ) No, ( ) Yes and several water right related depths
obtained, ( ) Yes and No water right related depths obtained because _____

_____

9. Manning's "n" obtained? - ( ) Yes, ( ) No because _____

_____

10. Property damage due to BW? - ( ) No ( ) Yes and elevation/property type
checked, ( ) Yes but elevation/property type not obtained because _____

_____

11. Environmental Hazards Present? - ( ) No, ( ) Yes, details obtained, ( ) Yes,
details not obtained because _____

_____

12. Ground Photos Taken? - ( ) Upstream floodplain and all property, ( ) Downstream floodplain and all property, ( ) Site looking from downstream, ( ) Site
looking from upstream, ( ) Channel Material w/scale, ( ) Evidence of channel
instability, ( ) Evidence of scour, ( ) Existing structure inlet/outlet, ( ) Other__

_____

13. Effective drainage area visually verified? ( ) Yes, ( ) No because _____

_____

IV. Post Inspection Survey Annotation

1. Section II Findings annotated on survey? - ( ) Yes, ( ) No and see section
    attached (attach typed explanation by site station and site name, and check list
    section and number).

2. Survey Originals and check lists forwarded to ( ) Municipality's Roadway
    Unit, _____ ea,   site's, and the ( ) Public Works or Engineering, _
    _____ea. site's for hydraulic design.

                        _____/s/_____
                             (Designer Making Inspection)

Chapter 7 Urban Hydrology

7.1 Introduction

Hydrology is generally defined as a science dealing with the interrelationship between water on and under the earth and in the atmosphere. For the purpose of this chapter, hydrology will deal with estimating flood magnitudes, volumes and time distributions as the result of precipitation. In the design of storm water management facilities, floods are usually considered in terms of peak runoff or discharge in cubic feet per second (cfs) and hydrographs as discharge per time. For structures which are designed to control volume of runoff, like detention storage facilities, or where flood routing through culverts is used, then the entire discharge hydrograph will be of interest.

There are a large number of hydrologic procedures covering topics such as snow melt, precipitation, water supply, infiltration, drought, evapotranspiration, etc. In basic urban hydrology, procedures commonly used are reduced to:

* precipitation and losses,
* peak flow determination,
* flow hydrograph or volume determination,
* hydrograph routing and combining, and
* storage routing.

The Open Channel and Storage Facilities chapters cover hydrograph routing and storage routine respectively. This chapter covers the other three major topics.

The analysis of the peak rate of runoff, volume of runoff, and time distribution of flow is fundamental to the design of drainage facilities. Errors in the estimates will result in a structure that is either undersized and causes storm water management problems or oversized and costs more than necessary. On the other hand, it must be realized that any hydrologic analysis is only an approximation. The relationship between the amount of precipitation on a drainage basin and the amount of runoff from the basin is complex, and too little data are available on the factors influencing the rural and urban rainfall-runoff relationship to expect exact solutions.

The type and source of information available for hydrologic analysis will vary from site to site and it is the responsibility of the designer to determine what information is available and applicable to a particular analysis.

Hydrologic analysis methods vary from simple formulas (e.g., Rational Method, Berkely Ziegler) to unit hydrograph methods (e.g., SCS Method, Santa Barbara Method) to regression equations based on local data (e.g., U. S. Geological Survey Equations) to complex continuous simulation computer models (e.g., Hydrocomp Simulation Model, EPA SWMM Model). Each of these methods has its place in municipal engineering for the analysis and design of different storm water management facilities. No one method is optimum for all analysis and design but several of the methods have severe limitations which should be accounted for. Recently, some municipalities and States have passed ordinances and legislation which further restricts the use of some hydrologic methods.

The purpose of this chapter is to present several hydrologic methods that have been widely used in municipalities throughout the United States. Other methods may also be applicable and methods specifically developed for particular locations, in most cases, will produce better results than general methods that have not been verified for local use. Since the number of hydrologic methods available is much too large to be covered in one chapter, the interested reader is encouraged to seek other sources for documentation of other methods and the availability of computer programs to assist in using these methods.

7.2 Concept Definitions

Following are definitions of concepts which will be important for understanding different hydrologic methods. These concepts will be used throughout the remainder of this book in dealing with different aspects of hydrologic studies.

Antecedent Moisture Conditions
: Antecedent soil moisture conditions are the soil moisture conditions of the watershed at the beginning of a storm. These conditions affect the volume of runoff generated by a particular storm event. Notably they affect the peak discharge only in the lower range of flood magnitudes -- say below about the 10-year event threshold. As the frequency of a flood event increases, antecedent moisture has a rapidly decreasing influence on runoff.

Depression Storage Frequency
: Depression storage is the natural depressions within a watershed which store runoff. Generally after the depression storage is filled runoff will commence. Frequency is the number of times a flood of a given magnitude can be expected to occur on an average over a long period of time. Frequency analysis is then the estimation of peak discharges for various recurrence intervals. Another way to express frequency is with probability. Probability analysis seeks to define the flood flow with a probability of being equalled or exceeded in any year. For example, a 25-year flood has the probability of occurrence of once every 25 years on the average, or a 4 percent chance of occurrence in any given year.

Hydraulic Roughness
: Hydraulic roughness is a composite of the physical characteristics which influence the flow of water across the earth's surface, whether natural or channelized. It affects both the time response of a watershed and drainage channel as well as the channel storage characteristics.

Hydrograph
: The hydrograph is a graph of the time distribution of runoff from a watershed.

Hyetograph
: The hyetograph is a graph of the time distribution of rainfall over a watershed.

Infiltration
: Infiltration is a complex process of allowing runoff to penetrate the ground surface and flow through the upper soil surface. The infiltration curve is a graph of the time distribution at which this occurs.

Interception
: Storage of rainfall on foliage and other intercepting surfaces during a rainfall event is called interception storage.

Lag Time   The lag time is defined as the time from the centroid of the excess rainfall to the peak of the runoff hydrograph.

Peak Discharge   The peak discharge, sometimes called peak flow, is the maximum rate of flow of water passing a given point during or after a rainfall event or snowmelt.

Rainfall Excess   After interception, depression storage and infiltration have been satisfied, if there is excess water available to produce runoff this is the rainfall excess.

Stage   The stage of a channel is the elevation of the water surface above some elevation datum.

Time Of   The time of concentration is the time required for water to flow from the most
Concentration   hydraulically remote point of the basin to the location being analyzed. Thus the time of concentration is the maximum time for water to travel through the watershed, which is not always the maximum distance from the outlet to any point in the watershed.

Unit   A unit hydrograph is the direct runoff hydrograph resulting from a rainfall event
Hydrograph   which has a specific temporal and spatial distribution and which lasts for a specific duration of time (thus there could be a 5-, 10-, 15-minute, etc., unit hydrograph for the same drainage area). The ordinates of the unit hydrograph are such that the volume of direct runoff represented by the area under the hydrograph is equal to one inch of runoff from the drainage area.

7.3 Hydrologic Design Policies

For all hydrologic analysis, the following factors should be evaluated and included when they will have a significant effect on the final results.

Drainage Basin Characteristics
Size and Shape
Slope
Ground Cover and Land Use
Geology
Soil Types
Surface Infiltration
Ponding and Storage
Watershed Development Potential
Antecedent Moisture Design Conditions
History of Urban Development
Other Characteristics

Stream Channel And Conveyance System Characteristics
Geometry and Configuration
Natural Controls
Artificial Controls
Channel Modifications

Aggradation - Degradation
Debris
Manning's "n"
Slope
Connectedness of Impervious Areas
Age, Condition and Type of Structures
Other Characteristics

Floodplain Characteristics
Slope
Vegetation
Alignment
Storage
Location of Structures and Development
Obstructions to Flow
Other Characteristics

Meteorological Characteristics
Time Rate, Geographical Distribution and Amounts of Precipitation
Historical Flood Heights
Other Characteristics

Many hydrologic methods are available. The methods included in this chapter are recommended for general municipal use and the circumstances for their use are included in the description of each method. Engineers and hydrologists will often argue the virtues of one hydrologic method over another but most methods will produce acceptable results if used within the limitations of the method and calibrated or verified for local use. Calibration is probably much more important than the method selected and will be further discussed later in this chapter. This is true because most methods have factors or parameters which can be adjusted within a wide range to reflect special situations or conditions. Sound engineering judgment is still necessary because urban hydrology is still as much art as science.

Municipalities should only accept those hydrologic methods that have been proven to produce acceptable results for their area and complete source documentation is submitted or available for approval. Consistent application of similar methods is important.

The methods in this chapter were selected for use based on several considerations, including the following.
- Verification of their accuracy in duplicating local hydrologic estimates in municipalities throughout the United States and for a range of design storms.
- Availability of equations, nomographs, and public domain computer programs.
- Use and familiarity with the methods by local municipalities and consulting engineers.
- Applicability to a wide range of geographical locations with different hydrologic characteristics.

It should be remembered that no method is universally applicable and that every method's accuracy will be compromised by attempting to apply it in situations where its basic assumptions or underlying bases are violated. Recognition of circumstances which may render a particular method inappropriate is important. It is also important to make an estimate of the relative accuracy of the predictions of the methods used either through sensitivity analysis or a less formal recognition of approximations and judgments employed. Urban drainage calculations are often performed with little or no true flow data; use hydrology which is a rough approximation; employ parameters with much leeway for judgment; and use assumptions on the stability of land use and flow conveyance conditions which are almost always violated.

7.4 Design Frequency And Risk

Risk

Since it is not economically nor often physically feasible to design storm water management structures for the maximum runoff a watershed is capable of producing, a design frequency is usually established. This use of a frequency implies an assumed risk of the failure of the structure or its overwhelming with an event of greater magnitude than that used for design. This area of consideration is often called risk-based analysis. While use of risk-based analysis is not common in urban storm water design, a knowledge of the risks and uncertainties involved is appropriate to any type of design and analysis.

The frequency with which a given flood can be expected to occur is the reciprocal of the probability or chance that the flood will be equaled or exceeded in a given year. The probability of occurrence of an event (say a flood of magnitude equal to or greater than X1) is expressed as:

$$P\{X1\} = N1/N \tag{7.1}$$

Where N1 is the number of occurrences of a flood greater than or equal to a certain event, and N is the total population of observations. P is always between zero and one. The return period is then the reciprocal:

$$Tr = 1/P\{X1\} \tag{7.2}$$

For example, the probability of a 5-year flood being equalled or exceeded in any one year has a probability of occurrence of 20 percent (0.2). Alternately, it would occur, on the average, once every five years for any given stream. In the same way a 5-year rainfall would tend to occur once every five years for any given spot on the ground. Across a municipality there may be many 5-year or greater storms each year since each spot in the city would tend to experience a 5-year or greater storm every five years. The probability of nonoccurrence of X1 is:

$$P\{\text{not } X1\} = 1-P\{X1\} = 1-1/Tr \tag{7.3}$$

The probability of nonoccurrence of an event X1 in "n" years is:

$$P\{\text{not } X1^n\} = (1-1/Tr)^n \tag{7.4}$$

Risk (R) is defined as the probability that an event X1 will occur in n years and is given as:

$$R = 1-(1-1/Tr)^n \tag{7.5}$$

Or rearranged:

$$Tr = 1/[1-(1-R)^{1/n}] \tag{7.6}$$

These equations provide the designer an estimate of the reasonableness of the design parameters. For example, what is the risk that the capacity of a culvert designed for a 25-year flood will be equalled or exceeded in the first ten years? Using equation 7.5 the risk is calculated to be about 33 percent as:

$$R = 1-(1-1/Tr)^n = 1-(1-1/25)^{10} = 0.335$$

Or, what should be the design flood return period (recurrence interval) to reduce the risk of overtopping of a roadway to five-percent within the fifty-year design life of a highway culvert? Using equation 7.6 the return period is:

$$Tr = 1/[1-(1-R)^{1/n}] = 1/[1-(1-.05)^{1/50}] = 975 \text{ years}$$

A flood with a recurrence interval of 975 years would be considered excessive for normal design purposes, and may still not protect the designer from liability even if it could be accurately estimated (Lewis, 1992).

Risk-Based Analysis

All estimates of design and economic parameters have inherent error and variability due to the complex physical, social and economic situations for even the most simple hydrologic and hydraulic analyses and projects. The designer typically develops a "most likely" estimate of a certain design parameter (for example, 10-year storm rainfall or Manning's roughness coefficient) and then uses sensitivity analysis to test the impact of variability in the parameter estimate on the final solution. However, there is nothing but a "gut feeling" on the actual likelihood of the parameter being correct or its actual likely range. "Engineering judgment" is relied on for all such answers.

Risk-based analysis was developed to attempt to estimate the actual amount of risk and uncertainty inherent in any design and to treat each key variable and parameter in probabilistic terms. It can seek specifically to quantify risk and uncertainty, taking into account the vagaries of the physical environment, the limitations of understanding of physical phenomena involved, cost and willingness to pay, institutional redgidities and social or political risk (Shabman, 1985). It can be used to quantify the likelihood of both physical and economic success of a project (COE, 1993). Risk-based design is often taking the place of more conventional analysis, especially in larger Federal flood control projects.

Risk is an estimated chance of an occurrence (such as flooding). Uncertainty is the error associated with key parameters or functions used in computing economic or reliability estimates. For flood control projects, economic variables of interest include: damage relationships, structure and contents values, structure elevations and types, and flood warning and evacuation times and effectiveness. Uncertainty arises due to errors in sampling, measurement, estimation and forecasting, and modeling. For hydrologic and hydraulic analysis, stage and discharge are of prime importance. Uncertainty in discharge is due to short or nonexistent flood records, inaccurate rainfall-runoff modeling, and inaccuracy in known flood flow regulation where it exists. Stage uncertainty comes from errors and unknowns in roughness, geometry, debris accumulation, ice effects, sediment effects, and others (COE, 1993).

An example of the difference in design approach is the determination of design height for a levee. Normal design approaches would be to determine the design flood, develop a backwater profile, design the levee to contain the design flood and add freeboard. "Enlightened" design would also plan to account for storms larger than the design storm by controlling the location of overtopping to reduce damage to the maximum extent practicable. Risk-based design takes a different approach. The end goal of risk-based design is to have a levee with a known reliability and performance. For example, it may be designed to contain the 100-year flood with a 95% reliability and the 500-year flood with a 50% reliability. It also seeks to manage events exceeding the design. The levees may look the same, but the design for the risk-based designed levee included a determined of the chances of failure even for floods less than the "design" flood. This is not always comforting to local citizens but it is more realistic.

The difference is represented in design by adding a probability density function around each estimated parameter. For example, each point on a simple stage-discharge curve has a bell-shaped or other shaped probability density function built in. The "most likely" stage for a discharge of, say, 100 cfs is elevation 520. But there is a 20 percent chance the elevation will be above 523, and so on.

Detailed discussion of risk-based analysis beyond this introduction is beyond the scope of this book. Readers are encouraged to explore the emerging literature on the subject or obtain information from the Corps of Engineers, hydrologic Engineering Center in Davis, California.

Design Frequency

A drainage facility should be designed to accommodate a discharge with a given return period(s). The design should be such that the backwater (termed the "headwater") caused by the structure for the design storm does not:
- increase the flood hazard significantly for property,
- dangerously overtop a street or highway, and/or
- exceed a certain depth on a street or highway embankment.

Different design storm frequencies or return-periods are important for different types of studies or structure sizes and uses:
- Large hydraulic structures are often designed for very infrequent floods such as some fraction of the Probable Maximum Flood (PMF) when considering structural integrity, and slightly less remote events when designing outlet structures (100-year or 500-year events).
- Major urban conveyance systems are often designed for the 100-year flood, particularly if they fall under the purview of the National Flood Insurance program administered by FEMA. The range of design frequencies is generally from the 25-year to 100-year events. Studies on flood damages have shown that if the 25-year and more frequent storms are controlled, the majority of the damages in urban flooding will be avoided (Johnson, 1985). The practice for some major structures, such as highway bridges, is to designed to pass the 50-year peak flow. Some agencies make allowance for debris buildup partial blockage (ST. Paul District, 1985). Very often channels are designed for a flow much less than the 100-year event but should pass the 100-year event in the floodplain without structural damage to adjacent properties.
- Smaller urban conveyance systems, storm drains and feeder streams (often termed the "convenience system") are designed for a range of flows generally centering on the 5-year to the 10-year storms. Although many municipalities allow for gutter and inlet design for design storms less than the 5-year storm, benefit cost analysis has shown that the 5-year storm is often the most cost effective standard to use for storm drain design (Casamayor and Rodgers, 1980).
- Concern for erosion control in channels usually centers around the 2-year to 5-year storms. This frequency of storm is often called the "channel forming" or "dominant" discharge. Most stable streams form naturally to pass this range of floods within their banks. Often partial duration frequency flow values are important here since the number of storms is also important (Chow, 1964).

- Pollution calculations are concerned with very frequent storms on the order of fractions of a year. This is because these storms are numerous and, cumulatively, convey the major portion of pollution to and through streams. Again, partial duration values are important here since the number of storms is important.

Following are some design frequency ranges that are used by many municipalities across the United States. These frequencies should be increased or decreased to account for local conditions, specific site conditions, importance of the storm water management facility, and other factors important to the proposed facilities and adjacent development.

Cross Drainage - Cross drainage facilities transport storm runoff under roadways. For many municipalities such drainage facilities are designed to accommodate a 25-year flood, though designs from the 10-year through the 100-year are not uncommon. Assuming a 25-year storm is chosen, the peak flows and hydrographs used for cross drainage design should be based on fully developed land use conditions. Thus the cross drainage should be designed so that the roadway is not overtopped for all floods that are equal to or less than the 25-year frequency event. Thus if a storm drainage system crosses under a roadway, then a 25-year flood must be routed through the system to show that the roadway will not be overtopped by this event. The excess storm runoff from events larger than the 25-year storm may be allowed to inundate the roadway or may be stored in areas other than on the roadway until the drainage system can accommodate the additional runoff. The final design should be checked using the 100-year flood to be sure structures are not flooded or increased damage does not occur to the highway or adjacent property for this design event.

Storm Drains - Storm drains are often designed to accommodate a 10-year flood. Frequency ranges from as low as the 5-year storm to as high as the 25-year storm are common. The design should be such that the storm runoff does not:
- increase the flood hazard significantly for property,
- encroach on to the street or highway so as to cause a significant hazard,
- limit traffic, emerging vehicles, or pedestrian movements to an unreasonable extent.

Based on these design criteria, a design involving temporary street or road inundation is acceptable practice for flood events greater than the design event but not for floods that are equal to or less than the design event.

Inlets - Inlets to storm drain systems should be designed for a 5- or 10-year flood depending on the roadway type and other factors at the site. Some cities allow for a much more frequent storm design for inlets (such as the 2-year storm) while sizing the storm drain which carries the flow from the inlets for the 5- or 10-year storm. See the storm drainage systems chapter for more details on such designs.

Detention And Retention Storage Facilities - All storage facilities should be designed to provide sufficient storage and release rates to accommodate a range of design storm events. The 2- and 10-year design storm events are common and cover the range of normal flooding. Larger regional ponds may be designed to accommodate events up to the 100-year storm. The design should be such that the storm runoff does not:
- increase the flood hazard significantly for adjacent, upstream, or downstream property as defined in the drainage ordinance, or
- cause any safety hazards associated with the facility.

Emergency spillway facilities should be provided to accommodate the 100-year storm. The final design should be checked to be sure that the downstream flood peaks for the storage discharge hydrographs have not increased for the 2- and 10-year floods. Dam Safety requirements may dictate emergency spillway size.

Review Frequency

After sizing a storm water management facility using a design event, a review frequency flood event should be used such as the 100-year flood. This is done to ensure that there are no unexpected flood hazards inherent in the proposed facilities. In some cases a flood event larger than the 100-year flood should be used to ensure the safety of the drainage structure and downstream development.

7.5 Hydrologic Procedure Selection

In performing hydrologic calculations it is important to have a systematic approach and to know:
* the types and accuracy of answers needed,
* the available methodologies to provide necessary information,
* the available data and information, and
* the assumptions inherent in and the limitations on selected methods.
   A general approach for hydrologic calculations could be outlined as the following.
   Step 1 - Determine requirements (e.g. peak flow, hydrograph etc.) and accuracy and select a design procedure.
   Step 2 - Collect necessary data.
   Step 3 - Identify design storm criteria and develop the design storm or rainfall.
   Step 4 - Compute time of concentration or other lag times required.
   Step 5 - Determine rainfall excess if appropriate to the methodology.
   Step 6 - Compute peak rate of runoff or flood hydrograph.
   Step 7 - Perform detention storage or channel routing if appropriate.
   Step 8 - Estimate or test sensitivity to engineering judgments and data error ranges. Adjust approach as appropriate.
   Step 9 - Document all estimates and calculations in detail.
   Streamflow measurements for determining a flood frequency relationship at a site are usually unavailable; in such cases, it is accepted practice to estimate peak runoff rates and hydrographs using statistical, empirical or physically based formula methods. In general results from using several methods should be compared, not averaged. If hydrologic procedures have been developed for the local area they will probably produce the best results and should be used. The accuracy of general hydrologic procedures can be greatly increased for a local area if they are calibrated for use within a specific area. The next section in this chapter discusses calibration.
   A consideration of peak runoff rates for design conditions is sometimes adequate for conveyance systems such as storm drains or open channels. However, if the design must include flood routing (e.g., storage basins or complex conveyance networks), a flood hydrograph is required. Many municipal ordinances now require that discharges from storm water management facilities be evaluated through a portion of the downstream drainage system. For such an analysis, a hydrograph and channel routing is necessary. Although the development of runoff hydrographs (typically more complex than estimating peak runoff rates) is usually accomplished using computer programs, some methods are adaptable to nomographs or other desktop procedures.
   Following are some of the commonly used categories or types of methods for generating peak flow estimates and/or hydrographs for rural and urban watersheds. Some of these methods are

available in the computer program HYDRAIN Drainage Design System and HEC-1. The hydrologic model within the HYDRAIN system is called HYDRO.

Analysis Of Stream Gage Data

Where reliable stream gage data are available they can be used to develop peak discharges and (less often) hydrographs. Data may be available from the regional U.S. Geological Survey (USGS), Corps of Engineers, the local municipality, or other agencies. These data can be used to generate very accurate hydrologic estimates.

A large number of types of analysis of gage data are available in the literature. For peak flow estimates log Pearson Type III frequency distribution analysis is considered to be one of the most reliable methods for estimating flood frequency relationships, though often data manipulation and adjustment is necessary (Water Resources Council, 1981).

Rules of thumb have been developed by the USGS relating the frequency estimate required and the length of record desirable as:

Design Return Period (years)	Desired Period of Record (years)
10	10
25	15
50	20
100	25

Unfortunately many urban stream are not gaged and such direct discharge data are seldom available. In these cases, data may be available for similar areas and can be transferred to the watershed in question through one of several methods of regional frequency analysis. Such methods are found in standard hydrology textbooks. Alternately, the gaged watershed could be modelled and the final modelling methodology transferred to the new watershed.

If urban development has been occurring in the watershed throughout the gaged period allowance must be made to adjust the record for such development. Many methods can be found in the literature for such adjustments (Sarma, et al., 1969, Dunne and Leopold, 1978, Rantz, 1971, Gundlach, 1978). While data on actual storms are scarce, information indicates that for every 1 percent increase in impervious area in an urban setting the peak flow increases between 1 and 1.25 percent. The former value is for more infrequent storms in the 100-year return period range and the later for return periods in the range of the 2-year storm. Development has a less percent impact on the larger storms due to the fact that for these storms the ground is normally saturated during the storm event and the impact of impervious area is lessened.

If urban regression equations are available for the area in question which have impervious area as one of the parameters the impact of development can be explicitly determined by differentiating the regression equation with respect to impervious area. This gives the change in peak flow for a change in impervious area. It may be necessary to consider a change in lag time or time-of-concentration too in this analysis since these times normally shorten considerably with increases in impervious area.

Heggen (1982) provides a nomographic method for using the SCS methodology for estimating changes in runoff volume for known changes in the SCS curve number. Based on an analysis of a number of methods for determining the impacts of development a general relationship can be derived as depicted in Figure 7-1 (McCuen, 1993). Using this figure the change in peak flow can be estimated in relation to undeveloped flows.

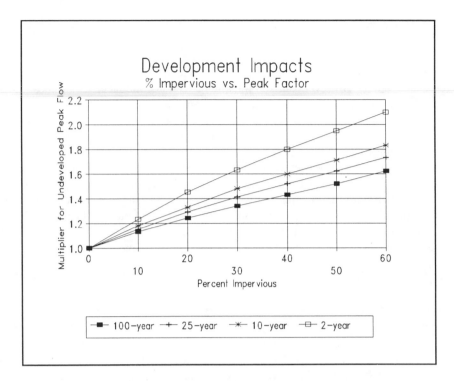

Figure 7-1 Impacts Of Urban Development On Peak Flow

Source: McCuen, R., Introduction to Chapter 9, ASCE TC on Hydrology, 1993.
Reproduced by permission of ASCE.

Regression Equations

If specific gage data are not available, peak flow, and sometimes whole hydrographs, can be calculated by using regression equations developed empirically for specific geographic regions. The equations are normally in the form of a log-log formula, where the dependent variable would be the peak flow or flow volume for a given frequency, and the independent variables may be parameters such as area, impervious area, slope, channel geometry, and other meteorological, physical or site specific data. Regression equations are available in some areas for both urban and rural streams, large and small. However, close attention should be paid to limits on data parameters and geographical applicability. The average error of estimate should be assessed as well as sensitivity to specific parameters. Regression results are often used for assistance in calibration as a "reality check", or for preliminary estimating in master planning.

A complete, though simulated, flood hydrograph can be computed using regression methods by applying lagtime, obtained from the proper regression equation, and peak discharge of a specific recurrence interval, to a dimensionless hydrograph. The coordinates of the runoff hydrograph can be computed by multiplying lagtime by the time ratios and peak discharge by the discharge ratios developed for different areas. Such a method has been developed and is being used by some municipalities in Georgia (Inman, 1986) and has been developed or verified by the USGS for other locations. Care must be taken that the definitions for lag time are equivalent as different agencies sometimes use different definitions.

Rational Method

A number of theoretically derived peak flow methods exist in the literature. Most provide a physically reasonable mathematical relationship between key parameters and peak flow much like a regression equation. However, unlike regression equations, the methods are meant to be used in general application across a wide range of geography and climate. The parameters are then estimated or calculated using local knowledge, standard values and engineering judgment. Great errors can occur when these methods are applied in ways or situations contrary to the assumptions inherent in the equations or when parameters are incorrectly estimated.

The well known Rational Method is one such general use methodology. It provides peak runoff rates for small urban and rural watersheds, often limited to 100 acres or much less, but is best suited to urban storm drain systems. The Rational Method should be used with caution if the time of concentration exceeds 30 minutes. Rainfall is a necessary input. Many municipalities will not allow the use of the Rational Method to estimate hydrographs or for detention storage designs. The Rational Method will be discussed in detail later in this chapter.

Unit Hydrograph Methods

Most standard hydrology books contain a derivation of the linear theory of hydrologic systems culminating in the unit hydrograph theory. A unit hydrograph is essentially the runoff, distributed correctly in time, from a unit of excess rainfall falling in a certain predetermined time period and applied uniformly over a watershed or sub-basin. Thus the five-minute unit hydrograph is the runoff hydrograph from, say, one inch of excess rainfall falling uniformly over a five-minute interval. Determination of an actual runoff hydrograph from a storm event (termed "convolution") is accomplished when measured five-minute blocks of rainfall (termed a rainfall hyetograph) are multiplied by the ordinates of this unit hydrograph, shifted in time by five-minute steps and added together. Use of unit hydrographs involves the determination of excess rainfall through the use of some sort of initial loss and infiltration methodology.

A number of unit hydrograph types are available in the literature. Two of the most common are described briefly here:

Synder's Unit Hydrograph - This method was developed (Snyder, 1938) for the Appalachian area watersheds ranging from 10 to 10,000 square miles. It has been applied to watersheds across the United States by the Corps of Engineers and is one of the methods found in the popular HEC-1 program. It provides a means of generating a synthetic unit hydrograph. It relies on the calculation of lag time and peak flow through two relationships involving area, length measurements and estimated parameters. Since it does not define the total hydrograph shape other relationships must be used with the Snyder method for such a definition. For example, HEC-1 uses the Clark relationship for such a definition along with empirically developed estimates of the hydrograph widths at the 50 and 75 percent of peak levels (HEC, 1990). Details of the method can be found in Chow (1964).

SCS Synthetic Unit Hydrograph - The Soil Conservation Service has developed a family of hydrologic procedures, one of which is a synthetic unit hydrograph procedure. It has been widely used for developing rural and urban hydrographs. The unit hydrograph used by the SCS method is based upon an analysis of a large number of natural unit hydrographs from a broad cross section of geographic locations and hydrologic regions. Rainfall is a necessary input. This method is discussed in detail later in this chapter.

Continuous Simulation Models

All of the methods described above are event models which use rainfall as the input and flood peaks and hydrographs as the output. This is because most municipal storm water management facilities are designed for a specified flood event and rainfall data are readily available throughout the country. A consequence of this approach is that the complex interaction between rainfall and resulting storm runoff must be estimated and only one or two flood events are used in the design. In contrast, continuous simulation models such as the EPA Storm Water Management Model (SWMM) or the Hydrocomp Simulation Model attempt to represent the entire hydrologic system on the computer so as to simulate the natural system. In this way the model simulates the runoff process including interception, infiltration, overland flow, channel flow, etc. This simulation is over a long period of time and is continuous so that both flood events and low flow events are simulated. If accurately simulated, the models will provide information on particular aspects of the runoff process, such as antecedent moisture, which is very important when estimating flood peaks and hydrographs.

Although these models have been available since the early 1970's, their application for municipal use has been limited. This probably results from the large amount of data needed and until quite recently the difficulty in using the models. Today, user friendly continuous simulation models are available and in time their use for municipal storm water management analysis and design will increase. The use of these models is especially useful for water quality simulation, storm movement analysis, and complex designs and analyses. Due to the complexity of these models and their present limited application, no further discussion of continuous simulation models will be presented in this book. The interested reader should consult the extensive literature related to these models and documentation that is readily available from the model developers and software distributors.

Summary

When sufficient streamflow data are available, the findings from a Log Pearson III method can be used for hydrological estimates. Often adjustments or regionalization must be performed to make the data useful.

If available, regression equations and regression analysis are an acceptable method of estimating peak flows and hydrographs for rural and urban watersheds. These equations and analysis have been shown to be accurate, reliable, and easy to use as well as providing consistent findings when applied by different hydraulic engineers (Newton and Herin, 1982). Regression equations are used to relate such things as the peak flow or some other flood characteristic at a specified recurrence interval to the watershed's physiographic, hydrologic and meteorological characteristics.

The major problem with using stream gage data or regression equations is their lack of availability within many municipalities. Also where available, limitations of using the equations related to the scope of the data base used for regression analysis limits the use in many locations. It is very common that these equations can only be used for large or undeveloped watersheds while most municipal hydrologic studies deal with small urbanized watersheds and site developments. If regression equations are available for specific municipalities the associated documentation should be used to determine their applicability and how to apply the equations.

The following discussion will focus on the two most popular hydrologic methods used within most municipalities: the Rational Method and the SCS Unit Hydrograph Method. Other unit hydrograph methods are available such as the Snyder Synthetic Unit Hydrograph, and for some areas local unit hydrographs have been developed such as the Colorado Unit Hydrograph Method. Those interested in the Snyder Synthetic Unit Hydrograph should consult Chow (1964) or the

AASHTO Model Drainage Manual, and for those municipalities where unit hydrograph methods have been developed, associated documentation should be used.

7.6 Calibration

Calibration is a process of varying the parameters or coefficients of a hydrologic method so that it will estimate peak discharges and hydrographs consistent with local rainfall and streamflow data. Most hydrologic procedures used for storm water management facility design contain general equations that have been developed for large geographic areas. These procedures cannot be expected to take into account local hydrologic conditions and as a result, unless calibrated, they may not produce acceptable estimates for analysis and design.

Following is an illustration of a hydrograph resulting from flow data as compared to a hydrograph resulting from using a non-calibrated and calibrated hydrologic procedure. It can be seen that the calibrated hydrograph, although not exactly duplicating the hydrograph from stream-flow data, is a much better representation of the streamflow hydrograph than the non-calibrated hydrograph.

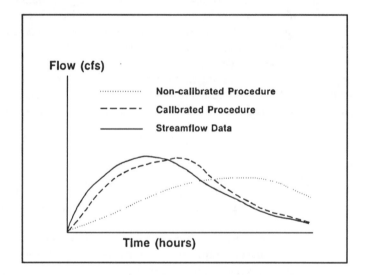

Figure 7-2 Hydrograph Calibration

The accuracy of the hydrologic estimates will have a major effect on the design of storm water management facilities. Although it might be argued that one hydrologic procedure is more accurate than another, practice has shown that all of the methods discussed in this chapter can, if calibrated and not used beyond their intended purposes, produce acceptable results consistent with observed or measured events. What should be emphasized is the need to calibrate the method for local and site specific conditions. This calibration process can result in much more accurate and consistent estimates of peak flows and hydrographs.

The calibration process can vary depending on the data or information available for a local area. Following are some general steps in the calibration process.

1. If streamflow data are available for an area, the hydrologic procedures can be calibrated to these data. The process would involve generating peak discharges and hydrographs for different input conditions (e.g., slope, area, antecedent soil moisture conditions) and comparing

these results to the gaged data. Changes in the procedures would then be made to improve the estimated values as compared to the measured values. It is typically a good idea to first match flow volumes through adjustment of losses, then match peaks and timing. Also, it is best to change one parameter at a time and to measure sensitivity at each change.

Note: When changing hydrologic procedures care should be exercised to be sure that the basic theory of the hydrologic procedures is not violated by changes made during the calibration process.

2. After changing the variables or parameters in the hydrologic procedure the results should be checked against another similar gaged stream or another portion of the streamflow data that were not used for calibration.

3. If some local agency has developed procedures or equations (for example a local regression equation for lag time or rainfall losses) for an area based on streamflow data, general hydrologic procedures can be calibrated to these local procedures. In this way the general hydrologic procedures can be used for a greater range of conditions (e.g., land uses, watershed size, slope).

4. The calibration process should only be undertaken by personnel highly qualified in hydrologic procedures and design. There are many pitfalls. For example, the use of too long a time step duration may give an artificially low peak flow simply because two time adjacent steps straddled the peak and did not capture it.

5. Should it be necessary to use unreasonable values for variables in order for the model to produce reasonable results, then the model should be considered suspect and its use carefully considered (e.g., having to use terrain variables that are obviously dissimilar to the geographic area in order to calibrate to measured discharges or hydrographs). However, the skilled modeler can often modify parameters to mimic conditions in nature not available in the model. For example, for small watersheds where absolute timing is not critical, artificially extending the lag time can "mimic" inadvertent storage without resorting to a series of very minor detention routings behind each undersized culvert within a subbasin.

7.7 Precipitation And Losses

Storm water management facilities are normally designed based on some flood frequency. Floods and their corresponding frequencies are not often available. Therefore, most synthetic hydrologic procedures use rainfall and rainfall frequency as the basic input, and derive flood estimates. To do this it is commonly assumed that, for example, the 10-year rainfall will produce the 10-year flood. Depending on antecedent soil moisture conditions, and other hydrologic parameters, for a specific storm, this may or may not be true.

Other methods, mostly regression type equations and gage analysis, rely on direct estimation of flood magnitudes without resorting to the use of rainfall.

Intensity-Duration-Frequency Curves

Rainfall data are available for most geographic areas. Such data can be obtained from the National Weather Service publications or from local sources if available. The two most common sources of such information for use in the 37 states east of the 105th meridian for durations from 5 minutes to 24 hours and return periods from 1- to 100-years are Technical Paper 40 (Hershfield, 1961) and NWS HYDRO-35 (NOAA, 1977). Other Technical Papers and local or regional reports, as well as digital data and information, are available from the National Oceanic and Atmospheric Administration, National Climatic Data Center, Asheville, NC (phone 704-259-0682) for other areas of the country and less frequent events. Information on the Probable Maximum Precipitation required for some dam safety studies is found in HMR-51 and HMR-52 (NOAA, 1978, 1982).

Since TP40 and HYDRO-35 are now quite dated it may be advisable, when generating a new rainfall intensity-duration-frequency (IDF) curve for an area, to obtain the latest rainfall information from NOAA in digital form and develop new IDF curves. From these publications or new data, IDF curves can be developed for the commonly used design frequencies. These IDF curves then become the basic input for the hydrologic procedures discussed in this chapter.

The procedure for developing an IDF curve for a particular return period from TP40 and HYDRO-35 is as follows:

STEP 1 - For periods one hour and greater in duration obtain point rainfall information by interpolating between curves from the maps in TP40. For periods one hour and less obtain information in a similar way from HYDRO-35.

STEP 2 - Because the values given are for partial duration series they must be converted for most uses to annual series data by multiplying the map derived information by factors obtained from the Table 7-1 below.

STEP 3 - Plot family of curves with duration on the horizontal axis and intensity on the vertical axis and perform minor manual smoothing to remove irregularities.

Table 7-1 Empirical Factors For Converting
Partial-Duration Series To Annual Series

Return Period	Conversion Factor
2-year	0.88
5-year	0.96
10-year	0.99

IDF data for any city can be easily fit by an equation of the form:

$$i = a/(t+b)^c \qquad (7.7)$$

Where: i =   the rainfall intensity, in/hr
        t =   the duration, min
        a, b, and c are curve fitting parameters

By taking the log of equation 7.7 the equation becomes:

$$\log i = \log a - c \log (t+b) \qquad (7.7a)$$

This is the form of the equation for a straight line with: c being the slope and log a being the y-intercept. By assuming trial values for b, a straight line can be fit through plots of log (t+b) versus log i. This can be done automatically using the linear regression fitting functions found in most common spreadsheet software.

Often c is set equal to 1, simplifying the equation without losing great accuracy. All IDF curves for a particular location can be approximated by adding the term Td to the numerator where T is the recurrance interval and d is a curve fitting parameter.

Table 7-2 is an example of rainfall intensity data available for the Fayetteville, North Carolina area. Similar data can be obtained for other municipalities.

Table 7-2 Rainfall Intensity Data - Fayetteville, North Carolina

Storm Duration		Rainfall Intensity (in./hr)					
hours	minutes	2	5	10	25	50	100
0	5	5.17	6.19	6.93	8.02	8.87	9.72
	10	4.33	5.26	5.94	6.91	7.68	8.44
	15	3.70	4.51	5.10	5.95	6.62	7.28
	30	2.58	3.35	3.87	4.62	5.20	5.78
1		1.67	2.24	2.63	3.17	3.58	4.00
2		0.93	1.25	1.47	1.78	2.01	2.25
3		0.68	0.92	1.08	1.31	1.49	1.66
6		0.42	0.57	0.68	0.82	0.93	1.04
12		0.24	0.34	0.40	0.48	0.55	0.61
24		0.14	0.19	0.23	0.28	0.32	0.35

Rainfall Durations and Time Distributions

When using the design storm method, the urban storm water designer must decide which design storm to use. Earlier sections talked about the design frequency applicable for the sizing of various components of the drainage system. But the choice of frequency is only part of the storm selection process. All design storms are made up of four main components: frequency, duration, rainfall distribution, and depth (assuming stationary storms). The storm frequency, depth and duration are inter-related through use of the intensity-duration-frequency curves.

Duration - The chosen duration of the storm is often determined by either the municipality's design criteria or by the type of design being performed. A rule-of-thumb often used for minor system peak flow design in urbanized areas using the Rational Method is to use the storm duration which is equal to the time-of-concentration of the site to the calculation point. However, it can be easily shown that this may not yield the highest peak if a portion of the site is grassy and the rest is paved. Often using only the paved portion with its higher runoff and shorter time of concentration (and thus higher rainfall intensity) gives the highest peak flow.

In developing a runoff hydrograph using unit hydrograph methods a storm duration significantly longer than the time of concentration is warranted. For example, it can be shown that the duration of the storm when using SCS methods for unit hydrograph development and a number of pulses of rainfall in the rainfall hyetograph, that the duration of the storm must be almost twice the time-of-concentration. For detention design the duration of the storm should be that which yields the highest storage requirement. This then becomes a function of the relative sizes of the pond, watershed and outlet but will, in any case, be much longer than the duration necessary for simple peak runoff estimation.

Distribution - The appropriate time distribution of the design rainfall has been a subject of some speculation among urban storm water designers. It should be remembered that the objective is to produce a runoff "event" of a particular frequency. It is normally assumed that a rainfall of a particular frequency will, on average, produce a runoff of the denticle frequency. Though each individual storm may not at all reproduce its associated frequency of runoff. While rainfall depths for a given duration have an associated calculated frequency, rainfall distributions generally do not. Thus, a lesser amount of rain, but falling in an atypical way, may produce a higher peak runoff than more rain depth but falling in a more uniformly distributed manner. This distribution selection becomes more important as the size of the watershed decreases and the imperviousness increases. Larger or less impervious watersheds have the opportunity to "dampen out" pulses of rainfall and smooth the runoff hydrographs through attenuation and runoff hydrograph timing combinations.

There are probably a dozen methods used around the United States to distribute rainfall and others in other countries. Four commonly used methods to distribute rainfall (other than a uniform distribution) are illustrated by an example for a 1-hour storm with five minute time increments.

Example Information - A Midwestern city has the following IDF curve information for the 10-year storm:

Table 7-3 IDF Information for the 10--Year Storm

Time (min)	5	10	15	30	60
Depth (in)	0.512	0.867	1.125	1.588	1.967
Intensity (in/hr)	6.145	5.201	4.498	3.176	1.967

Other method-specific data will be given for each approach.

The "Balanced Storm Approach" - This approach is described in COE (1982) and in the popular HEC-1 computer software manual (HEC, 1990). In this method the storm depth at any duration of storm is equal to the depth from the IDF curve. Therefore the total storm depth in the first five minutes, from this example, is 0.512 inches. The total depth after 10-minutes is 0.867 inches. The incremental depth from five to ten minutes is equal to the difference between the two or 0.867-0.512 = 0.355 inches. Continuing this process and using either linear interpolation or curve fitting for the five-minute increments for which there is no value leads to 12 incremental rainfall depths for each five minutes. These depth increments are then arranged so that the most intense increment is at the center of the storm. The second most intense is placed before it. The third is placed after it, the forth at the front, and so on, staggering the placement of all the increments until all 12 are placed alternately in front or at the end of the storm. Table 7-4 illustrates this procedure for the example given. The last two columns are the final storm increments and accumulated depth S-curve.

The Balanced Storm Method has the advantage in that it is relatively independent of storm duration. It will always maximize the rainfall depth for any duration for a given frequency. That is, due to the nesting of the depths, any duration chosen will yield the depth of rainfall for that frequency, it is "balanced". This has certain advantages when working with different sized areas with one storm. It has the disadvantage in that it is a theoretical frequency distribution, not observed in nature, and gives the most intense storm possible for any given frequency and duration.

And, in fact, comparing its S-curve to that of a series of S-curves derived by Huff (discussed below) indicates that the distribution has a less than 1 in 10 chance of occurring in nature. This has led to a concern that it may overpredict storm peaks for smaller urban areas where attenuation and timing effects do not dampen out the intense rainfall. It is the method used by HEC-1 when PH records are used (HEC, 1990).

Table 7-4 Balanced Storm Example

Time (min)	Intensity (in/hr)	Depth (in)	Increment (in)	Arranged (in)	Accum. (in)
0	0	0	0	0	0
5	6.144	0.512	0.512	0.039	0.039
10	5.200	0.867	0.355	0.054	0.094
15	4.497	1.124	0.258	0.079	0.172
20	3.955	1.318	0.194	0.119	0.291
25	3.525	1.469	0.150	0.194	0.485
30	3.175	1.588	0.119	0.355	0.840
35	2.886	1.684	0.096	0.512	1.352
40	2.643	1.762	0.079	0.258	1.610
45	2.436	1.827	0.065	0.150	1.760
50	2.258	1.882	0.054	0.096	1.856
55	2.103	1.928	0.046	0.065	1.921
60	1.967	1.967	0.039	0.046	1.967

SCS 24-Hour Storm - The Soil Conservation Service has developed several semi-dimensionless 24-hour storms which are considered typical of different locations within the United States. Section 7.9 in this chapter covers the different storm types, S-curve appearance and geographic applicability. The SCS storm is semi-dimensionless in that the horizontal axis (X-axis) of the storm S-curve covers a time period of 24-hours while the vertical axis (Y-axis) is a dimensionless ratio of the storm accumulation to that time divided by the storm total depth. The SCS 24-hour storm is a generalization of the Balanced Storm Approach and is therefore also theoretical in nature. It was developed by taking the IDF curve information from a number of municipalities and developing an average representative curve. Thus it shares all the advantages and disadvantages of the Balanced Storm Approach.

One additional disadvantage of the SCS storm approach is that it is cumbersome and inaccurate to use for short duration storms owing to the general unavailability of detailed tables of values for very short time increments. The SCS storm method was derived for larger areas where longer time increments are applicable (SCS, 1986). The general procedure for storms less than 24 hours is to take the central portion of the SCS curve equal to the duration necessary. So, for example, for a 1 hour storm the time period form 11.5 to 12.5 hours would be used. The total rainfall for one hour (1.967 inches) would be proportioned in a manner equivalent to the SCS curve. Since the total portion of the curve is 0.4518 (0.7351 - 0.2833) and 84% of that occurs between 11.5 and 12.0 hours of the curve, 84% of the 1.967 inch for a one hour storm total should also occur in the first half hour. Table 7-5 shows typical calculations for the example. Note that the actual resultant storm occurs in two blocks. This is due to the fact that the available table has time increments of only 0.5 hours and linear interpolation was used for the five minute values.

Table 7-5 SCS 24-Hour Storm Method Example

SCS Hour (hrs)	SCS Value of P/P24	Time (min)	Interp. Rainfall (in)	Increment. Rainfall (in)
11.5	0.2833	0	0.000	0.000
		5	0.276	0.276
		10	0.551	0.276
		15	0.827	0.276
		20	1.103	0.276
		25	1.378	0.276
12.0	0.6632	30	1.654	0.276
		35	1.706	0.052
		40	1.758	0.052
		45	1.810	0.052
		50	1.863	0.052
		55	1.915	0.052
12.5	0.7351	60	1.967	0.052

Huff Distributions - Huff (1967) looked at 261 storms in a 400 square mile network during the period 1955-1966 in East-central Illinois and assigned typical distributions to them according to whether they tended to have the bulk of the rainfall in the first, second, third or fourth quartile. These "Huff" curves are applicable to areas of from 50 to 400 square miles. The time distributions were expressed as percentages of total duration and total rainfall accumulation into a series of S-curves. For each of the quartile groupings the storms were analyzed and a family of curves was derived, one for each probability interval of accumulated rainfall plotted versus time. For the 90% curve, for example, at each point on the curve ninety percent of the storms had accumulated at least the given percentage of rainfall by that point in the storm's total duration. Table 7-6 gives the 50 percent (median) time distributions for each of the four quartile storms versus percentage of storm duration. Table 7-7 gives first-quartile storm information for the 10%, 50% and 90% storms. Later Huff and Vogel (1976) used Chicago data from 471 storms with depths greater than 0.5 inches to derive point data distributions following the same procedures as the East-central Illinois data. Huff (1990) compared the two sets of data to derive curves for areas from 0 to 10 square miles and 10 to 50 square miles. These values are given in Table 7-8.

According to Huff, it is recommended to use either first or second quartile storms for durations less than 12 hours, and forth quartile storms for durations greater than 24 hours. For most design purposes the median curves are probably the most applicable. They are more established than the more extreme curves at the 10% and 90% levels, which typify unusual storm distribution conditions. The storm distributions are thought to be applicable to all areas with storm climatology similar to the Midwest. Some have thought they should be applicable wherever the SCS Type II storm is applicable since, according to SCS records, these areas indicate similar storm distributions.

Table 7-6 Median Time Distributions of Heavy
Storm Rainfall in Areas of 50 to 400 Square Miles

Cum. Percent of Storm Time	First-quartile Storm	Second-quartile Storm	Third-quartile Storm	Forth-quartile Storm
5	8	2	2	2
10	17	4	4	3
15	34	8	7	5
20	50	12	10	7
25	63	21	12	9
30	71	31	14	10
35	76	42	16	12
40	80	53	19	14
45	83	64	22	16
50	86	73	29	19
55	88	80	39	21
60	90	86	54	25
65	92	89	68	29
70	93	92	79	35
75	95	94	87	43
80	96	96	92	54
85	97	97	95	75
90	98	98	97	92
95	99	99	99	97

Source: Huff, 1990.

Table 7-7 Median Time Distributions of Areal Mean Rainfall in First-quartile Storms at 10%, 50% and 90% Probability Levels

Cumulative Percent of Storm Time	Cumulative Percent of Storm Rainfall for Given Storm Probability		
	10%	50%	90%
5	24	8	2
10	50	17	4
15	71	34	13
20	84	50	28
25	89	63	39
30	92	71	46
35	94	76	49
40	95	80	52
45	96	83	55
50	97	86	57
55	98	88	60
60	98	90	63
65	98	92	67
70	99	93	72
75	99	95	76
80	99	96	82
85	99	97	89
90	99	98	94
95	99	99	97

Source: Huff, 1990.

Table 7-8 Median Time Distributions of Heavy Storm Rainfall
at a Point and on Areas of 10 to 50 Square Miles.

Cum. Percent of Storm Time	Cumulative Percent of Point Storm Rainfall for a Given Storm Type				Cumulative Percent of Storm Rainfall on Areas of 10 to 50 Square Miles for a Given Storm Type			
5	16	3	3	2	12	3	2	2
10	33	8	6	5	25	6	5	4
15	43	12	9	8	38	10	8	7
20	52	16	12	10	51	14	12	9
25	60	22	15	13	62	21	14	11
30	66	29	19	16	69	30	17	13
35	71	39	23	19	74	40	20	15
40	75	51	27	22	78	52	23	18
45	79	62	32	25	81	63	27	21
50	82	70	38	28	84	72	33	24
55	84	76	45	32	86	78	42	27
60	86	81	57	35	88	83	55	30
65	88	85	70	39	90	87	69	34
70	90	88	79	45	92	90	79	40
75	92	91	85	51	94	92	86	47
80	94	93	89	59	95	94	91	57
85	96	95	92	72	96	96	94	74
90	97	97	95	84	97	97	96	88
95	98	98	97	92	98	98	98	95

Source: Huff, 1990

The Huff distribution has the advantage in that it is derived from real observed data for heavy storms. Therefore the distribution in time should give a realistic picture of a typical storm. It should be realized that, for the median storm, half the storms observed will have steeper (and half milder) S-curve accumulations.

The Huff distributions have some significant disadvantages for the smaller watersheds. Huff used 30-minute data for his observations and thus missed the short-time rainfall "burst" information inherent in one-minute data. Thus for longer storm durations with short time increments the short duration bursts will be missed. If the method is used for these short times the actual rainfall accumulation for a short time period may be greater or less than the IDF curve would indicate for the same duration of storm. Secondly, graphical information has been shown to be difficult to read, interpolate accurately, and use (Ruthroff and Bodtmann, 1976). Thus information in this section has been expressed in Huff's tabular format to facilitate consistent use. Thirdly, since the distribution is duration independent, strict duration criteria must be set or the designer should be required to choose the duration which maximizes the peak, because different durations of the same storm can yield significantly different peak flows. Lastly, the method used combined a number of storms in each quartile averaging them at each percentage accumulation interval. This also tends to dampen out the variation effects of individual storms producing an averaged smooth curve unlike that found in nature. Huff's separating the data into quartiles and probabilities somewhat compensates for this and allows more flexibility in its use.

Table 7-9 illustrates the application of the Huff method to the example problem for the first and second quartile storms. The basic procedure is to scale the total storm duration into percentages of the total duration and find the dimensionless accumulation of the total storm rainfall from the appropriate table. Multiplying this by the total storm depth gives the accumulation up to that point in the storm. Subtracting subsequent values gives the incremental values of the storm. No arrangement is necessary since the shape of the S-curve gives the arrangement.

Table 7-9 Example Problem Using Huff Distributions

Time	Huff - 1st Quartile		Huff - 2nd Quartile	
	accum	inc	accum	inc
0	0.00	0.00	0.00	0.00
5	0.54	0.54	0.13	0.13
10	0.91	0.37	0.26	0.14
15	1.18	0.28	0.43	0.17
20	1.36	0.18	0.70	0.27
25	1.50	0.14	1.08	0.37
30	1.61	0.11	1.38	0.30
35	1.68	0.07	1.56	0.18
40	1.74	0.07	1.69	0.13
45	1.81	0.07	1.79	0.10
50	1.88	0.07	1.86	0.07
55	1.92	0.04	1.92	0.06
60	1.97	0.05	1.97	0.05

Yen and Chow's Method

Yen and Chow (1980) evaluated hourly data from over 7400 storms from Urbana, IL., Boston, MA., and Elizabeth City, NJ. to develop a "typical" dimensionless rainfall distribution for use in basins under 20 square miles. From these they developed a rainfall hyetograph in the shape of a triangle. The figure below depicts this hyetograph.

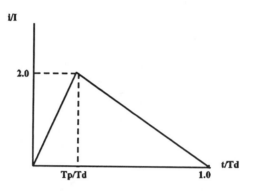

Figure 7-3 Yen and Chow's Dimensionless Hyetograph

In this procedure Tp is the time to peak, Td is the total storm duration, and I is the total storm rainfall. From a known total storm depth and duration the only unknown to totally define the hyetograph is the value of Tp/Td. Based on a further analysis of many thousand rainfall gages around the U.S., Yen and Chow determined typical values of Tp/Td (Yen and Chow, 1983).

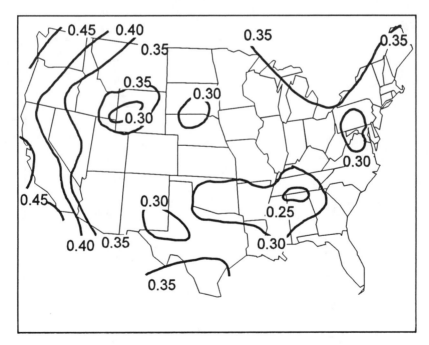

Figure 7-4 Tp/Td Values (after Yen and Chow, 1983)

Applying Yen and Chow's method to the example problem yields the following information:

Tp/Td = 0.32   I = 1.967 in   Td = 60 min

Since i/I = 2.0 the peak instantaneous rainfall rate (i) is 3.934 in/hr and Tp is 19.2 minutes. Then five minute blocks of rainfall are derived from the triangular shape by calculating the area under the triangle for each five minutes. The following storm results.

Table 7-10 Yen and Chow's Method

Time	Accum. Rain	Inc. Rain
0	0	0
5	0.043	0.043
10	0.171	0.128
15	0.384	0.213
20	0.681	0.297
25	0.983	0.301
30	1.244	0.261
35	1.465	0.221
40	1.646	0.181
45	1.786	0.141
50	1.887	0.1
55	1.947	0.06
60	1.967	0.02

The different storm S-curves are plotted for comparison in Figure 7-5.

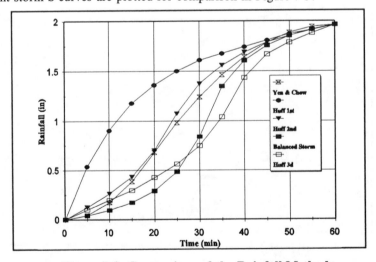

Figure 7-5 Comparison of the Rainfall Methods

The Huff third quartile storm is also plotted to compare it to the balanced storm approach. Notice that the balanced storm has the steepest curve of any of the methods given. The Yen and Chow method, for this example, is roughly equivalent to Huff's second quartile storm.

7.8 Rational Method

The Rational Method is a simple formula for the estimation of peak flow rates from small urban and rural watersheds. Its form is a simple ratio between runoff and rainfall rates. It is often recommended for estimating the design storm peak runoff for small urban and rural watersheds in drainage system design. Often a size limit of less than 100 acres is used and some regulations prohibit its use for areas larger than 20 acres (South Carolina, 1992). The method is best suited for estimating peak discharges for inlet and storm drainage system design.

Rational Formula

The Rational Formula estimates the peak rate of runoff at any location in a watershed as a function of the drainage area, runoff coefficient, and mean rainfall intensity for a duration equal to the time of concentration (the time required for water to flow from the most remote point of the basin to the location being analyzed). The Rational Formula is expressed as follows:

$$Q = CIA \tag{7.8}$$

Where: Q = maximum rate of runoff, cfs
C = runoff coefficient representing a ratio of runoff to rainfall
I = average rainfall intensity for a duration equal to the time of concentration, in./hr
A = drainage area contributing to the design location, acres

Note that the units of the Rational Formula are not consistent as the left side is in cfs and the right side is in acre-inches per hour. However the conversion from one set of units to the other is approximately equal to one, thus allowing for the discrepancy.

The Rational Method, while first introduced in 1889 (Kuichling), is derived from earlier work in both America and England. Even though it has frequently come under criticism for its simplistic approach, no other drainage design method has received such widespread use. This does not imply that it produces accurate hydrologic estimates. However, many systems have been designed with the Rational Method which have been tested by storms and found not to be grossly undersized. However, adequate data has not been collected with which to test the validity of the method in a variety of situations. It may, in fact, over or underdesign storm systems. Opinions in the literature are mixed. With the increased complexity of urban storm drainage systems and the need for hydrograph routing downstream from storm water management facilities the use of the Rational Method is being severely restricted in many municipalities. Proper application of the method and recognition of instances where standard application of the Rational Method does not apply are still the governing principles.

For example, in some situations it can be demonstrated that taking the peak from only the directly-connected impervious area of a development yields a higher peak flow estimate than taking the runoff from a combination of the grassy and paved areas, even though the area considered is smaller. Or, in backwater situations, the ability of the pipe to handle the calculated peak flow is reduced even though the peak may be calculated correctly.

Characteristics And Limits Of The Rational Method

When using the Rational Method some precautions should be considered in order to correctly apply the method.

1. The first step in applying the Rational Method is to obtain a good topographic map and define the boundaries of the drainage area in question. A field inspection of the area should also be made to determine if the natural drainage divides have been altered.

2. In determining the C value (land use) for the drainage area, thought should be given to future changes in land use that might occur during the service life of the proposed facility that could result in an inadequate drainage system.

3. Since the Rational Method uses a composite C value for the entire drainage area, if the distribution of land uses within the drainage basin will affect the results of hydrologic analysis, then the basin should be divided into two or more sub-drainage basins for analysis.

4. Restrictions to the natural flow such as highway crossings and dams that exist in the drainage area should be investigated to see how they affect the design flows. Also, the effects of upstream detention facilities may be taken into account.

5. The charts, graphs, and tables included in this section are given to assist the designer in applying the Rational Method. The designer should use good engineering judgment in applying these design aids and should make appropriate adjustments when specific site characteristics dictate that these adjustments are or are not appropriate.

Following are characteristics of, or assumptions inherent in, the Rational Method which may limit its use for analysis of storm water management facilities. A more detailed discussion can be found in Rossmiller (1980) or Chow (1964).

The rate of runoff resulting from any rainfall intensity is a maximum when the rainfall intensity lasts as long or longer than the time of concentration.

This assumption limits the size of the drainage basin that can be evaluated by the Rational Method. For large drainage areas, the time of concentration can be so large that constant rainfall intensities for such long periods do not occur and shorter more intense rainfalls can produce larger peak flows. Further, in semi arid and arid regions, storm cells are relatively small with extreme intensity variations thus making the Rational Method inappropriate for large watersheds. There is even some question about whether the time-of-concentration can be measured or is significant (Schaake, et al., 1967).

The frequency of peak discharges is the same as that of the rainfall intensity for the given time of concentration.

Frequencies of peak discharges depend on rainfall frequencies, antecedent moisture conditions in the watershed, and the response characteristics of the drainage system. For small and largely impervious areas, rainfall frequency is the dominant factor. For larger drainage basins, the physical characteristics of the drainage basin control. For drainage areas with few impervious surfaces, antecedent moisture conditions usually govern, especially for rainfall events with a return period of 10 years or less.

The fraction of rainfall that becomes runoff (C) is independent of rainfall intensity or volume.

The assumption is reasonable for impervious areas, such as streets, rooftops, and parking lots. For pervious areas, the fraction of runoff varies with rainfall intensity and the accumulated volume of rainfall (Schaake, et al., 1967). Thus, the art necessary for application of the Rational Method involves the selection of a coefficient that is appropriate for the storm, soil, and land use conditions. Many guidelines and tables have been established, but seldom have they been supported with empirical evidence.

The peak rate of runoff is sufficient information for the design.

Modern drainage practice often includes detention of urban storm runoff to reduce the peak rate of runoff downstream. No recommended hydrograph is available from the Rational Method to route through detention storage facilities. Lack of a reliable hydrograph severely limits the use of the Rational Method for the evaluation of design alternatives available in urban and in some instances, rural drainage design. Other hydrologic methods discussed in this chapter can be used to generate hydrographs.

One of the greatest drawbacks of the Rational Method is that its results are not usually replicable from application to application. A formula with the simplicity of the Rational Method lends itself to wide variations in application, judgment of the parameters, and use of various adjustment procedures (Rossmiller, 1980). Many municipalities have identified one or several adjustments considered important for their local environment. Complex adjustments which yield only minor changes in the final peak flow estimate are usually not warranted (except, perhaps for legal reasons). Despite all these concerns the Rational Method remains a mainstay of hydrologic design.

The Rational Method depends on four variables: time of concentration, rainfall, area and the C factor. The results of using the Rational Formula to estimate peak discharges are very sensitive to the variables that are used. Careful selection of the variables and recognition of unusual circumstances is important. Following is a discussion of the different variables used in this method.

Time Of Concentration

The time factor used in the Rational Method is simply a period within the total storm duration during which the maximum average rainfall intensity occurs. It is sometimes called the "rainfall intensity averaging time". Popular application of the Rational Method uses the time of concentration (t_c) for each design point within the drainage basin as the rainfall intensity averaging time.

The time of concentration is the time required for water to flow from the hydraulically most remote point of the drainage area to the point under investigation. The duration of rainfall is then set equal to the time of concentration and is used to estimate the design average rainfall intensity (I). The time of concentration used in the design consists of an inlet time to the point where the runoff enters the storm drain or channel inlet plus the time of flow in a closed conduit or open channel to the design point.

Two common errors should be avoided when calculating the time of concentration. First, in some cases runoff from a portion of the drainage area which is highly impervious may result in a greater peak discharge than would occur if the entire area were considered. In these cases, adjustments can be made to the drainage area by disregarding those areas where flow time is too slow to add to the peak discharge. Sometimes it is necessary to estimate several different times of concentration to determine the design flow that is critical for a particular application. Second, when designing a drainage system, the overland flow path is not necessarily the same before and after development and grading operations have been completed. Selecting overland flow paths in excess of 100 feet in urban areas and 300 feet in rural areas should be done only after careful consideration.

Figure 7-6 can be used to estimate inlet time for design conditions that do not involve complex drainage conditions. For each drainage area, the distance is determined from the inlet to the most remote point in the tributary area. From a topographic map, the average slope is determined for the same distance. The runoff coefficient (C) is determined by the procedure described in a subsequent section of this chapter.

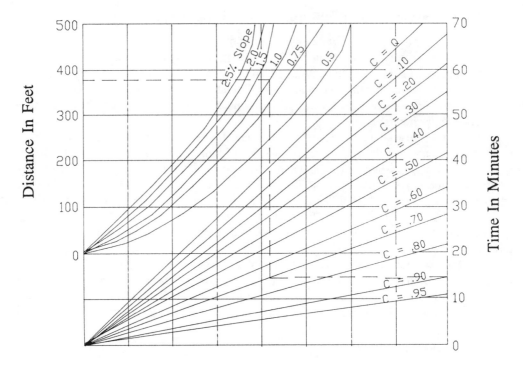

Figure 7-6 Overland Time Of Flow

Source: Airport Drainage, Federal Aviation Administration, 1965

  To obtain the total time of concentration, the pipe or open channel flow time must be calculated and added to the inlet time. After first determining the average flow velocity in the pipe or channel (using the Manning formula for example), the travel time is obtained by dividing velocity into the pipe or channel length. Velocity can be estimated by using the nomograph shown on Figure 7-7. Note: time of concentration cannot be less than 5 minutes.
  A large number of other methods are available for estimating inlet time. They fall into categories based on their form and independent variables. The travel time estimation method used for the SCS Method could be substituted for the above procedure or used to check the values obtained from the charts and nomographs described for use with the Rational Method. The SCS travel time estimation method is described under the SCS Unit Hydrograph method in this chapter.
  Kirpich (1940) developed a formula for inlet time based on data from six steeply sloped agricultural and forested watersheds in West Tennessee as:

$$t_c = 0.0078 \ (L/(S^{0.5}))^{0.077} \tag{7.9}$$

Where: t_c = time of concentration, min
    L = length of travel, ft
    S = slope of the flow path from the most remote part of the basin to the calculation
      point divided by the horizontal distance between the two points, ft/ft

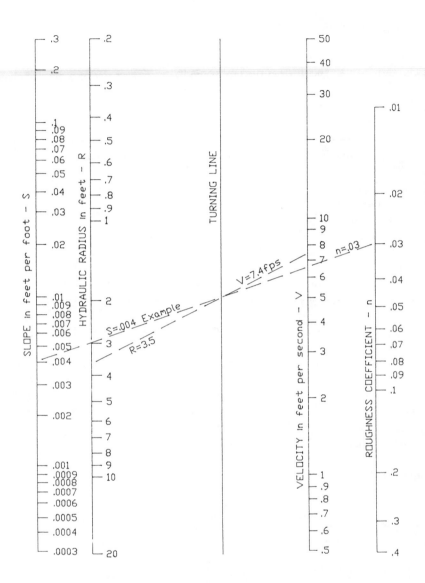

Figure 7-7 Manning Formula For Travel Time

Source: AASHTO Model Drainage Manual, 1991

Equation 7.9 is most applicable for natural basins with well defined channels, bare earth overland flow or flow in mowed channels. Rossmiller (1980) gives adjustment factors for other conditions based on literature values:

- For general overland flow and flow in natural grassed channels multiply t_c by 2.
- For concrete or asphalt surfaces multiply t_c by 0.4.
- For concrete channels multiply t_c by 0.2.

Kerby (1959) developed an estimate of the overland flow portion of the inlet time as:

$$t_c = 0.83(N_kL/(S^{0.5}))^{0.467}$$ (7.10)

Where: t_c = time of concentration, min
         L = straight line length from the top of the basin to the point of a defined channel, ft
         S = slope calculated similar to the Kirpich equation, ft/ft
         N_k = a coefficient of roughness given in Table 7-11.

Table 7-11 Values Of N_k For The Kerby Formula

N_k	Surface Type
0.02	smooth impervious surface
0.10	smooth bare packed soil, free of stones
0.20	poor grass, cultivated row crops or moderately rough bare surfaces
0.40	pasture or average grass cover
0.60	deciduous timberland
0.80	conifer timberland, deciduous timberland with deep forest litter or dense grass cover

Rainfall Intensity

The rainfall intensity (I) is the average rainfall rate in inches/hr for a duration equal to the time of concentration for a selected return period. Once a particular return period has been selected for design and a time of concentration calculated for the drainage area, the rainfall intensity can be determined from intensity-duration-frequency curves. Figure 7-8 gives an example of IDF curves for Raleigh, North Carolina. Rational Method calculations are made easier by fitting a curve of the form given in equation 7.7 to the IDF information.

Drainage Area

The choice of drainage area may, on the surface, appear simple. However there are a number of details to consider such as:
• existing and planned flow directions in key areas such as along lot lines, at street intersections, flow bypass points, etc.
• impacts of regrading on flow direction and speed;
• impacts of drainage structures on various flow frequencies;
• the use of both pervious and impervious areas in the area calculation or only directly connected imperious areas;
• the use of several or one land use composite calculation; and
• the flow directions and amounts for flows in excess of the design storm.

Runoff Coefficient

The runoff coefficient (C) is the variable of the Rational Method least susceptible to precise determination and requires judgment and understanding on the part of the designer. While engineering judgment will always be required in the selection of runoff coefficients, a typical coefficient represents the integrated effects of many drainage basin parameters. The following discuss-

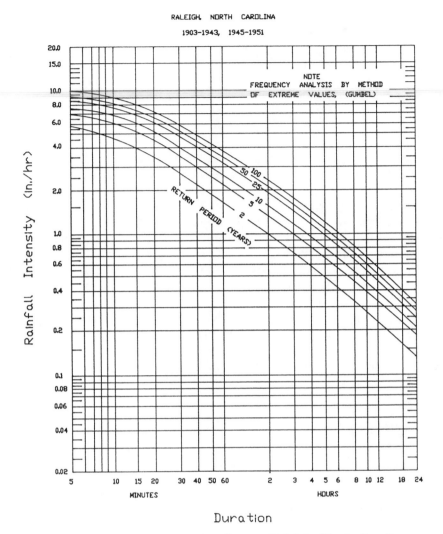

Figure 7-8 **Intensity Duration Curve - Raleigh, North Carolina**

Source: Weather Bureau Tech. Paper 25

ion considers C factor modifications based on the effects of design frequency, soil groups, land use, and average land slope.

The coefficients given in Table 7-15 are applicable for storms of 5-year to 10-year frequencies. Less frequent, higher intensity storms will require modification of the coefficient because infiltration and other losses have a proportionally smaller effect on runoff (Wright-McLaughlin, 1969). The adjustment of the Rational Method for use with major storms can be made by multiplying the right side of the Rational Formula by a frequency factor C_f. The Rational Formula now becomes:

$$Q = CC_f IA \qquad\qquad (7.11)$$

The C_f values are listed in Table 7-12. The product of C_f times C should not exceed 1.0.

Table 7-12 Frequency Factors For Rational Formula

Recurrence Interval (years)	C_f
25	1.1
50	1.2
100	1.25

Three methods for determining the runoff coefficient are presented based on soil groups and land slope (Tables 7-13 and 7-14), land use (Table 7-15), and a composite coefficient for complex watersheds (Table 7-16).

Table 7-14 gives the recommended coefficient (C) of runoff for pervious surfaces by selected hydrologic soil groupings and slope ranges. From this table the C values for non-urban areas such as forest land, agricultural land, and open space can be determined. Soil properties influence the relationship between runoff and rainfall since soils have differing rates of infiltration. Infiltration is the movement of water through the soil surface into the soil. Based on infiltration rates, the Soil Conservation Service (SCS) has divided soils into four hydrologic soil groups as follows:

Group A   Soils having a low runoff potential due to high infiltration rates. These soils consist primarily of deep, well drained sands and gravels.

Group B   Soils having a moderately low runoff potential due to moderate infiltration rates. These soils consist primarily of moderately deep to deep, moderately well to well drained soils with moderately fine to moderately coarse textures.

Group C   Soils having a moderately high runoff potential due to slow infiltration rates. These soils consist primarily of soils in which a layer exists near the surface that impedes the downward movement of water or soils with moderately fine to fine texture.

Group D   Soils having a high runoff potential due to very slow infiltration rates. These soils consist primarily of clays with high swelling potential, soils with permanently high water tables, soils with a claypan or clay layer at or near the surface, and shallow soils over nearly impervious parent material.

As an example, a list of soils for Orange County, North Carolina and their hydrologic classification is presented in Table 7-13.

As the slope of the drainage basin increases, the selected C value should also increase. This is caused by the fact that as the slope of the drainage area increases, the velocity of overland and channel flow will increase allowing less opportunity for water to infiltrate the ground surface. Thus, more of the rainfall will become runoff from the drainage area.

It is often desirable to develop a composite runoff coefficient based on the percentage of different types of surface in the drainage area. Composites can be made with Tables 7-14 and 7-15. At a more detailed level composites can be made with Table 7-14 and the coefficients with respect to surface type given in Table 7-16. The composite procedure can be applied to an entire drainage area or to typical "sample" blocks as a guide to selection of reasonable values of the coefficient for an entire area.

Table 7-13 Hydrologic Soils Groups For Orange County, North Carolina

Series Name	Hydrologic Groups	Series Name	Hydrologic Groups
Altavista	C	Herndon	B
Appling	B	Hiwassee	B
Cecil	B	Iredell	D
Chewacla	C	Lignum	C
Congaree	B	Louisburg	B
Creedmoor	C	Orange	D
Enon	C	Tatum	C
Georgeville	B	Vance	C
Goldston	C	Wedowee	D

Source: Soil Conservation Service

Table 7-14 Recommended Coefficient Of Runoff For Pervious Surfaces By Selected Hydrologic Soil Groupings And Slope Ranges

Slope	A	B	C	D
Flat (0-1%)	0.04-0.09	0.07-0.12	0.11-0.16	0.15-0.20
Average (2-6%)	0.09-0.14	0.12-0.17	0.16-0.21	0.20-0.25
Steep	0.13-0.18	0.18-0.24	0.23-0.31	0.28-0.38

Source: Storm Drainage Design Manual, Erie and Niagara Counties Regional Planning Board.

It should be remembered that the Rational Method assumes that all land uses within a drainage area are uniformly distributed throughout the area. If it is important to locate a specific land use within the drainage area then another hydrologic method should be used where hydrographs can be generated and routed through the drainage system.

Perhaps the most complex formulation for C factor estimation is given by Rossmiller (1980) as:

$$C = 7.2(10)^{-7}CN^3 Cr(Cf)^{Cs}(Cn)^{Ci}(Cm) \qquad (7.12)$$

Where:
$Cr = RI^{0.05}$
$Cs = (-S)^{0.2}$
$Ci = 0.15\text{-}i$
$CN = $ SCS Curve Number
$S = $ Average watershed slope, %, (4% = 4)
$Imp = $ Watershed imperviousness as a decimal (20% = 0.20)
$Cf = (0.01CN)^{0.6}$
$Cn = (0.001CN)^{1.48}$
$Cm = ((Imp+1)/2)^{0.7}$
$RI = $ Recurrence interval, yrs
$i = $ Rainfall intensity, in./hr

The equation takes into account, respectively, frequency, slope, rainfall intensity, modifier for slope factor, modifier for intensity factor, and surface roughness.

Table 7-15 Recommended Coefficient Of Runoff Values For Various Selected Land Uses

Description of Area		Runoff Coefficients
Business:	Downtown areas	0.70-0.95
Neighborhood areas		0.50-0.70
Residential:	Single-family areas	0.30-0.50
	Multi units, detached	0.40-0.60
	Multi units, attached	0.60-0.75
	Suburban	0.25-0.40
Residential (1/2 acre lots or more)		0.30-0.45
Apartment dwelling areas		0.50-0.70
Industrial:	Light areas	0.50-0.80
	Heavy areas	0.60-0.90
Parks, cemeteries		0.10-0.25
Playgrounds		0.20-0.40
Railroad yard areas		0.20-0.40
Unimproved areas		0.10-0.30

Source: Hydrology, Federal Highway Administration, HEC No. 19, 1984

Table 7-16 Coefficients For Composite Runoff Analysis

Surface		Runoff Coefficients
Street:	Asphalt	0.70-0.95
	Concrete	0.80-0.95
Drives and walks		0.75-0.85
Roofs		0.75-0.95

Source: Hydrology, Federal Highway Administration, HEC No. 19, 1984

Example Problem - Rational Method

Following is an example problem which illustrates the application of the Rational Method to estimate peak discharges. Preliminary estimates of the maximum rate of runoff are needed at the inlet to a culvert for a 25-yr and 100-yr return period. From a topographic map and field survey, the area of the drainage basin upstream from the point in question was measured to be 18 acres. In addition the following data were measured:

Average overland slope = 2.0%   Length of overland flow = 50 ft
Length of main basin channel = 1300 ft   Slope of channel - .018 ft/ft = 1.8%
Roughness coefficient (n) of channel was estimated to be 0.090

From existing land use maps, land use for the drainage basin was estimated to be:

Residential (single family) = 80%   Graded - sandy soil, 3% slope = 20%

From existing land use maps, the land use for the overland flow area at the head of the basin was estimated to be:

Lawn - sandy soil, 2% slope = 100%

A runoff coefficient (C) for the overland flow area is determined from Table 7-15 to be .15.

From Figure 7-6 with an overland flow length of 50 ft, slope of 2.0 % and a C of .15, the overland flow time is 10 min. Channel flow velocity is determined from Figure 7-7 to be 3.5 ft/s (n = 0.090, R = 1.97 and S = .018). Therefore,

Flow Time = (1300 ft)/[(3.5 ft/s)(60 s/min)] = 6.2 min and t_c = 10+6.2 = 16.2 min = 16 min

From Table 7-2 with duration equal to 16 min (values obtained by linear interpolation between values for 15 and 30 minutes),

I_{25} (25-yr return period)   =   5.86 in./hr I_{100} (100-yr return period)   =   7.18 in./hr

A weighted runoff coefficient (C) for the total drainage area is determined in the following table by utilizing the values from Table 7-15.

Land Use	(1) Percent of Total Land Area	(2) Runoff Coefficient	(3) Weighted Runoff Coefficient*
Residential (single family)	.80	.50	.40
Graded area	.20	.30	.06
Total Weighted Runoff Coefficient			.46

* Column 3 equals column 1 multiplied by column 2.
From the Rational Method equation:

Q_{25} = C_fCIA = 1.1 X .46 X 5.86 X 18 = 53 cfs
Q_{100} = C_fCIA = 1.25 X .46 X 7.18 X 18 = 74 cfs

These are the estimates of peak runoff for a 25-yr and 100-yr design storm for the given drainage basin.

7.9 SCS Hydrologic Methods

The grouping of methods known as the SCS methodology were derived and modified throughout the last half century (Rallison, 1980). The techniques developed by the U. S. Soil Conservation Service for calculating rates of runoff require the same basic data as the Rational Method: drainage area, a runoff factor, time of concentration, and rainfall. The SCS method, however, is more sophisticated in that it considers also the time distribution of the rainfall, the initial rainfall losses to interception and depression storage, and an infiltration rate that decreases during the course of a storm. With the SCS method, the direct runoff can be calculated for any storm, either real or fabricated, by subtracting infiltration and other losses from the rainfall to obtain the precipitation excess. Details of the methodology can be found in the <u>SCS National Engineering Handbook, Section 4.</u>

The SCS method can be used to find peak flow, outflow hydrographs, or unit hydrographs for use with rainfall excess hyetographs also generated from a modification of the SCS method. The SCS method includes the following basic equations and concepts:

- Runoff volume can be found using the basic SCS rainfall-runoff equation.
- "Curve Numbers" used in determination of rainfall losses are developed based on soils, land uses and antecedent moisture assumptions within the drainage area.
- Standard dimensionless storm distributions are used based on geographic location throughout the United States.
- A lag time is determined based on calculation of the time of concentration to the study point (similar to the Rational Method).

From some or all of this basic information peaks, hydrographs, storms and unit hydrographs can be found:

- Peak flows can be estimated from the SCS peak flow methods.
- Total outflow hydrographs can be developed through the SCS Tabular method or any of a number of computer models utilizing the SCS unit hydrograph.
- Using one of the rainfall distributions and known rainfall totals, total and excess rainfall amounts are determined using the SCS rainfall-runoff relationship.
- Using the curvilinear (or triangular approximation) dimensionless unit hydrographs, dimensional unit hydrographs are developed for the drainage area based on drainage area and lag time calculations.

Basic Equations And Concepts

The following discussion outlines the equations and basic concepts used in the SCS method.

<u>Rainfall-Runoff Equation</u> - A relationship between accumulated rainfall and accumulated runoff was derived by SCS from experimental plots for numerous soils and vegetative cover conditions. Data for land-treatment measures, such as contouring and terracing, from experimental watersheds were included. The equation was developed mainly for small rural watersheds for which only daily rainfall and watershed data are ordinarily available. It was developed from recorded storm data that included the total amount of rainfall in a calendar day but not its distribution with respect to time. The SCS runoff equation is therefore primarily a method of estimating direct runoff from 24-hour or 1-day storm rainfall, though shorter durations are routinely used. The basic proportionality relationship, the center of the methodology, is:

$$F/S = Q/(P-I_a) \qquad\qquad (7.13)$$

Where: Q =   accumulated direct runoff, in.
$\quad\quad\quad P$ =   accumulated rainfall (potential maximum runoff), in.
$\quad\quad\quad I_a$ =   initial abstraction including surface storage, interception, and infiltration prior to runoff, in.
$\quad\quad\quad S$ =   potential maximum retention, in.
$\quad\quad\quad F$ =   retained rainfall volume, in.

By substituting $F = (P-I_a) - Q$ into equation 7.13 the following rainfall-runoff relationship emerges:

$$Q = (P - I_a)^2 \ / \ (P - I_a) + S \qquad\qquad (7.14)$$

A relationship between I_a and S was developed from experimental watershed data, which removes the necessity for estimating I_a for common usage. The empirical relationship used in the SCS runoff equation is:

$$I_a = 0.2S \qquad\qquad (7.15)$$

Research has shown that the value of $0.2S$ is not correct for all circumstances. There is much scatter in the data. Others have suggested $0.1S$ better fits even the SCS data. Other values can be used for $0.2S$ though care must be taken to use equations for which this assumption has not been included. Bosznay (1989) avoided the need for the $0.2S$ assumption (and the use of curve numbers) by seeking to determine S directly and plotting Q directly as a function of S for various assumed values of $(P - I_a)$.

Substituting $0.2S$ for I_a in equation 7.14, the SCS rainfall-runoff equation becomes:

$$Q = (P - 0.2S)^2 \ / \ (P + 0.9S) \qquad\qquad (7.16)$$

SCS has also defined a physically based "curve number" such that:

$$CN = 1000/(10+S) \ \text{ or} \qquad\qquad (7.17)$$
$$S = 1000/CN - 10 \qquad\qquad (7.18)$$

Curve numbers vary between about 40 and 99, the higher the curve number the more runoff per unit of rainfall. The curve numbers reflect all physical aspects of the land use and antecedent moisture within the basin. Curve numbers are discussed in more detail below.

Figure 7-10 shows a graphical solution of equation 7-16 which enables the precipitation excess from a storm to be obtained if the total rainfall and watershed curve number are known. For example, 4.8 inches of direct runoff would result if 6.5 inches of rainfall occurs on a watershed with a curve number of 85.

The family of curves in Figure 7-10 can be collapsed to a single curve through manipulation of equation 7.16 as (Hawkins, 1979):

$$Q/S = (P/S-0.2)^2/(P/S+0.8) \qquad\qquad (7.19)$$

It has been stated that this equation has little basis in physical reality but it seems to work well for a variety of applications throughout the United States except in portions of the arid southwest (Hjelmfelt, 1991). For very dry climates the SCS methodology has often not appeared applicable and often gives too high a runoff for large rainfall events. In these conditions reliance may be placed on regression type equations available for much of the area (Daly, 1981).

Boughton (1987) has demonstrated that the equation is a special case of a spatially varied, saturation overflow model and has derived curves similar in form to those in Figure 7-9 from purely physical reasoning. Figure 7-9 illustrates this reasoning. The watershed is considered to be a series of "buckets" each of which produce runoff after it has become filled. The buckets are of different sizes. Each size range of buckets begins to produce runoff at a different point in the storm. When all the buckets are full runoff equals rainfall, and the rainfall-runoff line in Figure 7-9 approaches 45 degrees. For the simple case of two buckets as illustrated in the figure, line A results when bucket A is full and begins to produce runoff. When bucket B is full the runoff begins to equal the rainfall and the whole watershed is producing runoff.

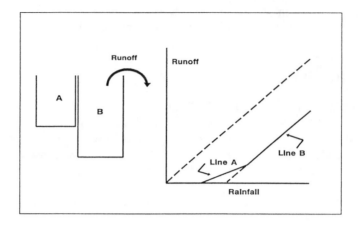

Figure 7-9 Spatial Variability Explanation of SCS Curves
Source: Boughton, W. C., "Evaluation of Partial Areas of Watershed Runoff", J. of Ir. and Drain. Engr., ASCE, Vol. 113, No. 3, August, 1987. Reproduced by permission of ASCE.

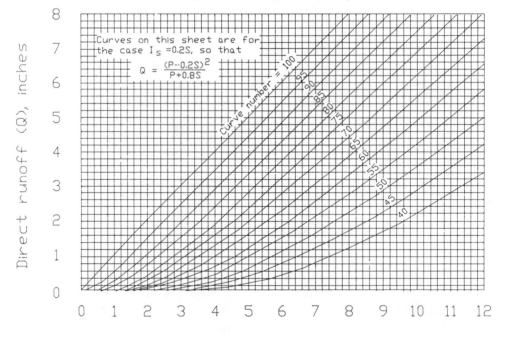

Figure 7-10 SCS Solution Of The Runoff Equation

Source: SCS, TR-55, 1986

For a real watershed there are millions of "buckets" of all different sizes. Some runoff begins almost immediately. The watershed response line curves slowly toward the 45 degree dotted line as more and more rainfall occurs. The line shifts to the right for dryer initial conditions, for greater initial losses assumed or for more porous soil. For many different soil and land use conditions (curve numbers) a family of curves would result eventuating in a figure similar to Figure 7-10, the SCS Solution to the Runoff Equation.

Hjelmfelt (1980) has shown that the equation is an effective "frequency transformer"; transforming rainfall to runoff of the same frequency. Regardless of any controversy or questions surrounding the method, it has received wide attention and application across the United States and foreign countries.

Curve Number - In hydrograph applications, runoff is often referred to as rainfall excess or effective rainfall - all defined as the amount by which rainfall exceeds the capability of the land to infiltrate or otherwise retain the rain water. The principal physical watershed characteristics affecting the relationship between rainfall and runoff are land use, land treatment, soil types, and land slope.

Land use is the watershed cover, and it includes both agricultural and nonagricultural uses. Items such as type of vegetation, water surfaces, roads, roofs, etc. are all part of the land use. Land treatment applies mainly to agricultural land use, and it includes mechanical practices such as contouring or terracing and management practices such as rotation of crops.

The SCS method uses a combination of soil conditions and land-use (ground cover) to assign a runoff factor to an area. These runoff factors, called runoff curve numbers (CN), indicate the runoff potential of an area when the soil is not frozen. The higher the CN, the higher is the runoff potential.

Soil properties influence the relationship between rainfall and runoff by affecting the rate of infiltration. The SCS has divided soils into four hydrologic soil groups based on infiltration rates (Groups A, B, C, and D). These groups were previously described for the Rational Method.

Consideration should be given to the effects of urbanization on the natural hydrologic soil group. If heavy equipment can be expected to compact the soil during construction or if grading will mix the surface and subsurface soils, appropriate changes should be made in the soil group selected. Also runoff curve numbers vary with the antecedent soil moisture conditions, defined as the amount of rainfall occurring in a selected period preceding a given storm. In general, the greater the antecedent rainfall, the more direct runoff there is from a given storm. A five-day period is used as the minimum for estimating antecedent moisture conditions. Antecedent soil moisture conditions also vary during a storm; heavy rain falling on a dry soil can change the soil moisture condition from dry to average to wet during the storm period.

The following pages give a series of tables related to runoff factors. The first tables (Tables 7-18 - 7-21) give curve numbers for various land uses. These tables are based on an average antecedent moisture condition i.e., soils that are neither very wet nor very dry when the design storm begins. Curve numbers should be selected only after a field inspection of the watershed and a review of zoning and soil maps.

Several factors, such as the percentage of impervious area and the means of conveying runoff from impervious areas to the drainage system, should be considered in computing CN for urban areas. For example, do the impervious areas connect directly to the drainage system, or do they outlet onto lawns or other pervious areas where infiltration can occur? The curve number values given in Table 7-18 are based on directly connected impervious area. An impervious area is considered directly connected if runoff from it flows directly into the drainage system. It is also considered directly connected if runoff from it occurs as concentrated shallow flow that runs over pervious areas and then into a drainage system. It is possible that curve number values from urban areas could be reduced by not directly connecting impervious surfaces to the drainage system.

Antecedent Moisture Conditions - Most of the curve numbers given are for what might be termed "normal" conditions. Dryer conditions will work to increase the amount of available storage effectively sliding the curve in Figure 7-10 to the right (i.e. lower curve number) while wetter conditions will use up some of the available storage resulting in a higher curve number for a specific storm. Table 7-22 gives conversion factors to convert average curve numbers to wet and dry curve numbers. The SCS has defined three wetness conditions. Table 7-23 gives a common definition of the antecedent conditions for the three classifications.

If S is treated as a random variable then it has been shown (Hjelmfelt, 1991) that the dry and wet conditions correspond roughly to the 10 and 90 percent exceedance frequency. Or they correspond roughly to the lower and upper envelope CN curves in a plot of actual rainfall versus runoff values. The normal moisture condition would plot through the middle of the scatter of points for any given watershed.

The actual CN for any storm event can be calculated from the quadratic formulation of the rainfall-runoff equation (Hawkins, 1979):

$$CN = 100/(1 + 1/2(P+2Q-(4Q^2 + 5PQ)^{1/2}))$$ (7.20)

The following discussion will give some guidance for adjusting curve numbers for different types of impervious areas. Imperviousness can be estimated by laying a grid over an aerial photograph and selecting a minimum of 200 sampling points at grid intersections (Cochran, 1963).

Connected Impervious Areas - Urban CN's given in Table 7-18 were developed for typical land use relationships based on specific assumed percentages of impervious area. These CN values were developed on the assumptions that:
(1) pervious urban areas are equivalent to pasture in good hydrologic condition, and
(2) impervious areas have a CN of 98 and are directly connected to the drainage system.
Some assumed percentages of impervious area are shown in Table 7-18.

If all of the impervious area is directly connected to the drainage system, but the impervious area percentages or the pervious land use assumptions in Table 7-18 are not applicable, use Figure 7-11 to compute a composite CN. For example, Table 7-18 gives a CN of 70 for a 1/2-acre lot in hydrologic soil group B, with an assumed impervious area of 25 percent. However, if the lot has 20 percent impervious area and a pervious area CN of 61, the composite CN obtained from Figure 7-11 is 68. The CN difference between 70 and 68 reflects the difference in percent impervious area.

Unconnected Impervious Areas - Runoff from these areas is spread over a pervious area as sheet flow. To determine CN when all or part of the impervious area is not directly connected to the drainage system, (1) use Figure 7-12 if the total impervious area is less then 30 percent or (2) use Figure 7-11 if the total impervious area is equal to or greater than 30 percent, because the absorptive capacity of the remaining pervious areas will not significantly affect runoff.

When impervious area is less than 30 percent, obtain the composite CN by entering the right half of Figure 7-12 with the percentage of total impervious area and the ratio of total unconnected impervious area to total impervious area. Then move left to the appropriate pervious CN and read down to find the composite CN. For example, for a 1/2-acre lot with 20 percent total impervious area (75 percent of which is unconnected) and pervious CN of 61, the composite CN from Figure 7-12 is 66. If all of the impervious area is connected, the CN (from Figure 7-11) would be 68.

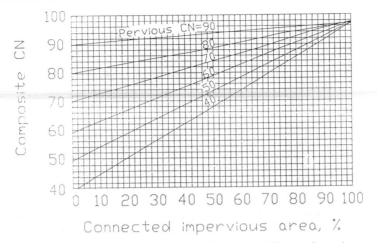

Figure 7-11 Composite CN With Connected Impervious Areas

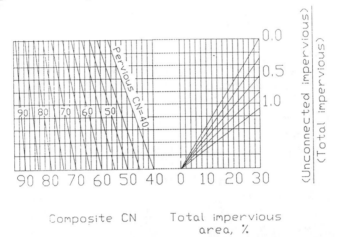

Figure 7-12 Composite CN With Unconnected Impervious Areas
(Total Impervious Area Less Than 30%)

Source: SCS, TR-55, 1986

Composite Curve Numbers - When a drainage area has more than one land use, a composite curve number can be calculated and used in the analysis. It should be noted that when composite curve numbers are used, the analysis does not take into account the location of the specific land uses but sees the drainage area as a uniform land use represented by the composite curve number.

Composite curve numbers for a drainage area can be calculated by entering the required data into a table such as the one presented in Table 7-17.

Table 7-17 Composite Curve Number Calculations

(1) Land Use	(2) Curve Number	(3) Area	(4) % of Total Area	(5) Composite Curve Number (Col 2 X Col 4)

Table 7-18 Runoff Curve Numbers For Urban Areas[1]

Cover description		Curve numbers for hydrologic soil groups			

Cover type and hydrologic condition		A	B	C	D
Fully developed urban areas (vegetation established)					
Open space (lawns, parks, golf courses, cemeteries, etc.)[3]					
Poor condition (grass cover < 50%)		68	79	86	89
Fair condition (grass cover 50% to 75%)		49	69	79	84
Good condition (grass cover > 75%)		39	61	74	80
Impervious areas:					
Paved parking lots, roofs, driveways, etc. (excluding right-of-way)		98	98	98	98
Streets and roads:					
Paved; curbs and storm drains (excluding right-of-way)		98	98	98	98
Paved; open ditches (including right-of-way)		83	89	92	93
Gravel (including right-of-way)		76	85	89	91
Dirt (including right-of-way)		72	82	87	89
Western desert urban areas:					
Natural desert landscaping (pervious areas only)		63	77	85	88
Artificial desert landscaping (impervious weed barrier, desert shrub with 1- to 2-inch sand or gravel mulch and basin borders)		96	96	96	96
Urban districts:	Average Percent impervious area[2]				
Commercial and business	85	89	92	94	95
Industrial	72	81	88	91	93
Residential districts by average lot size:					
1/8 acre or less (town houses)	65	77	85	90	92
1/4 acre	38	61	75	83	87
1/3 acre	30	57	72	81	86
1/2 acre	25	54	70	80	85
1 acre	20	51	68	79	84
2 acres	12	46	65	77	82
Developing urban areas					
Newly graded areas (pervious areas only, no vegetation)		77	86	91	94
Idle lands (CN's are determined using cover types similar to those in Table 7-20).					

Source: SCS, TR-55, 1986

[1] Average runoff condition, and $I_a = 0.2S$

[2] The average percent impervious area shown was used to develop the composite CN's. Other assumptions are as follows: impervious areas are directly connected to the drainage system, impervious areas have a CN of 98, and pervious areas are considered equivalent to open space in good hydrologic condition. If the impervious area is not connected, the SCS method has an adjustment to reduce the effect.

[3] CN's shown are equivalent to those of pasture. Composite CN's may be computed for other combinations of open space cover type.

Table 7-19 Cultivated Agricultural Land[1]

Cover description			Curve numbers for hydrologic soil group			
Cover type	Treatment[2]	Hydrologic condition[3]	A	B	C	D
Fallow	Bare soil	-	77	86	91	94
	Crop residue	Poor	76	85	90	93
	cover (CR)	Good	74	83	88	90
Row Crops	Straight row (SR)	Poor	72	81	88	91
		Good	67	78	85	89
	SR + CR	Poor	71	80	87	90
		Good	64	75	82	85
	Contoured (C)	Poor	70	79	84	88
		Good	65	75	82	86
	C + CR	Poor	69	78	83	87
		Good	64	74	81	85
	Contoured &	Poor	66	74	80	82
	terraced (C & T)	Good	62	71	78	81
	C&T + CR	Poor	65	73	79	81
		Good	61	70	77	80
	Small grain SR	Poor	65	76	84	88
		Good	63	75	83	87
	SR + CR	Poor	64	75	83	86
		Good	60	72	80	84
	C	Poor	63	74	82	85
		Good	61	73	81	84
	C + CR	Poor	62	73	81	84
		Good	60	72	80	83
	C&T	Poor	61	72	79	82
		Good	59	70	78	81
	C&T + CR	Poor	60	71	78	81
		Good	58	69	77	80
	Close-seeded SR	Poor	66	77	85	89
	or broadcast	Good	58	72	81	85
	Legumes or C	Poor	64	75	83	85

Rotation	Good	55	69	78	83
Meadow C&T	Poor	63	73	80	83
	Good	51	67	76	80

Source: SCS, TR-55, 1986

[1] Average runoff condition, and $I_a = 0.2S$.

[2] Crop residue cover applies only if residue is on at least 5% of the surface throughout the year.

[3]Hydrologic condition is based on a combination of factors that affect infiltration and runoff, including (a) density and canopy of vegetative areas, (b) amount of year-round cover, (c) amount of grass or closed-seeded legumes in rotations, (d) percent of residue cover on the land surface (good > 20%), and (e) degree of roughness.

Poor: Factors impair infiltration and tend to increase runoff.

Good: Factors encourage average and better than average infiltration and tend to decrease runoff.

Table 7-20 Other Agricultural Lands[1]

Cover description		Curve numbers for hydrologic soil group			
Cover type	Hydrologic condition	A	B	C	D
Pasture, grassland, or	Poor	68	79	86	89
range-continuous for-	Fair	49	69	79	84
age for grazing[2]	Good	39	61	74	80
Meadow--continuous grass, protected from grazing and generally mowed for hay	--	30	58	71	78
Brush--brush-weed-grass	Poor	48	67	77	83
mixture with brush the	Fair	35	56	70	77
major element[3]	Good	[4]30	48	65	73
Woods--grass	Poor	57	73	82	86
combination (orchard	Fair	43	65	76	82
or tree farm)[5]	Good	32	58	72	79
Woods[6]	Poor	45	66	77	83
	Fair	36	60	73	79
	Good	[4]30	55	70	77
Farmsteads--buildings, lanes,driveways, and surrounding lots	--	59	74	82	86

Source: SCS, TR-55, 1986

[1] Average runoff condition, and $I_a = 0.2S$

[2] Poor: < 50% ground cover or heavily grazed with no mulch

Fair: 50 to 75% ground cover and not heavily grazed
Good: > 75% ground cover and lightly or only occasionally grazed
[3] Poor: < 50% ground cover
Fair: 50 to 75% ground cover
Good: > 75% ground cover
[4] Actual curve number is less than 30; use CN = 30 for runoff computations.
[5] CN's shown were computed for areas with 50% grass (pasture) cover. Other combinations of conditions may be computed from CN's for woods and pasture.
[6] Poor: Forest litter, small trees, and brush are destroyed by heavy grazing or regular burning.
Fair: Woods grazed but not burned, and some forest litter covers the soil.
Good: Woods protected from grazing, litter and brush adequately cover soil.

Table 7-21 Arid And Semiarid Rangelands[1]

Cover type	Hydrologic condition[2]	A[3]	B	C	D
Herbaceous--mixture of grass,	Poor		80	87	93
weeds, and low-growing brush,	Fair		71	81	89
with brush the minor element	Good		62	74	85
Oak-aspen--mountain brush	Poor		66	74	79
mixture of oak brush, aspen,	Fair		48	57	63
mountain mahogany, bitter	Good		30	41	48
brush, maple, and other brush					
Pinyon-juniper--pinyon,	Poor		75	85	89
juniper, or both; grass	Fair		58	73	80
understory	Good		41	61	71
Sagebrush with grass	Poor		67	80	85
understory	Fair		51	63	70
	Good		35	47	55
Desert shrub--major plants	Poor	63	77	85	88
include saltbush, greasewood,	Fair	55	72	81	86
creosote-bush, blackbrush,	Good	49	68	79	84
bursage, paloverde, mesquite, and cactus					

Source: SCS, TR-55, 1986

[1] Average runoff condition, and $I_a = 0.2S$
[2] Poor: < 30% ground cover (litter, grass, and brush overstory)
Fair: 30 to 70% ground cover Good: > 70% ground cover
[3] Curve numbers for Group A have been developed only for desert shrub

**Table 7-22 Conversion From Average Antecedent Moisture Conditions
To Dry And Wet Conditions**

CN For Average Conditions	Corresponding CN's For	
	Dry	Wet
100	100	100
95	87	98
90	78	96
85	70	94
80	63	91
75	57	88
70	51	85
65	45	82
60	40	78
55	35	74
50	31	70
45	26	65
40	22	60
35	18	55
30	15	50
25	12	43
15	6	30
5	2	13

Source: SCS, TP-149, 1973.

**Table 7-23 Rainfall Groups For Antecedent Soil Moisture Conditions
During Growing And Dormant Seasons**

Antecedent Condition	Growing Season Five-Day Antecedent Rainfall	Dormant Season Five-Day Antecedent Rainfall
Dry[1]	Less than 1.4 inches	Less than 0.5 inches
Average[2]	1.4 to 2.1 inches	0.5 to 1.1 inches
Wet[3]	Over 2.1 inches	Over 1.1 inches

[1] An optimum condition of watershed soils, where soils are dry but not to the wilting point, and when satisfactory plowing or cultivation takes place.
[2] The average case for annual floods.
[3] When a heavy rainfall, or light rainfall and low temperatures, have occurred during the five days previous to a given storm.
Source: SCS, TP-149, 1973.

Rainfall - The SCS method is based on a 24-hour rainfall distribution which has a shape dependent on the location within the United States. These storm time distributions have been prepared by the SCS from analysis of numerous rainfall records. Figures 7-13 and 7-14 show the distributions presently available for the SCS method. Table 7-24 gives the tabular values for the SCS Types I, IA, II and III distribution.

Shorter durations can be used by taking the most steeply sloped portion of the S-curve corresponding to the desired duration and distributing the rainfall accordingly. For example, a six-hour storm would take the center 6-hour portion (hours/values 9 - 0.1467 through 15 - 0.8538) of the 24-hour rainfall distribution.

**Figure 7-13 Approximate Geographic Boundaries For
SCS Rainfall Distributions**

Source: SCS, TR-55, 1986

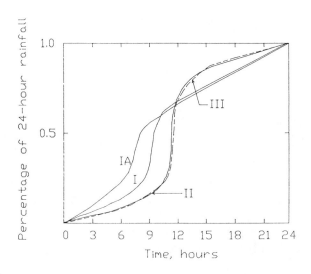

Figure 7-14 SCS 24-Hour Rainfall Distributions

Source: SCS, TR-55, 1986

Table 7-24 Ordinates Of The SCS
Type I, IA, II, Type III Precipitation Distributions

Storm Time (Hours)	I	AI	II	III	Storm Time (Hours)	I	IA	II	II
0.0	.000	.000	.0000	.000					
0.5	.008	.010	.0053	.005	12.5	.705	.683	.7351	.702
1.0	.017	.022	.0108	.010	13.0	.727	.701	.7724	.751
1.5	.026	.036	.0164	.015	13.5	.748	.719	.7989	.785
2.0	.035	.051	.0223	.020	14.0	.767	.736	.8179	.811
2.5	.045	.067	.0284	.026	14.5	.784	.753	.8380	.830
3.0	.055	.083	.0347	.032	15.0	.800	.769	.8538	.848
3.5	.065	.099	.0414	.037	15.5	.816	.785	.8676	.867
4.0	.076	.116	.0483	.043	16.0	.830	.800	.8801	.886
4.5	.087	.135	.0555	.050	16.5	.844	.815	.8914	.895
5.0	.099	.156	.0632	.057	17.0	.857	.830	.9019	.904
5.5	.122	.179	.0712	.065	17.5	.870	.844	.9115	.913
6.0	.125	.204	.0797	.072	18.0	.882	.858	.9206	.922
6.5	.140	.233	.0887	.081	18.5	.893	.871	.9291	.930
7.0	.156	.268	.0984	.089	19.0	.905	.884	.9371	.939
7.5	.174	.310	.1089	.102	19.5	.916	.896	.9446	.948
8.0	.194	.425	.1203	.115	20.0	.926	.908	.9519	.957
8.5	.219	.480	.1328	.130	20.5	.936	.920	.9588	.962
9.0	.254	.520	.1467	.148	21.0	.946	.932	.9653	.968
9.5	.303	.550	.1625	.167	21.5	.955	.944	.9719	.973
10.0	.515	.577	.1808	.189	22.0	.965	.956	.9777	.979
10.5	.583	.601	.2042	.216	22.5	.974	.967	.9836	.984
11.0	.624	.623	.2351	.250	23.0	.983	.978	.9892	.989
11.5	.654	.644	.2833	.298	23.5	.992	.989	.9947	.995
12.0	.682	.664	.6632	.500	24.0	1	1	1	1

Source: McCuen, 1982, Akan, 1993

Calculation Of Lag Time And Time Of Concentration - There are many ways to consider definitions of lag time and time of concentration. SCS warns against a simplified consideration of each of them as simply a calculation of a "theoretical velocity of a segment of water moving through a hydraulic system" (SCS, 1985). The lag time (T_l) is really a weighted time of concentration for each segment of the watershed. In hydrograph analysis it is the time from the center of mass of the rainfall to the peak of the outflow. (Note that the USGS defines lag time in most of its regression equations differently as the time from the center of mass of the rainfall to the center of mass of the outflow. This will lead to errors in calculation if methods are mixed unless the time is adjusted based on the shape of the outflow hydrograph). Time of concentration (T_c) is defined as the time of travel for water falling on the hydraulically most distant point in the watershed to the outlet.

The average slope within the watershed together with the overall length and retardance of overland flow are the major factors affecting the runoff rate through the watershed. However, differing storage characteristics for different storm events, seasonal changes and urbanization will all affect the lag time. Thus average values can sometimes lead to gross errors. Nevertheless, without adequate rainfall runoff information to estimate basin lag time, regression or synthetic methods are the only available recourse.

Travel Time And Time Of Concentration Empirical Equation - Lag (T_l) can be considered as a weighted time of concentration and is related to the physical properties of a watershed, such as area, length, and slope. The SCS derived the following empirical relationship between lag and time of concentration:

$$T_l = 0.6 \ T_c \tag{7.21}$$

The SCS equation to estimate lag is:

$$T_l = (l^{0.8} \ (S + 1)^{0.7}) \ / \ (1900 \ Y^{0.5}) \tag{7.22}$$

Where: T_l = lag, hrs
       l = length of mainstream to farthest divide, ft
       Y = average slope of watershed, %
       S = 1000/CN - 10
       CN = SCS curve number

The lag time can be corrected for the effects of urbanization by using Figures 7-15 and 7-16. The amount of modifications to the hydraulic flow length usually must be determined from topographic maps or aerial photographs following a field inspection of the area. The modification to the hydraulic flow length not only includes pipes and channels but also the length of flow in streets and driveways.

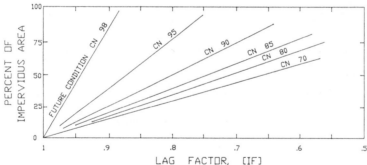

Figure 7-15 Factors For Adjusting Lag For Impervious Areas Within The Watershed

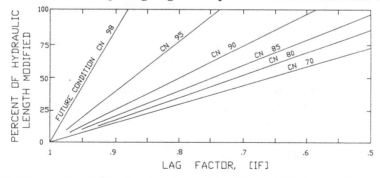

Figure 7-16 Factors For Adjusting Lag For Main Channel Hydraulically Improvement
Source: HEC-19

In small urban areas (less than 2000 acres), a curve number method can be used to estimate the time of concentration from watershed lag. In this method the lag for the runoff from an increment of excess rainfall can be considered as the time between the center of mass of the excess rainfall increment and the peak of its incremental outflow hydrograph.

This equation and the urbanization correction factors should be tested to determine if satisfactory results are obtained for local conditions.

After the lag time is adjusted for the effects of urbanization, equation 7.21 can be used to calculate the time of concentration for use in the SCS method.

If lag and/or time of concentration equations have been developed for use within your municipality, these equations can be substituted for equations 7.21 and 7.22 for use with the SCS Unit Hydrograph Method. Also, variations of the Manning's equation can be used to estimate travel time and time of concentration for the SCS Unit Hydrograph Method. The following describes the procedures and equations for using the Manning's equation.

Travel Time And Time Of Concentration Manning's Equation - In using the Manning's equation, travel time (T_t) is the time it takes water to travel from one location to another within a watershed, through the various components of the drainage system. Time of concentration (T_c) is computed by summing all the travel times for consecutive components of the drainage conveyance system from the hydraulically most distant point of the watershed to the point of interest within the watershed. Following is a discussion of related procedures and equations (TR-55, 1986).

Water moves through a watershed as sheet flow, shallow concentrated flow, open channel flow, or some combination of these. The type that occurs is a function of the conveyance system and is best determined by field inspection. The travel time is the ratio of flow length to flow velocity:

$$T_t = L/(3600V) \tag{7.23}$$

Where: T_t = travel time, hr
  L = flow length, ft
  V = average velocity, ft/s
  3600 = conversion factor from seconds to hours

The time of concentration is the sum of T_t values for the various consecutive flow segments along the path extending from the hydraulically most distant point in the watershed to the point of interest.

$$T_c = T_{t1} + T_{t2} + ... T_{tm} \tag{7.24}$$

Where: T_c = time of concentration, hr
  m = number of flow segments

Sheet flow is flow over plane surfaces and it usually occurs in the headwater of streams. With sheet flow, the friction value (Manning's n) is an effective roughness coefficient that includes the effect of raindrop impact; drag over the plane surface; obstacles such as litter, crop ridges, and rocks; and erosion and transportation of sediment. These n values are for very shallow flow depths of about 0.1 foot or so. For sheet flow of less than 300 feet, use Manning's kinematic solution (Overton and Meadows, 1976) to compute T_t:

$$T_t = [0.42 \ (nL)^{0.8} \ / \ (P_2)^{0.5}(S)^{0.4}] \tag{7.25}$$

Where: T_t = travel time, min
  n = Manning roughness coefficient

   L =   flow length, ft
   P_2 =   2-year, 24-hour rainfall = 3.6 in.
   S =   slope of hydraulic grade line (land slope), ft/ft
   This simplified form of the Manning's kinematic solution is based on the following:
* shallow steady uniform flow,
* constant intensity of rainfall excess (rain available for runoff),
* rainfall duration of 24 hours, and
* minor effect of infiltration on travel time.

   Another approach is to use the kinematic wave equation. For details on using this equation consult the publication by R. M. Regan, A Nomograph Based On Kinematic Wave Theory For Determining Time Of Concentration For Overland Flow, Report Number 44, Civil Engineering Department, University of Maryland at College Park, 1971.

Table 7-25 Roughness Coefficients (Manning's n) For Sheet Flow

Surface Description		n
Smooth surfaces (concrete, asphalt, gravel, or bare soil)		0.011
Fallow (no residue)		0.05
Cultivated soils:	Residue cover $\leq$ 20%	0.06
	Residue cover > 20%	0.17
Grass:	Short grass prairie	0.15
	Dense grasses[1]	0.24
	Bermudagrass[2]	0.41
Range (natural)		0.13
Woods:[3]	Light underbrush	0.40
	Dense underbrush	0.80

[1]The n values are a composite of information by Engman (1986).
[2]Includes species such as weeping lovegrass, bluegrass, buffalo grass, blue grama grass, and native grass mixtures.
[3]When selecting n, consider cover to a height of about 0.1 ft. This is the only part of the plant cover that will obstruct sheet flow.

Source: SCS, TR-55, 1986

   After a maximum of 300 feet (100 feet in urban areas), sheet flow usually becomes shallow concentrated flow. The average velocity for this flow can be determined from Figure 7-17, in which average velocity is a function of watercourse slope and type of channel.
   Average velocities for estimating travel time for shallow concentrated flow can be computed from using Figure 7-17, or the following equations. These equations can also be used for slopes less then 0.005 ft/ft.

Unpaved	$V = 16.1345(S)^{0.5}$	(7.26)
Paved	$V = 20.3282(S)^{0.5}$	(7.27)

Where: V =   average velocity, ft/s
          S =   slope of hydraulic grade line (watercourse slope), ft/ft

These two equations are based on the solution of Manning's equation with different assumptions for n (Manning's roughness coefficient) and r (hydraulic radius, ft) for unpaved areas, n is 0.05 and r is 0.4; for paved areas, n is 0.025 and r is 0.2.

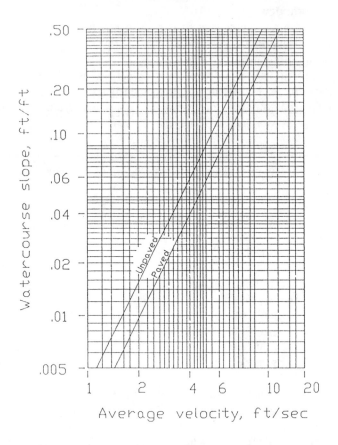

Figure 7-17 Average Velocities - Shallow Concentrated Flow

Source: SCS, TR-55, 1986

After determining average velocity using Figure 7-17 or equations 7.26 or 7.27, use equation 7.23 to estimate travel time for the shallow concentrated flow segment.

Open channels are assumed to begin where surveyed cross section information has been obtained, where channels are visible on aerial photographs, or where blue lines (indicating streams) appear on United States Geological Survey (USGS) quadrangle sheets. Manning's equation or water surface profile information can be used to estimate average flow velocity. Average flow velocity is usually determined for bank-full elevation.

$$\textbf{Manning's equation is } V = [1.49 \ (r)^{2/3} \ (s)^{1/2}] \ / \ n \qquad\qquad \textbf{(7.28)}$$

Where: V = average velocity, ft/s
           r = hydraulic radius (equal to a/p_w), ft
           a = cross sectional flow area, ft^2

p_w =   wetted perimeter, ft
s =   slope of the hydraulic grade line, ft/ft
n =   Manning's roughness coefficient for open channel flow

After average velocity is computed using equation 7.28, T_t for the channel segment can be estimated using equation 7.23. Velocity in channels should be calculated from the Manning equation. Cross sections from all channels that have been field checked should be used in the calculations. This is particularly true of areas below dams or other flow control structures.

Sometimes it is necessary to estimate the velocity of flow through a reservoir or lake at the outlet of a watershed. This travel time is normally very small and can be assumed as zero. If the travel time through the reservoir or lake is important to the analysis then the hydrograph should be routed through the storage facility. A reservoir can have an impact in reducing peak flows which can be accounted for by routing.

Following are some limitations of this procedure which should be taken into account.

- Manning's kinematic solution should not be used for sheet flow longer than 300 feet (100 feet in urban areas).
- In watersheds with storm drains, carefully identify the appropriate hydraulic flow path to estimate T_c. Storm drains generally handle only a small portion of a large event. The rest of the peak flow travels by streets, lawns, and so on, to the outlet.
- A culvert or bridge can act as a reservoir outlet if there is significant storage behind it. Detailed storage routing procedures should be used to determine the outflow through the culvert.

Drainage Area - The drainage area of a watershed is determined from topographic maps and field surveys. For large drainage areas it might be necessary to divide the area into sub-drainage areas to account for major land use changes, the location of storm water management facilities, to obtain analysis results at different points within the drainage area, and to route flows to points of interest. Also a field inspection of the existing or proposed drainage systems should be made to determine if the natural drainage divides have been altered.

Procedures, Tables And Figures

The following sections provide procedures for rainfall excess, peak flow calculation, hydrograph calculation and development of an SCS unit hydrograph. The procedures herein were taken directly or developed from the SCS Technical Release 55 (TR-55), and the national Engineering Handbook, Chapter 4 (NEH4) which present simplified procedures to calculate storm runoff volume, peak rate of discharges and hydrographs.

SCS Peak Discharges - The SCS peak discharge method is applicable for estimating the peak run-off rate from watersheds with homogeneous land uses. The following method is based on the results of computer analyses performed using TR-20, "Computer Program for Project Formulation - Hydrology" (SCS, 1983). The peak discharge equation is:

$$Q_p = q_u AQF_p \tag{7.29}$$

Where: Q_p =   peak discharge, cfs
q_u =   unit peak discharge, cfs/mi^2/in.
A =   drainage area, mi^2
Q =   runoff, in.
F_p =   pond and swamp adjustment factor

The input requirements for this method are as follows:
- T_c - hours
- Drainage area - mi^2
- Rainfall distribution type
- 24-hour design rainfall
- CN value
- Pond and Swamp adjustment factor (If pond and swamp areas are spread throughout the watershed and are not considered in the T_c computation, an adjustment is needed.)

Computations - Computations for the peak discharge method, using a Type II rainfall distribution, proceed as follows:

1. The 24-hour rainfall depth is determined from the following table for the selected return frequency.

Frequency	24-hour Rainfall
2-year	3.34 in.
5-year	4.62 in.
10-year	5.48 in.
25-year	6.67 in.
50-year	7.59 in.
100-year	8.50 in

2. The runoff curve number, CN, is estimated from Table 7-18 and direct runoff, Q, is calculated using equation 7.29.
3. The CN value is used to determine the initial abstraction, I_a, from Table 7-27, and the ratio I_a/P is then computed. (P = accumulated rainfall or potential maximum runoff.)
4. The watershed time of concentration is computed and is used with the ratio I_a/P to obtain the unit peak discharge, q_u, from Figure 7-20. If the ratio I_a/P lies outside the range shown in Figure 7-20, either the limiting values or another peak discharge method should be used.
5. The pond and swamp adjustment factor, F_p, is estimated from below:

Table 7-26 Adjustment Factor F_p For Pond And Swamp Areas That Are Spread Throughout The Watershed

Pond & Swamp Areas (%)	F_p
0	1.00
0.2	0.97
1.0	0.87
3.0	0.75
5.0	0.72

Source: SCS, TR-55, 1986

---

6. The peak runoff rate is computed using equation 7.29.

Table 7-27 I_a Values For Runoff Curve Numbers

Curve Number	I_a (in)	Curve Number	I_a (in)
40	3.000	70	.857
41	2.878	71	.817
42	2.762	72	.778
43	2.651	73	.740
44	2.545	74	.703
45	2.444	75	.667
46	2.348	76	.632
47	2.255	77	.597
48	2.167	78	.564
49	2.082	79	.532
50	2.000	80	.500
51	1.922	81	.469
52	1.846	82	.439
53	1.774	83	.410
54	1.704	84	.381
55	1.636	85	.353
56	1.571	86	.326
57	1.509	87	.299
58	1.448	88	.273
59	1.390	89	.247
60	1.333	90	.222
61	1.279	91	.198
62	1.226	92	.174
63	1.175	93	.151
64	1.125	94	.128
65	1.077	95	.105
66	1.030	96	.083
67	.985	97	.062
68	.941	98	.041
69	.899		

Source: SCS, TR-55, 1986

Limitations - The accuracy of the peak discharge method is subject to specific limitations, including the following.
1. The watershed must be hydrologically homogeneous and describable by a single CN value.
2. The watershed can have only one main stream, or if more than one, the individual branches must have nearly equal time of concentrations.
3. Hydrologic routing cannot be considered.
4. F_p applies only to areas located away from the main flow path.
5. Accuracy is reduced if the ratio I_a/P is outside the range given in Figures 7-18 - 7-21.
6. The weighted CN value must be greater than or equal to 40 and less than or equal to 98.
7. The same procedure should be used to estimate pre- and post-development time of concentration when computing pre- and post-development peak discharge.
8. The watershed time of concentration must be between 0.1 and 10 hours.

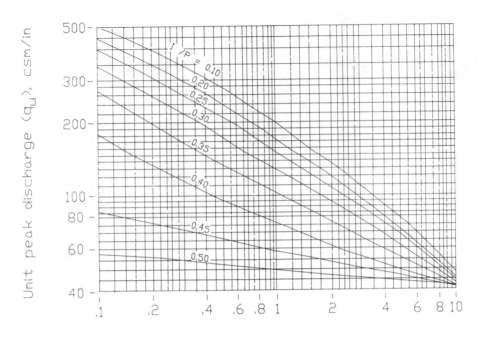

Figure 7-18 SCS Type I Unit Peak Discharge Graph

Source: SCS, TR-55, 1986

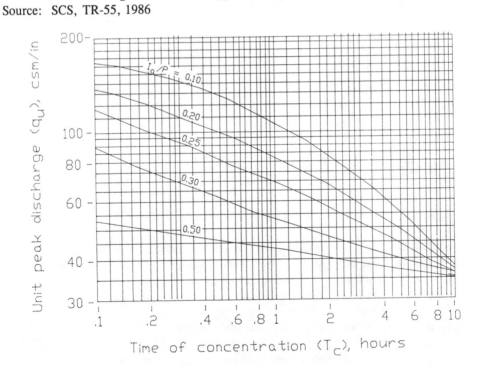

Figure 7-19 SCS Type IA Unit Peak Discharge Graph

Source: SCS, TR-55, 1986

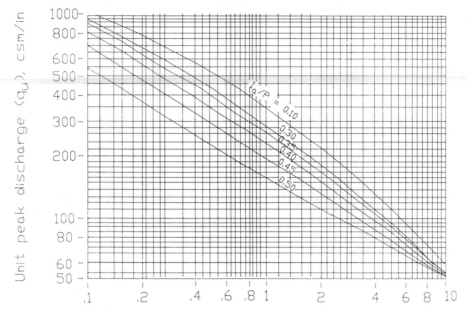

Figure 7-20 SCS Type II Unit Peak Discharge Graph
Source: SCS, TR-55, 1986

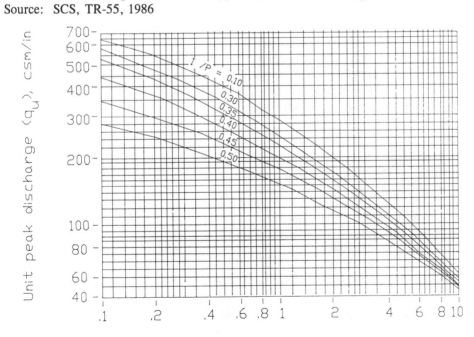

Figure 7-21 SCS Type III Unit Peak Discharge Graph
Source: SCS, TR-55, 1986

SCS Peak Discharge Example - Compute the 25-year peak discharge for a 50-acre wooded watershed which will be developed as follows:
- (1) Forest land - poor cover (hydrologic soil group B) = 10 acres
- (2) Forest land - poor cover (hydrologic soil group C) = 10 acres
- (3) Town house residential (hydrologic soil group B) = 20 acres
- (4) Industrial development (hydrological soil group C) = 10 acres
- Other data include: percentage of pond and swamp area = 0

1. Calculate rainfall excess:
    - The 25-year, 24-hour rainfall is 6.67 inches.
    - Composite weighted runoff coefficient is:

Dev. #	CN	% Total	Area	Composite CN
1	55	.20	10 ac	11.0
2	70	.20	10 ac	14.0
3	85	.40	20 ac	34.0
4	91	.20	10 ac	18.2
Total		1.00	50 ac	77.2 use 77

    - From Figure 7-10, Q = 4.0 in.
2. Calculate time of concentration
    - The hydrologic flow path for this watershed = 2,000 ft

Segment	Type of Flow	Length	Slope (%)
1	Overland n = .45	70 ft	2.0 %
2	Shallow channel	750 ft	1.7 %
3	Main channel*	1100 ft	0.20 %

* For the main channel, n = .025, width = 10 ft, depth = 2 ft, rectangular channel.
    - Segment 1 - Travel time from equation 7.25 with P_2 = 3.34 in.
        $T_t = [0.42(0.45 \times 60)^{.8}] / [(3.34)^{.5} (.035)^{.4}]$
        T_t = 12.3 min
    - Segment 2 - Travel time from Figure 7-17 and equation 7.23
        V = 2.6 ft/s (from Figure 7-17 or equation 7.23)
        T_t = 750 / 60 (2.6) = 4.8 min
    - Segment 3 - Using equation 7.28 and 7.23
        V = $(1.49/.025) (1.43)^{.67} (.002)^{.5}$ = 3.4 ft/s
        T_t = 1100 / 60 (3.4) = 5.4 min
    - T_c = 12.3 + 4.8 + 5.4 = 22.5 min (.38 hrs)
3. Calculate I_a/P for CN = 77 (from Step 1), I_a = .597 (Table 7-27)
        I_a/P = (.597 / 6.67) = .09
        (Note: Use I_a/P = .10 to facilitate use of Figures 7-18 - 7-21. Straight line interpolation could also be used.)
4. Estimate unit discharge q_u from Figure 7-20 = 600 cfs
5. Calculate peak discharge with F_p = 1 using equation 7.29
        Q_{25} = 600 (50/640) (4.0) (1) = 188 cfs.

SCS Hydrograph Generation

In addition to estimating the peak discharge, the SCS method can be used to estimate the entire hydrograph. The Soil Conservation Service has developed a Tabular Hydrograph procedure which can be used to generate the hydrograph for small drainage areas. The Tabular Hydrograph procedure uses unit discharge hydrographs which have been generated for a series of time of concentrations. SCS has developed tables which give the unit discharges (csm/in.) for different times of concentration for the four SCS rainfall distributions (USDA, TR-55,1986). Straight line interpolation can be used for time of concentrations and travel times between the values given in the tables. Since these tables do not account for reservoir routing through storage facilities and are more difficult to use if the drainage area is not homogeneously developed, their application within most municipal applications is limited. Due to this limitation and the availability of more appropriate hydrologic methods for municipal application, the tables and details of their use are not presented in this book but can be obtained from the SCS publication.

For drainage areas which are not homogeneous where hydrographs need to be generated from sub-areas and then routed and combined at a point downstream, the designer is referred to the procedures outlined by the SCS in the 1986 version of TR-55 available from the National Technical Information Service in Springfield, Virginia 22161. The catalog number for TR-55, "Urban Hydrology for Small Watersheds," is PB87-101580. These procedures do involve some channel and reservoir routing assumptions which may affect the accuracy of the hydrologic estimates. What would be more appropriate for most municipalities hydrologic studies which involve channel and/or reservoir routing would be to use one of the many hydrologic computer models available (e.g., HEC-1, TR-55)

Rainfall Excess Using SCS Methods

Calculation of rainfall excess using the SCS methodology involves a step-wise procedure. Combining equations 7.13 and 7.14 yields:

$$F = S(P - I_a)/(P - I_a + S) \tag{7.30}$$

Differentiating each side with respect to time yields a loss rate (dF/dt) and a rainfall intensity (dP/dt). Using a difference approximation yields:

$$(F_{i+1} - F_i)/d = S_2/(P_A - I_a + S)^2 * (P_{i+1} - P_i)/d \tag{7.31}$$

$$P > I_a$$

Where: F_i = loss at time step i
$\quad\quad\quad$ F_{i+1} = loss at time step i+1
$\quad\quad\quad$ d = time step
$\quad\quad\quad$ P_A = Average rainfall during time step = $(P_i + P_{i+1})/2$
$\quad\quad\quad$ P_i = rainfall rate at time step i
$\quad\quad\quad$ P_{i+1} = rainfall rate at time step i+1

The rainfall excess can then be calculated for each time step (a necessary step in the application of unit hydrographs) through the use of a table with column headings of: time, cumulative rainfall, cumulative abstractions $(I_a + F)$ from equation 7.30, cumulative excess rainfall, and excess rainfall hyetograph.

The steps in the development of a rainfall excess hyetograph are:

1. Determine the SCS Curve number for the area.

    Divide the area into suitable sub-basins as appropriate. Determine the antecedent moisture condition necessary for the design. While some agencies require use of wet conditions for less frequent storm designs, comparison of this approach with gage data often indicates an overprediction of peak flows (especially when using longer duration storms). Remember that the goal is to reproduce a storm "event" to test the drainage system, not to develop an overly conservative hypothetical storm whose nominal frequency is significantly different from its actual "event" frequency.

2. Determine the frequency and duration of the design storm.

    For smaller watersheds use of a long duration is not warranted nor, in some instances, realistic. For smaller, more impervious watersheds a shorter more intense storm will sometimes yield a higher peak flow (Aron and Kibler, 1990). This is especially true if there is a lot of connected imperviousness which may act almost like a separate watershed. The median storm duration east of the Mississippi is about six hours. Choice of a duration somewhat longer than the time of concentration is usually sufficient.

3. Calculate S, I_a, T_c, T_l, and d from the equations in the unit hydrograph section.

    Make sure d is within the limits set above. It is convenient to round it to the nearest minute and to see if a multiple of 60 will fit within the limits of 0.2 and 0.25 T_l.

4. Distribute the total rainfall according to the proportions from the proper SCS S-curve.

    The difference between the end and the beginning of the storm duration interval chosen is the total portion of the S-curve contained in the storm duration. This can be done by hand using Figure 7-14 or Table 7-24. Then for each time increment a proportion can be set up as:

$$P_i/P = R_i/R \qquad\qquad\qquad (7.32)$$

    Where: P_i =   the increment of rainfall at time interval d for use with the unit hydrograph
            R_i =   the increment of the S-curve at time interval d
            R =   total S-curve rainfall dimensionless rainfall at the storm duration

5. Set up a calculation table (see example) below.

    The table should have column headings for time, S-curve accumulated values (R), accumulated rainfall (P), initial abstraction (I_a), infiltration and other losses (F), rainfall excess (Q), and hyetograph values. This format works well with spreadsheet applications. Aaron (1992) provides a linear filter to correct for certain defects in the infiltration assumptions inherent in this SCS methodology.

Rainfall Excess Example

Step 1: Curve Number = 70
Step 2: 10-Year, 6-Hour storm for Charlotte, NC = 3.67 in.
        SCS Type II rainfall distribution
Step 3: A = 700 acres T_c = 40 min (calculated)
        T_l = 24 min T_p = 27 min
        d = 6 min (equals 0.25 T_l)   S = 1000/70 - 10 = 4.29 in.
        I_a = 0.2S = 0.0858 in.
Steps 4 and 5:
        The time actually started in hour 9 of the SCS Type 2 curve. A description of each
        column follows:
        (1)   Time ordinate, in this example starting with hour 1.0 and going through hour 7.0
                for a total of six hours. The time increment is 0.1 hours (6 minutes) equal to d.
        (2)   The total dimensionless time period is hour 15.0 minus hour 9.0 from the SCS
                Type II storm distribution table or 0.8538 - 0.1467 = 0.7071. Each dimensionless
                rainfall increment was linearly interpolated at 0.1 hour intervals between the half-
                hour interval values given in the table.
        (3)   The precipitation increments were calculated from equation 7.32 as:
                P_i = 3.67 * ((R_i - R_{i-1})/0.7071) + P_{i-1}
                Note that some rounding has occurred in the table presentation which did not take
                place in the spreadsheet calculation.
        (4)   The initial abstraction was deducted from the precipitation until the total of 0.2S
                or 0.0858 in.
        (5)   Values taken from a solution of equation 7.30 using the accumulated values from
                Column 3.
        (6)   Col. 3 minus Col. 4 and Col. 5.
        (7)   Incremental excess from Col. 6.

Table 7-28 Example Problem Solution Table

(1) Time Ord. i (hrs)	(2) Cum. Precip. R_i (in.)	(3) SCS Init. P_i (in.)	(4) Cum. Abs. Ia_i (in.)	(5) Cum. Losses F_i (in.)	(6) Cum. Excess Q_i (in.)	(7) Cum. Excess Hyetograph (in.)
1.0	0.147	0.000	0.000	0.000	0.00	0.00
1.1	0.150	0.016	0.016	0.000	0.00	0.00
1.2	0.153	0.033	0.033	0.000	0.00	0.00
1.3	0.156	0.049	0.049	0.000	0.00	0.00
1.4	0.159	0.066	0.066	0.000	0.00	0.00
1.5	0.163	0.082	0.082	0.000	0.00	0.00
1.6	0.166	0.101	0.101	0.000	0.00	0.00
1.7	0.170	0.120	0.120	0.000	0.00	0.00
1.8	0.173	0.139	0.139	0.000	0.00	0.00
1.9	0.177	0.158	0.158	0.000	0.00	0.00
2.0	0.181	0.177	0.177	0.000	0.00	0.00
2.1	0.185	0.201	0.201	0.000	0.00	0.00
2.2	0.190	0.226	0.226	0.000	0.00	0.00

2.3	0.195	0.250	0.250	0.000	0.00	0.00
2.4	0.200	0.274	0.274	0.000	0.00	0.00
2.5	0.204	0.298	0.298	0.000	0.00	0.00
2.6	0.210	0.331	0.331	0.000	0.00	0.00
2.7	0.217	0.363	0.363	0.000	0.00	0.00
2.8	0.223	0.395	0.395	0.000	0.00	0.00
2.9	0.229	0.427	0.427	0.000	0.00	0.00
3.0	0.235	0.459	0.459	0.000	0.00	0.00
3.1	0.245	0.509	0.509	0.000	0.00	0.00
3.2	0.254	0.559	0.559	0.000	0.00	0.00
3.3	0.264	0.609	0.609	0.000	0.00	0.00
3.4	0.274	0.659	0.659	0.000	0.00	0.00
3.5	0.283	0.709	0.709	0.000	0.00	0.00
3.6	0.359	1.103	0.858	0.232	0.01	0.01
3.7	0.435	1.498	0.858	0.557	0.08	0.07
3.8	0.511	1.892	0.858	0.833	0.20	0.12
3.9	0.587	2.286	0.858	1.072	0.36	0.16
4.0	0.663	2.681	0.858	1.279	0.54	0.19
4.1	0.678	2.755	0.858	1.316	0.58	0.04
4.2	0.692	2.830	0.858	1.351	0.62	0.04
4.3	0.706	2.905	0.858	1.386	0.66	0.04
4.4	0.721	2.979	0.858	1.419	0.70	0.04
4.5	0.735	3.054	0.858	1.452	0.74	0.04
4.6	0.743	3.093	0.858	1.469	0.77	0.02
4.7	0.750	3.131	0.858	1.486	0.79	0.02
4.8	0.757	3.170	0.858	1.502	0.81	0.02
4.9	0.765	3.209	0.858	1.519	0.83	0.02
5.0	0.772	3.248	0.858	1.535	0.85	0.02
5.1	0.778	3.275	0.858	1.546	0.87	0.02
5.2	0.783	3.303	0.858	1.557	0.89	0.02
5.3	0.788	3.330	0.858	1.568	0.90	0.02
5.4	0.794	3.358	0.858	1.579	0.92	0.02
5.5	0.799	3.385	0.858	1.590	0.94	0.02
5.6	0.803	3.407	0.858	1.599	0.95	0.01
5.7	0.807	3.428	0.858	1.607	0.96	0.01
5.8	0.811	3.450	0.858	1.616	0.98	0.01
5.9	0.816	3.471	0.858	1.624	0.99	0.01
6.0	0.820	3.493	0.858	1.632	1.00	0.01
6.1	0.823	3.512	0.858	1.640	1.01	0.01
6.2	0.827	3.531	0.858	1.647	1.03	0.01
6.3	0.831	3.550	0.858	1.654	1.04	0.01
6.4	0.834	3.569	0.858	1.661	1.05	0.01
6.5	0.838	3.588	0.858	1.668	1.06	0.01
6.6	0.841	3.604	0.858	1.674	1.07	0.01
6.7	0.844	3.621	0.858	1.681	1.08	0.01
6.8	0.847	3.637	0.858	1.687	1.09	0.01
6.9	0.851	3.654	0.858	1.693	1.10	0.01
7.0	0.854	3.670	0.858	1.699	1.11	0.01

Figure 7-22 depicts the accumulated rainfall losses and excess (I_a, F and P) from Table 7-28. Figure 7-23 shows the excess rainfall hyetograph. The step-like appearance and slight increase in rainfall comes from the half-hour linear interpolation increments and the fact that while equal values of rainfall are falling for a half hour in six minute increments the infiltration (F) is decreasing throughout that time period.

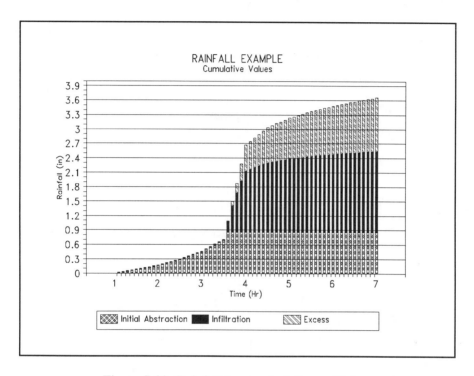

Figure 7-22 Rainfall Losses And Excess Values

SCS Dimensionless Unit Hydrographs

Unit hydrograph hydrological methods can be employed using SCS methods and information. The SCS methods use, as a basis, a dimensionless unit hydrograph whose shape is an average compilation of extensive analysis of measured data. Details of the derivation of the unit hydrograph can be found in the SCS National Engineering Handbook, Section 4 (SCS, 1985).

Dimensionless Unit Hydrograph Discussion - Figure 7-24 depicts the average dimensionless curvilinear unit hydrograph while the discharge ratios are tabulated in Table 7-29. The horizontal axis is a ratio of the time to the time to peak (T_p) while the vertical axis is a ratio of discharge to the peak discharge (q_p) or accumulated volume to total flow volume (Q).

It can be seen from this figure that the time base is approximately five times the time to peak and about 3/8 of the total volume occurred prior to the time of the peak.

The triangular hydrograph is a practical approximation of the curvilinear unit hydrograph. Its geometric makeup can be easily described mathematically, which makes it very useful in the process of estimating discharge rates, and produces results that are sufficiently accurate for most storm water management facility designs. Figure 7-25 depicts the triangular dimensionless unit

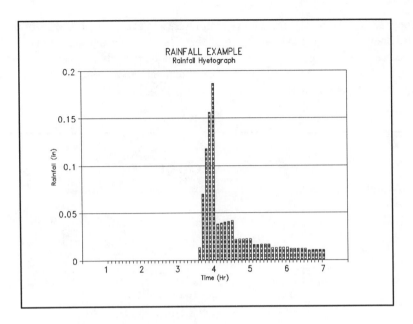

Figure 7-23　Rainfall Hyetograph

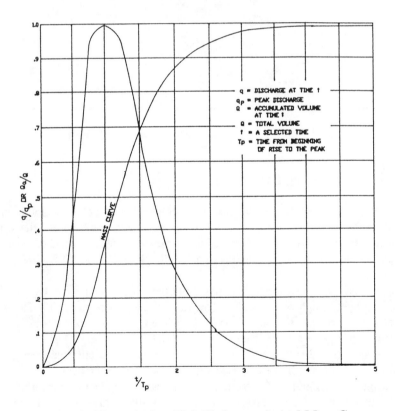

Figure 7-24　Dimensionless Unit Hydrograph And Mass Curve
Source: SCS, 1985

hydrograph and the curvilinear hydrograph. The areas under the rising limbs of the two hydrographs are the same as is the total volume. The SCS developed the following equation to estimate the peak rate of discharge for the dimensionless unit hydrograph:

$$q_p = 484 \ A(Q/T_p) \tag{7.33}$$

Where: q_p = peak rate of discharge, cfs
A = area, mi^2
Q = storm runoff volume, in.
T_p = time to peak, hr
484 includes appropriate conversion factors for the units

Table 7-29 Ratios For Dimensionless Unit Hydrograph And Mass Curve

Time Ratios (t/T_p)	Discharge Ratios (q/q_p)	Mass Curve Ratios (Qa/Q)
0	.000	.000
.1	.030	.001
.2	.100	.006
.3	.190	.017
.4	.310	.035
.5	.470	.065
.6	.660	.107
.7	.820	.163
.8	.930	.228
.9	.990	.300
1.0	1.000	.375
1.1	.990	.450
1.2	.930	.522
1.3	.860	.589
1.4	.780	.650
1.5	.680	.705
1.6	.560	.751
1.7	.460	.790
1.8	.390	.822
1.9	.330	.849
2.0	.280	.871
2.2	.207	.908
2.4	.147	.934
2.6	.107	.953
2.8	.077	.967
3.0	.055	.977
3.2	.040	.984
3.4	.029	.989
3.6	.021	.993
3.8	.015	.995
4.0	.011	.997
4.5	.005	.999
5.0	.000	1.000

Source: SCS, 1985

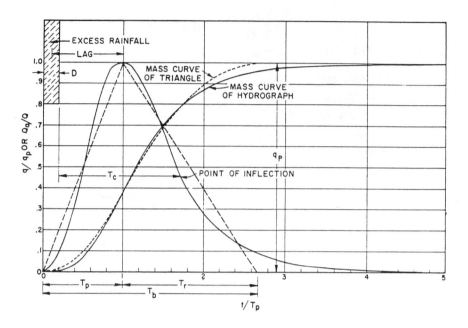

**Figure 7-25 Dimensionless Curvilinear Unit Hydrograph And
Equivalent Triangular Hydrograph**

Source: SCS, 1985

Additional relationships (in consistent units) which can be derived from a geometric consideration of the SCS dimensionless unit hydrographs are:

$$T_L = 0.6 \ T_c \tag{7.34}$$
$$T_p = d/2 + T_L \tag{7.35}$$
$$T_b = 2.67 \ T_p \ \text{(Triangular hydrograph)} \tag{7.36}$$

Where: T_L = watershed lag time, hr
  T_b = time of base, hr
According to SCS, d must be between 0.2 and 0.25 T_p. If d = 0.2 T_p, then:

$$d = 1.11 \ T_p = 0.133 \ T_c \tag{7.37}$$
$$T_p = 0.667 \ T_c \tag{7.38}$$

Where: d = rainfall time increment, hr

The constant 484, or peak rate factor (PR), is valid for the average SCS dimensionless unit hydrograph. Any change in the dimensionless unit hydrograph reflecting a change in the percent of volume under the rising side would cause a corresponding change in the shape factor associated with the triangular hydrograph and therefore a change in the constant 484. This constant has been known to vary from about 600 in steep terrain to 300 in very flat swampy country.

Characteristics of the dimensionless hydrograph vary with the size, shape, and slope of the tributary drainage area. The most significant characteristics affecting the dimensionless hydrograph shape are the basin lag and the peak discharge for a given rainfall. Basin lag in this method is defined as the time from the center of mass of rainfall excess to the hydrograph peak. Steep slopes, compact shape, and an efficient drainage network tend to make lag time short and peaks high; flat slopes, elongated shape, and an inefficient drainage network tend to make lag time long and peaks low for the same rainfall. Any change in the 484 factor would need to also be reflected in a change in the shape of the hydrograph to ensure a unit of volume remains under the curve. Special hydrographs have been developed for specific applications.

Meadows, et al., (1992) provide a methodology for the modification of the SCS curvilinear hydrograph for other peak rate factors. For the curvilinear hydrograph, a two parameter gamma function was developed which fits the curvilinear form closely:

$$q = (q_p*(t/tp) * e^{(1-(t/tp))})^{(N-1)} \tag{7.39}$$

N can be found from the fit equation:

$$N = 0.8679 * e^{(0.00353*PR)} \tag{7.40}$$

Where PR is the peak rate factor chosen (484 being the normal value for an N of 4.7). The procedure then is to choose a shape factor and calculate q_p and tp. For various values of t/tp, q and finally q/q_p can be calculated.

The dimensionless unit hydrograph with peak rate factor of 484 can also be fitted with an exponential equation as:

$$q/q_p = t/T_p^{3.5} * e^x \tag{7.41}$$

Where: $x = -3.5*(t/T_p-1)$

Unit Hydrograph Applications - Steps in the application of the unit hydrograph method are:
1. Measure basin area and parameters necessary to calculate the time of concentration.
2. Calculate t_p and d from the relationship between lag time, time of concentration and time step for rainfall increments. Ensure the time step for rainfall increments falls in the acceptable range.
3. Calculate q_p from equation 7.33 for a peak rate factor of 484 or from equation 7.39 for a chosen peak rate factor setting t equal to t_p.
4. Use Figure 7-24, Table 7-29 or the Meadows equation 7.39 to determine the flow rates for various time increments equal to d to define the unit hydrograph.
5. Convolute the hydrograph using the excess rainfall hyetograph generated using the SCS or other appropriate rainfall methods.

Unit Hydrograph Example - The rainfall hyetograph from the previous example is used with an SCS unit hydrograph to develop a runoff hydrograph. Figure 7-26 illustrates both the triangular and curvilinear hydrographs as well as the accumulated volume of outflow S-curve. The first few iterations of the convolution process are given in Table 7-30 below and are illustrated in Figure 7-27.

Step 1: A = 700 acres = 1.09 sq. mi.
Step 2: T_c = 40 min (calculated)
         T_l = 24 min
         T_p = 27 min = 0.45 hrs.
         d = 6 min (equals 0.25 T_l)

Step 3: Q = 1.0 for a unit hydrograph

q_p = 484AQ/T_p = (484)(1.09)(1.0)/(0.45) = 1,172 cfs

Table 7-30 Unit Hydrograph Convolution Example

RAINFALL --> HYETOGRAPH			0.01	0.07	0.12	0.16	0.19

CURVILINEAR UNIT HYDROGRAPH			TOTAL HYDROGRAPH				
TIME (HRS)	FLOW (CFS)	FLOW (CFS)					
0.0	0.00	0.00	0.00				
0.1	92.59	1.23	1.23	0.00			
0.2	481.25	12.84	6.39	6.46	0.00		
0.3	913.92	56.60	12.13	33.56	10.91	0.00	
0.4	1149.24	150.13	15.25	63.73	56.70	14.44	0.00
0.5	1152.95	295.47	15.30	80.14	107.68	75.06	17.29
0.6	1002.67	461.52	13.31	80.40	135.41	142.54	89.86
0.7	790.11	566.14	10.49	69.92	135.85	179.24	170.65
0.8	579.26	575.34	7.69	55.10	118.14	179.82	214.59
0.9	401.91	510.49	5.33	40.40	93.09	156.38	215.28
1.0	266.99	410.27	3.54	28.03	68.25	123.23	187.22
1.1	171.23	306.12	2.27	18.62	47.35	90.34	147.53
1.2	106.67	215.66	1.42	11.94	31.46	62.68	108.16
1.3	64.85	145.16	0.86	7.44	20.18	41.64	75.05
1.4	38.62	94.16	0.51	4.52	12.57	26.71	49.85
1.5	22.59	59.24	0.30	2.69	7.64	16.64	31.97
1.6	13.01	36.33	0.17	1.58	4.55	10.12	19.92
1.7	7.39	21.80	0.10	0.91	2.66	6.02	12.11
1.8	4.15	12.84	0.06	0.52	1.53	3.52	7.21
1.9	2.30	7.44	0.03	0.29	0.87	2.03	4.22
2.0	1.27	4.25	0.02	0.16	0.49	1.15	2.43
2.1	0.69	2.39	0.01	0.09	0.27	0.65	1.38
2.2	0.37	1.34	0.00	0.05	0.15	0.36	0.77
2.3	0.00	0.73	0.00	0.03	0.08	0.20	0.43
2.4		0.39		0.00	0.04	0.11	0.24
2.5		0.19			0.00	0.60	0.13
2.6		0.07				0.00	0.07
2.7		0.00					0.00

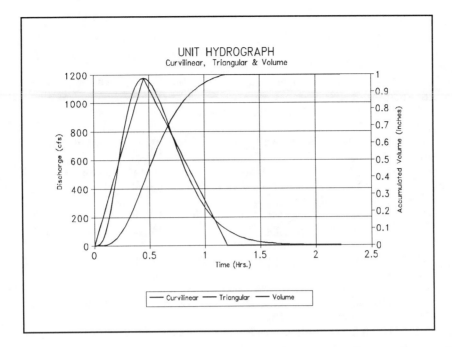

Figure 7-26   Unit Hydrographs And Accumulated Runoff Volume

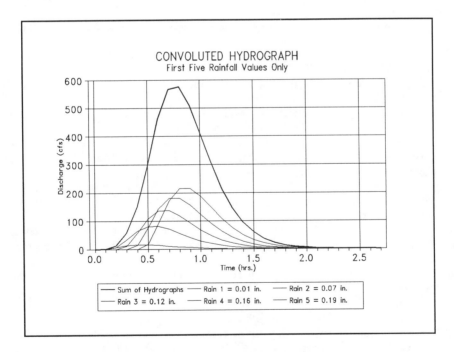

Figure 7-27   Convoluted Hydrograph - First Five Values

References

Aaron, G. "Adaptations of Horton and SCS Infiltration Equations to Complex Storms", J. of Ir. and Drain. Engr., ASCE, Vol. 118, No. 2, March/April, 1992.

Aron, G. and D. F. Kibler, "Pond Sizing for Rational Formula Hydrographs", AWRA Water Resources Bull., Vol. 26, No. 2, April, 1990.

Bosznay, M., "Generalization of SCS Curve Number Method", J. of Ir. and Drain. Engr., ASCE, Vol. 115, No. 1, February, 1989.

Boughton, W. C., "Evaluation of Partial Areas of Watershed Runoff", J. of Ir. and Drain. Engr., ASCE, Vol. 113, No. 3, August, 1987.

Casamayor, J. E., and J. R. Rodgers, Cost Analysis of 2-year, 5-year and 10-year Stormwater Protection, Hydraulics Technical Group, San Antonio, TX, Texas Section, ASCE.

Chow, V. T. ed, Handbook of Hydrology, p 8-21ff, McGraw-Hill, 1964.

Cochran, W. G., Sampling Techniques, John Wiley and Sons, Inc., 1963.

Dunne, T., and L. B. Leopold, Water in Environmental Planning, W. H. Freeman & Co., San Francisco, 1978.

Daly, M., "Discussion of Empirical Investigation of Curve Number Technique", J. of Hydraulic Div., ASCE, HY5, May, 1981, p. 651.

Federal Highway Administration, HYDRAIN Documentation, 1991

Gundlach, D. L., Adjustment of Peak Discharge Rates for Urbanization, U.S. Army Corps of Engineers Hydrologic Engineering Center, Tech. Paper No. 54., 1978.

Hawkins, R. H., "Runoff Curve Numbers from Partial Area Watersheds", J. of Ir. and Drain. Engr., ASCE, Vol. 105, No. IR4, December, 1979.

Hershfield, D. M., "Rainfall Frequency Atlas of the United States", Technical Paper 40, 1961.

Hjelmfelt, A. T., "Investigation of Curve Number Procedure", J. of Hydraulic Div., ASCE, Vol. 117, No. 6, June, 1991.

Hjelmfelt, A. T., "An Empirical Investigation of the Curve Number Technique", J. of Hydraulic Div., ASCE, Vol. 106, No. 9, Sept., 1980.

Huff, F. A., "Time Distributions of Heavy Rainstorms in Illinois,", State of Illinois, Water Survey, ISWS/CIR-173/90, Circular 173, 1990.

Huff, F. A., "Time Distribution of Rainfall in Heavy Storms,", Water Resources Research, Vol. 3, No. 4, 1967.

Huff, F. A. and J. L. Vogel, "Hydrometeorology of Heavy Storms in Chicago and Northeastern Illinois," Illinois State Water Survey Rpt. of Investigation 82, 1976.

Hydrologic Engineering Center, HEC-1 Flood Hydrograph Package, Davis California, September, 1990

Hydrologic Engineering Center, Hydrologic Analysis of Ungaged Watersheds Using HEC-1, Davis California, Training Document No. 15, April, 1982.

Hydrologic Engineering Center, U.S. Army Corps of Engineers, HEC-1 Flood Hydrograph Package, 1990.

Johnson, William K., "Significance of Location in Computing Flood Damage", ASCE J. of Wat. Res. Planning and Mgmt., Vol 111, No.1, Jan., pp 65-81, 1985.

Kerby, W. S., "Time of Concentration for Overland Flow", Civil Engineering, Vol. 29, March, 1959.

Kirpich, Z. P., "Time of Concentration from Small Agricultural Watersheds", Civil Engineering, Vol. 10, No. 6, June 1940.

Kuichling, E., The Relation Between the Rainfall and the Discharge of Sewers in Populous Districts, Trans. Am. Soc. of Civ. Eng., vol. 20, pp. 1-56, 1889.

Lewis, G.L., "Jury Verdict: Frequency Versus Risk-Based Culvert Design", ASCE J. of Water Res. Plan. & Mgmt., Vol. 118, No. 2, Mar/April, 1992, pp. 166-185.

McCuen, R. H., "A Guide to Hydrologic Analysis Using SCS Methods", Prentice-Hall, 1982.

McCuen, R., Introduction to Chapter 9, ASCE TC on Hydrology, DRAFT, 1993.

Meadows, M. E., K. B. Morris, W. E. Spearmen, "Improved Runoff Estimation Methods and Single Outlet Detention Performance Curves Applicable to Urban Watersheds in the Midlands of South Carolina", Supplemental Rpt., Dept. of Civil Engineering, U. of South Carolina and South Carolina Land Resources Conservation Commission, March, 1992.

NOAA, "Five- to 60-Minute Precipitation Frequency for the Eastern and Central United States", NOAA Tech. Memo. NWS HYDRO-35, 1977.

NOAA and the Army Corps of Engineers, Hydrometeorological Report No. 51, Probable Maximum Precipitation Estimates, US East of the 105th Meridian, June 1978.

NOAA and the Army Corps of Engineers, Hydrometeorological Report No. 52, Application of the Probable Maximum Precipitation Estimates - US East of the 105th Meridian August, 1982.

Overton, D. E. and M. E. Meadows, Storm Water Modeling, Academic Press. New York, N.Y. pp. 58-88, 1976.

Rantz, S. E., Suggested Criteria for Hydrologic Design of Storm-Drainage Facilities in the San Francisco Bay Region, California., USGS Open File Report, Menlo Park, CA., 1971.

Rossmiller, R. L., The Rational Formula Revisited, Proc. Int. Symp. on Urb. Storm Runoff, University of Kentucky, Lexington, KY, July 28-31, 1980.

Ruthroff, C. L. and W. F. Bodtmann, "Computing Derivatives from Equally Spaced Data," J. of Applied Meteorology, Vol. 15, No. 11, Nov., 1976.

Sarma, P. G. S., J. W. Delleur, and A. R. Rao, A Program in Urban Hydrology, Part II, Technical Report No. 99, Purdue University, Water Resources Research Center, 1969.

Schaake, J. C., J. C. Geyer, and J. W. Knapp, "Experimental Examination of the Rational Method", J. of the Hydraulics Division, ASCE, Vol. 93, No. HY6, pp. 353-370, Nov. 1967.

Shabman, L., Water Project Design and Safety: Prospects for Use of Risk Analysis in Public Sector Organizations, in Risk-Based Decision Making in Water Resources, ASCE Conf. Proc., Nov 3-5, 1985, pp. 16-29, 1985.

Snyder, F. F., Synthetic Unit Hydrographs. Trans. Am. Geophysical Union, Vol. 19, Part 1, 1938, pp. 447-454.

Soil Conservation Service, Urban Hydrology for Small Watersheds, Technical Release No. 55, Washington, D.C., 1986.

South Carolina Stormwater Management and Sediment Reduction Regulations, South Carolina Land Resources Conservation Commission, Division of Engineering, Columbia, South Carolina, 1992.

U. S. Army Corps of Engineers, St. Paul District, Analysis for the Determination of Percentage of Plugging for Hydraulic Analysis of Bridges. St. Paul, MN, Hydraulic Design Section.

U. S. Department of Agriculture, Soil Conservation Service, Engineering Division, Urban Hydrology For Small Watersheds, Technical Release 55 (TR-55), 1986.

U. S. Department of Agriculture, Soil Conservation Service, Engineering Division, National Engineering Handbook, Section 4, Hydrology, 1985.

U. S. Department of Transportation, Federal Highway Administration, Hydrology, Hydraulic Engineering Circular No. 19, 1984.

Water Resources Council Bulletin 17B, Guidelines For Determining Flood Flow Frequency, 1981.

Wright-McLaughlin Engineers, Urban Storm Drainage Criteria Manual, Vol. I and II, Prepared for the Denver Regional Council of Governments, Denver, Colorado, 1969.

Yen, B. C., and V. T. Chow, "Design Hyetographs for Small Drainage Structures,", ASCE J. of Hydraulics, Vol. 106, No. HY6, June, 1980.

Yen, B. C., and V. T. Chow, "Local Design Storm Vol. II," Federal Highway Administration, Rept. No. FHWA/RD-82-064, 1983.

Chapter 8 Storm Drainage Systems

8.1 Introduction

Storm drainage facilities collect storm water runoff and convey it away from structures and through the roadway right-of-way in a manner which adequately drains sites and roadways and minimizes the potential for flooding and erosion to properties. Storm drainage facilities consist of curbs, gutters, storm drains, channels, ditches and culverts. The placement and hydraulic capacities of storm drainage structures and conveyances should be designed to take into consideration damage to adjacent property and to secure as low a degree of risk of traffic interruption by flooding as is consistent with the importance of the road, the design traffic service requirements, and available funds.

8.2 Concept Definitions

Following are definitions of concepts important in storm drain analysis and design as used in this chapter

Bypass	Flow which bypasses an inlet on grade and is carried in the street or channel to the next inlet downgrade. Inlets can be designed to allow a certain amount of bypass. Also, inlets may be designed to allow a certain amount of bypass for one design storm and larger or smaller amounts for other design storms.
Combination Inlet	A drainage inlet usually composed of a curb-opening and a grate inlet.
Curb-Opening Inlet	A drainage inlet consisting of an opening in the roadway curb.
Drop Inlet	A drainage inlet with a horizontal or nearly horizontal opening.
Equivalent Cross Slope	An imaginary continuous cross slope having conveyance capacity equal to that of the given compound cross slope.
Flanking Inlet	Inlets placed upstream and on either side of an inlet at the low point in a sag vertical curve. The purpose of these inlets is to intercept debris as the slope decreases and to act in relief of the inlet at the low point.
Frontal Flow	The portion of the flow which passes over the upstream side of a grate.

Grate Inlet   A drainage inlet composed of a grate in the roadway section or at the roadside
                       in a low point, swale or channel.

Grate Perimeter   The sum of the lengths of all sides of a grate, except that any side adjacent to
                       a curb is not considered a part of the perimeter in weir flow computations.

Gutter   That portion of the roadway section adjacent to the curb which is utilized to
                       convey storm runoff water. It may include a portion or all of a traveled lane,
                       shoulder or parking lane, and a limited width adjacent to the curb may be of
                       different materials and have a different cross slope.

Hydraulic   The hydraulic grade line is the locus of elevations to which the water would rise
Grade Line   in successive piezometer tubes if the tubes were installed along a pipe run.

Inlet Efficiency   The ratio of flow intercepted by an inlet to total flow in the gutter.

Pressure Head   Pressure head is the height of a column of water that would exert a unit pressure
                       equal to the pressure of the water.

Scupper   A vertical hole through a bridge deck for the purpose of deck drainage. Some-
                       times, a horizontal opening in the curb or barrier is called a scupper.

Side-Flow   Flow which is intercepted along the side of a grate inlet, as opposed to frontal
Interception   interception.

Slotted Drain   A drainage inlet composed of a continuous slot built into the top of a pipe which
Inlet   serves to intercept, collect and transport the flow.

Splash-Over   Portion of the frontal flow at a grate which skips or splashes over the grate and
                       is not intercepted.

Spread   The width of flow measured laterally from the roadway curb.

Velocity Head   Velocity head is a quantity proportional to the kinetic energy of flowing water
                       expressed as a height or head of water.

8.3 Pavement Drainage

Design Steps

There are many details to consider in the design and specification of storm drain systems.
ASCE Manuals of Engineering Practice (1960, 1982, 1983) as well as other trade and vendor
publications provide construction and specification details beyond the scope of this text. Typical
steps in the hydraulic design of a storm drain system include the following.

1. Design Problem And Design Criteria Specification - Determine the design type to be done
and the design criteria to be followed for the specific locality. Where criteria are not adequate
or complete establish internal criteria using other sources including those given in this text.

2. System Drainage Area Definition And Preliminary Layout - Based on street layout or other criteria, identify the total drainage area to be handled (including off site drainage). With reference to design criteria, layout the preliminary drainage routes. Consider bypass flow locations for storms in excess of design. If parcels or lots are involved layout bypass locations on parcel boundaries where possible. Look for an ability to introduce environmental features and for integration with other neighborhood amenities. Locate quantity and quality control structures.

3. Field And Office Data Collection - Make a field visit to determine feasibility of preliminary layout and to look for site specific problems such as: rock outcrops, large trees, environmentally sensitive areas, utility locations, unmapped structures, etc.

4. System Layout - Perform "final" layout in coordination with others on the design team to ensure minimization of design conflicts and an ability to take advantage of any multi-objective use opportunities of the drainage system. Locate all ditches, inlets, manholes, mains, laterals, culverts, etc. Determine flow direction from grading plan or map of site and mark with arrows on site map.

5. Hydrologic Calculations - Outline drainage area for each inlet or ditch start. Develop flow estimates for the design frequency throughout the system.

6. Street Flow - Develop flow and spread calculations for streets or roadways and determine permissible maximum spread. Section 8.5 provides details of spread calculations and 8.6 for gutter flow.

7. Inlet Spacing And Layout - Beginning at upstream end determine the location of the upstream inlet based on trial drainage area delineations. Size inlet and calculate bypass. Continue downstream locating inlets and determining bypass amounts. Accommodate special situations and point sources of inflow. Seek to visualize the flow of water to determine potential problems. For example, placing reliance on an inlet at the end of a downhill cul-de-sac often allows excess water to bypass the inlet into a residential driveway or parking lot if it should become clogged. Extra inlets should be placed at all street crossings to eliminate flow bypass. Sections 8.6 through 8.9 give details for inlet design.

8. Pipe Sizing And Layout - Beginning at the upstream end size pipes for the calculated flow and the given design criteria. Sometimes pipe size criteria are different from the street flow criteria. Velocity should not appreciable decrease at inlets. Pipe sizes should never decrease downstream even for flow on a steeper slope. Take great care in allowing flow to enter inlet boxes from laterals and tapping culverts with pipes larger than 50 percent of the minimum box dimension. Provide for maintenance and other utility spacing minimums. Minimum pipe sizes may apply in some locations. Sections 8.10 and 8.11 provide details of pipe sizing and hydraulic gradient calculations. For ditches the additional consideration of erosion and bank protection is a concern. Smooth flow lines and gradual transitions should be planned.

9. Hydraulic Gradeline - Follow design criteria for hydraulic gradeline calculations in pipes. Begin at the downstream end using a suitable flow depth in the receiving stream if a free outfall cannot be assumed. Sections 8.10 and 8.11 provide details of hydraulic gradeline calculations.

Design Factors

Design factors to be considered for drainage calculations include:

- Return period
- Spread
- Inlet types and spacing
- Longitudinal slope
- Cross slope
- Curb and gutter sections
- Roadside and median channels

- Bridge decks
- Shoulder gutter
- Median/Median barriers
- Storm Drains
- Detention Storage
- Cost
- Erosion

Following is a summary discussion of each of these factors. Most factors are covered in detail in later sections of this chapter. Where appropriate, some typical municipal criteria are given.

Return Period - The design storm return period for pavement drainage should be consistent with the frequency selected for other components of the drainage system. See Section 8.5 for further discussion of design storm return period (frequency) criteria.

Spread - For multi-lane curb and gutter or guttered roadways with no parking, it is not practical to avoid travel lane flooding when grades are flat (0.2 to one percent). However, flooding should never exceed the lane adjacent to the gutter (or shoulder) for design conditions. Municipal bridges with curb and gutter should also use this criterion. For single-lane roadways at least 8 feet of roadway should remain unflooded for design conditions.

Inlet Types And Spacing - Inlet types should be selected from locally available inlets. Municipalities may want to specify the manufacturer and specific inlet types that are acceptable, and where performance information is available for engineering analysis and design. Inlets should be located or spaced in such a manner that the design curb flow does not exceed the spread limitations. Flow should not be allowed to cross intersecting streets unless approved by the municipal Engineering Department. Some percentage of lockage should be planned for if appropriate.

Longitudinal Slope - A minimum longitudinal gradient is more important for a curbed pavement, since it is susceptible to storm water spread. Flat gradients on uncurbed pavements can lead to a spread problem if vegetation is allowed to build up along the pavement edge.

Desirable gutter grades should not be less than 0.3 percent for curbed pavements, and with a minimum 0.2 percent in very flat terrain. Minimum grades can be maintained in very flat terrain by use of a sawtooth profile.

To provide adequate drainage in sag vertical curves, a minimum slope of 0.3 percent should be maintained within 50 feet of the level point in the curve. This is accomplished where the length of the curve divided by the algebraic difference in grades is equal to or less than 167. Special gutter profiles should be developed to maintain a minimum slope of 0.2 percent up to the inlet. Although ponding is not usually a problem at crest vertical curves, on extremely flat curves a similar minimum gradient should be provided to facilitate drainage.

Cross Slope - The design of pavement cross slope is often a compromise between the need for reasonably steep cross slopes for drainage and relatively flat cross slopes for driver comfort. The USDOT, FHWA (FHWA-RD-79-30, 31, 1979) reports that cross slopes of 2 percent have little effect on driver effort in steering, especially with power steering, or on friction demand for vehicle stability. Use of a cross slope steeper than 2 percent on pavements with a central crown line is not desirable. In areas of intense rainfall, a somewhat steeper cross slope may be necessary to facilitate drainage. In such areas, the cross slope could be increased to 2.5 percent.

When three or more lanes are inclined in the same direction on multi-lane pavements, it is desirable that each successive pair of lanes, or the portion of the outside lanes from the first two lanes from the crown line, have an increased slope. The two lanes adjacent to the crown line should be pitched at the normal slope, and successive lane pairs, or portions of the outside lanes, should be increased by about 0.5 to one percent. Where three or more lanes are provided in each direction, the maximum pavement cross slope should be limited to 4 percent.

Median areas should not be drained across traveled lanes and shoulders should generally be sloped to drain away from the pavement, except with raised, narrow medians.

Curb And Gutter Sections - Curbing at the outside edge of pavements is normal practice for low-speed, urban highway facilities. Gutters may be 1.0 feet to 3.5 feet wide. Standard curb and gutter has a width of 1.5 feet, and may be integral with the curb. Gutters are on the same cross slope as the pavement on the high side and depressed with a steeper cross slope on the low side, usually one inch per foot. Typical practice is to place curbs at the outside edge of shoulders or parking lanes on low speed facilities. The gutter width may be included as a part of the parking lane. Shoulder gutter is not required adjacent to barrier walls on high fills.

Curbed highway sections are relatively inefficient at conveying water, and the area tributary to the gutter section should be kept to a minimum to reduce the hazard from water on the pavement. Where practicable, the flow from major areas draining toward curbed highway pavements should be intercepted by channels as appropriate.

Roadside And Median Channels - Roadside channels are commonly used with uncurbed roadway sections to convey runoff from the highway pavement and from areas which drain toward the highway. Due to right-of-way limitations, roadside channels cannot be used on most urban arterials. They can be used in cut sections, depressed sections, and other locations where sufficient right-of-way is available and driveways or intersections are infrequent.

It is preferable to slope median areas and inside shoulders to a center swale, to prevent drainage from the median area from running across the pavement. This is particularly important for high-speed facilities, and for facilities with more than two lanes of traffic in each direction. If used, temporary storage in shallow medians must be carefully engineered to handle high intensity rainfall.

Bridge Decks - Drainage of bridge decks is similar to other curbed roadway sections. It is often less efficient, because cross slopes are flatter, parapets collect large amounts of debris, and small drainage inlets on scuppers have a higher potential for clogging by debris. Because of the difficulties in providing and maintaining adequate deck drainage systems, gutter flow from roadways should be intercepted before it reaches a bridge. In many cases, deck drainage must be carried several spans to the bridge end for disposal.

Short, continuous span bridges, particularly over-passes, may be built without inlets. The water from the bridge roadway should be carried downslope by open or closed chutes near the end of the bridge structure. Some type of bridge end drainage must be provided at all bridges.

Zero gradients and sag vertical curves should be avoided on bridges. The minimum desirable longitudinal slope for bridge deck drainage should be 0.2 percent. When bridges are placed at a vertical curve and the longitudinal slope is less than 0.2 percent, the gutter spread should be checked to ensure a safe, reasonable design.

Scuppers are the recommended method of deck drainage because they can reduce the problems of transporting a relatively large concentration of runoff in an area of generally limited right-of-way. However, the use of scuppers should be evaluated for site-specific concerns. Scuppers should not be located over embankments, slope protection, navigation channels, driving lanes, or railroad tracks. Runoff collected and transported to the end of the bridge should generally be collected by inlets and down drains, although sod flumes may be used for extremely minor flows in some areas. Downspouts, where required, should be made of rigid corrosion-resistant material not less than 6 inches in least dimension and should be provided with cleanouts.

The details of deck drains should be such as to prevent the discharge of drainage water against any portion of the structure or on moving traffic below, and to prevent erosion at the outlet of the downspout. Deck drainage may be connected to conduits leading to storm water outfalls at ground level. Overhanging portions of concrete decks should be provided with a drip bead or notch. Water in a roadway gutter section should be intercepted prior to the bridge.

The placement of bridges over environmentally sensitive areas should be avoided if possible. Where not possible precautions should be taken to minimize the adverse environmental impact caused by roadway runoff from bridge decks. Precautions include: restricting the use of scuppers and downspouts, and directing water through overland treatment systems, detention ponds or designed wetlands prior to entering the water course (EPA, 1993, VERSAR, 1985, Woodward-Clyde, 1989). For situations where traffic under the bridge or environmental concerns prevent the use of scuppers, grated bridge drains could also be used.

Shoulder Gutters - Shoulder gutters may be appropriate to protect fill slopes from erosion caused by water from the roadway pavement. Shoulder gutter is required on fill slopes higher than 20 feet - standard slopes are 2:1. It is also required on fill slopes higher than 10 feet - standard slopes are 6:1 and 3:1 if the roadway grade is greater than 2 percent. In areas where permanent vegetation cannot be established, a shoulder gutter is required on fill slopes higher than 10 feet regardless of the grade. Inspection of the existing/proposed site conditions and contact with maintenance and construction personnel should be made by the designer to determine if vegetation will survive.

Shoulder gutters may be appropriate at bridge ends where concentrated flow from the bridge deck would otherwise run down the fill slope. This section of gutter should be long enough to include the transitions. Shoulder gutters are not required on the high side of super-elevated sections or adjacent to barrier walls on high fills.

Median/Median Barriers - Weep holes are often used to prevent ponding of water against barriers (especially on superelevated curves). In order to minimize flow across traveled lanes, it is preferable to collect the water into a subsurface system connected to the main storm drain system.

Storm Drains - Storm drains are used to convey water from the inlets to an acceptable outlet. Storm drains comprise a drainage system usually consisting of one or more pipes connecting two or more drop inlets. Cross storm drains "hydraulically designed" to function as a culvert or culverts are an exception to that statement. Storm drains should have adequate capacity so that they can accommodate runoff that enters the system. The storm drainage system for sag vertical curves should have a higher level of flood protection to decrease the depth of ponding on the roadway and bridges.

Storm drains should be designed to protect the roadway from flooding at the appropriate return period. Reserve capacity should be available at critical locations such as vertical curve sags and at bridge approaches. Where feasible, the storm drains should be designed to avoid existing utilities. Attention should be given to the storm drain outfalls to ensure that the potential for erosion is minimized.

Detention Storage - Reduction of peak flows can be achieved by the storage of runoff in detention basins, storm drains, swales and channels, and other detention storage facilities. Storm water is then released to the downstream conveyance facility at a reduced flow rate. The concept should be considered for use in highway drainage design where existing downstream conveyance facilities are inadequate to handle peak flow rates from highway storm drainage facilities, where environmental concerns are present, where the highway would contribute to increased peak flow rates and aggravate downstream flooding problems, and as a technique to reduce the right-of-way, construction, and operation costs of outfalls from highway storm drainage facilities.

Costs - The cost of drainage is neither incidental nor minor on most roads. Careful attention to requirements for adequate drainage and protection of the roadway from storm water in all phases of location and design will prove to be effective in reducing costs in both construction and maintenance. Unless drainage is properly accommodated, maintenance costs will be unduly high.

Construction costs too can be minimized through proper layout and design. For example, it is cost effective to avoid deep cuts, rock blasting, numerous junctions, utility relocation and pumping of storm water. Storm drain design should anticipate upstream development which will use the system and seek to accommodate such development in system sizing.

8.4 Storm Water Inlet Overview

The primary aim of drainage design is to limit the amount of water flowing along the gutters or ponding at the sags to quantities which will not interfere with the passage of traffic for the design frequency. This is accomplished by placing inlets at such points and at such intervals to intercept flows and control spread.

In this chapter, guidelines are given for evaluating roadway features and design criteria as they relate to gutter and inlet hydraulics and storm drain design. Procedures for performing gutter flow calculations are based on a modification of Manning's Equation. Inlet capacity calculations for grate and combination inlets are based on information contained in HEC-12 (USDOT, FHWA, 1984). Storm drain design is based on the use of the rational formula.

Drainage inlets are sized and located to limit the spread on traffic lanes to tolerable widths for the design storm. Because grates may become blocked by trash accumulation, curb openings, or combination inlets with both grate and curb openings, are advantageous for urban conditions. Grate inlets and depressions of curb-opening inlets should be located outside the through-traffic lanes. Inlet grates should safely accommodate bicycle and pedestrian traffic where appropriate.

Inlets should be selected, sized and located to prevent silt and debris carried in suspension from being deposited on the traveled way where the longitudinal gradient is decreased.

Inlets at vertical curve sags in the roadway grade should also be capable of limiting the spread to tolerable limits. The width of water spread on the pavement should not be substantially greater than the width of spread encountered on continuous grades. At high discharges this can only be accomplished by the use of inlets just upstream of the sag inlet on either or both sides of the sag. These additional inlets, often referred to as flanking inlets, serve to pick up the runoff before it reaches the sag and also limit the spread in the event that the sag inlet is clogged by debris. Where there is a danger of damage to adjacent property by runoff overtopping the curb in a sag, flanking inlets should be used and the location checked to ensure that the curb is not overtopped due to insufficient inlet capacity.

Inlets should be located so that concentrated flow and heavy sheet flow will not cross traffic lanes. Where pavement surfaces are warped, as at cross streets or ramps, surface water should be intercepted just before the change in cross slope. Also, inlets should be located just upgrade of pedestrian crossings.

Inlets should be placed upstream of locations where the pavement cross slope reverses, such as on the high side of superelevated horizontal curves, to avoid concentrated flows crossing the roadway. Special care should be given to inlet placement to ensure adequate capacity at bridge approaches and at sag vertical curves where ponding deeper than the curb height could occur.

The design of a drainage system for a roadway traversing an urbanized region is generally a more complex problem than for roadways traversing sparsely settled rural areas. This is often due to:

- the wide roadway sections, flat grades, both in longitudinal and transverse directions, shallow water courses, absence of side channels;
- the more costly property damages which may occur from ponding of water or from flow of water through built-up areas;
- the fact that the roadway section must carry traffic, but also act as a channel to convey the water to a disposal point. Unless proper precautions are taken, this flow of water along the roadway will interfere with or possibly halt the passage of highway traffic.

There are four storm water inlet categories:

- curb opening inlets
- combination inlets
- grated inlets
- multiple inlets

In addition, inlets may be classified as being on a continuous grade or in a sump. The term "continuous grade" refers to an inlet located on the street with a continuous slope past the inlet with water entering from one direction. The "sump" condition exists when the inlet is located at a low point and water enters from both directions.

Although design storm criteria differ from one municipality to another, the 2-year, 5-year, and 10-year design storms are the most frequent design storms used for storm water inlet design. In addition to design storm criteria, following are some typical spread limit criteria used by municipalities.

- Maximum spread of 6 feet in a travel lane.
- For a street with a valley gutter, another foot for the gutter is allowed with a total maximum spread of 7 feet.
- For a street with a standard 2 feet 6 inch curb and gutter, an additional 2 feet is allowed with a total maximum spread of 8 feet from the face of the curb.

8.5 Design Frequency And Spread

Following are some recommended criteria for design storm frequency and spread. These criteria should be analyzed and modified to fit local conditions.

The design storm frequency for pavement drainage should be consistent with the frequency selected for other components of the drainage system. However, for a full-shoulder bridge condition, the spread should be contained within the shoulder for a 10-year design storm.

For multi-lane curb and gutter, or guttered roadways with no parking, it is not practical to avoid travel lane flooding when longitudinal grades are flat (0.2 to one percent). However, flooding should never exceed the lane adjacent to the gutter (or shoulder) for design conditions. Municipal bridges with curb and gutter should also use this criterion. For single-lane roadways at least 8 feet of roadway should remain unflooded for design conditions.

The major considerations for selecting a design frequency and spread is highway classification. Ponding should be minimized on the traffic lanes of high-speed, high-volume highways, where it is not expected by the public to occur, whereas for local streets some ponding is common.

Highway speed is another major consideration, because at speeds greater than 45 miles per hour, even a shallow depth of water on the pavement can cause hydroplaning. Design speed is recommended for use in evaluating hydroplaning potential. When the design speed is selected, consideration should be given to the likelihood that legal posted speeds may be exceeded. It is clearly unreasonable to provide the same level of protection for low speed facilities as for high speed facilities.

Other considerations include inconvenience, hazards, and nuisances to pedestrian traffic and buildings adjacent to roadways which are located within the splash zone. These considerations should not be minimized and, in some locations (such as commercial pedestrian areas), may assume major importance.

Table 8-1 shows recommended design criteria for frequency and spread.

Table 8-1 Frequency And Spread Design Criteria

Road Classification		Design Frequency	Design Spread
High Volume	< 45 mph	10-year	Shoulder + 3 feet
	> 45 mph	10-year	Shoulder
	sag point	50-year	Shoulder + 3 feet
Collector	< 45 mph	10-year	1/2 driving lane
	> 45 mph	10-year	Shoulder
	sag point	10-year	1/2 driving lane
Local Streets	low ADT	5-year	1/2 driving lane
	high ADT	10-year	1/2 driving lane
	sag point	10-year	1/2 driving lane

8.6 Gutter Flow Calculations

The following form of Manning's Equation can be used to evaluate gutter flow hydraulics:

$$Q = [0.56 / n] \ S_x^{5/3} \ \ S^{1/2} \ \ T^{8/3} \tag{8.1}$$

Where: Q = gutter flow rate, cfs
  n = Manning's roughness coefficient
  S_x = pavement cross slope, ft/ft
  S = longitudinal slope, ft/ft
  T = width of flow or spread, ft

A nomograph for solving equation 8.1 is presented in Figure 8-1. Manning's n values for various pavement surfaces are given in Table 8-2.

Uniform Cross Section

The nomograph in Figure 8-1 is used with the following procedures to find gutter capacity for uniform cross slopes:

Condition 1: Find spread, given gutter flow.

1. Determine input parameters, including longitudinal slope (S), cross slope (S_x), gutter flow (Q), and Manning's n.
2. Draw a line between the S and S_x scales and note where it intersects the turning line.
3. Draw a line between the intersection point from Step 2 and the appropriate gutter flow value on the capacity scale. If Manning's n is 0.016, use Q from Step 1; if not, use the product of Q and n.
4. Read the value of the spread (T) at the intersection of the line from Step 3 and the spread scale.

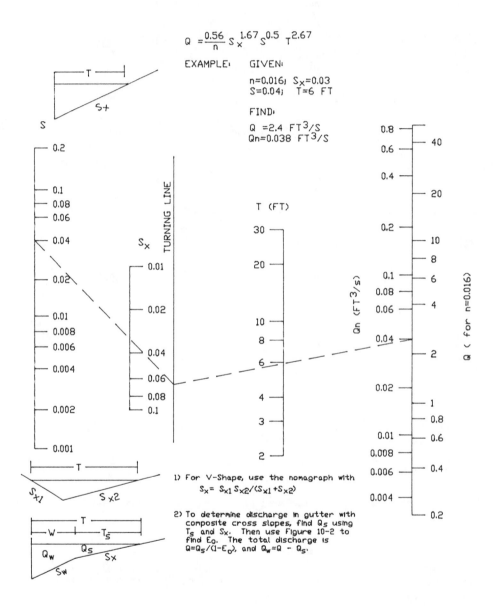

$$Q = \frac{0.56}{n} S_x^{1.67} S^{0.5} T^{2.67}$$

EXAMPLE: GIVEN:

n=0.016; S_x=0.03
S=0.04;   T=6 FT

FIND:

Q =2.4 FT^3/S
Qn=0.038 FT^3/S

1) For V-Shape, use the nomagraph with
$S_x = S_{x1} S_{x2}/(S_{x1}+S_{x2})$

2) To determine discharge in gutter with
composite cross slopes, find Q_s using
T_s and S_x. Then use Figure 10-2 to
find E_0. The total discharge is
$Q=Q_s/(1-E_0)$, and $Q_w=Q - Q_s$.

Figure 8-1 Flow In Triangular Gutter Sections

Source: AASHTO Model Drainage Manual, 1991

Table 8-2 Manning's n Value For Street And Pavement Gutters

Type of Gutter or Pavement	Range of Manning's n
Concrete gutter, troweled finish	0.012
Asphalt pavement:	
Smooth texture	0.013
Rough texture	0.016
Concrete gutter with asphalt pavement:	
Smooth	0.013
Rough	0.015
Concrete pavement:	
Float finish	0.014
Broom finish	0.016
For gutters with small slopes, where sediment	
may accumulate, increase above values of n by	0.002

Note: Estimates are by the Federal Highway Administration
Reference: USDOT, FHWA, HDS-3 (1961).

Condition 2: Find gutter flow, given spread.

1. Determine input parameters, including longitudinal slope (S), cross slope (S_x), spread (T), and Manning's n.
2. Draw a line between the S and S_x scales and note where it intersects the turning line.
3. Draw a line between the intersection point from Step 2 and the appropriate value on the T scale. Read the value of Q or Qn from the intersection of that line on the capacity scale.
4. For Manning's n values of 0.016, the gutter capacity Q from Step 3 is selected. For other Manning's n values, the gutter capacity times n (Qn) is selected from Step 3 and divided by the appropriate n value to give the gutter capacity.

Composite Gutter Sections

Figure 8-2 in combination with Figure 8-1 can be used to find the flow in a gutter with width (W) less than the total spread (T). Such calculations are generally used for evaluating composite gutter sections or frontal flow for grate inlets.

Figure 8-3 provides a direct solution of gutter flow in a composite gutter section. The flow rate at a given spread or the spread at a known flow rate can be found from this figure. Figure 8-3 involves a complex graphical solution of the equation for flow in a composite gutter section. Typical of graphical solutions, extreme care in using the figure is necessary to obtain accurate results.

Condition 1: Find spread, given gutter flow.

1. Determine input parameters, including longitudinal slope (S), cross slope (S_x), depressed section slope (S_w), depressed section width (W), Manning's n, gutter flow (Q), and a trial value of the gutter capacity above the depressed section (Q_s).

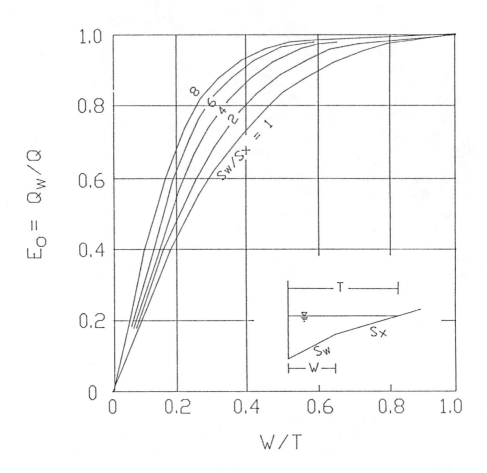

Figure 8-2 Ratio Of Frontal Flow To Total Gutter Flow

Source: AASHTO Model Drainage Manual, 1991

2. Calculate the gutter flow in W (Q_w), using the equation:

$$Q_w = Q - Q_s \qquad\qquad (8.2)$$

3. Calculate the ratios Q_w/Q or E_o and S_w/S_x and use Figure 8-2 to find an appropriate value of W/T.
4. Calculate the spread (T) by dividing the depressed section width (W) by the value of W/T from Step 3.
5. Find the spread above the depressed section (T_s) by subtracting W from the value of T obtained in Step 4.

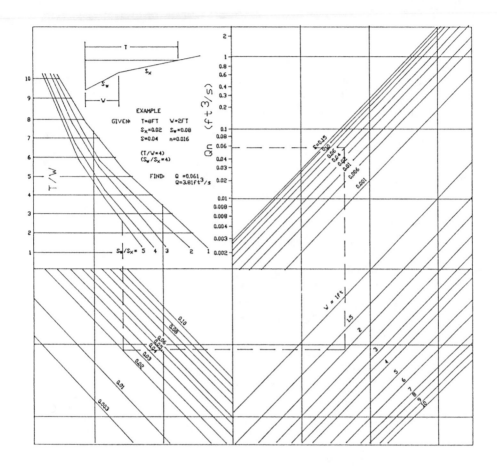

Figure 8-3 Flow In Composite Gutter Sections

Source: AASHTO Model Drainage Manual, 1991

6. Use the value of T_s from Step 5 along with Manning's n, S, and S_x to find the actual value of Q_s from Figure 8-1.

7. Compare the value of Q_s from Step 6 to the trial value from Step 1. If values are not comparable, select a new value of Q_s and return to Step 1.

Condition 2: Find gutter flow, given spread.

1. Determine input parameters, including spread (T), spread above the depressed section (T_s), cross slope (S_x), longitudinal slope (S), depressed section slope (S_w), depressed section width (W), Manning's n, and depth of gutter flow (d).

2. Use Figure 8-1 to determine the capacity of the gutter section above the depressed section (Q_s). Use the procedure for uniform cross slopes (Condition 2), substituting T_s for T.
3. Calculate the ratios W/T and S_w/S_x, and, from Figure 8-2, find the appropriate value of E_o (the ratio of Q_w/Q).
4. Calculate the total gutter flow using the equation:

$$Q = Q_s / (1 - E_o)$$ (8.3)

Where: Q = gutter flow rate, cfs
       Q_s = flow capacity of the gutter section above the depressed section, cfs
       E_o = ratio of frontal flow to total gutter flow, Q_w/Q
5. Calculate the gutter flow in width (W), using equation 8.2.

Example Problems

Example 1 Given: T = 8 ft   n = 0.015
                       S_x = 0.025 ft/ft   S = 0.01 ft/ft

Find: (1) Flow in gutter at design spread
           (2) Flow in width (W = 2 ft) adjacent to the curb

Solution: (1) From Figure 8-1, Qn = 0.03
                Q = Qn/n = 0.03/0.015 = 2.0 cfs
           (2) T = 8 - 2 = 6 ft
                $(Qn)_2$ = 0.014 (Figure 8-1) (flow in 6 ft width outside of width W)
                Q = 0.014/0.015 = 0.9 cfs
                Q_w = 2.0 - 0.9 = 1.1 cfs
                Flow in the first 2 ft adjacent to the curb is 1.1 cfs and 0.9 cfs in the remainder of the gutter.

Example 2 Given: T = 6 ft S_w = 0.0833 ft/ft
                       T_s = 6 - 1.5 = 4.5 ft   W = 1.5 ft
                       S_x = 0.03 ft/ft   n = 0.014
                       S = 0.04 ft/ft

Find: Flow in the composite gutter

Solution: (1) Use Figure 8-1 to find the gutter section capacity above the depressed section.
                Q_sn = 0.038
                Q_s = 0.038/0.014 = 2.7 cfs
           (2) Calculate W/T = 1.5/6 = 0.25 and
                S_w/S_x = 0.0833/0.03 = 2.78
                Use Figure 8-2 to find E_o = 0.64
           (3) Calculate the gutter flow using equation 8.3:
                Q = 2.7/(1 - 0.64) = 7.5 cfs
           (4) Calculate the gutter flow in width, W, using equation 8.2:
                Q_w = 7.5 - 2.7 = 4.8 cfs

8.7 Grate Inlet Design

Inlets are drainage structures utilized to collect surface water through grate or curb openings and convey it to storm drains or direct outlet to culverts. Grate inlets subject to traffic should be bicycle safe and be load bearing adequate. Appropriate frames should be provided.

Inlets used for the drainage of highway surfaces can be divided into three major classes. These classes are the following.

1. Grate Inlets - These inlets include grate inlets consisting of an opening in the gutter covered by one or more grates, and slotted inlets consisting of a pipe cut along the longitudinal axis with a grate of spacer bars to form slot openings.
2. Curb-Opening Inlets - These inlets are vertical openings in the curb covered by a top slab.
3. Combination Inlets - These inlets usually consist of both a curb-opening inlet and a grate inlet placed in a side-by-side configuration, but the curb opening may be located in part upstream of the grate.

In addition, where significant ponding can occur, in locations such as underpasses and in sag vertical curves in depressed sections, it is good engineering practice to place flanking inlets on each side of the inlet at the low point in the sag. The flanking inlets should be placed so that they will limit spread on low gradient approaches to the level point and act in relief of the inlet at the low point if it should become clogged or if the design spread is exceeded.

The design of grate inlets will be discussed in this section, curb inlet design in Section 8.8, and combination inlets in Section 8.9.

Grate Inlets On Grade

The capacity of an inlet depends upon its geometry and the cross slope, longitudinal slope, total gutter flow, depth of flow and pavement roughness. The depth of water next to the curb is the major factor in the interception capacity of both gutter inlets and curb opening inlets. At low velocities, all of the water flowing in the section of gutter occupied by the grate, called frontal flow, is intercepted by grate inlets, and a small portion of the flow along the length of the grate, termed side flow, is intercepted. On steep slopes, only a portion of the frontal flow will be intercepted if the velocity is high or the grate is short and splash-over occur. For grates less than 2 feet long, intercepted flow is small.

Inlet interception capacity has been investigated by agencies and manufacturers of grates. For inlet efficiency data for various sizes and shapes of grates, refer to Hydraulic Engineering Circular No. 12, Federal Highway Administration and inlet grate capacity charts prepared by grate manufacturers.

A parallel bar grate is the most efficient type of gutter inlet; however, when crossbars are added for bicycle safety, the efficiency is greatly reduced. Where bicycle traffic is a design consideration, the curved vane grate and the tilt bar grate are recommended for both their hydraulic capacity and bicycle safety features. They also handle debris better than other grate inlets but the vanes of the grate must be turned in the proper direction.

Where debris is a problem, consideration should be given to debris handling efficiency rankings of grate inlets from laboratory tests in which an attempt was make to qualitatively simulate field conditions. Table 8-3 presents the results of debris handling efficiencies of several grates.

The ratio of frontal flow to total gutter flow, E_o, for straight cross slope is expressed by the following equation:

$$E_o = Q_w/Q = 1 - (1 - W/T)^{2.67} \tag{8.4}$$

Table 8-3 Grate Debris Handling Efficiencies

Rank	Grate	Longitudinal Slope (0.005)	(0.04)
1	CV - 3-1/4 - 4-1/4	46	61
2	30 - 3-1/4 - 4	44	55
3	45 - 3-1/4 - 4	43	48
4	P - 1-7/8	32	32
5	P - 1-7/8 - 4	18	28
6	45 - 2-1/4 - 4	16	23
7	Reticuline	12	16
8	P - 1-1/8	9	20

Source: HEC-12, 1984

Where: Q =   total gutter flow, cfs
        Q_w = flow in width W, cfs
        W =   width of depressed gutter or grate, ft
        T =   total spread of water in the gutter, ft

Figure 8-2 provides a graphical solution of E_o for either depressed gutter sections or straight cross slopes.

The ratio of side flow, Q_s, to total gutter flow is:

$$Q_s/Q = 1 - Q_w/Q = 1 - E_o \qquad (8.5)$$

The ratio of frontal flow intercepted to total frontal flow, R_f, is expressed by the following equation:

$$R_f = 1 - 0.09 \ (V - V_0) \qquad (8.6)$$

Where: V =   velocity of flow in the gutter, ft/s
        V_o = gutter velocity where splash-over first occurs, ft/s

This ratio is equivalent to frontal flow interception efficiency. Figure 8-4 provides a solution of equation 8.6 which takes into account grate length, bar configuration and gutter velocity at which splash-over occurs. The gutter velocity needed to use Figure 8-4 is total gutter flow divided by the area of flow.

The ratio of side flow intercepted to total side flow, R_s, or side flow interception efficiency, is expressed by:

$$R_s = 1 \ / \ [1 + (0.15V^{1.8}/S_xL^{2.3})] \qquad (8.7)$$

Where: L = length of the grate, ft
        Figure 8-5 provides a solution to equation 8.7.
        The efficiency, E, of a grate is expressed as:

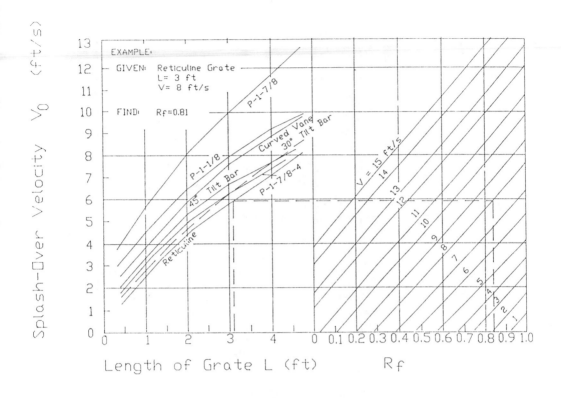

Figure 8-4 Grate Inlet Frontal Flow Interception Efficiency

Source: HEC-12, 1984

$$E = R_fE_o + R_s(1 - E_o) \qquad (8.8)$$

The interception capacity of a grate inlet on grade is equal to the efficiency of the grate multiplied by the total gutter flow:

$$Q_i = EQ = Q[R_fE_o + R_s(1 - E_o)] \qquad (8.9)$$

Example Problem

The following example illustrates the use of this procedure.

Given:

$W = 2$ ft	$T = 8$ ft
$S_x = 0.025$ ft/ft	$S = 0.01$ ft/ft
$E_o = 0.69$	$Q = 3.0$ cfs
$V = 3.1$ ft/s	Gutter depression = 2 in.

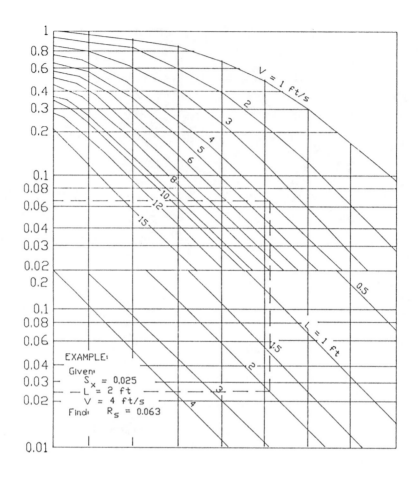

Figure 8-5 Grate Inlet Side Flow Interception Efficiency

Source: HEC-12, 1984

Find: Interception capacity of:
             (1) a curved vane grate, and
             (2) a reticuline grate 2-ft long and 2-ft wide

Solution: From Figure 8-4 for Curved Vane Grate, $R_f = 1.0$
             From Figure 8-4 for Reticuline Grate, $R_f = 1.0$
             From Figure 8-5 $R_s = 0.1$ for both grates.
             From equation 8.9:
             $Q_i = 3.0[1.0 * 0.69 + 0.1(1 - 0.69)] = 2.2$ cfs

For this example the interception capacity of a curved vane grate is the same as that for a reticuline grate for the sited conditions.

Grate Inlet In Sag

A grate inlet in a sag operates as a weir up to a certain depth dependent on the bar configuration and size of the grate and as an orifice at greater depths. For a standard gutter inlet grate, weir operation continues to a depth of about 0.4 foot above the top of grate and when depth of water exceeds about 1.4 feet, the grate begins to operate as an orifice. Between depths of about 0.4 foot and about 1.4 feet, a transition from weir to orifice flow occurs.

The capacity of grate inlets operating as a weir is:

$$Q_i = CPd^{1.5} \tag{8.10}$$

Where: P = perimeter of grate excluding bar widths and the side against the curb, ft
        C = 3.0
        d = depth of water above grate, ft

The capacity of grate inlets operating as an orifice is:

$$Q_i = CA(2gd)^{0.5} \tag{8.11}$$

Where: C = 0.67 orifice coefficient
        A = clear opening area of the grate, ft^2
        g = 32.2 ft/s^2

Figure 8-6 is a plot of equations 8.10 and 8.11 for various grate sizes. The effects of grate size on the depth at which a grate operates as an orifice is apparent from the chart. Transition from weir to orifice flow results in interception capacity less than that computed by either the weir or the orifice equation. This capacity can be approximated by drawing in a curve between the lines representing the perimeter and net area of the grate to be used.

Example Problem

The following example illustrates the use of Figure 8-6.

Given: A symmetrical sag vertical curve with equal bypass from inlets upgrade of the low point; allow for 50% clogging of the grate.

           Q_b = 3.6 cfs   Q = 8 cfs, 10-year storm (design)
           Q_b = 4.4 cfs   Q = 11 cfs, 25-year storm (check)
           T = 10 ft, design S_x = 0.05 ft/ft
           d = TS_x = 0.5 ft

Find: Grate size for design Q and depth at curb for check Q. Check spread at S = 0.003 on approaches to the low point.

Solution: From Figure 8-6, a grate must have a perimeter of 8 ft to intercept 8 cfs at a depth of 0.5 ft. Some assumptions must be made regarding the nature of the clogging in order to compute the capacity of a partially clogged grate. If the area of a grate is 50 percent covered by debris so that the debris-covered portion does not con-tribute to interception, the effective perimeter will be reduced by a lesser amount than 50 percent. For example if a 2-ft x 4-ft grate is clogged so that the effective width is 1-ft, then the perimeter, P = 1 + 4 + 1 = 6 ft, rather than 8 ft, the total perimeter, or

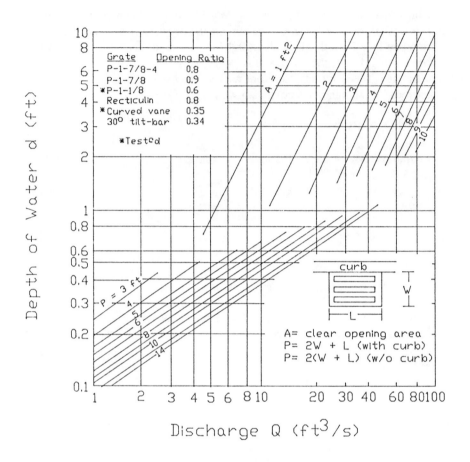

Figure 8-6 Grate Inlet Capacity In Sump Conditions

Source: HEC-12, 1984

4 ft, half of the total perimeter. The area of the opening would be reduced by 50 percent and the perimeter by 25 percent. Therefore, assuming 50 percent clogging along the length of the grate, a 4 x 4, a 2 x 6, or a 3 x 5 grate would meet requirements of an 8-ft perimeter 50 percent clogged.

Assuming that the installation chosen to meet the design is a double 2 x 3 ft grate, for 50 percent clogged conditions:

$P = 1 + 6 + 1 = 8$ ft

For 10-year flow: $d = 0.5$ ft (from Figure 8-6)
For 25-year flow: $d = 0.6$ ft (from Figure 8-6),
                        $T = 12.0$ ft

At the check flow rate, ponding will extend 2 ft into a traffic lane if the grate is 50 percent clogged in the manner assumed.

The American Society of State Highway and Transportation Officials (AASHTO) geometric policy recommends a gradient of 0.3 percent within 50 ft of the level point in a sag vertical curve.

Check T at S = 0.003 for the design and check flow:
Q = 3.6 cfs, T = 8.2 ft (10-year storm) (Figure 8-1)
Q = 4.4 cfs, T = 9 ft (25-year storm) (Figure 8-1)

Thus a double 2 x 3-ft grate 50 percent clogged is adequate to intercept the design flow at a spread which does not exceed design spread and spread on the approaches to the low point will not exceed design spread. However, the tendency of grate inlets to clog completely warrants consideration of a combination inlet, or curb-opening inlet in a sag where ponding can occur, and flanking inlets on the low gradient approaches.

8.8 Curb Inlet Design

Curb Inlets On Grade

Following is a discussion of the procedures for the design of curb inlets on grade which is followed by curb inlets in a sump.

Curb-opening inlets are effective in the drainage of highway pavements where flow depth at the curb is sufficient for the inlet to perform efficiently. Curb openings are relatively free of clogging tendencies and offer little interference to traffic operation. They are a viable alternative to grates in many locations where grates would be in traffic lanes or would be hazardous for pedestrians or bicyclists. The length of curb-opening inlet required for total interception of gutter flow on a pavement section with a straight cross slope is determined using Figure 8-7. The efficiency of curb-opening inlets shorter than the length required for total interception is determined using Figure 8-8.

The length of inlet required for total interception by depressed curb-opening inlets or curb-openings in depressed gutter sections can be found by the use of an equivalent cross slope, S_e, in the following equation.

$$S_e = S_x + S'_w E_o \qquad (8.12)$$

Where: E_o = ratio of flow in the depressed section to total gutter flow
S'_w = cross slope of the gutter measured from the cross slope of the pavement, ft/ft
S_x = (a/12W), ft/ft
a = gutter depression, in.

It is apparent from examination of Figure 8-7 that the length of curb opening required for total interception can be significantly reduced by increasing the cross slope or the equivalent cross slope. The equivalent cross slope can be increased by use of a continuously depressed gutter section or a locally depressed gutter section.

Steps for using Figures 8-7 and 8-8 in the design of curb inlets on grade are given below.
(1) Determine the following input parameters:
Cross slope = S_x (ft/ft)   Longitudinal slope = S (ft/ft)
Gutter flow rate = Q (cfs)   Manning's n = n
Spread of water on pavement = T (ft) from Figure 8-1

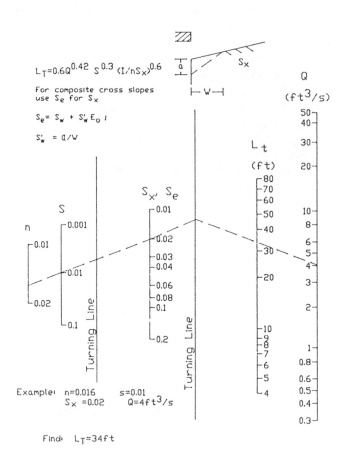

Figure 8-7 Curb-Opening And Slotted Drain Inlet Length For Total Interception

Source: HEC-12, 1984

(2) Enter Figure 8-7 using the two vertical lines on the left side labeled n and S. Locate the value for Manning's n and longitudinal slope and draw a line connecting these points and extend this line to the first turning line.

(3) Locate the value for the cross slope (or equivalent cross slope) and draw a line from the point on the first turning line through the cross slope value and extend this line to the second turning line.

(4) Using the far right vertical line labeled Q locate the gutter flow rate. Draw a line from this value to the point on the second turning line. Read the length required from the vertical line labeled L_T.

(5) If the curb-opening inlet is shorter than the value obtained in step 4, Figure 8-8 can be used to calculate the efficiency. Enter the x-axis with the L/L_T ratio and draw a vertical line upward to the E curve. From the point of intersection, draw a line horizontally to the intersection with the y-axis and read the efficiency value.

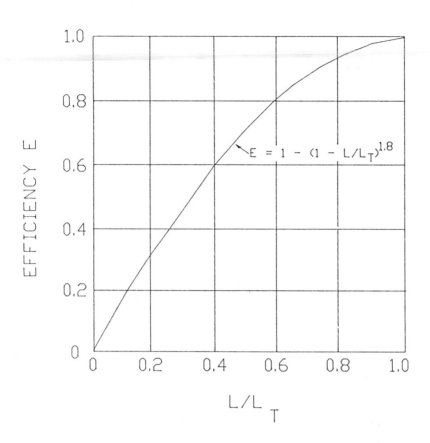

Figure 8-8 Curb-Opening And Slotted Drain Inlet Interception Efficiency

Source: HEC-12, 1984

Example Problem

Given: S_x = 0.03 ft/ft   n = 0.016
           S = 0.035 ft/ft S'_w = 0.083
           Q = 5 cfs

Find: (1) Q_i for a 10-ft curb-opening inlet
           (2) Q_i for a depressed 10-ft curb-opening inlet with a = 2 in., W = 2 ft, T = 8 ft
                 (Figure 8-1)

Solution: (1)   From Figure 8-7, L_T = 41 ft
                 L/L_T = 10/41 = 0.24
                 From Figure 8-8, E = 0.39
                 Q_i = EQ = 0.39 x 5 = 1.95 cfs = 2 cfs

(2) $Qn = 5.0 \times 0.016 = 0.08$ cfs
$S_w/S_x = (0.03 + 0.083)/0/03 = 3.77$
$T/W = 3.5$ (from Figure 8-3)
$T = 3.5 \times 2 = 7$ ft
$W/T = 2/7 = 0.29$ ft
$E_o = 0.72$ (from Figure 8-2)
Therefore, $S_e = S_x + S'_w E_o$
$S_e = 0.03 + 0.083(0.72) = 0.09$
From Figure 8-7, $L_T = 23$ ft
$L/L_T = 10/23 = 0.43$
From Figure 8-8, $E = 0.64$
$Q_i = 0.64 \times 5 = 3.2$ cfs

The depressed curb-opening inlet will intercept 1.6 times the flow intercepted by the un-depressed curb opening and over 60 percent of the total flow.

Curb Inlets In Sump

For the design of a curb-opening inlet in a sump location, the inlet operates as a weir to depths equal to the curb opening height and as an orifice at depths greater than 1.4 times the opening height. At depths between 1.0 and 1.4 times the opening height, flow is in a transition stage.

The capacity of curb-opening inlets in a sump location can be determined from Figure 8-9 which accounts for the operation of the inlet as a weir and as an orifice at depths greater than 1.4h. This figure is applicable to depressed curb-opening inlets and the depth at the inlet includes any gutter depression. The height (h) in the Figure assumes a vertical orifice opening (see sketch on Figure 8-9). The weir portion of Figure 8-9 is valid for a depressed curb-opening inlet when d $\leq$ (h + a/12).

The capacity of curb-opening inlets in a sump location with a vertical orifice opening but without any depression can be determined from Figure 8-10. The capacity of curb-opening inlets in a sump location with other than vertical orifice openings can be determined by using Figure 8-11.

Steps for using Figures 8-9, 8-10, and 8-11 in the design of curb-opening inlets in sump locations are given below.

(1) Determine the following input parameters:
Cross slope = S_x (ft/ft)
Spread of water on pavement = T (ft) from Figure 8-1
Gutter flow rate = Q (cfs) or dimensions of curb-opening inlet [L (ft) and H (in.)]
Dimensions of depression if any [a (in.) and W (ft)]

(2) To determine discharge given the other input parameters, select the appropriate Figure (8-9, 8-10, or 8-11 depending whether the inlet is in a depression and if the orifice opening is vertical).

(3) To determine the discharge (Q), given the water depth (d), locate the water depth value on the y-axis and draw a horizontal line to the appropriate perimeter (p), height (h), length (L), or width times length (hL) line. At this intersection draw a vertical line down to the x-axis and read the discharge value.

(4) To determine the water depth given the discharge, use the procedure described in step 3 except you enter the figure at the value for the discharge on the x-axis.

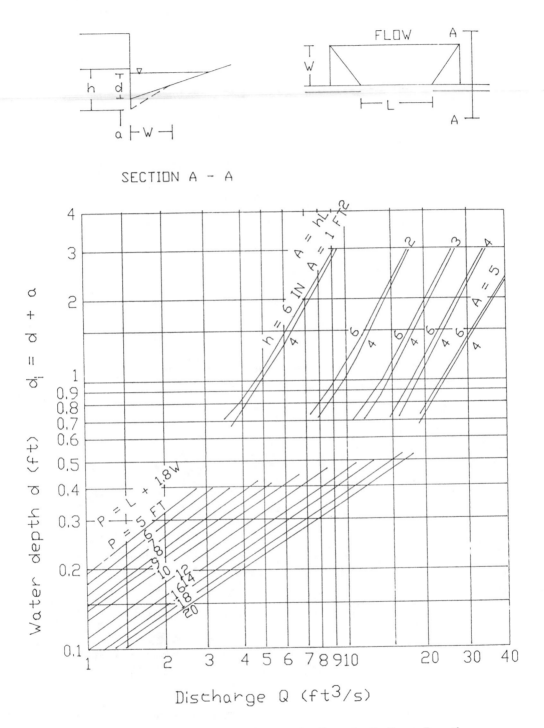

SECTION A - A

Figure 8-9 Depressed Curb-Opening Inlet Capacity In Sump Locations

Source: AASHTO Model Drainage Manual, 1991

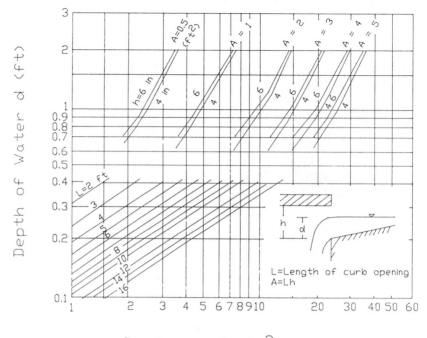

Figure 8-10 Curb-Opening Inlet Capacity In Sump Locations
Source: AASHTO Model Drainage Manual, 1991

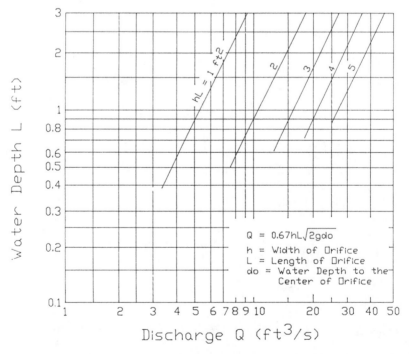

Figure 8-11 Curb-Opening Inlet Orifice Capacity - Inclined And Vertical Orifice Throats
Source: AASHTO Model Drainage Manual, 1991

Example Problem

Given: Curb-opening inlet in a sump location
           L = 5 ft   h = 5 in.
           (1)   Undepressed curb opening
                 S_x = 0.05 ft/ft
                 T = 8 ft
           (2)   Depressed curb opening
                 S_x = 0.05 ft/ft
                 a = 2 in.
                 W = 2 ft, T = 8 ft

Find: Discharge Q_i

Solution: (1)   d = TS_x = 8 x 0.05 = 0.4 ft
                 d < h
                 From Figure 8-10, Q_i = 3.8 cfs
           (2)   d = 0.4 ft
                 h + a/12 = (5 + 2/12)/12 = 0.43 ft
                 since d < 0.43 the weir portion of Figure 8-9 is applicable (lower portion of the Figure).
                 P = L + 1.8W = 5 + 3.6 = 8.6 ft
                 From Figure 8-9, Q_i = 5 cfs
                 At d = 0.4 ft, the depressed curb-opening inlet has about 30 percent more capacity than an inlet without depression.

8.9 Combination Inlets

On a continuous grade, the capacity of an unclogged combination inlet with the curb opening located adjacent to the grate is approximately equal to the capacity of the grate inlet alone. Thus capacity is computed by neglecting the curb opening inlet and the design procedures should be followed based on the use of Figures 8-4, 8-5 and 8-6.

All debris carried by storm water runoff that is not intercepted by upstream inlets will be concentrated at the inlet located at the low point, or sump. Because this will increase the probability of clogging for grated inlets, it is generally appropriate to estimate the capacity of a combination inlet at a sump by neglecting the grate inlet capacity. Assuming complete clogging of the grate, Figures 8-9, 8-10, and 8-11 for curb-opening inlets should be used for design.

8.10 Energy Losses In A Pipe System

Following are the equations needed to calculate the energy losses for the hydraulic grade line calculations for storm drainage systems.

Friction Losses

Energy losses from pipe friction may be determined by rewriting the Manning Equation.

$$S_f = [Qn/1.486\ A(R^{2/3})]^2 \tag{8.13}$$

Then the head losses due to friction may be determined by the formula:

$$H_f = S_f L \tag{8.14}$$

Where: H_f = friction head loss, ft
 S_f = friction slope, ft/ft
 L = length of outflow pipe, ft

Velocity Head Losses

From the time storm water first enters the storm drain system at the inlet until it discharges at the outlet, it will encounter a variety of hydraulic structures such as inlets, manholes, junctions, bends, contractions, enlargements and transitions, which will cause velocity head losses. Velocity losses may be expressed in a general form derived from the Bernoulli and Darcy-Weisback Equations.

$$H = KV^2/2g \tag{8.15}$$

Where: H = velocity head loss, ft
 K = loss coefficient for the particular structure
 V = velocity of flow, ft/s
 g = acceleration due to gravity, 32.2 ft/s^2

Entrance Losses

Following are the equations used for entrance losses for beginning flows.

$$H_{tm} = V^2/2g \tag{8.16}$$
$$H_e = KV^2/2g \tag{8.17}$$

Where: H_{tm} = terminal (beginning of run) loss, ft
 H_e = entrance loss for outlet structure, ft
 K = 0.5 (assuming square-edge)
 Other terms defined above.

Junction Losses - Incoming Opposing Flows

The head loss at a junction, H_{j1} for two almost equal and opposing flows meeting head on with the outlet direction perpendicular to both incoming directions, is considered as the total velocity head of outgoing flow.

$$H_{j1} = (V_3^2) \text{ (outflow)}/2g \tag{8.18}$$

Where: H_{j1} = junction losses, ft
 Other terms are defined above.

Junction Losses - Changes In Direction Of Flow

When main storm drain pipes or lateral lines meet in a junction, velocity is reduced within the chamber and specific head increases to develop the velocity needed in the outlet pipe. The sharper the bend (approaching 90°) the more severe this energy loss becomes. When the outlet conduit is sized, determine the velocity and compute head loss in the chamber by the formula:

$$H_b = K(V^2) \text{ (outlet)}/2g \qquad\qquad (8.19)$$

Where: H_b = bend head loss, ft
   K = junction loss coefficient
Table 8-4 lists the values of K for various junction angles.

Table 8-4 Values Of K For Change In Direction Of Flow In Lateral

K	Degree of Turn (In Junction)
0.19	15
0.35	30
0.47	45
0.56	60
0.64	75
0.70	90 and greater

K values for other degree of turns can be obtained by interpolating between values in Table 8-4.

Junction Losses - Several Entering Flows

The computation of losses in a junction with several entering flows utilizes the principle of conservation of energy. For a junction with several entering flows, the energy content of the inflows is equal to the energy content of outflows plus additional energy required by the collision and turbulence of flows passing through the junction. The total junction losses at the sketched intersection are as follows.

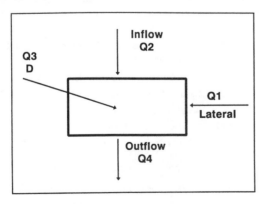

The following equation can be used to calculate junction losses:

$$H_{j2} = [(Q_4V_4^2)-(Q_1V_1^2)-(Q_2V_2^2)+(KQ_1V_1^2)]/(2gQ_4)] \tag{8.20}$$

Where: H_{j2} = junction losses, ft
           Q = discharges, cfs
           V = horizontal velocities, ft/s (V_3 is assumed to be zero)
           g = acceleration due to gravity, 32.2 ft/s^2
           K = bend loss factor
Where subscript nomenclature is as follows:
           Q_1 = 90° lateral, cfs
           Q_2 = straight through inflow, cfs
           Q_3 = vertical dropped-in flow from an inlet, cfs
           Q_4 = main outfall = total computed discharge, cfs
           V_1,V_2,V_3,V_4 are the horizontal velocities of foregoing flows, respectively in feet per second

Assume: $H_b = K(V_1^2)/2g$ for change in direction. Also, no velocity head of an incoming line is greater than the velocity head of the outgoing line. Water surface of inflow and outflow pipes in junction to be level.

When losses are computed for any junction condition for the same or a lesser number of inflows, the above equation will be used with zero quantities for those conditions not present. If more directions or quantities are at the junction, additional terms will be inserted with consideration given to the relative magnitudes of flow and the coefficient of velocity head for directions other than straight through.

The final step in designing a storm drain system is to check the hydraulic grade line (HGL) as described in the next section of this chapter. Computing the HGL will determine the elevation, under design conditions, to which water will rise in various inlets, manholes, junctions, etc.

In Figure 8-12 is a summary of energy losses which should be considered. Following this in Figure 8-13 is a sketch showing the proper and improper use of energy losses in developing a storm drain system.

8.11 Storm Drains

After the tentative locations of inlets, storm drains, and outfalls with tail-waters have been determined and the inlets sized, the next logical step is the computation of the rate of discharge to be carried by each storm drain and the determination of the size and gradient of pipe required to accommodate this discharge. This is done by proceeding in steps from upstream of a line to downstream to the point at which the line connects with other lines or the outfall, whichever is applicable. The discharge for a run is calculated, the storm drain serving that discharge is sized, and the process is repeated for the next run downstream. It should be recognized that the rate of discharge to be carried by any particular section of storm drain is not necessarily the sum of the inlet design discharge rates of all inlets above that section of pipe, but as a general rule is somewhat less than this total. It is useful to understand that the time of concentration is most influential and as the time of concentration grows larger, the proper rainfall intensity to be used in the design grows smaller.

For ordinary conditions, storm drains should be sized on the assumption that they will flow full or practically full under the design discharge but will not be placed under pressure head. The Manning Formula is recommended for capacity calculations.

$$H_{tm} = \frac{V^2}{2g}$$

TERMINAL JUNCTION LOSSES
(at beginning of run)

Where g = gravitational constant
32.2 feet per second
per second

$$H_e = 0.5 \frac{V^2}{2g}$$

ENTRANCE LOSSES
(for structure at beginning of run)
Assuming square edge

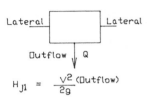

$$H_{j1} = \frac{V^2}{2g} \text{(Outflow)}$$

JUNCTION LOSSES

Use only where flows are
identical to above, otherwise
use H_{j2} Equation.

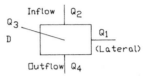

$$H_{j2} = \frac{Q_4 V_4{}^2 - Q_1 V_1{}^2 - Q_2 V_2{}^2 + K Q_1 V_1{}^2}{2g Q_4}$$

JUNCTION LOSSES
(After FHWA)

Total losses to include H_{j2} plus losses
for change in direction of less than 90
(H_b).

Where   K = Bend loss factor
Q_3 = Vertical dropped-in flow from
an inlet
V_3 = Assumed to be zero

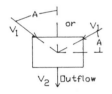

$$H_b = \frac{K V_1{}^2}{2g}$$

BEND LOSSES
(changes in direction of flow)

FRICTION LOSS (H_f)

$$H_f = S_f \times L$$

Where H_f = friction head
S_f = friction slope
L   = length of conduit

Where K	Degree of Turn in Junction
0.19	15
0.35	30
0.47	45
0.56	60
0.64	75
0.70	90

$$S = \left(\frac{Qn}{1.486 A R^{\frac{2}{3}}} \right)$$

Where   Q = discharge of conduit
n = Mannings coefficient of
roughness (use 0.013
for R. C. Pipes)
A = area of conduit
R = hydraulic radius of conduit
(D/4 for round pipe)

TOTAL ENERGY LOSSES AT EACH JUNCTION

$$H_T = H_{tm} + H_e + H_{ji} \quad \text{or} \quad H_{j2} + H_b + H_f$$

Figure 8-12 Summary Of Energy Losses

Source: AASHTO Model Drainage Manual, 1991

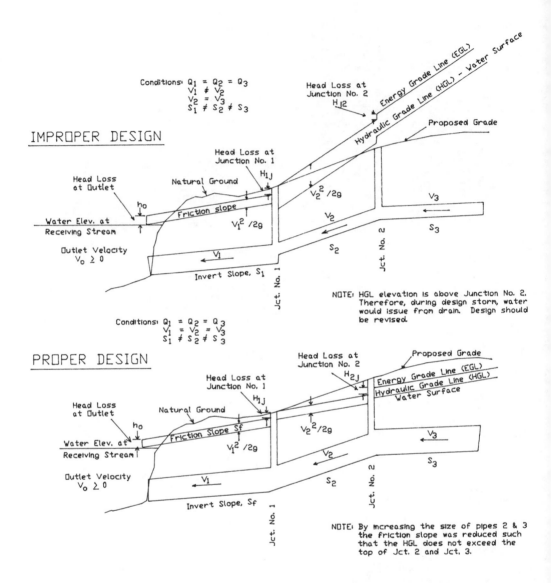

Figure 8-13 Energy And Hydraulic Grade Lines For Storm Drains Under Constant Discharge

Source: AASHTO Model Drainage Manual, 1991

The standard recommended maximum and minimum slopes for storm drains should conform to the following criteria:
1. The maximum hydraulic gradient should not produce a velocity that exceeds 20 feet per second.
2. The minimum desirable physical slope should be 0.5 percent or the slope which will produce a velocity of 2.5 feet per second when the storm drain is flowing full, whichever is greater.

Systems should generally be designed for non-pressure conditions. When hydraulic calculations do not consider minor energy losses such as expansion, contraction, bend, junction, and manhole losses, the elevation of the hydraulic gradient for design flood conditions should be at least 1.0 foot below the ground elevation. As a general rule, minor losses should be considered when the velocity exceeds 6 feet per second (lower if flooding could cause critical problems). If all minor energy losses are accounted for, it is usually acceptable for the hydraulic gradient to reach the gutter elevation or the bottom of the casting (Nashville, 1988).

Formulas For Gravity And Pressure Flow

The most widely used formula for determining the hydraulic capacity of storm drain pipes for gravity and pressure flows is the Manning Formula and it is expressed by the following equation:

$$V = [1.486 \ R^{2/3}S^{1/2}]/n \tag{8.21}$$

Where: V = mean velocity of flow, ft/s
  R = the hydraulic radius, ft - defined as the area of flow divided by the wetted flow surface or wetted perimeter (A/WP)
  S = the slope of hydraulic grade line, ft/ft
  n = Manning's roughness coefficient
In terms of discharge, the above formula becomes:

$$Q = [1.486 \ AR^{2/3}S^{1/2}]n \tag{8.22}$$

Where: Q = rate of flow, cfs
  A = cross sectional area of flow, ft^2
For pipes flowing full, the above equations become:

$$V = [0.590 \ D^{2/3}S^{1/2}]/n \tag{8.23}$$
$$Q = [0.463 \ D^{8/3}S^{1/2}]/n \tag{8.24}$$

Where: D = diameter of pipe, ft

The Manning's equation can be written to determine friction losses for storm drain pipes as:

$$H_f = [2.87 \ n^2V^2L]/[S^{4/3}] \tag{8.25}$$
$$H_f = [29 \ n^2LV^2]/[(R^{4/3})(2g) \tag{8.26}$$

Where: H_f = total head loss due to friction, ft
  n = Manning's roughness coefficient
  D = diameter of pipe, ft
  L = length of pipe, ft
  V = mean velocity, ft/s
  R = hydraulic radius, ft
  g = acceleration of gravity, 32.2 ft/s^2
The nomograph solution of Manning's formula for full flow in circular storm drain pipes which is shown on Figures 8-14 - 8-16, and Figure 8-17, has been provided to solve the Manning's equation for part full flow in storm drains.

Saatci (1990) provides a direct solution for depth and velocity for the partial flow situation if discharge (Q), slope (S) and Manning's n are known. The procedure, originally derived for metric units but converted for application here, lends itself to spreadsheet calculations for storm sewer design. The following equations, derived from Manning equations and geometric considerations, are used.

$$K = 0.673 \; Qn \; D^{-8/3} \; S^{-1/2} \tag{8.27}$$
$$\theta = 3\pi/2 \; \{1- [1- (\pi K)^{1/2}]^{1/2}\}^{1/2} \tag{8.28}$$
$$h/D = 1/2 \; [1 - \cos (\theta/2)] \tag{8.29}$$
$$A = D^2 \; [(\theta - \sin \theta)/8] \tag{8.30}$$

where: Q = the discharge (m³/sec)
       K = a constant
       h = the depth, ft)
       D = the pipe diameter, ft
       n = Manning's roughness
       S = the pipe slope, ft/ft
       θ = the central angle

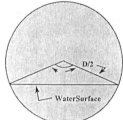

Equation 8.28 is fit to the curves of K versus θ. This method is valid for values of θ from 0 to 265 degrees. The procedure is illustrated by an example. A 2.5 ft diameter pipe has a flow of 2 cfs, an n value of 0.013 and a slope of 0.0022. Find the depth and velocity of flow.
From equation 8.27:

$$K = 0.673 Qn D^{-8/3} S^{-1/2} = (0.673)(2)(0.013)(2.5)^{-8/3}(0.0022)^{-1/2} = 0.0324$$

Substituting K into equation 8.28 gives:

$$\theta = 3\pi/2 \; \{1- [1- (\pi K)^{1/2}]^{1/2}\}^{1/2} \; = 3\pi/2 \; \{1-[1-(\pi \; 0.0324)^{1/2}]^{1/2}\}^{1/2} =$$
1.9702 rad = 112.9 deg which is less than 265 deg (ok).

Substituting into equation 8.29 gives:

$$h/D = 1/2 \; [1 - \cos (\theta/2)]$$
$$h/D = 1/2 \; [1 - \cos (112.9/2)] = 0.224$$

Then h = (0.224)(2.5) = 0.560 ft. The flow area is calculated from equation 8.30 as:

$$A = D^2 \; [(\theta - \sin \theta)/8] = (2.5)^2 \; [(1.9702 - \sin 112.9)/8 = 0.875 \; ft^2$$

And the velocity is Q/A = 2/0.875 = 2.286 ft/s.

Hydraulic Grade Line

The total energy in a pipe is equal to:

$$V^2/2g + P/\gamma + Z \tag{8.31}$$

Where: P = Pressure head, lb/ft²
       γ = specific weight of water, 62.4 lb/ft³

V = average velocity of flow, ft/s
g = acceleration of gravity, 32.2 ft/s²
Z = elevation relative to some datum, ft

Each term in this equation has the dimension of length. The quantity $P/\gamma + Z$ establishes the elevation of the hydraulic grade line. For non-pressure flow the term P/γ is just the flow depth. The hydraulic grade line is the elevation to which water would rise if not for the fact that it is contained in a pipe under pressure.

The general design procedure is to establish a downstream control elevation. From that elevation calculations proceed upstream from junction to junction or manhole to manhole. At the lower end of each junction the pipe friction losses from the downstream section, expressed in terms of feet of loss, are added to the downstream hydraulic gradeline elevation. At the upstream end of each junction the minor losses through the junction are added. If the transition is to a ditch section the losses added are the entrance losses to the pipe system. This then establishes the starting elevation of the hydraulic gradeline for the next upstream leg of the calculations.

Hydraulic Grade Line Design Procedure

The head losses are calculated beginning from the downstream control point to the first junction and the procedure is repeated for the next junction. The computation for an outlet control may be tabulated on Figure 8-18 using the following procedure:

1. Enter in Col. 1 the station for the junction immediately upstream of the outflow pipe. Hydraulic grade line computations begin at the outfall and are worked upstream taking each junction into consideration.
2. Enter in Col. 2 the outlet water surface elevation if the outlet will be submerged during the design storm or 0.8 diameter plus invert out elevation of the outflow pipe, whichever is greater.
3. Enter in Col. 3 the diameter (D_o) of the outflow pipe.
4. Enter in Col. 4 the design discharge (Q_o) for the outflow pipe.
5. Enter in Col. 5 the length (L_o) of the outflow pipe.
6. Enter in Col. 6 the friction slope (S_f) in ft/ft of the outflow pipe. This can be determined by using the following formula:

$$S_f = (Q^2)/K \qquad (8.32)$$

Where: S_f = friction slope, ft/ft
K = $[1.486\ AR^{2/3}]/n$, ft³

7. Multiply the friction slope (S_f) in Col. 6 by the length (L_o) in Col. 5 and enter the friction loss (H_f) in Col. 7. On curved alignments, calculate curve losses by using the formula H_c = 0.002 (Δ)($V_o^2/2g$), where Δ = angle of curvature in degrees and add to the friction loss.
8. Enter in Col. 8 the velocity of the flow (V_o) of the outflow pipe.
9. Enter in Col. 9 the contraction loss (H_o) by using the formula H_o = $[0.25\ (V_o^2)]/2g$, where g = 32.2 ft/s².
10. Enter in Col. 10 the design discharge (Q_i) for each pipe flowing into the junction. Neglect lateral pipes with inflows of less than ten percent of the mainline outflow. Inflow must be adjusted to the mainline outflow duration time before a comparison is made.
11. Enter in Col. 11 the velocity of flow (V_i) for each pipe flowing into the junction (for exception see Step 10).

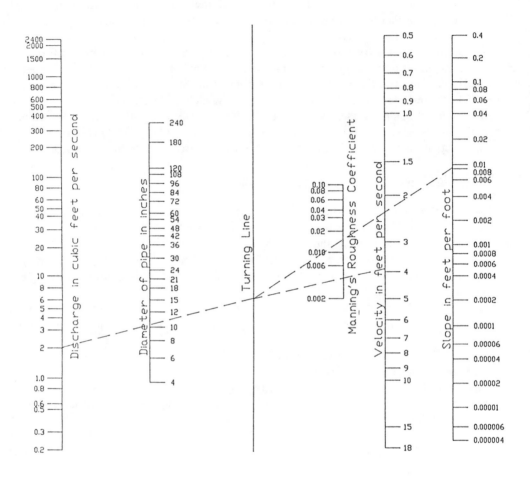

Figure 8-14 Nomograph For Solution Of Manning's Formula For Flow In Storm Drains

Source: AASHTO Model Drainage Manual, 1991

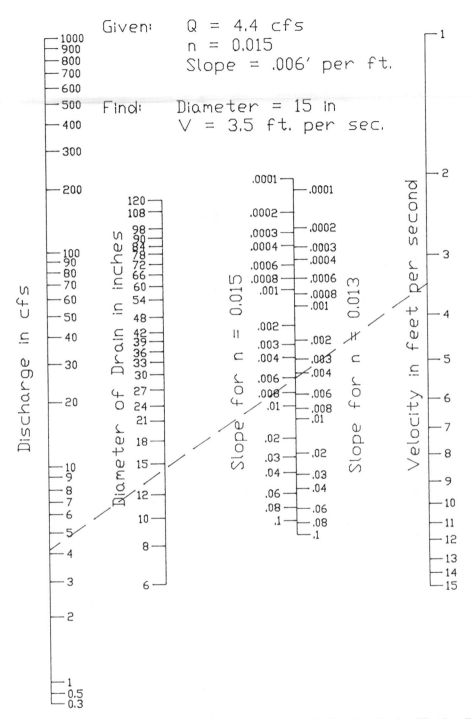

Given: Q = 4.4 cfs
n = 0.015
Slope = .006′ per ft.

Find: Diameter = 15 in
V = 3.5 ft. per sec.

Figure 8-15 Nomograph For Computing Required Size Of Circular Drain, Flowing Full n = 0.013 or 0.015

Source: AASHTO Model Drainage Manual, 1991

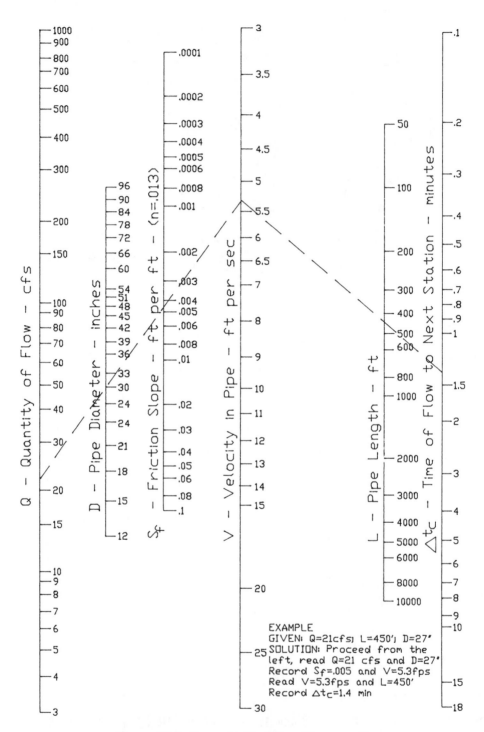

Figure 8-16 Concrete Pipe Flow Nomograph

Source: AASHTO Model Drainage Manual, 1991

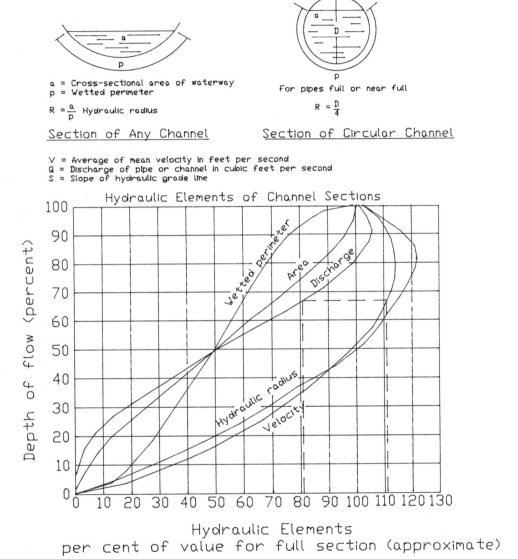

a = Cross-sectional area of waterway
p = Wetted perimeter

$R = \frac{a}{p}$ Hydraulic radius

Section of Any Channel

For pipes full or near full

$R = \frac{D}{4}$

Section of Circular Channel

V = Average of mean velocity in feet per second
Q = Discharge of pipe or channel in cubic feet per second
S = Slope of hydraulic grade line

Figure 8-17 Values Of Various Elements Of Circular Section For Various Depths Of Flow

Source: AASHTO Model Drainage Manual, 1991

12. Enter in Col. 12 the product of Q_i times V_i for each inflowing pipe. When several pipes inflow into a junction, the line producing the greatest Q_i times V_i product is the line which will produce the greatest expansion loss (H_i). (For exception, see Step 10).

13. Enter in Col. 13 the controlling expansion loss (H_i) using the formula $H_i = [0.35\ (V_i^2)]/2g$.

14. Enter in Col. 14 the angle of skew of each inflowing pipe to the outflow pipe (for exception, see Step 10).

15. Enter in Col. 15 the greatest bend loss (H_Δ) calculated by using the formula $H_\Delta = [KV_i^2]/2g$ where $K =$ the bend loss coefficient corresponding to the various angles of skew of the inflowing pipes.

16. Enter in Col. 16 the total head loss (H_t) by summing the values in Col. 9 (H_o), Col. 13 (H_i), and Col. 15 (H_Δ).

17. If the junction incorporates adjusted surface inflow of ten percent or more of the mainline outflow, i.e., drop inlet, increase H_t by 30 percent and enter the adjusted H_t in Col. 17.

18. If the junction incorporates full diameter inlet shaping, such as standard manholes, reduce the value of H_t by 50 percent and enter the adjusted value in Col. 18.

19. Enter in Col. 19 the FINAL H, the sum of H_f and H_t, where H_t is the final adjusted value of the H_t.

20. Enter in Col. 20 the sum of the elevation in Col. 2 and the Final H in Col. 19. This elevation is the potential water surface elevation for the junction under design conditions.

21. Enter in Col. 21 the rim elevation or the gutter flow line, whichever is lowest, of the junction under consideration in Col. 20. If the potential water surface elevation exceeds the rim elevation or the gutter flow line, whichever is lowest, adjustments are needed in the system to reduce the elevation of the hydraulic grade line (H.G.L.).

22. Repeat the procedure starting with Step 1 for the next junction upstream.

Figure 8-18 can be used to summarize the hydraulic grade line computations.

All storm drains should be designed such that velocities of flow will not be less than 2.5 feet per second at design flow or lower, with a minimum slope of 0.5 percent. For very flat flow lines the general practice is to design components so that flow velocities will increase progressively throughout the length of the pipe system. Upper reaches of a storm drain system should have flatter slopes than the slopes of lower reaches. Progressively increasing slopes keep solids moving toward the outlet and deters settling of particles due to steadily increasing flow streams.

The slopes are calculated by the modified Manning formula (term previously defined):

$$S = [(nV)^2]/[2.208\ R^{4/3}] \tag{8.33}$$

If downstream drainage facilities are undersized for the design flow, an above- or below-ground detention structure may be needed to reduce the possibility of flooding. The required storage volume can be provided by using larger than needed storm drain pipes sizes and restrictors to control the release rates at manholes and/or junction boxes in the storm drain system. The same design criteria for sizing the detention basin is used to determine the storage volume required in the system. Figure 8-19 can be used to summarize the storm drain design computations.

Example Problem

The following example will illustrate the hydrologic calculations needed for storm drain design using the rational formula (see Hydrology Chapter for Rational Method description and procedures). Figure 8-20 shows a hypothetical storm drain system that will be used in this example. Following is a tabulation of the data needed to use the rational equation to calculate inlet flow rate for the seven inlets shown in the system layout.

PROJECT

	HYDRAULIC GRADE LINE								JUNCTION LOSS											
STATION	Outlet Water Surf. Elev.	D_o	Q_o	L_o	Sf_o	H_f	V_o	H_o	Q_i	V_i	$Q_i V_i \frac{V_i^2}{2g}$	H_i	ANGLE	H_Δ	H_t	1.3 H_t	0.5 H_t	FINAL H	Inlet Water Surf. Elev.	Rim. Elev.
(1)	(2)	(3)	(4)	(5)	(6)	(7)	(8)	(9)	(10)	(11)	(12)	(13)	(14)	(15)	(16)	(17)	(18)	(19)	(20)	(21)

$$H_i = 0.35 \frac{V_i^2}{2g} \qquad H_o = 0.25 \frac{V_o^2}{2g} \qquad H_\Delta = K \frac{V_i^2}{2g}$$

FINAL $H = H_f + H_t$

$H_t = H_o + H_i + H_\Delta$

90° K = 0.70	50° K = 0.47	20° K = 0.16
80° K = 0.66	40° K = 0.38	15° K = 0.10
70° K = 0.61	30° K = 0.28	
60° K = 0.55	25° K = 0.22	

Figure 8-18 Hydraulic Grade Line Computation Form

Source: AASHTO Model Drainage Manual, 1991

Table 8-5 Hydrologic Data

Inlet[a]	Drainage Area	Time of Concentration	Rainfall Intensity	Runoff Coefficient	Inlet Flow[b] cfs
1	2.0	8	6.3	.9	11.3
2	3.0	10	5.9	.9	15.8
3	2.5	9	6.1	.9	13.6
4	2.5	9	6.1	.9	13.6
5	2.0	8	6.3	.9	11.3
6	2.5	9	6.1	.9	13.6
7	2.0	8	6.3	.9	11.3

[a] Inlet and storm drain system configuration are shown in Figure 8-20
[b] Calculated using the Rational Equation (see Hydrology Chapter).

The following table shows the data and results of the calculations needed to determine the design flow rate in each segment of the hypothetical storm drain system.

Table 8-6 Storm Drain System Calculations

Storm Drain Segment	Drainage Area (acres)	Time of Concentration (minutes)	Rainfall Intensity (inches/hr)	Runoff Coefficient	Design Flow Rate (cfs)
I_1-M^1	2.0	8	6.3	.9	11.3
I_2-M_1	3.0	10	5.9	.9	15.8
M_1-M_2	5.0	10.5	5.8	.9	25.9
I_3-M_2	2.5	9	6.1	.9	13.6
I_4-M_2	2.5	9	6.1	.9	13.6
M_2-M_3	10.0	11.5	5.6	.9	50.2
I_5-M_3	2.0	8	6.3	.9	11.3
I_6-M_3	2.5	9	6.1	.9	13.6
M_3-M_4	14.5	13.5	5.3	.9	68.6
I_7-M_4	2.0	8	6.3	.9	11.3
M_4-O	16.5	14.7	5.1	.9	75.4

8.12 Environmental Design Considerations

Roadway Pollution

In 1972 Sartor and Boyd developed data on the accumulation of pollutants in streets and roads. Well over 75 percent of all pollutants accumulate within three to five feet of the gutter. The types of pollutants coming off roads include solids, metals, nutrients, oil and grease, bacteria and other pollutants. Specific sources include both vehicle related pollutants and a myriad of other contributors. Some of the more common are (Kobinger, 1982):

COMPUTED ____________ DATE ____________

CHECKED ____________ DATE ____________

ROUTE ____________

SECTION ____________

COUNTY ____________

STATION		LENGTH FEET	DRAINAGE AREA 'A' (ACRES)		RUNOFF COEFFICIENT 'C'	'A' 'C'		FLOW TIME MIN.		RAINFALL INTENSITY 'I' in/hour	TOTAL RUNOFF C.I.A.= Q (c.f.s.)	DIAMETER PIPE inches	CAPACITY FULL c.f.s.	VELOCITY (f.p.s.)		INVERT ELEV		MANHOLE INVERT DROP	SLOPE OF DRAIN ft/ft
FROM	TO		INCREMENT	TOTAL		INCREMENT	TOTAL	TO UPPER END	IN SECTION					FLOWING FULL	DESIGN FULL	UPPER END	LOWER END		

Figure 8-19 Storm Drain Computation Form

Source: AASHTO Model Drainage Manual, 1991

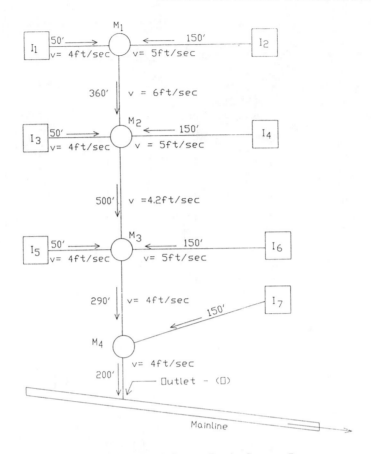

Figure 8-20 Hypothetical Storm Drain System Layout

Source: AASHTO Model Drainage Manual, 1991

- exhaust emissions containing nitrogen oxides, hydrocarbons, carbon monoxide, metals and salts which can settle over the municipality;
- particulants from pavement wear and windblown deposits;
- nutrients and pesticides from roadside maintenance programs;
- sediment from bank and shoulder erosion;
- hydrocarbons from leaks and spills;
- bacteria from fecal deposition and animal carcasses;
- thermal impacts from solar heating of pavement;
- a number of trace metals from rust, tire wear, brake lining wear, engine wear, plating deterioration, etc., and
- salts from deicing activities.

Studies have also shown that actual physical impacts of these pollutants are very site specific and hard to quantify. Both acute (first flush) and chronic impacts have been reviewed in the literature. Sediment sampling from commercial and residential areas has indicated the deposition of metals and PCB's downstream from outfall points (Shaheen, 1975). Many of the metals go through several changes in species making them less available for later uptake. Average density of traffic (ADT) is the greatest predictor of pollution with a value of 30,000 being the normal dividing point between a rural and urban roadway. Specific land use and traffic type also heavily influences the pollutant type and intensity.

A number of state DOT's have instituted programs to investigate storm water pollution from their highways most notably California, Florida, Washington and Virginia. The Federal Highway Administration has an ongoing program of study and research in this area and has published a number of reports on the subject most notably three separate studies: (1) characterizing the pollution sources and types, (2) developing possible best management practices (BMPs) and (3) developing a design and analysis methodology. Attempts have been made to predict pollutant discharge using both physically based modelling and regression equations. In-stream aquatic sampling has been used as a screening technique to prioritize management programs with some success in Washington State (Horner & Mar, 1985).

BMP measures to treat highway runoff fall into several categories including: source control, post-deposition treatment, and post-runoff treatment. Source controls include: traffic regulation, litter control, chemical use controls. Post-deposition treatment includes street cleaning, debris removal, spill cleanup, and curb or barrier elimination. Post-runoff treatment includes: vegetative controls, wet detention, infiltration and wetlands.

Design of streets to take into account sensitive wetland or other water areas can have a great long-term effect in reduction of pollutants. In many situations there is ample room along road shoulders or within interchange areas to develop highway controls. At the least, vegetative controls and erosion protection can be built into many designs. Some specific structural BMPs include:

- velocity controls,
- retention and infiltration ponds,
- streets without curbs,
- vegetated filter strips and overland flow,
- street designs for sumped catch basins,
- coalescing plate oil/water separator,
- first flush devices,
- vegetated pavement blocks (parking).

- direct discharge reduction,
- infiltration/exfiltration trenches,
- grassed swales,
- screened catch basins,
- exfiltrating curb & gutter,
- oil/grit separators,
- porous pavement (parking), and

There are also some low cost non-structural operations, maintenance, and management measures which can be effective and can be incorporated into ongoing programs throughout a municipality. EPA has recommended that municipalities look at: snow maintenance, vegetation management, road repair, transportation planning, structural retrofitting, catchbasin cleaning, litter control and targeted street sweeping for general pollution reduction.

Retrofitting streets during roadway rehabilitation is also an option in design. It is recommended that specific screening and studies be undertaken prior to investing in more costly retrofitting of structural measures. Certain rules-of-thumb have been developed for such screening and sampling techniques have been researched and demonstrated (Versar, 1985). Detailed design guidelines are available (Woodward-Clyde, 1989).

Structural Roadway BMP Design Overview

Design of highway and roadway BMPs is similar to general BMP design in that pollutant loads, runoff volumes and impacts to receiving water body must be estimated; a BMP sized to achieve a particular goal or objective; and the details of maintenance, placement, materials, etc. worked out. Highway impacts on waterbodies immediately downstream are normally short-term and acute in nature. Therefore, exceedance frequency type design is normally appropriate. Woodward-Clyde (1989) provides a detailed design procedure while Versar (1989) provides specific BMP design guidelines for vegetative control, wet ponds, infiltration systems and wetlands. Relatively ineffective structural BMPs include sumped catch basins and detention ponds. Performance of these can be improved somewhat with proper design and a strong maintenance program. See the Structural Best Management Practices, Chapter 13 for more details on BMP design.

Vegetative Measures - include grassed channels and overland flow buffer strips. The effectiveness is dependent on the density of the stand and the amount of flow short-circuiting the strip or flowing with depth in the channel. Vegetative BMPs reduce nutrients through soil and vegetative processes; breakdown hydrocarbons by bacterial degradation; filter suspended solids; cause settling of suspended solids through velocity reduction; and complex metals through soil adsorption and biological assimilation.

Retention Basins - Performance of wet ponds ranges from poor to excellent depending on the size of the pond relative to the drainage area. The primary removal mechanism is sedimentation, though soluble nutrients reduction often takes place through biological uptake. This can be enhanced with the construction of a bench along the shore for emergent vegetation growth. Often a partnership can be worked out wherein the roadway and other development share the pond for pollution treatment.

Infiltration Systems - Infiltration systems temporarily store roadway runoff allowing it to infiltrate into the ground. This type of system can be very effective in the removal of pollutants, though construction quality control and maintenance practices are essential for its overall effectiveness. Pretreatment of runoff for removal of oil and sediment greatly lengthens the life of an infiltration system (Galli, 1992a, 1992b). Low water tables (3 feet from trench or pond bottom) and permeable soil (more than 1/2" per hour) are requirements for the use of such a system.

Wetlands - Wetlands are really a combination of the above methods with a thriving and complex ecosystem, a wider range of chemical reactions and dense growth of marsh plants. Removal capabilities are high for many types of pollutants. Constructed wetlands for the purpose of storm water treatment may not have the complex ecological system of a natural or enhanced wetland (MWCOG, 1992). The three important components in wetland creation are water, soil and vegetation (Hammer, 1992). Sufficient baseflow must be available to maintain water levels which may limit urban applications to locations where joint use can be made of the location. The soils in natural wetlands are hydric in nature and anaerobic which leads to a wide potential for chemical reaction within the soil. Wetland vegetation is capable of surviving up to five days inundated. In areas where cold weather shortens the growing season pollution removal may be drastically reduced or even negative. A wide variety of plants are recommended to provide a balance in chemical and biological reaction. Maintenance costs in the first few years are relatively high until the marsh vegetation is established.

8.13 Computer Programs

To assist with storm drain system design many microcomputer software models have been developed for the computation of hydraulic gradeline. A public domain computer model has been attached to the program HYDRA, which has been adopted by the Federal Highway Administration organized Pooled Fund Study on Integrated Drainage System, as a recommended program for storm drain design and analysis. The model developed in this study, called HYGRD, allows a user to check design adequacy and also to analyze the performance of a storm drain system under assumed inflow conditions.

The HYDRA computer program is integrated into the Federal Highway Administration's HYDRAIN computer model system which is available from McTrans Software, University of Florida, 512 Weil Hall, Gainesville, Florida 32611. The program HYDRA is also available in private domain versions which interface with GIS or CADD software to provide a complete design management system. While not perfect, the system does allow for a hierarchical treatment of storm drain design and analysis using either a Rational Method peak flow design, inflow hydrographs or a full dynamic routing based on the SWMM EXTRAN program block in some versions. Note that the program is not without "bugs" so check output carefully.

Appendix A shows examples of using HYDRA for storm drain design.

References

American Association Of State Highway And Transportation Officials, Model Drainage Manual, Suite 225, 444 North Capital Street, N.W., Washington, D.C., 20001, 1991.

American Society of Civil Engineers, Manual of Practice No. 37, "Design and Construction of Sanitary and Storm Sewers", ASCE, 1960.

American Society of Civil Engineers, Manual of Practice No. 60, "Gravity Sanitary Sewer Design and Construction", ASCE, 1982.

American Society of Civil Engineers, Manual of Practice No. 62, "Existing Sewer Evaluation and Rehabilitation", ASCE, 1983.

Bellevue, WA, Standards and Specifications, Ch. 4, Storm Drainage and Streams, Drawing No. 34, 1988.

Environmental Protection Agency, "Guidance Specifying Management Measures for Sources of Nonpoint Pollution in Coastal Waters", EPA Ofc. of Water 840-B-92-002, Jan., 1993.

Environmental Protection Agency, "Detailed Guidance For Part 2 Municipal NPDES Permit Application", 1993.

Federal Aviation Administration, Manual On Airport Drainage, Advisory Circular 150-5320-5B, Washington, DC, GPO, 1984.

Federal Highway Administration, HYDRAIN Documentation, 1991.

Galli, F.J., "Preliminary Analysis of the Performance and Longevity of Urban BMPs installed in Prince George's County, Maryland", Dept. of Env. Resources, Prince George's County, 1992a.

Galli, F.J., "Analysis of Urban BMP Performance and Longevity in Prince George County, Maryland", Metor. Wash. COG, August, 1992b.

GKY & Assoc., HYDRAIN: Integrated Drainage Design Computer System, 5411-E Backlick Rd., Springfield, VA 22151, 1990.

Horner, R. R., and Mar, B. W., " Assessing the Impacts of Operating Highways on Aquatic Ecosystems", Trans. Res. Brd. Rpt. 1017, 1985.

Kobinger, N.P. "Sources and Migration of Highway Runoff Pollutants (Four Volumes)", Rpt. No. FHWA-/RD-84/057-060, Fed. Hwy. Admin., Wash. D.C., 1982.

Metropolitan Washington Council of Governments, "A Current Assessment of Urban Best Management Practices", March, 1992.

Saatci, A., "Velocity and Depth of Flow Calculations in Partially Filled Pipes", ASCE Journal of Environmental Engineering., Vol. 116, No. 6, November/December, 1990.

Sartor, J. D., and Boyd, G. B., "Water Pollution Aspects of Street Surface Contaminants", EPA-R2-72-081, November, 1972.

Shaheen, D. G., "Contributions of Urban Roadway Usage to Water Pollution", Rpt. EPA-600/2-75-004, April, 1975.

U. S. Department of Transportation, Federal Highway Administration, Drainage Of Highway Pavements, Hydraulic Engineering Circular No. 12, 1984.

Versar, Inc., Springfield, VA. Management Practices for Mitigation of Highway Stormwater Runoff Pollution Volumes 1-4., 1985.

Woodward-Clyde Consultants. Pollutant Loadings and Impacts From Highway Stormwater Runoff Volume 1-4. Design Procedure, 1989.

Appendix A - HYDRA Storm Drain Calculations

Example Applications

HYDRA is a storm drain and sanitary sewer analysis program. This section will limit itself to a discussion of the Rational Method storm drain design and analysis capabilities of HYDRA. Some additional related capabilities include: cost estimating, hydrograph routing, and inlet computations. The reader is directed to the HYDRAIN user documentation available from MCTRANS for a complete description of these and other capabilities of the model.

The program will select pipe size, slope and invert elevations given certain design information. Alternately it will analyze a given pipe and channel network given pertinent information. It will also work in combination within a single run sizing and analyzing pipe systems. In the analysis mode it will indicate which pipes are overloaded or surcharged and will suggest a flow removal quantity which will render the system capable of handling the remaining flow.

HYDRA requires the creation of an input file either within the HYDRAIN shell or through the use of any word processor capable of creating ASCII files. The input file consists of commands which describe the system in a logical sequence working from upstream to downstream. It is possible that several command sequences can produce the same result.

After a brief discussion of the techniques used by HYDRA for design and analysis in the Rational Method a design example is presented which illustrates the capabilities of HYDRA.

Storm Flow: The Rational Method

The Rational Method is a widely used method for the sizing of storm water systems to handle peak flow. The method does not give any information on flow hydrographs or routing. However, for smaller areas and for simple designs where pipe storage is not a consideration it has proved effective. The Rational Method takes advantage of the fact that the conversion from acre-inches per hour (on the right side of the equation) to cubic feet per second is about 1.00. The Rational Method equation is:

$$Q = C I A \tag{8.A.1}$$

Where: Q = peak flow, cfs
C = dimensionless runoff coefficient
I = rainfall intensity for design storm, in/hr
A = drainage area, acres

The rainfall intensity is chosen from a local intensity-duration-frequency (IDF) curve for the chosen design frequency and for a duration equivalent to the time of concentration of the drainage area. The C factor is chosen as an area weighted composite of the land use. Proper selection of C values and use of the Rational Method for sizing of storm drains is found in the Hydrology chapter of this book, as well as most hydrology textbooks.

To produce flows HYDRA multiplies CA times the rainfall interpolated from the input IDF curve, for the input or computed time of concentration. At any point in the system HYDRA calculates the longest time of concentration which will be either the time for that increment of drainage area at that point (inlet time) or the sum of the longest combination of inlet time plus pipe flow time to that junction point. In the particular case where the individual area at that point is greater than the sum of all the other areas from upstream, HYDRA makes a correction to the flow.

Time of concentration can either be input as a single number, as a gutter flow value with sheet flow calculated, or with both sheet flow and gutter flow calculated. Overland flow is calculated from the Federal Aviation Administration equation as:

$$tc = (1.8 \ (1.1-C) \ L^{0.5}) \ / \ S^{0.33} \tag{8.A.2}$$

Where: C = dimensionless runoff coefficient
         L = distance traveled, ft
         S = slope, percent
For gutter flow a second formula is used:

$$tc = L/(K \ S^{0.5}) \tag{8.A.3}$$

Where: L = distance traveled, ft
         S = slope, ft/ft
         K = an empirical coefficient equal to 32 ft/s
Rainfall is input on RAI records with pairs of time; rainfall intensity is taken directly from the local IDF curve. Alternately rainfall can be read from a file with an .idf suffix in similar format by simply providing the filename after the record header. To avoid extrapolation difficulties it is recommended that the first time be zero (with the first rainfall equal to the second rainfall value) and the last rainfall value be set equal to the second-to-last value to avoid the possibility of negative rainfall values.

Rational Method information for each drainage area is given on a STO record in the form of: area, C factor, and time of concentration value(s).

Flow Conveyance: The Manning Equation

Flow generated by the Rational Method input is transported through pipes and channels. The basic equation employed by HYDRA for the sizing of pipes and channels is the Manning equation given here combined with the continuity equation:

$$Q = 1.486/n \ A \ R_h^{0.667} \ S^{0.5} \tag{8.A.4}$$

Where: Q = flow discharge, cfs
         n = Manning friction coefficient or roughness factor
         A = flow area, ft^2
         R_h = the hydraulic radius of the flow area, ft
         S = pipe slope, ft/ft

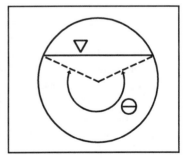

The Izzard equation for flow in roadway gutters is also available. Information on the Izzard equation is available in the HYDRAIN user's manual.

In the design case the equation is solved for pipe diameter or channel flow depth given the known flow. HYDRA will choose the minimum pipe required to carry the design flow but in no case will it suggest a pipe smaller than the minimum size specified on the PIP record or 12 inches (whichever is greater). For sizes less than 48 inches it rounds up to the nearest 3 inches. For larger sizes it rounds up to the nearest 6 inches in diameter.

Hydraulic radius in pipes is a function of depth. In most cases the pipes will not be flowing full. The hydraulic radius is calculated from:

$$R_h = D/4 \ (1 - \sin 2(\Theta)/2(\Theta)) \tag{8.A.5}$$

Where: D = pipe diameter, in.
$\quad\quad\quad \Theta$ = angle in radians as indicated in the figure above.
Information for pipe flow is input on the PIP record and for channels on the CHA record.

Hydraulic Grade Line Calculations: The Energy Equation

The user has the option, through the use of the HGL command, of initiating the calculation of the hydraulic gradeline through the drainage system. HYDRA uses information supplied on the PIP and PNC records on elevations, connectivity and junction types and angles to calculate the hydraulic gradeline. Calculations begin at the outfall and proceed upstream to each of the terminal nodes. Major and minor losses are included. Major losses are friction losses while minor losses are those incurred as the flow enters and leaves a junction.

Flow depth at the outfall is determined internally by HYDRA or can be input manually as a tailwater through the use of the TWE record. Note that the most downstream PNC record must have a zero in the manhole width field for HYDRA to detect that it is the outfall.

To determine the hydraulic gradeline the energy equation is solved. The energy equation states that at any two points the total energy upstream equals the total energy downstream plus any energy losses. It is expressed for pipe flow as (i=inflow of pipe, o=outflow of pipe):

$$Vi^2/2g + Pi/\Gamma + Zi = Vo^2/2g + Po/\Gamma + Zo + \Sigma_\blacktriangle E \tag{8.A.6}$$

Where: V =   velocity, ft/s
$\quad\quad\quad g$ =   acceleration due to gravity, 32.2 ft/s^2
$\quad\quad\quad P$ =   pressure, lbs/ft^2
$\quad\quad\quad \Gamma$ =   specific weight of water, 62.4 lbs/ft^3
$\quad\quad\quad Z$ =   water surface elevation, ft
$\quad\quad\quad \blacktriangle E$ = head loss, ft

Minor losses and head losses are calculated as proportional to the velocity head. Additional information on these losses can be found in the HYDRAIN user manual and in the literature.

HYDRA begins with the tailwater and checks to see if it is above the pipe crown on the downstream end. If it is, HYDRA assumes the pipe is surcharged (operating under pressure conditions) and computes a friction slope (using Manning's equation) necessary to achieve the calculated flow. The slope is multiplied by the length and any user supplied losses (using BEN or LOS records) are added to obtain an upstream elevation. This elevation is compared with the summation of the flow depth at the upstream end plus the upstream invert elevation. The greater of the two values is assumed to be the upstream hydraulic gradeline elevation. If the tailwater is below the pipe crown the upstream depth plus invert elevation is assumed to be the gradeline.

Minor losses are then added to the gradeline. This is the starting "tailwater elevation" for the next pipe upstream and the calculations begin again as previously explained.

HYDRA Commands

The commands used for the Rational Method analysis fall into several categories. The paragraphs below give a brief description of the commands applicable to the Rational Method design of storm drain systems. There are a number of other commands. Reference should be made to the HYDRAIN user manual for a complete listing and description of the commands.

Job Control Information

JOB -   Initiates JOB and enters Job title. Up to 50 alphanumeric characters for "title header" for each page. Use only once.

SWI -   Sets SWItch for determining method of storm/sanitary flow analysis. Set to "2" for Rational Method design.

HGL -   Signals that Hydraulic Gradeline computations should be made. Include for the computations and exclude if not wanted.

CRI -   Determines whether inverts or crowns are to be matched   (CRIteria). "0" (default) inverts of pipes will be matched in free design. "1" crowns will be matched. Can use any time.

REM -   Allows a line for REMarks or comments.

END -   ENDs a command string. Last command in the file.

Logic Flow Information

HOL -   HOLds system flow at the lower end of a lateral. Input is a number from 1 to 25 which is a holding register. Can reuse same number but will overwrite previous information in that register. REC record retrieves data from the register it was HOL'd to.

NEW -   Loads NEW lateral name. Up to 20 alphanumeric characters.

REC -   RECalls flows previously stored using the HOL or DIV commands. See HOL. Can recall up to five held numbers at one node.

Rainfall And Peak Flow Generation

RAI -   Sets the values on a RAInfall intensity versus duration curve. Pairs of time, intensity values (minutes, inches per hour) from a local IDF curve. It is wise to make the first time "0" with the first two intensities equal and to make the last two intensities equal for different times so rainfall will never go to zero or be extrapolated incorrectly for a very short time of concentration.

STO -   Enters sub-basin data for determining STOrm water design flow for the Rational Method. Structure is: acres, C, time of concentration (min). Alternate structure employs equations to calculate times. See HYDRAIN user manual for these alternate methods.

Pipe Or Channel Information

CHA -   Allows you to define an open CHAnnel or ditch. Structure is: length (ft), upstream invert elevation, downstream invert elevation (if number is less than 1 but greater than zero HYDRA will read it as a slope), Manning n, left side slope (foot horizontal per 1 foot vertical), bottom width (ft), right side slope, OPTIONAL equation choice (enter 1 to use Izzard's equation, blank for Manning).

PDA -   Establishes basic Pipe design DAta to be used throughout the system unless over ridden. Structure is: Manning n, minimum diameter for free design (inches), minimum depth, minimum cover (ft), minimum velocity (ft/s), minimum slope (ft/ft), and OPTIONAL maximum diameter. Must be placed prior to first PIP record and can be reused anyplace.

PIP -   Moves water from one point to another in a circular PIPe. Structure is: length (ft), ground elevation upstream, ground elevation downstream, OPTIONAL invert elevation upstream, OPTIONAL invert elevation downstream, minimum diameter (use a <u>negative number</u> to specify an existing pipe of known diameter). Note that if the last three values are omitted HYDRA will switch to free design mode and calculate these values for this pipe.

PNC -   Specifies Pipe-Node Connections for hydraulic gradeline computation. Each PNC record must immediately follow a PIP record and gives information about the downstream node of the PIP it follows. Structure is: upstream node number of PIPe, downstream node number of PIPe, manhole or node width of downstream end (set to zero for outfall node) (ft), angle between pipe entering node and pipe leaving node (not the water deflection angle) (degrees), benching code (0=flat bench (default), 1=1/2 bench, 2=full bench, 3=improved bench (see HYDRAIN user manual)).

<u>Losses And Miscellaneous</u>

BEN -   Specifies Pipe BENd data such as angle and radius for curved pipe. Placed after PNC record. Structure is: bend radius (ft), bend angle (normally between 0 and 120 degrees).

LOS -   Allows input of additional pipe LOSes. Will be included in the hydraulic gradeline computation. Input after the PNC record. Structure: loss (ft).

TWE -   Allows for the input of a tailwater elevation at the system outfall. Structure: tailwater elevation.

Examples

In this section two examples are given. The figure below gives the input information. The first demonstrates a "free design" of a system. The second example is an analysis of the same system as HYDRA designs in example one but with one pipe deliberately undersized. This undersized pipe can then be analyzed by HYDRA by simply leaving the pipe field in the PIP record blank. This triggers HYDRA to design a pipe to fill the blank. In this way HYDRA can be used to first analyze a system and then design replacement pipes.

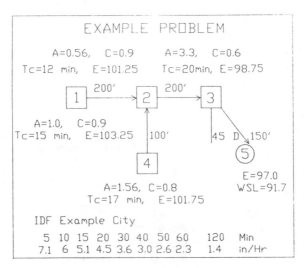

Figure 8-A-1 Example Problem

JOB EXAMPLE 1 - FREE DESIGN

```
SWI 2
HGL 1
PDA 0.013 12 3 2 2 0.002
RAI 0 7.1 5 7.1 10 6 15 5.1 20 4.5 30 3.6 40 3 50 2.6 60 2.3
    120 1.4 160 1.4
NEW 1-2
STO 1 .9 15
PIP 200 103.25 101.25
PNC 1 2 2 180
HOL 1
NEW 3-2
STO 1.56 .8 17
PIP 100 101.75 101.25
PNC 3 2 2 90
REM COMPUTATION FROM 2-4
REC 1
STO .56 .9 12
PIP 200 101.25 98.75
PNC 2 4 2 135
REM COMPUTATION FROM 4-5
STO 3.3 .6 20
PIP 150 98.75 97
PNC 4 5 0 180
END
```

JOB EXAMPLE 2 - PIPE ANALYSIS

```
SWI 2
HGL 1
PDA 0.013 12 3 2 2 0.002
RAI 0 7.1 5 7.1 10 6 15 5.1 20 4.5 30 3.6 40 3 50 2.6 60 2.3
    120 1.4 160 1.4
NEW 1-2
STO 1 .9 15
PIP 200 103.25 101.25 99.6 97.6 -15
PNC 1 2 2 180
NEW 3-2
STO 1.56 .8 17
PIP 100 101.75 101.25 98.13 97.63 -18
PNC 3 2 2 90
REM COMPUTATION FROM 2-4
REC 1
STO .56 .9 12
REM ********************************************
REM ******** SUBST HERE 15" FOR THE 21" PIPE
REM ********************************************
```

PIP 200 101.25 98.75 96.73 94.85 -15
PNC 2 4 2 135
REM COMPUTATION FROM 4-5
STO 3.3 .6 20
PIP 150 98.75 97 93.41 92.56 -27
PNC 4 5 0 180
END

EXAMPLE 1 - FREE DESIGN

****** HYDRA ******* (Version 3.0) *****

EXAMPLE 1 - FREE DESIGN

Commands Read From File C:\HYDRA\EX1.HDA
  JOB
  SWI2
  HGL1
  PDA0.013 12 3 2 2 0.002
  RAI0 7.1 5 7.1 10 6 15 5.1 20 4.5 30 3.6 40 3 50 2.6 60 2.3
  120 1.4 160 1.4

                        IDF CURVE

.71E+01*.*...
      .   .
      .   .
      .   *   .
.53E+01.   .
      .   *   .
      .   .
      .   *   .
      .   .
.35E+01.   *   .
      .   .
      .   *   .
      .   *   .
      .   *   .
.18E+01.   .
      .   *   *   .
      .   .
      .   .
      .   .
.00E+00*...
      0.   23.   46.   69.   91.   114.   137.   160.

                PLOT-DATA (TIME Vs.VALUE)

----
   0.   7.10   20.   4.50   60.   2.30   0.   .00   0.   .00
   5.   7.10   30.   3.60   120.   1.40   0.   .00   0.   .00
  10.   6.00   40.   3.00   160.   1.40   0.   .00   0.   .00
  15.   5.10   50.   2.60   0.   .00   0.   .00   0.   .00

  NEW1-2
  STO1 .9 15
  PIP200 103.25 101.25

```
*** Tc =  15.0 MINUTES
*** CA =   .9
 PNC1 2 2 180
 HOL1
 NEW3-2
 STO1.56 .8 17
 PIP100 101.75 101.25
*** Tc =  17.0 MINUTES
*** CA =   1.2
 PNC3 2 2 90
 REMCOMPUTATION FROM 2-4
 REC1
 STO.56 .9 12
 PIP200 101.25 98.75
*** Tc =  17.4 MINUTES
*** CA =   2.7
 PNC2 4 2 135
 REMCOMPUTATION FROM 4-5
 STO3.3 .6 20
 PIP150 98.75 97
*** Tc =  20.0 MINUTES
*** CA =   4.6
 PNC4 5 0 180
 END
 END OF RUN.
```

****** HYDRA ******* (Version 3.0) *****

EXAMPLE 1 - FREE DESIGN

***   Pipe Design

Link	Length (ft)	Diam (in.)	Invert Up/Dn (ft)	Slope (ft/ft)	Depth Up/Dn (ft)	Min. Cover (ft)	Velocity Act/Full (ft/s)	--Flow-- Act/Full (cfs)	Estimated Cost ($)
1	200	15	99.90 / 97.90	.01000	3.4 / 3.4	2.0	5.7 / 5.3	4.59 / 6.48	0.
2	100	18	98.13 / 97.63	.00500	3.6 / 3.6	2.0	4.7 / 4.2	6.07 / 7.45	0.
3	200	21	96.73 / 94.85	.00938	4.5 / 3.9	2.0	7.2 / 6.4	12.78 / 15.38	0.
4	150	27	93.42 / 92.56	.00569	5.3 / 4.4	2.0	6.7 / 5.9	20.84 / 23.43	0.

```
     LENGTH =   450.  TOTAL LENGTH =   650.
     COST   =    0.  TOTAL COST   =    0.
```

****** HYDRA ******* (Version 3.0) *****

Pipe #	Downstream Node #	Hydraulic Gradeline Elevation	Crown Elevation	Possible Surcharge	Ground Elevation
1	2	98.7	99.1	N	101.3
2	2	98.7	99.1	N	101.3
3	4	96.1	96.6	N	98.8
4	5	94.2	94.8	N	97.0

Pipe #	Terminal Node #	Hydraulic Gradeline Elevation	Ground Elevation
1	1	101.4	103.3
2	3	99.7	101.8

EXAMPLE 2 - PIPE ANALYSIS

****** HYDRA ******* (Version 3.0) *****

EXAMPLE 2 - ANALYSIS

Commands Read From File C:\HYDRA\EX2.HDA
  JOB
  SWI2
  HGL1
  PDA0.013 12 3 2 2 0.002
  RAI0 7.1 5 7.1 10 6 15 5.1 20 4.5 30 3.6 40 3 50 2.6 60 2.3
  120 1.4 160 1.4

IDF CURVE

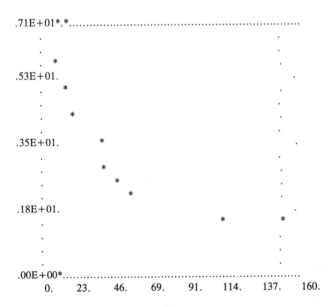

STORM DRAINAGE SYSTEMS

313

PLOT-DATA (TIME Vs.VALUE)

0.	7.10	20.	4.50	60.	2.30	0.	.00	0.	.00
5.	7.10	30.	3.60	120.	1.40	0.	.00	0.	.00
10.	6.00	40.	3.00	160.	1.40	0.	.00	0.	.00
15.	5.10	50.	2.60	0.	.00	0.	.00	0.	.00

```
NEW1-2
STO1 .9 15
PIP200 103.25 101.25 99.6 97.6 -15
*** Tc =  15.0 MINUTES
*** CA =   .9
PNC1 2 2 180
HOL1
       ****** HYDRA ******* (Version 3.0) *****

EXAMPLE 2 - ANALYSIS
NEW3-2
STO1.56 .8 17
PIP100 101.75 101.25 98.13 97.63 -18
*** Tc =  17.0 MINUTES
*** CA =   1.2
PNC3 2 2 90
REMCOMPUTATION FROM 2-4
REC1
STO.56 .9 12
REM**********************************************
REM******** SUBST HERE
REM**********************************************
PIP200 101.25 98.75 96.73 94.85 -15
*** Tc =  17.4 MINUTES
*** CA =   2.7
PNC2 4 2 135
REMCOMPUTATION FROM 4-5
STO3.3 .6 20
PIP150 98.75 97 93.41 92.56 -27
*** Tc =  20.0 MINUTES
*** CA =   4.6
PNC4 5 0 180
END
END OF RUN.
       ****** HYDRA ******* (Version 3.0) *****
```

EXAMPLE 2 - ANALYSIS
***   Analysis of Existing Pipes

Link	Length (ft)	Diam (in.)	Invert Up/Dn (ft)	Slope (ft/ft)	Depth Up/Dn (ft)	Cover Up/Dn (ft)	Velocity Act/Full (ft/s)	--Flow-- Act/Full (cfs)	Load (%)	Remove (cfs)	Diam (in.)
1	200	15	99.60	.01000	3.7	2.3	5.7	4.59	71		
			97.60		3.7	2.3	5.3	6.48			
2	100	18	98.13	.00500	3.6	2.0	4.7	6.07	81		
			97.63		3.6	2.0	4.2	7.45			
3	200	15	96.73	.00940	4.5	3.2	10.4	12.78	203	6.50	18
			94.85		3.9	2.5	5.1	6.28			

4   150   27   93.41   .00567 5.3   2.9   6.6   20.84   89
                 92.56   4.4   2.0   5.9   23.38

LENGTH =   450.   TOTAL LENGTH =   650.

****** HYDRA ******* (Version 3.0) *****

Page No   4

Pipe #	Downstream Node #	Hydraulic Gradeline Elevation	Crown Elevation	Possible Surcharge	Ground Elevation
1	2	98.4	98.8	N	101.3
2	2	98.7	99.1	N	101.3
3	4	96.1	96.1	N	98.8
4	5	94.2	94.8	N	97.0

Pipe #	Terminal Node #	Hydraulic Gradeline Elevation	Ground Elevation
1	1	101.1	103.3
2	3	99.7	101.8

Chapter 9 Design Of Culverts

9.1 Introduction

A culvert is defined as the following.
- A structure used to convey surface runoff through embankments.
- A structure which is usually designed hydraulically to take advantage of submergence to increase hydraulic capacity.
- A structure, as distinguished from bridges, which is usually covered with embankment and is composed of structural material around the entire perimeter, although some are supported on spread footings with the streambed serving as the bottom of the culvert.
- A structure which is 20 feet or less in centerline length between extreme ends of openings for multiple boxes. However, a structure designed hydraulically as a culvert is treated in this chapter, regardless of its total length.
- For economy and hydraulic efficiency, culverts should be designed to operate with the inlet submerged during flood flows, if conditions permit.
- Cross-drains are those culverts and pipes that are used to convey runoff from one side of a highway to another.

Culvert flow is one of the most complex forms of hydraulics an engineer will encounter. Easy to understand nomographs and "cookbook" procedures can belie the reality that control of flow in a culvert can shift dramatically and unpredictably between inlet control (weir or orifice), barrel control and outlet control causing relatively sudden rises in headwater and, in the worst case in larger culverts, structural damage or failure due to sudden pressure surges. Certain types of culverts on certain slopes can exhibit a looped rating curve depending on whether the flow is rising or falling. Therefore, the most critical aspect of culvert design is to determine stable and predictable performance for all expected flow levels. When the type of flow is known the well known equations for orifice, weir, or pipe flow and backwater profiles can be applied to determine the relationships between head and discharge (Blaisdell, 1966). The good news is that modern culvert design nomographs and instructions are based both on sound theory and extensive laboratory and field studies, and can be relied on in all but the most uncommon situations.

The design of a culvert is influenced by cost, hydraulic efficiency, purpose, and the topography at the proposed culvert site. Thus physical data must be integrated with engineering and economic considerations. The information contained in this chapter should give the designer the ability to design culverts taking into account the factors that influence their design and selection.

The primary purpose of a culvert is to convey surface water, but properly designed it may also be used to restrict flow and reduce downstream storm runoff peaks. In addition to the hydraulic function, a culvert must also support the embankment and roadway for traffic conveyance, and protect the traveling public and adjacent property owners from flood hazards to the extent practicable and in a reasonable and prudent manner.

Primary considerations for the final selection of any drainage structure are that its design be based upon appropriate hydraulic principles, economy, and minimized effects on adjacent property by the resultant headwater depth and outlet velocity. The allowable headwater elevation is that elevation above which damage may be caused to adjacent property and/or a highway. It is this allowable headwater depth that is the primary basis for sizing a culvert.

To ensure safety during major flood events, access and egress routes to developed areas should be checked for a more infrequent flood than the design flood. Often the 100-year flood is checked to determine if these streets will provide safe access for emergency vehicles and local residents. Thus, an analysis of how the culvert will function during a less frequent flood is an important part of culvert design. Some municipalities have different culvert design standards for streets of different classifications.

At many sites, either a bridge or a culvert will fulfill the structural and hydraulic requirements. The structural choice should be based on:
• risk of property damage,
• construction and maintenance cost,
• traffic safety,
• environmental considerations,
• risk of failure, and
• aesthetic considerations.

Performance curves should be developed for all culverts for evaluating the hydraulic capacity of a culvert for various headwaters. These will display the consequence of high flow rates at the site and any possible hazards. Sometimes a small increase in flow rate can affect a culvert design. If only the design peak discharge is used in the design, the designer cannot assess what effects increases in the estimated design discharge will have on the culvert design.

For culverts with significant headwater storage the site should be treated as a detention design and flow should be routed using techniques found in the Storage Facilities, Chapter 11.

9.2 Concept Definitions

Following are definitions of concepts which will be important for understanding different culvert design procedures. These concepts are used throughout the remainder of this book in dealing with different aspects of culvert designs.

Critical Depth   Critical depth can best be illustrated as the depth at which water flows over a weir, this depth being attained automatically where no other backwater forces are involved. For a given discharge and cross-section geometry there is only one critical depth. Appendix B at the end of this chapter gives a series of critical depth figures for the different culvert shapes.

Cross-Drainage   Cross-drainage culverts extend under a roadway and transport runoff from one
Culverts   side of a roadway and discharge the runoff at the other side of the roadway.

Energy Grade   The energy grade line represents the total energy at any point along the culvert
Line   barrel.

Free Outlets   Free outlets are outlets whose tailwater is equal to or lower than critical depth. For culverts having free outlets, lowering of the tailwater has no effect on the discharge or the backwater profile upstream of the tailwater.

Hydraulic Grade Line	The hydraulic grade line is the depth to which water would rise in vertical tubes connected to the sides of the culvert barrel. In full flow, the energy grade line and the hydraulic grade line are parallel lines separated by the velocity head except at the inlet and the outlet.
Invert	Invert refers to the inside bottom of the culvert.
Improved Inlets	Flared, improved, or tapered inlets indicate a special entrance condition which decreases the amount of energy needed to pass the flow through the inlet and thus increases the capacity of culverts at the inlet.
Normal Flow	Normal flow occurs in a channel reach when the discharge, velocity and depth of flow do not change throughout the reach. The water surface profile and channel bottom slope will be parallel. This type of flow will be approximated in a culvert operating on a steep slope provided the culvert is sufficiently long.
Soffit	Soffit refers to the inside top of the culvert. The soffit is also referred to as the crown of the culvert.
Steep And Mild Slope	A steep slope culvert operation is where the computed critical depth is greater than the computed uniform depth. A mild slope culvert operation is where critical depth is less than uniform depth.
Submerged Inlets	Submerged inlets are those inlets having a headwater greater than about 1.5 times the diameter.
Submerged Outlets	Partially submerged outlets are outlets whose tailwater is higher than critical depth and lower than the height of the culvert. Submerged outlets are outlets having a tailwater elevation higher than the soffit of the culvert.
Uniform Flow	Uniform flow is flow in a prismatic channel of constant cross section having a constant discharge, velocity and depth of flow throughout the reach. This type of flow will exist in a culvert operating on a steep slope provided the culvert is sufficiently long.

9.3   Culvert Design Steps And Criteria

Design Steps

1.Determine And Analyze Site Characteristics - Site characteristics include the generalized shape of the highway embankment, bottom elevations and cross sections along the stream bed, the approximate length of the culvert, and the allowable headwater elevation. In determining the allowable headwater elevation, roadway elevations and the elevation of upstream property should be considered. The consequences of exceeding the allowable headwater elevation should be evaluated and kept in mind throughout the design process.

Culvert design is actually a trail-and-error procedure because the length of the barrel cannot be accurately determined until the size is known, and the size cannot be precisely determined until the length is known. In most cases, however, a reasonable estimate of length will be accurate enough to determine the culvert size.

A field visit is essential to determine any site characteristics that might affect the culvert design and the effect the culvert might have on the natural drainage system.

2. Perform Hydrologic Analysis - Outline the drainage area above the culvert site. Develop flow estimates for the design frequencies and determine if sufficient storage will be available at the culvert inlet to justify routing flows through the culvert for design purposes. Design frequencies should include the design flow rate and a larger flood (check flood) which is usually the 100-year flow rate. The probable accuracy of the estimate should be kept in mind as the design proceeds. The accuracy is dependent on the method used to define the flow rate, the available data on which it is based, etc. If routing through the culvert is needed, a hydrograph must be developed.

3. Perform Outlet Control Calculations And Select Culvert - These calculations are performed before inlet control calculations in order to select the smallest feasible barrel which can be used without the required headwater elevation in outlet control exceeding the allowable headwater elevation. The full flow outlet control performance curve for a given culvert (size, inlet edge, shape, material) defines its maximum performance. Therefore, the inlet improvements beyond the beveled edge or changes in inlet invert elevation will not reduce the required outlet control headwater elevation. This makes the outlet control performance curve an ideal limit for improved inlet design. The results of these calculations should be the outlet control performance curve. In addition to considering the allowable headwater elevation, the velocity of flow at the exit to the culvert should be checked to determine if downstream erosion problems will be created.

4. Perform Inlet Control Calculations For Conventional And Beveled Edge Culvert Inlets - Perform the inlet control calculations to develop the inlet control performance curve to determine if the culvert design selected will be on inlet or outlet control for the design and check flood frequencies. A fall may be incorporated upstream of the culvert to increase the flow through the culvert.

5. (Optional) Perform Throat Control Calculations For Side- And Slope-Tapered Inlets - The same concepts are involved here as with a conventional or beveled edge culvert design.

6. (Optional) Analyze The Effect of FALLS On Inlet Control Section Performance - The purpose of this step is to determine if having a FALL before the inlet of the culvert would increase the capacity of the culvert and if a FALL can be justified from a cost perspective and site characteristics.

7. (Optional) Design Side- And/Or Slope-Tapered Inlet - Side- and slope-tapered inlets can be used to significantly increase the capacity of many culvert designs. Develop performance curves based on side- and/or slope-tapered inlets and determine from a cost perspective and site characteristics if such a design would be justified.

8. Complete File Documentation - As outlined in the Documentation, Chapter 16, complete the documentation file for the final design selected.

Engineering And Technical Design Criteria

The design of a culvert should take into account many different engineering and technical aspects at the culvert site and adjacent areas. The following design criteria should be considered for all culvert designs as applicable.

Engineering Aspects
- Flood frequency
- Velocity limitations
- Buoyancy protection

Site criteria
- Length and slope

	• Debris control
	• Ice buildup
Design limitations	• Headwater limitations
	• Tailwater conditions
	• Storage - temporary or permanent
Design options	• Culvert inlets
	• Inlets with headwalls
	• Wingwalls and aprons
	• Improved inlets
	• Material selection
	• Culvert skews
	• Culvert sizes and shapes
Related designs	• Weep holes
	• Outlet protection
	• Erosion and sediment control
	• Environmental considerations
	• Safety considerations

Some culvert designs are relatively simple involving a straight-forward determination of culvert size and length. Other designs are more complex where structural, hydraulic, environmental, or other considerations must be evaluated and provided for in the final design. The designer must incorporate personal experience and judgment to determine which criteria must be evaluated and how to design the final culvert installation. Following is a discussion of each of the above criteria as it relates to culvert siting and design.

Flood Frequency

The appropriate flood frequency for determining the flood carrying capacity of a culvert is dependent on the roadway at the crossing and the level of risk associated with failure of the culvert crossing as well as the level of risk associated with increasing the flood hazard to upstream (back-water) or downstream (redirection of floodwaters) property. It is recommended that culverts be designed to accommodate the following minimum flood frequencies:
• All cross-drainage culverts - 25-year frequency
• Other culverts - 10-year frequency

Future development of contributing watersheds and floodplains that have been zoned or delineated should be considered in determining the design flood frequency. In addition, the 100-year frequency storm should be routed through all culverts to be sure structures are not flooded or increased damage does not occur to the roadway or adjacent property for this design event.

An economic or risk analysis (Young, et al, 1970) may justify a design to pass floods greater than those noted above where potential damage to adjacent property, to human life, or heavy financial loss due to flooding as determined by a flood assessment or analysis, commensurate with the site. Even designing for a very infrequent flood may not protect the designer or owner from potential liability should the culvert back water onto private property at any flow level (Lewis, 1992). Risk concepts may be used in design.

Also, in compliance with the National Flood Insurance Program it is necessary to consider the 100-year frequency flood at locations identified as being special flood hazard areas. This does not necessitate that the culvert be sized to pass the 100-year flood, provided the capacity of the drainage system including the culvert plus flow by-passing the culvert is sufficient to accommodate the 100-year flood without raising the water surface elevation 1 foot or if raised greater than 1 foot mitigation is provided. The designer should review the municipal floodway regulations for more information related to floodplain regulations.

Velocity Limitations

Both minimum and maximum velocities should be considered when designing a culvert. The maximum velocity should be consistent with channel stability requirements at the culvert outlet. As outlet velocities increase, the need for channel stabilization at the culvert outlet increases. If velocities exceed permissible velocities for the various types of nonstructural outlet lining material available, the installation of structural energy dissipators is appropriate.

The maximum allowable velocity for corrugated metal pipe is 10 feet per second. There is no specified maximum allowable velocity for reinforced concrete pipe, but outlet protection should be provided where discharge velocities will cause erosion problems.

A minimum velocity of 2.5 feet per second when the culvert is flowing partially full is recommended to ensure a self-cleaning condition during partial depth flow. When velocities below this minimum are anticipated to cause unacceptable sedimentation, the installation of a sediment trap upstream of the culvert should be considered. For streams carrying heavy sediment loads, higher velocities and lower headwater depths may be required to ensure sediment passage and avoid accumulation of sediment near the culvert entrance.

Buoyancy Protection

Headwalls, endwalls, slope paving or other means of anchoring to provide buoyancy protection should be considered for all flexible culverts. Buoyancy is more serious with steepness of the culvert slope, depth of the potential headwater (debris blockage may increase), flatness of the upstream fill slope, height of the fill, large culvert skews, or mitered ends.

Length And Slope

Since the capacity of culverts on outlet control will be affected by the length of the culvert, their length should be kept to a minimum and existing facilities should not be extended without determining the decrease in capacity that will occur. In addition, the culvert length and slope should be chosen to approximate existing topography, and to the degree practicable, the culvert invert should be aligned with the channel bottom and the skew angle of the stream, and the culvert entrance should match the geometry of the roadway embankment.

Debris Control

In designing debris control structures it is recommended that the Hydraulic Engineering Circular No. 9 entitled "Debris - Control Structures" be consulted. Debris control should be considered:
- where experience or physical evidence indicates the watercourse will transport a heavy volume of controllable debris,
- for culverts located in mountainous or steep regions,
- for culverts that are under high fills, and
- where clean out access is limited. However, access must be available to clean out the debris control device.

Ice Buildup

Ice buildup should be mitigated as necessary by:
* increasing the culvert height one foot above the total of the maximum observed ice buildup plus any winter flow depth, and
* increasing the culvert width to encompass the observed channel's static ice width plus 10 percent where appropriate to prevent property damage.

Headwater Limitations

The allowable headwater elevation is determined from an evaluation of land use upstream of the culvert and the proposed or existing roadway elevation. Headwater is the depth of water above the culvert invert at the entrance end of the culvert. In general the constraint which gives the lowest allowable headwater elevation establishes the criteria for the hydraulic calculations.

The following criteria related to headwater should be considered.
* The allowable headwater for design frequency conditions should allow for or consider the following upstream controls.
  * Reasonable freeboard.
  * Upstream property damage.
  * Elevations established to delineate floodplain zoning.
  * Low point in the road grade that is not at the culvert location.
  * Ditch elevation of the terrain that will permit flow to divert around culvert.
  * Following are some recommended HW/D design criteria -
    1. For drainage facilities with cross-section area equal to or less than 30 sq ft - HW/D = to or < 1.5.
    2. For drainage facilities with cross-section area greater than 30 sq ft - HW/D = to or < 1.2.
* The headwater should be checked for the 100-year flood to ensure compliance with floodplain management criteria and for most facilities the culvert should be sized to maintain flood-free conditions on major thoroughfares for one-half lane of two-lane facilities and one lane of multi-lane facilities.
* The maximum acceptable outlet velocity should be identified. Either the headwater should be set to produce acceptable velocities or stabilization or energy dissipation should be provided where these velocities are exceeded.

After determining the allowable headwater elevation, tailwater elevation, and approximate length, invert elevations must be established. Scour can be minimized if the culvert has the same slope as the channel. In addition, the flow conditions and velocity in the channel upstream from the culvert should be investigated to determine if scour will occur.

If there is insufficient headwater elevation available to convey the required discharge, it will be necessary to either use a larger culvert, lower the inlet invert, use an irregular cross section, use an improved inlet if in inlet control, use multiple barrels, or use a combination of these measures. If the inlet invert is lowered, special consideration must be given to scour and sedimentation at the entrance.

Tailwater Conditions

The hydraulic conditions downstream of the culvert site must be evaluated to determine a tailwater depth for a range of discharges. At times there may be a need for calculating backwater curves to establish the tailwater conditions. If the culvert outlet is operating with a free outfall, the critical depth and equivalent hydraulic grade line should be determined. For culverts which discharge to an open channel, the stage-discharge curve for the channel must be determined. See the Open Channel, Chapter 10, in this book.

If an upstream culvert outlet is located near a downstream culvert inlet or other control, the headwater elevation of the downstream control may establish the design tailwater depth for the upstream culvert. If the culvert discharges to a lake, pond, or other major water body, the expected high water elevation of the particular water body may establish the culvert tailwater.

Storage - Temporary Or Permanent

If storage is being assumed upstream of the culvert, consideration should be given to:
- the total area of flooding,
- the average time that bankfull stage is exceeded for the design flood up to 48 hours in rural areas or 6 hours in urban areas, and
- ensuring that the storage area will remain available for the life of the culvert through the purchase of right-of-way or easement.

Culvert Inlets

Selection of the type of inlet is an important part of culvert design - particularly with inlet control. Hydraulic efficiency and cost can be significantly affected by inlet conditions. The inlet coefficient K_e is a measure of the hydraulic efficiency of the inlet, with lower values indicating greater efficiency. All methods described in this chapter, directly or indirectly, use inlet coefficients. Recommended inlet coefficients are given in Table 9-1.

Following are some considerations that should be used to determine the type inlet to be used for a particular installation.

Projecting Inlets or Outlets
- Extend beyond the embankment of the roadway.
- Have low construction cost.
- Are susceptible to damage during roadway maintenance and automobile accidents.
- Have poor hydraulic efficiency for thin materials.
- Should be mitered to fill slope.
- Used predominantly with metal pipe.
- Should always include anchoring the inlet to concrete slope paving to strengthen the weak leading edge.

Headwalls with Bevels
- Increase the efficiency of metal pipe.
- Provide embankment stability and erosion protection.
- Provide protection from buoyancy.
- Shorten the required structure length.
- Reduce maintenance damage.

Improved Inlets
- Should be considered for culverts which will operate in inlet control.
- Can increase the hydraulic performance of the culvert, but may also add to the total culvert cost. Therefore, they should only be used if practicable.

Commercial End Sections
- Are available for both corrugated metal and concrete pipe.
- They retard embankment erosion and incur less damage from maintenance.
- May improve projecting metal pipe entrances by increasing hydraulic efficiency, reducing the accident hazard, and improving appearance.
- They are hydraulically equal to a headwall, but can be equal to a bevelled or side- tapered entrance if a flared, enclosed transition takes place before the barrel.

Table 9-1 Inlet Coefficients

Type of Structure and Design of Entrance	Coefficient K_e
Pipe, Concrete	
Projecting from fill, socket end (groove-end)	0.2
Projecting from fill, square cut end	0.5
Headwall or headwall and wingwalls	
Socket end of pipe (groove-end)	0.2
Square-edge	0.5
Rounded [radius = 1/12(D)]	0.2
Mitered to conform to fill slope	0.7
*End-Section conforming to fill slope	0.5
Beveled edges, 33.7° or 45° bevels	0.2
Side- or slope-tapered inlet	0.2
Pipe, or Pipe-Arch, Corrugated Metal	
Projecting from fill (no headwall)	0.9
Headwall or headwall and wingwalls square-edge	0.5
Mitered to fill slope, paved or unpaved slope	0.7
*End-Section conforming to fill slope	0.5
Beveled edges, 33.7° or 45° bevels	0.2
Side- or slope-tapered inlet	0.2
Box, Reinforced Concrete	
Headwall parallel to embankment (no wingwalls)	
Square-edged on 3 edges	0.5
Rounded on 3 edges to radius of [1/12(D)] or beveled edges on 3 sides	0.2
Wingwalls at 30° to 75° to barrel	
Square-edged at crown	0.4
Crown edge rounded to radius of [1/12(D)] or beveled top edge	0.2
Wingwalls at 10° or 25° to barrel	
Square-edged at crown	0.5
Wingwalls parallel (extension of sides)	
Square-edged at crown	0.7
Side- or slope-tapered inlet	0.2

* Note: End Section conforming to fill slope, made of either metal or concrete, are the sections commonly available from manufacturers. From limited hydraulic tests they are equivalent in operation to a headwall in both inlet and outlet control. Some end sections, incorporating a closed taper in their design, have a superior hydraulic performance.

Inlets With Headwalls

Headwalls may be used for a variety of reasons:
- increasing the efficiency of the inlet,
- providing embankment stability,
- providing embankment protection against erosion,
- providing protection from buoyancy, and
- shorten the length of the required structure.

The relative efficiency of the inlet depends on the pipe material. Headwalls are usually required for all metal culverts and where buoyancy protection is necessary. Corrugated metal pipe in a headwall is essentially square-edged with an inlet coefficient of about 0.5. For tongue and groove, or bell and concrete pipe, little increase in hydraulic efficiency is realized by adding a headwall.

Wingwalls And Aprons

Wingwalls are used where the side slopes of the channel adjacent to the entrance are unstable or where the culvert is skewed to the normal channel flow. Little increase in hydraulic efficiency is realized with the use of normal wingwalls, regardless of the pipe material used and, therefore, the use should be justified for other reasons. Wingwalls can be used to increase hydraulic efficiency if designed as a side-tapered inlet.

If high headwater depths are to be encountered, or the approach velocity in the channel will cause scour, a short channel apron should be provided at the toe of the headwall. This apron should extend at least one pipe diameter upstream from the entrance, and the top of the apron should not protrude above the normal streambed elevation.

Improved Inlets

Where inlet conditions control the amount of flow that can pass through the culvert, improved inlets can greatly increase the hydraulic performance at the culvert. For these designs refer to Section 9.8 which describes the design of improved inlets.

Material Selection

The material selection should consider replacement cost and difficulty of construction as well as traffic delay. The material selected should be based on a comparison of the total cost of alternate materials over the design life of the structure which is dependent upon the following:
- durability (service life),
- structural strength,
- hydraulic roughness,
- bedding conditions,
- abrasion and corrosion resistance, and
- water tightness requirements.

There are a number of different types of culvert pipe materials on the market, and a number of variations within each basic material type. There is significant competition among pipe manufacturers concerning which material is superior generally and for specific applications. Cost and longevity data from other than unbiased sources is not always trustworthy as the myriad of assumptions which go into the estimates can slant the results in any direction.

Generally material costs are comparable for the smaller diameters among the three major categories: concrete, steel and plastic, with installation costs being higher (normally in the range of 10 to 30 percent) for plastic and steel because of the greater bedding care than must be taken due to the pipe's flexibility. As the size increases above 36-inches in diameter concrete pipe begins to be more expensive, by as much as 40 percent for 72-inch pipes. This factor is balanced by the fact that concrete pipes have an expected life of 60 to 100 years or more while steel pipes have an expected life of 25 to 75 years depending heavily on: coating type(s) and thickness, the abrasive nature of the sediments, the acidity of the flow and soil, the use of street salts, groundwater resistivity and chemical content, and the soil conductivity. Plastic pipes have not had the track record to estimate longevity though it is generally thought to fall between concrete and steel pipes. Fire hazards may be an issue with plastic pipe.

Often bituminous coatings and pavings are used with corrugated metal pipe to prolong life by providing resistance to corrosion and abrasion. When coated, the layer should be a minimum of 0.05 inch thick. Culverts are either hot or cold dipped. Bottom paved pipe normally has the bottom 1/4 covered with a minimum thickness of 1/8 inch above the corrugations. Pipe arches have the bottom 40 percent covered. Polymer coatings and fiber-bonding are also used and should be applied in accordance with AASHTO and ASTM standards. Corrosion is repaired with aluminum-filled epoxy or aluminum-filled and fibered asphaltic coatings.

On balance, most cities require the use of concrete pipe for storm drains and sewers and for culverts placed in critical areas or within the public right-of-way. Often other types of pipe are allowed for less important applications such as driveways or low traffic parking lot entrances. For larger pipes or pipes on slopes too steep to allow easy concrete pipe installation, steel pipes are sometimes specified. The basic pipe materials and applicable testing standards from various sources are given in Table 9-2.

Table 9-2 Culvert Pipe Material Types and Standards

Pipe Material	Standards
Aluminum-alloy Structural Plate	- ASTM B790, B746 - culvert manufacture - ASTM B209 - alloy type
Galvanized Structural Plate	- ASTM A761, A796 - culvert manufacture - ASTM A444 - pipe material - AASHTO M243 - field coatings of asphaltic mastic or tar base material for shallow buried pipe

Corrugated Metal Pipe	- Joined with either rivet lap joint construction (annular corrugations) or continuous lock or welded seam (helical corrugations) and wrapped in nonwoven filter fabric or "o" ring gaskets. - Aluminum alloy corrugated pipe fabrication - ASTM B745, ASTM B209 - alloy material - Aluminum coated steel type 2 fabrication - ASTM A760, with ASTM A819 coupling bands - Galvanized steel fabrication - ASTM A760, with ASTM A444 material - Installation according to AASHTO Standard Specifications of Highway Bridges Section, 12 and 23, ASTM A796, A798
High Density Polyethylene Pipe (HDPE)	- Male and female ends with gasketed joints, ASTM D3212, or external coupling band joints in accordance with AASHTO standards - Precast fittings (wyes, tees, etc.) not normally accepted in lieu of precast storm sewer manholes - Corrugated HDPE - manufacture AASHTO M294, cell class minimum 324420C, ASTM D3350 with flexibility factor less than 0.095 - Ribbed HDPE - manufacture ASTM F894, cell class minimum 334433C, ASTM D3350 - Smoothwall HDPE - manufacture ASTM F714, cell class minimum 35434C, ASTM D3350 - Installation according to AASHTO Standard Specifications of Highway Bridges Section 18, ASTM D2321.
Polyvinyl Chloride Pipe (PVC)	- Storm sewer pipe, profile wall should be the bell and spigot type with elastomeric joints and smooth inner walls in accordance with AASHTO M304, minimum cell class 12454C or 12364C, ASTM D1784 - Smooth wall PVC manufacture ASTM F679 or AASHTO M278 with same cell class minimum - Precast fittings (wyes, tees, etc.) not normally accepted in lieu of precast storm sewer manholes - Installation according to AASHTO Standard Specifications of Highway Bridges Section 18, ASTM D2321.
Reinforced Concrete Pipe (RCP)	- Bell or groove and spigot or tongue pipe required with rubber gasket, ASTM C443 - Mastic type joints are acceptable for non-critical pipes if joints are wrapped with a one foot wide nonwoven geotextile fabric. - Material class III, IV or V, ASTM C76 - Elliptical RCP minimum class HE-II, ASTM C507 - Installation according to AASHTO Standard Specifications of Highway Bridges Sections 17 and 28.

Reinforced Concrete Box Sections	- Material standard, ASTM C789 with compressive strength tested prior to shipping - Precast box sections, ASTM C850 - Joints should be smooth to allow for a continuous line, sealed with butyl rubber or asphaltic mastic, and wrapped with a one foot wide nonwoven geotextile fabric. - Steel reinforcement minimum outside cover of 1-inch extending into the male and female ends. Steel wire should be between 1/2" and 2" from the ends of the pipe.

Table 9-3 gives recommended Manning's n values for different materials used for culverts.

Table 9-3 Manning's n Values

Type of Conduit	Wall & Joint Description	Manning's n
Concrete Pipe	Good joints, smooth walls	0.011-0.013
	Good joints, rough walls	0.014-0.016
	Poor joints, rough walls	0.016-0.017
	Badly Spalled	0.015-0.020
Concrete Box	Good joints, smooth finished walls	0.014-0.018
	Poor joints, rough, unfinished walls	0.014-0.018
Corrugated	2 2/3 by 1/2 inch corrugations	0.027-0.022
Metal Pipes and	6 by 1 inch corrugations	0.025-0.022
Boxes Annular	5 by 1 inch corrugations	0.026-0.025
Corrugations	3 by 1 inch corrugations	0.028-0.027
	6 by 2 inch structural plate	0.035-0.033
	9 by 2 1/2 inch structural plate	0.037-0.033
Corrugated	2 2/3 by 1/2 inch corrugated	
Metal Pipes,	24 inch plate width	0.024-0.012
Helical Corrugations, Full Circular		
Flow Spiral Rib Metal Pipe	3/4 by 3/4 in recesses at 12 inch spacing, good joints	0.012-0.013

Note: For further information concerning Manning n values for selected conduits consult Hydraulic Design of Highway Culverts, Federal Highway Administration, HDS No. 5, page 163.

Culvert Skews

Culvert skews should not exceed 45 degrees as measured from a line perpendicular to the roadway centerline without approval from the Municipal Engineering Department.

Culvert Sizes And Shapes

The culvert size and shape selected should be based on engineering and economic criteria related to site conditions. The following minimum sizes should be used to avoid maintenance prob-

lems, clogging and provide sufficient capacity:
- 24 inches for Interstate System,
- 18 inches for other systems, and
- 12 inches for a side-drain or drive.

Land use requirements can dictate a larger or different barrel geometry than required for hydraulic considerations. It is also recommended to only use arch or oval shapes if required by hydraulic limitations, site characteristics, structural criteria, or environmental criteria.

Multiple barrel culverts should fit within the natural dominant channel with minor widening of the channel so as to avoid conveyance loss through sediment deposition in some of the barrels. Multiple barrel culverts should be avoided where:
- the approach flow is high velocity, particularly if supercritical (these sites require either a single barrel or special inlet treatment to avoid adverse hydraulic jump effects),
- irrigation canals or ditches are present unless approved by the canal or ditch owner, and
- fish passage is required unless special treatment is provided to ensure adequate low flows (commonly one barrel is lowered).

Weep Holes

Weep holes are sometimes used to relieve uplift pressure. Filter materials should be used in conjunction with the weep holes in order to intercept the flow and prevent the formation of piping channels. The filter materials should be designed as underdrain filter so that it will not become clogged and so that piping cannot occur through the pervious material and the weep hole. Plastic woven filter cloth would be placed over the weep hole in order to keep the pervious material from being carried into the culvert. If weep holes are used to relieve uplift pressure, they should be designed in a manner similar to underdrain systems.

Outlet Protection

See Energy Dissipation, Chapter 12, for information on the design of outlet protection. In general scour holes at culvert outlets provide efficient energy dissipators. As such, outlet protection for the selected culvert design flood should only be provided where the outlet scour hole depth computations indicate:
- the scour hole will undermine the culvert outlet,
- the expected scour hole may cause costly property damage,
- the scour hole causes a nuisance effect (most common in urban areas),
- the scour hole blocks fish passage, and
- the scour hole will restrict land use requirements.

Erosion And Sediment Control

For erosion and sediment control, the use of silt boxes, brush silt barriers, temporary silt fence and filter cloth, and check dams may be appropriate. These measures should be utilized as necessary during construction to minimize pollution of streams and damages to wetlands.

Environmental Considerations

In addition to controlling erosion, siltation and debris at the culvert site, care must be exercised in selecting the location of the culvert site. Environmental considerations are a very important aspect of culvert design. Where compatible with good hydraulic engineering, a site should be selected that will permit the culvert to be constructed to cause the least impact on the stream or wetlands. This selection must consider the entire site, including any necessary lead channels.

Safety Considerations

Traffic should be protected from culvert ends as follows.
- Small culverts (30 in. diameter or less) should use an end section or a sloped headwall.
- Culverts greater than 30 in. diameter should receive one of the following treatments.
    a. Extended to the appropriate "clear zone" distance per AASHTO Roadside Design Guide.
    b. Safety treated with a grate if the consequences of clogging and causing a potential flooding hazard is less than the hazard of vehicles impacting an unprotected end. If a grate is used, an open area should be provided between the bars of 1.5 to 3.0 times the area of the culvert entrance.
    c. Shielded with a traffic barrier if the culvert is very large, cannot be extended, has a channel which cannot be safely traversed by a vehicle, or has a significant flooding hazard with a grate.
- Periodically inspect each site to determine if safety problems exist for traffic or for the structural safety of the culvert and embankment.
- Grating, if used for human safety, should be designed at an angle to cause the flow to force someone up above the culvert and not against the grate. See discussion in the Storage Facilities, Chapter 11.

9.4 Culvert Flow Controls And Equations

An exact theoretical analysis of culvert flow is extremely complex because the following is required:
- analyzing nonuniform flow with regions of both gradually varying and rapidly varying flow,
- determining how the flow type changes as the flow rate and tailwater elevations change,
- applying backwater and drawdown calculations,
- applying energy and momentum balance,
- applying the results of hydraulic model studies, and
- determining if hydraulic jumps occur and if they are inside or downstream of the culvert barrel.
The design procedures contained in this chapter are for the design of culverts for a constant discharge, considering inlet and outlet control. Generally, the hydraulic control in a culvert will be at the culvert outlet if the culvert is operating on a mild slope. Entrance control usually occurs if the culvert is operating on a steep slope.

For outlet control, the head losses due to tailwater and barrel friction are predominant in controlling the headwater of the culvert. The entrance will allow the water to enter the culvert faster than the backwater effects of the tailwater and barrel friction will allow it to flow through the culvert.

For inlet control, the entrance characteristics of the culvert are such that entrance head losses are predominant in determining the headwater of the culvert. The barrel will carry water through the culvert more efficiently than the water can enter the culvert. Proper culvert design and analysis requires checking for both inlet and outlet control to determine which will govern particular culvert designs. For more information on inlet and outlet control see the Federal Highway Administration publication entitled - Hydraulic Design Of Highway Culverts, HDS-5, 1985.

Design Equations

The following diagram illustrates the terms and dimensions used in the culvert headwater equations.

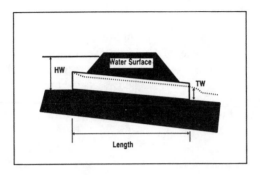

Figure 9-1 Culvert Terms And Dimensions

There are many combinations of conditions which may classify a particular culvert's hydraulic operation. By consideration of a succession of parameters, the designer may arrive at the appropriate calculation procedure. The most common types of culvert operations for any barrel type are classified as follows.

Critical Depth - Outlet Control - The entrance is unsubmerged (HW $\leq$ 1.5D), the critical depth is less than uniform depth at the design discharge ($d_c < d_u$) and the tailwater is less than or equal to critical depth (TW $\leq d_c$). This condition is a common occurrence where the natural channels are on flat grades and have wide, flat floodplains. The control is critical depth at the outlet.

$$HW = d_c + V_c^2/(2g) + H_e + H_f - SL \tag{9.1}$$

Where: HW =   headwater depth, ft
     d_c =   critical depth, ft
     V_c =   critical velocity, ft/s
     g =   32.2 ft/sec^2
     H_e =   entrance headloss, ft
     H_f =   friction headloss, ft
     S =   slope of culvert, ft/ft
     L =   length of culvert, ft

Tailwater Depth - Outlet Control - The entrance is unsubmerged (HW $\leq$ 1.5D), the critical depth is less than uniform depth at design discharge ($d_c < d_u$) and TW is greater than critical depth (TW $> d_c$) and TW is less than D (TW $<$ D). This condition is a common occurrence where the channel is deep, narrow, and well defined. The control is tailwater at the culvert outlet. The outlet velocity is the discharge divided by the area of flow in the culvert at tailwater depth.

$$HW = TW + V^2/(2g) + H_e + H_f - SL \tag{9.2}$$

Where: HW =   headwater depth, ft
     TW =   tailwater at the outlet, ft
     V =   velocity based on tailwater depth, ft/s

g = 32.2 ft/sec^2
H_e = entrance headloss, ft
H_f = friction headloss, ft
S = slope of culvert, ft/ft
L = length of culvert, ft

Tailwater Depth > Barrel Depth - Outlet Control - This condition will exist if the critical depth is less than uniform depth at the design discharge ($d_c < d_u$) and TW depth is greater than D (TW > D), or the critical depth is greater than the uniform depth at the design discharge ($d_c > d_u$) and TW is greater than (SL + D), [TW > (SL + D)]. The HW may or may not be greater than 1.5D, though often it is greater. If the critical depth of flow is determined to be greater than the barrel depth (only possible for rectangular culvert barrels), then this operation will govern. Outlet velocity is based on full flow at the outlet.

$$HW = H + TW - SL \qquad (9.3)$$

Where: HW = headwater depth, ft
H = total head loss of discharge through culvert, ft
TW = tailwater depth, ft
SL = culvert slope times length of culvert, ft

Tailwater < Barrel - Outlet Control - The entrance is submerged (HW > 1.5D) and the tailwater depth is less than D (TW < D). Normally, the designer should arrive at this type of operation only after previous consideration of the operations depth covered when the critical depth, tailwater depth, or "slug" flow controls the flow in outlet control conditions. On occasion, it may be found that (HW ≥ 1.5D) for the three previously outlined conditions but (HW < 1.5D) for equation 9.4. If so, the higher HW should be used. Outlet velocity is based on critical depth if TW depth is less than critical depth. If TW depth is greater than critical depth, outlet velocity is based on TW depth.

$$HW = H + P - SL \qquad (9.4)$$

Where: HW = headwater depth, ft
H = total head loss of discharge through culvert, ft
P = empirical approximation of equivalent hydraulic grade line, ft. P = (d_c + D)/2 if TW depth is less than critical depth at design discharge. If TW is greater than critical depth, then P = TW.
SL = culvert slope times length of culvert, ft

Tailwater Insignificant - Inlet Control - The entrance may be submerged or unsubmerged, the critical depth is greater than uniform depth at the design discharge ($d_c > d_u$), TW depth is less than SL (tailwater elevation is lower than the upstream flowline). Tailwater depth with respect to the diameter of the culvert is inconsequential as long as the above conditions are met. This condition is a common occurrence for culverts in rolling or hilly country. The control is critical depth at the entrance for HW values up to about 1.5D. Control is the entrance geometry for HW values over about 1.5D. HW is determined from empirical curves in the form of nomographs that are discussed later in this chapter. If TW is greater than D, outlet velocity is based on full flow at the outlet. If TW is less than D, outlet velocity is based on uniform depth for the culvert.

Inlet Or Outlet Control - For "slug" flow operation the entrance may be submerged or unsub-merged, critical depth is greater than uniform depth at the design discharge ($d_c > d_u$), TW depth is greater than ($SL + d_c$) (TW elevation is above the critical depth at the entrance), and TW depth is less than $SL + D$ (TW elevation is below the upstream soffit). TW depth with respect to D alone is inconsequential as long as the above conditions are met. This condition is a common occurrence for culverts in rolling or hilly country. The control for this type of operation may be at the entrance or the outlet or control may transfer back and forth between the two (commonly called "slug" flow). For this reason, it is recommended that HW be determined for both entrance control and outlet control and the higher of the two determinations be used. Entrance control HW is determined from the inlet control nomographs and outlet control HW is determined by equations 9.3 or 9.4 or the outlet control nomographs.

If TW depth is less than D, outlet velocity should be based on TW depth. If TW depth is greater than D, outlet velocity should be based on outlet full flow.

9.5   Design Procedures

The following design procedure provides a convenient and organized method for designing culverts for constant discharge, considering inlet and outlet control. There are two procedures presented for designing culverts: (1) the manual use of inlet and outlet control nomographs and (2) the use of a personal computer system HYDRAIN.

It is recommended that the HYDRAIN computer model be used for culvert design since it will allow the designer to easily develop performance curves rather than only examining one design situation. The personal computer system HYDRAIN uses the theoretical basis for the nomographs to size a culvert. In addition, this system can evaluate improved inlets, route hydrographs, consider road overtopping, and evaluate outlet streambed scour. By using water surface profiles, this proce-dure is more accurate in predicting backwater effects and outlet scour.

The following will outline the design procedures for use of the nomographs. The use of the computer model is described in Section 9.11.

Tailwater Elevations

In some cases culverts fail to perform as intended because of tailwater elevations high enough to create backwater. The problem is more severe in areas where gradients are very flat, and in some cases in areas with moderate slopes. Thus, as part of the design process, the normal depth of flow in the downstream channel at discharges equal to those being considered should be computed.

If the tailwater computation leads to water surface elevations below the invert of the culvert exit, there are obviously no problems; if elevations above the culvert invert are computed, the culvert capacity will be somewhat less than assumed. The tailwater computation can be simple, and on steep slopes requires little more than the determination of a cross section downstream where normal flow can be assumed, and a Manning equation calculation. (See Open Channel, Chapter 10, for more information on open channel analysis.) Conversely, with sensitive flood hazard sites, if the slopes are flat, or natural and man-made obstructions exist downstream, a water surface profile analysis reaching beyond these obstructions may be required.

Use Of Inlet And Outlet Control Nomographs

The use of nomographs requires a trial and error solution. The solution is quite easy and provides reliable designs for many applications. It should be remembered that velocity, hydrograph routing, roadway overtopping, and outlet scour require additional, separate computations beyond what can be obtained from the nomographs.

Figures 9-2 and 9-3 show examples of an inlet control and outlet control nomograph that can be used to design concrete pipe culverts. For culvert designs not covered by these nomographs, refer to the complete set of nomographs given in Appendix C at the end of this chapter. Following is the design procedure which requires the use of inlet and outlet control nomographs.

Step	Action
(1)	List design data:

        Q = discharge, cfs L = culvert length, ft
        S = culvert slope, ft/ft K_e = inlet loss coefficient
        V = velocity, ft/s TW = tailwater depth, ft
        HW = allowable headwater depth for the design storm, ft

(2)   Determine trial culvert size by assuming a trial velocity 3 to 5 ft/s and computing the culvert area, $A = Q/V$. Determine the culvert diameter (inches).

(3)   Find the actual HW for the trial size culvert for both inlet and outlet control.

- For inlet control, enter inlet control nomograph with D and Q and find HW/D for the proper entrance type.
- Compute HW and, if too large or two small, try another culvert size before computing HW for outlet control.
- For outlet control enter the outlet control nomograph with the culvert length, entrance loss coefficient, and trial culvert diameter.
- To compute HW, connect the length scale for the type of entrance condition and culvert diameter scale with a straight line, pivot on the turning line, and draw a straight line from the design discharge through the turning point to the head loss scale H. Compute the headwater elevation HW from the equation:

$$HW = H + h_o - LS \tag{9.5}$$

where: $h_o = 1/2$ (critical depth + D), or tailwater depth, whichever is greater.

(4)   Compare the computed headwaters and use the higher HW nomograph to determine if the culvert is under inlet or outlet control. If outlet control governs and the HW is unacceptable, select a larger trial size and find another HW with the outlet control nomographs. Since the smaller size of culvert had been selected for allowable HW by the inlet control nomographs, the inlet control for the larger pipe need not be checked.

(5)   Calculate exit velocity and expected streambed scour to determine if an energy dissipator is needed.

A performance curve for any culvert can be obtained from the nomographs by repeating the steps outlined above for a range of discharges that are of interest for that particular culvert design. A graph is then plotted of headwater vs. discharge with sufficient points so that a curve can be drawn through the range of interest. These curves are applicable through a range of headwater, velocities, and scour depths versus discharges for a length and type of culvert. Usually curves with length intervals of 25 to 50 feet are satisfactory for design purposes. Such computations are made much easier by available computer programs.

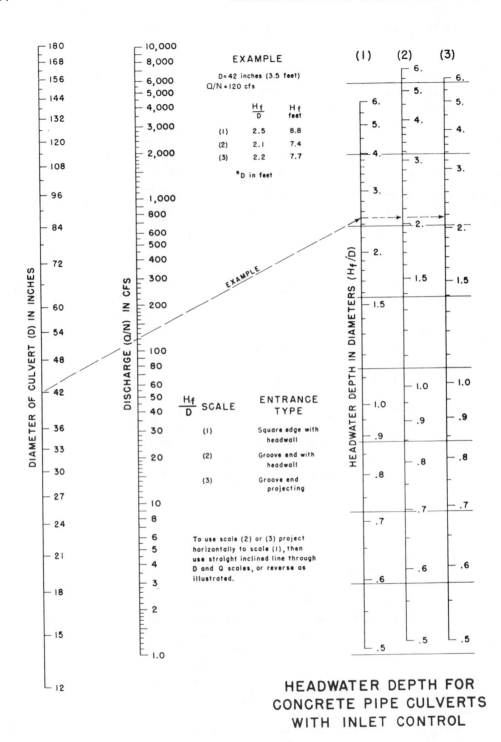

EXAMPLE

D= 42 inches (3.5 feet)
Q/N = 120 cfs

	$\frac{H_f}{D}$	H_f feet
(1)	2.5	8.8
(2)	2.1	7.4
(3)	2.2	7.7

*D in feet

$\frac{H_f}{D}$ SCALE   ENTRANCE TYPE

(1)   Square edge with headwall

(2)   Groove end with headwall

(3)   Groove end projecting

To use scale (2) or (3) project horizontally to scale (1), then use straight inclined line through D and Q scales, or reverse as illustrated.

HEADWATER DEPTH FOR
CONCRETE PIPE CULVERTS
WITH INLET CONTROL

Figure 9-2 Inlet Control Nomograph

Source: FHWA, 1973

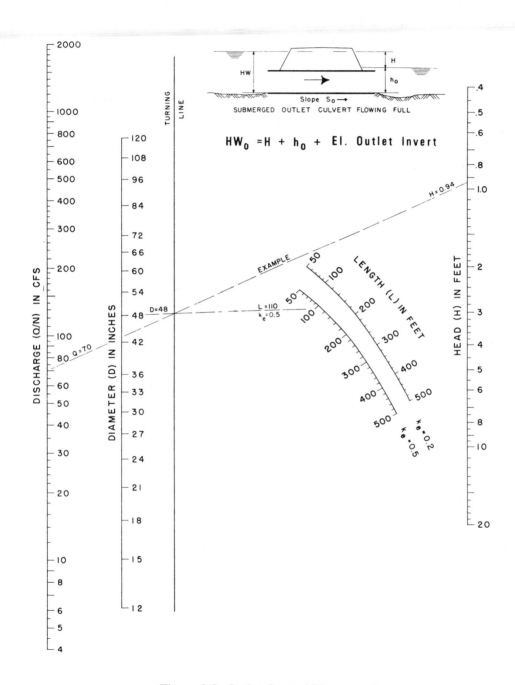

Figure 9-3 Outlet Control Nomograph

Source: Bureau Of Public Roads, 1963

To complete the culvert design, roadway overtopping should be analyzed. A performance curve showing the culvert flow as well as the flow across the roadway is a useful analysis tool. Rather than using a trial and error procedure to determine the flow division between the overtopping flow and the culvert flow, an overall performance curve can be developed.

The overall performance curve can be determined as follows:

Step   Action

(1)   Select a range of flow rates and determine the corresponding headwater elevations for the culvert flow alone. The flow rates should fall above and below the design discharge and cover the entire flow range of interest. Both inlet and outlet control headwaters should be calculated.

(2)   Combine the inlet and outlet control performance curves to define a single performance curve for the culvert.

(3)   When the culvert headwater elevations exceed the roadway crest elevation, overtopping will begin. Calculate the equivalent upstream water surface depth above the roadway (crest of weir) for each selected flow rate. Use these water surface depths and equation 9.6 to calculate flow rates across the roadway.

$$Q = C_d L (HW)^{1.5} \hspace{4cm} (9.6)$$

Where: $Q =$ overtopping flow rate, cfs
$C_d =$ overtopping discharge coefficient
$L =$ length of roadway, ft
$HW =$ upstream depth, measured from the roadway crest to the water surface upstream of the weir drawdown, ft

(4)   See Figure 9-4 for guidance in determining a value for C_d. For more information on calculating overtopping flow rates see pages 39 - 42 in HDS No. 5.

(5)   Add the culvert flow and the roadway overtopping flow at the corresponding headwater elevations to obtain the overall culvert performance curve.

Storage Routing

A significant storage capacity behind a highway embankment attenuates a flood hydrograph. Because of the reduction of the peak discharge associated with this attenuation, the required capacity of the culvert, and its size, may be reduced considerably. If significant storage is anticipated behind a culvert, the design should be checked by routing the design hydrographs through the culvert to determine the discharge and stage behind the culvert. Routing procedures are outlined in Hydraulic Design of Highway Culverts, Section V - Storage Routing, HDS No. 5, Federal Highway Administration. Section 9.10 in this chapter gives some basic information on flood routing and culvert design.

9.6 Culvert Design Example

The following example problem illustrates the procedures to be used in designing culverts using the nomographs. Example - size a culvert given the following design conditions which were determined by physical limitations at the culvert site and hydraulic procedures described elsewhere in this book.

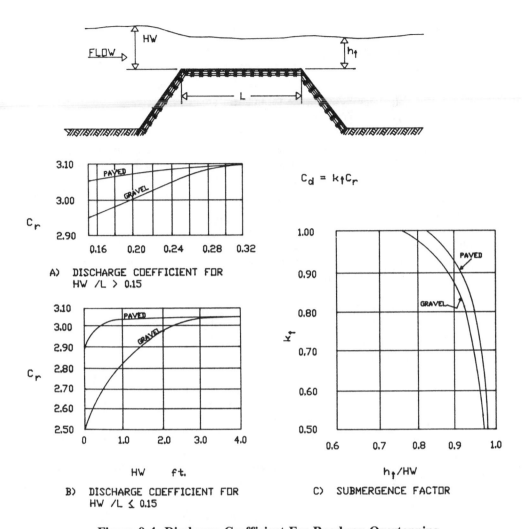

$$C_d = k_t C_r$$

A) DISCHARGE COEFFICIENT FOR
    HW /L > 0.15

B) DISCHARGE COEFFICIENT FOR
    HW /L ≤ 0.15

C) SUBMERGENCE FACTOR

Figure 9-4 Discharge Coefficient For Roadway Overtopping

Source: HDS No. 5, 1985

Input Data

Discharge for 10-yr flood = 70 cfs
Discharge for 100-yr flood = 176 cfs
Allowable H_w for 10-yr discharge = 4.5 ft
Allowable H_w for 100-yr discharge = 7.0 ft
Length of culvert = 100 ft
Natural channel invert elevations - inlet = 15.50 ft, outlet = 15.35 ft
Culvert slope = 0.0015 ft/ft
Tailwater depth for 10-yr discharge = 3.0 ft
Tailwater depth for 100-yr discharge = 4.0 ft
Tailwater depth is the normal depth in downstream channel
Entrance type = Groove end with headwall

Steps	Computation

1.   Assume a culvert velocity of 5 ft/s
       Required flow area = 70 cfs/5 ft/s = 14 ft^2 (for the 10-yr flood).
2.   The corresponding culvert diameter is about 48 in.
       This can be calculated by using the formula for area of a circle:
       Area = (3.14D^2)/4 or D = (Area times 4/3.14)$^{0.5}$
       Therefore: D = ((14 sq ft x 4)/3.14)$^{0.5}$ x 12 in./ft = 50.7 in.
3.   A grooved end culvert with a headwall is selected for the design. Using the inlet control
       nomograph (Figure 9-2), with a pipe diameter of 48 in. and a discharge of 70 cfs; read
       a HW/D value of 0.93.
4.   The depth of headwater (HW) is (0.93) x (4) = 3.72 ft which is less than the allowable
       headwater of 4.5 ft.
5.   The culvert is checked for outlet control by using Figure 9-3.
       With an entrance loss coefficient K_e of 0.20, a culvert length of 100 ft, and a pipe
       diameter of 48 in., an H value of 0.77 ft is determined. The headwater for outlet control
       is computed by the equation:
       $$HW = H + h_o - LS$$
       For the tailwater depth lower than the top of culvert,
       h_o = T_w or 1/2 (critical depth in culvert + D) whichever is greater.
       h_o = 3.0 ft or h_o = 1/2 (2.55 + 4.0) = 3.28 ft
       The headwater depth for outlet control is:
       $$HW = H + h_o - LS$$
       $$HW = 0.77 + 3.28 - (100) \text{ x } (0.0015) = 3.90 \text{ ft}$$
6.   Since HW for outlet control (3.90 ft) is greater than the HW for inlet control (3.72 ft),
       outlet control governs the culvert design.
       Thus, the maximum headwater expected for a 10-yr recurrence flood is 3.90 ft, which
       is less than the allowable headwater of 4.5 ft.
7.   The performance of the culvert is checked for the 100-yr discharge. The allowable
       headwater for a 100-yr discharge is 7 ft; critical depth in the 48 in. diameter culvert for
       the 100-yr discharge is 3.96 ft.
       For outlet control, an H value of 5.2 ft is read from the outlet control nomograph. The
       maximum headwater is:
       $$HW = H + h_o - LS$$
       $$HW = 5.2 + 4.0 - (100) \text{ x } (0.0015) = 9.05 \text{ ft}$$
       This depth is greater than the allowable depth of 7 ft, thus a larger size culvert must be
       selected.
8.   A 54 in. diameter culvert is tried and found to have a maximum headwater depth of 3.74
       ft for the 10-yr discharge and of 6.97 ft for the 100-yr discharge. These values are
       acceptable for the design conditions.
9.   Estimate outlet exit velocity. Since this culvert is on outlet control and discharges into
       an open channel downstream, the culvert will be flowing full at the flow depth in the
       channel. Using the 100-year design peak discharge of 176 cfs and the area of a 54 in.
       or 4.5 ft diameter culvert the exit velocity will be: Q = VA.
              Therefore: V = 176 / (π(4.5)^2)/4 = 11.8 ft/s.
       With this high velocity some energy dissipator may be needed downstream from this
       culvert for streambank protection. It will first be necessary to compute a scour hole depth
       and then decide if protection is needed. See Energy Dissipation, Chapter 12, for design
       procedures related to energy dissipators.

10.   The designer should check minimum velocities for low frequency flows if the larger storm
        event (100-year) controls culvert design. Note: Figure 9-5 provides a convenient form
        to organize culvert design calculations. For an example of a design which incorporates
        roadway overtopping, see Appendix A - example application of the HY8 Culvert Analysis
        Microcomputer Program.

9.7 Long Span Culvert

Long span culverts are better defined on the basis of structural design aspects than on the basis
of hydraulic considerations. According to the AASHTO Specifications for Highway Bridges, long
span structural plate structures: (1) exceed certain defined maximum sizes for pipes, pipe-arches,
and arches, or (2) may be special shapes of any size that involve a long radius of curvature in the
crown or side plates. Special shapes include vertical and horizontal ellipses, underpasses, low and
high profile arches, and inverted pear shapes. Generally, the spans of long span culverts range
from 20 ft to 40 ft.

Long span culverts depend on interaction with the earth embankment for structural stability.
Therefore, proper bedding and selection and compaction of backfill are of utmost importance.
For multiple barrel structures, care must be taken to avoid unbalanced loads during backfilling.
Anchorage of the ends of long span culverts is required to prevent flotation or damage due to high
velocities at the inlet. This is especially true for mitered inlets. Severe miters and skews are not
recommended.

Long span culverts generally are hydraulically short (low length to equivalent diameter ratio)
and flow partly full at the design discharge. The same hydraulic principles apply to the design
of long span culverts as to other culverts. However, due to their large size and variety of shapes,
it is very possible that design nomographs are not available for the barrel shape of interest. For
these cases, dimensionless inlet control design curves have been prepared. For these nomographs
and design curves consult the publication, Hydraulic Design of Highway Culverts, Federal Highway
Administration, HDS No. 5.

For outlet control, backwater calculations are usually appropriate, since design headwaters
exceeding the crowns of these conduits are rare. The bridge design techniques of HDS No. 1,
Hydraulics of Bridge Waterways, are appropriate for the design of most long span culverts.

9.8 Design Of Improved Inlets

A culvert operates in either inlet or outlet control. As previously discussed under outlet
control, headwater depth, tailwater depth, entrance configuration, and barrel characteristics all
influence a culvert's capacity. The entrance configuration is defined by the barrel cross sectional
area, shape, and edge condition, while the barrel characteristics are area, shape, slope, length, and
roughness.

The flow condition for outlet control may be full or partly full for all or part of the culvert
length. The design discharge usually results in full flow. Inlet improvements in these culverts
reduce the entrance losses, which are only a small portion of the total headwater requirements.
Therefore, only minor modifications of the inlet geometry which results in little additional cost
are justified.

In inlet control, only entrance configuration and headwater depth determine the culvert's
hydraulic capacity. Barrel characteristics and tailwater depth are of no consequence. These
culverts usually lie on relatively steep slopes and flow only partly full. Entrance improvements
can result in full, or nearly full flow, thereby increasing culvert capacity significantly.

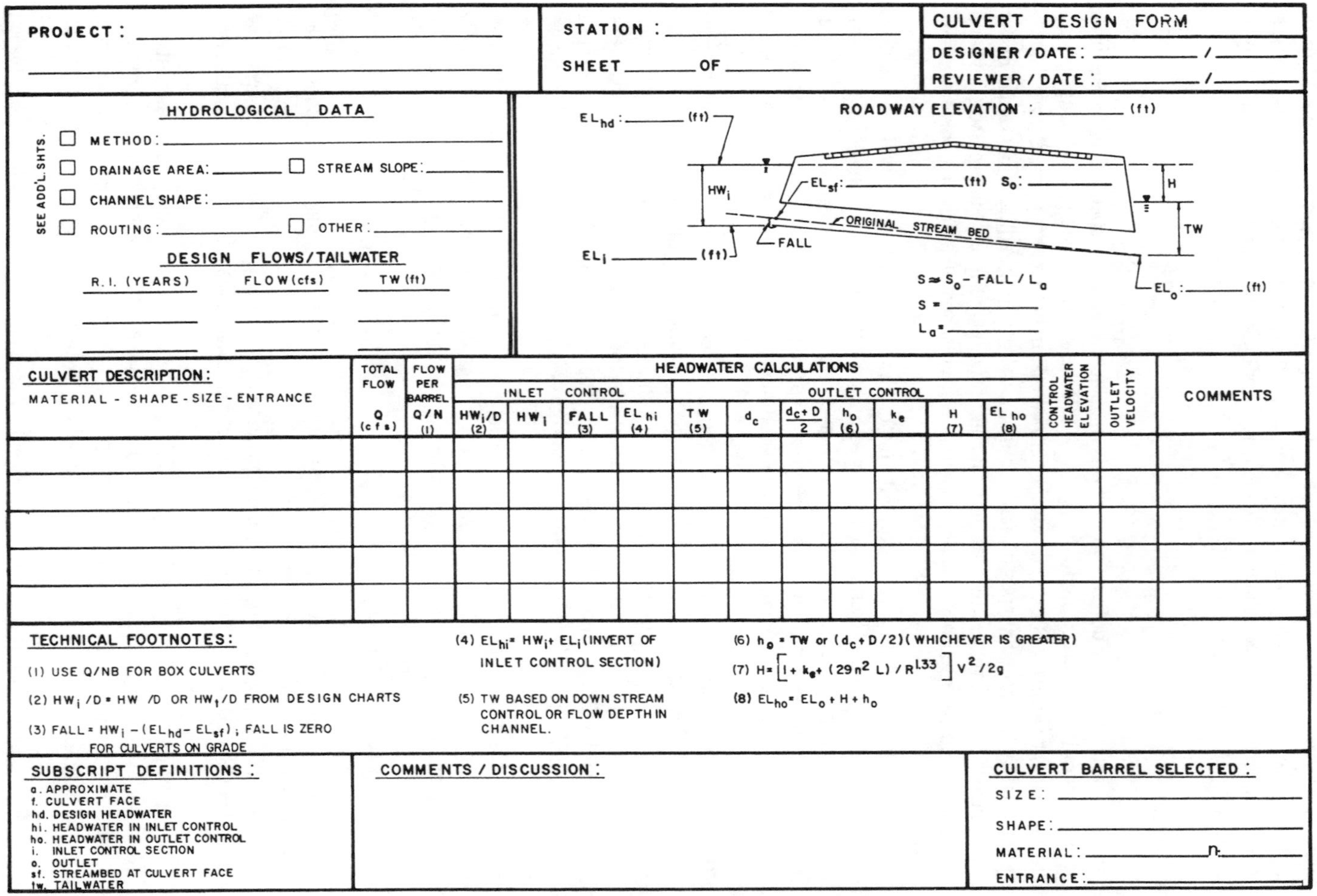

PROJECT: ______

STATION: ______ SHEET ______ OF ______

CULVERT DESIGN FORM

DESIGNER / DATE: ______ / ______

REVIEWER / DATE: ______ / ______

HYDROLOGICAL DATA

SEE ADD'L. SHTS.

- ☐ METHOD: ______
- ☐ DRAINAGE AREA: ______ ☐ STREAM SLOPE: ______
- ☐ CHANNEL SHAPE: ______
- ☐ ROUTING: ______ ☐ OTHER: ______

DESIGN FLOWS/TAILWATER

R.I. (YEARS)	FLOW (cfs)	TW (ft)

CULVERT DESCRIPTION: MATERIAL - SHAPE - SIZE - ENTRANCE	TOTAL FLOW Q (cfs)	FLOW PER BARREL Q/N (1)	HEADWATER CALCULATIONS: INLET CONTROL HW_i/D (2)	HW_i	FALL (3)	EL_{hi} (4)	OUTLET CONTROL TW (5)	d_c	$\frac{d_c+D}{2}$	h_o (6)	k_e	H (7)	EL_{ho} (8)	CONTROL HEADWATER ELEVATION	OUTLET VELOCITY	COMMENTS

TECHNICAL FOOTNOTES:

(1) USE Q/NB FOR BOX CULVERTS

(2) $HW_i/D = HW/D$ OR HW_i/D FROM DESIGN CHARTS

(3) $FALL = HW_i - (EL_{hd} - EL_{sf})$; FALL IS ZERO FOR CULVERTS ON GRADE

(4) $EL_{hi} = HW_i + EL_i$ (INVERT OF INLET CONTROL SECTION)

(5) TW BASED ON DOWN STREAM CONTROL OR FLOW DEPTH IN CHANNEL.

(6) $h_o = TW$ or $(d_c + D/2)$ (WHICHEVER IS GREATER)

(7) $H = \left[1 + k_e + (29n^2 L)/R^{1.33}\right] V^2/2g$

(8) $EL_{ho} = EL_o + H + h_o$

SUBSCRIPT DEFINITIONS:

- a. APPROXIMATE
- f. CULVERT FACE
- hd. DESIGN HEADWATER
- hi. HEADWATER IN INLET CONTROL
- ho. HEADWATER IN OUTLET CONTROL
- i. INLET CONTROL SECTION
- o. OUTLET
- sf. STREAMBED AT CULVERT FACE
- tw. TAILWATER

COMMENTS / DISCUSSION:

CULVERT BARREL SELECTED:

SIZE: ______

SHAPE: ______

MATERIAL: ______ n: ______

ENTRANCE: ______

Figure 9-5 Culvert Design Calculation Form

Source: HDS-5, 1985

Figure 9-6 illustrates the performance of a 30-inch circular culvert in inlet control with three commonly used entrances: thin-edged projecting, square-edged, and groove-edged.

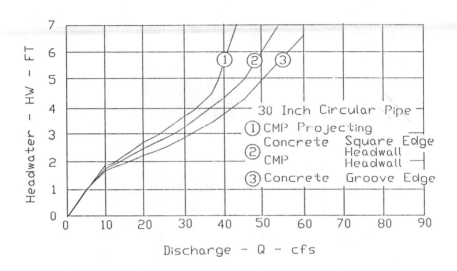

Figure 9-6 Performance Curves - Inlet Control

Source: HDS-5, 1985

Improved inlets include inlet geometry refinements beyond those normally used in conventional culvert design practice. Several degrees of improvements are possible, including bevel-edged, side-tapered, and slope-tapered inlets.

It is clear that inlet type and headwater depth determine the capacities of many culverts. For a given headwater, a groove-edged inlet has a greater capacity than a square-edged inlet, which in turn out performs a thin-edged projecting inlet.

The performance of each inlet type is related to the degree of flow contraction. A high degree of contraction requires more energy, or headwater, to convey a given discharge than a low degree of contraction. Figure 9-7 shows schematically the flow contractions of the three inlet types.

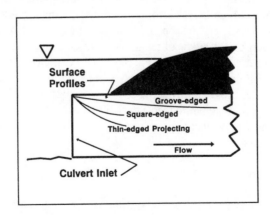

Figure 9-7 Schematic Flow Contractions For Conventional Culvert Inlets

Bevel-Edged Inlet

The first degree of inlet improvement is a beveled edge (Figure 9-8). The bevel is proportioned based on the culvert barrel or face dimension and operates by decreasing the flow contraction at the inlet. A bevel is similar to a chamfer except that a chamfer is smaller and is generally used to prevent damage to sharp concrete edges during construction.

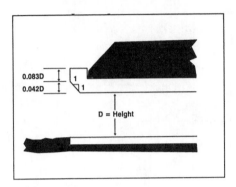

0.083D ⫶ 1
0.042D ⫶ 1

D = Height

Figure 9-8 Bevel-Edged Inlet

Adding bevels to a conventional culvert design with a square-edged inlet increases culvert capacity by 5 to 20 percent. The higher increase results from comparing a bevel-edged inlet with a square-edged inlet at high headwaters. The lower increase is the result of comparing inlets with bevels, with structures having wingwalls of 30 to 45 degrees. Although the bevels referred to in this book are plane surfaces, rounded edges which approximate the bevels are also acceptable.

As a minimum, bevels should be used on all culverts which operate in inlet control, both conventional and improved inlet types. The exception to this is circular concrete culverts where the socket end performs much the same as a beveled edge.

Culverts flowing in outlet control cannot be improved as much as those in inlet control, but the entrance loss coefficient, k_e, is reduced from 0.5 for a square edge to 0.2 for beveled edges. It is recommended that bevels be used on all culvert entrances if little additional cost is involved.

Side-Tapered Inlet

The second degree of improvement is a side-tapered inlet. This inlet has an enlarged face area with the transition to the culvert barrel accomplished by tapering the sidewalls. The inlet face has the same height as the barrel, and its top and bottom are extensions of the top and bottom of the barrel. The intersection of the sidewall tapers and barrel is defined as the throat section. If a headwall and wingwalls are going to be used at the culvert entrance, side-tapered inlets should add little if any to the overall cost while significantly increasing hydraulic efficiency.

The side-tapered inlet provides an increase in flow capacity of 25 to 40 percent over that of a conventional culvert with a square edged inlet. Whenever increased inlet efficiency is needed or when a headwall and wingwalls are planned to be used for a culvert installation, a side-tapered inlet should be considered.

Slope-Tapered Inlet

A slope-tapered inlet is the third degree of improvement. Its advantage over the side-tapered inlet without a depression is that more head is available at the inlet. This is accomplished by incorporating a fall in the enclosed entrance section. The slope-tapered inlet can have over a 100 percent greater capacity than a conventional culvert with square edges. The degree of increased capacity depends largely upon the amount of fall available. Since this fall may vary, a range of increased capacities is possible.

Side- and slope-tapered inlets should be used in culvert design when they can economically be used to increase the inlet efficiency over a conventional design. For a complete discussion of tapered inlets including figures and illustrations, see pages 65-93, FHWA, HDS-5, 1985.

Improved Inlet Performance

Table 9-4 compares the inlet control performance of the different inlet types. The top part of the table shows the increase in discharge that is possible for a headwater depth of 8 feet. The bevel-edged inlet, side-tapered inlet and slope-tapered inlet show increases in discharge over the square-edged inlet of 16.7, 30.4 and 55.6 percent, respectively. It should be noted that the slope-tapered inlet incorporates only a minimum fall. Greater increases in capacity are often possible if a larger fall is used. The bottom part of the table depicts the reduction in headwater that is possible for a discharge of 500 cfs. The headwater varies from 12.5 ft for the square-edged inlet to 7.6 ft for the slope-tapered inlet. This is a 39.2 percent reduction in required headwater.

9.9 Design Procedures For Beveled-Edged Inlets

This section will outline the procedures and figures to use when incorporating bevel-edged inlets in the design of culverts. For the design of side- and slope-tapered inlets consult the detailed design criteria and example designs outlined in the U. S. Department of Transportation publication Hydraulic Engineering Circular No. 5 entitled Hydraulic Design of Highway Culverts.

Table 9-4

Comparison Of Inlet Performance At Constant Headwater For 6 ft x 6 ft Concrete Box Culvert

Inlet Type	Headwater	Discharge	%Improvement
Square-edge	8.0 feet	336 cfs	0
Bevel-edge	8.0 feet	392 cfs	16.7
Side-tapered	8.0 feet	438 cfs	30.4
Slope-tapered*	8.0 feet	523 cfs	55.6

Comparison Of Inlet Performance At Constant Discharge For 6 ft x 6 ft Concrete Box Culvert

Inlet Type	Discharge	Headwater	% Improvement
Square-edge	500 cfs	12.5 feet	0
Bevel-edge	500 cfs	10.1 feet	19.2
Side-tapered	500 cfs	8.8 feet	29.6
Slope-tapered*	500 cfs	7.6 feet	39.2

* Minimum fall in inlet = D/4 = 6/4 = 1.5 ft

Four inlet control figures for culverts with beveled edges are included in Appendix D at the end of this chapter. Following is a list of the figures and their use in culvert design.

Figure   Use for -

1   90° headwalls (same for 90° wingwalls)
2   skewed headwalls
3   wingwalls with flare angles of 18 to 45 degrees
4   circular pipe culverts with beveled rings
             Note that Figures 2, 3, and 4 apply only to bevels having either a 33° angle (1.5:1) or a 45° angle (1:1).

The figures for bevel-edged inlets are used for design in the same manner as the conventional inlet design nomographs discussed earlier. For box culverts the dimensions of the bevels to be used are based on the culvert dimensions. The top bevel dimension is determined by multiplying the height of the culvert by a factor. The side bevel dimensions are determined by multiplying the width of the culvert by a factor. For a 1:1 bevel, the factor is 1/2 in./ft. For a 1.5:1 bevel the factor is 1 in./ft.

For example the minimum bevel dimensions for a 8 ft x 6 ft box culvert with 1:1 bevels would be:

Top Bevel $= d = 6$ ft x 1/2 in./ft $= 3$ in.

Side Bevel $= b = 8$ ft x 1/2 in./ft $= 4$ in.

For a 1.5:1 bevel computations would result in $d = 6$ in. and $b = 8$ in.

Design Figure Limits

The improved inlet design figures are based on research results from culvert models with barrel width, B, to depth, D, ratios of from 0.5:1 to 2:1.

For box culverts with more than one barrel, the figures are used in the same manner as for a single barrel, except that the bevels must be sized on the basis of the total clear opening rather than on individual barrel size. For example, in a double 8 ft by 8 ft box culvert:

Top Bevel - is proportioned based on the height of 8 ft which results in a bevel of 4 in. for the 1:1 bevel and 8 in. for the 1.5:1 bevel.

Side Bevel - is proportioned based on the clear width of 16 ft which results in a bevel of 8 in. for the 1:1 bevel and 16 in. for the 1.5:1 bevel.

The ratio of the inlet face area to the barrel area remains the same as for a single barrel culvert. Multibarrel pipe culverts should be designed as a series of single barrel installations since each pipe requires a separate bevel.

For multibarrel installations exceeding a 3:1 width to depth ratio, the side bevels become excessively large when proportioned on the basis of the total clear width. For these structures, it is recommended that the side bevel be sized in proportion to the total clear width, B, or three times the height, whichever is smaller. The top bevel dimension should always be based on the culvert height. The shape of the upstream edge of the intermediate walls of multibarrel installations is not as important to the hydraulic performance of a culvert as the edge condition of the top and sides. Therefore, the edges of these walls may be square, rounded with a radius of one-half their thickness, chamfered, or beveled. The intermediate walls may also project from the face and slope downward to the channel bottom to help direct debris through the culvert.

It is recommended that Figure 3 for skewed inlets not be used for multiple barrel installations, as the intermediate wall could cause an extreme contraction in the downstream barrels. This would result in under design due to a greatly reduced capacity. Skewed inlets should be avoided whenever possible, and should not be used with side- or slope-tapered inlets.

9.10 Flood Routing And Culvert Design

Flood routing through a culvert is a practice that evaluates the effect of temporary upstream ponding caused by the culvert's backwater. By not considering flood routing it is possible that the findings from culvert analyses will be conservative. If the selected allowable headwater is considered acceptable without flood routing, then costly overdesign of both the culvert and outlet protection may result, depending on the amount of temporary storage involved.

There are many ramifications associated with culvert flood routing.
- Ownership or easement of the upstream property may be required.
- A perceived loss of a subjective safety factor.
- Credibility, both in court as well as in technical negotiations.
- Evaluating environmental concerns.
- Realistic assessments of potential flood hazards.
- Estimation of sediment problems.

Ignoring temporary storage effects on reducing the selected design flood magnitude by assuming that this provides a factor of safety is not recommended. This practice results in inconsistent factors of safety at culvert sites as it is dependent on the amount of temporary storage at each site. Further, with little or no temporary storage at a site the factor of safety would be unity thereby precluding a factor of safety. If a factor of safety is desired, it is essential that flood routing practices be used to ensure consistent and defensible factors of safety are used at all culvert sites.

Improved hydrology methods or changed watershed conditions are factors that can cause an older, existing culvert to be inadequate. A culvert analysis that relies on findings that ignore any available temporary storage may be misleading. A flood routing analysis may show that what was thought to be an inadequate existing culvert is, in fact, adequate.

Often existing culverts require replacement due to corrosion or abrasion. This can be very costly, particularly where a high fill is involved. A less costly alternative is to place a smaller culvert inside the existing culvert. A flood routing analysis may, where there is sufficient storage, demonstrate that this is acceptable in that no increase in flood hazard results.

With legal proceedings or in resolving conflicting design findings it is essential that creditable and defensible practices be used. By ignoring flood routing where significant storage occurs, findings may be discredited. With legal proceedings, claims of design negligence may result depending on the nature of the case.

With culvert flood routing a more realistic assessment can be made where environmental concerns are important. The temporary time of upstream ponding can be easily identified. This allows environmental specialists to assess whether such ponding is beneficial or harmful to localized environmental features such as fisheries, beaver ponds, wetlands and uplands.

Potential flood hazards increase whenever a culvert increases the natural flood stage. Some of these hazards can conservatively be assessed without flood routing. However, some damages associated with culvert backwater are time dependent and thus require an estimate of depth versus duration of inundation. Some vegetation and commercial crops can tolerate longer periods of inundation than others, and to greater depths. Such considerations become even more important when litigation is involved.

Complex culvert sediment deposition ("silting") problems require the application of a sediment routing practice. This practice requires a time-flood discharge relationship, or hydrograph. This flood hydrograph must be coupled to a flood-discharge and sediment-discharge relationship in order to route the sediment through the culvert site.

There are situations where culvert sizes and velocities obtained through flood routing will not differ significantly from those obtained by designing to the selected peak discharge and ignoring any temporary upstream storage. This occurs when:
- there is no significant temporary pond storage available (as in deep incised channels),
- the culvert must pass the design discharge with no increase in the natural channel's flood stage, and
- runoff hydrographs last for long periods of time such as with snowmelt runoff or irrigation flows.

Design Procedure

The design procedure for flood routing through a culvert is the same as for reservoir routing, see Storage Facilities, Chapter 10. The site data and roadway geometry are obtained and the hydrology analysis completed to include estimating a hydrograph. Once this essential information is available, the culvert can be designed.

Flood routing through a culvert can be time consuming. It is recommended that the HYDRAIN system be used as it contains software that very quickly routes floods through a culvert to evaluate an existing culvert (review), or to select a culvert size that satisfies given criteria (design). However, the designer should be familiar with the culvert flood routing design process. This familiarization is necessary in order to:
- recognize and test suspected software malfunctions,
- circumvent any software limitations,
- flood route manually where the software is limited, and
- understand and discuss culvert flood routing in a creditable manner.

9.11 HYDRAIN Culvert Computer Program

The HYDRAIN culvert analysis microcomputer program will perform the calculation for the following:
1. culvert analysis (including independent multiple barrel sizing)
2. hydrograph generation
3. hydrograph routing
4. roadway overtopping
5. outlet scour estimates

The example problem in Appendix A provides the user with analysis approaches to be used with the culvert analysis portion of the program. The example provides instruction in data entry, file modification and culvert performance analysis.

The culvert alternatives for this example were chosen to illustrate the features of the software and do not necessarily represent cost effective designs. Since the program is still being developed, some of the screens shown in the examples may differ slightly from the version you obtain.

It is recommended that the user should work through the example on the computer while following the text so as to become familiar with the program. New users should consult the README file that accompanies the program for further help and directions.

HYDRAIN's culvert analysis program has several user-friendly features which permit easy data entry, editing and comparison of several design alternatives. Data are entered by selecting options on a menu or by entering numeric data at prompts. These data are periodically summarized in tables. Any incorrect entry can be changed, and design variations can be quickly analyzed. Another feature of HYDRAIN's culvert analysis program is that plots of irregular cross sections, channel rating curves and culvert performance curves can be obtained if the terminal has graphics capabilities.

References

American Association of State Highway and Transportation Officials, Highway Drainage Guidelines, 1982.

American Association of State Highway and Transportation Officials, Model Drainage Manual, 1992.

American Concrete Pipe Association, "Culvert Durability Study", No. 02-902, 1982.

Blaisdell, F. W., "Flow in Culverts and Related Design Philosophies", Proc. ASCE J. of Hy., Vol. 92, No. HY2, March, 1966.

City of Indianapolis, IN, "Specifications for Construction of Stormwater Drainage Improvements", Dept. of Capital Asset Management, April, 1994.

Federal Highway Administration, Hydraulics of Bridge Waterways, Hydraulic Design Series No. 1, 1978.

Federal Highway Administration, Hydraulic Design of Highway Culverts, Hydraulic Design Series No. 5, 1985.

Federal Highway Administration, Debris-Control Structures, Hydraulic Engineering Circular No. 9, 1971.

Federal Highway Administration, HY8 Culvert Analysis Microcomputer Program Applications Guide, Hydraulic Microcomputer Program HY8, 1987.

HYDRAIN Culvert Computer Program (HY8), Available from McTrans Software, University of Florida, 512 Weil Hall, Gainesville, Florida 32611.

Lewis, G.L., "Jury Verdict: Frequency Versus Risk-Based Culvert Design", ASCE J. of Water Res. Plan. & Mgmt., Vol. 118, No. 2, Mar/April, 1992, pp. 166-185.

U. S. Department of Interior, Design Of Small Canal Structures, 1983.

Young, G. K., R. S. Taylor and L. S. Costello, "Evaluation of the Flood Risk Factor in the Design of Box Culverts", Rpt. No. FHWA-RD-74-11, Sept., 1970.

Appendix A - Microcomputer Solution - HY8 Culvert Analysis

Introduction

The HYDRAIN Culvert Analysis microcomputer program HY8 consists of three main options:
   (1) Culvert Analysis,
   (2) Hydrograph Generation, and
   (3) Routing.
The purpose of this Appendix is to provide the user with analysis approaches to use with the Culvert Analysis portion of the program.

HY8 has several user-friendly features which permit easy data entry, editing and comparison of several design alternatives. Pressing the F1 key will summon a Help screen that will provide additional information on how to enter data. Pressing the F5 key will stop execution of the program and return the user to the operating system.

Data are entered by selecting options on a menu or by entering numeric data at prompts. These data are periodically summarized in tables. Any incorrect entry can be changed, and design variations can be quickly analyzed.

Another feature of HY8 is that plots of irregular cross sections, rating curves and performance curves can be obtained. The scale interval and range of values can be modified by advancing to the screen following the plot and making the desired changes.

Example Problem

The following example has been produced using the HY8 Culvert Analysis Microcomputer Program Version 3.0. The screens shown may not match exactly the version of HY8 that you are using since some editorial changes were made so that the screens would fit in this text.

After creating a file, the user will be prompted for the discharge range, site data and culvert shape, size, material and inlet type. The discharge range for this example will be from 0 to 500 cfs with a design discharge of 400 cfs.. The site data are entered by providing culvert invert data. If embankment data points are input, the program will fit the culvert in the fill and subtract the appropriate length.

Culvert Data

As an initial size estimate, try a 5 ft x 5 ft concrete box culvert. For the culvert assume that a conventional inlet with 1:1 bevels and 45 degree wingwalls will be used. As each group of data is entered the user is allowed to edit any incorrect entries. The following is how the screen that summarizes the culvert information will look.

```
     CULVERT ANALYSIS 3.0                         DATE:
     CULVERT FILE:EXAMPLE1                         CULVERT NO. 1

         ITEM                                SELECTED CULVERT

     (1) BARREL SHAPE:                       BOX 5.00 FT X 5.00FT
     (2) BARREL MATERIAL:                    CONCRETE
     (3) MANNING'S N:                        .012
     (4) INLET TYPE:                         CONVENTIONAL
     (5) INLET EDGE AND WALL:                1:1 BEVEL (45 DEG. FLARE)
     (6) INLET DEPRESSION:                   NONE
         TYPE ITEM NO. TO EDIT ITEM:
         <ENTER> TO CONTINUE DATA LISTING

     1HELP  2     3     4     5END  6    7    8     9SHELL  10
```

Channel Data

Next the program will prompt for data pertaining to the channel so that tailwater elevations can be determined. For this example problem, the channel is irregularly shaped and can be described by the 8 coordinates listed in the table below. After opening the irregular channel file the user will be prompted for channel slope (.05), number of cross-section coordinates (8) and subchannel option. The subchannel option in this case would be option (2), left and right overbanks (n = .08) and main channel (n = .03).

IRREGULAR CHANNEL CROSS-SECTION

CROSS-SECTION COORD. NO.	X (FT)	Y (FT)
1	12.00	180.00
2	22.00	175.00
3	32.00	174.50
4	34.00	172.50
5	39.00	172.50
6	41.00	174.50
7	51.00	175.00
8	61.00	180.00

TYPE COORD. NO. TO EDIT COORD.
<I> OR <D> TO INSERT OR DELETE   <ENTER> TO CONTINUE
<P> TO PLOT CROSS-SECTION
1HELP   2   3   4   5END   6   7   8   9SHELL 10

The next prompt, for channel boundaries, refers to the number of the coordinate pair defining the left subchannel boundary and the number of the coordinate pair defining the right subchannel boundary. The boundaries for this example are the 3rd and 6th coordinates. After this is input, the program prompts for channel coordinates. Once these are entered, pressing (P) will cause the computer to display the channel cross-section shown below. The user can easily identify any input errors by glancing at the plot. To return to the data input screens, press any key. If data are correct press (return). You can then enter the roughness data for the main channel and overbanks.

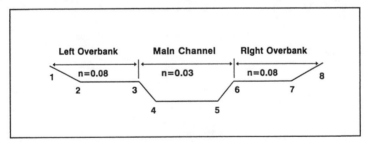

The program now has enough information to develop a uniform flow rating curve for the channel and provide the user with a list of options. Selecting option (T) on the Irregular Channel Data Menu will make the program compute the rating curve data and display the following table. Selecting option (I) will permit the user to interpolate data between calculated points.

CULVERT ANALYSIS 3.0   DATE:
CULVERT FILE:EXAMPLE1   CULVERT NO. 1
    TAILWATER RATING CURVE
    IRREGULAR CHANNEL FILE: EXAMPLE1

NO.	FLOW(CFS)	T.W.E.(FT)	VEL.(Ft/s)	SHEAR(PSF)
1	0.00	172.50	0.00	0.00
2	50.00	173.45	9.02	2.29

3	100.00	173.91	11.13	3.14
4	150.00	174.28	12.52	3.75
5	200.00	174.57	13.70	4.29
6	250.00	174.80	14.93	4.88
7	300.00	174.99	15.94	5.39
8	350.00	175.15	16.77	5.81
9	400.00	175.30	17.51	6.20
10	450.00	175.44	18.18	6.56
11	500.00	175.56	18.80	6.90

TYPE: <P> TO PLOT RATING CURVE
      <ENTER> TO CONTINUE
      <ESC> FOR TAILWATER MENU
   1HELP 2   3   4   5END 6   7   8   9SHELL 10

The Tailwater Rating Curve Table consists of tailwater elevation (T.W.E.) at normal depth, natural channel velocity (Vel.) in feet per second, and the shear stress in pounds per square foot at the bottom of the channel for various flow rates. At the design discharge of 400 cfs, the tailwater elevation will be 175.30 feet. The channel velocity will be 17.51 ft/s, and the shear will be 6.20 psf. This information will be useful in the design of channel linings if they are needed. Entering (P) will cause the computer to display the rating curve for the channel. This curve, shown below, is a plot of tailwater elevation vs. flow rate at the exit of the culvert.

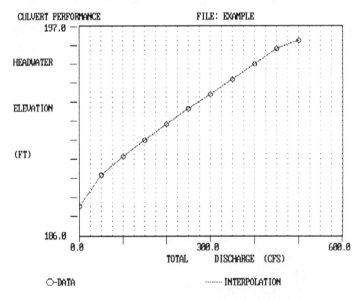

The next prompts are for the roadway profile, so that an overtopping analysis can be performed. For this example problem, the roadway profile is a sag vertical curve, which will require nine coordinates to define. Once these coordinates are input, the profile will be displayed when (P) is entered, as illustrated below. The other data required for overtopping analysis are roadway surface or weir coefficient and the embankment top width. For this example, the roadway is paved with an embankment width of 50 feet.

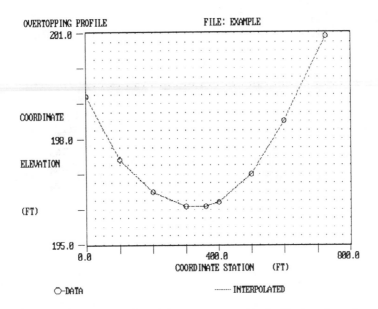

OVERTOPPING PROFILE   FILE: EXAMPLE

O—DATA   ------ INTERPOLATED

All the data has now been entered and the summary table is displayed as shown below. At this point any of the data can be changed or the user can continue by pressing (Return), which will bring up the Culvert Program Options Menu.

CULVERT ANALYSIS 3.0   DATE:
CULVERT FILE:EXAMPLE1   SUMMARY TABLE

C	A - SITE DATA			B - CULVERT SHAPE, MATERIAL, INLET				
U								
L	INLET	OUTLET	CULVERT	BARRELS	SPAN	RISE	MANN.	INLET
V	ELEV.	ELEV.	LENGTH	SHAPE			N	TYPE
NO.	(FT)	(FT)	(FT)	MATERIAL	(FT)	(FT)		
1	187.50	172.50	300	1 - RCB	5.00	5.00	.012	CONV.
2								
3								

TO REVIEW DATA PRESS
<S> FOR SITE DATA
<C> FOR CULVERT SHAPE, MATERIAL, OR INLET DATA
<D> FOR DISCHARGE RANGE
<T> FOR TAILWATER RATING CURVE
<O> FOR OVERTOPPING DATA
<A> TO ADD OR DELETE CULVERTS
<M> TO MINIMIZE CULVERT SPAN
<ENTER> TO CONTINUE ANALYSIS
1HELP   2   3   4   5END   6   7   8   9SHELL 10

This feature, "Minimize Culvert Size", is intended to allow the user to use HY8 Version 3.0 as a tool to perform culvert design for circular, box, elliptical, and arch shape culverts based on a user's defined allowable headwater elevation. This feature can be activated by selecting letter "M". Once this letter is selected it enables the user to input the allowable headwater elevation. That elevation will be the basis for adjusting the user's defined culvert size for the design discharge. The program will adjust the culvert span by increasing or decreasing by 0.5 foot increments. It will compute the headwater elevation for the span, and compare it with the user's defined allowable headwater. If the computed headwater elevation is lower than or equal to the defined allowable headwater elevation the minimization routine will stop, and the adjusted culvert can be used for the remainder of the program. Several hydraulic parameters are also computed while performing the minimization routine. These hydraulic parameters which are part of the output of the minimization routine table, must be printed from this screen because they are not printed with the output listing routine.

CULVERT ANALYSIS 3.0   DATE:
CULVERT FILE:EXAMPLE1   SUMMARY TABLE

C   A - SITE DATA   B - CULVERT SHAPE, MATERIAL, INLET
U
L INLET OUTLET CULVERT BARRELS SPAN RISE MANN. INLET
V ELEV. ELEV.   LENGTH SHAPE   N   TYPE
NO.(FT) (FT)   (FT)   MATERIAL (FT) (FT)
1 187.50 172.50   300   1 - RCB   6.50 6.00 .012   CONV.
2
3

HEADWATER ELEVATION   FLOW VELOCITY   FLOW DEPTHS
ENTER ALLOW. = 196.00   CULVERT = 31.71   CULVERT = 1.94
CONTROLLING = 195.50   CHANNEL = 17.51   CHANNEL = 2.80
INLET CONTROL = 195.50   DISCHARGE = 400.00   NORMAL = 1.94
OUTLET CONTROL = 180.32   SLOPE = 0.0500   CRITICAL = 4.91

MAX. HEADWATER <ENTER> TO CHANGE HEADWATER
    <ANY KEY> TO CONT.

1HELP 2   3   4   5END   6   7   8   9SHELL 10

This feature proves to be a time saver for users because it avoids the need for repetitively editing a culvert size to obtain a controlling headwater elevation.

At this point the data file can be saved or renamed by selecting option (S). The culvert performance curve table can be obtained by selecting option (N). If (N) is selected before (S) and an error occurs, the file can be retrieved by loading "current". When option (N) is selected, the program will compute the performance curve table without considering overtopping in the analysis. Since this 5 ft x 5 ft culvert is a preliminary estimate, the performance without considering overtopping is calculated and is shown in the next table.

This table indicates the controlling headwater elevation (HW), the tailwater elevation and the headwater elevations associated with all the possible control sections of the culvert. It is apparent from the table that at 400 cfs the headwater (HW) is 199.62 ft, which exceeds the design headwater of 195 feet. Consequently, the 5 ft x 5 ft box culvert is inadequate for the site conditions. The plot of inlet and outlet control headwaters can be obtained by entering (P). In this example, the culvert is operating in inlet control (the upper curve) throughout the discharge range.

CULVERT # 1 PERFORMANCE CURVE
FOR 1 BARREL(S)

Q	HWE	TWE	ICH	OCH	CCE	FCE	TCE	VO
(cfs)	(ft)	(ft)	(ft)	(ft)	(ft)	(ft)	(ft)	(ft/s)
0	187.50	172.50	0.00	-15.00	0.00	187.50	0.00	0.00
50	189.54	173.45	2.04	-12.67	0.00	0.00	0.00	17.29
100	190.91	173.91	3.41	-11.30	0.00	0.00	0.00	21.63
150	192.11	174.28	4.61	-10.15	0.00	0.00	0.00	24.57
200	193.30	174.57	5.80	-9.12	0.00	0.00	0.00	26.61
250	194.59	174.80	7.09	-8.18	0.00	0.00	0.00	28.45
300	196.04	174.99	8.54	-7.29	0.00	0.00	0.00	29.83
350	197.71	175.15	10.21	-6.35	0.00	0.00	0.00	31.11
400	199.62	175.30	12.12	-5.23	0.00	0.00	0.00	32.15
450	201.77	175.44	14.27	-3.96	0.00	0.00	0.00	33.12
500	204.16	175.56	16.66	-2.55	0.00	0.00	0.00	33.94

El. inlet face invert   187.50 ft   El. outlet invert   172.50 ft
El. inlet throat invert   0.00 ft   El. inlet crest   0.00 ft
PRESS <P> TO PLOT   PRESS <ENTER> TO CONTINUE
1   2   3   4   5END   6   7   8   9   10

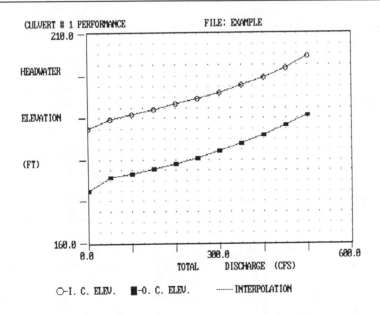

The user can easily modify the existing program file to analyze a larger barrel. Suppose a 6 ft x 6 ft culvert is tried. Hit any key to return to the Culvert Data Summary Table and enter (B) to modify culvert shape. The prompts will be the same as they were for the 5 ft x 5 ft culvert, and the user will be returned to the Culvert Data Summary Table directly without going through the tailwater and overtopping menus again. Pressing <enter> will bring up the Culvert Program Options menu, with which the file can be renamed and saved. The performance of this culvert can be checked by selecting option (N) for no overtopping. The following table appears.

CULVERT # 1 PERFORMANCE CURVE FOR 1 BARREL(S)

Q	HWE	TWE	ICH	OCH	CCE	FCE	TCE	VO
(cfs)	(ft)	(ft)	(ft)	(ft)	(ft)	(ft)	(ft)	(ft/s)
0	187.50	172.50	0.00	-15.00	0.00	187.50	0.00	0.00
50	189.26	173.45	1.76	-12.93	0.00	0.00	0.00	17.48
100	190.42	173.91	2.92	-11.72	0.00	0.00	0.00	21.11
150	191.44	174.28	3.94	-10.70	0.00	0.00	0.00	23.92
200	192.37	174.57	4.87	-9.79	0.00	0.00	0.00	25.99
250	193.27	174.80	5.77	-8.96	0.00	0.00	0.00	27.95
300	194.18	174.99	6.68	-8.18	0.00	0.00	0.00	29.42
350	195.12	175.15	7.62	-7.44	0.00	0.00	0.00	30.75
400	196.12	175.30	8.62	-6.73	0.00	0.00	0.00	31.98
450	197.21	175.43	9.71	-6.06	0.00	0.00	0.00	33.01
500	198.40	175.56	10.90	-5.41	0.00	0.00	0.00	33.93

El. inlet face invert   187.50 ft   El. outlet invert   172.50 ft
El. inlet throat invert   0.00 ft   El. inlet crest   0.00 ft
PRESS <P> TO PLOT   PRESS <ENTER> TO CONTINUE
1   2   3   4   5END   6   7   8   9   10

Since the design headwater criterion has still not been met, another size must be selected. Try a 7 ft x 6 ft culvert, and modify the file accordingly. The resulting performance table shown below indicates that the design headwater will not be exceeded at 400 cfs. However, the headwater elevation of 196.74 feet at 500 cfs indicates that some overtopping will occur due to the 100-year storm.

CULVERT # 1 PERFORMANCE CURVE FOR 1 BARREL(S)

Q	HWE	TWE	ICH	OCH	CCE	FCE	TCE	VO
(cfs)	(ft)	(ft)	(ft)	(ft)	(ft)	(ft)	(ft)	(ft/s)
0	187.50	172.50	0.00	-15.00	0.00	187.50	0.00	0.00
50	189.08	173.45	1.58	-13.14	0.00	0.00	0.00	17.76
100	190.11	173.91	2.61	-12.04	0.00	0.00	0.00	20.19
150	191.02	174.28	3.52	-11.12	0.00	0.00	0.00	23.31
200	191.85	174.57	4.35	-10.30	0.00	0.00	0.00	25.26
250	192.63	174.80	5.13	-9.55	0.00	0.00	0.00	27.16
300	193.40	174.99	5.90	-8.84	0.00	0.00	0.00	28.85

350	194.18	175.15	6.68	-8.18	0.00	0.00	0.00	30.20
400	194.98	175.30	7.48	-7.54	0.00	0.00	0.00	31.34
450	195.83	175.43	8.33	-6.93	0.00	0.00	0.00	32.55
500	196.74	175.56	9.24	-6.34	0.00	0.00	0.00	33.58

El. inlet face invert   187.50 ft   El. outlet invert   172.50 ft
El. inlet throat invert   0.00 ft   El. inlet crest   0.00 ft
PRESS <P> TO PLOT   PRESS <ENTER> TO CONTINUE
1   2   3   4   5END   6   7   8   9   10

To determine the amount of overtopping and the actual headwater, press (return), and then select (O) for overtopping. A Summary of Culvert Flows will appear on the screen, as shown below:

SUMMARY OF CULVERT FLOWS (CFS) FILE: EXAMPLE1   DATE:

ELEV(FT)	TOTAL	1	2	3	4	5	6	OT	ITER
187.50	0	0	0	0	0	0	0	0	0
189.08	50	50	0	0	0	0	0	0	2
190.11	100	100	0	0	0	0	0	0	2
191.02	150	150	0	0	0	0	0	0	2
191.85	200	200	0	0	0	0	0	0	2
192.63	250	200	0	0	0	0	0	0	2
193.40	300	300	0	0	0	0	0	0	2
194.18	350	350	0	0	0	0	0	0	2
194.98	400	400	0	0	0	0	0	0	2
195.83	450	450	0	0	0	0	0	0	2
196.25	500	473	0	0	0	0	0	37	5

PRESS:
<1> TO PLOT TOTAL RATING CURVE
<2> TO DETERMINE SPECIFIC INFORMATION ABOUT EACH CULVERT
<3> TO SEE MULTIPLE CULVERT COMPUTATIONAL ERROR TABLE
<4> TO PRINT CULVERT SUMMARY
<ENTER> TO RETURN FOR NEW RUN OR EXIT
1HELP   2   3   4   5END   6   7   8   9SHELL   10

This computation table is used when overtopping and/or multiple culvert barrels are used. It shows the headwater, total flow rate, the flow through each barrel and overtopping flow, and the number of iterations it took to balance the flows. From this information a total (culvert and overtopping) performance curve, shown on the next page, can be obtained by selecting option (1). This curve is a plot of the headwater elevation vs. the total flow rate which indicates how the culvert or group of culverts will perform over the selected range of discharges. It is especially useful for comparing the effects of various combinations of culverts.

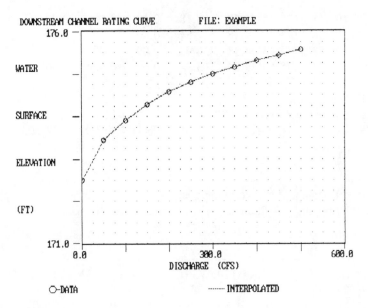

From the Summary table, when the total flow is 500 cfs, 473 cfs passes through the culvert and 37 cfs flows over the road. The headwater elevation will be 196.25 feet. Assume that in this case overtopping at 100-year frequency can be tolerated, and the 7 ft x 6 ft culvert will be used. Referring back to the performance curve data, the outlet velocity at 400 cfs is 31.34 ft/s. Since the tailwater rating curve generated previously indicates that the natural channel velocity at 400 cfs is 17.51 ft/s, an energy dissipator is warranted.

When overtopping occurs, the performance of the culvert will differ from that without overtopping. By selecting option (2), the culvert performance data can be obtained. The user also has the option to plot these data.

CULVERT #1 PERFORMANCE CURVE FOR 1 BARREL(S)								
Q	HWE	TWE	ICH	OCH	CCE	FCE	TCE	VO
(cfs)	(ft)	(ft)	(ft)	(ft)	(ft)	(ft)	(ft)	(ft/s)
0	187.50	172.50	0.00	-15.00	0.00	187.50	0.00	0.00
50	189.08	173.45	1.58	-13.14	0.00	0.00	0.00	17.76
100	190.11	173.91	2.61	-12.04	0.00	0.00	0.00	20.19
150	191.02	174.28	3.52	-11.12	0.00	0.00	0.00	23.31
200	191.85	174.57	4.35	-10.30	0.00	0.00	0.00	25.26
250	192.63	174.80	5.13	-9.55	0.00	0.00	0.00	27.16
300	193.40	174.99	5.90	-8.84	0.00	0.00	0.00	28.85
350	194.18	175.15	6.68	-8.18	0.00	0.00	0.00	30.20
400	194.98	175.30	7.48	-7.54	0.00	0.00	0.00	31.34
450	195.83	175.43	8.33	-6.93	0.00	0.00	0.00	32.55
473	196.24	175.56	8.74	-6.66	0.00	0.00	0.00	33.05

El. inlet face invert   187.50 ft   El. outlet invert   172.50 ft
El. inlet throat invert   0.00 ft   El. inlet crest   0.00 ft
PRESS <P> TO PLOT   PRESS <ENTER> TO CONTINUE
1HELP 2   3   4   5END   6   7   8   9SHELL 10

By pressing <enter> the program returns to the Summary of Culvert Flows menu. Selecting option (3), a Summary of Iterative Solution Errors is produced. This table, shown below, lists the amount of error present in the solution for a discharge of 500 cfs as 10 cfs.

SUMMARY OF ITERATIVE SOLUTION ERRORS FILE: EXAMPLE1 DATE:

HEAD ELEV(FT)	HEAD ERROR(FT)	TOTAL FLOW(CFS)	FLOW ERROR(CFS)	% FLOW ERROR
187.50	0.00	0	0	0.00
189.08	0.00	50	0	0.00
190.11	0.00	100	0	0.00
191.02	0.00	150	0	0.00
191.85	0.00	200	0	0.00
192.63	0.00	250	0	0.00
193.40	0.00	300	0	0.00
194.18	0.00	350	0	0.00
194.98	0.00	400	0	0.00
195.83	0.00	450	0	0.00
196.25	0.00	500	10	1.96

PRESS <ENTER> TO CONTINUE

1HELP 2   3   4   5END 6   7   8   9SHELL 10

Appendix B - Critical Depth Figures (Source: HEC No. 13, 1972)

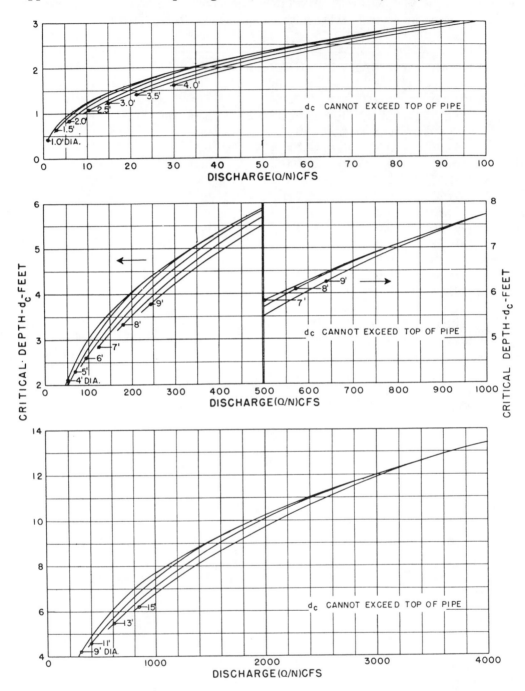

CRITICAL DEPTH
CIRCULAR PIPE

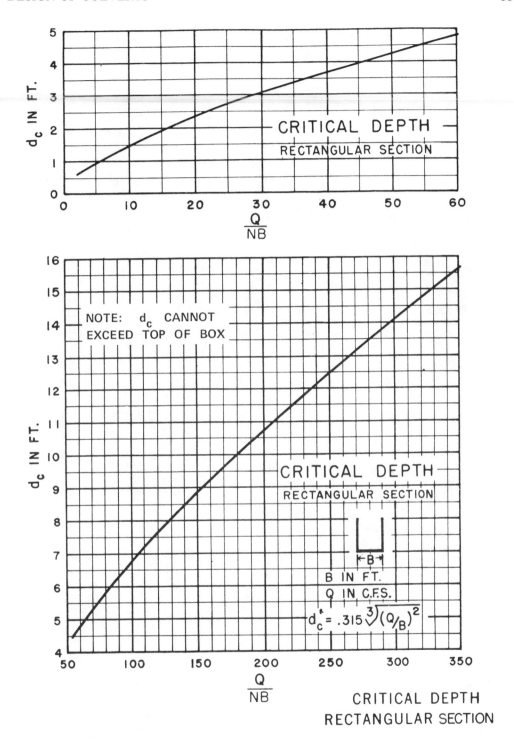

CRITICAL DEPTH
RECTANGULAR SECTION

Appendix C - Conventional Nomographs (Source: HEC No. 13, 1972)

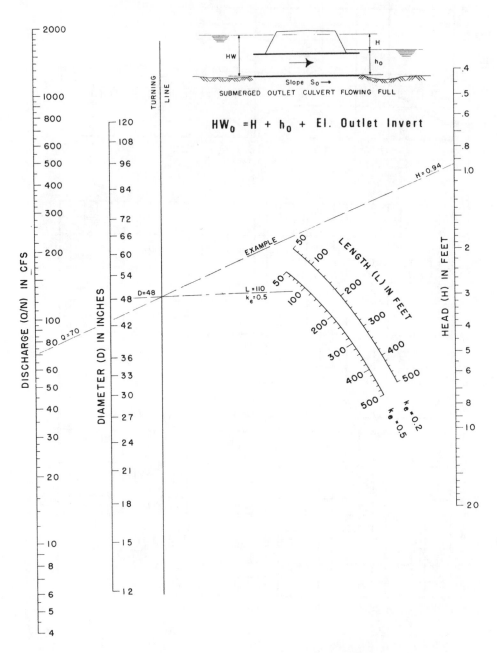

HEAD FOR
CONCRETE PIPE CULVERTS
FLOWING FULL
n = 0.012

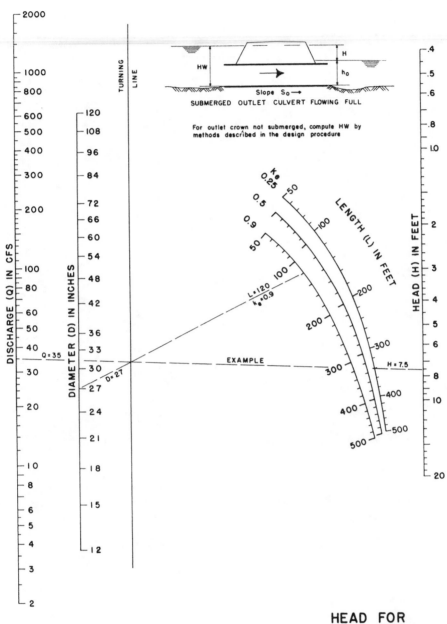

SUBMERGED OUTLET CULVERT FLOWING FULL

For outlet crown not submerged, compute HW by
methods described in the design procedure

EXAMPLE

HEAD FOR
STANDARD
C. M. PIPE CULVERTS
FLOWING FULL
n = 0.024

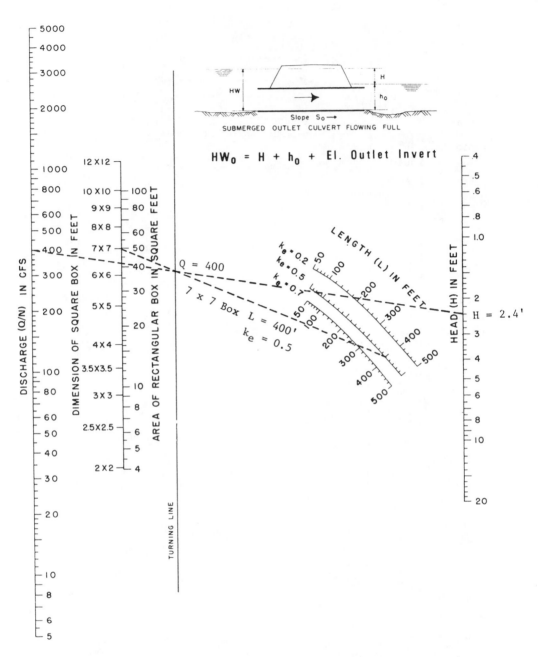

HEAD FOR
CONCRETE BOX CULVERTS
FLOWING FULL
n = 0.012

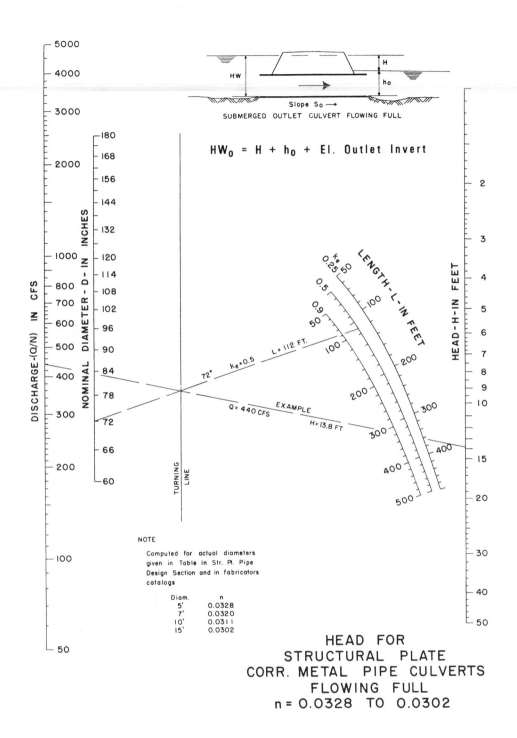

HEAD FOR
STRUCTURAL PLATE
CORR. METAL PIPE CULVERTS
FLOWING FULL
$n = 0.0328$ TO 0.0302

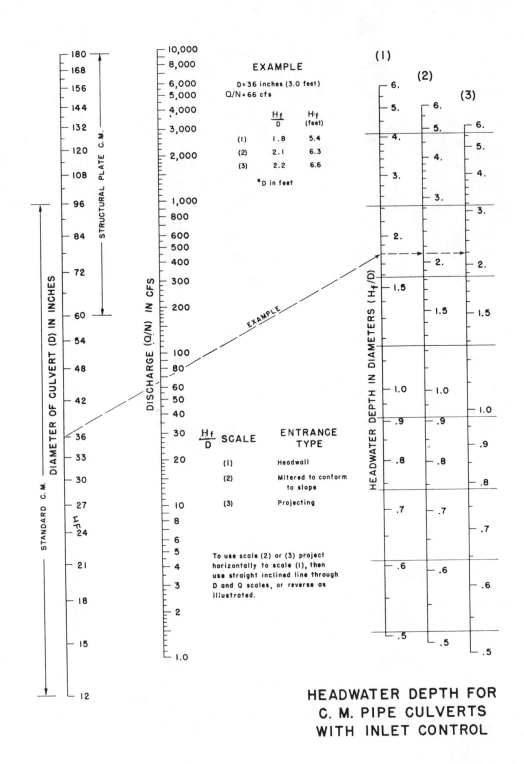

EXAMPLE

D = 36 inches (3.0 feet)
Q/N = 66 cfs

	$\frac{H_f}{D}$	H_f^* (feet)
(1)	1.8	5.4
(2)	2.1	6.3
(3)	2.2	6.6

*D in feet

$\frac{H_f}{D}$ SCALE   ENTRANCE TYPE

(1)   Headwall

(2)   Mitered to conform to slope

(3)   Projecting

To use scale (2) or (3) project horizontally to scale (1), then use straight inclined line through D and Q scales, or reverse as illustrated.

DIAMETER OF CULVERT (D) IN INCHES

STRUCTURAL PLATE C.M.

STANDARD C.M.

DISCHARGE (Q/N) IN CFS

HEADWATER DEPTH IN DIAMETERS (H_f/D)

(1) (2) (3)

HEADWATER DEPTH FOR
C. M. PIPE CULVERTS
WITH INLET CONTROL

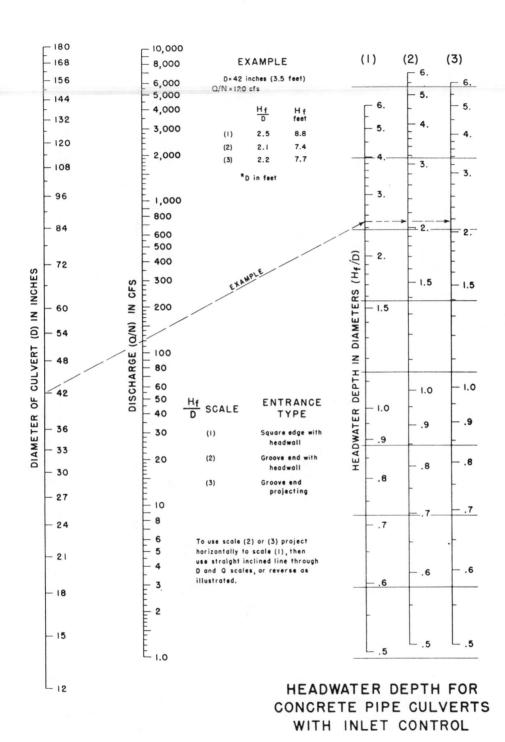

HEADWATER DEPTH FOR
CONCRETE PIPE CULVERTS
WITH INLET CONTROL

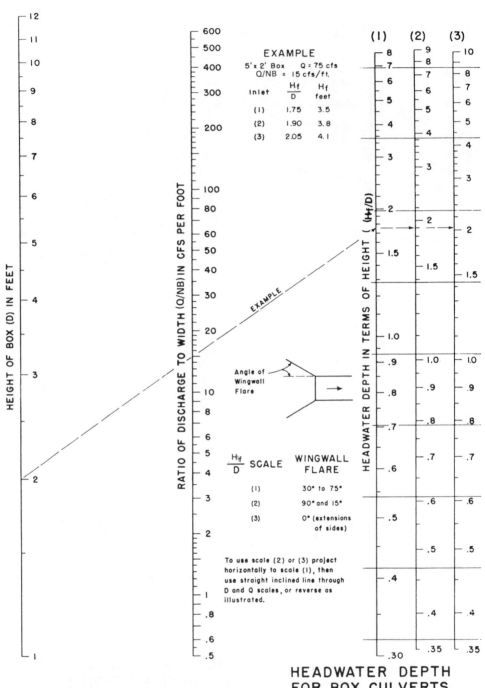

HEIGHT OF BOX (D) IN FEET

RATIO OF DISCHARGE TO WIDTH (Q/NB) IN CFS PER FOOT

HEADWATER DEPTH IN TERMS OF HEIGHT (H₊/D)

EXAMPLE

5' x 2' Box   Q = 75 cfs
Q/NB = 15 cfs/ft.

Inlet	$\frac{H_f}{D}$	H_f feet
(1)	1.75	3.5
(2)	1.90	3.8
(3)	2.05	4.1

EXAMPLE

Angle of
Wingwall
Flare

$\frac{H_{if}}{D}$ SCALE	WINGWALL FLARE
(1)	30° to 75°
(2)	90° and 15°
(3)	0° (extensions of sides)

To use scale (2) or (3) project
horizontally to scale (1), then
use straight inclined line through
D and Q scales, or reverse as
illustrated.

HEADWATER DEPTH
FOR BOX CULVERTS
WITH INLET CONTROL

Appendix D - Improved Inlet Figures (Bevel-Edge) (Source: HEC No. 13, 1972

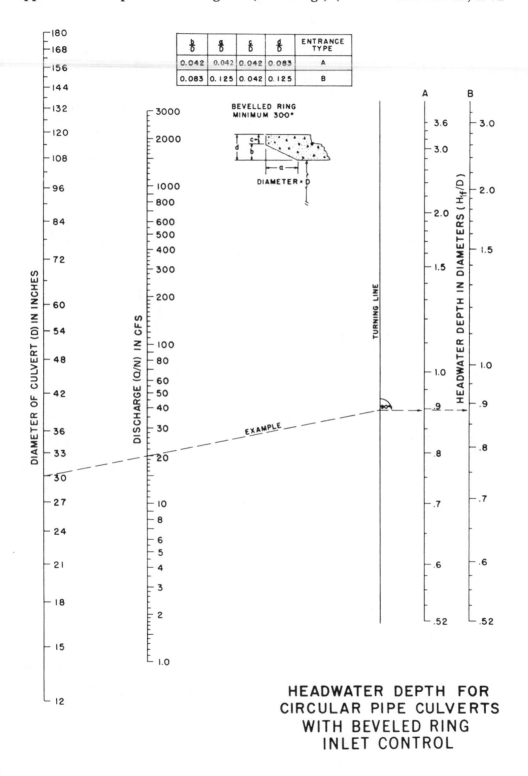

$\frac{b}{D}$	$\frac{a}{D}$	$\frac{c}{D}$	$\frac{d}{D}$	ENTRANCE TYPE
0.042	0.042	0.042	0.083	A
0.083	0.125	0.042	0.125	B

HEADWATER DEPTH FOR
CIRCULAR PIPE CULVERTS
WITH BEVELED RING
INLET CONTROL

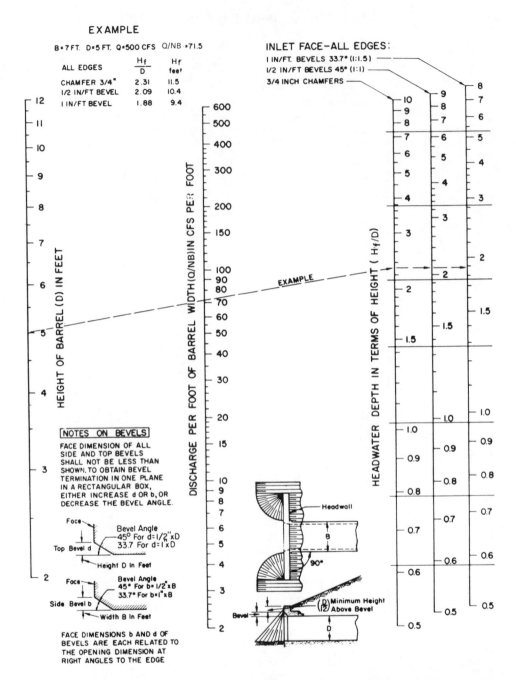

EXAMPLE

B = 7 FT.   D = 5 FT.   Q = 500 CFS   Q/NB = 71.5

ALL EDGES	$\frac{H_f}{D}$	H_f feet
CHAMFER 3/4"	2.31	11.5
1/2 IN/FT BEVEL	2.09	10.4
1 IN/FT BEVEL	1.88	9.4

INLET FACE—ALL EDGES:
1 IN/FT. BEVELS 33.7° (1:1.5)
1/2 IN/FT BEVELS 45° (1:1)
3/4 INCH CHAMFERS

HEIGHT OF BARREL (D) IN FEET

DISCHARGE PER FOOT OF BARREL WIDTH (Q/NB) IN CFS PER FOOT

HEADWATER DEPTH IN TERMS OF HEIGHT (H_f/D)

EXAMPLE

NOTES ON BEVELS

FACE DIMENSION OF ALL SIDE AND TOP BEVELS SHALL NOT BE LESS THAN SHOWN. TO OBTAIN BEVEL TERMINATION IN ONE PLANE IN A RECTANGULAR BOX, EITHER INCREASE d OR b, OR DECREASE THE BEVEL ANGLE.

Face
Top Bevel d
Bevel Angle
45° For d = 1/2"xD
33.7 For d = 1"xD
Height D In Feet

Face
Side Bevel b
Bevel Angle
45° For b = 1/2"xB
33.7° For b = 1"xB
Width B In Feet

FACE DIMENSIONS b AND d OF BEVELS ARE EACH RELATED TO THE OPENING DIMENSION AT RIGHT ANGLES TO THE EDGE

Headwall
B
90°
Bevel
(D/12) Minimum Height Above Bevel
D

HEADWATER DEPTH FOR INLET CONTROL
RECTANGULAR BOX CULVERTS
90° HEADWALL
CHAMFERED OR BEVELED INLET EDGES

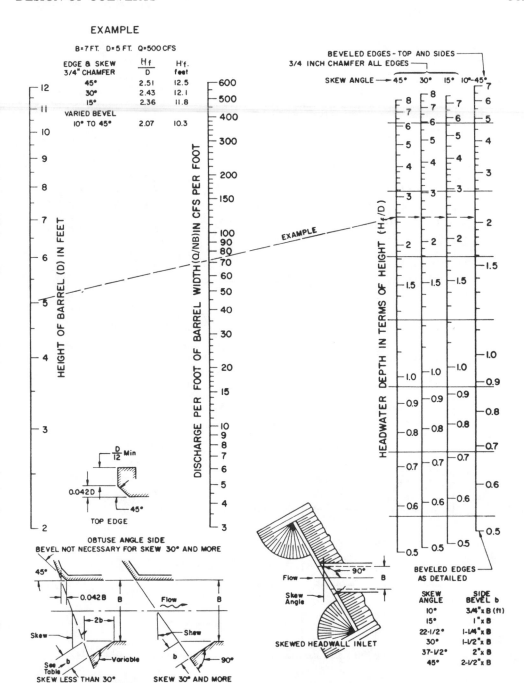

EXAMPLE

B=7 FT. D=5 FT. Q=500 CFS

EDGE & SKEW 3/4" CHAMFER	$\frac{H_f}{D}$	H'_f feet
45°	2.51	12.5
30°	2.43	12.1
15°	2.36	11.8
VARIED BEVEL 10° TO 45°	2.07	10.3

TOP EDGE

OBTUSE ANGLE SIDE
BEVEL NOT NECESSARY FOR SKEW 30° AND MORE

ACUTE ANGLE SIDE
BEVELED INLET EDGES
DESIGNED FOR SAME CAPACITY AT ANY SKEW

SKEWED HEADWALL INLET

SKEW ANGLE	SIDE BEVEL b
10°	3/4" x B (ft)
15°	1" x B
22-1/2°	1-1/4" x B
30°	1-1/2" x B
37-1/2°	2" x B
45°	2-1/2" x B

BEVELED EDGES AS DETAILED

**HEADWATER DEPTH FOR INLET CONTROL
SINGLE BARREL BOX CULVERTS
SKEWED HEADWALLS
CHAMFERED OR BEVELED INLET EDGES**

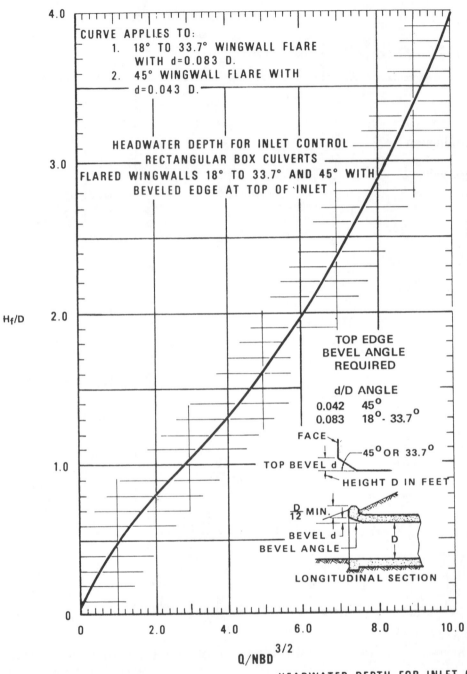

HEADWATER DEPTH FOR INLET CONTROL
RECTANGULAR BOX CULVERTS
FLARED WINGWALLS 18° TO 33.7° AND 45°
WITH BEVELED EDGE AT TOP OF INLET

Chapter 10 Open Channel Design

10.1 Introduction

A consideration of open channel hydraulics is an integral part of projects in which artificial channels and improvements to or analysis of natural channels are a primary concern. The design of channels in urban environments is complicated by a number of factors including: changing hydrology due to watershed land use changes, near bank encroachments, numerous channel crossings and modifications, intermittent upstream construction with sediment surges, and difficult or unknown environmental considerations.

In many municipalities urban stream corridors have become unattractive abandoned fields full of weeds and trash. However, many other municipalities have made the decision to consider urban streams a resource rather than a nuisance and have taken steps to make their corridors attractive to both wildlife and man. While it is nearly impossible to have a pristine and fully "natural" stream in an urban environment, it is very possible to have an attractive and functional "urban" stream. Examples abound of greenways, bike paths, walking trails, boardwalk wetlands, multiobjective parks and recreational areas... even urban salmon spawning areas.

But the beginning of any channel design or modification is to understand both the hydrology and hydraulics of the stream and watershed. This chapter emphasizes procedures for performing uniform flow calculations that aid in the selection or evaluation of appropriate channel linings, depths, and grades for natural or man-made channels. Allowable velocities are provided, along with procedures for evaluating channel capacity using Manning's equation. Additional sections discuss briefly greenways, stable channel design and other environmental considerations.

10.2 Design Criteria

Channel Types

The three main classifications of open channel types according to channel linings are: vegetative, flexible, and rigid. Vegetative linings include grass with mulch, sod, and lapped sod. Rock riprap and some forms of flexible man-made linings or gabions are examples of flexible linings, while rigid linings are generally concrete or rigid block.

Vegetative Linings - Vegetation, where practical, is the most desirable lining for an artificial channel. It stabilizes the channel body, consolidates the soil mass of the bed, checks erosion on the channel surface, provides habitat and controls the movement of soil particles along the channel bottom. Conditions under which vegetation may not be acceptable, however, include but are not limited to:
- standing or continuous flowing water,
- lack of the regular maintenance necessary to prevent domination by taller or woody vegetation,
- lack of nutrients and inadequate topsoil,

- excessive shade, and
- high velocities.

Proper seeding, mulching, and soil preparation are required during construction to assure establishment of a healthy growth of grass. Soil testing should be performed and the results evaluated by an agronomist or local agricultural extension office to determine soil treatment requirements for pH, nitrogen, phosphorus, potassium, and other factors. In many cases, temporary erosion control measures (such as jute or straw matting or spray tacking substances) are required to provide time for the seeding to establish a viable vegetative lining (UDFCD, 1991). Sodding should be staggered, to avoid seams in the direction of flow. Lapped or shingle sod should be staggered and overlapped by approximately 25 percent with the top overlap portion facing downstream to prevent sod "rollup" during high flows. Staked sod is usually only necessary for use on steeper slopes to prevent sliding.

Flexible Linings - Rock riprap, including rubble, is the most common type of flexible lining. It presents a rough surface that can dissipate energy and mitigate increases in erosive velocity. These linings are usually less expensive than rigid linings and have self-healing qualities that reduce maintenance. Depending on the soil underneath they may require use of filter fabric or gravel layers and allow the infiltration and exfiltration of water. The growth of grass and weeds through the lining may present maintenance problems though vegetation can help stabilize the channel if it is not woody and of large enough diameter to cause turbulence and riprap displacement. The use of a flexible lining may be restricted where right-of-way is limited, since the higher roughness values create larger cross sections.

Rigid Linings - Rigid linings are generally constructed of concrete and used where smoothness offers a higher capacity for a given cross-sectional area. Higher velocities, however, create the potential for scour at channel lining transitions. A rigid lining can be destroyed by flow undercutting or eroding under the lining, soil support loss at improperly constructed or maintained joints, channel headcutting, erosion along the tops of channels from improperly accounted side flow, or the buildup of hydrostatic pressure behind the rigid surfaces. Filter fabric may be required to prevent soil loss through pavement cracks. Soil cement can also be used profitably in areas where sand materials low in organic materials are available (Reese, 1987). Such channels in Arizona have withstood high velocities of sediment laden flow without abrasion or structural failure.

Experience in Albuquerque, NM and the Corps of Engineers in Southern California have produced design tips (Blair, 1984, USCOE, 1978) for concrete channels:

- eliminate contraction joints by using continuous reinforcing;
- use new joint sealants such as ethylene vinyl acetate foam or two-part urethane;
- use a staggered or stepped joint (see Blair, 1984) to eliminate joint failure;
- use 0.3 to 0.5 percent steel longitudinally and 0.2 to 0.27 percent transversely and place it in the upper portion of the middle third of thickness (longitudinal bar on top);
- use side ditches along tops of channels to intercept sheet flow and convey it to chutes or "dip" inlets;
- form a low flow or trickle channel in the bottom of the channel (see UDFCD, 1991);
- use 7 to 8 inch concrete thickness;
- use underdrains of 6 inch perforated pipe in gravel or coarse sand when experiencing temporary or permanent high water tables;
- limit joints and weepholes when possible;
- use expansion joints only when abutting other structures; and
- ensure proper compaction of soils under and adjacent to the concrete.

Rigid Low Flow Channels - Under continuous base flow conditions when a vegetative lining alone would not be appropriate, a small concrete (normally V shaped) pilot or trickle channel could be used to handle the continuous low flows. Vegetation could then be maintained for handling larger flows. The trickle channel allows the remainder of the channel to remain dry and easy to maintain; reduces erosion caused by a meandering low flow channel; and allows for sediment transport at low flows due to the higher velocities in the trickle channel (UDFCD, 1991). Sometimes rock lined channels are used for trickle channels; however they are more aesthetically pleasing but require more maintenance and can encourage sediment deposition. Rock imbedded in concrete can obtain the best of both designs... but at greater cost. A low flow pipe has also been used effectively if it is designed with a minimum diameter of 24 in.; maintains a minimum of 3 ft/s velocity at half-full conditions; and provides access manholes every 300 to 500 feet.

General Design Criteria

In general, the following criteria should be used for open channel design and analysis.
- Channels with bottom widths greater than 10 feet should be designed with a minimum bottom cross slope of 12 to 1.
- Low flow and high flow sections should be considered in the design of channels with large cross sections (Q > 50 cfs).
- Channel side slopes should be stable throughout the entire length and side slope should depend on the channel material. A maximum of 3:1 is usually allowed for vegetation and 2:1 for riprap, unless otherwise justified by calculations.
- Superelevation of the water surface at horizontal curves should be accounted for by increased freeboard.
- Trapezoidal or parabolic cross sections are preferred tp triangular shapes.
- If relocation of a stream channel is unavoidable, the cross-sectional shape, meander pattern, roughness, sediment transport capacity, and slope should conform to the existing conditions insofar as practicable. Some means of energy dissipation may be necessary when existing conditions cannot be duplicated.
- Streambank stabilization should be provided, when appropriate, as a result of any stream disturbance and should include both upstream and downstream banks as well as the local site.
- Open channel drainage systems are often sized to handle a 10-year design storm. The 100-year design storm should be routed through the channel flood plain system to determine if the 100-year plus applicable freeboard flood elevation restrictions are exceeded, structures are flooded, or flood damages increased.
- Where complex sections are used the low flow channel is often sized to handle the "dominant discharge" (normally between the 2-year and 5-year flow) while the high flow channel handles less frequent flows depending on overbank flooding circumstances. Low flow channels are also designed to handle base flows in larger streams.
- Sediment transport requirements must be considered for conditions of flow below the design frequency. A low flow channel component within a larger channel can reduce maintenance by improving sediment transport in the channel.

Channel Transitions - The following criteria should be considered at channel transitions in order for the channel system to operate as designed.
- Transition to channel sections should be smooth and gradual.
- A straight line connecting flow lines at the two ends of the transition should not make an angle greater than 12.5 degrees with the axis of the main channel.

- Transition sections should be designed to provide a gradual transition to avoid turbulence and eddies.
- Energy losses in transitions should be accounted for as part of the water surface profile calculations.
- Scour downstream from rigid-to-natural and steep-to-mild slope transition sections should be accounted for through velocity slowing and energy dissipating devices.

10.3 Hydraulic Terms And Equations

Design analysis of both natural and artificial channels proceeds according to the basic principles of open channel flow (see Chow, 1970; Henderson, 1966). The basic principles of fluid mechanics -- continuity, momentum, and energy -- can be applied to open channel flow with the additional complication that the position of the free surface is usually one of the unknown variables. The determination of this unknown is one of the principle problems of open channel flow analysis and it depends on quantification of the flow resistance. Natural channels display a much wider range of roughness values than artificial channels.

Flow Classification

The classification of one-dimensional open channel flow can be summarized as follows.
    Steady Flow
        1. Uniform Flow
        2. Nonuniform Flow
           a. Gradually Varied Flow
           b. Rapidly Varied Flow
    Unsteady Flow
        1. Unsteady Uniform Flow (rare)
        2. Unsteady Nonuniform Flow
           a. Gradually Varied Unsteady Flow
           b. Rapidly Varied Unsteady Flow
The steady uniform flow case and the steady nonuniform flow case are the most fundamental types of flow treated in municipal engineering hydraulics. Rapidly varying flow is handled through special design procedures such as inlet or weir equations.

Hydraulic Terms And Definitions

Total Energy Head - The total energy head is the specific energy head plus the elevation of the channel bottom with respect to some datum. The locus of the energy head from one cross section to the next defines the energy grade line.

Specific Energy - Specific energy (E) is defined as the energy head relative to the channel bottom. If the channel is not too steep (slope less than 10 percent) and the streamlines are nearly straight and parallel (so that the hydrostatic assumption holds), the specific energy E becomes the sum of the depth and velocity head:

$$E = y + \alpha \ (V^2/2g) \tag{10.1}$$

Where: y = depth, ft

α = kinetic energy correction coefficient
V = mean velocity, ft/s
g = gravitational acceleration, 32.2 ft/s^2

The kinetic energy correction coefficient is taken to have a value of one for turbulent flow in prismatic channels but may be significantly different from one in natural channels.

Kinetic Energy Coefficient - As the velocity distribution in a river varies from a maximum at the design portion of the channel to essentially zero along the banks, the average velocity head, computed as $V^2/2g$ for the stream at a section, does not give a true measure of the kinetic energy of the flow, but rather normally underpredicts the actual value. A weighted average value of the kinetic energy is obtained by multiplying the average velocity head, above, by a kinetic energy coefficient, α, defined as:

$$\alpha = [\Sigma(q_i v_i^2)/(QV^2] = [\Sigma(a_i v_i^3)/(QV^3)] \tag{10.2}$$

Where: v_i = average velocity in subsection i, ft/s
q_i = discharge in same subsection i, cfs
a_i = flow area of subsection i, ft^2
Q = total discharge in river, cfs
V = average velocity in river at section or Q/A, ft/s

Kolupaila (1956) has given typical values of α for regular channels between 1.1 and 1.2 (with an average value of 1.15); and for natural channels between 1.15 and 1.5 (with an average value of 1.3). For natural channels with flooded overbanks the value can range between 1.5 and 2.0 with an average value of 1.75.

Steady And Unsteady Flow - A steady flow is where the discharge passing a given cross-section is constant with respect to time. The maintenance of steady flow requires that the rates of inflow and outflow be constant and equal. When the discharge varies with time, the flow is unsteady.

Uniform Flow And Nonuniform Flow - A nonuniform flow is one in which the velocity and depth vary in the direction of motion, while they remain constant in uniform flow. Uniform flow can only occur in a prismatic channel, which is a channel of constant cross section, roughness, and slope in the flow direction; however, nonuniform flow can occur either in a prismatic channel or in a natural channel with variable properties.

Gradually-Varied And Rapidly-Varied - A nonuniform flow, in which the depth and velocity change gradually enough in the flow direction that vertical accelerations can be neglected, is referred to as a gradually-varied flow; otherwise, it is considered to be rapidly-varied.

Froude Number - The Froude number is an important dimensionless parameter in open channel flow. It represents the ratio of inertia forces to gravity forces and is defined by:

$$F = V/(gd)^{.5} \tag{10.3}$$

Where: V = mean velocity = Q/A, ft/s
g = acceleration of gravity, ft/s^2
d = hydraulic depth = A/B, ft
A = cross-sectional area of flow, ft^2
B = channel topwidth at the water surface, ft

This expression for Froude number applies to any single section channel of nonrectangular shape.

Critical Flow - The variation of specific energy with depth at a constant discharge shows a minimum in the specific energy at a depth called critical depth at which the Froude number has a value of one. Critical depth is also the depth of maximum discharge when the specific energy is held constant.

Subcritical Flow - Depths greater than critical occur in subcritical flow and the Froude number is less than one. In this state of flow, small water surface disturbances can travel both upstream and downstream, and the control is always located downstream.

Supercritical Flow - Depths less than critical depth occur in supercritical flow and the Froude number is greater than one. Small water surface disturbances are always swept downstream in supercritical flow, and the location of the flow control is always upstream.

Hydraulic Jump - Hydraulic jumps occur at abrupt transitions from supercritical to subcritical flow in the flow direction. There are significant changes in depth and velocity in the jump, and energy is dissipated. For this reason, the hydraulic jump is often employed to dissipate energy and control erosion at storm water management structures.

Equations

The following equations are those most commonly used to analyze open channel flow.

Manning's Equation - For a given channel geometry, slope, and roughness, and a specified value of discharge Q, a unique value of depth occurs in steady uniform flow. It is called the normal depth. The normal depth is used to design artificial channels in steady, uniform flow and is computed from Manning's equation:

$$Q = [(1.49/n)AR^{2/3}S^{1/2}] \qquad (10.4)$$

Where: Q = discharge, cfs
        n = Manning's roughness coefficient
        A = cross-sectional area of flow, ft^2
        R = hydraulic radius = A/P, ft
        P = wetted perimeter, ft
        S = channel slope, ft/ft

The selection of Manning's n is generally based on observation; however, considerable experience is essential in selecting appropriate n values. See Section 10.4 for a discussion of Manning's n values and selection tables.

If the normal depth computed from Manning's equation is greater than critical depth, the slope is classified as a mild slope, while on a steep slope, the normal depth is less than critical depth. Thus, uniform flow is subcritical on a mild slope and supercritical on a steep slope.

In channel analysis, it is often convenient to group the channel properties in a single term called the channel conveyance K:

$$K = (1.49/n)AR^{2/3} \qquad (10.5)$$

and then Manning's equation can be written as:

$$Q = KS^{1/2} \tag{10.6}$$

The conveyance represents the carrying capacity of a stream cross-section based upon its geometry and roughness characteristics alone and is independent of the streambed slope.

The concept of channel conveyance is useful when computing the distribution of overbank flood flows in the stream cross section and the flow distribution through the opening in a proposed stream crossing. Manning's equation should not be used for determining highwater elevations in a bridge opening.

Continuity Equation - The continuity equation is the statement of conservation of mass in fluid mechanics. For the special case of steady flow of an incompressible fluid, it assumes the simple form:

$$Q = A_1V_1 = A_2V_2 \tag{10.7}$$

Where: Q = discharge, cfs
        A = flow cross-sectional area, ft^2
        V = mean cross-sectional velocity, ft/s (which is perpendicular to the cross section)

The subscripts 1 and 2 refer to successive cross sections along the flow path. The continuity equation can be used together with Manning's equation to obtain the steady uniform flow velocity as:

$$V = Q/A = [(1.49/n)R^{2/3}S^{1/2}] \tag{10.8}$$

Energy Equation - The energy equation expresses conservation of energy in open channel flow expressed as energy per unit weight of fluid which has dimensions of length and is therefore called energy head. The energy head is composed of potential energy head (elevation head), pressure head, and kinetic energy head (velocity head). These energy heads are scalar quantities which give the total energy head at any cross section when added. Written between an upstream open channel cross section designated 1 and a downstream cross section designated 2, the energy equation is:

$$h_1 + \alpha_1(V_1^2/2g) = h_2 + \alpha_2(V_2^2/2g) + h_L \tag{10.9}$$

Where: h_1 and h_2 are the upstream and downstream stages, respectively, ft
        α = kinetic energy correction coefficient
        V = mean velocity, ft/s
        h_L = head loss due to local cross-sectional changes (minor loss) as well as boundary resistance, ft

The stage h is the sum of the elevation head z at the channel bottom and the pressure head, or depth of flow y, i.e. h=z+y. The terms in the energy equation are illustrated graphically in Figure 10-1. The energy equation states that the total energy head at an upstream cross section is equal to the energy head at a downstream section plus the intervening energy head loss. The energy equation can only be applied between two cross sections at which the streamlines are nearly straight and parallel so that vertical accelerations can be neglected.

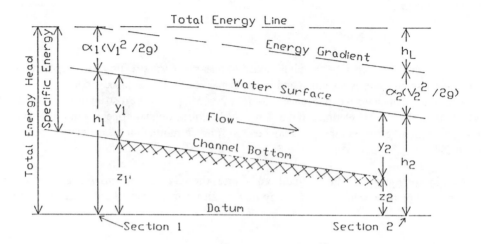

Figure 10-1 Terms In The Energy Equation

Source: FHWA, 1990

10.4 Manning's n Values

The Manning's n value is an important variable in open channel flow computations. Variation in this variable can significantly affect discharge, depth, and velocity estimates. Since Manning's n values depend on many different physical characteristics of natural and man-made channels, care and good engineering judgment must be exercised in the selection process.

The following general factors should be considered when selecting the value of Manning's n.

1. The physical roughness of the bottom and sides of the channel. Fine particle soils on smooth, uniform surfaces result in relatively low values of n. Coarse materials, such as gravel or boulders, or pronounced surface irregularity cause higher values of n.

2. The value of n depends on the height, density, type of vegetation, and how the vegetation affects the flow through the channel reach. The n value will increase in the spring and summer, as vegetation grows and foliage develops, and diminish in the fall, as the dormant season approaches.

3. Channel shape variations, such as abrupt changes in channel cross sections or alternating small and large cross sections, will require somewhat larger n values than normal. These variations in channel cross section become particularly important if they cause the flow to meander from side to side.

4. A significant increase in the value of n is possible if severe meandering occurs in the alignment of a channel. Meandering becomes particularly important when frequent changes in the direction of curvature occur with relatively small radii of curvature.

5. Active channel erosion or sedimentation will tend to increase the value of n, since these processes may cause variations in the shape of a channel. The potential for future erosion or sedimentation in the channel should also be considered.

6. Obstructions such as log jams or deposits of debris will increase the value of n. The level of this increase will depend on the number, type, and size of obstructions.

7. To be conservative, it is better to use a higher resistance for capacity calculations and a lower resistance for stability calculations. Sensitivity studies may be important.

8. Due to floodplain vegetation, n may vary with depth of flow.

All of these factors should be considered with respect to type of channel, degree of maintenance, seasonal requirements, and other considerations as a basis for making a determination of an appropriate design n value. The probable condition of the channel when the design event is anticipated should be considered. Values representative of a freshly constructed channel are rarely appropriate as a basis for design capacity calculations.

Recommended Manning's n values for channels which are either excavated or dredged and natural are given in Table 10-1. Recommended Manning's n values for artificial channels with rigid, unlined, temporary, and riprap linings are given in Table 10-2. For more information refer to the publication Guide For Selecting Manning's Roughness Coefficients For Natural Channels And Flood Plains, FHWA-TS-84-204, 1984.

Table 10-1 Uniform Flow Values Of Roughness Coefficient - n

Type Of Channel And Description	Minimum	Normal	Maximum
EXCAVATED OR DREDGED			
a. Earth, straight and uniform	0.016	0.018	0.020
1. Clean, recently completed	0.018	0.022	0.025
2. Clean, after weathering	0.022	0.025	0.030
3. Gravel, uniform section, clean	0.022	0.027	0.033
b. Earth, winding and sluggish			
1. No vegetation	0.023	0.025	0.030
2. Grass, some weeds	0.025	0.030	0.033
3. Dense weeds/plants in deep channels	0.030	0.035	0.040
4. Earth bottom and rubble sides	0.025	0.030	0.035
5. Stony bottom and weedy sides	0.025	0.035	0.045
6. Cobble bottom and clean sides	0.030	0.040	0.050
c. Dragline-excavated or dredged			
1. No vegetation	0.025	0.028	0.033
2. Light brush on banks	0.035	0.050	0.060
d. Rock cuts			
1. Smooth and uniform	0.025	0.035	0.040
2. Jagged and irregular	0.035	0.040	0.050
e. Channels not maintained, weeds and brush uncut			
1. Dense weeds, high as flow depth	0.050	0.080	0.120
2. Clean bottom, brush on sides	0.040	0.050	0.080
3. Same, highest stage of flow	0.045	0.070	0.110
4. Dense brush, high stage	0.080	0.100	0.140
NATURAL STREAMS			
Minor streams (top width at flood stage < 100 ft)			
a. Streams on Plain			
1. Clean, straight, full stage, no rifts or deep pools	0.025	0.030	0.033

2. Same as above, but with more stones and weeds	0.030	0.035	0.040
3. Clean, winding, some pools and shoals	0.033	0.040	0.045
4. Same as above, but with some weeds and some stones	0.035	0.045	0.050
5. Same as above, lower stages, more ineffective slopes and sections	0.040	0.048	0.055
6. Same as 4, but with more stones	0.045	0.050	0.060
7. Sluggish reaches, weedy, deep pools	0.050	0.070	0.080
8. Very weedy reaches, deep pools, or floodways with heavy stand of timber and underbrush	0.075	0.100	0.150
b. Mountain streams, no vegetation in channel, banks usually steep, trees and brush along banks submerged at high stages			
1. Bottom: gravels, cobbles, few bolders	0.030	0.040	0.050
2. Bottom: cobbles with large bolders	0.040	0.050	0.070
Floodplains			
a. Pasture, no brush			
1. Short grass	0.025	0.030	0.035
2. High grass	0.030	0.035	0.050
b. Cultivated area			
1. No crop	0.020	0.030	0.040
2. Mature row crops	0.025	0.035	0.045
3. Mature field crops	0.030	0.040	0.050
c. Brush			
1. Scattered brush, heavy weeds	0.035	0.050	0.070
2. Light brush and trees in winter	0.035	0.050	0.060
3. Light brush and trees, in summer	0.040	0.060	0.080
4. Medium to dense brush, in winter	0.045	0.070	0.110
5. Medium to dense brush, in summer	0.070	0.100	0.160
d. Trees			
1. Dense willows, summer, straight	1.110	0.150	0.200
2. Cleared land, tree stumps, no sprouts	0.030	0.040	0.050
3. Same as above, but with heavy growth of sprouts	0.050	0.060	0.080
4. Heavy stand of timber, a few down trees, little undergrowth, flood stage below branches	0.080	0.100	0.120
5. Same as above, but with flood stage reaching branches	0.100	0.120	0.160
Major Streams (top width at flood stage > 100 ft). The n value is less than that for minor streams of similar description, because banks offer less effective resistance.			
a. Regular section with no boulders or brush	0.025	.....	0.060
b. Irregular and rough section	0.035	.....	0.100

Table 10-2 Manning's Roughness Coefficients For Artificial Channels - n

Lining Category	Lining Type	0-0.5 ft	0.5-2.0 ft	> 2.0 ft
Rigid	Concrete	0.015	0.013	0.013
	Grouted Riprap	0.040	0.030	0.028
	Stone Masonry	0.042	0.032	0.030
	Soil Cement	0.025	0.022	0.020
	Asphalt	0.018	0.016	0.016
Unlined	Bare Soil	0.023	0.020	0.020
	Rock Cut	0.045	0.035	0.025
Temporary*	Woven Paper Net	0.016	0.015	0.015
	Jute Net	0.028	0.022	0.019
	Fiberglass Roving	0.028	0.022	0.019
	Straw With Net	0.065	0.033	0.025
	Curled Wood Mat	0.066	0.035	0.028
	Synthetic Mat	0.036	0.025	0.021
Gravel Riprap	1-inch D_{50}	0.044	0.033	0.030
	2-inch D_{50}	0.066	0.041	0.034
Rock Riprap	6-inch D_{50}	0.104	0.069	0.035
	12-inch D_{50}	----	0.078	0.040

Note: Values listed are representative values for the respective depth ranges. Manning's roughness coefficients, n, vary with the flow depth.
*Some "temporary" linings become permanent when buried.

Source: HEC-15, 1988.

Natural Channels - There will be times when the determination of Manning's n value is difficult in compound channels or for special circumstances. The partitioning of roughness among its several components is not new and has been used for sediment transport for years. Cowan (1956) has developed a method for the partitioning of the various components that make up Manning's n as:

$$n = (n_0 + n_1 + n_2 + n_3 + n_4) \, m_5 \qquad (10.10)$$

Where: n = Manning roughness coefficient for natural channel
n_0 = coefficient associated with lining material type
n_1 = coefficient associated with degree of irregularity
n_2 = coefficient associated with variations of the channel cross section
n_3 = coefficient associated with channel obstructions
n_4 = coefficient associated with channel vegetation
m_5 = coefficient associated with channel meandering

Table 10-3 gives pertinent information for the determination of each of the coefficients. For channels flowing in alluvial materials the roughness and flow rate are interdependent and are determined by both the bed grain size distribution and the bedforms of the alluvium (ripples, dunes, anti-dunes, etc.). There have been many analyses to determine roughness and stag-discharge relationships. For example see ASCE, 1977, Simons and Senturk, 1976.

Table 10-3 Cowan's Coefficients For Computing Manning's n

Material	Specific Condition	Value
Bank Material	earth	0.020
n_0	rock cut	0.025
	fine gravel	0.024
	coarse gravel	0.028
Degree of Irregularity	smooth	0.000
n_1	minor	0.005
	moderate	0.010
	severe	0.020
Cross Section Variations	gradual	0.000
n_2	alternating some	0.005
	alternating much	0.010 to 0.015
Obstructions	negligible	0.000
n_3	minor	0.010 to 0.015
	appreciable	0.020 to 0.030
	severe	0.040 to 0.060
Vegetation	low	0.005 to 0.010
n_4	medium	0.010 to 0.025
	high	0.025 to 0.050
	very high	0.050 to 0.100
Meandering	minor	1.000
m_5	appreciable	1.150
	severe	1.300

Source: Chow, 1959

10.5 Best Hydraulic Section

The cost of an open channel, other than environmental or regulatory costs, is comprised of: (1) the acquisition of land; (2) excavation; and (3) lining. Land cost varies with the top width of the channel, excavation cost with the cross sectional area, and lining with the wetted perimeter. In cases where land cost predominates, the most economical design is a buried facility. In other cases it can be shown the most economical design is one which is comprised of the so called "best hydraulic section". For a given discharge, slope and channel roughness, maximum velocity implies minimum cross sectional area. From Manning's equation, if velocity is maximized and area is minimized, wetted perimeter will also be minimized. The best hydraulic section therefore, simultaneously minimizes area and wetted perimeter.

The best of all prismatic sections is the semi-circle. Similarly, the best rectangular section is half of a square (i.e., width equal twice the depth) and the best triangular section is an isosceles right triangle, etc. Of course the semi-circular section would be impractical because of the labor involved in excavation and forming the lining. Since rectangular and triangular sections are special cases of trapezoidal sections, it follows that the best hydraulic section for trapezoidal sections is the one of primary practical interest.

Given that the desired side slope, M to one, has been selected for a given channel, the minimum wetted perimeter (P) exists when:

$$P = 4y \ (1 + M^2)^{0.5} - 2My \qquad (10.11)$$

(Figure 10-2 below shows a definition of variables.)

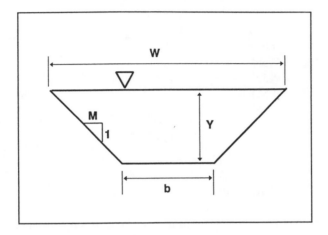

Figure 10-2 Trapezoidal Channel - Definition Of Variables

From the geometry of the channel cross-section and the Manning equation, design equations can be developed for determining the dimensions of the best hydraulic section for a trapezoidal channel.

The depth of the best hydraulic section is defined by:

$$y = C_M \ (Qn/(S^{1/2}))^{3/8} \qquad (10.12)$$

Where:
$$C_M = \frac{[(k + 2(M^2 + 1)^{1/2})^{2/3}]^{3/8}}{[1.49 \ (k + M)^{5/3}]^{3/8}} \qquad (10.13)$$

The associated bottom width is:

$$B = ky \qquad (10.14)$$

The cross-sectional area of the resulting channel is:

$$A = By + My^2 \qquad (10.15)$$

Table 10-4 lists values of C_M and K for various values of M.

**Table 10-4 Values Of C_M And k For Determining Bottom Width
And Depth Of Best Hydraulic Section**

M	C_M	k	Remarks
0/1	0.790	2.00	Vertical Sides
0.5/1	0.833	1.236	
0.577/1	0.833	1.155	60^0 Side Slopes
1.0/1	0.817	0.828	
1.5/1	0.775	0.606	
2.0/1	0.729	0.472	
2.5/1	0.688	0.385	
3.0/1	0.653	0.325	
3.5/1	0.622	0.280	
4.0/1	0.595	0.246	
5.0/1	0.522	0.198	
6.0/1	0.518	0.166	
8.0/1	0.467	0.125	
10.0/1	0.430	0.100	
12.0/1	0.402	0.083	

10.6 Uniform Flow Calculations

Following is a discussion of the equations that can be used for the design and analysis of open channel flow. Manning's equation, presented in three forms below, is recommended for evaluating uniform flow conditions in open channels:

$$v = (1.49/n) \ R^{2/3} \ S^{1/2} \tag{10.16}$$
$$Q = (1.49/n) \ A \ R^{2/3} \ S^{1/2} \tag{10.17}$$
$$S = [Qn/(1.49 \ A \ R^{2/3})]^2 \tag{10.18}$$

where: v = average channel velocity, ft/s
         Q = discharge rate for design conditions, cfs
         n = Manning's roughness coefficient
         A = cross-sectional area, ft^2
         R = hydraulic radius A/P, ft
         P = wetted perimeter, ft
         S = slope of the energy grade line, ft/ft

For prismatic channels, in the absence of backwater conditions, the slope of the energy grade line, water surface and channel bottom are equal. Area, wetted perimeter, hydraulic radius, and channel top width for standard channel cross-sections can be calculated from geometric dimensions.

Direct Solution

When the hydraulic radius, cross-sectional area, and roughness coefficient and slope are known, discharge can be calculated directly from equation 10.17. The slope can be calculated using equation 10.18 when the discharge, roughness coefficient, area, and hydraulic radius are known.

Nomographs for obtaining direct solutions to Manning's equation are presented in Figures 10-4 and 10-5. Figure 10-3 provides a general solution for the velocity form of Manning's equation, while Figure 10-4 provides a solution of Manning's equation for trapezoidal channels.

General Solution Nomograph - The following steps are used for the general solution using the nomograph in Figure 10-3.
1. Determine open channel data, including slope in ft/ft, hydraulic radius in ft, and Manning's n value.
2. Connect a line between the Manning's n scale and slope scale and note the point of intersection on the turning line.
3. Connect a line from the hydraulic radius to the point of intersection obtained in Step 2.
4. Extend the line from Step 3 to the velocity scale to obtain the velocity in ft/s.

Trapezoidal Solution Nomograph - The trapezoidal channel nomograph solution to Manning's equation in Figure 10-4 can be used to find the depth of flow if the design discharge is known or the design discharge if the depth of flow is known.
1. Determine input data, including slope in ft/ft, Manning's n value, bottom width in ft, and side slope in ft/ft.
2. a. Given the design discharge, find the product of Q times n, connect a line from the slope scale to the Qn scale, and find the point of intersection on the turning line.
   b. Connect a line from the turning point from Step 2a to the b scale and find the intersection with the z = 0 scale.
   c. Project horizontally from the point located in Step 2b to the appropriate z value and find the value of d/b.
   d. Multiply the value of d/b obtained in Step 2c by the bottom width b to find the depth of uniform flow, d.
3. a. Given the depth of flow, find the ratio d divided by b and project a horizontal line from the d/b ratio at the appropriate side slope, z, to the z = 0 scale.
   b. Connect a line from the point located in Step 3a to the b scale and find the intersection with the turning line.
   c. Connect a line from the point located in Step 3b to the slope scale and find the intersection with the Qn scale.
   d. Divide the value of Qn obtained in Step 3c by the n value to find the design discharge, Q.

Trial And Error Solution

A trial and error procedure for solving Manning's equation is used to compute the normal depth of flow in a uniform channel when the channel shape, slope, roughness, and design discharge are known. For purposes of the trial and error process, Manning's equation can be arranged as:

$$AR^{2/3} = (Qn)/(1.49\ S^{1/2}) \qquad\qquad (10.19)$$

where: A = cross-sectional area, ft
        R = hydraulic radius, ft
        Q = discharge rate for design conditions, cfs
        n = Manning's roughness coefficient
        S = slope of the energy grade line, ft/ft

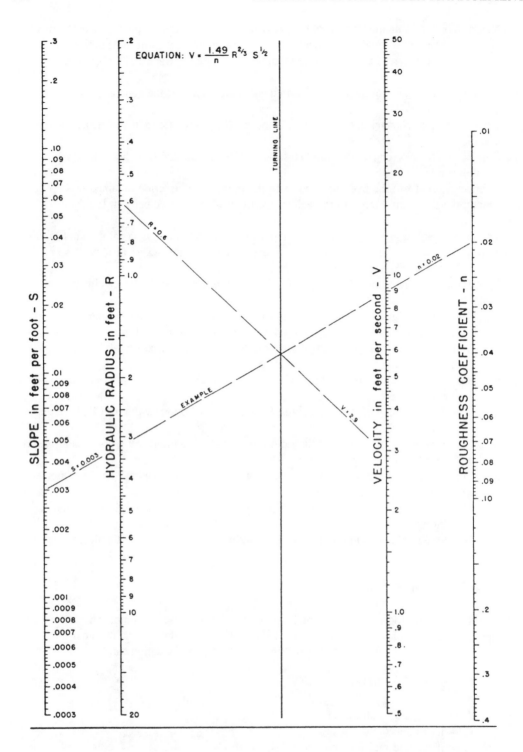

Figure 10-3 Nomograph For The Solution Of Manning's Equation

Source: USDOT, FHWA, HDS-3, 1961

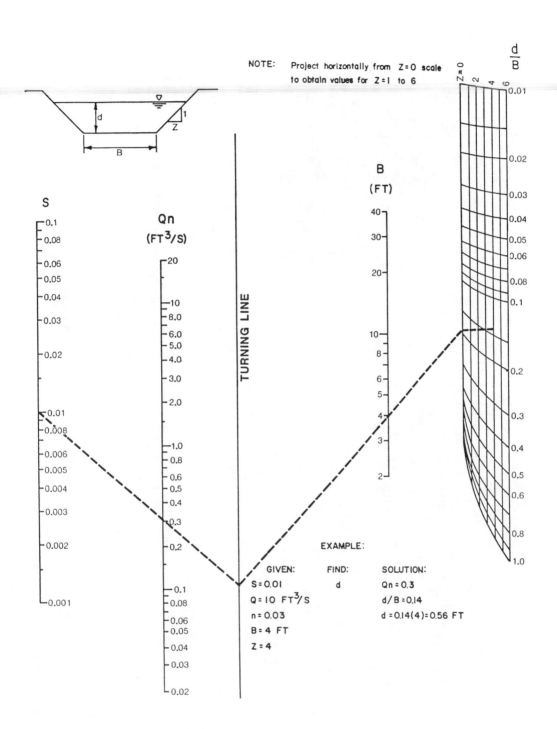

Figure 10-4 Solution Of Manning's Equation For Trapezoidal Channels

Source: USDOT, FHWA, HEC-15, 1986

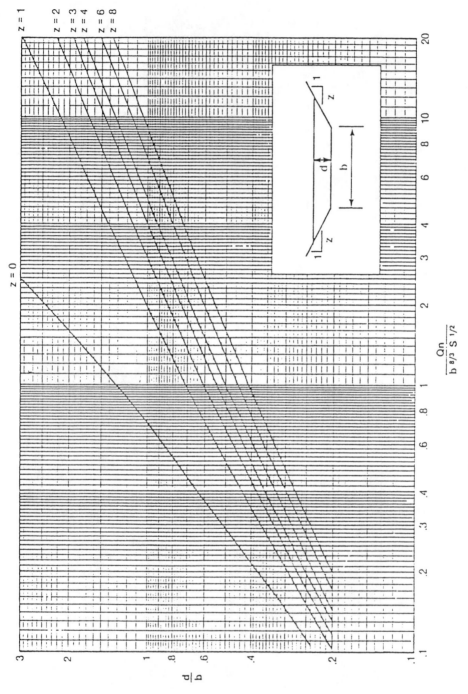

Figure 10-5 Trapezoidal Channel Capacity

Source: Storm Water Management Manual, Nashville, 1988

To determine the normal depth of flow in a channel by the trial and error process, trial values of depth are used to determine A, P, and R for the given channel cross section. Trial values of $AR^{2/3}$ are computed until the equality of equation 10.19 is satisfied such that the design flow is conveyed for the slope and selected channel cross section. Graphical procedures for simplifying trial and error solutions are presented in Figure 10-5 for trapezoidal channels.

1. Determine input data, including design discharge, Q, Manning's n value, channel bottom width, b, channel slope, S, and channel side slope, z.
2. Calculate the trapezoidal conveyance factor using the equation:

$$K_T = (Qn)/(b^{8/3}S^{1/2}) \qquad\qquad (10.20)$$

where: K_T = trapezoidal open channel conveyance factor
         Q = discharge rate for design conditions, cfs
         n = Manning's roughness coefficient
         b = bottom width, ft
         S = slope of the energy grade line, ft/ft

3. Enter the x-axis of Figure 10-5 with the value of K_T calculated in Step 2 and draw a line vertically to the curve corresponding to the appropriate z value from Step 1.
4. From the point of intersection obtained in Step 3, draw a horizontal line to the y-axis and read the value of the normal depth of flow over the bottom width, d/b.
5. Multiply the d/b value from Step 4 by b to obtain the normal depth of flow.

Irregular Channels

The calculation of uniform flow in irregular channels, either in shape or roughness, is a common problem for urban storm water designers. There are several techniques to handle it depending on the accuracy required and the shape of the channel (COE, 1970, Chow, 1959).

Average Roughness - For channels with simple concave geometry which have different roughnesses, for example, on the bed and banks, an equivalent Manning's n can be calculated and applied to the whole channel (Chow, 1959). If the channel is broken into N sub-sections then the equivalent Manning's n value can be expressed as:

$$n_e = (P_1 n_1^x + P_2 n_2^x + ... + P_N n_N^x)^y / (P)^y \qquad\qquad (10.21)$$

where: n_e =   equivalent Manning's n
         x,y =   coefficients depending on basic assumptions

Other variables are as defined previously. For the assumption that the mean velocity in each sub-section is equal to each other and the mean velocity in the channel, x = 3/2 and y = 2/3. For the assumption that the sum of the forces resisting the flow equal the total force resisting the flow, x = 2 and y = 1/2. For this type of channel using a single Manning's equation and an equivalent Manning's n is preferable to using a subdivided channel as the subdivision process leads to an overestimation of channel conveyance by as much as 7 to 10 percent (Garbrecht and Brown, 1991 and Bensen and Dalrymple, 1967)

Irregular Channel Calculations - For more complex convex channel shapes (flow with overbanks) or width to depth ratios greater than 10, uniform flow can be calculated by breaking the channel into N sub-sections, projecting vertical lines from the perimeter break point upward to the surface. Then for each sub-section, i:

$$v_i = (K_i/A_i)\ S^{1/2} \tag{10.22}$$

where: v_i = mean velocity in subsection i, ft/s
  K_i = conveyance in subsection i, ft^2 (see equation 10.5)
  A_i = flow area in subsection i, ft^2
and from the continuity equation:

$$Q = VA = \Sigma(v_i A_i) = \Sigma(K_i)\ S^{1/2} \tag{10.23}$$

where it should be noted that K_i the sectional conveyance equals $(1.49/n_i)A_i R_i^{2/3}$.

When computing the hydraulic radius of the subsections, the water depth common to the two adjacent subsections is not counted as wetted perimeter. The procedure then is to guess at a depth of flow; calculate conveyance at each subsection for the various flow areas, wetted perimeters and Manning's n values; and then find the total discharge. By plotting a stage-discharge curve the depth for the known Q can be readily determined through interpolation. This method lends itself to spreadsheet calculations. The slope is normally assumed equal in each section.

Alternately a backwater computer model can be used to quickly determine depth of flow by inputting the complex section and simply changing the vertical values by the product of slope times distance and moving the section an arbitrary distance upstream for several cross sections.

10.7 Critical Flow Calculations

Critical depth depends only on the discharge rate and channel geometry. The general equation for determining critical depth is:

$$Q^2/g = A^3/T \tag{10.24}$$

where: Q = discharge rate for design conditions, cfs
  g = acceleration due to gravity, 32.2 ft/s^2
  A = cross-sectional area, ft^2
  T = top width of water surface, ft
A trial and error procedure is needed to solve equation 10.24.

Semi-empirical equations (as presented in Table 10-5) can be used to simplify trial and error critical depth calculations. The following equation from Chow (1959) is used to determine critical depth with the critical flow section factor, Z:

$$Z = Q/(g^{0.5}) \tag{10.25}$$

where: Z = critical flow section factor
  Q = discharge rate for design conditions, cfs
  g = acceleration due to gravity, 32.2 ft/s^2
The following guidelines are presented for evaluating critical flow conditions of open channel flow:
1. A normal depth of uniform flow within about 10 percent of critical depth is unstable (relatively large depth changes are likely for small changes in roughness, cross sectional area or slope) and should be avoided in design, if possible.
2. If the velocity head is less than one-half the mean depth of flow, the flow is subcritical.
3. If the velocity head is equal to one-half the mean depth of flow, the flow is critical.

4. If the velocity head is greater than one-half the mean depth of flow, the flow is supercritical.
5. If an unstable critical depth cannot be avoided in design, the least favorable type of flow should be assumed for the design.

Table 10-5 Critical Depth For Uniform Flow In Selected Channel Types

Channel Type	Semi-Empirical Equation[1] For Estimating Critical Depth	Range of Applicability
Rectangular[2]	$d_c = [Q^2/(gb^2)]^{1/3}$	N/A
Trapezoidal[2]	$d_c = 0.81[Q^2/(gz^{0.75}b^{1.25})]^{0.27} - b/30z$	$.01 < 0.5522Q/b^{2.5} < 0.4$ For $0.5522Q/b^{2.5} < 0.1$, use rectangular equa.
Triangular[2]	$d_c = [(2Q^2)/(gz^2)]^{1/5}$	N/A
Circular[3]	$d_c = 0.325(Q/D)^{2/3} + 0.083D$	$0.3 < d_c/D < 0.9$
General[4]	$(A^3/T) = (Q^2/g)$	N/A

Where: c_c = critical depth, ft
Q = design discharge, cfs
g = acceleration due to gravity, 32.2 ft/s^2
b = bottom width of channel, ft
z = side slopes of a channel (horizontal to vertical)
D = diameter of circular conduit, ft
T = top width of water surface, ft

[1]Assumes uniform flow with the kinetic energy coefficient equal to 1.0
[2]Reference: French, 1985
[3]Reference: USDOT, FHWA, HDS-4, 1965
[4]Reference: Brater and King, 1976

The Froude number, Fr, calculated by the following equation, is useful for evaluating the type of flow conditions in an open channel:

$$Fr = v/(gA/T)^{0.5} \qquad (10.26)$$

where: Fr = Froude number (dimensionless)
v = velocity of flow, ft/s
g = acceleration of gravity, 32.2 ft/s^2
A = cross-sectional area of flow, ft^2
T = top width of flow, ft

If Fr is greater than 1.0, flow is supercritical; if it is under 1.0, flow is subcritical. Fr is 1.0 for critical flow conditions.

In compound channels (i.e. with both a high and low flow channel) or with flow in the main channel and overbanks, there can be more than one depth at which critical depth will occur. There is no readily accepted method for analysis of this phenomena. However, more details can be found in Petryk and Grant (1978), Blalock and Strum (1981), Schoellhamer, et al., (1985), and Chaudhry and Bhallamudi (1988).

10.8 Vegetative Design

A two-part procedure, adapted from Chow (1959), Temple, et al., (1987) and Temple (1979) and presented below, is recommended for the design of temporary and vegetative channel linings. Part 1, the design stability component, involves determining channel dimensions for low vegetative retardance conditions, using Class D as defined in Table 10-6. Part 2, the design capacity component, involves determining the depth increase necessary to maintain capacity for higher vegetative retardance conditions, using Class C as defined in Table 10-6. If temporary lining is to be used during construction, vegetative retardance Class E should be used for the design stability calculations.

If the channel slope exceeds 10 percent, or a combination of channel linings will be used, additional procedures not presented below are required. References include HEC-15 (USDOT, FHWA, 1986) and HEC-14 (USDOT, FHWA, 1983).

Table 10-6 Classification Of Vegetal Covers To Degrees Of Retardancy

Retardance	Cover	Condition
A	Weeping Lovegrass	Excellent stand, tall (average 30 in.)
	Yellow Bluestem	
	Ischaemum	Excellent stand, tall (average 36 in.)
B	Kudzu	Very dense growth, uncut
	Bermuda grass	Good stand, tall (average 12 in.)
	Native grass mixture	
	little bluestem, bluestem,	
	blue gamma other short	
	and long stem	
	Midwest grasses	Good stand, unmowed
	Weeping lovegrass	Good stand, tall (average 24 in.)
	Laspedeza sericea	Good stand, not woody, tall (average 19 in.)
	Alfalfa	Good stand, unmowed (average 11 in.)
	Weeping lovegrass	Good stand, unmowed (average 13 in.)
	Kudzu	Dense growth, uncut
	Blue gamma	Good stand, uncut (average 13 in.)
C	Crabgrass	Fair stand, uncut (10 - 48 in.)
	Bermuda grass	Good stand, mowed (average 6 in.)
	Common lespedeza	Good stand, uncut (average 11 in.)
	Grass-legume mix:[1]	Good stand, uncut (6 - 8 in.)
	Centipedegrass	Very dense cover (average 6 in.)
	Kentucky bluegrass	Good stand, headed (6 - 12 in.)
D	Bermuda grass	Good stand, cut to 2.5 in.
	Common lespedeza	Excellent stand, uncut (average 4.5 in.)
	Buffalo grass	Good stand, uncut (3 - 6 in.)
	Grass-legume mix:[2]	Good stand, uncut (4 - 5 in.)

	Lespedeza sericea	After cutting to 2 in. (good before cutting)
E	Bermuda grass	Good stand, cut to 1.5 in.
	Bermuda grass	Burned stubble

[1]summer (orchard grass, redtop, Italian ryegrass, and common lespedeza)
[2]fall, spring (orchard grass, redtop, Italian ryegrass, and common lespedeza)
Note: Covers classified were tested in experimental channel and were green and generally uniform.

Design Stability

The following are the steps for design stability calculations.
1. Determine appropriate design variables, including discharge, Q, bottom slope, S, cross section parameters, and vegetation type.
2. Use Table 10-15 to assign a maximum velocity, $v_{m'}$, based on vegetation type and slope range.
3. Assume a value of n and determine the corresponding value of vR from the n versus vR curves in Figure 10-6. Use retardance Class D for permanent vegetation and E for temporary construction. The method of Kouwen and Li (1980), Kouwen (1988) can be used for vegetal Manning's n values for channels with tall weeds. Resistance in wetlands areas can be obtained from Kadlec (1990). Alternately n can be calculated by the following equation (Gwinn & Ree, 1980, Green & Garton, 1983):

$$n = \exp \{[0.01329C(\ln R_v)^2] - [0.09543C(\ln R_v)] + [0.2971C] - 4.16\} \quad (10.27)$$

Where: R_v = vR/ν
v = channel velocity, ft/s
R = channel hydraulic radius, ft
ν = kinematic viscosity of water, ft^2/s
C = 10.0, 7.643, 5.601, 4.436 and 2.876 for retardance classes A, B, C, D, and E respectively

4. Calculate the hydraulic radius using the equation:

$$R = (vR)/v_m \quad (10.28)$$

where: R = hydraulic radius of flow, ft
vR = value obtained from Figure 10-6 in Step 2
v_m = maximum velocity from Step 2, ft/s

5. Use the following form of Manning's equation to calculate the value of vR:

$$vR = (1.49 R^{5/3} S^{1/2})/n \quad (10.29)$$

where: vR = calculated value of vR product
R = hydraulic radius value from Step 4, ft
S = channel bottom slope, ft/ft
n = Manning's n value assumed in Step 3

6. Compare the vR product value obtained in Step 5 to the value obtained from Figure 10-6 for the assumed n value in Step 3. If the values are not reasonably close, return to Step 3 and repeat the calculations using a new assumed n value.

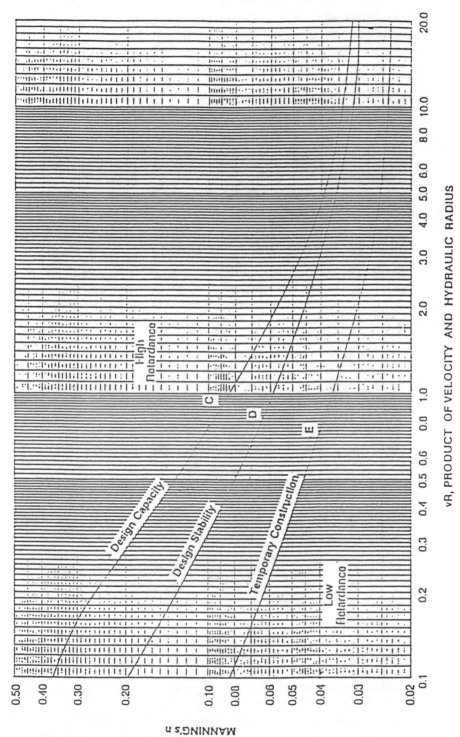

vR, PRODUCT OF VELOCITY AND HYDRAULIC RADIUS

Figure 10-6 Manning's n Values For Vegetated Channels

Source: USDA, TP-61, 1947

7. For trapezoidal channels, find the flow depth using Figures 10-4 or 10-5, as described in Section 10.6. The depth of flow for other channel shapes can be evaluated using the trial and error procedure described in Section 10.6.

8. If bends are considered, calculate the length of downstream protection, L_p, for the bend using Figure 10-7. Provide additional protection, such as gravel or riprap in the bend and extending downstream for length, L_p.

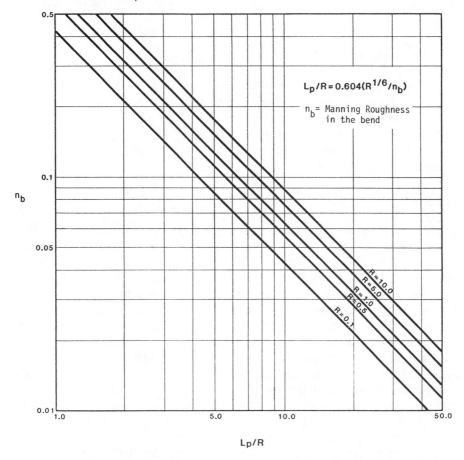

$$L_p/R = 0.604(R^{1/6}/n_b)$$

n_b = Manning Roughness in the bend

Figure 10-7 Protection Length, L_p, Downstream Of Channel Bend

Source: USDOT, FHWA, HEC-15, 1986

Design Capacity

The following are the steps for design capacity calculations.

1. Assume a depth of flow greater than the value from Step 7 above and compute the waterway area and hydraulic radius.

2. Divide the design flow rate, obtained using appropriate procedures from the Urban Hydrology, Chapter 7, by the waterway area from Step 1 to find the velocity.

3. Multiply the velocity from Step 2 by the hydraulic radius from Step 1 to find the value of vR.

4. Use Figure 10-6 to find a Manning's n value for retardance Class C based on the vR value from Step 3.

5. Use Manning's equation (equation 10.16) or Figure 10-3 to find the velocity using the hydraulic radius from Step 1, Manning's n value from Step 4, and appropriate bottom slope.
6. Compare the velocity values from Steps 2 and 5. If the values are not reasonably close, return to Step 1 and repeat the calculations.
7. Add an appropriate freeboard to the final depth from Step 6. Generally, 20 percent is adequate.
8. If bends are considered, calculate superelevation of the water surface profile at the bend using the equation:

$$\mathbf{delta\ d} = (v^2 T)/(gR_c) \qquad\qquad\qquad (10.30)$$

where: delta d = Superelevation of water surface profile due to bend, ft
          v =   average velocity from Step 6, ft/s
          T =   top width of flow, ft
          g =   acceleration of gravity, 32.2 ft/s^2
          R_c = mean radius of the bend, ft
Add freeboard consistent with the calculated delta d.

Erosion Control

Practice has shown that complete protection of grassed channels from erosion is costly (Wright-McLaughlin, 1969). It is far better to provide reasonable erosion-free design with the intention to take additional erosion control measures and corrective steps after the first year of operation. However, the use of erosion control cut-off walls at regular intervals in a grassed channel is often desirable. Such cut-off walls will safeguard a channel from serious erosion in case of a large storm flow prior to the grass developing a good root system.

Erosion control cut-off walls are usually of reinforced concrete, approximately 8 in. thick and from 18 in. to 2 ft deep, extending across the entire bottom of the channel. They can be shaped to fit a slightly sloped bottom to help direct water to the trickle channel. Grass will not grow under bridges, and therefore the erosion tendency is large. A cut-off wall and/or the use of riprap at the downstream edge of a bridge is good practice in some alluvial bed environments.

At bends in the channel special erosion control measures may be needed. However, once a good growth of grass is established and if the design velocities, depths, and curvatures are satisfactory, normally erosion at bends will not be a problem.

In maintaining the appropriate channel slope, the designer may find it necessary to use frequent drops. Erosion tends to occur at the edges immediately downstream of a drop even though it may be only 6 in. to 18 in. high (Wright-McLaughlin, 1969). Proper use of riprap and gabions at these drops may be necessary.

10.9 Approximate Flood Limits

For small streams and tributaries not included in the floodplain studies, analysis may be required to identify the 100-year flood elevation and to evaluate floodplain encroachment as required. For such cases, when the designer can demonstrate that a complete backwater analysis is unwarranted, approximate methods may be used.

100-Year Flood Elevation

A generally accepted method for approximating the 100-year flood elevation is outlined in the following steps (Powell, et al., 1980).
1. Divide the stream or tributary into reaches that may be approximated using average slopes, cross sections, and roughness coefficients for each reach.
2. Estimate the 100-year peak discharge for each reach using an appropriate hydrologic method from the Urban Hydrology, Chapter 7.
3. Compute normal depth for uniform flow in each reach using Manning's equation for the reach characteristics from Step 1 and peak discharge from Step 2.
4. Use the normal depths computed in Step 3 to approximate the 100-year flood elevation in each reach. The 100-year flood elevation is then used to delineate the floodplain.
    This approximate method is based on several assumptions, including the following.
*   A channel reach is accurately approximated by average characteristics throughout its length.
*   The cross-sectional geometry, including area, wetted perimeter, and hydraulic radius, of a reach may be approximated using typical geometric properties that can be used in Manning's equation to solve for normal depth.
*   Uniform flow can be established and backwater effects are negligible between reaches.
*   Expansion and contraction effects are negligible.
    As indicated, the approximate method is based on a number of restrictive assumptions that may limit the accuracy of the approximation and applicability of the method. The designer is responsible for appropriate application of this method.

Setback Limits

After the 100-year flood elevation and floodplain are established, floodway setback limits may be approximated by requiring conveyance of the encroached section, including any allowable flood elevation increases, to equal the conveyance of the non-encroached section. From Manning's equation, the conveyance is given as follows:

$$K = (1.49/n) \ A \ R^{2/3} \qquad\qquad (10.31)$$
$$Q = KS^{1/2} \qquad\qquad (10.32)$$

where: K = channel conveyance
           A = cross-sectional area, ft^2
           R = hydraulic radius A/P, ft
           P = wetted perimeter, ft
           Q = discharge, cfs
           S = slope of the energy grade line, ft/ft
The following procedure may be used to approximate setback limits for a stream.
1. Divide the stream cross section into segments for which the geometric properties may be easily solved and estimate a Manning's n value for each segment.
2. Compute the area, hydraulic radius, and conveyance (equation 10.31) of each segment for both the encroached and non-encroached segment.
3. Sum the conveyance for each cross-sectional segment to obtain the total conveyance for both the encroached and non-encroached conditions.
4. Set the total conveyance of the encroached cross section equal to the total conveyance of the non-encroached section and solve for the allowable encroachment by trial and error.

This method for approximating the allowable encroachment is based on the assumptions that the 100-year flood elevation has been established or can be approximated and that the energy grade line of the encroached and non-encroached sections remains unchanged. The accuracy of the results obtained using this method may be highly subject to the accuracy of the flood elevation used. In addition, since the method assumes no change in the energy grade line, the method should not be used near bridges or similar contraction-expansion areas.

For typical natural channel cross sections, the procedure may result in an equality that is very difficult to solve for the allowable encroachment dimensions. Morris (1984) provides a series of dimensionless graphs that are solved for the allowable encroachment as a percentage of the non-encroached overbank width. These graphs are based on an allowable flood elevation increase of 1 foot and assume a symmetrical cross section with triangular overbanks and equal encroachment on both overbanks. The limitations listed above for the general procedure also apply.

Because of the simplifying assumptions required, this approximate method will have limited applicability. Generally, only very small streams will satisfy the assumptions and the designer should use extreme caution to avoid misapplication.

10.10 Uniform Flow - Example Problems

Following are some example problems using the direct solution of the Manning's equation, for irregular channels and for grassed lined channels designed for stability and capacity.

Direct Solution Of Manning's Equation

Use Manning's equation to find the velocity, v, for an open channel with a hydraulic radius value of 0.6 ft, an n value of 0.020, and slope of 0.003 ft/ft. Solve using Figure 10-4.
1. Connect a line between the slope scale at 0.003 and the roughness scale at 0.020 and note the intersection point on the turning line.
2. Connect a line between that intersection point and the hydraulic radius scale at 0.6 ft and read the velocity of 2.9 ft/s from the velocity scale.

Irregular Channel (Example 1)

A channel consists of four subsections with the properties contained in the table on the next page. A discharge of about 3,500 cfs flows in the channel. Trial depths were developed and all parameters per subsection were computed based on the trial depth. The discharge was calculated and compared to the desired discharge. This was continued until the values matched within a set tolerance. Based on this table the maximum depth was found to be about 5.0 feet.

Grassed Channel Design Stability (Example 2)

A trapezoidal channel is required to carry 50 cfs at a bottom slope of 0.015 ft/ft. Find the channel dimensions required for design stability criteria (retardance Class D) for a grass mixture.
1. From Table 10-15, the maximum velocity, v_m, for a grass mixture with a bottom slope less than 5 percent is 4 ft/s.
2. Assume an n value of 0.035 and find the value of vR from Figure 10-6.
       vR = 5.4
3. Use equation 10.28 to calculate the value of R: R = 5.4/4 = 1.35 ft

	Sec. 1	Sec. 2	Sec. 3	Sec. 4
Topwidth, ft	25	75	15	75
Avg. Depth, ft	2.5	4.5	3	1
Area, ft^2	62.50	337.50	45.00	75.00
Wet. Perim., ft	25.5	75	15.1	75
Hyd. Radius, ft	2.45	4.50	2.98	1.00
Manning n	.045	.035	.035	.050
Conveyance	3,762	39,181	3,969	2,235

Total Conveyance	49,147
Total Area, ft^2	520
Discharge (slope = 0.005), cfs	3,489
Mean Velocity, ft/s	6.71

4. Use equation 10.29 to calculate the value of vR:
   $$vR = [1.49\ (1.35)^{5/3}\ (0.015)^{1/2}]/0.035 = 8.60$$
5. Since the vR value calculated in Step 4 is higher than the value obtained from Step 2, a higher n value is required and calculations are repeated. The results from each trial of calculations are presented below:

Assumed n Value	vR (Figure 10-6)	R (equation 10.28)	vR (equation 10.29)
0.035	5.40	1.35	8.60
0.038	3.8	0.95	4.41
0.039	3.4	0.85	3.57
0.040	3.2	0.80	3.15

   Select n = 0.040 for stability criteria.
6. Use Figure 10-4 to select channel dimensions for a trapezoidal shape with 3:1 side slopes.
   $$Qn = (50)\ (0.040) = 2.0 \qquad\qquad S = 0.015$$
   For b = 10 ft,   d = (10) (0.098) = 0.98 ft
   For b = 8 ft,   d = (8) (0.14) = 1.12 ft

   Select: b = 10 ft, such that R is approximately 0.80 ft
           z = 3   d = 1 ft
           v = 3.9 ft/s (equation 10.8)   Fr = 0.76 (equation 10.26)
           Flow is subcritical

Design capacity calculations for this channel are presented in Example 3 below.

Grassed Channel Design Capacity (Example 3)

Use a 10-ft bottom width and 3:1 side-slopes for the trapezoidal channel sized in Example 2 and find the depth of flow for retardance Class C.

1. Assume a depth of 1.0 ft and calculate the following (see Figure 10-6):

$$A = (b + zd) d = [10 + (3) (1)] (1)$$
$$A = 13.0 \text{ ft}^2$$
$$R = \{[b + zd] d\}/\{b + [2d (1 + z^2)^{0.5}]\}$$
$$R = \{[10 + (3)(1)] (1)\}/\{10 + [(2)(1)(1 + 3^2)^{0.5}]\} = 0.796 \text{ ft}$$

2. Find the velocity.

$$v = Q/A = 50/13.0 = 3.85 \text{ ft/s}$$

3. Find the value of vR.

$$vR = (3.85) (0.796) = 3.06$$

4. Using the vR product from Step 3, find Manning's n from Figure 10-6 for retardance Class C.

$$n = 0.047$$

5. Use Figure 10-3 or equation 10.16 to find the velocity for S = 0.015, R = 0.796, and n = 0.047.

$$v = 3.34 \text{ ft/s}$$

6. Since 3.34 ft/s is less than 3.85 ft/s, a higher depth is required and calculations are repeated. Results from each trial of calculations are presented below:

Assumed Depth (ft)	Area (ft²)	R (ft)	Velocity Q/A (ft/s)	vR	Manning's n (Fig. 10-6)	Velocity (Eq. 10.16)
1.0	13.00	0.796	3.85	3.06	0.047	3.34
1.05	13.81	0.830	3.62	3.00	0.047	3.39
1.1	14.63	0.863	3.42	2.95	0.048	3.45
1.2	16.32	0.928	3.06	2.84	0.049	3.54

7. Select a depth of 1.1 with an n value of 0.048 for design capacity requirements. Add at least 0.2 ft for freeboard to give a design depth of 1.3 ft. Design data for the trapezoidal channel are summarized as follows:

Vegetation lining = grass mixture, v_m = 4 ft/s
Q = 50 cfs
b = 10 ft, d = 1.3 ft, z = 3, S = 0.015 ft/ft
Top width = (10) + (2) (3) (1.3) = 17.8 ft
n (stability) = 0.040, d = 1.0 ft, v = 3.9 ft/s, Froude number = 0.76 (equation 10.26)
n (capacity) = 0.048, d = 1.1 ft, v = 3.45 ft/s, Froude number = 0.64 (equation 10.26)

10.11 Gradually Varied Flow

In reality all flow varies gradually or rapidly. The most common occurrence of gradually varied flow in storm drainage is the backwater created by some downstream control culverts, storm drain inlets, channel constrictions or the drawdown from a weir, dam or other overflow point. For these conditions, the flow depth will be different from normal depth in the channel and the water surface profile should be computed using backwater techniques.

Many computer programs are available for computation of backwater curves. The most general and widely used one dimensional flow programs are, HEC-2, developed by the U.S. Army Corps of Engineers (1982), and Bridge Waterways Analysis Model (WSPRO), developed for the Federal Highway Administration. These programs can be used to compute water surface profiles for both natural and artificial channels.

For prismatic channels, the backwater calculation can be computed manually using the direct step method, as presented by Chow (1959). For an irregular nonuniform channel, the standard step method is recommended, although it is a more tedious and iterative process. The use of HEC-2 is recommended for standard step calculations.

Cross sections for water surface profile calculations should be normal to the direction of flow. The number of sections required will depend on the irregularity of the stream and floodplain. In general, a cross section should be obtained at each location where there are significant changes in stream width, shape, or vegetal patterns. Sections should usually be no more than 4 to 5 channel widths apart or 100 ft apart for ditches or streams and 500 ft apart for floodplains, unless the channel is very regular.

Direct Step Method

The direct step method is limited to prismatic channels. A form for recording the calculations described below is presented in Table 10-7 (Chow, 1959).
1. Record the following parameters across the top of Table 10-7:

    Q = design flow, cfs n = Manning's n value
    S_o = channel bottom slope, ft/ft α = energy coefficient
    y_c = critical depth, ft y_n = normal depth, ft
2. Using the desired range of flow depths, y, recorded in column 1, compute the cross-sectional area, A, the hydraulic radius, R, and average velocity, v, and record results in columns 2, 3, and 4, respectively.
3. Compute the velocity head, $\alpha v^2/2g$, in ft, and record the result in column 5.
4. Compute specific energy, delta E, in ft, by summing the velocity head in column 5 and the depth of flow in column 1. Record the result in column 6.
5. Compute the change in specific energy, delta E, between the current and previous flow depths, record the result in column 7 (not applicable for row 1).
6. Compute the friction slope using the equation:

$$S_f = (n^2 \ v^2 \ )/(2.22 \ R^{4/3})   \text{(10.33)}$$

where: S_f =   friction slope, ft/ft
            n =   Manning's n value
            v =   average velocity, ft/s
            R =   hydraulic radius, ft

Record the result in column 8.
7. Determine the average of the friction slope between this depth and the previous depth (not applicable for row 1). Record the result in column 9.
8. Determine the difference between the bottom slope, S_o, and the average friction slope, $\overline{S}_f$, from column 9 (not applicable for row 1). Record the result in column 10.
9. Compute the length of channel between consecutive rows or depths of flow using the equation:

$$\text{delta } x = \text{delta } E/(S_o - S_f) = \text{Col. 7/Col. 10}   \text{(10.34)}$$

| Cross Section No. | Water Surface Elevation | | Area | Hydraulic Radius R | $R^{2/3}$ | n | K | $\bar{K}_t$ | $\dfrac{1000}{\bar{S}_f}$ | L | h_f | K^3/A^2 | α | V | $\alpha V^2/2g$ | $\Delta(\alpha V^2/2g)$ | h_o | Δ Water Surface Elevation |
	Assumed	Computed																
(1)	(2)	(3)	(4)	(5)	(6)	(7)	(8)	(9)	(10)	(11)	(12)	(13)	(14)	(15)	(16)	(17)	(18)	(19)

Table 10-7 Water Surface Profile Computation Form For The Direct Step Method

where: delta x = length of channel between consecutive depths of flow, ft
           delta E = change in specific energy, ft
           S_o =   bottom slope, ft/ft
           S_f =   friction slope, ft/ft
Record the result in column 11.

10. Sum the distances from the starting point to give cumulative distances, x, for each depth in column 1 and record the result in column 12.

Standard Step Method

The standard step method is a trial and error procedure applicable to both natural and prismatic channels. The step computations are arranged in tabular form, as shown in Table 10-8 and described below (Chow, 1959):

1. Record the following parameters across the top of Table 10-8:

   Q = design flow, cfs   n = Manning's n value
   S_o = channel bottom slope, ft/ft   α = energy coefficient
   k_e = eddy head loss coefficient, ft y_c = critical depth, ft
   y_n = normal depth, ft

2. Record the location of the measured channel cross sections and the trial water surface elevation, z, for each section in columns 1 and 2. The trial elevation will be verified or rejected based on computations of the step method.

3. Determine the depth of flow, y, based on trial elevation and channel section data. Record the result in column 3.

4. Using the depth from Step 3 and section data, compute the cross-sectional area, A, in ft², and hydraulic radius, R, in ft. Record results in columns 4 and 5.

5. Divide the design discharge by the cross-sectional area from Step 4 to compute the average velocity, v, in ft/s. Record the result in column 6.

6. Compute the velocity head, $\alpha v^2/2g$, in ft, and record the result in column 7.

7. Compute the total head, H, in ft, by summing the water surface elevation, z, in column 2 and velocity head in column 7. Record the result in column 8.

8. Compute the friction slope, S_f, using equation 10.33 and record the result in column 9.

9. Determine the average friction slope, $\overline{S}_f$, between the sections in each step (not applicable for row 1). Record the result in column 10.

10. Determine the distance between sections, delta x, and record the result in column 11.

11. Multiply the average friction slope, $\overline{S}_f$ (column 10), by the reach length, delta x (column 11), to give the friction loss in the reach, h_f. Record the result in column 12.

12. Compute the eddy loss using the equation:

$$h_e = (k_e\ v^2)/2g \qquad\qquad (10.35)$$

where: h_e =   eddy head loss, ft
           k_e =   eddy head loss coefficient, ft (for prismatic and regular channels, k_e = 0; for gradually converging and diverging channels, k_e = 0 to 0.1 or 0.2; for abrupt expansions and contractions, k_e = 0.5)
           v =   average velocity, ft/s (column 6)
           g =   acceleration due to gravity, 32.2 ft/s²

13. Compute the elevation of the total head, H, by adding the values of h_f and h_e (columns 12 and 13) to the elevation at the lower end of the reach, which is found in column 14 of the previous reach. Record the result in column 14.

Location _____

Q = ___ n = ___ S_o = ___ n = ___ k_o = ___ y_c = ___ y_n = ___													
Station (1)	z (2)	y (3)	A (4)	R (5)	v (6)	$nv^2/2g$ (7)	H (8)	$\bar{S}_l$ (9)	S_l (10)	Δx (11)	h_1 (12)	h_e (13)	H (14)
1.													
2.													
3.													
4.													
5.													
6.													
7.													
8.													
9.													
10.													
11.													
12.													
13.													
14.													
15.													
16.													
17.													
18.													
19.													
20.													
21.													
22.													

Table 10-8 Water Surface Profile Computation Form For The Standard Step Method

14. If the value of H computed above does not agree closely with that entered in column 8, a new trial value of the water surface elevation is used in column 2 and calculations are repeated until agreement is obtained. The computation may then proceed to the next step or section reported in column 1.

10.12 Gradually Varied Flow - Example Problems

Direct Step Method

Use the direct step method to compute a water surface profile for a trapezoidal channel using the following data:

Q = 400 cfs	B = 20 ft	z = 2
S = 0.0016 ft/ft	n = 0.025	α = 1.10

A dam backs up water to a depth of 5 ft immediately behind the dam. The upstream end of the profile is assumed to have a depth 1 percent greater than normal depth.

Results of calculations, as obtained from Chow (1959), are reported in Table 10-9. Values in each column of the table are briefly explained below.

1. Depth of flow, in ft, arbitrarily assigned values ranging from 5 to 3.4 ft.
2. Water area, in ft^2, corresponding to the depth, y, in column 1.
3. Hydraulic radius, in ft, corresponding to y in column 1.
4. Mean velocity, in ft/s, obtained by dividing 400 cfs by the water area in column 2.
5. Velocity head, in ft, calculated using the mean velocity from column 4 and an α value of 1.1.
6. Specific energy, E, in ft, obtained by adding the velocity head in column 5 to the depth of flow in column 1.
7. Change of specific energy, delta E, in ft, equal to the difference between the E value in column 6 and that of the previous step.
8. Friction slope, S_f, computed by equation 10.33, with n = 0.025, v as given in column 4, and R as given in column 3.
9. Average friction slope between the steps, $\overline{S}_f$, equal to the arithmetic mean of the friction slope computed in column 8 and that of the previous step.
10. Difference between the bottom slope, S_o, 0.0016 and the average friction slope, $\overline{S}_f$, in column 9.
11. Length of the reach, delta x, in ft, between the consecutive steps computed by equation 10.34 or by dividing the value of delta E in column 7 by the value of S_o - $\overline{S}_f$ in column 10.
12. Distance from the section under consideration to the dam site. This is equal to the cumulative sum of the values in column 11 computed for previous steps.

Standard Step Method

Use the standard step method to compute a water surface profile for the channel data and stations considered in the previous example. Assume the elevation at the dam site is 600 ft.

Results of the calculations, after Chow (1959), are reported in Table 10-10. Values in each column of the table are briefly explained below:

1. Section identified by station number such as "station 1 + 55." The locations of the stations are fixed at the distances determined in the previous example to compare the procedure with that of the direct step method.
2. Water surface elevation, z, at the station. A trial value is first entered in this column; this will be verified or rejected on the basis of the computations made in the remaining columns

y (1)	A (2)	R (3)	v (4)	αv²/2g (5)	E (6)	ΔE (7)	S_f (8)	S̄_f (9)	S_o − S̄_f (10)	Δx (11)	x (12)
5.00	150.00	3.54	2.667	0.1217	5.1217	--	0.000370	--	--	--	--
4.80	142.00	3.43	2.819	0.1356	4.9356	0.1861	0.000433	0.000402	0.001198	155	155
4.60	134.32	3.31	2.979	0.1517	4.7517	0.1839	0.000507	0.000470	0.001130	163	318
4.40	126.72	3.19	3.156	0.1706	4.5706	0.1811	0.000590	0.000553	0.001047	173	491
4.20	119.20	3.08	3.354	0.1925	4.3925	0.1781	0.000705	0.000652	0.000948	188	679
4.00	112.00	2.96	3.572	0.2184	4.2184	0.1741	0.000850	0.000778	0.000822	212	891
3.80	104.80	2.84	3.814	0.2490	4.0490	0.1694	0.001020	0.000935	0.000665	255	1,146
3.70	101.30	2.77	3.948	0.2664	3.9664	0.0826	0.001132	0.001076	0.000524	150	1,304
3.60	97.92	2.71	4.085	0.2856	3.8856	0.0808	0.001244	0.001180	0.000412	196	1,500
3.55	96.21	2.65	4.150	0.2950	3.8450	0.0390	0.001310	0.001277	0.000323	123	1,623
3.50	94.50	2.63	4.233	0.3067	3.8067	0.0391	0.001302	0.001346	0.000254	154	1,777
3.47	93.40	2.63	4.270	0.3131	3.7831	0.0236	0.001427	0.001405	0.000195	121	1,898
3.44	92.45	2.61	4.326	0.3202	3.7602	0.0229	0.001471	0.001449	0.000151	152	2,050
3.42	91.80	2.60	4.357	0.3246	3.7446	0.0156	0.001500	0.001486	0.000114	137	2,187
3.40	91.12	2.59	4.390	0.3292	3.7292	0.0154	0.001535	0.001510	0.000082	100	2,375

Note: $Q = 400$ cfs $n = 0.025$ $S_o = 0.0016$ $\alpha = 1.10$ $y_c = 2.22$ ft $y_n = 3.36$ ft

Reference: Chow (1959)

Table 10-9 Direct Step Method Results For Example

of the table. For the first step, this elevation must be given or assumed. Since the elevation of the dam site is 600 ft and the height of the dam is 5 ft, the first entry is 605.00 ft. When the trial value in the second step has been verified, it becomes the basis for the verification of the trial value in the next step, and the process continues.

3. Depth of flow, y, in ft, corresponding to the water surface elevation in column 2. For instance, the depth of flow at station 1 + 55 is equal to the water surface elevation minus the elevation at the dam site minus the distance from the dam site times bed slope.

$$605.048 - 600.00 - (155)(0.0016) = 4.80 \text{ ft}$$

4. Water area, A, in ft^2, corresponding to y in column 3.
5. Hydraulic radius, R, in ft, corresponding to y in column 3.
6. Mean velocity, v, equal to the discharge, 400 cfs, divided by the water area in column 4.
7. Velocity head, in ft, corresponding to the velocity in column 6 and an α value of 1.1.
8. Total head, H, equal to the sum of z in column 2 and the velocity head in column 7.
9. Friction slope, S_f, computed by equation 10.33, with n = 0.025, v from column 6, and R from column 5.
10. Average friction slope through the reach, $\overline{S}_f$, between the sections in each step, approximately equal to the arithmetic mean of the friction slope just computed in column 9 and that of the previous step.
11. Length of the reach between the sections, delta x, equal to the difference in station numbers between the stations.
12. Friction loss in the reach, h_f, equal to the product of the values in columns 10 and 11.
13. Eddy loss in the reach, h_e, equal to zero.
14. Elevation of the total head, H, in ft, computed by adding the values of h_f and h_e in columns 12 and 13 to the elevation at the lower end of the reach, which is found in column 14 of the previous reach. If the value obtained does not agree closely with that entered in column 8, a new trial value of the water surface elevation is assumed until agreement is obtained. The value that leads to agreement is the correct water surface elevation. The computation may then proceed to the next step.

10.13 Hydraulic Jump

A hydraulic jump can occur when flow passes rapidly from supercritical to subcritical depth. The evaluation of the hydraulic jump should consider the high energy loss and erosive forces that are associated with the jump. For rigid-lined facilities such as pipes or concrete channels, the forces and the change in energy can affect the structural stability or the hydraulic capacity. For grass-lined channels, unless the erosive forces are controlled, serious damage can result. Control of jump location is usually obtained by check dams or grade control structures that confine the erosive forces to a protected area. Flexible material such as riprap, or rubble usually affords the most effective protection.

The analysis of the hydraulic jump inside storm drains must be approximate, because of the lack of data for circular, elliptical, or arch sections. The jump can be approximately located by intersecting the energy grade line of the supercritical and subcritical flow reaches. The primary concerns are whether the pipe can withstand the forces, which may separate the joint or damage the pipe wall, and whether the jump will affect the hydraulic characteristics. The effect on pipe capacity can be determined by evaluating the energy grade line, taking into account the energy lost by the jump. In general, for Froude numbers less than 2.0, the loss of energy is less than 10 percent. French (1985) provides semi-empirical procedures to evaluate the hydraulic jump in circular and other non-rectangular channel sections.

Station (1)	z (2)	y (3)	A (4)	R (5)	v (6)	$\alpha v^2/2g$ (7)	H (8)	S_f (9)	$\bar{S}_f$ (10)	Δx (11)	h_f (12)	h_e (13)	H (14)
0 + 00	605.000	5.00	150.00	3.54	2.667	0.1217	605.122	0.000370	--	--	--	--	605.122
1 + 55	605.048	4.80	142.08	3.43	2.819	0.1356	605.184	0.000433	0.000402	155	0.062	0	605.184
3 + 10	605.109	4.60	134.32	3.31	2.979	0.1517	605.261	0.000507	0.000470	163	0.077	0	605.261
4 + 91	605.186	4.40	126.72	3.19	3.156	0.1706	605.357	0.000590	0.000553	173	0.096	0	605.357
6 + 79	605.286	4.20	119.28	3.08	3.354	0.1925	605.479	0.000705	0.000652	188	0.122	0	605.479
8 + 91	605.426	4.00	112.00	2.96	3.572	0.2184	605.644	0.000850	0.000778	212	0.165	0	605.644
11 + 46	605.633	3.80	104.80	2.84	3.814	0.2490	605.882	0.001020	0.000935	255	0.238	0	605.882
13 + 04	605.786	3.70	101.30	2.77	3.948	0.2664	606.052	0.001132	0.001076	158	0.170	0	606.052
15 + 00	605.999	3.60	97.92	2.71	4.085	0.2856	606.285	0.001244	0.001188	196	0.233	0	606.285
16 + 23	606.147	3.55	96.21	2.60	4.158	0.2950	606.442	0.001310	0.001277	123	0.157	0	606.442
17 + 77	606.343	3.50	94.50	2.65	4.233	0.3067	606.650	0.001382	0.001346	154	0.208	0	606.650
18 + 98	606.507	3.47	93.40	2.63	4.278	0.3131	606.820	0.001427	0.001405	121	0.170	0	606.820
20 + 50	606.720	3.44	92.45	2.61	4.326	0.3202	607.040	0.001471	0.001449	152	0.220	0	607.040
21 + 07	606.919	3.42	91.80	2.60	4.357	0.3246	607.244	0.001500	0.001486	137	0.204	0	607.244
23 + 75	607.201	3.40	91.12	2.59	4.380	0.3292	607.530	0.001535	0.001518	188	0.285	0	607.530

Note: Q = 400 cfs   n = 0.025 S_o = 0.0016 α = 1.10 h_e = 0 y_c = 2.22 ft y_n = 3.36 ft

Reference: Chow (1959)

Table 10-10 Standard Step Method Results For Example

For long box culverts with a concrete bottom, the concerns about jump are the same as for storm drains. However, the jump can be adequately defined for box culverts/drains and for spillways using the jump characteristics of rectangular sections.

The relationship between variables for a hydraulic jump in rectangular sections can be expressed as:

$$d_2 = -(d_1/2) + [(d_1^2/4) + (2v_1^2 d_1/g)]^{1/2} \qquad (10.36)$$

where: d_2 = depth below jump, ft
         d_1 = depth above jump, ft
         v_1 = velocity above jump, ft/s
         g = acceleration due to gravity, 32.2 ft/s^2

A nomograph for solving equation 10.36 is presented in Figure 10-8. Additional details on hydraulic jumps can be found in HEC-14 (1983), Chow (1959), Peterka (1978), and French (1985).

10.14 Construction And Maintenance Considerations

An important step in the design process involves identifying whether special provisions are warranted to properly construct or maintain proposed facilities. Open channels rapidly lose hydraulic capacity without adequate maintenance. Maintenance may include repairing erosion damage, mowing grass, cutting brush, and removing sediment or debris. Brush, sediment, or debris can reduce design capacity and can harm or kill vegetative linings, thus creating the potential for erosion damage during large storm events. Maintenance of vegetation should include the appropriate application of fertilizer, irrigation during dry periods, and reseeding or resodding to restore the viability of damaged areas. Implementation of a successful maintenance program is directly related to the accessibility of the channel system and the easements necessary for maintenance activities.

10.15 Stability Of Channels

Overview

In addition to being an integral component of storm drainage systems, urban streams provide important environmental benefits. Natural, or undisturbed, streams possess numerous environmental benefits that are derived from the diversity and stability of the stream and its associated riparian ecosystem. Urbanization is associated with processes that disturb natural streams, increasing velocity and volume of runoff resulting in increased erosion and sediment problems within the stream system. The purpose of this section is to provide some guidelines for ensuring the stability of streams and for designing drainage projects and streambank protection measures with environmental benefits.

Channels in both natural and man made states are not naturally stable and unchanging. All channels attempt to adjust to the ever changing combination of flows, sediment loads, topography, climate and bed and bank conditions imposed on them. Any change in these variables will result in some channel modification. If a channel is designed without due regard for a more stable configuration that channel will be stable only as long as rigid bank linings and bed stabilizers hold it in check and/or channel dredging removes excess sediment. Channels have three degrees of freedom or three ways of becoming unstable: width changes, depth changes and planform or horizontal instability (horizontal migration for example). There are many detailed treatments of

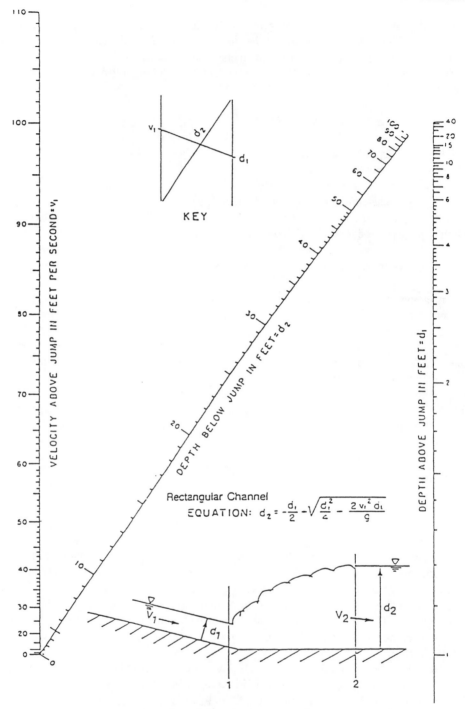

**Figure 10-8 Nomograph For Solving The Rectangular Channel
Hydraulic Jump Equation**

Source: U.S. Dept. Of Interior, 1973

stream sediment transport and stable channel design from both a river engineering and geomorphic approach, and a detailed treatment is beyond the scope of this book. See for example SLA (1982), ASCE (1977), Simons and Senturk (1976), USCOE (1989), or Schumm (1977). This section will summarize some general ways to consider and recognize stream stability problems, considerations for stable channel design, and will refer to other sources of detailed information.

Channel Response To Change

Lane (1955), using methods from fluvial geomorphology, has provided a simple proportion which allows for a qualitative "feel" for channel response to changes.

$$QS \propto Q_s D_{50} \tag{10.37}$$

where: $Q =$ discharge
        $S =$ channel slope
        $Q_s =$ sediment discharge
        $D_{50} =$ median grain size

In a proportion if one side increases or decreases the other must do likewise. Consider, for example the response of a stream or channel to upstream construction and greatly increased sediment yields. Q_s increases but Q is constant. Therefore, to maintain the proportionality the slope (S) must increase and D_{50}, the median sediment size transported, must decrease. The result is channel aggradation (sedimentation) near the inflow point of the sediment until such time as the slope is steep enough to transport the excess sediment.

Urban development increases Q. To adjust, S must decrease or Q_s or D_{50} or both must increase. The result is channel degradation (erosion), widening, deepening (depending on the relative soils and vegetation in bed and banks) and eventual slope flattening. If a sinuous channel is straightened the slope is increased, and a response similar to flow increase will normally result. Uncontrolled urban development normally both shortens channels and increases discharge.

If a large retention pond is built in line with a stream carrying significant sediment the result is channel degradation below the dam due to decreased sediment transport (Q_s) and the responding slope decrease. If a main channel is dredged to make it deeper the base level is dropped for each of the tributaries. The result of this base level lowering is an increased gradient (S) in the lower portions of the tributaries. To balance this impact Q_s increases inducing headcutting or rapid erosion of the tributaries.

Simons and Senturk (1976) give a compilation of channel response proportions based on the work and experience of many authors:
- depth, D, is proportional to discharge, Q;
- channel width, W, is proportional to both Q and Q_s;
- channel shape, expressed as width to depth ratio (W/D), is directly proportional to Q_s; and
- sinuosity, the measure of tendency to meander, is proportional to valley slope, S_v, and inversely proportional to Q_s.

The best way to avoid instability problems in urban stream channels and to maximize environmental benefits is to maintain streams in as natural a condition as possible, and when channel modification is necessary, to avoid altering channel dimensions, channel alignment, and channel slope as much as possible. When channel modification is necessary, the following set of guidelines should be followed to minimize erosion problems and maximize environmental benefits.
- Avoid channel enlargement whenever possible.
- When channels must be enlarged, avoid streambed excavation that would significantly increase streambed slope or streambank height.

- When channel bottom widths are increased more than 25 percent, provide for a low flow channel to concentrate flows during critical low flow periods.
- Avoid channel realignment whenever possible.

Assessing Streams

Whether existing erosion on streams needs to be repaired and protected from additional erosion depends on several factors, including type, extent, severity, and location of the erosion. Stream-banks that appear to be well-vegetated and stable usually are, but often banks that appear to be unstable and eroding may be relatively stable. Differentiating between stable and eroding banks involves observing and evaluating evidence (that may at times be contradictory) about bank conditions. General conditions frequently associated with stable and eroding banks are summarized in Table 10-11 (Nunnally and Keller, 1979).

Table 10-11 General Conditions For Stable And Eroding Banks

Characteristics	Stable Bank	Eroding Bank
bank slope	not vertical; may be compound with vegetated berm at toe	often vertical or near vertical; may have moss or sod or other failed material at toe
bank cover	may have variety of vegetation growing on slope, including ferns or moss	general absence of vegetation
trees	often has trees growing on bank or on the bed at toe	standing live or dead trees inside the bank line, often leaning toward channel, fallen trees may obstruct flow
bankline	relatively uniform or smoothly curing	irregular, sometimes with scalloped appearance
sediment	sediment located in bars, bars may be partially stabilized by vegetation, especially along bank toe	entire bed may be covered with sediment; bars not stabilized

---

Judging the severity of streambank erosion is a somewhat subjective task that may involve the amount or the rate of erosion, relative location of the problem, and even the type of erosion involved. Erosion is a natural process and several feet of lateral streambank erosion in a bend, or meander, might be no cause for concern. On the other hand, if the erosion threatens a culvert installation it might be considered severe. Table 10-12 provides typical basic guidance for establishing severity, in midwestern and eastern climates, but it should be used with considerable discretion.

Various modifications to streams have both good and adverse impacts. Table 10-13 gives a summary of these.

Table 10-12 Degree Of Erosion Severity

Degree of Erosion	Characteristics
stable	little or no evidence of erosion; if eroding banks are present, they are small in extent (less than 5 linear feet or 50 square feet) and rates are modest (less than 1/2 foot per year); greater erosion may be tolerated at bends if it causes no associated problems.
moderate	extent of problem or rate of erosion exceeds criteria for stable class, but is less than severe.
severe	erosion covers large area of bank and is occurring at a rate in excess of one foot per year or a rate that is unacceptable for safety, environmental, or economical reasons.

Stable Channel Design Approaches

There are four main approaches to consider stability in channel design: maximum permissible velocity approach, tractive stress approach, sediment transport approach and regime equations (SCS, 1977). A detailed treatment of all of these methods is beyond the scope of this book. The SCS (1977) or Simons and Senturk (1976) should be consulted for details of applications. Use of any of these approaches should be tempered with engineering judgment and experience. There is no substitute for understanding the stream types in the area and for knowing what has happened to similar streams when they have been designed, impacted by development or modified. The sediment transport method will not be covered in this book because the level of skill involved in understanding the methods and the complexities of sediment transport are beyond the needed skills of the normal urban storm water manager.

Maximum Permissible Velocity Method

The final design of artificial open channels should be consistent with the velocity limitations for the selected channel lining. Maximum velocity values for selected lining categories for bare earth channels are presented in Table 10-14. Seeding and mulch should only be used when the design value does not exceed the allowable value for bare soil. Velocity limitations for vegetative linings are reported in Table 10-15. Vegetative lining calculations are presented in Section 10.8 and riprap procedures are presented in Section 10.16. If lining costs are expensive there are methods available to design channels to minimize such costs (Trout, 1982 and Loganathan, 1991).

For D_{75} grain sizes greater than or equal to 1 mm, which act as discrete particles, the permissible velocities for both sediment laden water and sediment free water can be found from the following equations fit to SCS (1977) information. Sediment laden water is defined as having fine suspensions of sediment carried in the flow at concentrations greater than 20,000 ppm by weight. Flows with concentrations of fine sediment less than 1,000 ppm by weight are considered sediment free. Permissible velocities for flows in between can be found by linear interpolation between the two values found in equations 10.38 and 10.39.

Table 10-13 Consideration Of Channel Modifications (Source: COE, 1992)

POSITIVE IMPACTS	NEGATIVE IMPACTS
RESERVOIRS, PONDS AND DETENTION	
• flood flow and stage reduction at flood flow • downstream damage reduction • source of water for multi-uses • recreation uses	• loss of land • filling of reservoir with sediment • loss of wetlands • modification of stream flow regime • conveyance encroachment and risk • water quality changes, thermal impacts • evaporation losses in arid climates • elimination in fish spawning and movement
LEVEES AND FLOOD WALLS	
• no flooding from exterior until exceedance flows reached • protection of property values	• may induce flooding up- and downstream of site • potential of sudden large loss if exceeded • interior flooding and pumping needed • maintenance and observation
CHANNELIZATION AND CHANNEL WIDENING	
• flood stage reductions for all flows • local damage reduction through project reach	• potential impact on fish spawning and habitat • changed sediment transport • increased maintenance • induced flooding downstream if loss of floodplain storage
DIVERSIONS	
• flood stage reductions for all flows • local damage reduction in project reach	• increased maintenance • induced flooding downstream if loss of floodplain storage
NONSTRUCTURAL EFFORTS	
• individual structures protected • few environmental impacts • low cost	• high residual damage to infrastructure • emergency responses required for specific events • individual maintenance required

Source: COE, 1992

Table 10-14 Maximum Velocities For Comparing Lining Materials

Material	Clear Water	Water with Colloidal Silt	Water with Non-colloidal Silt, Sand or Gravel
Fine Sand (colloidal)	1.5	2.5	1.5
Sand Loam (noncolloidal)	1.45	2.5	2.0
Silt Loam (noncolloidal)	2.0	3.0	2.0
Alluvial Silt (noncolloidal)	2.0	3.5	2.0
Alluvial Silt (colloidal)	3.75	5.0	3.0
Firm Loam	2.5	3.5	2.25
Volcanic Ash	2.5	3.5	2.0
Fine Gravel	2.5	5.0	3.75
Stiff Clay (very colloidal)	3.75	5.0	3.0
Graded Loam to Cobbles (noncolloidal)	3.75	5.0	5.0
Graded Silt to Cobbles (colloidal)	3.75	5.0	3.0
Coarse Gravel	4.0	6.0	6.5
Cobbles and Shingles	5.0	5.5	6.5
Shales and Hard Pans	6.0	6.0	5.0

Source: Fortier and Scoby, 1926.

$$V_s = 2.6356\ D_{75}{}^{0.30596} \qquad D_{75} > 0.4\ mm \qquad\qquad (10.38)$$
$$V_c = 1.4653\ D_{75}{}^{0.37990} \qquad D_{75} > 2.0\ mm \qquad\qquad (10.39)$$

where: V_s = sediment laden water permissible velocity, ft/s
$\quad\quad\ V_c$ = sediment free water permissible velocity, ft/s
$\quad\quad\ D_{75}$ = sediment particle size for which 75% of particles by weight are smaller, mm

For sizes less than the indicated limits which act as discrete particles use 2.0 ft/s for the maximum permissible velocity without adjustments. For other sizes and types of soil the following adjustments should be made:

• adjust all channels types for both alignment and depth;
• adjust channels with soils which behave as discrete particles for side slope;
• apply a frequency adjustment factor for storms whose design frequency is less frequent than the 10 percent flood (more than 10-year return period) and the bed or bank materials do not operate as discrete particles; and
• apply a density correction to channels for all soil types except clean sands and gravels with less than 5 percent passing the #200 sieve.

For deeper channels the velocity can be multiplied by 1.1 for 4 foot depth and 1.4 for 18 foot depth. Linear interpolation between these depths is acceptable. If the channel is curved the

Table 10-15 Maximum Velocities For Vegetative Channel Linings

Vegetation Type	Slope Range (%)[1]	Maximum Velocity[2] (ft/s)	
		Erosion Resistant Soils	Easily Eroded Soils
Bermudagrass	0-5	8	6
	5-10	7	5
	>10	6	4
Kentucky bluegrass	0-5	7	5
Buffalo grass	5-10	6	4
	>10	5	3
Grass mixture	0-5[1]	5	4
	5-10	4	3
Lespedeza sericea Kudzu, alfalfa	0-5[3]	3.5	2.5
Annuals[4]	0-5	3.5	2.5
Sod		4.0	4.0
Lapped sod		5.5	5.5

Source: USDA, TP-61, 1954.

[1]   Do not use on slopes steeper than 10 percent except for side-slope in combination channel.
[2]   Use velocities exceeding 5 ft/s only where good stands can be established and maintained.
[3]   Do not use on slopes steeper than 5 percent except for side-slope in combination channel.
[4]   Annuals - used on mild slopes or as temporary protection until permanent covers are established.

permissible velocity should be multiplied by a factor based on a measured radius of curvature (to the channel centerline) and channel width. Table 10-16 gives appropriate values.

For bank materials which behave as discrete particles (noncolloidal) an adjustment in permissible velocity should be made based on side-slope of the bank. Table 10-17 gives appropriate values.

Frequency factor corrections for channels whose bed or bank materials do not act as discrete particles, and the design discharge is less frequent than the 10 percent storm, can be made using the following equation:

$$C_f = 1.705 - 0.3085 \ln (F) \qquad\qquad (10.40)$$

where: C_f = factor to multiply by permissible velocity
        F = flood frequency in percent chance (100 year storm = 1.0)

Density corrections should be made for all soils except clean sands and gravels with less than 5 percent material passing the #200 sieve. Corrections can be made on the basis of void ratio of the soil using the following equations:

SM, SC, GM and GC soils	$C_d = -0.617e + 1.417$	(10.41)
CL and ML soils	$C_d = -0.567e + 1.480$	(10.42)
CH and MH soils	$C_d = -0.358e + 1.361$	(10.43)

where: C_d = correction factor multiplied by the permissible velocity
         e = soils void ratio = $G(62.4/\Gamma_d) - 1$
         G = specific gravity = specific weight of material/specific weight of water
         Γ_d = dry density of solid material and voids, lb/ft^3

Table 10-16 Adjustment Factors For Channel Alignment

Curve Radius/Top Width	Correction Factor
16	1.00
14	0.99
12	0.96
10	0.93
8	0.89
6	0.81
5	0.70

Source: SCS TR 25, 1977

Table 10-17 Adjustment Factors For Channel Side-Slope

Cotangent of Slope Angle (x:1)	Correction Factor
1.5	0.50
2.0	0.72
2.5	0.82
3.0	0.86
4.0	0.90
>4.0	1.00

Source: SCS TR 25, 1977

Application of this method involves determination of design discharge and approximation of sediment concentration. Based on the bed and bank materials and the channel configuration, determine if the permissible velocity method is applicable. Compare the design velocities with the permissible velocities. If the design velocities are greater than the permissible velocities either reconfigure the channel or provide bed and bank protection as applicable.

Tractive Stress Method

Channels become unstable when the ability of the material in the bed and banks is overcome by the shearing force of the water. The theoretical point shear stress of the water on the boundary of an infinitely wide channel can be expressed as:

$$\tau = 62.4 \, D \, S_t \tag{10.44}$$

where: τ = average boundary shear stress, lbs/ft^2
        D = flow depth, ft
        S_t = friction slope due to particle resistance, ft/ft
        62.4 is the specific weight of water, lbs/ft^3.

S_t is calculated by "partitioning" the friction loss among the various components that go into Manning's n. The base value of Manning's n, based only on particle size, n_t, can be found from:

$$n_t = D_{75}{}^{1/6}/39 \qquad (10.45)$$

D_{75} is in inches in this equation. Then S_t can be found from a proportion as:

$$S_t = (n_t/n)^2 \, S \qquad (10.46)$$

where: n and S are the channel total friction slope and Manning's n

The maximum shear force on the bed of the channel is about 0.97τ while the maximum on the side slope of a trapezoidal channel is about 0.75τ. Shear coefficients can get as high as 2.5 on the outside of sharp bends (Ippen & Drinker, 1962). An approximate correction for the actual maximum shear stress on the downstream outer bank of a curved reach can be obtained from the empirical equation:

$$\tau_b = [-0.133 \, (R_c/W) + 2.224] \, \tau \qquad (10.47)$$

where: τ_b = shear stress in the bend, lbs/ft^2
        R_c = bend radius to channel centerline, ft
        W = topwidth, ft

For turbulent flood conditions in an urban stream the ability of a non-cohesive soil particle (in the size range $0.25" < D_{75} < 5.0"$) to resist movement depends on its density, position and size. A "critical" shear stress for particles on the bed can be defined for these conditions as:

$$\tau_c = 0.4 \, D_{75} \qquad (10.48)$$

where: τ_c = critical shear stress, lbs/ft^2
        D_{75} = sediment particle size for which 75% of particles by weight are smaller, inches
               ($0.25" < D_{75} < 5.0"$)

If the particle or stone is on a side slope of angle Θ and its natural angle of repose (the angle of a dumped pile of such stone) is Φ then the critical shear stress for a particle on a side slope is given as:

$$\tau_{ss} = \tau_c \, [(z^2 - \cot^2\Phi)/(1+z^2)]^{1/2} = \tau_c \, K_{ss} \qquad (10.49)$$

where: τ_{ss} = critical shear stress on a side slope, lbs/ft^2
        Θ = side slope angle, deg.
        Φ = natural angle of repose of sediment, deg.

The angle of repose (Φ) for particles in this size range can be found from the set of equations below for descriptive shape factors of the particles:

Very Rounded $\Phi = 33.82 - 2.09 \, D_{75} + 21.83 \log (D_{75})$ (10.50)
Moderately Rounded $\Phi = 34.96 - 1.83 \, D_{75} + 18.95 \log (D_{75})$ (10.51)
Slightly Rounded $\Phi = 35.14 - 0.76 \, D_{75} + 12.80 \log (D_{75})$ (10.52)

Slightly Angular	$\Phi = 37.63 - 1.71\ D_{75} + 15.22\ \log\ (D_{75})$	(10.53)
Moderately Angular	$\Phi = 39.05 - 1.73\ D_{75} + 13.59\ \log\ (D_{75})$	(10.54)
Very Angular	$\Phi = 40.14 - 1.51\ D_{75} + 11.11\ \log\ (D_{75})$	(10.55)

where: Φ = natural angle of repose of sediment, deg.

D_{75} = sediment particle size for which 75% of particles by weight are smaller, inches $(0.25" < D_{75} < 5.0")$

If the density is different from 160 lb/ft^3 the critical tractive stress should be multiplied by:

$$T = (\Gamma_s - 62.4)/97.6 \qquad (10.56)$$

where: T = correction factor

Γ_s = specific weight of particle, lb/ft^3

For grain sizes finer than $D_{75} = 0.25"$ the TP-25 (SCS, 1977) reference should be consulted.

Regime Equations For Channel Proportions

Regime methods are based on the development of empirical equations from observed stream data. These methods can yield unrealistic answers when blindly applied to streams not similar to the observed streams of the database. Because regime equations do not directly account for physical processes in their regressions and are not based on physical principles which can then be both relied upon and extrapolated with more comfort, they have been often criticized. However they have the advantage of being based on true observations rather than theory and thus bring a certain comfort level.

Determination of stable channel widths or width to depth ratios can be difficult. Flow area functions are generally more accurate. Local measurements on similarly situated channels are the best source of data. Regime equations can be useful in helping to determining channel proportions when other information is not available and the channel does not have rigid boundaries.

Regime equations are developed from observations of data and are often combined with physical relationships. They can be very helpful in stability analysis on urban streams and rivers as the designer considers the potential impacts of stream development and modifications. They typically apply to alluvial channels like those found most often in the west and southwest. They can be applied to any channel whose boundary contains a layer of loose material of the same type that is moved along the bed (termed the "bed load" or "bed material load"). There are a large number of regime equations and since they are primarily empirical in nature they cannot be readily applied beyond the conditions for which they were derived. Some of the better examples and discussions from the literature include: Lacey (1930), Blench (1966), Ackers and White (1973), Hey, et al., (1982), Ackers (1972), White, et al., (1982), Chien (1955, 1957) and Inglis (1948).

Simons And Albertson Equations - The equations provided below were developed by Simons and Albertson and apply generally to alluvial channels though most of the data were derived in the west. According to Simons and Albertson the Froude number F should be less than 0.3 for stable channels in alluvium. F is expressed as $V/(gD)^{1/2}$.

$D = 1.23\ R$ (R from 1 to 7 feet)	(10.57)
$D = 2.11 + 0.934\ R$ (R from 7 to 12 feet)	(10.58)
$W = 0.9\ P$	(10.59)
$W = 0.92\ T - 2.0$	(10.60)

C_1 - C_5 Coefficients by Channel Type

	A	B	C	
$P = C_1 Q^{0.512}$	3.30	2.51	2.12	(10.61)
$R = C_2 Q^{0.361}$	0.37	0.43	0.51	(10.62)
$A = C_3 Q^{0.873}$	1.22	1.08	1.08	(10.63)
$V = C_4 (R^2 S)^{1/3}$	13.9	16.1	16.0	(10.64)
$W/D = C_5 Q^{0.151}$	6.5	4.3	3.0	(10.65)

where: P = wetted perimeter, ft
        R = hydraulic radius, ft
        A = flow area, ft^2
        V = mean channel velocity, ft/s
        D = mean depth, ft.
        W = mean channel width, ft
        T = channel top width, ft
        C_i = coefficients by different channel types defined as:
               A = sand bed and sand banks
               B = sand bed and cohesive banks
               C = cohesive bed and banks (cohesive PI > 7)

Gravel Bed Equations - The following set of equations are based on hydraulic geometry relationships and are most applicable to gravel-bed streams. They are based on data on 70 streams in the Alberta, Canada area (Hey, et al., 1982, Bray, 1975):

$$B = 2.38 \, Q_2^{0.527} \qquad\qquad (10.66)$$
$$A = 0.632 \, Q_2^{0.860} \qquad\qquad (10.67)$$
$$d = 0.266 \, Q_2^{0.333} \qquad\qquad (10.68)$$
$$v = 1.58 \, Q_2^{0.140} \qquad\qquad (10.69)$$

and additionally:

$$B/d = 8.95 \, Q_2^{0.194} \qquad\qquad (10.70)$$
$$S = 0.0354 \, Q_2^{-0.342} \qquad\qquad (10.71)$$

Planform estimates can be made. Slopes falling above the line indicated by: $S = 0.06 \, Q_2^{-0.44}$ will tend to be braided rather than meandering.

Neill Method - Neill (1984) has developed a hybrid approach to regime analysis combining both physical considerations and regime supplements. His approach can be used for trapezoidal channel design or stability checking. It works best for channels in the low transport range and which are neither aggrading nor degrading actively. High sediment transport can cause the slope to increase greatly and adjustments in the procedure for this are only approximate. Braided or multi-channeled streams would have a higher slope and be wider than indicated. The method should not be applied to these type of streams (COE, 1990).

The basic approach is to: (1) determine a stable slope, (2) determine a channel width, (3) determine a flow depth, and (4) check the results against other stable channel methods. It is normally an iterative procedure.

The design discharge should be one that can remain stable. Therefore using design discharges above about the 10-year flood will result in a channel too wide and flat to be stable and result in sub-meandering within the channel bottom. If a larger capacity is desired, Neill recommends using the greater of 10-year or 50 percent of the design discharge and providing overbank storage and high flow channels for the remainder of the design flow.

Channel slope is based on both critical tractive force and regime relationships blended together. Figure 10-9 gives the chart for slope selection based on dominant discharge. The slope value should be multiplied by 1.7 for moderate sediment transport and 3.0 for high sediment transport. Moderate transport includes: active bedload several times per year, channel bars and some basin gullying of sandy soil. High transport includes: active bed for a substantial portion of the year, frequent bars and stream splitting or braiding, and generally sheet and rill erosion in the basin or caving banks.

Channel width is based on regime concepts alone. It is very rough and its results should be compared to other local streams. The width equation is:

$$W = C_w \, Q^{1/2} \tag{10.72}$$

Where: W = width, ft

Q = dominant discharge, cfs

C = 1.6 for high bank resistance, 2.1 for moderate bank resistance, 2.7 for low bank resistance

Channel depth is taken from Figure 10-10 on a preliminary basis. More accurate determination of channel cross sectional geometry can be obtained through the use of various applicable resistance equations. This figure can be used for preliminary selection and quick comparison of alternatives and different Q values.

Checking of the relationships can be done through the use of three charts prepared by Neill for: allowable mean velocity (Figure 10-11), allowable shear stress on cohesive materials (Figure 10-12) and limiting slope (Figure 10-13).

Stable Channel Design Examples

Permissible Velocity Example - A natural stream channel is to be modified to convey the 2 percent chance flood (50-year return period). Hydraulic analysis indicates a trapezoidal channel with 2:1 side slopes, a 40 foot bottom width, a design flow depth of 8.7 feet and design velocity of 5.45 ft/s. The material is glacial outwash, clean sandy gravel soil with a D_{75} of 2.25 inches. Sediment yield estimates and visual inspection of the channel indicate the transport concentration is less than about 500 ppm. The channel will be excavated as a straight reach with one bend of 600 foot radius of curvature. Determine the allowable velocity and reach stability.

1. The D_{75} size in mm is 57.15. From equation 10.73 for clear flow conditions:

$$V_c = 1.4653 \, D_{75}{}^{0.3799} \quad D_{75} > 2.0 \text{ mm} \tag{10.73}$$

$V_c = 1.4653 \, (57.15)^{0.3799}$

$V_c = 6.8$ ft/s

2. Correction factors are applied:
   - Depth correction factor = 1.22
   - Side slope correction factor = 0.72
   - Alignment correction factor

     curve radius/topwidth = (600)/74.8 = 8.02

     correction factor = 0.89
   - Frequency and density correction factors do not apply

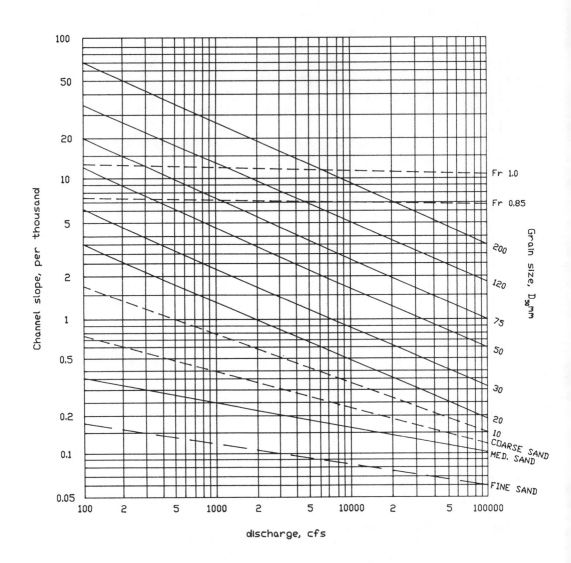

Figure 10-9 Slopes for Stable Channels (Neill, 1984)

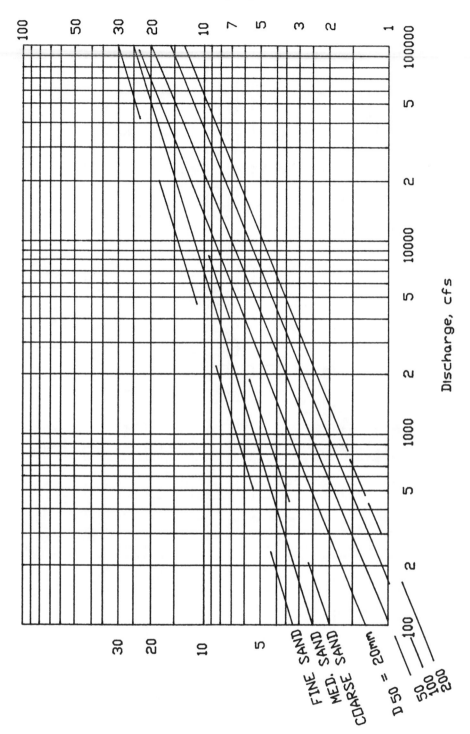

Figure 10-10 Preliminary Estimate Of Channel Depth (Neill, 1984)

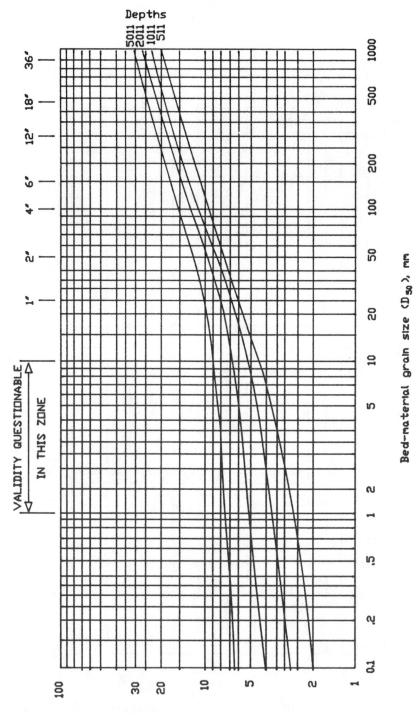

Figure 10-11 Allowable Mean Velocity For Stable Channels (Neill, 1984)

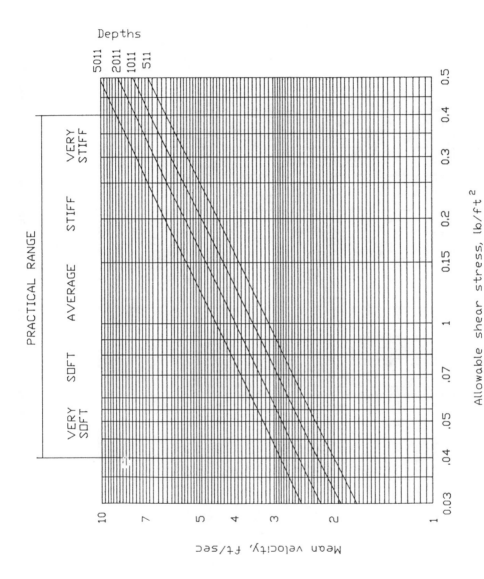

Figure 10-12 Allowable Mean Velocity And Shear Stress For Cohesive Soils And Stable Channels (Neill, 1984)

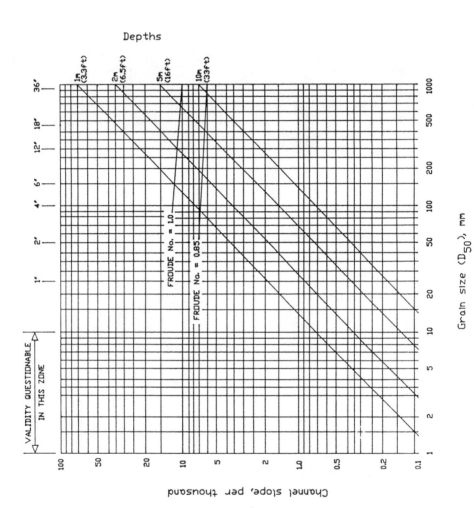

Figure 10-13 Allowable Limiting Slope For Stable Channels (Neill, 1984)

3. Straight reach permissible velocity
    = (6.8)(1.22) = 8.31 ft/s bed velocity
    = (6.8)(1.22)(0.72) = 5.88 ft/s side slope
4. Curved reach permissible velocity
    = (8.31)(0.89) = 7.40 ft/s bed velocity
    = (5.88)(0.89) = 5.24 ft/s side slope

Since the curve reach side slope permissible velocity is less than the actual mean velocity (5.24 versus 5.45 ft/s) there should be a check on bank protection methods that are available. Since the values are so close it may be appropriate to construct the channel with vegetated protection and watch the bend to determine its stability.

Note that for the sediment laden condition the permissible velocity increases to 9.08 ft/s uncorrected and 7.1 ft/s in the curved reach on the side slope. This increase is due to the fact that the sediment load acts as an exchange mechanism for the bed and banks, both depositing and scouring particles. It may also deposit a "protective" coating of colloidal clay effectively armoring the channel from normal erosion.

Tractive Stress Design Example A channel is to be constructed in an urban subdivision. The design flow is 262 cfs at a depth of 3.5 feet and velocity of 3.23 ft/s. The slope of the energy grade line and the channel, assuming uniform flow, is 0.0026 ft/ft. The bottom width of the channel is 18 feet with side slopes of 1 1/2:1. The mature channel total Manning's n value is expected to be 0.045. The channel is going to be lined with very angular gravel (GM) with a D_{75} of 0.90 inches (22.9 mm). Determine the channel's stability.

1. Since $D_{75} > 0.25$ inch the large particle method can be applied. The particle resistance factor then is:

$$n_t = D_{75}^{1/6}/39 \qquad (10.74)$$
$$= (0.9)^{1/6}/39 = 0.0252$$

The friction slope due to the gravel is:
$$S_t = (n_t/n)^2 \, S = (0.0252/0.045)^2 \,(0.0026 = 0.00082)$$

And point shear stress is:
$$\tau = 62.4 \, D \, S_t = (62.4)(3.5)(0.00082) = 0.179 \text{ lbs/ft}^2$$

2. The shear on the side slope and bed is taken as 75% and 97% respectively as:
$$\tau \text{ (side slope)} = 0.75 \, \tau = 0.134 \text{ lbs/ft}^2$$
$$\tau \text{ (bed)} = 0.97 \, \tau = .174 \text{ lbs/ft}^2$$

3. The shear in the bend can be calculated from the ratio of the radius of curvature and the bottom width as: radius/bottom width = 150/18 = 8.33

$$\tau_b = [-0.133 \, (R_c/W) + 2.224] \, \tau \qquad (10.75)$$
$$\tau_b = [-0.133 \, (8.33) + 2.224] \, (.174) = 0.194 \text{ lbs/ft}^2 \text{ for the bed}$$
$$\tau_b = [-0.133 \, (8.33) + 2.224] \, (.134) = 0.150 \text{ lbs/ft}^2 \text{ for the side slope*}$$
        * Some designers do not calculate a separate shear for side slopes in bends.

4. The allowable tractive stress can be found from:
$$\tau_c = 0.4 \, D_{75} = (0.4)(0.90) = 0.36 \text{ lbs/ft}^2 \text{ for the bed}$$

Adjusting for the side slope:
$$\Phi = 40.14\text{-}1.51 \, D_{75} + 11.11 \log (D_{75}) \qquad (10.76)$$
$$= 40.14\text{-}1.51 \, (0.90) + 11.11 \log (0.90 = 38.3 \text{ deg})$$
$$\tau_{ss} = \tau_c \, [(z^2 - \cot^2\Phi)/(1+z^2)]^{1/2} \qquad (10.77)$$
$$= 0.36 \, [(1.5^2 - \cot^2 38.3)/(1+1.5^2)]^{1/2} = 0.161 \text{ lbs/ft}^2$$

Comparing actual and allowable, the channel will be stable for both bed and side slope.

Regime Equation Example - A type B channel is to be designed to convey 600 cfs at bankfull with 2:1 side slopes and n=0.022.

1. $P = 2.51\ Q^{0.512} = 2.15 * 600^{0.512} = 66.4$ ft
2. $R = C_2\ Q^{0.361} = 0.43 * 600^{0.361} = 4.33$ ft
3. $A = C_3\ Q^{0.873} = 1.08 * 600^{0.873} = 288\ ft^2$ (equals PR)
4. $V = Q/A = 600/288 = 2.08$ ft/s
5. $D = 1.23R = 1.23 * 4.33 = 5.3$ ft
6. Froude number $= V/[gd]^{1/2} = 2.08/[32.2 * 5.3]^{1/2} = 0.159 < 0.3$
7. $W = 0.9\ P = 59.76$ ft
   $W = 0.92\ T - 2.0$ $T = 67.1$ ft.
   For 2:1 side slope bottom width $B = 67.1-(4)(5.3) = 45.9$ ft
8. Find the regime equation channel slope needed from:
   $V = C_4\ (R^2S)^{1/3} = 16.1\ [(4.33)^2\ S]^{1/3} = 2.08$ ft/s Then: $S = 0.000114$ ft/ft
9. Assuming Manning's equation, find the slope necessary to provide conveyance capacity for the channel dimensions given.
   $V = 2.08 = 1.49/n\ R^{2/3}\ S^{1/2} = 1.49/0.022\ 4.33^{2/3}\ S^{1/2}$
   $S = 0.00013$ difference of 14% in slope

One can choose either slope or one in the middle depending on the topography of the site. If a slope flatter than the Manning's slope is provided, the channel conveyance needs to be increased to account for this.

Neill Relations - Given a dominant discharge of 500 cfs, S (natural) $= 0.008$, n $= 0.025$, D_{50} $= 0.5$ mm and cross sectional shape to be trapezoidal with side slopes of 2:1, find stable channel dimensions and slope. Low transport condition applies (Neill, 1984).

Checking Figure 10-9 a slope of 0.006 is indicated. From the width relationships for moderate bank resistance the width is found to be 47 feet. A rough approximation of depth from Figure 10-10 is 1.65 feet. Based on these values the Manning equation is solved giving a velocity of about 6.45 ft/s. Checking this against Figure 10-11 gives a satisfactory result. Checking against Figure 10-13 gives a slope in about the right range though the chart does not go to such low depths.

It should be noted that conventional design might have chosen a greater depth and thus a higher velocity and narrower channel. While this channel would have carried the flow adequately it would probably have needed significant bank protection and constant maintenance.

Grade Control Structures

Grade control structures are used to prevent streambed degradation. This is accomplished in two ways. First, the structures provide local base levels that prevent bed erosion and subsequent slope increases. Second, some structures provide controlled dissipation of energy between the upstream and downstream sides of the structure. Structure choice depends on the type of existing or anticipated erosion, cost, and environmental objectives. The best source of design guidance is the National Engineering Handbook, Section 11, Drop Spillways and Section 14, Chute Spillways and UDFCD (1991). All grade control consists of a control section, an adjacent protection section and an energy dissipation section.

The general design procedure for grade control structures is to:
- determine the total fall through the design reach;
- for the selected channel determine the stable channel gradient;
- determine the amount of fall to be controlled by the grade control or drop structures (from the tailwater of the upstream structure to the head on the next structure downstream should be a stable slope); and

• select the size, location and type of structures to be used.

Grade control design will be improved if: entrance walls are rounded, end walls are straight rather than curved or flared, exit channel is larger than the control section, a preformed scour hole is supplied if the exit channel is 1.5 times the control section crest length, and the riprap section length downstream should be 100 times the depth of flow (Biedenharn, 1987).

Sills - A sill (or stabilizer) is a structure that extends across a channel and has a surface that is flush with the channel invert or that extends a foot or two above the invert. Because sills are intended to prevent scouring of the bed, they should be placed close enough together to control the energy gradeline and prevent scour between structures.

Drop Structures, Chutes, And Flumes - Drop structures provide for a vertical drop in the channel invert between the upstream and downstream sides, whereas chutes and flumes provide for a more gradual change in invert elevation. Because of the high energies that must be dissipated, pre-formed scour holes or plunge pools are required below these structures. When the ratio of the drop to critical depth in the channel is less than one, the CIT type drop structure is often used (Vanoni and Pollak, 1959, Murphy, 1967). When it is greater than one the SAF type is normally used (WES, 1988). Figures 10-14 and 10-15 illustrate these two types of drop structures. The other most common type of structure used extensively in the rural southeast is a sheet pile type shown in Figure 10-16 (Little and Murphy, 1982). A very simple drop structure can be formed from a culvert section as illustrated in Figure 10-17 (Biedenharn, 1987). Figure 10-18 shows a typical gabion control structure for a trapezoidal channel. The equation of discharge is:

$$q = 2.63 \, H^{1.62} \tag{10.78}$$

Where: q = unit discharge, cfs/ft
      H = total head on crest, ft
This type of structure shows submergence at a submergence ratio of about 80 percent. It will be stable as long as the unit discharge is less than the limit given as :

$$q \le 38.5[(H_c + h)/H_c]^{-3.5} \tag{10.79}$$

Where: H_c = crest height, ft
      h = tailwater height relative to the crest, ft

Environmental Considerations - Sills may be notched at the thalweg location to concentrate low flows to improve aquatic habitat and water quality or for aesthetic reasons. In highly visible locations, sills extending above the channel invert may be constructed of, or faced with, materials such as natural stone that create an attractive appearance. Sills may be modified to allow for passage of boats or fish, if desired.

10.16 Channel Bank Protection

The protection of the channel bed and banks is both an art and a science. Rarely will the designer find a situation which includes a straight uniform channel with uniform soils and no extenuating circumstances. Therefore the designer must choose among a wide variety of possible bank protection methods to meet the demands of space restrictions, rapid transitions, cost limitations, soil variability, aesthetic demands, environmental and habitat concerns, regulatory demands,

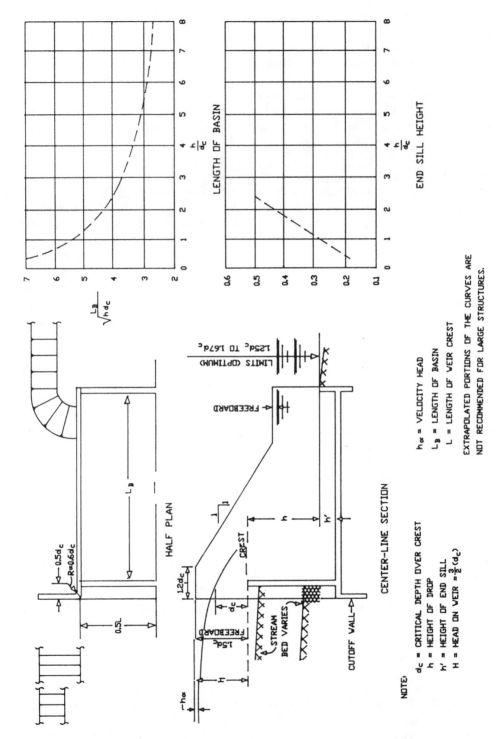

Figure 10-14 CIT Type Drop Structure (Vanoni and Pollak, 1959)

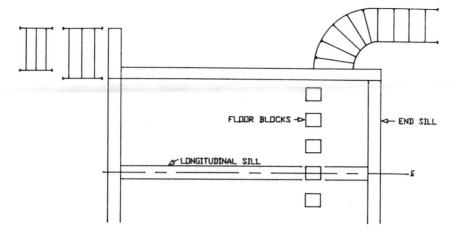

HALF PLAN

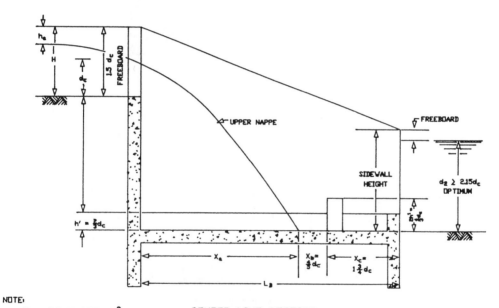

CENTER-LINE SECTION

NOTE:

H = HEAD ON WEIR = $\frac{3}{2}(d_c)$
h_a = VELOCITY HEAD
d_2 = TAILWATER DEPTH
d_c = CRITICAL DEPTH OVER CREST
h = HEIGHT OF DROP
h' = HEIGHT OF END SILL
L_B = LENGTH OF STILLING BASIN = $X_a + X_b + X_c$
X_a = HORIZONTAL DISTANCE FROM CREST TO
    INTERSECTION OF UPPER NAPPE AND
    STILLING BASIN FLOOR
X_b = HORIZONTAL DISTANCE FROM INTERSECTION OF
    UPPER NAPPE AND STILLING BASIN FLOOR TO
    UPSTREAM FACE OF FLOOR BLOCKS
X_c = HORIZONTAL DISTANCE FROM UPSTREAM FACE
    OF FLOOR BLOCKS TO END OF STILLING BASIN

Figure 10-15 SAF Type Drop Structure (WES, 1988)

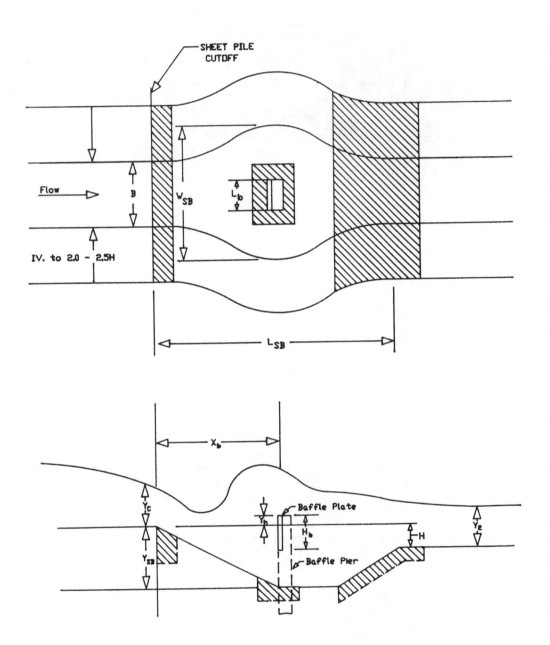

Figure 10-16 Low Drop Grade Control Structure

Source: Little, W. C. and J. B. Murphy, "Model Study of Low Drop Grade Structures," ASCE
J. of Hydrau., Vol. 108, No. HY10, Oct. 1982. Reproduced by permission of ASCE.

Figure 10-17   Culvert Section Drop Structure

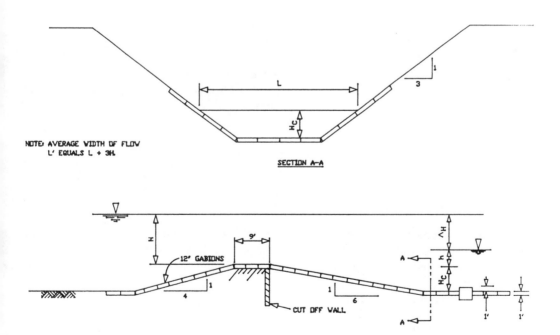

Figure 10-18   Typical Gabion Trapezoidal Channel Control Structure
(Source: COE, 1974)

and much more. Several of the more common types of bank protection used are described below. For all but the generic types the designer should consult additional literature to ensure a complete understanding of the limitations and details of design methods.

Streambanks subject to erosion are protected by stabilizing eroding soils, planting vegetation, covering the banks with various materials, or building structures to deflect stream currents away from the bank. Placement and type of bank protection vary, depending on the cause of erosion, environmental objectives, and cost. Table 10-18 identifies streambank protection measures appropriate for different problems and ranks them according to overall environmental benefits.

Table 10-18 Streambank Protection Measures

Problems	Appropriate Protection Measures
Toe erosion and upper bank failure	1. Earth core dikes with vegetation 2. Stone toe protection with vegetation 3. Cribwalls with vegetation 4. Gabions 5. Bulkheads
Local scour	1. Woody vegetation 2. Grass 3. Riprap and Cribwalls with vegetation 4. Conventional riprap 5. Gabions
General scour	1. Woody vegetation 2. Grass 3. Riprap with vegetation 4. Conventional riprap 5. Gabions
Mass failure	1. Rock toe protection with vegetation 2. Cribwalls with vegetation 3. Bulkheads
Overbank runoff and piping	1. Flow diversion 2. Drop inlets 3. Drop structures

Riprap Design

The following procedure is based on results and analysis of laboratory and field data (Maynord, 1987; Reese, 1984; Reese, 1988). This procedure applies to riprap placement in both natural and prismatic channels and has the following assumptions and limitations:

- minimum riprap thickness equal to d_{100},
- the value of d_{85}/d_{15} less than 4.6,
- Froude number less than 1.2,
- side slopes up to 2:1,
- a safety factor of 1.2, and maximum velocity less than 18 ft/s.

If significant turbulence is caused by boundary irregularities, such as installations near obstructions or structures, this procedure is not applicable.

Guidance for riprap design in a typical small stream urban setting is from a combination of sources including Reese (1984), Maynord (1987), Reese (1988) and guidance from several Federal agencies. Equation 10.80 gives the D_{50} size of stone (in inches) for riprap placed in a channel with average velocity "V" and depth "d".

$$D_{50} = 0.0136 \ V^3/d^{0.5}K^{1.5.} \tag{10.80}$$

K is the side slope correction factor and can be found from equation 10.81. It should be used for all side slope placement on slopes steeper then 1V:4H. For other placement K can be taken equal to one (1.0). θ is equal to the bank angle with the horizontal (e.g. a 1V:3H slope has a θ value of 18.43 degrees). The method assumes a typical riprap angle of repose of 39 degrees. If stone is very flat or very rounded the 0.396 term should be replaced by the Sin^2 of the estimated riprap angle of repose.

$$K = [1- (Sin^2\theta/0.396)]^{0.5} \tag{10.81}$$

Equation 10.80 is based on a safety factor of 1.2 and a stone specific weight of 165 lbs/ft³. For situations other than a uniform straight channel the D_{50} size from equation 10.80 should be multiplied by a Stability Correction Factor found below and used in equation 10.82.

Condition	Stability Factor (SF)
Uniform flow; straight or mildly curving reach (curve radius/channel topwidth ($R_c/T >$ 30)); little impact from wave action and floating debris; little uncertainty in design parameters.	1.0 - 1.2
Gradually varied flow; moderate bend curvature (30 > R_c/T > 10); moderate impact from waves or debris; moderate uncertainty in design parameters.	1.3 - 1.6
Approaching rapidly varied flow; sharp bend curvature (10 > R_c/T); significant impact from waves or debris; high flow turbulence; significant uncertainty in design parameters.	1.6 - 2.0

$$C_{SF} = (SF/1.2)^{1.5} \tag{10.82}$$

If the rock density is significantly different from 165 lbs/ft³ the D_{50} size found in equation 10.80 should be multiplied by a specific weight correction factor (C_w) found in equation 10.83. S_w is the specific weight of the stone.

$$C_w = [102.6 / (S_w -62.4)]^{1.5} \tag{10.83}$$

The riprap layer thickness should be a minimum of D_{100}, and the D_{85}/D_{15} value should be less than 4.6. Stone should be angular in shape. Riprap should be placed so as not to be flanked by the flow. The end of the protected section should be keyed into the bank to prevent scouring failure. For riprap blanket thicknesses greater than D_{100} the following reductions in D_{50} stone size are allowed:
   • for blanket thickness equal to 1.5 D_{100} the D_{50} size can be reduced 25 percent.

- for blanket thickness equal to 2.0 D_{100} the D_{50} size can be reduced 40 percent.

Channel design must account for riprap thickness in channel excavation. If the complete channel is to be designed as a riprap channel, the riprap itself controls the velocity and an iterative process between riprap size and average channel velocity due to riprap roughness is necessary. Channel roughness for riprap lined channels can be evaluated from (D_{50} in feet):

$$n = 0.0395 \ (D_{50})^{1/6}$$ (10.84)

The stone weight for the selected diameter can be calculated from:

$$W = 0.5236 \ S_w \ D_{50}{}^3$$ (10.85)

Where: W = stone weight, lbs
        S_w = specific weight of the stone, lb/ft^3

Normally an apron of riprap is placed and keyed into the channel bottom. Figure 10-20 gives details of the placement by several methods. Gradation of riprap material should meet the following rules for stone weight (W) for the upper and lower limit curves for gradation (COE, 1970).

- W_{50} lower limit not less than size indicated by procedure above
- W_{100} lower limit $\geq W_{50}$ lower limit
- upper limit $W_{100} \leq 5$ lower limit W_{50}
- lower limit $W_{15} \geq 1/16 \ W_{100}$

The relationship between the weight (W in pounds) of the stone and its equivalent spherical diameter (D in feet) is:

$$D = \left(\frac{6W}{\pi \gamma_s}\right)^{1/3}$$ (10.86)

Where: γ_s = the specific weight of the stone (about 165 lbs/ft^3)

The methods in Figure 10-20 should be used under the following circumstances

- Method A - excavation made in the dry, extend below the bed to anticipated scour depth.
- Method B - if toe excavation is wet the horizontal rock should be extended horizontally at a depth of 3 to 5 feet for medium sized channels, thickness "b" not less than the layer thickness, c $\geq$ a.
- Method C - for underwater placement with low velocities, toe at existing channel bottom and a = 1.5 thickness and c = 5 thickness.
- Method D - underwater placement with expected erosion, a thickened rock toe should be placed in trench and a = 3 thickness and c = 5 thickness.
- Method E - channel bottom is in rock, the layer should be keyed into the rock.

Riprap Design Example

A natural channel has a calculated velocity of 8.5 ft/s and a depth of 3 feet. Riprap is to be placed on the outside of a bend on a 1V:3H side slope. Moderate turbulence is expected (1.5 safety factor assumed). The side-slope correction factor can be found from:

K = [1- (Sin$^2\theta$/0.396)]$^{0.5}$ = [1- (Sin^{2}18.43/0.396)]$^{0.5}$ = 0.865

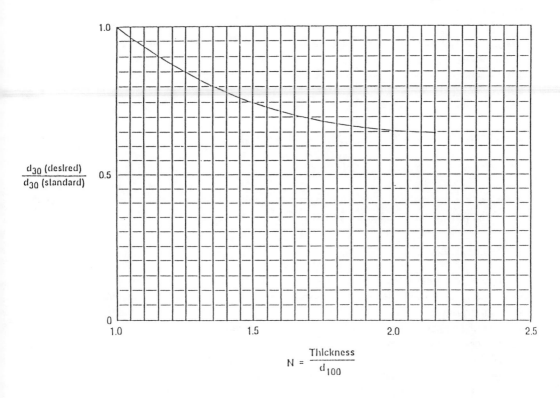

Figure 10-19 Riprap Lining Thickness Adjustment For d_{85}/d_{15} = 2.0 to 2.3

Source: Maynord, 1987

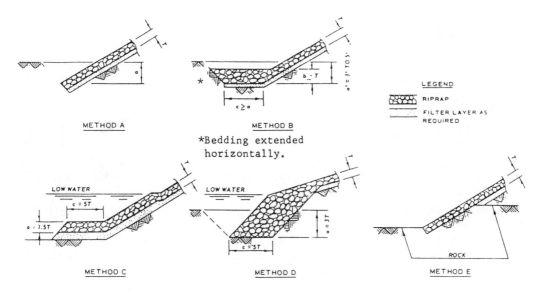

Figure 10-20 Typical Riprap Placement (COE, 1971)

Then the D_{50} is:

$$D_{50} = 0.0136 \ V^3/d^{0.5}K^{1.5.} = 0.0136 \ (8.5)^3/(3)^{0.5}(0.865)^{1.5} = 6 \text{ inches}$$

Applying the stability correction factor of 1.5:

$$C_{SF} = (SF/1.2)^{1.5} = (1.5/1.2)^{1.5} = 1.397$$

Then the corrected D_{50} size is:

$$1.397 * 6 = 8.4 \text{ inches (use 8 1/2 inches)}$$

The design is then finished by determining the available gradation, use of filter cloth, placement means, thickness, extent, and end treatment.

Grouted Riprap

Grouted riprap has advantages over loose riprap in that it: can be placed in steeper areas, is more stable for low flow channels, can take advantage of smaller stone sizes, and reduces weed growth and trash accumulation. Because it is no longer a flexible lining it loses the "self healing" properties of dumped stone. Grout should contain air entrainment, have a 28 day strength of at least 2400 psi and a high slump (5 to 7 inches) in order to penetrate at least two feet into the riprap layer (UDFCD, 1991). Grout penetration may be accomplished by rodding, vibrating or pumping. The fines from a typical gradation of riprap can be removed to allow for better penetration. The appearance can be improved by leaving the outer rocks exposed (similar to exposed aggregate). Weep holes should be provided to allow for reduction in lift forces and hydrostatic pressure buildup behind the stone.

Gabions And Rock Mattresses

Wire enclosed rock mattresses and blocks (gabions) can be used in most places where concrete would otherwise be specified because riprap would be unstable. Gabions are used routinely for vertical channel banks, rock fill drops and weirs, mattress type bank and shore protection, and other miscellaneous water resource uses (Jacobs, 1984). However, gabion wire is susceptible to chemical degradation from high sulfate soils or acidic pollution (plastic coated wire can reduce this impact) and from the abrasive action of cobbles, rocks and angular sand transported in the channel. The normal service life of wire enclosed rock is considered to be 15 years if there are no extraordinary circumstances to shorten this expectancy (UDFCD, 1991, Stephenson, 1979, Bekaert, 1979). Maintenance requirements include periodic inspection to patch broken or cut wires. Vandalism can be a problem with wire enclosed rock. In some urban areas mattresses should be buried under several inches of grassed soil to protect them.

Tests conducted on the stability of such mattresses found that the thickness of the wire enclosed rock could be one-third that of riprap while maintaining the same resistance to erosion, and that the stability of the rocks was twice that of non-enclosed rocks (Simons, et al., 1984). Field experience from studies by the US Army and in a few sites in Wyoming also prove the same finding. Many of the failures of gabions come from bedding failures not of the mattress itself. Standard sizes of gabions are given in Table 10-19. Figure 10-21 shows a typical installation. Eleven gage wire is used for typical mesh with 13.5 gage for ties.

Table 10-19 U.S.A. Gabion Dimensions Maccaferri (8 X 10 mesh)

Letter Code	Length (ft)	Width (ft)	Height (ft)	Number of Diaphragms	Capacity (yd³)	Min. Rock Dimensions (in.)
A	6	3	3	1	2.0	4
B	9	3	3	2	3.0	4
C	12	3	3	3	4.0	4
D	6	3	1.5	1	1.0	4
E	9	3	1.5	2	1.5	4
F	12	3	1.5	3	2.0	4
G	6	3	1	1	0.66	4
H	9	3	1	2	1.0	4
I	12	3	1	3	1.33	4

SLOPE (RENO) MATTRESSES (6 X 8 mesh)						Area (yd²)
Letter Code	Length (ft)	Width (ft)	Thickness (in.)	Number of Cells	Capacity (yd³)	
Q	9	6	6"	3	1.0	6
R	12	6	6"	4	1.33	8
T	9	6	9"	3	1.5	6
U	12	6	9"	4	2.0	8

Source: Maccaferri, 1990

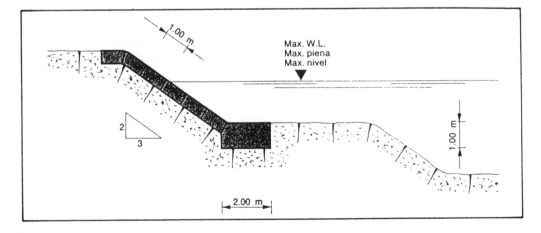

Figure 10-21 Typical Mattress Installation Schematic (Source: Maccaferri, 1979)

Another type of wire enclosed member is termed a "rock sausage" or "sack gabion" due to its round and elongated shape. Originally pioneered centuries ago by the Chinese (in bamboo and rope cages) rock sausages have found wide use in the far east. Little design information is available (Posey, 1973).

Considerations for design of mattresses of wire enclosed rock (similar to the proprietary Reno Mattress, Maccaferri, 1976, 1979) for placement in channels in lieu of tested local information and experience (Simons, et al., 1984) are:
- Manning roughness can be found from:

$$n_b = d_{90}^{1/6}/31.69 \qquad\qquad (10.87)$$

Where: n_b = Manning roughness coefficient
d_{90} = size fraction for which 90 percent is finer, ft
- Bed, maximum bank, critical and bank critical shear stress are found from (for a chosen rock median diameter):

$$\tau = \gamma DS \qquad\qquad (10.88)$$
$$\tau_m = 0.75\tau \qquad\qquad (10.89)$$
$$\tau_c = 0.1\,(\gamma_s - \gamma)\,d_m \qquad\qquad (10.90)$$

$$\tau_s = \tau_c \sqrt{1 - \frac{Sin^2\theta}{Sin^2\phi}} \qquad\qquad (10.91)$$

Where: τ = bed shear stress, lbs per ft²
τ_m = maximum shear on the banks, lbs per ft²
τ_c = critical shear stress, lbs per ft²
τ_s = critical shear stress on banks, lbs per ft²
γ = specific weight of water, 62.4 lbs/ft³
D = flow depth, ft
d_m = median rock diameter, ft
S = channel slope, ft/ft
θ = bank slope, degrees
ϕ = angle of repose of rock, about 41 degrees

For bed placement compare τ with τ_c. For bank placement compare τ_m with τ_s. If the former is greater than the latter there will be some displacement or deformation. If the former is less than 20 percent greater than the latter there will only be deformation, the mattresses will remain stable.
- Mattress thickness can be computed from:

$$T_m = 6.67\,\tau_c - 16.67 \qquad\qquad (10.92)$$

Where: T_m = mattress thickness, in.
- For velocities over 15 to 25 ft/s consider grouting the mattresses with sand asphalt mastic rather than using a considerably thicker mattress. Percentages of sand, filler and bitumen are about 70:15:15 with slightly more sand and bitumen for underwater placement. Application rates are 20 to 35 lbs/ft² for consolidating and 35 to 50 for sealing. (Maccaferri, 1976). Grouting modifies the Manning n value to 0.016.
- For smaller velocities use a geotextile filter under the mattresses. For larger velocities design a gravel filter for placement under the mattresses. Geotextiles are often used in combination

with granular material to provide drainage (Jacobs, 1984, UDFCD, 1991). Filter fabrics can tear at steeper slopes and should be limited to placement on 2.5:1 slopes. In fine silts and clays clogging of the pores can also take place and a granular filter is recommended. Compute the velocity at the mattress-filter and filter-bank interfaces using Manning equation with n = 0.02 for filter fabric and n = 0.025 for gravel.

$$V_f = 1.486/n_f \ (d_m/2)^{2/3} \ S^{1/2} \qquad\qquad\qquad\qquad (10.93)$$
$$V_b = 0.5 \ V_b \qquad\qquad\qquad\qquad\qquad\qquad\qquad (10.94)$$

Where: V_f = velocity at mattress-filter interface, ft/s
$\qquad\quad V_b$ = velocity at filter-bank soil interface, ft/s
Compare with erosive velocity of bank soil from either permissible velocity method above or for noncohesive and cohesive soils respectively:

$$V_e \ = 1.67 \ d_s^{1/2} \qquad\qquad\qquad\qquad\qquad\qquad (10.95)$$
$$V_e \ = 24.92 \ \tau_e^{1/2} \qquad\qquad\qquad\qquad\qquad\qquad (10.96)$$

τ_e can be computed from maximum permissible velocities given earlier or from Figure 10-22.

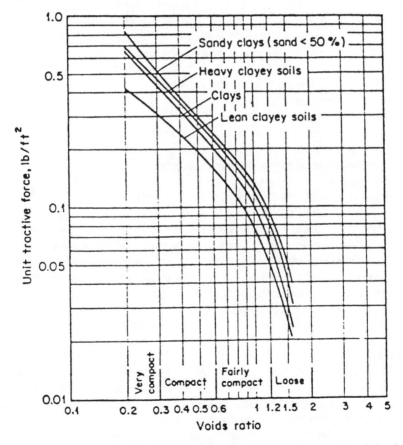

Figure 10-22 Permissible Unit Tractive Force For Channel In Cohesive Material From USSR Data (Source: USSR, 1936)

- If $V_b < V_e$ use geotextile filter fabric.
  If $V_b > V_e$ design a gravel filter whose thickness is not less than 6 to 9 inches using the following gradation:
      d50 (filter)/d50 (base) < 40
      5 < d15 (filter)/d15 (base) < 40
      d15 (filter)/d85 (base) < 5
- The filter thickness can be computed from the following equation. Use 6 inches or L (in the units of d_m) whichever is greater.

$$L = 4d_m\left[1 - \left(\frac{V_e}{V_f}\right)^2\right]$$ (10.97)

Soil Cement

Soil-cement has been used successfully and cost effectively in a number of applications throughout the United States and the world. It has been especially successful in arid climates where the naturally sandy soil, low in organic content, and wide unstable streams are conducive for soil cement use. Soil-cement is a material formed by blending, compacting and curing a mixture of soil/aggregate, portland cement, and water and allowing it to harden in place (Reese, 1987).

The type of soils conducive for use of soil cement has the following characteristics (PCA, 1975):
- sands with at least 55 percent passing No. 4 sieve
- no material retained on the No. 2 sieve
- 5 to 35 percent passing the No. 200 sieve
- plasticity index (PI) less than 8 (Pima County recommends 3, 1986)
- less than 2 percent organic material

Violation of the fines standard may require more cement to make up the strength loss. Violation of the plasticity standard will require hydrated lime to improve friability. Soils with too much organic material can be treated with calcium chloride to hasten hydration. Gravel content can be increased to improve abrasion resistance for bank protection in high sediment transport situations.

Pima County, Arizona has experienced long duration flows with high sediment content and velocities over 20 ft/s against soil-cement bank protection without noticeable loss of material. Other instances of abrasion and wear have been investigated with mixed results (Reese, 1987). The most common cause of failure was flanking and undermining of the soil-cement monolithic structure. Soil-cement emplacements for bank protection should be tied well into the bank and toed well into the soil below expected scour depths. A Manning n value of 0.022 is normally assigned to soil cement, trimmed and with a smooth appearance.

Cement content by dry weight is normally in the 7 to 14 percent range (bags of cement weigh 94 pounds) (PCA, 1976). ASTM cement type I and II are normally used with type V used for high sulfate content situations in the soil or water (> 0.2 percent in soil and > 1500 ppm in the water). Compressive strength tests from ASTM are available for soil-cement as D1632, D1633 and D2901 as well as wet-dry testing and freeze-thaw testing. Density of soil-cement can normally be taken as 120 lbs per cubic foot.

Soil cement for river bank protection and dam construction is normally made from soil excavated and mixed on site and placed in lifts by road construction equipment (Nussbaum and Colley, 1971). Lifts are kept moist between passes and can be dusted with cement to improve bonding lift to lift. The relationship between lift height, slope and width is given by:

$$W = T (S^2 + 1)^{1/2} SL \qquad\qquad (10.98)$$

Where: W = width of each lift, ft
        T = thickness of the layer perpendicular to the bank slope, ft
        S = bank slope horizontal, SH:1V
        L = thickness of each compacted lift, ft

Most lifts are eight feet wide due to construction equipment limitations. Lifts as little as four feet wide would be structurally stable if they could be placed easily with modified equipment. Slope is dictated by both structural failure if the slope is not steep enough and soil failure if it is too steep. Many bank placements are at 1:1 slopes. Nowatski and Reely (1986) and Nussbaum and Colley (1971) provide details of slope stability testing. Lift thicknesses are between 4 and 8 inches. Figure 10-23 shows a typical placement of this type.

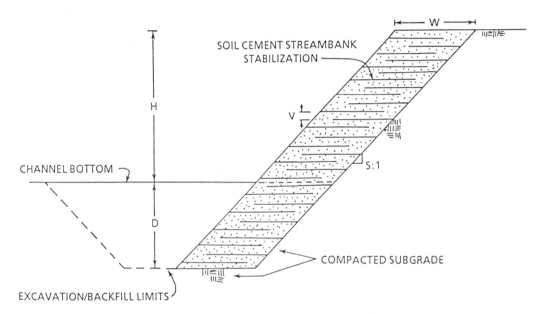

Figure 10-23 Typical Soil Cement Placement (Source: Reese, 1987)

Facings of soil-cement are generally 6 to 12 inches thick. Pima County specified 500 psi compressive strength plus two percent additional cement for a series of channels which underwent record flooding without loss. Compaction up to 3:1 slopes can be accomplished with a bulldozer with street pads, or a roller compactor can be let down the side of the channel by rope and winch.

Bioengineering

Soil bioengineering is an applied science combining mechanical, biological and ecological concepts to construct living structures for bank and slope protection. Biological methods have been used in the United States for a number of years but without the scientific background afforded by the bioengineering science. Soil bioengineering, pioneered in Europe, is gaining in popularity in the United States and has been used for various applications of bank and slope protection (Sotir and Gray, 1989, Gray and Leiser, 1982, and Schiechtl, 1980).

Bioengineering methods use structural support to hold live plantings in place while the root structure grows and the plants are established. This is done through the use of sprigging, live crib walls (see Figure 10-24 for a typical example), cut brush layers, live fascines, live stakes, and other methods.

Description	Notes
Six foot willow switches are wired together to form a mat which is then secured to the bank by stakes, fascines, poles or rock fill. The toe of the slope is reinforced with a brushlayer anchored by a live fascine or rip-rap. Mattress should lie perpendicular to the water.	1. The mat establishes a complete cover for the bank. Entire layer must be slightly covered with soil or fill. 2. Captures sediment and rebuilds bank. Plants provide long term durability and erosion control. 3. Establishes dense riparian growth. Allows for invasion of surrounding riparian vegetation. 4. An economical technique which provides protection soon after it is established. 5. Applicable on banks with a uniform gradient not exceeding 2:1. The toe of the slope is reinforced with a brushlayer anchored by a live fascine. 6. Mattress should lie perpendicular to the water. 7. Site should have moderate water level fluctuations.

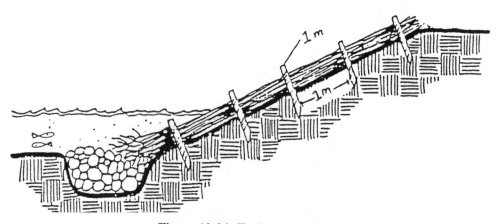

Figure 10-24 Typical Live Crib Wall

Advantages of soil bioengineering include the strength of intertwining root masses, self healing properties, natural appearance, habitat enrichment, faster recovery of ecosystems after disturbance, and resistance to slope failure. Disadvantages include labor intense installations, need for stability control until roots are established, and dependence on ability of materials to root and grow. Many of these disadvantages can be countered with both use of native vegetation, engineering stability analyses, skilled designers and the integration of structural members with the plantings as necessary.

Detailed knowledge of plant species (grasses, legumes, shrubs and trees) and ability to withstand inundation in bank and terrace zones is necessary for this method to be successful (Logan, 1979). Often an interdisciplinary team is helpful. Nurseries need notice if they will be counted on to supply various types of vegetation stock. Willows can be cut from other stream banks in the area or supplied by maintenance crews. Planting by phase is often done to facilitate best mixes of vegetation, availability periods and growing seasons. In any case close monitoring is necessary until the growth is established and assured.

Manufactured Bank Protection Methods

There are a wide variety of manufactured bank protection products. They include: mats and coverings, crib membranes, filter cloths, blocks and articulated concrete matting, rigid grid blocks, structural wall modules of concrete and various materials, sand filled mattresses, concrete filled mattresses, etc.

Manufactured materials have the advantage of uniformity, consistency, targeted strengths for specific applications, ability for steep or vertical banks, nice appearance in developed urban settings, tested stability, and technical support from suppliers. Disadvantages include unit cost, labor intense installation and the possibility of lack of availability should additional or matching materials be needed and the supplier is no longer in business.

10.17 Floodplain Management

One of the main duties of the urban storm water manager is to manage the floodplain as it interacts with the streams and open channels. This duty is assigned both by virtue of need and Federal, state and local regulations. The official regulatory floodplain is defined as depicted in Figure 10-25. The Floodplain Management Act of 1978 requires the floodplain manager to perform a number of duties. These, combined with other duties, spell out the purposes and programs of floodplain management. Table 10-20 lists typical and atypical duties of a local floodplain manager.

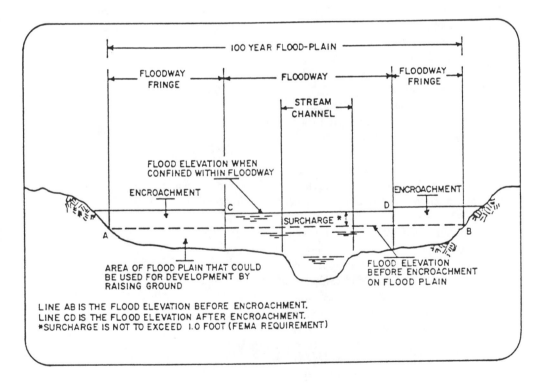

Figure 10-25 The Floodplain And Related Definitions

Table 10-20 Duties And Tools Of A Floodplain Manager

1. Administer the National Flood Insurance Program
    *   maintain flood maps
    *   maintains records
    *   administer flood map changes
    *   provide information on home locations
    *   maintain elevation and anchoring certificates
    *   update maps periodically
    *   coordinate with Federal and state agencies
2. Administer the local Community Rating System
    *   apply for rating annually
    *   administrate measures for which points can be given
3. Adopt and administer local floodplain ordinances and regulations
    *   develops design standards and criteria
    *   reviews plans
    *   monitors compliance
    *   perform inspections
    *   provide information
    *   issue permits
    *   hear appeals
    *   administer related regulations such as zoning, subdivision, or building codes
4. Provide technical assistance to the community in floodplain and flooding matters
    *   assistance on flood insurance
    *   assistance on development design
    *   assistance on flood reduction and avoidance measures
5. Perform floodplain planning
    *   perform technical master planning
    *   identify and seek to mitigate local flooding problems
    *   eliminate repeated loss situations
    *   seek to preserve and enhance natural values of floodplain
    *   look into integrated uses and parks planning
    *   work with Federal and state agencies on flood reduction and stream restoration measures
6. Plan and manage flood fighting and emergency preparedness
    *   develop preparedness plans
    *   maintain flood warning and mitigation systems
    *   flood response
    *   stockpile equipment and supplies
    *   manage or assist in flood response and recovery operations
    *   maintain flood protection measures
    *   maintain flooding records and statistics
7. Educate self and the public
    *   keep up with new developments in the program
    *   keep up with new technical approaches
    *   coordinate with Federal and state agencies
    *   coordinate with other local agencies
    *   conduct public education program
    *   keep staff and political leaders educated and informed

- maintain information
- develop self-help programs
- assist in disaster assistance programs
- provide individual technical and regulatory assistance
- coordinate with realtors, developers, engineers and bankers

---

There are literally hundreds (maybe thousands) of publications, brochures and papers on how to manage various aspects of the floodplain. FEMA (1986) provides an overview of all the pertinent details of floodplain management, and provides a conceptual framework for floodplain management which stands on four important legs:

- reduce flood losses and threats to health and safety;
- preserve and restore the natural and beneficial uses of floodplains;
- take a balanced view that minimizes exposure to loss rather than promoting floodplain abandonment on the one hand, or intense floodplain development on the other; and
- develop and use the tools available to provide careful and technically sound consideration of all information and alternative uses of floodplains.

In terms of flooding, floodplain management seeks to: reduce human and property susceptibility to flooding, reduce the impacts of flooding, and reduce the actual flood levels and peaks. Reducing human susceptibility includes: restrictions on the mode and time of day/season of occupancy of floodplains; in the ways and means of access; in the pattern, density and elevation of structures; in the materials out of which structures are constructed and their designs; and in the infrastructure and landscaping in flood prone areas. This is done through regulations, disaster preparedness and assistance, floodproofing, and flood warning/ mitigation measures.

Reducing the impacts of floods on individuals includes the actions of: flood insurance, education, tax adjustments as incentives for wise development, flood emergency response measures and post flood recovery operations. Reducing floods themselves include all the aspects discussed in Table 10-21 (positive and negative aspects). Nonstructural alternatives should be pursued aggressively especially when taking into account nonpoint source control measures.

Floodplain management, when integrated with the overall storm water management program, provides a regulatory means to improve the surface water system throughout the municipality. Flooding does not stop at the edge of the regulatory floodplain; neither does proper floodplain management. Flood prone areas should be identified throughout the municipality both upstream from and adjacent to regulatory floodplains. Life and property losses due to persons entering floodplain areas should be dealt with just as strongly as those from persons living within floodplain areas. With proper, and patient, application of sound principles of floodplain management, great and positive changes have been demonstrated in various municipalities, around the country.

Flood Insurance Programs

The Federal Emergency Management Agency (FEMA), through the National Flood Insurance Program (NFIP), offers a wide range of services and requirements. NFIP was created in 1968 by the National Flood Insurance Act and further defined and modified by several subsequent acts. The program has two major thrusts: to shift the burden of costs for flood losses from taxpayers at large to floodplain occupants, and to reduce losses due to flooding through floodplain regulation. The two major vehicles to do this are flood insurance availability to all homeowners and the required locally passed floodplain regulations and program necessary for a municipality to retain eligibility for flood insurance.

Based on studies of the flooding in the Midwest in 1993, there is expected to be many basic changes on how the Federal government looks at managing floodplains in the future, and the types of programs and support it will offer. It can be expected that the NFIP will also change. The Report of the Interagency Floodplain Management Review Committee (1994) recommends fundamental changes in both philosophy and implementation including: a stronger and more coordinated floodplain management strategy with the three thrusts of avoiding, minimizing and mitigating flood damages at all levels of government and the private sector; establishment of a National Floodplain Management Program by legislative mandate; and revitalization of the Federal Water Resources Council.

The vehicle to accomplish flood prone designation is through the use of the 100-year floodplain delineation (termed the base flood) as conveyed in a series of maps and reports for each municipality in the program. Originally municipalities entered the "emergency phase" of the program after the development of Flood Hazard Boundary Maps (FHBM) was complete. These maps do not show flood elevations and thus are not as accurate as recently completed studies. Insurance rates are higher for municipalities still in the emergency phase. They can convert to the regular program through a restudy of the municipality's streams and the adoption of stricter flood plain regulations after a model regulation published by FEMA.

As shown in Figure 10-25, FEMA has designated an area immediately adjacent to the stream as the regulatory floodway. This is an area of higher velocity and deeper flooding. It is determined in engineering hydraulic backwater profile studies by mathematically encroaching in the floodplain form both sides until the water surface rises a maximum of one foot. The area between the edge of the floodplain and the floodway line is termed the floodway fringe and is depicted in Figure 10-25. Development outside the regulatory floodplain is unconstrained by flooding considerations from a Federal regulatory standpoint. Development within the floodway fringe must be elevated or flood-proofed to a minimum of one foot above the flood elevation (this elevation is termed the flood protection elevation). Development or fill within the floodway is prohibited unless a hydraulic study is done and submitted to FEMA showing that the fill or development does not raise upstream flood elevations. This is normally difficult to do without redefining channel dimensions, roughness or proving a flaw in the original study.

Prior to 1988 FEMA used three vehicles to convey the floodplain information.

● The Flood Insurance Rate Map (FIRM) shows 100-year and 500-year floodplain boundaries and flood elevations and an insurance designation.

● The Flood Boundary and Floodway Map (FBFM) shows the same 100-year and 500-year boundaries but without elevations. Instead it shows the key cross sections used in developing the detailed study of the stream. These cross sections are keyed to the Federal Insurance Study (FIS).

● The FIS contains complete information on the municipality, the study methods and results, and other floodplain management information. Included are Floodway Data Tables and, generally, tables of elevations and river mileage of each cross section.

FEMA also used a wide range of designations to show various flooding depths. Due primarily to resistance by insurance companies unwilling to write many different policies for each of the designations and the apparent redundancy in the FIRM and FBFM, FEMA has changed both the designations and the way results are reported. Since 1988 one map depicts all necessary flood information found on the FIRM and the FBFM. And, since 1988, the flood designations are as follows.

A   Areas within the 100-year floodplain boundary determined by approximate methods. Development within these areas requires a detailed hydraulic and hydrologic study submitted to the municipality.

AE   Areas within the 100-year floodplain boundary determined by detailed methods.

AH   Areas of 100-year shallow flooding with constant water surface elevation and average depths form one to three feet.

AO   Areas of shallow flooding (usually sheet flow) where average depths are one to three feet.

A99   Areas currently within the 100-year floodplain to be protected by a Federally authorized flood protection system to be constructed.

X   Areas outside the regulatory 100-year floodplain which are flood prone (normally areas upstream from the one-mile cutoff for most detailed studies), protected from the 100-year flood by a levee; areas of sheet flow less than one foot in depth.

D   Unstudied areas where flooding is possible but undetermined.

The program has had several drawbacks which have been minimized by some municipalities. The FIS program allows for encroachment sufficient to raise the flood elevation a maximum of one foot. While this rarely happens in practice (theoretically a levee located at the floodway boundary would be necessary to accomplish a one-foot rise), some municipalities have a stricter definition of the flood protection elevation requiring as much as five feet freeboard (though most are under three feet). Others allow encroachments at some lesser level. For example, Indiana allows only a 0.1 foot rise due to encroachment.

Another drawback is that the floodplain boundary is determined based on hydrology derived from current land use conditions. As development occurs greater flows result rendering the floodplain boundary inadequate. Some municipalities have established full build-out boundaries or a combination of assumed partially reduced peaks due to detention requirements and full-build-out. Other municipalities seek to keep their flood maps current through frequent and mandatory use of FEMA's method for flood map updating: Letters of Map Revision (LOMRs), Conditional Letters of Map Revision (CLOMARs) and Physical Map Revisions (PMRs).

10.18 Environmental Features Of Channel Projects

Floodplains and surface water resources provide many environmentally related benefits to a local urban municipality. Natural areas, ponds and lakes and linear parks can be sources of municipal recreation. Habitat areas and flood control areas can be in and adjacent to water sources. Water quality can be improved using the overbank and wetland areas adjacent to streams and rivers. This section covers both instream and floodplain environmental features and activities.

Instream Mitigation Features

A consideration of the environmental impacts of channel projects is, or should be, an important part of every channel design and alteration project. When properly designed, channels can both maintain flood control objectives and habitat preservation or restoration objectives. This type of consideration has not always been the case. Thaxton and Sneed (1982) have reviewed the use of environmental features in Corps of Engineers projects and McCarley, et al., have reviewed the use for flood control projects (1990). Both have found some successful but not widespread integration of environmental features in the past. The Corps of Engineers has surveyed hundreds of environmental measurement and assessment techniques in an attempt to find quantitative measurement techniques for both impact and improvement of streams and rivers (Henderson, 1982).

The beneficial and adverse environmental impacts of various types of channel projects have been detailed by Nunnally and Shields (1985) and are summarized in Table 10-21. Impacts are interrelated in complex ways but can be categorized as: aesthetics, recreation, water quality, terrestrial habitat, and aquatic habitat. They can also be seen as primary, secondary and tertiary. For example, a channel straightening project has the primary effect of rapid bed and bank erosion, the secondary effect of reduced water quality due to increased suspended sediment levels, and the tertiary effect of impacts on recreation and aquatic habitat.

Table 10-21 Possible Adverse Environmental Impacts Of Channel Related Projects

Negative Environmental Impacts	Positive Environmental Impacts

Channel Brush Snagging

Negative Environmental Impacts	Positive Environmental Impacts
• reduction of macroinvertebrate food supply • reduction of fish habitat	• aesthetic improvements • recreational improvements

Channel Bank Clearing

Negative Environmental Impacts	Positive Environmental Impacts
• reduced shade, thermal impacts • destroyed terrestrial habitat • reduced aquatic habitat • reduced food supply to food chain base • increased in-channel photosynthesis, shift toward autotrophic state • reduced bank stability, increased erosion	• limited improvement of aesthetics • limited recreational improvement

Channel Excavation

Negative Environmental Impacts	Positive Environmental Impacts

• same as clearing
• channel instability, erosion
    and sedimentation
• physical destruction and reduction of
    aquatic habitat
• can reduce recreational fishing
• inadvertent wetlands draining
	• can show landscaped aesthetic appeal
	• can improve some types of recreation

Negative Environmental Impacts	Positive Environmental Impacts

Channel Paving

Negative Environmental Impacts	Positive Environmental Impacts
• same as clearing and excavating • reduced aesthetic appeal • temperature fluctuations • loss of all habitat • block access if sides are steep and smooth	• possible improved aesthetics if original channel was very poor

Streambank Protection

Negative Environmental Impacts	Positive Environmental Impacts

• loss of near bank habitat
• recreation can be impaired if access or
    use are reduced
	• reduced erosion
	• possible improved aesthetics for
	    urban environment

- normally impaired aesthetics
- water quality effects same as clearing
- terrestrial habitat reduction
- same aquatic habitat as clearing

- recreation can be improved if access is increased
- stable banks provide habitat for small creatures and macroinvertebrates
- projecting bank protection can create instream habitat and slack water habitat

Detention Control And Drops

- blockage of fish passage
- natural sediment reduction
- can block recreational use
- can cause hazard

- sediment reduction if flow is reduced
- can be aesthetically pleasing

Levees And Floodwalls

- creation of drier floodplain conditions
- modified flow regime may impact spawning
- floodplain habitat reduction due to converted land use
- normally aesthetically dull

- can provide recreational overlooks, trails, access points, etc.

General environmental guidelines employed in successful channel projects include (COE, 1985):
- minimize structural modifications and channel alterations;
- channel instability is one of the largest impacts, so pay close attention to geomorphic and stable channel analysis; and
- layout channel and modifications in great detail showing all existing and created habitat and habitat (such as specific trees) to be saved.

In addition, the following aspects should be considered.
- Low flow channels are subchannels constructed within main channels. They can concentrate flow for biological habitat restoration, recreation, aesthetics and water quality benefits.
- High flow channels are used to pass larger flood flows. Benefit can be accrued if only one side of the channel is excavated - single side construction.
- Natural channels have a periodic meander with natural pools and riffles. An artificially but correctly sized meander, pool and riffle sequence will both help the channel to remain stable and provide fish habitat. Five to seven channel widths is the normal spacing between pools.
- Flow through design for high flows to enter cutoff and oxbow lakes preserve habitat and allow for flood control diversions.
- Instream habitat structures include sills, large rocks, ledges and overhangs, deflectors, bank embedded pipes with large holes cut in them, etc.
- Paved channels can be improved if fish are present by providing cover, tree shade and resting places.
- Construction timing and scheduling for non-sensitive seasons is important especially in areas that experience fish migration.

- For migrating fish, culvert outlets should have resting pools, low entrance jumps and sufficient flows. Baffles can reduce velocities. One municipality has even installed lights for especially long culverts to be used during the spawning season.
- Interior drainage areas and borrow pits can be redesigned as ponds for habitat as well as flood control.

Environmental Consideration Of Floodplains

Surface waters, floodplains and the watersheds draining to them must be seen as one interrelated ecological system. One part of that system is the open channels and the immediate floodplain areas. These areas can provide (FEMA, 1986):
- water resource values including natural moderation of flood flows, water quality benefits, and groundwater recharge;
- living resource values such as animal habitat, plant habitat and human-nature interaction areas;
- cultural resource values such as archeological sites, historical sites, scientific, cultural and recreational sites as well as areas for agriculture, forestry and aquaculture.

Floodplain Environmental Activities - Examples of management of floodplains for enhancement and preservation of natural values are included in Table 10-22 (FEMA, 1986).

Table 10-22 Environmental Management of Floodplains

1. Natural Flood Storage and conveyance
   - minimize floodplain filling
   - maintain wetlands
   - minimize grading and regrading to minimize soil compaction
   - relocate nonconforming structures out of the floodplain
   - return sites to natural contours
   - preserve natural drainage when constructing infrastructure (such as utility lines, bridges, sanitary lines) in the floodplain
   - prevent intrusion into sensitive ecosystem areas
2. Water Quality Maintenance
   - maintain wetland and vegetative buffers
   - use sound agricultural practices
   - control urban runoff quality
   - minimize erosion around streams and lakes
   - restrict sources of potential toxics and pathogens from the floodplain
   - restrict transportation of toxics and pathogens from routes over or adjacent to sensitive waterways
   - plan for emergency response to spills
3. Groundwater Recharge
   - require pervious surfaces near floodplains for low traffic areas
   - design for runoff infiltration
   - dispose of waste materials in noncontaminating ways
4. Living Resources
   - identify and protect wildlife habitat and ecologically sensitive areas from destruction
   - require topsoil protection programs
   - restrict wetland drainage

- • reestablish floodplain ecosystems
- • minimize tree cutting and vegetation removal
- • design floodgates and seawalls for natural flow
5. Cultural Resources
- • provide public access to and near water resources
- • locate and preserve historical and cultural resources and integrate with recreational areas
6. Agricultural Resources
- • minimize soil erosion on cropped areas
- • control chemical and fertilizer use
- • strengthen "carrot" type incentive programs locally
- • minimize irrigation return flows
- • eliminate feedlot operations and cattle access to stream banks
- • discourage new agricultural production areas
- • encourage sound agricultural practices in productive areas remote from flood problems
7. Aquacultural Resources
- • construct impoundments which minimize impact on natural drainage
- • limit exotic species not natural to the area
- • discourage large scale mechanical operations
- • use caution in animal waste disposal
8. Forestry Resources
- • control clear cutting
- • encourage enforcement state and Federal laws
- • include fire management in overall management plans
- • require erosion control on all logging operations

Source: FEMA, 1986

Greenway Planning - The use of linear parks and greenways in urban floodplain and channel management is gaining wide public acceptance around the United States. Any program seeking to enhance the environment of open channels within a local urban area could benefit from a greenway program. Greenway programs are often combined with an overall stream restoration effort. Objectives of a number of urban stream corridor programs are included in Table 10-23.

The National Parks Service provides a successful Federally assisted program for stream corridor restoration, Rivers and Trails Conservation Assistance Program, though many municipalities find it easier and faster to work independently. Other state conservation interests can be found in the River Conservation Directory (1990). Other guidance can be found in the "Riverworkbook" (1988), the Illinois Stream Preservation Handbook (1983), and Greenways (Fink and Searns, 1993). The steps in the development of a corridor restoration include: identification of issues, resources and problems; development of public involvement program; setting of goals; consideration of alternatives; setting a plan of action; and taking action.

Eugster (1988), Flanagan (1991) and the National Park Service (1983) lay out greenway planning steps once the planning team is put together.
- • Define role and function of the proposed greenway.
- • Determine project goals and orientation including early land acquisition/donation and development of regulations for land use practices.
- • Initiate the project through approaching each stakeholder with the proposal and "selling it".
- • Involve the public in hearings, workshops, etc.
- • Assess resources and land value, often through the use of GIS and a point rating overlay system.

Table 10-23 Stream Corridor Program Objectives

	Austin, TX	Boulder, CO	Chaska, MN	Chicago, IL	Eagan, MN	Fort Collins, CO	Jackson County, OR	Johnson County, KS	Kissimmee River, FL	Lexington, KY	Lincoln, NE	North Richmond, CA	Raleigh, NC	Salem, OR	San Antonio, TX	Scottsdale, AZ	Tucson, AZ	Tulsa, OK	Wabash River Corridor, IN
Flood Control/Stormwater Management										●		●				●		●	
Flood Insurance Rate Reductions						●				●			●						
Water Quality Protection							●												●
Satisfying NPDES Objectives						●				●									
Recreation		●							●				●			●		●	
Alternative Transportation										●	●								
Habitat Protection						●	●		●			●							
Community Acceptance/Involvement										●		●						●	
Cost Effectiveness										●	●					●			
Economic Development															●				
Public Education									●										●

Source: Ogden Environmental, 1993

- Analyze the issues including resource use conflicts and public attitudes.
- Explore regulatory and administrative alternatives to take advantage of programs and possible financial resources.
- Assess the political situation.
- Develop an implementation strategy.

References

American Society of Civil Engineers, "Sedimentation Engineering", ASCE Manuals and Reports on Engineering Practice No. 54, 1977.

Ackers, P, and White, W. R., "Sediment Transport: New Approach and Analysis", ASCE J. of Hy. Div., Vol. 99, No. HY11, Nov. 1973.

Ackers, P., "River Regime: Research and Application", J. Inst. of Wat. Engrs., Vol. 26, No. 5, 1972.

Bekaert, "Gabions", Terra Aqua Corp., Riverdale Station, NY, 1979.

Bensen, M. A. and T. Dalrymple, USGS Techniques of Water Resources Investigations, Book 3, 1967.

Biedenharn, D. S., "Grade Control Structures", US COE Training Course Notes, USAEWES, Nov. 1987.

Blair, L. A., "Trapezoidal Concrete Flood Control Channels in Albuquerque: Lessons Learned", Proc. 1984 Int. Symp. on Urb. Hydrology, Hydraulics and Sed. Control, Lexington, KY, 1984.

Blalock, M. E., and T. W. Strumm, "Minimum Specific Energy in Compound Channel", ASCE J. Hyd. Vol. 107, pp. 699-717, 1981.

Blench, T., "Mobile-Bed Fluviology", Dept of Tech. Svc., U. of Alberta, Edmonton, Alberta, CA, 1966.

Bray, D. I., "Representative Discharges for Gravel-Bed Rivers in Alberta, CA", J. of Hydrology, Vol. 27, 1975.

Chaudhry, M. H., and S. M. Bhallamudi, "Computation of Critical Depth in Symmetrical Compound Channels", J. Hyd. Research, Vol. 26, No. 4, 1988.

Chien, N., "A Concept of Lacey's Regime Theory", Proc. ASCE, Irr. & Drain Div., Feb. 1955.

Chien, N, "A Concept of the Regime Theory", ASCE Trans., paper 2884, with Discussions, 1957.

Chow, V. T., ed., Open Channel Hydraulics, McGraw Hill Book Company, New York, 1959.

Cowan, W. L., "Estimating Hydraulic Roughness Coefficients", Agriculture Engrg., Vol. 37, No. 7, 1956.

Denver Urban Drainage and Flood Control District, "Urban Storm Drainage Criteria Manual", with changes dated 1991.

Denver Urban Drainage and Flood Control District (UDFCD), Flood Hazard News, September, 1980.

Eugster, J. G., "Steps in State and Local Greenway Planning", Nat. Wetlands Newsletter, Sept. - Oct., 1988.

Federal Emergency Management Agency, "A Unified National Program for Floodplain Management", FEMA 100/ March, 1986.

Federal Highway Administration, Bridge Waterways Analysis Model (WSPRO), Users Manual, FHWA IP-89-027, 1989.

Flanagan, R. D., "Multi-purpose Planning for Greenway Corridors", ASFPM National Conf. Handout, 1991.

Flink, C.A. and R.M. Searns, "Greenways", Island Press, 1993.

Fortier, S. and F. C. Scoby, "Permissible Canal Velocities", Trans. ASCE, Vol. 89, pp. 940-956, 1926.

French, R. H., Open Channel Hydraulics, McGraw Hill Book Company, New York, 1985.

Garbrecht, J. and G. O. Brown, "Calculation of Total Conveyance in Natural Channels", ASCE J. of Hydraulics, Vol. 117, No. 6, June, 1991.

Gray, D. T. and A. T. Leiser, "Biotechnical Slope Protection and Erosion Control", Van Nostrand Reinhold, 1982.

Green, J. E. P. and J. E. Garton, "Vegetation Lined Channel Design Procedures", Trans. ASAE, 1983.

Gwinn, W. R. and Ree, W. O., "Maintenance Effects on the Hydraulic Properties of a Vegetation-Lined Channel", Trans. ASAE, 1980.

Harza Engineering Company, Storm Drainage Design Manual, Prepared for the Erie and Niagara Counties Regional Planning Board, Harza Engineering Company, Grand Island, N. Y., 1972.

Henderson, J. E., "Handbook of Environmental Quality Measurement and Assessment: Methods and Techniques", IR E-82-2, USAEWES, 1982.

Hey, R. D., J. C. Bathurst and C. R. Thorne, Gravel-bed Rivers, Chpt. 19 by D. I. Bray, "Regime Equations in Gravel-Bed Rivers", John Wiley & Sons, 1982.

Inglis, C. C., "Historical Note on Empirical Equations, Developed by Engineers in India for Flow of Water in Sand and Alluvial Channels", Proc. Int. Assoc. for Hy. Res., Second Meeting, Stockholm, 1948.

Interagency Floodplain Management Review Committee, "Sharing the Challenge: Floodplain Management into the 21st Century", Executive Office of the President, Washington DC, June, 1994.

Ippen, A. T. and P. A. Drinker, "Boundary Shear Stress in Curved Trapezoidal Channels", Proc. ASCE, HY5, September, 1962.

Jacobs, E. L., "Gabions: Flexible Solutions for Urban Erosion", Proc. Symp. on Urb. Hydrology, Hyd., and Sed. Cont., Lexington, KY, 1984.

Kadlec, R. H., "Overland Flow in Wetlands: Vegetation Resistance", ASCE J. of Hydraulic Engrg., Vol. 116, No. 5, 1990.

Kolupaila, K. "Methods of Determination of the Kinetic energy Factor", The Port Engineer, Calcutta, India, Vol. 5, no. 1, January, 1956.

Kouwen, N., "Field Estimation of the Biomechanical Properties of Grass", J. of Hy. Res., Vol. 26, No. 5, 1988.

Kouwen, N. and R, Li., "Biomechanics of Vegetative Channel Linings", ASCE J. of HY, Vol. 106, No. HY6, June, 1980.

Lacey, G., "Stable Channels in Alluvium", Proc. Inst. for Civ. Eng., 229, 259-292, 1930.

Lane, E. W., "The Importance of Fluvial Morphology in Hydraulic Engineering", Proc. ASCE, Vol. 21, No. 745, 1955.

Little, W. C. and J. B. Murphy, "Model Study of Low Drop Grade Structures," ASCE J. of Hydrau., Vol. 108, No. HY10, Oct. 1982.

Logan, "Vegetation and Mechanical Systems", USDA Forest Service, Missoula, Montana, Feb., 1979.

Loganathan, G. V., "Optimal Design of Parabolic Canals", ASCE J. Ir. & Drainage, Vol. 117, No. 5, 1991.

Maccaferri, "Flexible Linings in Reno Mattress and Gabions for Canals and Canalized Water Courses", Bologna, Italy, 1985.

Maccaferri, "Maccaferri Gabions Channeling Works", Bologna, Italy, 1989.

Maccaferri, "Maccaferri Gabions for River Training, Earth Control and Soil Conservation", Bologna, Italy, 1990.

Maynord, S. T., Stable Riprap Size for Open Channel Flows, Ph.D. Dissertation, Colorado State University, Fort Collins, Colorado, 1987.

McCarley, R. W., Ingram, J. J., Brown, B. J. and Reese, A. J., "Flood-control Channel National Inventory", MP HL-90-10, USAEWES, 1990.

Morris, J. R., A Method of Estimating Floodway Setback Limits in Areas of Approximate Study, In Proceedings of 1984 International Symposium on Urban Hydrology, Hydraulics and Sediment Control, Lexington, Kentucky: University of Kentucky, 1984.

Murphy, T. E., "Drop Structures for the Gering Valley Project, Scotsbluff County, NE", USAEWES Tech. Rpt. 2-760, Feb. 1967.

National Park Service, "Greenway Planning", U.S. Dept. of the Interior, Nat. Park. Svc., Mid-Atlantic Regional Office, Sept., 1983.

National Park Service, "Riverworkbook", Dept. of Interior, Mid-Atlantic Regional Office, 1988.

National Park Service, "River Conservation Directory", Dept. of Interior, National Assoc. for State River Conservation Programs, 1990.

Neill, C., "Hydraulic Design of Stable Flood Control Channels -Draft Guidelines for Preliminary Design", Seattle District COE, Northwest Hydraulic Consultants, Jan. 1984.

Nowatzki, E. A., and B. T. Reely, "Effects of Flyash Content on Strength and Durability Characteristics of Pantano Soil-Cement Mixes", Dept. of Civ. Engineering, U. of Arizona, Tucson, 1986.

Nunnally, N. R. and Shields, F. D., "Incorporation of Environmental Features in Flood Control Projects", TR E-85-3, USAEWES, 1985.

Nunnally, N. R. and E. Keller, "Use of Fluvial Processes to Minimize Adverse Effects of Stream Channelization", Wat. Res. Research Inst., UNC, July, 1979.

Nussbaum, P. J. and B. E. Colley, "Dam Construction and Facing with Soil-Cement" Portland Cement Assoc., R&D Bull., 1971.

Ogden Environmental, "A Nationwide Inventory of Stream Corridor Planning Programs", Ogden Environmental, Nashville, TN, 1993.

Peterska, A. J., Hydraulic Design of Stilling Basins and Energy Dissipators, Engineering Monograph No. 25., U. S. Department of Interior, Bureau of Reclamation, Washington, D. C., 1978.

Petryk, S. and E. U. Grant, "Critical Flow in Rivers with Flood Plains", ASCE J. Hydrau., May, pp. 583-594, 1978.

Pima County Dept. of Trans. and Flood Control Dist., "Soil-cement Applications and Use in Pima County for Flood Control Projects", Tucson, AZ, 1986.

Portland Cement Association (PCA), "Soil-cement Slope Protection for Embankments: Planning and Design", 1975.

Portland Cement Association (PCA), "Soil-cement for Water Control -Laboratory Tests", 1976.

Posey, C. J., "Stability of Rock Sausages", U. of Conn. Inst. of Water Resources, Rpt. No. 19, Nov. 1973.

Powell, R. F., L. D. James and D. E. Jones, "Approximate Method for Quick Flood Plain Mapping", ASCE J. Water Resources, Vol. 106, No. WR1, March, 1980.

Reese, A. J., Riprap Sizing, Four Methods, In Proceedings of ASCE Conference on Water for Resource Development, Hydraulics Division, ASCE, David L. Schreiber, ed., 1984.

Reese, A.J., "Use of Soil-Cement for Bank Protection," DRAFT ETL 1110-2, U.S. Army Corps of Engineers, Waterways Experiment Station, March, 1987.

Reese, A. J., Nomographic Riprap Design, Miscellaneous Paper HL 88-2. Vicksburg, Mississippi: U. S. Army Engineers, Waterways Experiment Station, 1988.

Schiechtl, H., "Bioengineering for Land Reclamation and Conservation", U. of Alberta Press, Edmonton Alberta, CA, 1980.

Schoellhamer, D. H., J. C. Peters and B. E. Larock, "Subdivision Froude Number", ASCE J. Hyd. Vol. 111, pp. 1099-1104, 1985.

Schumm, S. A., "The Fluvial System", Wiley & Sons, Inc., New York, 1977.

Simons, D. B., and F. Senturk, "Sediment Transport Technology", Water Resources Press, Ft. Collins, CO, 1976.

Simons, D. B. and M. L. Albertson, "Uniform Water Conveyance Channels in Alluvial Material", Tranc. ASCE, Vol. 128, Part 1, Paper No. 3399, 1963.

Simons, Li & Assoc., (SLA), "Engineering Analysis of Fluvial Systems", SLA, Ft. Collins, CO, 1982.

Sotir, R. B. and Gray, D. H., "Fill Slope Repair Using Soil Bioengineering Systems", Public Works, Dec. 1989.

State of Illinois, "Stream Preservation Handbook", IDOT, 1983.

Stephenson, D., "Rockfill and Gabions for Erosion Control", Civil Engr. in S. Africa, Sept., 1979.

Temple, D. M., "Tractive Force Design of Vegetated Channels", Proc. ASAE-CSAE Joint Meeting, U. of Manitoba, Winnipeg, CA, June, 1979.

Temple, D. M., K. M. Robonson, R. M. Ahring and A. G. Davis, "Stability Design of Grass-Lined Open Channels", USDA ARS Handbook # 667, 1987.

Thackston, E. L. and R. B. Sneed, "Review of Environmental Consequences of Waterway Design", TR E-82-4, USAEWES, 1982.

Trout, T. J., "Channel Design to Minimize Lining Material Costs", ASCE J. Ir. & Drainage, Vol. 108 No. IR4, Dec., 1982.

U. S. Army Corps of Engineers, "Hydraulic Design of Flood Control Channels", EM 1110-2-1601, 1970.

U. S. Army Corps of Engineers, "Gabion Channel Control Structures", ETL 1110-2-194, 1974.

U. S. Army Corps of Engineers, "Design Criteria - Paved Concrete Flood Control Channels", ETL 1110-2-236, June, 1978.

U. S. Army Corps of Engineers, "Environmental Engineering for Flood Control Channels", DRAFT, 1985.

U. S. Army Corps of Engineers, Waterways Experiment Station (WES), "Hydraulic Design Criteria", Vicksburg, MS, 1988.

U. S. Army Corps of Engineers, "Stability of Flood Control Channels", DRAFT, 1990.

U. S. Army Corps of Engineers, "Hydrologic Engineering Analysis Concepts for Cost-shared Flood Damage Reduction Studies", EP 1110-2-6005, Dec. 1992.

U. S. Department of Agriculture, SCS, "Handbook of Channel Design for Soil and Water Conservation", TP-61, 1954.

U. S. Department of Agriculture, Soil Conservation Service (SCS), "Design of Open Channels", Tech. Release 25, October, 1977.

U. S. Department of Transportation, Federal Highway Administration, Design Charts For Open Channel Flow, Hydraulic Design Series No. 3, Washington, D.C., 1973.

U. S. Department of Transportation, Federal Highway Administration, Hydraulic Design of Energy Dissipators for Culverts and Channels, Hydraulic Engineering Circular No. 14, Washington, D. C., 1982.

U. S. Department of Transportation, Federal Highway Administration, Guide for Selecting Manning's Roughness Coefficients For Natural Channels and Flood Plains, FHWA-TS-84-204, Washington, D. C., 1984.

U. S. Department of Transportation, Federal Highway Administration, Design of Stable Channels with Flexible Linings, Hydraulic Engineering Circular No. 15, Washington, D. C., 1986.

USSR, Maximum Permissible Velocity In Open Channels, Hydro-Technical Construction, Number 5, Moscow, 1936.

Vanoni, V. A., and R. E. Pollak, "Experimental Design of Low Rectangular Drops for Alluvial Flood Channels", Rpt. E-82, Calf. Inst. of Tech., Pasadena, CA, 1959.

White, W. R., R. Bettess and E. Paris, "Analytical Approach to River Regime", ASCE J. of Hy, Vol. 108, No. HY10, 1982.

Wright-McLaughlin Engineers, Urban Storm Drainage Criteria Manual, Vol. 2, Prepared for the Denver Regional Council of Governments, Wright-McLaughlin Engineers, Denver, Col., 1969.

Chapter 11 Storage And Detention Facilities

11.1 Introduction

The traditional design of storm drainage systems has been to collect and convey storm runoff as rapidly as possible to a suitable location where it can be discharged. As areas urbanize this type of design may result in major drainage and flooding problems downstream. The impacts of urban development include: faster and higher peaks, more flow volume, higher velocities, higher temperatures, lower base flows during non-storm conditions, and greater levels of pollution. This chapter deals with flood flow peak reduction, timing strategies and design. A later chapter will address storm water quality (Chapter 15, Storm Water Quality Management Plans).

11.2 Uses And Types Of Storage Facilities

Under favorable conditions, the temporary storage of some of the storm runoff can decrease downstream flows and often the cost of the downstream conveyance system. Detention storage facilities can range from small facilities contained in parking lots or other on-site facilities to large lakes and reservoirs. They can be:
- single purpose or dual purpose;
- on-line or off-line;
- regional or site specific;
- temporary or permanent;
- reduce runoff or delay runoff;
- single objective or multi-objective;
- single outlet or multi-outlet;
- above ground or under ground;
- wet ponds or dry detention;
- integrated with the surroundings or separate from them; and
- structural in nature or developed from natural materials.

This chapter provides general design criteria for detention/retention storage facilities as well as procedures for performing preliminary and final sizing and reservoir routing calculations.

It should be noted that the location of storage facilities is very important as it relates to the effectiveness of these facilities to control downstream flooding. Small facilities will only have minimal flood control benefits and these benefits will quickly diminish as the flood wave travels downstream. Multiple storage facilities located in the same drainage basin will affect the timing of the runoff through the conveyance system which could decrease or increase flood peaks in different downstream locations. Thus it is important for the designer to design storage facilities as a storm water management structure that both controls runoff from a defined area and interacts with other storm water management structures within the drainage basin. Effective storm water management must be coordinated on a regional or basin-wide planning basis. The municipality should encourage and participate in such planning.

Urban storm water storage facilities are often referred to as either detention or retention facilities. For the purposes of this chapter, detention facilities are those that are designed to reduce the peak discharge and only detain runoff for some short period of time. These facilities are designed to completely drain after the design storm has passed. Recharge basins are a special type of detention facility designed to drain into the groundwater table. Retention facilities are designed to contain a permanent pool of water while discharging runoff at a controlled rate. Since most of the basic design procedures are the same for detention and retention facilities, the term storage facilities will be used in this chapter to include detention and retention facilities. If special procedures are needed for detention or retention facilities these will be specified.

Routing calculations needed to design storage facilities, although not extremely complex, are time consuming and repetitive. To assist with these calculations there are many available reservoir routing computer programs. Storage facilities should be designed and analyzed using reservoir routing calculations. An approximate hand calculation method is presented in this chapter which closely approximates reservoir routing for smaller facilities. For very small facilities in highly impervious areas, with accurate target flow reduction criteria, approximate methods may be appropriate. Sometimes retention facilities are designed with no outlet. For this type of facility a long-term water balance is important rather than routing.

Uses Of Storage Facilities

The use of storage facilities for storm water management has increased dramatically in recent years. The benefits of storage facilities can be divided into two major control categories of quality and quantity.

Controlling the quantity of storm water using storage facilities can provide the following potential benefits:
* prevention or reduction of peak runoff rate increases caused by urban development,
* mitigation of downstream drainage capacity problems,
* recharge of groundwater resources, and
* reduction or elimination of the need for downstream outfall improvements.
  Control of storm water quality using storage facilities offers the following potential benefits:
* decrease downstream channel erosion (with proper design) through velocity control and flow reduction,
* reduction of pollution loading through deposition, chemical reaction and biological uptake mechanisms,
* improved base flow conditions,
* aesthetic and ecological habitat benefits at multi-objective sites,
* control of sediment deposition, and
* improved water quality through storm water filtration (wet ponds only).
  The objectives for managing storm water quantity by storage facilities are typically based on limiting peak runoff rates to match one or more of the following values:
* historic rates for specific design conditions (i.e., post-development peak equals pre-development peak for a particular frequency of occurrence),
* non hazardous discharge capacity of the downstream drainage system, and
* a specified value for allowable discharge set by a regulatory jurisdiction.

Types Of Storage Facilities

Table 11-1 gives a number of applications of ways to reduce or delay runoff. It is provided to encourage the urban storm water manager to "think outside the box" when it comes to storm water management design. Urban development will have impacts. Streams in urban areas will never again be rural streams. But these streams do not have to become eyesores nor sterile linear stretches of eroded urban blight. Urban storm water managers are in a position to bring about a change in thinking about how precipitation is handled from the time it first hits a surface until it leaves the municipal jurisdiction. Just as pollution in urban settings is the accumulated product of many minor acts of careless or overt polluting so urban flooding and erosion is the result of many minor (and a few major) development choices. Many municipalities and counties have learned how to handle the problem and to bring about the changes necessary to consider rain as a resource and surface streams as positive neighborhood enhancements.

Table 11-1 Advantages And Disadvantages Of Measures For Reducing Or Delaying Runoff

Measure	Advantages	Disadvantages
Cisterns and Covered Ponds	Alternate uses for the water stored Occupy small areas Land above has alternate uses Water conservation	Expensive to install Not large capacity Restricted maintenance access
Open Space and Grassed Areas	Aesthetically pleasing Cost effective Pollution removal	Land availability Public acceptance problems if not well done Safety issues
Blue-green Storage	Multi-purpose capabilities High public acceptance Aesthetically pleasing Often cost effective	Difficulty finding suitable sites Maintenance special difficulties May be safety hazard
Rooftop Ponding	Runoff delay Cooling effect on building Possible fire protection	Structural loading Clogging possibility if trees overhang Freezing problems during winter Roof leakage
Surface Ponds	Ability to control large areas Can be aesthetically pleasing Multi-purpose capability Aquatic habitat provider Can increase land value Pollution reduction	Requires large areas Possible pollution and eutrophication Pest breeding area Can become urban eyesore Safety hazard Maintenance problems possible
Increased Roughness on Roof - ripples and gravel	Runoff delay	Cost and structural loading

Table 11-1 Advantages And Disadvantages Of
Measures For Reducing Or Delaying Runoff

Measure	Advantages	Disadvantages
Porous Pavement, Gravel and Paving Blocks	Runoff reduction potential Groundwater recharge Gravel may be cheaper Pollution reduction	Cost and maintenance Clogging or earth compaction possibility Groundwater pollution Frostheave Grass and weeds grow through
Grassed Channels and Ditches, Vegetative Strips	Runoff delay Some runoff reduction Aesthetically pleasing Pollution reduction	Land loss Maintenance increase
Ponding on Impervious Areas	Runoff delay Pollution reduction	Restricts other uses when raining Damage due to wetness and freeze-thaw Dirt and debris collection in depressions
Infiltration Devices	Runoff reduction Groundwater recharge Little evaporation loss Pollution reduction Water conservation	Clogging Initial expense Groundwater contamination
"Micro-landscaping", Xeroscaping, Terracing, Flow Routing, Urban Forestry	Runoff reduction and delay Aesthetically pleasing Pollution reduction Water conservation	More expensive to design, construct and maintain

11.3 Design Criteria

Storage may be concentrated in large basin-wide or regional facilities or distributed throughout an urban drainage system. Possible dispersed or on-site storage may be developed in depressed areas in parking lots, road embankments and freeway interchanges, parks and other recreation areas, and small lakes, ponds and depressions within urban developments. The utility of any storage facility depends on the amount of storage, its location within the system, and its operational characteristics. An analysis of such storage facilities should consist of comparing the design flow(s) at a point or points downstream of the proposed storage site with and without storage. In addition to the design flow, other flows in excess of the design flow that might be expected to pass through the storage facility should be included in the analysis (i.e., 100-year flood). The design criteria for storage facilities should include:
- release rate,
- storage volume,
- grading and depth requirements,

- safety considerations and landscaping,
- environmental impacts and multi-objective use,
- outlet works, and
- location.

Commonly, control structure release rates are designed to approximate pre-developed peak runoff rates for the 2-year through 10-year storms, with emergency overflow capable of handling the 100-year discharge. The point of this comparison is either at the outlet of the pond, at the property boundary, or at some point downstream where the developed property makes up some proportion of the total drainage to that point. Design calculations are required to demonstrate that runoff from the 2- and 10-year design storms are controlled. If so, runoff from intermediate storm return periods can be assumed to be adequately controlled. Multi-stage control structures may be required to control both runoff from the 2- and 10-year storms. Other municipalities' basic criteria concentrate on the 2-year and 25-year storms or even control of the 100-year storm. This has been done in an effort to reduce downstream effects of urban development (caused by the increase of volume of flow as much as the peak) to an extent that often cannot be accomplished when the criteria is to meet pre-developed conditions.

If sedimentation during construction causes loss of detention volume, design dimensions should be restored before completion of the project, as built requirements or post construction surveys are often used to satisfy this requirement. For detention facilities, all detention volume should be drained within 72 hours. For a detailed discussion of the many types of detention and design specifications for many types of outlet works see Stahre and Urbonas (1990).

Following is a discussion of the general grading and depth criteria for "typical" storage facilities followed by criteria related to detention and retention facilities.

General Design Criteria

The construction of storage facilities usually requires excavation or placement of earthen embankments to obtain sufficient storage volume. Vegetated embankments should be less than 20 feet in height and should have side slopes no steeper than 3:1 (horizontal to vertical). Riprap-protected embankments should be no steeper than 2:1. Geotechnical slope stability analysis is recommended for embankments greater than 10 feet in height and is mandatory for embankment slopes steeper than those given above. Procedures for performing slope stability evaluations can be found in most soil engineering textbooks, including those by Spangler and Handy (1982) and Sowers and Sowers (1970).

A minimum freeboard of 1 foot above the 100-year design storm high water elevation is often provided for impoundment depths between about 5 to 20 feet. Smaller impoundments may or may not have freeboard requirements. Impoundment depths greater than 20 feet are subject to the requirements of the Safe Dams Act (see Section 11.4) unless the facility is excavated to this depth.

Other considerations when setting depths include flood elevation requirements, public safety, land availability, land value, present and future land use, water table fluctuations, soil characteristics, maintenance requirements, and required freeboard. Aesthetically pleasing features are also important in urbanizing areas.

In designing the detention the designer should gain information on some or all of the following:
- design policy, standards and criteria;
- hydrologic characteristics of the tributary basin or area;
- future development potential and resulting hydrology;
- the downstream conditions;
- risks of design flow exceedance;
- precision and accuracy requirements;

- local practice in design types and methods;
- regulatory and legal requirements;
- backwater impacts;
- low flow requirements;
- the aquatic ecosystem of the stream and potential detention pond impacts;
- dual purpose requirements;
- constructability and value engineering;
- availability of materials and land;
- physical suitability of the site utilities, easements, pedestrian access, building location, open space reservation, etc.;
- soils suitability;
- groundwater tables and locations of wells and other potable water supplies;
- special factors such as archeological sites, historical sites, endangered species, wetlands, rock formations, etc.;
- maintenance requirements;
- cost and schedule limitations;
- aesthetic and recreational requirements;
- human interaction with the site; and
- institutional, social and political concerns.

Specific Detention Design Criteria

Areas above the normal high water elevations of storage facilities should be sloped at a minimum of 5 percent toward the facilities to allow drainage and to prevent standing water. Careful finish grading is required to avoid creation of upland surface depressions that may retain runoff. The bottom area of storage facilities should be graded toward the outlet to prevent standing water conditions. A minimum 2 percent bottom slope is recommended. A low flow or pilot channel constructed across the facility bottom from the inlet to the outlet is recommended to convey low flows, and prevent standing water conditions. Often a sediment collection forebay is provided with easy maintenance access.

Specific Retention Design Criteria

The maximum depth of permanent storage facilities will be determined by site conditions, design constraints, climatic conditions, and environmental needs. In general, if the facility provides a permanent pool of water, a depth sufficient to discourage growth of weeds, without creating undue potential for anaerobic bottom conditions, should be considered. A depth of 6 to 8 feet is generally reasonable unless fishery requirements dictate otherwise. Aeration may be required in permanent pools to prevent anaerobic conditions. Where aquatic habitat is required the cognizant wildlife experts should be contacted for site specific criteria relating to such things as depth, habitat, and bottom and shore geometry. In some cases a shallow bench along the perimeter is constructed to encourage emergent vegetation growth to enhance the pollution reduction capabilities, safety or aesthetics of the pond.

Outlet Works Design Criteria

Outlet works selected for storage facilities typically include a principal spillway and an emergency overflow, and must be able to accomplish the design functions of the facility. Outlet works can take the form of combinations of drop inlets, pipes, weirs, and orifices. The principal spillway is intended to convey the design storm(s) without allowing flow to enter an emergency outlet. For large storage facilities, selecting a flood magnitude for sizing the emergency outlet should be consistent with the potential threat to downstream life and property if the facility embankment were to fail. The minimum flood to be used to size the emergency outlet for a more major system pond is the 100-year flood. For smaller ponds in some municipalities lesser floods are used for the emergency overflow spillway. The sizing of a particular outlet works should be based on results of hydrologic routing calculations.

Outlet works are designed to fit a specific design standard or to meet a specific need. Often the standard for storm water quantity control (see Structural Best Management Practices, Chapter 13, and Storm Water Quality Management Plans, Chapters 15, for a discussion of dual quality-quantity designs) is to control several different storm frequencies with one pond. Sometimes a specific range of frequencies is of particular interest, or a certain location within the watershed must be specifically protected. All outlet works should strive to meet stated goals while considering safety, aesthetics, maintenance minimization, longevity, debris handling, cost and simplicity of design.

Location Design Criteria

Stahre and Urbonas (1990) classify storage as either source control or downstream control. Source control consists of a larger number of smaller basins located high up in the drainage or conveyance system. They might include local percolation, injection or infiltration basins, smaller inlet control basins and on-site detention. Downstream controls are normally on- or off-line larger open pond type basins or large dry detention ponds. Source controls have the advantage in that they are lower cost to build, can be targeted to a certain area or pollution source and type, and are very flexible in design types. Maintenance and regulation are more difficult and costly for many source controls providing the same level of protection as fewer and larger downstream controls, and source controls are not normally effective for controlling flooding very far downstream. Downstream controls offer greater opportunities for larger scale multiobjective site use. On the other hand downstream controls require coordination for funding and development (perhaps master planning), require more upfront capital funding, may have greater environmental impacts in the forms of "abandonment" of reaches upstream from the pond and impact on flow interruption and riparian wetlands. Thus there is a trade-off between source and downstream controls. Usually, through master planning, a mix of both is found to be most appropriate.

In addition to controlling the peak discharge from the outlet works, storage facilities will change the timing of the entire hydrograph. If several storage facilities are located within a particular basin it is important to determine what effects a particular facility may have on combined hydrographs in downstream locations. Great cost savings can be realized by combining detention with channel improvements, oversizing detention in one area to compensate for no detention in another, and using different criteria for the sizing of the ponds based on the needs of the site or watershed (Hartigan, 1986, Hartigan and George, 1989, Jones, 1990, Maryland, 1986, 1987, James, et al., 1987, Tyrpak, 1990, McCuen and Moglen, 1987).

Detention can be located within floodplains and still effectively control flooding through the use of timing calculations. In this situation the flood peak coming down the stream rarely coincides with local on-site flooding. It is often advantageous to allow the on-site water to pass with simple erosion control and a properly sized conveyance system. Then locate the detention pond to "skim" the peak from the oncoming flood hydrograph through the use of a side-channel weir or a simple flow-through depression along the banks (Smith and Cook, 1987).

For all storage facilities, channel routing calculations should proceed downstream to a suitable location (normally to the point where the controlled land area is less than ten percent of the total drainage to that point). At this point the effect of the hydrograph routed through the proposed storage facility on the downstream hydrograph should be assessed and shown not to have detrimental effects on downstream areas.

11.4 Safe Dams Act

National responsibility for the promotion and coordination of dam safety lies with the Federal Emergency Management Agency (FEMA). In addition, state agencies have some responsibility for administration of the provisions of the Federal Dam Safety Act. Under the federal regulations, a dam is an artificial barrier that does or may impound water and that is 20 feet or greater in height or has a maximum storage volume of 30 acre-feet or more. Detailed engineering requirements are given in the regulations for new dams and these regulations should be consulted for all engineering requirements. A number of exemptions are allowed from the Safe Dams Act and the cognizant state office should be contacted to resolve questions.

Dams are classified as either new or existing, by hazard potential, and by size. Hazard potential categories are listed below.

Category 1 dams are located where failure would probably result in:
* loss of human life,
* excessive economic loss due to damage of downstream properties,
* public hazard, or
* public inconvenience due to loss of impoundment and/or damage to roads or any public or private utilities.

Category 2 dams are located where failure may damage downstream private or public property, but such damage would be relatively minor and within the general financial capabilities of the dam owner. Public hazard or inconvenience due to loss of roads or any public or private utilities would be minor and of short duration. Chances of loss of human life would be possible but remote.

Category 3 dams are located where failure may damage uninhabitable structures or land but such damage would probably be confined to the dam owner's property. No loss of human life would be expected.

Dams are also classified as either small, intermediate or large depending on their storage capacity and height. Table 11-2 gives the criteria for this classification.

Table 11-2 Size Categories For Dam Classification

Category	Storage (acre-feet)	Height (ft)
Small	30 to < 1,000	20 to < 41
Intermediate	1,000 to 50,000	41 to 100
Large	> 50,000	> 100

11.5 General Design Procedure For Storage Routing

The following data will be needed to complete storage design and routing calculations.
* Inflow hydrograph for all selected design storms.
* Stage-storage curve for proposed storage facility (see Figure 11-1 below for an example).
* Stage-discharge curve for all outlet control structures (see Figure 11-2 below for an example).

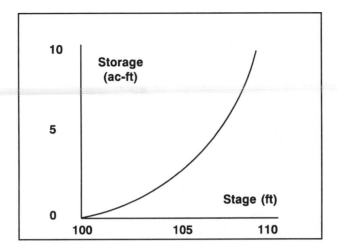

Figure 11-1 Example Stage-Storage Curve

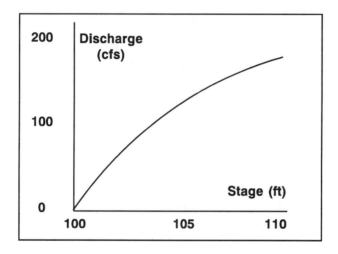

Figure 11-2 Example Stage-Discharge Curve

Using these data a design procedure is used to route the inflow hydrograph through the storage facility with different basin and outlet geometry until the desired outflow hydrograph is achieved.

Stage-Storage Curve

A stage-storage curve defines the relationship between the depth of water and storage volume in a reservoir. The data for this type of curve are usually developed using a topographic map and the double-end area, frustum of a pyramid, prismoidal formulas or circular conic section. The double-end area formula is expressed as:

$$V_{1,2} = [(A_1 + A_2)/2]d \qquad (11.1)$$

Where: $V_{1,2}$ = storage volume, ft^3, between elevations 1 and 2

A_1 =   surface area at elevation 1, ft^2
A_2 =   surface area at elevation 2, ft^2
d =   change in elevation between points 1 and 2, ft

The frustum of a pyramid is expressed as:

$$V = [d/3] [A_1 + (A_1 \times A_2)^{0.5} + A_2]/3 \tag{11.2}$$

Where: V =   volume of frustum of a pyramid, ft^3
Other terms as defined above.

The prismoidal formula for trapezoidal basins is expressed as:

$$V = LWD + (L + W) ZD^2 + 4/3 Z^2 D^3 \tag{11.3}$$

Where: V =   volume of trapezoidal basin, ft^3
L =   length of basin at base, ft
W =   width of basin at base, ft
D =   depth of basin, ft
Z =   side slope factor, ratio of horizontal to vertical

The circular conic section formula is:

$$V = 1.047 D (R_1^2 + R_2^2 + R_1R_2) \tag{11.4}$$
$$V = 1.047 D (3R_1^2 + 3ZDR_1 + Z_2D^2) \tag{11.5}$$

Where: R_1 and R_2 = bottom and surface radii of the conic section, ft
D =   depth of basin, ft
Z =   side slope factor, ratio of horizontal to vertical

Often the stage-storage curve can take the form of a power curve where storage is a function of a constant times stage to some power. This can be easily done through the use of computerized spreadsheets by taking the logarithm and fitting a straight line to sets of area-elevation or volume-elevation points. In fitting a dam to an existing site, areas or volumes can be determined by planimeter. For areas where detention will be excavated one of the formulas above can be used and size, depth, shape and side-slope adjusted to obtain the necessary storage volume.

Stage-Discharge Curve

A stage-discharge curve defines the relationship between the depth of water and the discharge or outflow from a storage facility. A typical storage facility has two spillways: principal and emergency. The principal spillway is designed with a capacity sufficient to convey the design flood without allowing flow to enter the emergency spillway. A pipe culvert, weir, or other appropriate outlet can be used for the principal spillway or outlet. The emergency spillway is sized to provide a bypass for floodwater during a flood that exceeds the design capacity of the principal spillway. This spillway should be designed taking into account the potential threat to downstream life and property if the storage facility were to fail. The stage-discharge curve should take into account the discharge characteristics of both the principal and emergency spillways.

General Procedure

Following are the steps in a general procedure for using the above data in the design of storage facilities.

Step 1   Compute inflow hydrograph for runoff from the 2-, 10-, and 100-year design storms using the procedures outlined in the Urban Hydrology, Chapter 7 (or use the local design criteria). Both pre- and post-development hydrographs are required for the 2- and 10-year design storms. Only the post-development hydrograph is required for runoff from the 100-year design storm.

Step 2   Perform preliminary calculations to evaluate detention storage requirements for the hydrographs from Step 1 (see Section 11.7). If storage requirements are satisfied for runoff from the 2- and 10-year design storms, runoff from intermediate storms is assumed to be controlled.

Step 3   Determine the physical dimensions necessary to hold the estimated volume from Step 2, including freeboard. The maximum storage requirement calculated from Step 2 should be used. From the selected shape determine the maximum depth in the pond.

Step 4   Select the type of outlet and size the outlet structure. The estimated peak stage will occur for the estimated volume from Step 2. The outlet structure should be sized to convey the allowable discharge at this stage.

Step 5   Perform routing calculations using inflow hydrographs from Step 1 to check the preliminary design using the storage routing equations. If the routed post-development peak discharges from the 2- and 10-year design storms exceed the pre-development peak discharges, or if the peak stage varies significantly from the estimated peak stage from Step 4, then revise the estimated volume and return to step 3.

Step 6   Consider emergency overflow from runoff due to the 100-year (or specified) design storm and established freeboard requirements.

Step 7   Evaluate the downstream effects of detention outflow to ensure that the routed hydrograph does not cause downstream flooding problems. The exit hydrograph from the storage facility should be routed though the downstream channel system until a confluence point is reached. Example criteria for locating the confluence point is where the drainage area being analyzed represents ten percent of the total drainage area (Debo & Reese, 1992).

Step 8   Evaluate the control structure outlet velocity and provide channel and bank stabilization if the velocity will cause erosion problems downstream.

This procedure can involve a significant number of reservoir routing calculations to obtain the desired results. Because there are almost limitless methods for design of outlet structures no generic guidance will cover every situation. But, for outlet structures which must control multiple storm frequencies there are several approaches. Normally, for the 2-year and 10-year storm scenario, the 10-year storm volume will determine the size of the pond while the 2-year storm peak control need will determine the size of a minimum outlet structure. Thus the pond is sized for the 10-year storm. Then the 2-year storm is routed through the pond and the 2-year outlet structure is sized. The depth of the storage when the 2-year storm is routed through the pond then normally becomes the elevation of the invert or weir for the 10-year storm outlet. The 10-year storm is then routed through both outlets.

11.6 Outlet Hydraulics

The outlet is the heart of the detention and retention design. The outlet meters the water through the pond at a controlled rate and in a controlled way. There are many types and combinations of outlet works. Most of these outlet works consist of a combination of weirs and orifices in conjunction with outlet pipes or conduits. Basic equations for weirs, orifices and pipes are given in the Data Sources And Collection, Chapter 6, in the flow measurement section. If culverts are used as outlets works or HW/D criteria are not met, such that orifices act as culverts, procedures presented in the Design Of Culverts, Chapter 9, should be used to develop stage-discharge data. Combinations of various components and special use outlet works are discussed in this chapter.

It is important to obtain accuracy in the design of outlet works through proper estimation of discharge coefficients, submergence, rating curves, control shifts, etc. But it should be remembered that the accuracy of hydrologic estimates is often very poor (McCuen, 1983), and therefore over-zealous optimization of detention is hardly warranted (ASCE, 1985).

The decision to use off-line storage versus on-line storage will effect the type of outlet structure used and its size. Nix and Tsay (1988) have shown that the use of off-line storage can reduce the size of the basin required by as much as 20 to 40 percent. This kind of storage can be used, for example, with a channel diversion such as a side-channel weir to "skim" higher flows into the detention structure at its upstream end for later release at the detention structure's downstream end. The structure then lies parallel to the flow channel.

Combination Outlets

Combinations of weirs, pipes and orifices can be put together to provide a variable control stage-discharge curve suitable for control of multiple storm flows. They are generally of two types: shared outlet control and separate outlet controls. Shared outlet control is typically a number of individual outlet openings, weirs or drops at different elevations on a riser pipe or box which all flow to a common larger conduit or pipe. Separate outlet controls are less common and normally consist of a single opening through the dam of a detention facility in combination with an overflow spillway for emergency use.

The most common shared outlet control practice is to place a low flow outlet at the base of a riser pipe. The riser pipe can then serve as an emergency overflow spillway and could also have openings, slots or perforations in it for variable discharge by stage. In this case separate stage-discharge curves are computed for each element of the outflow and for each type of possible control. The curve which gives the highest stage for a given flow is expected to control. The total stage-discharge curve is the upper stage envelope of all possible shared outlet stage-discharge curves. Figure 11-3 illustrates this practice and Figure 11-4 illustrates the various curves and the controlling curve. For the case of the overflow riser pipe diagrammed in Figure 11-3 the control can follow a three step process.

- Initially as the flow reaches and begins to overflow the riser pipe, weir flow controls and an equation of the weir form is applicable.
- As the stage rises the flow begins to interfere with itself around the circular opening and eventually the flow control transitions to that of orifice flow.
- If the flow is great and the outlet pipe is small in comparison, the barrel of the pipe may become the controlling factor and the outlet operates as a barrel control or outlet control culvert.

Perforated Risers

To extend detention times risers can be designed with perforations, though clogging potential is high for such designs. In that case the top of the riser is left open with a trash rack and vortex plate attached. The flow through the perforations acts like an orifice with the head measured from the center of the opening to the water surface. Holes should normally be a minimum of three diameters, center-to-center, apart. If the pipe is thick the discharge will be that of a short tube with appropriate adjustment in discharge coefficient. If it is corrugated metal pipe it will act as an orifice. Clogging can be reduced with the use of a jacket of gravel and wire mesh around the pipe. Various designs are provided by MWCOG, 1987, UDFCD, 1992 and Washington State Ecology, 1992. Figure 11-5 illustrates this practice and Table 11-3 gives hole spacing and sizes.

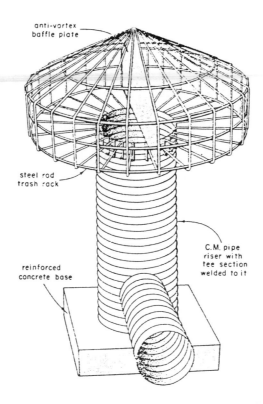

Figure 11-3 Corrugated Metal Pipe Riser With Trash Rack And Baffle

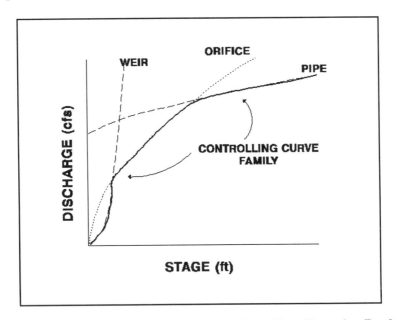

Figure 11-4 Stage-Discharge Curve For Riser Type Detention Pond

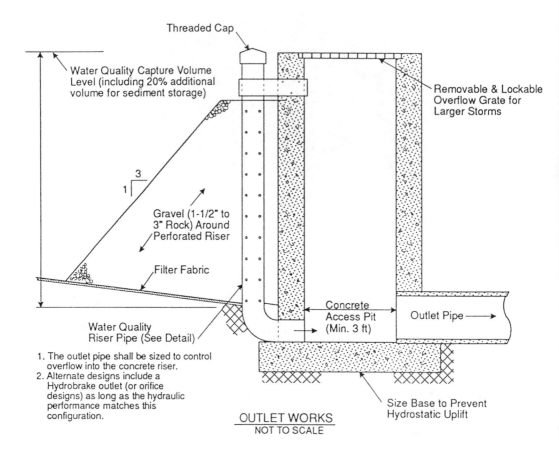

Figure 11-5 Typical Perforated Riser Outlet Design (UDFCD, 1992)

There is some question concerning the accuracy of using a simple orifice equation for such designs (MWCOG, 1987). McEnroe, et al., (1988) conducted experiments with perforated risers and developed an equation to predict flows through the perforations in the riser. The outlet capacity must be greater than the riser for the equation to retain validity.

$$Q = C_h \frac{2A_h}{3h_h} \sqrt{2g}\, h^{3/2}$$

(11.6)

Where: Q = discharge through all the riser holes, cfs
  C_h = discharge coefficient given as 0.611 by the author
  A_h = area of all the holes in the perforated section, ft^2
  h_h = distance from d/2 below the centerline of the lowest hole to d/2 above the center-line of the highest hole, ft (where d is the vertical) spacing between holes, ft
  h = head on riser pipe measured from d/2 below the centerline of the lowest hole, ft

Table 11-3 Maximum Number Of Perforated Columns (UDFCD, 1992)

Riser Diameter (in.)	Hole Diameter (in.)			
	1/4"	1/2"	3/4"	1"
4	8	8	-	-
6	12	12	9	-
8	16	16	12	8
10	20	20	14	10
12	24	24	18	12

Hole Diameter (in)	Area of Hole (in^2)
1/8"	0.013
1/4"	0.049
3/8"	0.110
1/2"	0.196
5/8"	0.307
3/4"	0.442
7/8"	0.601
1"	0.785

Two-Way Drop Inlets

A two-way drop inlet is a combination inlet/outlet structure to a conduit or channel in which the flow discharges over one of two weirs, which are parallel to the axis of the outlet channel or conduit. Figure 11-6 gives details of the design and construction (HDC, 1988). As in other combination inlets the flow transitions through three stages. The Corps of Engineers recommends that for larger applications the inlet be designed to avoid the unstable, slug flow orifice flow regime. The basic procedure is to assume a weir crest length and to plot each of the rating curves and ensure orifice flow never controls. The conduit control curve is plotted with two friction factors, high and low, to establish a sensitivity range. When a satisfactory plot is achieved the anti-vortex plate is set one foot above the pool elevation required to establish conduit controlled flow. The following design information applies (SCS, 1965).

- Initially determine the minimum pond water surface elevation to set the weir crest elevations.
- Determine the conduit invert, diameter and length from flow calculations.
- Determine, from structural considerations, dimensions T and E (Figure 11-6).
- Weir flow discharge coefficient has been determined to be 3.8 when the weir is rounded to a radius of half the thickness of the weir walls (See Figure 11-6).
- Orifice flow calculations are done using the following equations:

$$Q = C_o \, A_o \, (2gH_w)^{1/2} \tag{11.7}$$
$$A_o = (1/2) \, L_w \, (D-E) \tag{11.8}$$

Where: Q = discharge, cfs

C_o = orifice discharge coefficient
A_o = area of the orifice, ft^2
H_w = head on the weir, ft
D = conduit diameter, ft
E = separation wall thickness, ft
L_w = length of the weir, ft

C_o can be determined from:

$$C_o = C_1 \left(\frac{E}{D}\right)^{0.083} \left(\frac{L_W}{2d}\right)^{-0.2934}$$

(11.9)

$$C_1 = -15.6993 \left(\frac{T}{D}\right)^2 + 11.3136 \left(\frac{T}{D}\right) - 0.2032$$

(11.10)

T = thickness of the wall, ft

• Conduit control is developed when full pressure flow exists throughout the inlet structure and is determined from:

$$Q = A \sqrt{\frac{2gH_c}{K}}$$

(11.11)

Where: A = conduit cross sectional area, ft^2
g = gravitational constant, ft/s^2
H_c = head on pipe equal to the difference between the upper pool elevation and the hydraulic gradeline elevation at the exit portal, ft (use the outlet pipe center elevation for freeflow conditions and tailwater elevation for submerged conditions).
K = total loss coefficient on velocity head (i.e. KV^2/2g) equal to entrance loss coefficient (0.2 for this design) plus friction loss in pipe plus 1.0 exit loss.

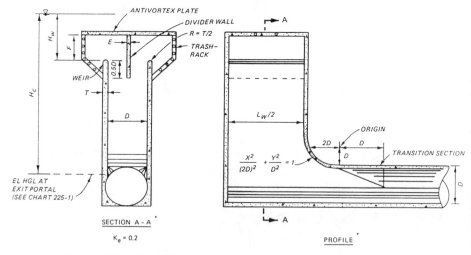

Figure 11-6 Two Way Drop Inlet (COE, 1988)

Drop type inlets can also be placed at different elevations as in Figure 11-7 to gain control over several different flow frequencies (Rossmiller, 1982). In this figure the lower inlet pipe was sized to control the 2-year flow while the two drop inlets were sized for the 5-year through 10-year and 25-year through 100-year flows respectively.

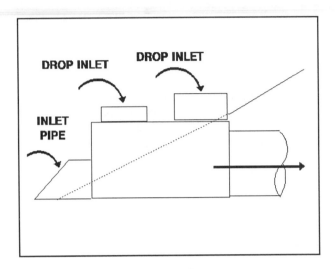

Figure 11-7 Multiple Drop Inlets

Weir, Orifice And Drop Inlet Boxes

Figure 11-8 illustrates three different types of box type inlet structures for control of multiple discharges (after Rossmiller, 1992).

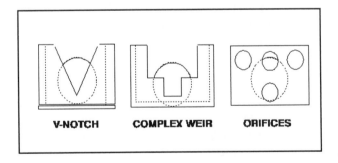

Figure 11-8 Inlet Boxes For Detention Control

In the first of these a V-notch weir is used as the controlling mechanism. As flow increases the ability to pass that flow also increases. The spread of the V-notch is sized to pass flow at the correct rate for several design flows. The flow through the notch needs to be adjusted for submergence effects based on the inlet head on the conduit needed to pass the flow through the V-notch. At some point the flow through the notch will not be sufficient to pass the required discharge due to submergence. At that point the top of the box elevation is set and flow begins to pass over the box and through the conduit.

The second of the figures is a complex weir. In this case a more narrow weir lower in elevation is used for the more frequent storm. A wider weir is then placed at the maximum elevation of the more frequent flood to handle necessary attenuation for less frequent storms. The 100-year storm is then passed over the top of the first two weirs and the flow box. The stage discharge curve is calculated as the combination of the inner lower weir, and the head on it, and the two pieces of the upper outer weir, and the lesser stage on it. The inner weir is not contracted while the outer weir has end contractions as appropriate to the design. The flow coefficient is approximate since this type of weir has not been extensively model tested. This type of weir also works well as a spill-through type opening through a dam.

The third type of box type outlet consists of multiple orifices at different elevations passing through the box walls. The orifices are placed such that the lower orifice(s) control a more frequent flow while the upper orifice invert is placed at an elevation to begin to operate for a less frequent design storm. There is significant trial and error in the placement of orifices. It should also be remembered that orifice flow does not begin until the orifice is suitably submerged. Until that time the orifice will act as a weir and culvert design equations would be applicable, though calculations for such low flows are rarely warranted.

Side-Channel Weirs

Side channel spillways are located along channel banks with the spillway crest parallel to the alignment of the channel. As the water level in the channel or stream rises above the crest the excess is diverted into the side channel. This type of structure works well in urban settings for flood diversions, skimming surface waters or pollution from streams, irrigation purposes, or for detention ponds located along channels, CSO or other storm sewer overflows or in floodplains. In conjunction with debris or surface pollution removal a floating boom is often placed diagonally across the channel (USBR, 1978).

While the flow from a side weir is very complex and dependent on a large number of factors, according to the USBR (1978) for normal purposes where flow is relatively tranquil, a standard weir equation with discharge coefficient will suffice. The exit channel must be set to allow for free flow over the weir without submergence. Overtopping of the weir walls should be prevented with freeboard. In closed conduit situations, open channels with faster flows, or, where a grate is needed for accuracy, the complexities of side-channel weir design must be dealt with. The water surface profile encountered along the lateral side weir is varying as is the discharge over the weir. The normal approach is to determine the discharge per unit length of weir, attempting to account for a large number of factors, and then integrate along the weir using a momentum or constant specific energy approach. Various authors have approached the problem for different types of weirs and channel shapes: Hager - general discussion of factors (1987), Uyumaz and Smith - design procedure for circular and rectangular channels (1991), Uyumaz - triangular channels (1992), and Cheong (1991) or Ramamurthy, et al., - trapezoidal channels (1986).

Metcalf and Eddy (1972) provide a description of the three types of situations encountered in side-channel weir design. Figure 11-9 shows possible water surface profiles for: (a) steeply sloping supercritical flow situation, (b) subcritical flow situation with the crest elevation above critical depth, and (c) subcritical flow situation with the crest elevation below critical depth. In a and c the water surface is falling along the axis of the weir crest. In b it is actually rising, if small amounts of water are withdrawn. If the ratio of the height of the weir to the specific energy is less than about 0.6, a falling water surface profile is likely (Metcalf & Eddy, 1972). Approximate solutions for the falling water surface profile and rising water surface profile respectively are:

$$Q_w = 0.67 \ L^{\ 0.72} \ E_w^{\ 1.645} \tag{11.12}$$
$$Q_w = 3.32 \ L^{0.83} \ E_w^{\ 1.67} \tag{11.13}$$

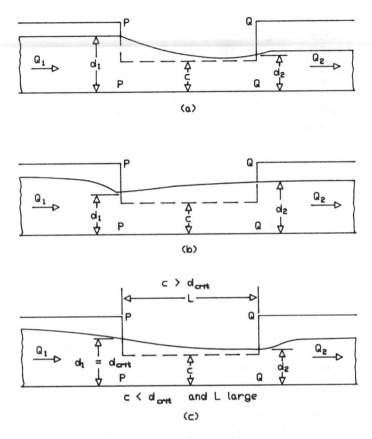

Figure 11-9 Side-Channel Weir Water Surface Profiles
(Source: Metcalf & Eddy, 1972)

Based on the work of Akers (1957), Metcalf & Eddy (1972), De Marchi (1934) and Uyumaz and Smith (1991) the following design procedure can be used with the assumption that specific energy is constant along the weir (found to be true within ± 3 percent).

The variables used in the following design procedure are:

B,D   = channel width or diameter, ft
c,p   = height of weir above channel, ft
m   = discharge coefficient of weir
$h_{1,2}$ = up- and downstream heads on the weir, ft
D   = hydraulic depth, cross sectional area/free surface width, ft
$d_{1,2}$ = depths of flow up- and downstream, ft
E   = specific energy relative to the channel bottom, ft
E_w = specific energy relative to the weir crest elevation, ft
$F_{1,2}$ = up- and downstream Froude numbers at ends of weir
L   = length of weir, ft
$L_{1,2}$ = length of weir in rising water surface case, ft

n_2 = ratio of h_1/h_2
$Q_{1,2}$ = discharge up- and downstream, cfs
$V_{1,2}$ = velocity up- and downstream, ft/s
$y_{1,2}$ = flow depths up- and downstream ends of weir, ft
$\alpha_{1,2}$ = velocity coefficient up- and downstream
α' = pressure-head correction
Θ = channel slope angle
Φ = varied flow function

The normal situation is to have an existing or designed channel with a known flow and a desired reduction in that flow, or flow level, to achieve some goal. The side weir is then designed to remove the desired amount of flow from the channel when it is flowing at the design flow.

1. Calculate the upstream depth, velocity and flow type.
- Upstream depth is calculated from the known flow and flow geometry using the Manning equation.
- Upstream velocity is calculated from the known flow and flow geometry using the Manning equation.
- Calculate the upstream Froude number. According to Metcalf & Eddy (1979) α may be taken as 1.2 upstream and 1.4 downstream.

$$F_1 = \frac{V_1}{\sqrt{g D_1 \cos\theta / \alpha_1}}$$ (11.14)

If the Froude number is greater than 1 the flow is supercritical (upper profile in Figure 11-9). If it is less than 1 the flow is subcritical (the lower two water surface profiles in Figure 11-9).

2. Compute critical depth in the channel and set the trial weir height.
- By setting the Froude number equal to 1 in equation 11.14, critical depth is determined.
- Set the weir height above the channel bottom. For subcritical flow, if the weir height is set above critical depth the water surface will be rising. By setting the weir height below critical depth the water surface will be falling. For supercritical flow the water surface will be falling. Another check is given by Metcalf & Eddy (1979), if the ratio of the weir height to the specific energy is less than about 0.6 the water surface is falling.

Note that there may be a number of factors which go into setting the weir height including: low flow maintenance restrictions, physical geometry of the channel, need for tail channel slope or free overfall, need to minimize weir length, etc.

3. Compute the specific energy.
- The specific energy relative to the weir crest elevation may be computed as (α' is taken as 1.0 upstream and 0.95 downstream):

$$E_w = \alpha \frac{V^2}{2g} + \alpha'(d-c)$$ (11.15)

- The head on the weir is defined as the depth of flow minus the weir height in feet.

4. Find the weir length.
- Compute c/E_w
- Find the weir length from (Akers, 1957):

$$L = 2.03\ B\ (5.28 - 2.63\ c/E_w)$$ (11.16)

Note: In this equation the ratio of the up- and downstream heads on the weir is set to 10.0
5. Check length for flow target (optional).
•   Solve for the downstream head on the weir. The flow goes through critical depth at the upstream end. At critical flow:

$$\alpha V_1^2/2g = \alpha' h_1/2 \tag{11.17}$$

Then the specific energy is:

$$E_w = \alpha' h_1/2 + \alpha' h_1 \tag{11.18}$$
$$h_1 = E_w/(1.5\alpha') \text{ and } h_2 = 0.1 h_1$$

•   Substitute h_2 into the specific energy equation to find V_2.
•   From the continuity equation $Q = VA$ find Q_2 and compare it to the target.
For the rising water case the determination of weir length and flow over the weir is handled in a different way. The problem becomes one of calculating the channel discharge at the beginning and ending of an assumed weir length, the difference being the discharge over the side weir. Knowing conditions at section 1 (Q, y_1), Φ_2 and thus Q_2 can be found. Alternately a Q_2 value can be set and the L value determined. See Chow (1959), Subramanya and Awasthy (1972), Metcalf & Eddy (1972), and Uyumaz and Smith (1991) for more details. The general equations are:

$$L_{1,2} = 3/2 \ B/m \ (\Phi_2 - \Phi_1) \tag{11.19}$$

$$\Phi = \frac{2E-3c}{E-c} \sqrt{\frac{E-y}{y-c}} - 3\sin^{-1}\sqrt{\frac{E-y}{E-c}} \tag{11.20}$$

The discharge coefficient is given as:
•   for rectangular channels $F_1 < 1$

$$m = \frac{2}{3} 0.611 \sqrt{1 - \left(\frac{3F_1^2}{F_1^2 + 2}\right)} \tag{11.21}$$

•   for rectangular channels $F_1 > 2$

$$m = 2/3 \ (0.036 - 0.008 \ F_1) \tag{11.22}$$

Discharge coefficients for other types of channels and weirs should be taken from the above referenced literature.

Example Of Side Channel Weir

A rectangular channel is 10 feet wide and flows 5 feet deep. The slope is 0.001 and Manning's n is taken as 0.014. A side channel weir is desired to reduce the depth to about 3.7 feet.
•   From the Manning equation the upstream normal depth flow characteristics are:
    $Q_1 = 310$ cfs $V_1 = 6.20$ ft/s

- The desired flow characteristics downstream from the weir are:

   $Q_2 = 206$ cfs $V_2 = 5.56$ ft/s

- The Froude number in the approaching flow is:

   $F = V/(gD \cos\theta/\alpha)^{1/2} = (6.20)/[(32.2 * 5 * 1)/1.2]^{1/2} = 0.035$ (subcritical)

- At critical depth $F = 1$ and:

   $1 = Q/(B\ D_c^{3/2}\ g^{1/2}) = 310/(10 * D_c^{3/2} * 32.2^{1/2})$ therefore, $D_c = 3.1$ ft

- Set weir height below critical depth (after several trials it is set at 1.2 ft)

- Specific energy is:

   $E_w = \alpha V^2/2g + \alpha'(y\text{-}c) = (1.2 * 6.2)^2/(2 * 32.2) + (5\text{-}1.2) = 4.516$ ft

- $c/E_w = 1.2/4.516 = 0.266$ (falling water surface profile)

- Weir length is calculated as:

   $L = 2.03\ B\ (5.28 - 2.63\ c/E_w) = 2.03*10 * (5.28 - 2.63*0.266) = 93$ ft

- Checking flow target, h_1 is at critical flow upstream where depth equals twice the velocity head and substituting into the specific energy equation for the velocity head:

   $h_2 = 0.1\ h_1 = (0.1)*(2/3)*(4.516) = 0.301$ ft

   $E_w = \alpha V^2/2g + \alpha'(h_2) = (1.4* V_2)^2/(2*32.2)+(0.95*0.301) = 4.516$ ft

   $V_2 = 13.95$ ft/s

   $y_2 = h_2 + c = 0.301 + 1.2 = 1.501$ ft

   $Q_2 = AV = (10)(1.501)(13.95) = 209$ cfs. OK

Rooftop Detention

Rooftop detention is often a viable option for larger commercial and industrial facilities. Water weighs 62.4 pounds per cubic foot. Six inches of standing water places a loading of 31.2 pounds per square foot. This is roughly equivalent to the snow loading requirement in many municipalities, though most building codes are uniformly silent on rooftop detention. Most municipalities limit storage depth on rooftops to about three inches. To help prevent clogging, designs often include the provision for overtopping into an unrestricted downspout and periodic inspection. Roof deflection may need to be considered in storage calculations (APWA, 1981). Often roofs for detention storage are dead level or have a pitch of 0.25 inch per foot.

Figure 11-10 gives the detail of a typical roof drain restrictor device. When water reaches the top of the ring it spills into the center and through the strainer section. Minimum hole spacing is 2 inches on center. Discharge capacity for this type of drain per set of holes (half hole plus full hole above, 1" in diameter) in cfs is:

- 0.0022 cfs for 1.5 in. depth
- 0.0046 cfs for 2.0 in. depth
- 0.0114 cfs for 2.5 in. depth

- 0.0137 cfs for 3.0 in. depth
- 0.0156 cfs for 3.5 in. depth
- 0.0179 cfs for 4.0 in. depth

For this type of restrictor the 6 inch diameter collar has 9 sets of holes and the 12 inch has 18 sets. Other devices offer: inverted parabolic openings for a linear depth inflow relationship, loading distribution pads built into the device, sumped designs, and durable plastic or aluminum construction.

Design and analysis is normally fairly simple often being more of a mass balance than a true routing. Water is stored as a wedge based on roof slope toward the drain line. Simple triangular, graphical or chainsaw routing, modified rational methods (Aaron and Kibler, 1990, Curry, 1974) or even approximate probability methods (Meredith, et al., 1990) can provide reasonable results for this type of application provided storm volumes are correct; however errors can be encountered which may result in inadequate designs and protection of downstream facilities (see later discussion of the modified rational method). Tables of required storage and allowable release rates based on roof top area can be prepared for a specific site and rainfall depth. An example of this type

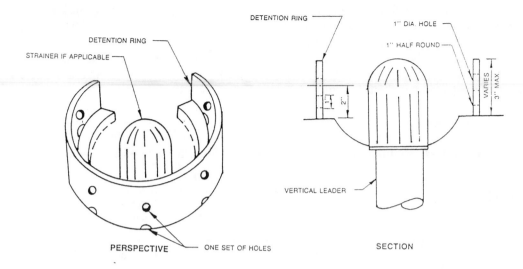

DETENTION RING

DETENTION RING

STRAINER IF APPLICABLE

1" DIA. HOLE

1" HALF ROUND

VARIES
3" MAX

VERTICAL LEADER

PERSPECTIVE   ONE SET OF HOLES   SECTION

Figure 11-10 Typical Roof Drain Restrictor Device (Source: Bellevue, 1988)

of analysis is given below. Care should be taken in gaining a clear understanding of assumptions in such a mass balance method and accounting for any possible underpredictions of resulting storage volumes. These examples are given for illustration purposes and the authors do not recommend the use of these methods. The following examples show the large discrepancy that can be obtained from using approximate methods and thus routing is recommended for the design of all storage and detention facilities.

Frequent inspection is one of the keys to successful rooftop detention. When rooftops are near trees leaves can clog the outlet devices allowing water to stand for long periods of time; sometimes leading to roof leaks (Poertner, 1974).

Example Rooftop Design

A rooftop with an area of 1.19 acres (200 ft by 260 ft) is to be designed to hold three inches of water (maximum) for the 25-year storm. Assuming virtually all the rainfall will runoff, a factor of 0.95 is used as a "C" factor and A is 1.19. The maximum allowable discharge from the roof is 1.5 cfs. Following most municipal practices, the minimum time of concentration is taken as 5 minutes.

Assuming the roof is dead level with parapets around the sides, the total allowable storage of the rooftop is: $260 * 200 * 3/12 = 13,000 \text{ ft}^3$

Modified Rational - Mass Balance Method - For any duration the rainfall is constant and equal to the intensity given in the IDF curve. This volume of rainfall accumulates on the roof at the rate indicated in Table 11-4. A release rate, assumed constant for simplicity, is then determined to limit the accumulated volume to the maximum allowable.

Application of the rainfall intensity information is:

$$Q = CIA = 0.95 * I * 1.19 = 1.13 \text{ I}$$

Table 11-4 gives information on application of the modified rational or mass balance method for this problem. For the outflow rate of 1.5 cfs the storage volume required is 6102 cubic feet at a storm duration of about 45 minutes.

Using this method and trial outflow rates, the lowest possible average outflow rate was determined to meet the requirements for less than 13,000 ft^3 of storage. That value was 0.30 cfs for a storm duration of about 6 hours and a storage of 12,960 ft^3.

Table 11-4 Modified Rational - Mass Balance For Rooftop Storage Example

(1) Duration (hr)	(2) Intensity (in/hr)	(3) Inflow (cfs)	(4) Volume (cfs-hr)	(5) Accumulated Volume (ft^3)	(6) Average Outflow Rate (cfs)	(7) Outflow Volume (cfs-hr)	(8) Storage Volume (ft^3)
0	0	0.00	0.00	0.00	0	0.00	0.00
0.083	8.00	9.04	0.75	2,700.00	1.5	0.12	2,268.00
0.167	6.60	7.46	1.25	4,500.00	1.5	0.25	3,600.00
0.25	5.50	6.22	1.56	5,616.00	1.5	0.38	4,248.00
0.50	4.00	4.52	2.26	8,136.00	1.5	0.75	5,436.00
0.75	3.33	3.76	2.82	10,160.00	1.5	1.13	6,102.00
1	2.66	3.01	3.01	10,836.00	1.5	1.50	5,436.00
2	1.68	1.90	3.80	13,680.00	1.5	3.00	2,880.00
3	1.28	1.45	4.35	15,660.00	1.5	4.50	-540.00
6	0.80	0.90	5.40	19,440.00	1.5	9.00	-12,960.00
12	0.46	0.52	6.24	22,464.00	1.5	18.00	-42,336.00

Columns 1 and 2 are the rainfall intensity-duration-frequency curve information for the 25-year storm for this municipality. Column 3 multiplies column 2 by 1.13. Column 4 is the product of columns 1 and 3 and column 5 is obtained by multiplying column 4 by 3600. Column 6 is a trial or target discharge rate and column 7 converts it to a volume by multiplying by column 1. Column 8 is the difference between columns 4 and 7 multiplied by 3600.

An average trial outflow rate of 0.30 cfs will handle the storm volume at the maximum 3 inches of depth. As mentioned previously, this design method, although used by some municipalities, is somewhat unrealistic in that the average outflow will vary with head on the outlet, and thus storage is somewhat underpredicted. Because of the unreliability of this method, some municipalities (i.e., within the Atlanta, Georgia Metropolitan Area) will not accept storage designs based on this method.

Drywells For Roof Drains

A more expensive alternate for roof detention is to take roof water to a drywell. Figure 11-11 shows an example. Drywells have the advantage of pollution reduction as well but may have maintenance difficulties if not well designed and discharging onto the surface or through well graded soils or gravel. Overflow Y's should be placed at the ground level and cleanouts supplied in case of clogging.

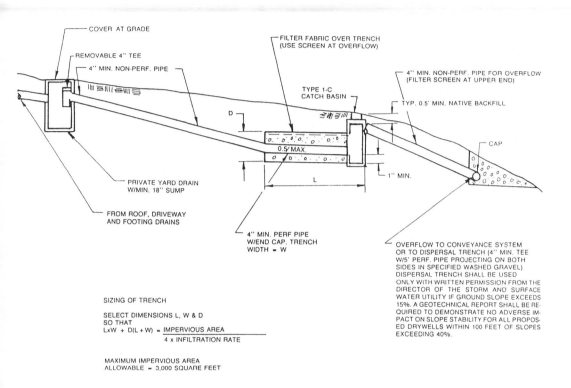

SIZING OF TRENCH

SELECT DIMENSIONS L, W & D
SO THAT

$$\frac{LxW + D(L + W)}{4 \times \text{INFILTRATION RATE}} = \text{IMPERVIOUS AREA}$$

MAXIMUM IMPERVIOUS AREA
ALLOWABLE = 3,000 SQUARE FEET

Figure 11-11 Typical Drywell (Bellevue, 1988)

Detention Chambers And Pipes

For very tight locations on smaller sites (typically areas with high land costs) it may be cost effective to store water in a chamber or oversized pipe or in an inlet chamber. For details see Urbonas and Stahre (1993).

Parking Lot Detention

Parking lot detention can be used when other sites are not available. Parking lot detention has been shown to be more effective than rooftop detention in reducing peak flow in intense urban developments (McCuen, 1975). Obviously such detention sites will share the same space as parked vehicles. Detention sites in parking lots should not be located where normal traffic or pedestrians will pass through. Maximum depths should be less than 7 to 12 inches and grades less than 7 percent (Poertner, 1974). Maximum ponding should occur no more frequently than every two years and sites should drain in 30 minutes or less. Often a 25 percent freeboard is added to the site.

Outlets should be as vandal (and owner) proof as possible. Figure 11-12 illustrates a type of parking lot detention outlet approved for Bellevue, Washington (1988). In this case the grate is sized for discharge restriction. Poertner (1974) describes a similar outlet device but with the addition of an orifice plate seated in the riser collar below the grate. Flow restriction is then not the responsibility of the grate (which acts as a trash rack) but of the orifice below the grate. Columbus, Ohio uses a similar design but with a lower orifice plate.

Other outlets include simple curb openings or curb inlets discharging directly to a ditch or stream. All outlets should provide for emergency overflows. An alternative to storing water on the lot is to construct the lot with grassy swales or gravel filled trenches as dividers and store water in the swales or trenches.

Porous Pavement

An alternative, discussed in more detail in the Structural Best Management Practices, Chapter 13, is porous pavement. It has a rather large storage capacity and, when considering the cost of both conventional pavement and the drainage facilities, is often more cost effective to install than conventional pavement and drainage facilities. Maintenance includes sweeping with vacuum assist. For low volume traffic areas with well drained subsoils, it can be a cost effective solution. Water storage in porous pavement for a combined surface and base thickness of 10 inches is about 2.4 inches. For a combined surface and base thickness of 25 inches it is nearly 7 inches of rainfall (Poertner, 1974). Exfiltration systems can be employed with porous pavement systems to improve performance and to allow for the acceptance of off site drainage (MWCOG, 1987). See MWCOG, 1987 for details.

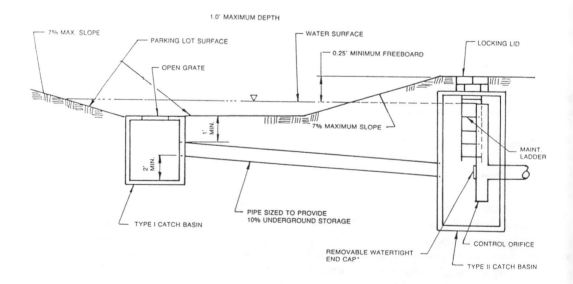

Figure 11-12 Typical Parking Lot Ponding Basin Overflow (Bellevue, 1988)

11.7 Preliminary Detention Calculations

A preliminary estimate of the storage volume required for peak flow attenuation may be obtained from a simplified design procedure that replaces the actual inflow and outflow hydrographs with the standard triangular shapes shown in Figure 11-13.

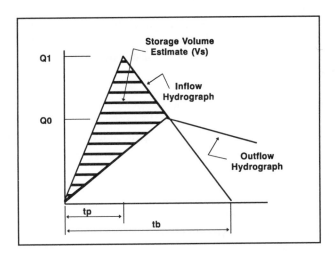

Figure 11-13 Triangular Shaped Hydrographs (Preliminary Analysis)

The required storage volume may be estimated from the area above the outflow hydrograph and inside the inflow hydrograph, expressed as:

$$V_S = 0.5T_i(Q_i - Q_o)$$ (11.23)

Where: V_S = storage volume estimate, ft^3
      Q_i = peak inflow rate, cfs
      Q_o = peak outflow rate, cfs
      T_i = duration of storage facility inflow, sec
Any consistent units may be used for equation 11.23.

Alternative Method

An alternative preliminary estimate of the storage volume required for a specified peak flow reduction can be obtained by the following regression equation procedure (Wycoff & Singh, 1986).
1. Determine input data, including the allowable peak outflow rate, Q_o, the peak flow rate of the inflow hydrograph, Q_i, the time base of the inflow hydrograph, t_b, and the time to peak of the inflow hydrograph, t_p.
2. Calculate a preliminary estimate of the ratio V_S/V_r using the input data from Step 1 and the following equation:

$$V_S/V_r = [1.291(1-Q_o/Q_i)^{0.753}]/[(t_b/t_p)^{0.411}]$$ (11.24)

Where: V_S =   volume of storage, in.
          V_r =   volume of runoff, in.
          Q_o =   outflow peak flow, cfs
          Q_i =   inflow peak flow, cfs
          t_b =   time base of the inflow hydrograph, hr (Determined as the time from the beginning of rise to a point on the recession limb where the flow is 5 percent of the peak.)
          t_p =   time to peak of the inflow hydrograph, hr

3. Multiply the peak flow rate of the inflow hydrograph, Q_i, times the potential peak flow reduction calculated in Step 2 to obtain the estimated peak outflow rate, Q_o, for the selected storage volume.

Peak Flow Reduction

A preliminary estimate of the potential peak flow reduction for a selected storage volume can be obtained by the following procedure.
1. Determine the following:
    *   volume of runoff, V_r
    *   peak flow rate of the inflow hydrograph, Q_i
    *   time base of the inflow hydrograph, t_b
    *   time to peak of the inflow hydrograph, t_p
    *   storage volume, V_S
2. Calculate a preliminary estimate of the potential peak flow reduction for the selected storage volume using the following equation (Singh, 1976):

$$Q_o/Q_i = 1 - 0.712(V_S/V_r)^{1.328}(t_b/t_p)^{0.546} \qquad (11.25)$$

Where: Q_o =   outflow peak flow, cfs
          Q_i =   inflow peak flow, cfs
          V_S =   volume of storage, in.
          V_r =   volume of runoff, in.
          t_b =   time base of the inflow hydrograph, hr (Determined as the time from the beginning of rise to a point on the recession limb where the flow is 5 percent of the peak.)
          t_p =   time to peak of the inflow hydrograph, hr

3. Multiply the peak flow rate of the inflow hydrograph, Q_i, times the potential peak flow reduction calculated from step 2 to obtain the estimated peak outflow rate, Q_o, for the selected storage volume.

11.8 Routing Calculations

The following procedure is used to perform routing through a reservoir or storage facility (Puls Method of storage routing).

Step 1   Develop an inflow hydrograph, stage-discharge curve, and stage-storage curve for the proposed storage facility. Example stage-storage and stage-discharge curves are shown in Figures 11-1 and 11-2.

Step 2   Select a routing time period, Δt, to provide at least five points on the rising limb of the inflow hydrograph.

Step 3   Use the storage-discharge data from Step 1 to develop storage characteristics curves that provide values of $S \pm (O/2)\Delta t$ versus stage. An example tabulation of storage characteristics curve data is shown in Table 11-5.

Table 11-5 Storage Characteristics

(1) Stage (ft)	(2) Storage[1] (ac-ft)	(3) Discharge[2] (cfs)	(4) Discharge[2] (ac-ft/hr)	(5) S-(O/2)Δt (ac-ft)	(6) S+(O/2)Δt (ac-ft)
100	0.05	0	0	0.05	0.05
101	0.3	15	1.24	0.20	0.40
102	0.8	35	2.89	0.56	1.04
103	1.6	63	5.21	1.17	2.03
104	2.8	95	7.85	2.15	3.45
105	4.4	143	11.82	3.41	5.39
106	6.6	200	16.53	5.22	7.98
107	10.0	275	22.73	8.11	11.89

[1] Obtained from the Stage-Storage Curve.
[2] Obtained from the Stage-Discharge Curve.
Note: t = 10 minutes = 0.167 hours and 1 cfs = 0.0826 ac-ft/hr.

Step 4   For a given time interval, I_1 and I_2 are known. Given the depth of storage or stage, H_1, at the beginning of that time interval, $S_1-(O_1/2)\Delta t$ can be determined from the appropriate storage characteristics curve (Figure 11-14).

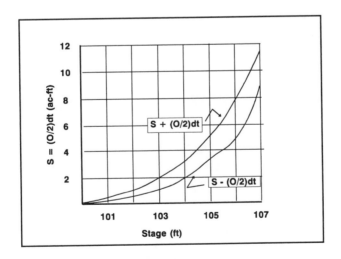

Figure 11-14 Storage Characteristic Curve

Step 5   Determine the value of $S_2 + (O_2/2)\Delta t$ from the following equation:

$$S_2 + (O_2/2)\ \Delta t = [S_1 - (O_1/2)\ \Delta t] + [(I_1 + I_2)/2\ \Delta t] \qquad (11.26)$$

Where: S_2 = storage volume at time 2, ft^3
O_2 = outflow rate at time 2, cfs
Δt = routing time period, sec
S_1 = storage volume at time 1, ft^3
O_1 = outflow rate at time 1, cfs
I_1 = inflow rate at time 1, cfs
I_2 = inflow rate at time 2, cfs

Other consistent units are equally appropriate.

Step 6   Enter the storage characteristics curve at the calculated value determined in step 5 of $S_2+(O_2/2)\Delta t$ and read off a new depth of water, H_2.

Step 7   Determine the value of O_2, which corresponds to a stage of H_2 determined in Step 6, using the stage-discharge curve.

Step 8   Repeat Steps 1 through 7 by setting new values of I_1, O_1, S_1, and H_1 equal to the previous I_2, O_2, S_2, and H_2, and using a new I_2 value. This process is continued until the entire inflow hydrograph has been routed through the storage facility.

11.9 Example Problem

This example demonstrates the application of the methodology presented in this chapter for the design of a typical detention storage facility. Example inflow hydrographs and associated peak discharges for both pre- and post-development conditions are assumed to have been developed using hydrologic methods from the Urban Hydrology, Chapter 7.

Storage facilities should be designed for runoff from both the 2- and 10-year design storms and an analysis done using the 100-year design storm runoff to ensure that the structure can accommodate runoff from this storm without damaging adjacent and downstream property and structures. Example peak discharges from the 2- and 10-year design storm events are as follows.

- Pre-developed 2-year peak discharge = 150 cfs
- Pre-developed 10-year peak discharge = 200 cfs
- Post-development 2-year peak discharge = 190 cfs
- Post-development 10-year peak discharge = 250 cfs

Since the post-development peak discharge must not exceed the predevelopment peak discharge, the allowable design discharges are 150 and 200 cfs for the 2- and 10-year storms, respectively. Example runoff hydrographs are shown in Table 11-6. Inflow durations from the post-development hydrographs are about 1.2 and 1.25 hours, respectively, for runoff from the 2- and 10-year storms.

Preliminary Volume Calculations

Preliminary estimates of required storage volumes are obtained using the simplified method outlined in Section 11.7. For runoff from the 2- and 10-year storms, the required storage volumes, V_S, are computed using equation 11.23:

$$V_S = 0.5T_i(Q_i - Q_o)$$

2-year storm: V_S = [0.5(1.2 x 3,600)(190 - 150)]/43,560 = 1.98 acre-feet
10-year storm: V_S = [0.5(1.25 x 3,600)(250 - 200)]/43,560 = 2.58 acre-feet

Table 11-6 Example Runoff Hydrographs

	Pre-Development Runoff		Post-Development Runoff	
(1)	(2)	(3)	(4)	(5)
Time	2-Year	10-Year	2-Year	10-Year
(Hours)	(cfs)	(cfs)	(cfs)	(cfs)
0	0	0	0	0
0.1	18	24	38	50
0.2	61	81	125	178
0.3	127	170	190 > 150	250 > 200
0.4	150	200	125	165
0.5	112	150	70	90
0.6	71	95	39	50
0.7	45	61	22	29
0.8	30	40	12	16
0.9	21	28	7	9
1.0	13	18	4	5
1.1	10	15	2	3
1.2	8	13	0	1

Design And Routing Calculations

Stage-discharge and stage-storage characteristics of a storage facility that should provide adequate peak flow attenuation for runoff from both the 2- and 10-year design storms are presented in Table 11-7. The storage-discharge relationship was developed by requiring the preliminary storage volume estimates of runoff for both the 2-and 10-year design storms to be provided when the corresponding allowable peak discharges occurred. Storage values were computed by solving the broad-crested weir equation for head, H, assuming a constant discharge coefficient of 3.1, a weir length of 4 feet, and no tailwater submergence. The capacity of storage relief structures was assumed to be negligible.

Storage routing was conducted for runoff from both the 2- and 10-year design storms to confirm the preliminary storage volume estimates and to establish design water surface elevations. Routing results using the Stage-Discharge-Storage Data given in Table 11-7 and the Storage Characteristics Curve given on Figure 11-14, and 0.1-hour time steps are given in Tables 11-8 and 11-9 for runoff from the 2- and 10- year design storms, respectively. The preliminary design provides adequate peak discharge attenuation for both the 2- and 10-year design storms.

For the routing calculations the following equation was used:

$$S_2 + (O_2/2) \Delta t = [S_1 - (O_1/2) \Delta t] + [(I_1 + I_2)/2 \Delta t]$$

Also, column 6 = column 3 + column 5

Since the routed peak discharge is lower than the maximum allowable peak discharges for both design storm events, the weir length could be increased or the storage decreased. If revisions are desired, routing calculations must be repeated.

Table 11-7 Stage-Discharge-Storage Data

(1) Stage (ft)	(2) Q (cfs)	(3) S (acre-feet)	(4) $S_1 - (O/2)\Delta t$ (acre-feet)	(5) $S_1 + (O/2)\Delta t$ (acre-feet)
0.0	0	0.00	0.00	0.00
0.9	10	0.26	0.30	0.22
1.4	20	0.42	0.50	0.33
1.8	30	0.56	0.68	0.43
2.2	40	0.69	0.85	0.52
2.5	50	0.81	1.02	0.60
2.9	60	0.93	1.18	0.68
3.2	70	1.05	1.34	0.76
3.5	80	1.17	1.50	0.84
3.7	90	1.28	1.66	0.92
4.0	100	1.40	1.81	0.99
4.5	120	1.63	2.13	1.14
4.8	130	1.75	2.29	1.21
5.0	140	1.87	2.44	1.29
5.3	150	1.98	2.60	1.36
5.5	160	2.10	2.76	1.44
5.7	170	2.22	2.92	1.52
6.0	180	2.34	3.08	1.60
6.4	200	2.58	3.41	1.76
6.8	220	2.83	3.74	1.92
7.0	230	2.95	3.90	2.00

Table 11-8 Storage Routing For The 2-Year Storm

(1) Time (hrs)	(2) Inflow (cfs)	(3) $[(I_1+I_2)]/2$ (acre-ft)	(4) H1 (ft)	(5) $S_1-(O_1/2)\Delta t$ (acre-ft) (6)-(8)	(6) $S_2+(O_2/2)\Delta t$ (acre-ft) (3)+(5)	(7) H (ft)	(8) Outflow (cfs)
0.0	0	0.00	0.00	0.00	0.00	0.00	0
0.1	38	0.16	0.00	0.00	0.16	0.43	3
0.2	125	0.67	0.43	0.10	0.77	2.03	36
0.3	190	1.30	2.03	0.50	1.80	4.00	99
0.4	125	1.30	4.00	0.99	2.29	4.80	130
							< 150 OK
0.5	70	0.81	4.80	1.21	2.02	4.40	114
0.6	39	0.45	4.40	1.12	1.57	3.60	85
0.7	22	0.25	3.60	0.87	1.12	2.70	55
0.8	12	0.14	2.70	0.65	0.79	2.02	37
0.9	7	0.08	2.08	0.50	0.58	1.70	27
1.0	4	0.05	1.70	0.42	0.47	1.03	18
1.1	2	0.02	1.30	0.32	0.34	1.00	12
1.2	0	0.01	1.00	0.25	0.26	0.70	7
1.3	0	0.00	0.70	0.15	0.15	0.40	3

Table 11-9 Storage Routing For The 10-Year Storm

(1) Time (hrs)	(2) Inflow (cfs)	(3) $[(I_1+I_2)]/2$ (acre-ft)	(4) H1 (ft)	(5) $S_1-(O_1/2)\Delta t$ (acre-ft) (6)-(8)	(6) $S_2+(O_2/2)\Delta t$ (acre-ft) (3)+(5)	(7) H (ft)	(8) Outflow (cfs)
0.0	0	0.00	0.00	0.00	0.00	0.00	0
0.1	50	0.21	0.21	0.00	0.21	0.40	3
0.2	178	0.94	0.40	0.08	1.02	2.50	49
0.3	250	1.77	2.50	0.60	2.37	4.90	134
0.4	165	1.71	4.90	1.26	2.97	2.97	173
						< 200 OK	
0.5	90	1.05	5.80	1.30	2.35	4.00	137
0.6	50	0.58	4.95	1.25	1.83	4.10	103
0.7	29	0.33	4.10	1.00	1.33	3.10	68
0.8	16	0.19	3.10	0.75	0.94	2.40	46
0.9	9	0.10	2.40	0.59	0.69	1.90	32
1.0	5	0.06	1.90	0.44	0.50	1.40	21
1.1	3	0.03	1.40	0.33	0.36	1.20	16
1.2	1	0.02	1.20	0.28	0.30	0.90	11
1.3	0	0.00	0.90	0.22	0.22	0.60	6

Although not shown for this example, runoff from the 100-year storm should be routed through the storage facility to establish freeboard requirements and to evaluate emergency overflow and stability requirements. In addition, the preliminary design provides hydraulic details only. Final design should consider site constraints such as depth of water, side slope stability and maintenance, grading to prevent standing water, and provisions for public safety.

Downstream Effects

An estimate of the potential downstream effects (i.e., increased peak flow rate and recession time) of detention storage facilities may be obtained by comparing hydrograph recession limbs from the pre-development and routed post-development runoff hydrographs. Example comparisons are shown in Figure 11-15 for the 10-year design storms.

Potential effects on downstream facilities should be minor when the maximum difference between the recession limbs of the pre-developed and routed outflow hydrographs is less than about 20 percent. As shown in Figure 11-15, the example results are well below 20 percent; downstream effects can thus be considered negligible and downstream flood routing omitted.

11.10 "Chainsaw Routing" Technique For Spreadsheet Application

Overview

The chainsaw routing procedure is a short-cut method of routing runoff hydrographs through storage facilities to approximate the outflow hydrograph (Malcom, 1987) . The same information is required for the chainsaw routing procedure as the storage indication method, as follows:

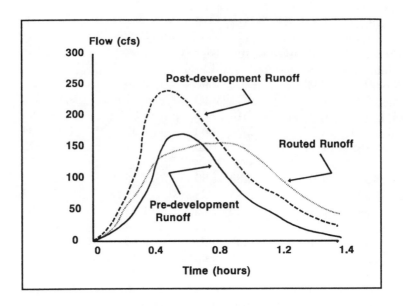

Figure 11-15 Runoff Hydrographs

- inflow hydrographs for all design storms,
- stage-storage curve for the proposed facility, and
- stage-discharge curve for all outlet control structures.

In this method the actual routing procedure has been simplified to allow the computations to be performed by hand or in a standard spreadsheet program. The normal routing equation is derived as:

$$dS/dt = I\text{-}O \tag{11.27}$$

which can be simplified as:

$$(S_j\text{-}S_i)/\Delta t = (I_i+I_j)/2 - (O_i+O_j)/2 \tag{11.28}$$

The chainsaw method simplifies it as:

$$\Delta S_{i,j} = (I_i - O_j)\,\Delta t_{i,j} \tag{11.29}$$

Where: S = storage, ft³
      I = inflow, cfs
      O = outflow, cfs
      Δt = time step, sec
      i,j = time step subscripts j following i in time

In this method the approximation is not considered a trapezoid but a parallelogram with $S_{i,j}$ representing storage across two time steps. In order to ensure accuracy, the computation interval should be less than or equal to 0.1 of the time to peak of the inflow hydrograph.

Stage-Storage

The stage-storage function represents the relation of accumulated storage volume to elevation within the basin. This relation can be expressed as a graph or as a function. However, writing the relation as a function is most beneficial for spreadsheet applications in the chainsaw routing method. The source of the data for the stage-storage function is typically a site plan or topographic map which illustrates the contours of the area proposed to be used for detention storage.

Apart from fitting a curve to a prism, discussed in a previous section, one method of writing a stage-storage function for natural basin shapes is accomplished by assuming that the stage-storage function can be approximated by a power relationship of the form:

$$S = aH^m \tag{11.30}$$

Where: a =   constant in the stage-storage power equation
   m =   coefficient in the stage-storage power equation

The steps for developing a stage-storage function for a natural area using the planimetered end-area method are the following.

1. Determine the elevations of interest within the storage volume of the detention basin and list them in increasing order in a spreadsheet or table. The units should be in feet.
2. Planimeter or measure the contour elevations for all stages within the detention basin storage volume and enter the areas in a column of a spreadsheet or table. The units should be in square feet.
3. Compute the incremental storage volume by using the average-end area method (upper contour area plus the lower contour area divided by two and multiplied by the difference in elevation).
4. Compute the accumulated volume by adding each incremental volume calculated in step 3.
5. Compute the relative stage by setting the bottom elevation of the pond equal to zero and adjusting all subsequent elevations appropriately.
6. Calculate the natural logarithm of the accumulated volume in Step 4. These values are assigned the variable name S_i.
7. Calculate the natural logarithm of the stage listed in step 5. These values are assigned the variable name H_i. Plot step five versus step six.
8. Calculate the exponent "m" in equation 11.30 by fitting a representative straight line through the points. Select two points on the line. The exponent "m" is then computed as:

$$m = [\ln (S_2 / S_1)] / [\ln (H_2 / H_1)] \tag{11.31}$$

9. Compute the variable "a" with equation 11.32.

$$a = S_2 / H_{S2}^{m} \tag{11.32}$$

Chainsaw Routing

The steps to be completed for the chainsaw routing are accomplished through a spreadsheet with the following columns:

Column 1   Time increment, seconds less than or equal to one-tenth of the time to peak of the inflow hydrograph.

Column 2   The inflow hydrograph which is generated using any of the appropriate methods described in the Urban Hydrology, Chapter 7 of this book or can be automatically generated using the procedure in the next section with an equation fit to the SCS shape.

Column 3   Storage in cubic feet.

Column 4   Stage computed from the storage in Column 3 and the computed stage-storage relationship.

Column 5   Total outflow based on the stage-discharge relationship set in succeeding columns.

Column 6-? Other stage-discharge relationships for the various portions of the outlet structure such as orifice and overflow weir, etc.

In order to begin the computations, the spreadsheet or table must be initialized by performing the following steps:

1. Set the initial inflow and outflow equal to zero.
2. Set the initial stage equal to the invert elevation of the outlet spillway or outlet pipe which would be at the pond surface for a wet pond.
3. Set the initial storage volume to its rightful value based on the starting stage. For a dry pond the value would be set to zero.

Refer to Table 11-10 for an example of several rows from a typical routing. Computations then take a time step from time "i" to time "j". The example will follow the computations from time 28 minutes to time 32 minutes:

Step 1.   Column 3 - The change in storage for time i-j is computed as the inflow from time step i minus the outflow from time step i times the time increment. This increment of storage is added to the storage total in Column 3 for a new total storage at time step j.

$(328-130)$ cfs * 4 min * 60 sec/min = 47,520 ft^3

$47,520 + 103,103 = 150,623$ ft^3

Step 2.   Column 4 - At time step j the new storage is calculated from the stage-storage relationship solved for stage.

new stage = 6.69 ft from stage-storage curve

Step 3.   Col 6-7 - Calculate the new discharge through each portion of the detention facility outlet works based on the stage in Column 4.

discharges = 117 cfs (culvert) and 25 cfs (weir)

Step 4.   Col 5 - Sum the total discharges from Columns 6 through the end for the new total discharge at time step j.

$117 + 25 = 142$ cfs outflow

Then begin STEP 1 again for the next time step and so on through the routing.

Table 11-10 Example Section Of Chainsaw Routing Table

1	2	3	4	5	Outflow Devices	
Time (min)	Inflow (cfs)	Storage (ft^3)	Stage (ft)	Outflow (cfs)	Culvert (cfs)	Weir (cfs)
28	328	103103	5.97	130	115	15
32	358	150623	6.69	142	117	25
36	368	202463	7.32	151	120	31

After: Malcom, 1987

And so on through the end of the routing.

Numerical Instability

In some cases (when the stage-discharge relation has high outflows relative to the stage-storage relation with low storage values) this method will result in unstable results. This becomes evident when the outflow exceeds the inflow, perhaps to the degree that negative storage will be computed. This error can be corrected by re-initializing the row at which this instability occurs, with the following procedure.

1. Set outflow (Column 5) equal to inflow (Column 2).
2. Set stage (Column 4) equal to outflow (Column 5) using the stage-discharge function.
3. Set storage (Column 3) equal to stage (Column 4), by using the stage-storage function.

11.11 Modified Rational Method Detention Design

The modified rational method is actually a collection of methods used in different parts of the country each employing the Rational Method to calculate a peak flow and then, after some fashion, building an outflow hydrograph around it. This hydrograph is then compared to a certain release rate or release hydrograph and the required storage volume determined graphically or numerically in tabular form.

Figure 11-16 depicts two common ways to perform a modified rational method design of detention. The top diagram in the figure is termed the Mass Balance method. The bottom diagram can be called the Graphical Hydrograph method. The cross-hatched portion is the estimated storage volume. The first is the most common method used in the Eastern United States. The second is slightly more conservative and generally gives more realistic answers when compared to fully routed applications.

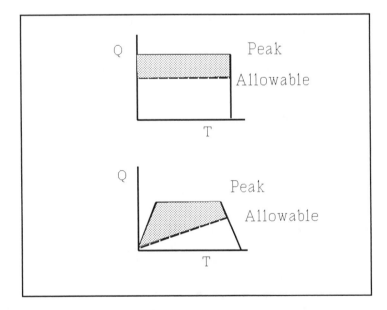

**Figure 11-16 The Mass Balance And Graphical Hydrograph Methods
For Modified Rational Method Detention Design**

In the Mass Balance method, shown in the top half of Figure 11-16, both a constant inflow hydrograph and constant release rate are used. It basically assumes a constant rainfall rate for a certain duration neglecting the flow buildup to peak. The total volume of inflow at any time is equal to 60CIAt. Where C, I and A are defined in the Rational Method and t is time in consistent units of measure. The release rate is a constant which can be expressed as R.

In the Graphical Hydrograph method, depicted in the bottom half for Figure 11-16, the inflow hydrograph from the watershed accumulates until the time-of-concentration is reached. Then it levels out until the storm duration is reached. The recession limb is assumed to be similar to the rising limb. The total inflow volume under each of the inflow hydrographs for any duration is equal, though the volume is accumulated at different overall rates.

The required storage volume equation for the Mass Balance method's volume is:

$$V = 60*(CiAt - Rt) \tag{11.33}$$

where: V = the required volume of the pond, ft^3
          C = the post-development C factor
          I = the rainfall intensity from the IDF curve, in./hr
          R = the allowable release rate, cfs
          t = the storm duration to maximize the volume, min

The required storage volume equation for the Graphical Hydrograph method is:

$$V = 60*[CiAt - R\ (t+t_c)/2] \tag{11.34}$$

where all variables are defined as before except t_c is the post-development time of concentration for the watershed in minutes. It can be seen that for a storm duration equal to the time-of-concentration ($t=t_c$) the two methods yield identical volume estimates. For durations longer than the time-of-concentration the second method yields a larger volume.

The normal method of solution is to set up a table for calculation of volume over a range of durations similar to Table 11-4. That duration which maximizes the required storage volume is chosen. It is normally a trial and error procedure. However, a closed-form solution can be found which gives the required durations and volumes without trial and error by using the following method. Both methods can be transformed to solve directly for either critical duration or maximum storage volume by: (1) substituting the equation:

$$I = a/(t+b) \tag{11.35}$$

where: I = the rainfall intensity, in./hr
          t = the time, min
          a and b are curve fitting constants

for rainfall intensity (I); (2) differentiating the resulting volume equation with respect to time; (3) setting the result equal to zero and solving for the critical duration; and (4) substituting the expression for critical duration found in step 3 into the original volume equation to find the maximum volume.

The values of a and b in equation 3 are easily found by curve fitting equation 3. Most IDF curves can be accurately expressed by such an equation for durations less than one hour. The logarithm of each side of the equation is taken as:

$$\log (I) = \log (a) - \log (t+b) \tag{11.36}$$

This is an equation of a straight line with slope equal to minus one and intercept equal to "log a". The "y" value is log (I) and the "x" value is log (t+b). The values of a and b are found using any standard spreadsheet's linear regression function. The table below illustrates the procedure. Column 1 is the time and column 2 is the intensity value from the municipality's IDF curve. Column 3 adds the trial value of b indicated at the top of the table to the time. Columns 4 and 5 are the logarithms of columns 3 and 2 respectively and are the X and Y values the spreadsheet preforms the regression on. Column 5 is the dependent variable. Columns 6 and 7 are optional showing, the accuracy. The procedure is to use trial values of b until the slope (from the regression output) of the regression line is close to -1.0 (in this case -1.0001). The value of a can then be calculated from the regression output. An alternate graphical method is found in Froehlich (1993) for sites without developed IDF curves.

Table 11-11 Example Curve Fitting

b =	18.48	a =	222.4	slope=	-1	
1	2	3	4	5	6	7
t (min)	I actual (in/hr)	t+b (min)	ln(t+b) --X--	ln(I) --Y--	i pred. (in/hr)	dif
5	9.692	23.48	3.156	2.271	9.468	-0.23
10	7.767	28.48	3.349	2.05	7.805	0.039
15	6.527	33.48	3.511	1.876	6.64	0.113
30	4.504	48.48	3.881	1.505	4.585	0.081
60	2.88	78.48	4.363	1.058	2.832	-0.05

When this is done the critical storm duration and maximum velocity for the Mass Balance method are found from equations 5 and 6 respectively. The critical duration and maximum volume are found for the Graphical Hydrograph method in equations 7 and 8 respectively. Note that the volume from the Graphical Hydrograph method is always greater than the Mass Balance. This difference is in the 20 percent range in most cases.

$$t = \sqrt{\frac{ACab}{R}} - b \qquad\qquad (11.37)$$

$$V_M = 60*[CaA - 2(CabAR)^{1/2} + Rb] \qquad\qquad (11.38)$$

$$t = \sqrt{\frac{2ACab}{R}} - b \qquad\qquad (11.39)$$

$$V_M = 60*[CaA - (2CabAR)^{1/2} + R/2 \ (b-t_c)] \qquad\qquad (11.40)$$

where: t =   the critical storm duration, min
       a =   municipal specific constant in the rainfall equation
       b =   municipal specific constant in the rainfall equation

C =   fully developed runoff factor in the Rational Method equation
A =   area, acres
R =   release rate, cfs
V_M =   maximum storage volume, ft^3
t_c =   fully developed time-of-concentration, min

Actual application in a municipality which has a predetermined release rate criteria (such as pre-developed conditions with a C factor of 0.1) can be developed graphically by dividing the volume equation by the area (A), giving a storage volume per unit area and a release rate effectively canceling A from the equation. Then a simple plot of unit volume versus C factor can be developed for the design storm and the required volume read directly, for curves of constant time-of-concentration. Equation 11.40 would become:

$$V_{MA} = 60*[Ca - (2CabR_A)^{1/2} + R_A/2 \ (b-t_c)] \tag{11.41}$$

where: V_{MA} =   the critical volume per unit drainage area, ft^3/acre
           R_A =   the release rate per unit drainage area, cfs/acre

There has been much discussion about the proper use of the modified rational method for detention design. Discussion centers on the question of whether the method properly sizes ponds, or for what limits it is applicable. Some engineers have found that it undersizes ponds when compared to other methods. Some have said it works well for smaller ponds with watershed size limits given in the range of from 2 acres on the low end to about 100 acres on the high end. There are several reasons for the undersizing statements.

The SCS method is often compared as a "standard". Depending on how the SCS method is employed it can show an undersize problem or it can show that the modified rational method works well for smaller basins. If, for example, a Rational Method C factor of 0.2 is used for pre-development conditions the resulting target discharge for detention reduction will be significantly higher than the corresponding pre-development peak predicted using the SCS method for many of the pre-development land uses. For a 10-year storm a C factor of 0.2 corresponds roughly to an SCS curve number in the low 60's. However the range of curve numbers for existing soils and cover conditions without development range from about 30 to as high as 89 depending on cover, soils and slope. If a high C factor is chosen but a low target value is set for the SCS method in comparison the Modified Rational method will greatly underpredict the required volume. Since the SCS method was derived originally for undeveloped land use conditions it is probably more accurate across a range of conditions than an arbitrary C factor. The answer is to lower the pre-development C factor estimate to correspond to cover, slope, storm frequency and soils conditions inherent in the SCS method. Table 7-7 in the Urban Hydrology, Chapter 7, can give guidance for estimating a more accurate undeveloped C factor. Forest and cultivated land will be at the low end of the range, meadow and lawns in the middle, and poorly covered pasture at the high end of the range.

Due to its theoretical worst-case rainfall distribution the SCS method has been thought to often overpredict volume and peak flows as imperviousness increases. As a small watershed runoff hydrograph moves from being generated by an infiltration dominated pervious surface to a mass transfer dominated impervious surface the basic assumptions of the Rational Method become more and more true. A block of rainfall is translated into runoff with little loss. If the SCS method is set, through reasonable adjustment of the curve number and time of concentration to mimic peak flows predicted by the Rational Method, for the more impervious cases, the volumes required begin to match more closely. In a series of tests for a Midwestern city the modified rational method consistently sized ponds adequately when compared to the SCS method when the pre-development and post-development outflow peaks were matched and the SCS outflow hydrograph was routed through the modified rational designed pond.

If the required amount of peak flow reduction increases the volume in the runoff hydrograph tails becomes increasingly important. The modified rational method does not reflect this in its averaging period approach and then tends to underpredict somewhat. Also, if a constant discharge is used for the modified rational method the discharge will be too large for a portion of the period and will also tend toward underprediction.

The same considerations apply for comparison of the modified rational method and routing using hydrographs generated using the Huff distribution. When the peak outflows from the Huff distribution and the rational method are approximately equal the volumes also tend to be approximately equal. But, unlike the SCS method, when the durations of the Huff storm are lengthened the peak decreases but the volume increases. If this duration storm is chosen the modified rational method will underpredict the required volume (according to the Huff method) by between 20 and 60 percent.

On balance the authors recommend that if the the modified rational method is to be used it should be used with caution and limited to highly impervious developments under about five acres for normal detention design. For extended detention times, methods which can account for longer storms and actual rainfall distributions should be used.

Example Problem

A 2 acre site is to be developed as a restaurant. The pre-development C factor is estimated to be 0.1 and the post development C factor is estimated to be 0.85. A 5 minute time of concentration is chosen. The local municipality requires that detention be provided such that the pre- and post-development 10-year storms outflow peaks match. The IDF curve and a and b values are given in the Table 11-11.

The pre-development peak flow is calculated to be:

$$Q = CIA = 0.1*9.69*2 = 1.94 \text{ cfs} = R$$

Using the mass Balance Method the required volume is:

$$V_M = 60*[CaA - 2(CabAR)^{1/2} + Rb]$$
$$= 60*[0.85*222.37*2-2*(0.85*222.37*18.48*2*1.94)^{1/2}+1.94*18.48] = 10,860 \text{ ft}^3$$

Using the Graphical Hydrograph method the required volume is:

$$V_M = 60*[CaA - (2CabAR)^{1/2} + R/2 (b-t_c)]$$
$$= 60*[0.85*222.37*2-(2*0.85*222.37*18.48*2*1.94)^{1/2}+1.94/2(18.48-5)]$$
$$= 13,590 \text{ ft}^3$$

Using the unit area formulation of the Graphical Hydrograph method gives:

$$V_{MA} = 60*[Ca - (2CabR_A)^{1/2} + R_A/2 (b-t_c)]$$
$$= 60*[0.85*222.37 - (2*0.85*222.37*18.48*1.94/2)^{1/2}+(1.94/2)/2*(18.48-5)]$$
$$= 6,795 \text{ ft}^3/\text{acre}$$

which is the actual storage volume divided by the area.

11.12 Hand Routing Method For Small Ponds

There are many approximate methods for estimating the volume of storage required for detention design including graphical estimation procedures and a number of variations involving the Rational Method. Each of these methods attempt to approximate the results which would be generated through a full storage-indication routing described in previous sections without actually performing the routing. The problem with these methods is that they often underpredict the detention volume required because the shortened triangular design storm does not contain sufficient volume.

The method provided here was developed based on the work of Horn (1987) and graphically approximates the routing of the volume of flow from any storm using a hydrograph shape equivalent to the SCS dimensionless unit hydrograph and approximated by the equations below. It will provide a volume which is slightly greater than the volume generated by unit hydrograph methods with convoluted outflow hydrographs.

Limitations

This method is approximately equivalent to a standard routing method and can be used with the following limitations:
- the method is subject to the limitations of the SCS unit hydrograph methodology and shape;
- the method should be limited to smaller applications where a single basin is modeled with the SCS dimensionless unit hydrograph shape;
- the method requires the use of a single outlet with a continuous stage-discharge curve which can be expressed as a power function of the form:

$$S = bH^n \tag{11.33}$$

Where: b = constant in power equation for either weir or orifice flow as (Lk_w) for weir flow or $(a_o k_o (2g)^{1/2})$ for orifices.

n = the coefficient in the stage-discharge equation is equal to 0.5 for orifice flow and 1.5 for weir flow
- outflow must begin when inflow begins (no dead storage capacity) and fits the power equation from the start of outflow;
- stage-storage can be expressed as a power function of the form:

$$S = aH^m \tag{11.34}$$

Where: a = constant in the stage-storage power equation
m = coefficient in the stage-storage power equation

Basic Approach Overview

Given that the inflow hydrograph can be approximated by the SCS dimensionless unit hydrograph and that both the stage-storage and stage-discharge functions can be expressed as simple power functions, the following dimensionless substitutions can be made to the hydrologic continuity equation used for storage routing and dimensionless parameters developed.

$$\alpha = m/n \tag{11.44}$$

$$K = a/b^\alpha \tag{11.45}$$
$$N_r = (KQ_i^{\alpha-1})/t_p \tag{11.46}$$
$$R = Q_o/Q_i \tag{11.47}$$

N_r is called the "routing number" and R is called the "attenuation ratio" which is the peak flow reduction target. Figures 11-16 and 11-17 show relationships between N_r and R for either spillway outlets or orifice outlets for a range of values of α. These figures constitute the "routing" in this method. For a known attenuation ratio the routing number can be read directly from the graph. Alternately, for a known N_r the attenuation ratio (R) can be read.

For design the basic approach is to use a known attenuation ratio target and determine N_r. From N_r, a known stage-storage function, known t_p, known Q_i, and K can be determined. From a known K, b can be found and thus orifice size or weir length.

For analysis the process works in reverse. From known physical site information N_r is calculated. From either Figure 11-16 or Figure 11-17 the attenuation ratio is read and the outflow peak value is calculated.

Emergency Spillway Approximation

Most municipalities require that all detention be able to pass safely some infrequent post-development peak flow through an emergency spillway. If this spillway is to be the only spillway for the detention facility then the orifice or weir and embankment or wall height need to be sized for this flow as well as the 2-year and 10-year or other design controlled peaks. If an additional overflow weir is to be added to the minor storage facility an approximate method will normally give adequate results.

The basic procedure is to distribute the additional flow from say the 50-year peak inflow between the previously sized orifice or weir and an overflow spillway. No routing is done and no peak attenuation assumed. It is assumed that the crest of the emergency overflow spillway is set at an elevation equal to the stage (H) of the peak 10-year storm outflow.

For orifice or weir flow for the main outlet and weir flow for the overflow emergency spillway the combined flow equation is:

$$Q_d = b*[(H + \blacktriangle H)^n - (H)^n] + k_{ew} * L_{ew} * \blacktriangle H^{1.5} \tag{11.48}$$

Where: Q_d = difference between the 50-year peak post-development inflow and the orifice or weir outflow (Q_p) at peak stage (H) (i.e. the flow to be proportioned between the orifice or weir and the emergency overflow spillway), cfs

k_{ew} = emergency overflow spillway discharge coefficient
L_{ew} = emergency overflow spillway effective length, ft
$\blacktriangle H$ = head on the emergency overflow spillway measured from H upward, ft
And b and n are the constant and coefficient in the stage-discharge equations for either orifice or weir flow of the principal outlet for the detention pond as described previously.

Details Of Approach

For design the following steps should be accomplished to size a small detention pond. By putting this procedure in a spreadsheet the design can e done quite easily and fast.

1. Data Input - Calculate or estimate C factors, Curve Numbers, and times of concentration for the pre- and post-development conditions. Determine location and approximate size limits of pond, vertically and horizontally. Choose an orifice or weir type and discharge coefficient (k_o or k_w).

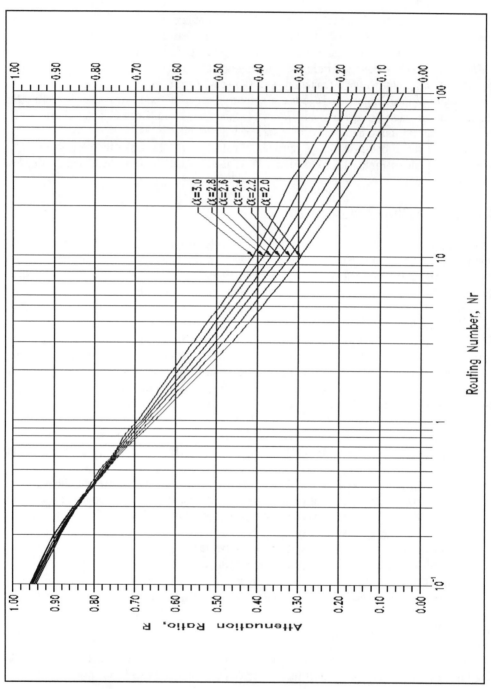

Figure 11-16 Attenuation Ratio For Orifices

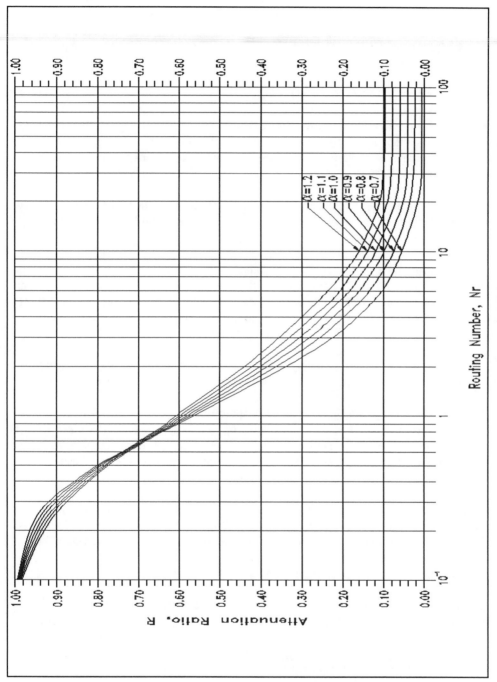

Figure 11-17 Attenuation Ratio For Weirs

2. Basic Calculations - Calculate peak flows for the design and emergency spillway discharges using Rational or other suitable method. Calculate runoff volume from SCS curve number equations. Calculate time-to-peak for the design storm using equations in the Urban Hydrology chapter for the SCS dimensionless unit hydrograph.

3. Volume Estimate and Detention Layout - Estimate design storm storage volume required from the following equation:

$$V_s = 1.39 * 60 * t_p * (Q_i - Q_o) \tag{11.49}$$

Estimate reasonable peak design storm stage at peak outflow using the design storm pre-development peak flow (or other municipal design criteria) as the flow target and the weir or orifice equation as appropriate. With known volume estimate and stage at peak outflow, perform pond layout and develop the stage-storage power function.

4. Find Routing Number - For the design storm calculate R and α from the equations above. Enter appropriate figure for either orifice or spillway flow (Figures 11-16 and 11-17 respectively) to read the routing number (N_r).

5. Find Preliminary Weir or Orifice Size and Other Data - From the routing number calculate K. From K calculate b and finally preliminary orifice or weir size. From orifice or weir size and peak outflow target calculate the actual head and storage from the stage-discharge and stage-storage equations. If the weir or orifice size is to be rounded to a next available size proceed through steps 6 and 7. If not, go to step 8 to size the emergency spillway.

6. (Optional) Recalculate Routing Number - From a chosen (rounded) weir or orifice size for the design storm flow calculate b and K. Finally calculate a new routing number (N_r).

7. (Optional) Recalculate R and Other Data - Enter appropriate figure for either orifice or spillway flow (Figures 11-16 and 11-17 respectively) with the new routing number (N_r) and find the attenuation number (R). From R calculate new Q_o and make sure it is less than the target. If so, calculate new stage and storage for this peak outflow for the design storm flow. This is the final size and stage data.

8. Size Emergency Spillway - Using the equation for the emergency spillway sizing, input an assumed $\blacktriangle H$ and calculate the emergency overflow spillway length (L_{ew}). Often one foot is used for $\blacktriangle H$, though site constraints and safety considerations should determine the assumed head.

Design Example - Sizing Of A Small Pond For Orifice Flow

A detention pond is to be sized for a small commercial development. The drainage area to the pond is 2.0 acres   Other data is as given. The design criteria for this community is that the 10-year storm peak for post development should be equal to the 10-year storm peak for pre-development using a C factor of 0.3 for pre-development conditions.   A 50-year storm must be used to size an emergency spillway. A 6-hour storm is to be used for design purposes.

Step 1 - Input Data

    Area = 2 acres $k_o = 0.6$

C pre-development = 0.2   C post-development = 0.8
CN pre-development = 62   CN post-development = 85
t_c pre-development = 5 min t_c post-development = 5 min
i_{10} = 7.03 in/hr P_{6hr} = 3.72 inches
i_{50} = 9.00 in/hr

Step 2 - Basic Calculations - Rational method peak flow and SCS method for volume and time-to-peak using a given rainfall IDF curve:

Q_{10} (predevelopment) = CiA = 0.2*7.03*2 = 2.81 cfs
Q_{10} (postdevelopment) = CiA = 0.8*7.03*2 = 11.25 cfs
Q_{50} (postdevelopment) = CiA = 0.8*9.00*2 = 14.40 cfs

The runoff volume is estimated from the SCS equations (see the hydrology chapter):

S = 1000/CN - 10 = 1000/85 - 10 = 1.765
Q_v = (P - 0.2S)²/(P+0.8S)
    = (3.72 - 0.2*1.765)²/(3.72 + 0.8*1.765) = 2.21 inches

The time to peak is estimated from the volume equation for the SCS unit hydrograph:

Q_v * 43560/12 * A = 1.39 * 60 * t_p * Q_{10}
2.21 * 43560/12 * 2 = 1.39 * 60 * t_p * 11.25

t_p = 17.10 min = 1,026 seconds

Step 3 - Preliminary Estimates - Based on a triangular hydrograph approximation a first estimate of volume (using the 10-year storm peaks and volume) required is:

V = 1.39 * 60 * t_p * (Q_i - Q_o) = 1.39*60*17.10*(11.25-2.81) = 12,036 ft³

Based on site characteristics the basin should produce a storage volume of about 12,000 cubic feet at a depth of about 5 feet. Using a simple spreadsheet program, a basin with bottom width of 25 feet and length of 55 feet with side-slope of 2.5:1 was chosen. A curve fit to these data yielded values for the stage-storage equation of:

a = 1535   m = 1.35

Step 4 - Routing Number - For an orifice n=0.5 (for a weir it is 1.5)

α = m/n = 1.35/0.5 = 2.7
R = Q_o/Q_i = 2.81/11.25 = 0.25

From Figure 11-17: N_r = 34

Step 5 - Orifice Size and Other Data - From equation 11.37, K is calculated (t_p in seconds), and then the orifice size is backed out:

K = N_r * t_p/ $Q_i^{\alpha-1}$ = 34 * 17.10 * 60/ (11.25$^{2.7-1}$) = 570

$b = (a/K)^{n/m} = (1535/570)^{0.37} = 1.44$
$a_o = b/(8.02*k_o) = 1.44/(8.02*0.6) = 0.30 \text{ ft}^2$
$D = [(4/\pi) * a_o]^{1/2} = 0.617 \text{ ft} = 7.4 \text{ in.}$

In this case it was desired to use a standard available size of 7.5 inches. To check this the calculations proceed backward. This is also the procedure to check an existing pond.

Step 6 (optional) - Recalculate Routing Number

$a_o = \pi * (D^2/4) = 0.307 \text{ ft}^2$
$b = a_o k_o (2g)^{1/2} = 1.48$
$\alpha = m/n = 1.35/0.5 = 2.7$
$K = a/b^\alpha = 1535/1.48^{2.7} = 533$
$N_r = KQ_i^{\alpha-1}/t_p = (533*11.25^{1.7})/(17.10*60) = 31.81$

Step 7 (optional) - Recalculate R and Other Data - From Figure 11-17 R = 0.255

$Q_o = RQ_i = 0.255 * 11.25 = 2.87 \text{cfs (Target 2.81 cfs) (OK)}$
$H = 1/(2g) * [Q_o/(k_o a_o)]^2 = 1/(2g) * [2.87/(0.6*0.307)]^2 = 3.91 \text{ ft}$
$S = 1535 H^{1.35} = 9{,}672 \text{ ft}^3$

Step 8 - Emergency Spillway Approximation - Assume a $\blacktriangle H$ of 1 foot and a k_{ew} of 3.0

$Q_d = b*[(H+\blacktriangle H)^n - (H)^n] + k_{ew} * L_{ew} * \blacktriangle H^{1.5}$
$14.4 - 2.87 = 1.48 * [(3.91 + 1)^{0.5} - (3.91)^{0.5}] + 3 * L_{ew} * 1^{1.5}$
$L_{ew} = 3.7 \text{ ft (OK)}$

Method Comparison

It should be noted that, for this problem, the required storage using the Graphical Hydrograph method (discussed in the Modified Rational section) is 8,090 ft³. The required volume by routing SCS method hydrographs through the designed pond using HEC-1 is 8,620 ft³. Figure XX below shows a comparison of the inflow hydrographs among the three methods when the peak flows are kept equal for both pre- and post-development conditions.

The fully routed hydrograph is the most "realistic" of the three in terms of both convolution of the hydrograph for the design storm and routing through the pond. In this example all three methods give equivalent volumes. This is not always the case. The Hand Routing method is more sensitive to large or small reductions in peak flow and may tend to underpredict required volumes for great peak flow reductions and overpredict required volumes for small peak flow reductions. Therefore it is recommended that it be used in the mid-range of the R values given in the figures.

Outflow hydrographs using the Huff rainfall distribution (discussed in the hydrology chapter) vary depending on the duration of storm chosen. If a duration is chosen to maximize the peak outflow the resulting volume required for storage is often equivalent or slightly less than that required for Modified Rational method design. However, if longer durations are chosen the peak flows will decrease but the required storage volumes will increase. In this case it is not uncommon, using the Huff distribution, for the required volume to be 20 to 60 percent higher than that necessary for the Modified Rational method.

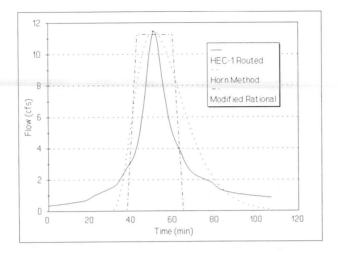

Figure 11-19. Comparison of Three Detention Design Methods

11.13 Land-Locked Retention

Watershed areas that drain to a central depression with no positive outlet (playa lakes) are typical of many topographic areas including karst topography, and can be evaluated using a mass flow routing procedure to estimate flood elevations. Although this procedure is fairly straightforward, the evaluation of storage facility outflow is a complex hydrogeologic phenomenon that requires good field measurements and a thorough understanding of local conditions. Since outflow rates for flooded conditions are difficult to calculate, field measurements are desirable.

The steps in the procedure presented below for the mass routing procedure are illustrated by the example graph given in Figure 11-18.

Step 1   Obtain cumulative rainfall data for the 100-year frequency, 10-day duration design event from Figure 11-19.

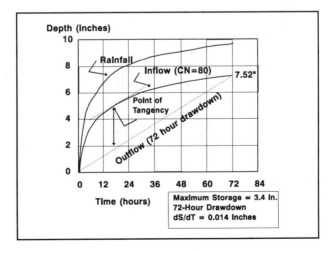

Figure 11-20 Mass Routing Curve

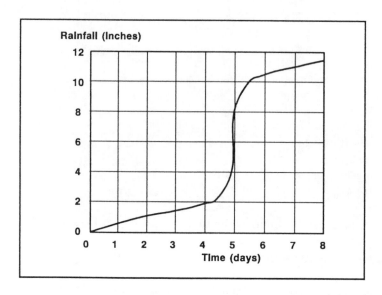

Figure 11-21 Cumulative Rainfall Data For 100-Year 10-Day Design Storm
(Note: Local data should be used to develop this figure.)

Step 2   Calculate the cumulative inflow to the land-locked retention facility using the rainfall data
         from Step 1 and runoff procedure from the Urban Hydrology, Chapter 7. Plot the mass
         inflow to the retention facility.

Step 3   Develop the facility outflow from field measurements of hydraulic conductivity, taking
         into consideration worst-case water table conditions. Hydraulic conductivity should be
         established using insitu test methods, then results compared to observed performance char-
         acteristics of the site. Plot the mass outflow as a straight line with a slope corresponding
         to worst-case outflow in inches/hour.

Step 4   Draw a line tangent to the mass inflow curve from Step 2, which has a slope parallel to
         the mass outflow line from Step 3.

Step 5   Locate the point of tangency between the mass inflow curve of step 2 and the tangent line
         drawn for Step 4. The distance from this point of tangency and the mass outflow line
         represents the maximum storage required for the design runoff.

Step 6   Determine the flood elevation associated with the maximum storage volume determined
         in Step 5. Use this flood elevation to evaluate flood protection requirements of the
         project. The zero volume elevation should be established as the normal wet season water
         surface or water table elevation or the pit bottom, whichever is highest.

Step 7   If runoff from the project area discharges into a drainage system tributary to the land-
         locked depression, detention storage facilities are required to comply with the pre-develop-
         ment discharge requirements for the project. Unless the storage facility is designed as
         a retention facility, including water budget calculations, environmental needs and provi-
         sions for preventing anaerobic conditions, relief structures should be provided to prevent
         standing water conditions.

11.14 Retention Storage Facilities

The use of retention storage facilities which have a permanent pool (wet ponds) is often discouraged because of the extensive maintenance that is sometimes required. Provisions for weed control and aeration for prevention of anaerobic conditions should be considered. Also, facilities should not be built that have the potential of becoming nuisances or health hazards. Note, wet ponds are required where water quality problems are to be addressed.

Water budget calculations are required for all permanent pool facilities and should consider performance for average annual conditions. The water budget should consider all significant inflows and outflows including, but not limited to, rainfall, runoff, infiltration, exfiltration, evaporation, and outflow.

Average annual runoff may be computed using a weighted runoff coefficient for the tributary drainage area multiplied by the average annual rainfall volume. Infiltration and exfiltration should be based on site-specific soils testing data. Evaporation may be approximated using the mean monthly pan evaporation or free water surface evaporation data appropriate for the given facility location.

11.15 Retention Facility Example Problem

A shallow retention facility with an average surface area of 3 acres and a bottom area of 2 acres is planned for construction at the outlet of a 100-acre watershed. The watershed is estimated to have a post-development runoff coefficient of 0.3. Site-specific soils testing indicates that the average infiltration rate is about 0.1 inch per hour. Determine for average annual conditions if the facility will function as a retention facility with a permanent pool.

1. From rainfall records, the average annual rainfall is about 50 in.
2. From local data the mean annual evaporation is 35 in.
3. The average annual runoff is estimated as:
   Runoff = (0.3) (50 in.) (100 acres) = 1,500 acre-in.
4. The average annual evaporation is estimated as:
   Evaporation = (35 in.) (3 acres) = 105 acre-in.
5. The average annual infiltration is estimated as:
   Infiltration = (0.1 in./hr) (24 hours/day) (365 days/yr) (2 acres)
   Infiltration = 1,752 acre-in.
6. Neglecting facility outflow and assuming no change in storage, the runoff (or inflow) less evaporation and infiltration losses is:
   Net Budget = 1,500 - 105 - 1,752 = -357 acre-in.
   Thus, the proposed facility will not function as a retention facility with a permanent pool.
7. Revise pool design as follows:
   Average surface area = 2 acres and bottom area = 1 acre
8. Recompute the evaporation and infiltration
   Evaporation = (35) (2) = 70 acre-in.
   Infiltration = (0.1) (24) (365) (1) = 876 acre-in.
9. The revised runoff less evaporation and infiltration losses is:
   Net Budget = 1,500 - 70 - 876 = +554 acre-in.
   The revised facility is assumed to function as a retention facility with a permanent pool.

11.16 Construction And Maintenance Considerations

An important step in the design process is identifying whether special provisions are warranted to properly construct or maintain proposed storage facilities. To assure acceptable performance and function, storage facilities that require extensive maintenance are discouraged. The following maintenance problems are typical of urban detention facilities and facilities should be designed to minimize these problems (See Poertner, 1974, MWCOG, 1987 and APWA, 1981 for more details):

- weed growth,
- grass and vegetation maintenance,
- sedimentation control,
- bank deterioration,
- standing water or soggy surfaces,
- mosquito control,
- blockage of outlet structures,
- litter accumulation, and
- maintenance of fences and perimeter plantings.

Proper design should focus on the elimination or reduction of maintenance requirements by addressing the potential for problems to develop. Following are some examples.

- Both weed growth and grass maintenance may be addressed by constructing side slopes that can be maintained using available power-driven equipment, such as tractor mowers.
- Sedimentation may be controlled by constructing traps to contain sediment for easy removal or low-flow channels to reduce erosion and sediment transport.
- Bank deterioration can be controlled with protective lining or by limiting bank slopes.
- Standing water or soggy surfaces may be eliminated by sloping basin bottoms toward the outlet, constructing low-flow pilot channels across basin bottoms from the inlet to the outlet, or by constructing underdrain facilities to lower water tables.
- In general, when the above problems are addressed, mosquito control will not be a major problem.
- Outlet structures should be selected to minimize the possibility of blockage (i.e., very small pipes tend to block quite easily and should be avoided).
- Finally, one way to deal with the maintenance associated with litter and damage to fences and perimeter plantings is to locate the facility for easy access where this maintenance can be conducted on a regular basis.

11.17 Protective Treatment

Protective treatment may be required to prevent entry to facilities that present a hazard to children and, to a lesser extent, all persons. Fences may be required for detention areas where one or more of the following conditions exist:

- Rapid stage increases would make escape practically impossible where small children frequent the area.
- Water depths either exceed 2.5 feet for more than 24 hours or are permanently wet and have side slopes steeper than 4:1.
- A low-flow watercourse or ditch passing through the detention area has a depth greater than 5 feet or a flow velocity greater than 5 ft/s.
- Side slopes equal or exceed 1.5:1.

Guards or grates may be appropriate for other conditions, debris accumulation may become a problem. In some cases, it may be advisable to fence the watercourse or ditch rather than the

Guards or grates may be appropriate for other conditions, debris accumulation may become a problem. In some cases, it may be advisable to fence the watercourse or ditch rather than the detention area. Fencing should be considered for dry retention areas with design depths in excess of 2.5 feet for 24 hours, unless the area is within a fenced, limited access facility.

Trash Racks And Safety Grates

Trash racks and safety grates serve several functions:
- they trap larger debris well away from the entrance to the outlet works where they will not clog the critical portions of the works;
- they trap debris in such a way that relatively easy removal is possible;
- they keep people and large animals out of confined conveyance and outlet areas;
- they provide a safety system whereby persons caught in them will be stopped prior to the very high velocity flows immediately at the entrance to outlet works and persons will be carried up and onto the outlet works allowing for an ability to climb to safety; and
- well designed trash racks can have an aesthetically pleasing appearance.

When designed well trash racks serve these purposes without interfering significantly with the hydraulic capacity of the outlet (or inlet in the case of conveyance structures) (ASCE, 1985, Allred-Coonrod, 1991). The location and size of the trash rack depends on a number of factors including: head losses through the rack, structural convenience, safety, and size of outlet.

Trash racks at entrances to pipes and conduits should be sloped at about 3H:1V to 5H:1V to allow trash to slide up the rack with flow pressure and rising water level, the slower the approach flow the flatter the angle. Rack opening rules-of-thumb abound in the literature. Figure 11-20 gives opening estimates based on outlet diameter (UDFCD, 1992). Judgment should be used in that an area with higher debris (e.g. a wooded area) may require more opening space.

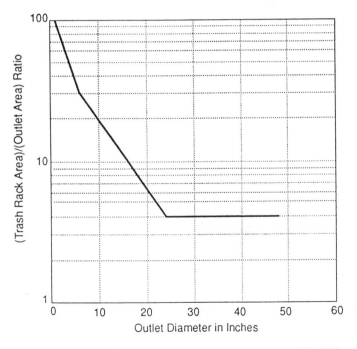

Figure 11-22 Minimum Rack Size vs. Outlet Diameter (UDCFD, 1992)

The bar opening space for small pipes should be less than the pipe diameter. For larger diameter pipes openings should be 6 inches or less. Collapsible racks have been used in some places if clogging becomes excessive or a person becomes pinned to the rack. Alternately debris for culvert openings can be caught upstream from the opening by using pipes placed in the ground or a chain safety net (USBR, 1978, UDFCD, 1991). Racks can be hinged on top to allow for easy opening and cleaning.

The control for the outlet should not shift to the grate. Nor should the grate cause the headwater to rise above planned levels. Therefore headlosses through the grate should be calculated. A number of empirical loss equations exist though many have difficult to estimate variables. Three will be given to allow for comparison. Metcalf & Eddy (1972) give the following equation (based on German experiments) for losses. Grate openings should be calculated assuming a certain percentage blockage as a worst case to determine losses and upstream head. Often 40 to 50 percent is chosen as a working assumption (Perham, 1987). Abt, et al., (1992) studied effects of blockage in supercritical flow and found at a 3H:1V or flatter slope debris tended to be pushed up and off the grate and that localized flooding could occur for a 40 percent blockage.

$$H_g = K_{g1} (w/x)^{4/3} (V_u^2/2g) \sin \theta_g \qquad (11.50)$$

Where: H_g = head loss through grate, ft
K_{g1} = bar shape factor:
  2.42 - sharp edged rectangular
  1.83 - rectangular bars with semicircular upstream faces
  1.79 - circular bars
  1.67 - rectangular bars with semicircular up- and downstream faces
w = maximum cross sectional bar width facing the flow, in.
x = minimum clear spacing between bars, in.
V_u = approach velocity, ft/s
θ = angle of the grate with respect to the horizontal, degrees

The Corps of Engineers (HDC, 1988) has developed curves for trash racks based on similar and additional tests. These curves are for vertical racks but presumably they can be adjusted, in a manner similar to the previous equation, through multiplication by the sine of the angle of the grate with respect to the horizontal.

$$H_g = K_{g2} V_u^2/2g \qquad (11.51)$$

Where K_{g2} is defined from a series of fit curves as:
• sharp edged rectangular (length/thickness = 10)
  $K_{g2} = 0.00158 - 0.03217 A_r + 7.1786 A_r^2$
• sharp edged rectangular (length/thickness = 5)
  $K_{g2} = -0.00731 + 0.69453 A_r + 7.0856 A_r^2$
• round edged rectangular (length/thickness = 10.9)
  $K_{g2} = -0.00101 + 0.02520 A_r + 6.0000 A_r^2$
• circular cross section
  $K_{g2} = 0.00866 + 0.13589 A_r + 6.0357 A_r^2$
and A_r is the ratio of the area of the bars to the area of the grate section.

Creager and Justin (1950) give:

$$K_{g2} = 1.45 - 0.45 A_o - A_r^2 \qquad (11.52)$$

Where: A_r = ratio of net open area (gross grate area less bar area) divided by gross area of the
              grate.

References

Aaron, G. and D. F. Kibler,"Pond Sizing for Rational Formula Hydrographs", Wat. Res. Bull., Vol. 26,
       No. 2, 1990.
Abt, S. R., T. E. Brisbane, D. M. Frick and C. A. McKnight, "Trash Rack Blockage in Supercritical Flow",
       ASCE J. of Hy. Engrg., Vol. 118, No. 12, Dec. 1992.
Allred-Coonrod, J. E., "Safety Grates in Open Channels", MS Prof. Paper, U. of New Mexico, Albuquerque,
       1991.
Akers, P., "A Theoretical Consideration of Side Weirs in Stormwater Overflows", Proc. Inst. of Civ. Engrs.,
       Vol. 6. No. 2, 1957.
American Public Works Assoc., "Urban Stormwater Management", Spec. Rpt. 49, 1981.
American Society of Civil Engineers, "Stormwater Detention Outlet Control Structures", Task Comm. on
       the Design of Outlet Control Structures, ASCE, New York, NY, 1985.
Bellevue, Washington, "Specifications Manual - Chpt. 4 Storm Drainage and Streams", Bellevue, Washington,
       1988.
Brater, E. F. and H. W. King, Handbook of Hydraulics, 6th ed, New York, McGraw Hill Book Company,
       1976.
Cheong, H., "Discharge Coefficient of Lateral Diversion From Trapezoidal Channel", ASCE J. of Irr. and
       Drain. Engrg., Vol. 117, No. 4, 1991.
Chow, C. N., Open Channel Hydraulics, New York, McGraw Hill Book Company, 1959.
Creager, W. P. and J. D. Justin, Hydroelectric Handbook, 2nd ed., John Wiley and Sons, Inc., NY, 1950.
Curry, L., "Relationship of Rational and Unit Graph Methods in Retention Basin Design", Proc. Nat. Symp.
       on Urban Rainfall and Runoff and Sediment Control, U. of Kentucky, 1974.
De Marchi, G., (quoted in Uyumaz and Smith, 1991) "Essay on the Performance of Lateral Weirs",
       L'Energia Electricia, Milan, Italy, Vol. 11, No. 11, Nov., 1934.
Debo, T. N. and A. J. Reese, "Determining Downstream Analysis Limits For Detention Facilities",
       Proceedings For NOVATECH 92, Inter. Conf. of Innovative Technologies In The Domain Of Urban
       Storm Water Drainage, Lyon, France, 1992.
Denver Urban Drainage and Flood Control District (UDFCD), "Urban Drainage Criteria Manual - Vol. 2",
       Denver UDFCD, 1969 with changes 1991.
Denver Urban Drainage and Flood Control District (UDFCD), "Urban Drainage Criteria Manual - Vol. 3",
       Denver UDFCD, 1992.
Froehlich, D. C., "Short-Duration-Rainfall Intensity Equations for Drainage Design," ASCE Journal of
       Irrigation and Drainage Engineering, Vol. 119, No. 5, 1993.
Hager, W. H., "Lateral Outflow Over Side Weirs", ASCE, J. of Hy. Engrg., Vol. 113, No. 4, 1987.
Hartigan, J. P., "Regional BMP Masterplans", Proc. ASCE Engrg. Foundation Conf., Urban Runoff
       Technology, Henniker, NH, 1986.
Hartigan, J. P. and George, "Optimizing the Performance of a Regional Stormwater Detention System",
       ASCE Civ. Engrg. Convention and Expo., 1989.
Horn, D. R., 1987, "Graphic Estimation of Peak Flow Reduction in Reservoirs", J. of Hydraulic Engineering,
       ASCE, Vol. 113, No. 11, pp. 1441-1450.
James, W. P., J. F. Bell and D. L. Leslie, "Size and Location of Detention Storage", ASCE J. of Wat. Res.
       Plan. and Mgmt., Vol. 113, No. 1, 1987.
Jones, J. E., "Multipurpose Stormwater Detention Ponds", Public Works, 1990.
Malcom, H. R., "Elements of Urban Stormwater Design", North Carolina State University.
McCuen, R. H., "Design Accuracy of Stormwater Detention Basins", Tech. Rpt., Dept. of Civil Engineering,
       U. of Maryland, 1983.
McCuen, R. H., "Flood Runoff From Urban Areas", Completion Rpt. A-025-Md, U. of Maryland, 1975.
McCuen, R. H. and G. E. Moglen, "Multicriterion Stormwater Management Methods", ASCE J. Wat. Res.
       Plan. and Mgmt., Vol. 114, No. 4, 1988.

McEnroe, B. M., J. M. Steichen and R. M. Schweiger, "Hydraulics of Perforated Riser Inlets for Under-ground-Outlet Terraces", Trans. ASAE, Vol. 31, No. 4, 1988.

Meredith, D. D., A. C. Middleton and J. R. Smith, "Design of Detention Basins for Industrial Sites", ASCE J. of Wat. Res. Plan. and Mgmt., Vol. 116, No. 4, 1990.

Metcalf & Eddy, Inc., "Wastewater Engineering", McGraw-Hill Book Company, 1972.

Metropolitan Washington Council of Governments (MWCOG), "Controlling Urban Runoff", 1987.

Nix, S. J. and T. Tsay, "Alternative Strategies for Stormwater Detention", Water Res. Bull., Vol. 24, No. 3, June, 1988.

Perham, R. E., "Floating Debris Control: A Literature Review", U.S. Army CREL, Hanover, NH, 13, 1987.

Poertner, H. G. "Practices in the Detention of Urban Stormwater Runoff", American Public Works Assoc., OWWR Proj. C-3380, 1974.

Ramamurthy, A. S., U. S. Tim, and L. B. Carballada, "Lateral Weirs in Trapezoidal Channels", ASCE J. of Irr. and Drain. Engrg., Vol. 112, No. 2, 1986.

Rossmiller, R. L., "Outlet Structure Hydraulics for Detention Facilities", Proc. Int. Symp. on Urban Hydrology, Hydraulics and Sediment Control, U. of Kentucky, Lexington, KY, 1982.

Sandvik, A., "Proportional Weirs for Stormwater Pond Outlets", Civil Engineering, ASCE, pp. 54-56, March, 1985.

Smith P. H. and J. S. Cook, "Stormwater Management Detention Pond Design Within Floodplain Areas", Trans. Res. Record 1017, 1987.

Soil Conservation Service, "Hydraulics of Two-way Covered Risers", TR No. 29, 1965.

Sowers, G. B. and G. F. Sowers, Introductory Soil Mechanics and Foundations, 3rd ed., New York, MacMillan Publishing Company, 1970.

Spangler, M. G. and R. L. Handy, Soil Engineering, 4th ed., New York, Harper & Row, 1982.

Stahre, P. and B. Urbonas, "Stormwater Detention", Prentice Hall, 1990.

State of Maryland, "The Effects of Alternative Stormwater Management Design Policy on Detention Basins", Dept. of Env., Sed. and Stormwater Div., 1986.

State of Maryland, " Design procedures for Stormwater Management Detention Structures", Dept. of Env., Sed. and Stormwater Div., 1987.

Stormwater Management Manual - Volume 2 Procedures, Metropolitan Government of Nashville and Davidson County, The EDGe Group, Inc. and CH2M Hill, July 1988.

Tyrpak, K. A., "Individualized Water Detention Saves Money", Public Works, Jan., 1990.

U.S. Army Corps of Engineers, "Hydraulic Design Criteria", USAE Waterways Experiment Station, Vicksburg, MS, 1988.

U.S. Bureau of Reclamation, "Design of Small Canal Structures", 1978.

Uyumaz, A., and R. H. Smith, "Design Procedure for Flow Over Side Weir", ASCE J. of Irr. and Drain. Engrg., Vol. 117, No. 1, 1991.

Uyumaz, A., "Side Weir Triangular Channel", ASCE J. of Irr. and Drain. Engrg., Vol. 118, No. 6, 1992.

Washington State Dept. of Ecology, "Stormwater Management Manual for the Puget Sound Basin", 1992.

Wycoff, R. L. and U. P. Singh, "Preliminary Hydrologic Design of Small Flood Detention Reservoirs, Water Resources Bulletin, Vol. 12, No. 2, pp 337-49, 1976.

Chapter 12   *Energy Dissipation*

12.1 Introduction

The failure or damage of many culverts and detention basin outlet structures can be traced to unchecked erosion. Erosive forces which are at work in the natural drainage network are often increased by the construction of highways or other urban developments. Interception and concentration of overland flow and constriction of natural waterways inevitably results in an increased erosion potential. To protect the culvert and adjacent areas, it is sometimes necessary to employ an energy dissipating device.

An energy dissipator is any device designed to protect downstream areas from erosion by reducing the velocity of flow to acceptable limits. Unlike bank protection which reduces the channel's susceptibility to erosion, energy dissipaters reduce the erosive force of the flow. These devices cover a wide range in complexity and cost and the particular type selected will depend on the assessment of the erosion hazard. This assessment includes determining the ability of the natural channel to withstand erosive forces and the scour potential represented by the super-imposed flow conditions. The purpose of this chapter is to aid in selecting and designing an energy dissipator capable to control erosive velocities that could damage channels and stream banks.

Generally, energy dissipators should be employed whenever the velocity of flow leaving a storm water management facility exceeds the erosion velocity of the downstream channel system. Several standard energy dissipator designs are in use including hydraulic jump, forced hydraulic jump, impact basins, drop structures, stilling wells, and riprap.

12.2 Recommended Energy Dissipators

Erosion problems at culvert or detention basin outlets are common. Determination of the flow conditions, scour potential, and channel erosion resistance should be standard procedure for all designs. Two types of scour can occur in the vicinity of culvert and other outlets: general channel degradation and local scour. Channel degradation may proceed in a fairly uniform manner over a long length, or may be evident in one or more abrupt drops progressing upstream with every runoff event. The abrupt drops, referred to as headcutting, can be detected by location surveys or by periodic maintenance following construction.

Local scour is the result of high-velocity flow at the culvert outlet, but its effect extends only a limited distance downstream. The highest outlet velocities will be produced by long, smooth-barrel culverts and channels on steep slopes. At most sites these cases will require protection of the outlet. However, protection is also often required for culverts and channels on mild slopes. For these culverts flowing full, the outlet velocity will be the critical velocity with low tailwater and the full barrel velocity for high tailwater.

Standard practice has been to use the same treatment at the culvert entrance and exit, however this does not always make sense. It is important to recognize that the inlet is designed to improve culvert capacity, improve structural stability, and/or reduce headloss while the outlet structure should provide a smooth flow transition back to the natural channel or into an energy dissipator.

For many designs, the following outlet protection and energy dissipators provide sufficient protection at a reasonable cost.
• Riprap apron
• Riprap outlet basins
• Baffled outlets
This chapter will focus primarily on these measures. The reader is referred to the Federal Highway Administration Hydraulic Engineering Circular No. 14 entitled, "Hydraulic Design Of Energy Dissipators For Culverts And Channels", for the design procedures of the other energy dissipators.

For special conditions that involve steep chutes or pipes in tight physical surrounding, with few debris problems or where debris can be caught on trash grates, an internal dissipator may be needed. The stilling well energy dissipator developed by the Corps of Engineers may be applicable (COE, 1988). For this dissipator, energy dissipation is accomplished by flow expansion in the stilling well, impact on the well and by momentum loss by the upward redirection of the flow. Interested designers should consult the Corps of Engineers' publication for further details.

12.3 Design Criteria

The dissipator type selected for a site must be appropriate to the location. In this chapter, the terms internal and external are used to indicate the location of the dissipator in relationship to the culvert. An external dissipator is located outside of the culvert and an internal dissipator is located within the culvert barrel.

Dissipator Type Selection

Internal dissipators are used where:
• the scour hole at the culvert outlet is unacceptable,
• the right-of way is limited,
• debris is not a problem, and
• moderate velocity reduction is needed.

Natural scour holes are used where:
• undermining of the culvert outlet will not occur or it is practicable to be checked by a cutoff wall,
• the expected scour hole will not cause costly property damage, and
• there is no nuisance effect.

External dissipators are used where:
• the outlet scour hole is not acceptable,
• moderate amount of debris is present, and
• the culvert outlet velocity (V_o) is moderate, Fr < 3.

Stilling Basins are used where:
• the outlet scour hole is not acceptable,
• debris is present, and
• the culvert outlet velocity (V_o) is high, Fr > 3

Ice Buildup

If ice buildup is a factor, it should be mitigated by:
• sizing the structure to not obstruct the winter low flow, and
• using external dissipators.

Debris Control

Debris control should be designed using FHWA Hydraulic Engineering Circular No. 9, "Debris-Control Structures" and should be considered:
• where clean out access is limited, and
• if the dissipator type selected cannot pass debris.

Flood Frequency

The flood frequency used should be the flood frequency used for the culvert design or a greater frequency, if justified by:
• low risk of failure of the crossing or development,
• substantial cost savings,
• limited or no adverse effect on the downstream channel, and
• limited or no adverse effect on downstream development.

Maximum Culvert Exit Velocity

The culvert exit velocity should be consistent with the maximum velocity in the natural channel or should be mitigated by using:
• normal channel stabilization,
• special riprap transition below the stilling basins, and
• energy dissipation.

Tailwater Relationship

The hydraulic conditions downstream should be evaluated to determine a tailwater depth and the maximum velocity for a range of discharges.
• Open channels (See Open Channel, Chapter 10).
• Lake, pond, or large water body should be evaluated using the high water elevation that has the same frequency as the design flood for the culvert if events are known to occur concurrently (statistically dependent). If statistically independent, evaluate the joint probability of flood magnitudes and use a likely combination. Use a reasonable worst case.
• Tidal conditions should be evaluated using the mean high tide, but should be checked using low tide.

Material Selection

The material selected for the dissipator should be based on a comparison of the total cost over the design life of alternate materials and should not be made using first cost as the only criteria. This comparison should consider replacement cost, maintenance cost, the difficulty of construction, and traffic and other delays.

Culvert Outlet Type

In choosing a dissipator, the selected culvert end treatment has the following implications.
- Culvert ends which are projecting or mitered to the fill slope offer no outlet protection.
- Headwalls provide embankment stability and erosion protection. They provide protection from buoyancy and reduce damage to the culvert.
- Commercial end sections add little cost to the culvert and may require less maintenance, retard embankment erosion, and incur less damage from maintenance.
- Aprons do not reduce outlet velocity, but if used should extend at least one culvert height downstream. They should not protrude above the normal streambed elevation.
- Wingwalls are used where the side slopes of the channel are unstable, where the culvert is skewed to the normal channel flow, to redirect outlet velocity, or to retain fill.

Safety Considerations

Traffic should be protected from external energy dissipators by locating them outside the appropriate "clear zone" distance per AASHTO Roadside Design Guide or shielding them with a traffic barrier.

Weep Holes

If weep holes are used to relieve uplift pressure, they should be designed in a manner similar to underdrain systems.

12.4 Design Procedure

Data Needs

The following data may be needed for the design and evaluation of energy dissipation facilities.
- Culvert and other terminal outlet structures
  - design capacity
  - type of control
  - barrel slope
  - outlet depth
  - outlet velocity
  - length
  - tailwater
  - Froude number

- Channel
  - capacity
  - bottom slope
  - cross section dimensions
  - normal depth
  - average velocity
  - allowable velocity
  - debris and bedload
  - soil plasticity index
  - saturated shear strength

- Allowable scourhole dimensions, based on site conditions
  - depth, h_s
  - width, W_s
  - length, L_s
  - volume, V_s

Procedure Outline

Following are the steps in the procedure that is generally applicable for outlet protection facilities for pipe and culvert type outlets.

Step 1   Compute local scourhole dimensions with the procedure in Section 12.5. A nonerodible layer (e.g., bedrock) may limit scourhole depth but only slightly affect scourhole width and length.

Step 2   Compare the local scourhole dimensions from Step 1 to the allowable scourhole dimensions from site data. If the allowable dimensions are exceeded, outlet protection is required.

Step 3   If outlet protection is required, choose an appropriate type. Suggested outlet protection facilities and applicable flow conditions (based on Froude number and dissipation velocity) are presented in Table 12-1. When outlet protection facilities are selected, appropriate design flow conditions and site-specific factors affecting erosion and scour potential, construction cost, and long-term durability should be considered. Consult the Federal Highway Administration Hydraulic Engineering Circular No. 14 (Federal Highway Administration, 1983) for other energy dissipators if the ones given in Table 12-1 are not applicable.

Step 4   Design downstream transition section if flow velocity leaving the end sill or end of riprap apron is higher than allowable.

Step 5   If outlet protection is not provided, energy dissipation will occur through formation of a local scourhole. A cutoff wall will be needed at the discharge outlet to prevent structural undermining. The wall depth should be slightly greater than the computed scourhole depth, h_s. The scourhole should then be stabilized. If the scourhole is of such size that it will present maintenance, safety, or aesthetic problems, other outlet protection will be needed.

Step 6   Evaluate the downstream channel stability and provide appropriate erosion protection if channel degradation is expected to occur.

Following is a discussion of applicable conditions for each outlet protection measure.

a.   Riprap aprons may be used when the outlet Froude number (Fr) is less than or equal to 2.5. In general, riprap aprons prove economical for transitions from culverts to overland sheet flow at terminal outlets, but may also be used for transitions from culvert sections to stable channel sections. Stability of the surface at the termination of the apron should be considered.

b.   Riprap outlet basins may also be used when the outlet Fr is less than or equal to 2.5. They are generally used for transitions from culverts to stable channels. Since riprap outlet basins function by creating a hydraulic jump to dissipate energy, performance is impacted by tailwater conditions.

c.   Baffled outlets have been used with outlet velocities up to 50 feet per second. Practical application typically requires an outlet Froude number between 1 and 9. Baffled outlets may be used at both terminal outlet and channel outlet transitions. They function by dissipating energy through impact and turbulence and are not significantly affected by tailwater conditions.

12.5 Local Scourhole Estimation

Estimates of erosion at culvert outlets must consider factors such as discharge, culvert diameter, soil type, duration of flow, and tailwater depth. In addition, the magnitude of the total erosion can consist of local scour and channel degradation. Blaisdell has also developed a design procedure for plunge pools from culverts which should be consulted when culverts discharge at some height above a streambed (Blaisdell and Anderson, 1988 a,b, 1991).

Table 12-1   Suggested Outlet Protection Type Based On Froude Number And Velocity

Outlet Protection	$Fr \leq 2.5$	Fr Between 2.5 and 4.5	$Fr \geq 4.5$ and $V < 50^a$	$V \geq 50^a$
Riprap Apron	X			
Riprap Outlet Basin	X			
Baffled Outlet	X^b	X^b	X^b	

Note: For more outlet protection types within the categories given above, refer to the FHWA Hydraulic Engineering Circular No. 14 for design procedures.

[a] Velocity is based on the energy to be dissipated. Theoretically, the dissipation velocity can be calculated using the equation:

$$V = (2gh)^{0.5} \tag{12.1}$$

where: V = theoretical dissipation velocity, ft/s
         g = acceleration due to gravity, 32.2 ft/s^2
         h = energy head to be dissipated, ft (can be approximated as the difference between channel invert elevations at the inlet and outlet)
[b]   Practical application requires that $1 \leq Fr \leq 9$.

Empirical equations for estimating the maximum dimensions of a local scourhole are presented in Table 12-2. These equations are based on test data obtained as part of a study conducted at Colorado State University (USDOT, FHWA, HEC-14, 1983). A form for recording the following local scourhole computations is presented in Figure 12-1. Following are the calculation steps.

1. Prepare input data, including:
   Q = design discharge, cfs
   For circular culvert, D = diameter, in. For other shapes, use the equivalent depth $Y_e = (A/2)^{0.5}$ (A = cross-sectional area)
   t = time of scour, min
   V_o = outlet mean velocity, ft/s

$$\tau_c = 0.0001 \ (S_v + 180) \ \tan \ (30 + 1.73 \ PI) \tag{12.2}$$

where: τ_c = critical tractive shear stress, pounds/in.^{2}
         S_v = saturated shear strength, pounds/in.^{2}
         PI = plasticity index from Atterburg limits

The time of scour should be based on a knowledge of peak flow duration. As a guideline, a time of 30 minutes is recommended. Tests indicate that approximately 2/3 to 3/4 of the maximum scour occurs in the first 30 minutes of the flow duration.

2. Based on the channel material, select the proper scour equations and coefficients from Table 12-2.

MATERIAL	NOMINAL GRAIN SIZE d50 (mm)	SCOUR EQUATION	DEPTH h_s				WIDTH W_s				LENGTH L_s				VOLUME V_s			
			α	β	θ	α_e	α	β	θ	α_e	α	β	θ	α_e	α	β	θ	α_e
Uniform Sand	0.20	V-1 or V-2	2.72	.375	0.10	2.79	11.73	0.92	.15	6.44	16.82	0.71	0.125	11.75	203.36	2.0	0.375	80.71
Uniform Sand	2.0	V-1 or V-2	1.86	0.45	0.09	1.76	8.44	0.57	0.06	6.94	18.28	0.51	0.17	16.10	101.48	1.41	0.34	79.62
Graded Sand	2.0	V-1 or V-2	1.22	0.85	0.07	.75	7.25	0.76	0.06	4.78	12.77	0.41	0.04	12.62	36.17	2.09	0.19	12.94
Uniform Gravel	8.0	V-1 or V-2	1.78	0.45	0.04	1.68	9.13	0.62	0.08	7.08	14.36	0.95	0.12	7.61	65.91	1.86	0.19	12.15
Graded Gravel	8.0	V-1 or V-2	1.49	0.50	0.03	1.33	8.76	0.89	0.10	4.97	13.09	0.62	0.07	10.15	42.31	2.28	0.17	32.82
Cohesive Sandy Clay 60% Sand PI 15	0.15	V-1 or V-2	1.86	0.57	0.10	1.53	8.63	0.35	0.07	9.14	15.30	0.43	0.09	14.78	79.73	1.42	0.23	61.84
Clay PI 5-16	Various	V-3 or V-4	0.86	0.18	0.10	1.37	3.55	0.17	0.07	5.63	2.82	0.33	0.09	4.48	0.62	0.93	0.23	2.48

EQUATIONS:

V-1. FOR CIRCULAR CULVERTS. Cohesionless material or the 0.15mm cohesive sandy clay

$$\left[\frac{h_s}{D}, \frac{W_s}{D}, \frac{L_s}{D}, \text{ or } \frac{V_s}{D^3} \right] = \alpha \left(\frac{Q}{\sqrt{g}\, D^{5/2}} \right)^\beta \left(\frac{t}{t_o} \right)^\theta$$

where: t_o = 316 min.

V-2. FOR OTHER CULVERTS SHAPES. Same material as above.

$$\left[\frac{h_s}{y_e}, \frac{W_s}{y_e}, \frac{L_s}{y_e}, \text{ or } \frac{V_s}{y_e^3} \right] = \alpha_e \left(\frac{q}{\sqrt{g}\, y_e^{5/2}} \right)^\beta \left(\frac{t}{t_o} \right)^\theta$$

where: t_o = 316 min.

V-3. FOR CIRCULAR CULVERTS. Cohesive sandy clay with PI = 5-16

$$\left[\frac{h_s}{D}, \frac{W_s}{D}, \frac{L_s}{D}, \text{ or } \frac{V_s}{D^3} \right] = \alpha \left(\frac{\rho v^2}{\tau_c} \right)^\beta \left(\frac{t}{t_o} \right)^\theta$$

where: t_o = 316 min.

V-4. FOR OTHER CULVERT SHAPES. Cohesive sandy clay with PI = 5-16

$$\left[\frac{h_s}{y_e}, \frac{W_s}{y_e}, \frac{L_s}{y_e}, \text{ or } \frac{V_s}{y_e^3} \right] = \alpha_e \left(\frac{\rho v^2}{\tau_c} \right)^\beta \left(\frac{t}{t_o} \right)^\theta$$

where: t_o = 316 min.

Table 12-2 Experimental Coefficients For Culvert Outlet Scour (Source: HEC-14, 1983)

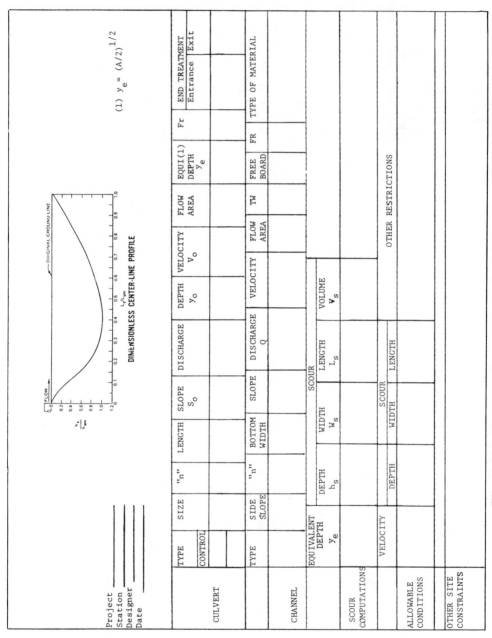

Figure 12-1 Culvert, Channel, Scour And Other Site Data Form

Source: HEC-14, 1983

3. Using the results from the equations selected in Step 2, compute the following scourhole dimensions:
   • Depth, h$_s$ • Length, L$_s$ • Width, W$_s$ • Volume, V$_s$

   Observations indicate that a nonerodible layer at a depth less than h$_s$ below the pipe outlet affects only scourhole depth. The width, and length, may be computed using the equations in Table 12-2.

12.6 Riprap Aprons

A flat riprap apron can be used to prevent erosion at the transition from a circular or box culvert outlet to a natural channel. Protection is provided primarily by having sufficient length and flare to dissipate energy by expanding the flow. Riprap aprons are appropriate when the culvert outlet Fr is less than or equal to 2.5.

Design Procedure

The design procedure presented in this section is taken from USDA, SCS, 1975. Two sets of curves, one for minimum and one for maximum tailwater conditions, are used to determine the apron size and the median riprap diameter, d$_{50}$. If tailwater conditions are unknown, or if both minimum and maximum conditions may occur, the apron should be designed to meet criteria for both. Although the design curves are based on round pipes flowing full, they can be used for partially full pipes and box culverts. The design procedure consists of the following steps.

Step 1 If possible, determine tailwater conditions for the channel. If tailwater is less than one-half the discharge flow depth (pipe diameter if flowing full), minimum tailwater conditions exist and the curves in Figure 12-2 apply. Otherwise, maximum tailwater conditions exist and the curves in Figure 12-3 should be used.

Step 2 Determine the correct apron length and median riprap diameter, d$_{50}$, using the appropriate curves from Figures 12-2 and 12-3. If tailwater conditions are uncertain, find the values for both minimum and maximum conditions and size the apron as shown in Figure 12-4.

a. For pipes flowing full: Use the depth of flow, d, which equals the pipe diameter, in feet, and design discharge, in cfs, to obtain the apron length, L$_a$, and median riprap diameter, d$_{50}$, from the appropriate curves.

b. For pipes flowing partially full: Use the depth of flow, d, in feet, and velocity, V, in ft/s. On the lower portion of the appropriate figure, find the intersection of the d and V curves, then find the riprap median diameter, d$_{50}$, from the scale on the right. From the lower d and V intersection point, move vertically to the upper curves until intersecting the curve for the correct flow depth, d. Find the minimum apron length, L$_a$, from the scale on the left.

c. For box culverts: Use the depth of flow, d, in feet, and velocity, V, in ft/s. On the lower portion of the appropriate figure, find the intersection of the d and V curves, then find the riprap median diameter, d$_{50}$, from the scale on the right. From the lower d and V intersection point, move vertically to the upper curve until intersecting the curve equal to the flow depth, d. Find the minimum apron length, L$_a$, using the left scale.

Step 3 If tailwater conditions are uncertain, the median riprap diameter should be the larger of the values for minimum and maximum conditions. The dimensions of the apron will be as shown in Figure 12-4. This will provide protection under either of the tailwater conditions.

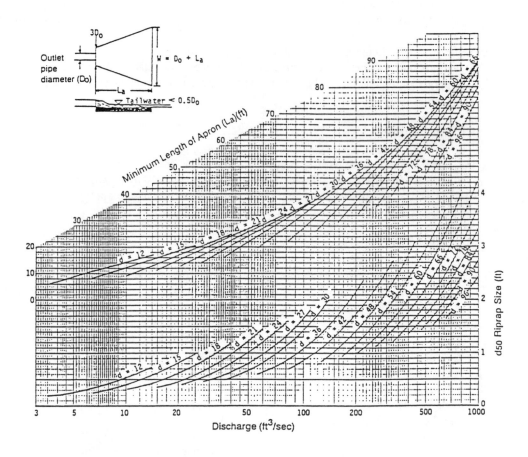

Curves may not be extrapolated.

Figure 12-2 Riprap Apron Under Minimum Tailwater Conditions

Source: USDA, SCS, 1975

Design Considerations

The following items should be considered during riprap apron design:
1. The maximum stone diameter should be 1.5 times the median riprap diameter.
   $$d_{max} = 1.5 \times d_{50}$$
   d_{50} = the median stone size in a well-graded riprap apron.
2. The riprap thickness should be 1.5 times the maximum stone diameter or 6 inches, whichever is greater.
   Apron thickness = $1.5 \times d_{max}$
   (Apron thickness may be reduced to $1.5 \times d_{50}$ when an appropriate filter fabric is used under the apron.)
3. The apron width at the discharge outlet should be at least equal to the pipe diameter or culvert width, d_w. Riprap should extend up both sides of the apron and around the end of the pipe or culvert at the discharge outlet at a maximum slope of 2:1 and a height not less than the pipe diameter or culvert height, and should taper to the flat surface at the end of the apron.

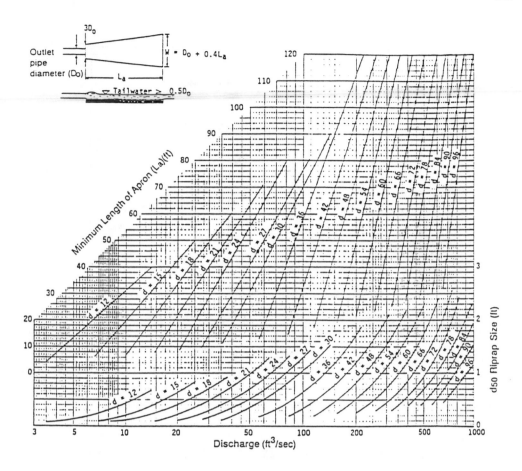

Figure 12-3 Riprap Apron Under Maximum Tailwater Conditions

Source: USDA, SCS, 1975

4. If there is a well-defined channel, the apron length should be extended as necessary so that the downstream apron width is equal to the channel width. The sidewalls of the channel should not be steeper than 2:1.
5. If the ground slope downstream of the apron is steep, channel erosion may occur. The apron should be extended as necessary until the slope is gentle enough to prevent further erosion.
6. The potential for vandalism should be considered if the rock is easy to carry. If vandalism is possible, the rock size may be increased or the rocks held in place using concrete or grout.

Example Problems

Example No. 1 - Riprap Apron Design for Minimum Tailwater Conditions

A flow of 280 cfs discharges from a 66-inch pipe with a tailwater of 2 ft above the pipe invert. Find the required design dimensions for a riprap apron.

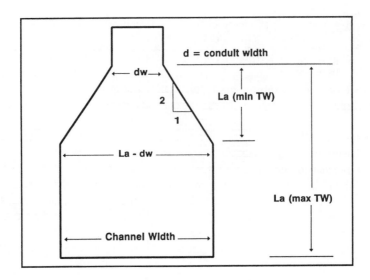

Figure 12-4 Riprap Apron Schematic For Uncertain Tailwater Conditions

Source: USDA, SCS, 1975

1. Minimum tailwater conditions = 0.5 d_o = 66 in. = 5.5 ft - therefore, 0.5 d_o = 2.75 ft.
2. Since TW = 2 ft, use Figure 12-2 for minimum tailwater conditions.
3. From Figure 12-2, the apron length, L_a, and median stone size, d_{50}, are 38 ft and 1.2 ft, respectively.
4. The downstream apron width equals the apron length plus the pipe diameter:
    W = d + L_a = 5.5 + 38 = 43.5 ft
5. Maximum riprap diameter is 1.5 times the median stone size:
    1.5 (d_{50}) = 1.5 (1.2) = 1.8 ft
6. Riprap depth = 1.5 (d_{max}) = 1.5 (1.8) = 2.7 ft.

Example No. 2 - Riprap Apron Design for Maximum Tailwater Conditions

    A concrete box culvert 5.5 ft high and 10 ft wide conveys a flow of 600 cfs at a depth of 5.0 ft. Tailwater depth is 5.0 ft above the culvert outlet invert. Find the design dimensions for a riprap apron.
1. Compute 0.5 d_o = 0.5 (5.0) = 2.5 ft.
2. Since TW = 5.0 ft is greater than 2.5 ft, use Figure 12-3 for maximum tailwater conditions.
    V = Q/A = [600/(5) (10)] = 12 ft/s
3. On Figure 12-3, at the intersection of the curve, d_o = 60 in. and V = 12 ft/s, d_{50} = 0.4 ft. Reading up to the intersection with d = 60 in, find L_a = 40 ft.
4. Apron width downstream = d_w + 0.4 L_a = 10 + 0.4 (40) = 26 ft.
5. Maximum stone diameter = 1.5 d_{50} = 1.5 (0.4) = 0.6 ft.
6. Riprap depth = 1.5 d_{max} = 1.5 (0.6) = 0.9 ft.

12.7 Riprap Basin Design

Outlet velocities from culverts and other drainage structures are one of the primary indicators of erosion potential. Different culvert and outlet designs should be investigated to limit the outlet velocity, but the degree of velocity reduction is, in most cases, limited. The continuity equation, $Q = VA$, can be utilized to compute channel exit velocities and culvert velocities. One method to reduce the exit velocities from outlets is to install a riprap basin. A riprap outlet basin is a preshaped scourhole lined with riprap that functions as an energy dissipator by forming a hydraulic jump.

General details of the basin recommended in this chapter are shown on Figure 12-5. Principal features of the basin are the following.

1. The basin is preshaped and lined with riprap of median size (d_{50}).
2. The floor of the riprap basin is constructed at an elevation of h_s below the culvert invert. The dimension h_s is the approximate depth of scour that would occur in a thick pad of riprap of size d_{50} if subjected to design discharge. The ratio of h_s to d_{50} of the material should be between 2 and 4.
3. The length of the energy dissipating pool is 10 x h_s or 3 x W_o whichever is larger. The basin overall length is 15 x h_s or 4 x W_o whichever is larger.

Design Procedure

Following are the steps in the procedure for the design of riprap basins.

Step 1   Estimate the flow properties at the brink (outlet) of the culvert. Establish the outlet invert elevation such that $TW/y_o \leq 0.75$ for the design discharge.

Step 2   For subcritical flow conditions (culvert on mild or horizontal slope) utilize Figures 12-6 or 12-7 to obtain y_o/D, then obtain V_o by dividing Q by the wetted area associated with y_o. D is the height of a box culvert. If the culvert is on a steep slope, V_o will be the normal velocity obtained by using the Manning equation for appropriate slope, section, and discharge.

Step 3   For channel protection, compute the Froude number for brink conditions with $y_e = (A/2)^{1.5}$. Select d_{50}/y_e appropriate for locally available riprap (usually the most satisfactory results will be obtained if $0.25 < d_{50}/y_e < 0.45$). Obtain h_s/y_e from Figure 12-8, and check to see that $2 < h_s/d_{50} < 4$. Recycle computations if h_s/d_{50} falls out of this range.

Step 4   Size basin as shown in Figure 12-5.

Step 5   Where allowable dissipator exit velocity is specified:
    a. Determine the average normal flow depth in the natural channel for the design discharge.
    b. Extend the length of the energy basin (if necessary) so that the width of the energy basin at section A-A, Figure 12-5, times the average normal flow depth in the natural channel is approximately equal to the design discharge divided by the specified exit velocity.

Step 6   In the exit region of the basin, the walls and apron of the basin should be warped (or transitioned) so that the cross section of the basin at the exit conforms to the cross section of the natural channel. Abrupt transition of surfaces should be avoided to minimize separation zones and eddies.

Step 7   If high tailwater is a possibility and erosion protection is necessary for the downstream channel, the following design procedure is suggested.
    • Design a conventional basin for low tailwater conditions in accordance with the instructions above.

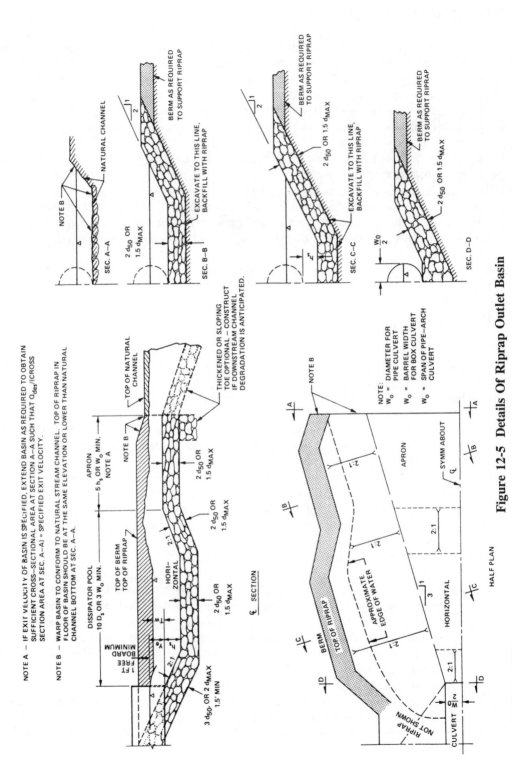

Figure 12-5 Details Of Riprap Outlet Basin

Source: HEC-14, 1983

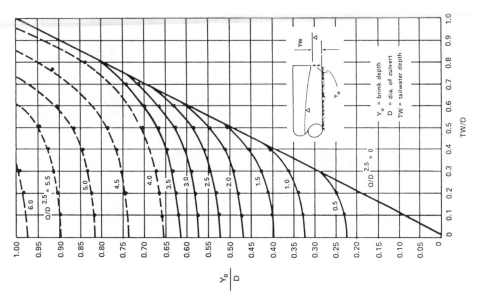

Figure 12-7 Circular Curves

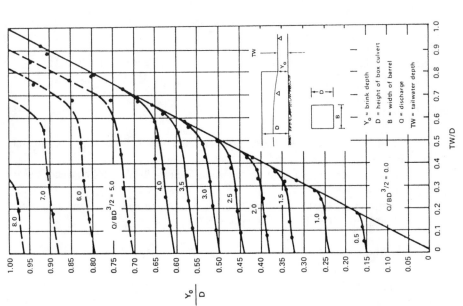

Dimensionless Rating Curves For Outlets On Horizontal And Mild Slopes

Figure 12-6 Rectangular Curves

Source: HEC-14, 1983

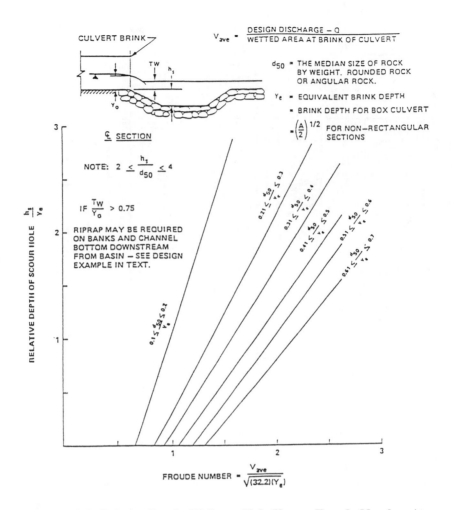

Figure 12-8 Relative Depth Of Scour Hole Versus Froude Number At Brink Of Culvert With Relative Size Of Riprap As A Third Variable

Source: HEC-14, 1983

- Estimate centerline velocity at a series of downstream cross sections using the information shown in Figure 12-9.
- Shape downstream channel and size riprap using Figure 12-12 and the stream velocities obtained above.

For materials, construction techniques, and design details consult the Federal Highway publication HEC-11 entitled Use of Riprap For Bank Protection.

Design Considerations

Riprap basin design should include consideration of the following.

1. The dimensions of a scourhole in a basin constructed with angular rock can be approximately the same as the dimensions of a scourhole in a basin constructed of rounded material when rock size and other variables are similar.

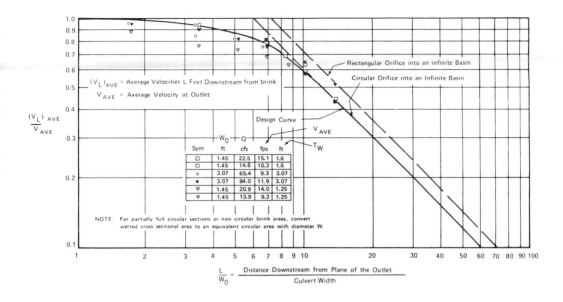

Figure 12-9 Distribution Of Centerline Velocity For Flow From Submerged Outlets To Be Used For Predicting Channel Velocities Downstream From Culvert Outlet Where High Tailwater Prevails

Source: HEC-14, 1983

2. When the ratio of tailwater depth to brink depth, TW/y_0, is less than 0.75 and the ratio of scour depth to size of riprap, h_s/d_{50}, is greater than 2.0, the scourhole should function very efficiently as an energy dissipator. The concentrated flow at the culvert brink plunges into the hole, a jump forms against the downstream extremity of the scourhole, and flow is generally well dispersed as it leaves the basin.

3. The mound of material formed on the bed downstream of the scourhole contributes to the dissipation of energy and reduces the size of the scourhole; that is, if the mound from a stable scoured basin is removed and the basin is again subjected to design flow, the scourhole will enlarge.

4. For high tailwater basins (TW/y_0 greater than 0.75), the high velocity core of water emerging from the culvert retains its jet-like character as it passes through the basin and diffuses, similarly to a concentrated jet diffusing in a large body of water. As a result, the scourhole is much shallower and generally longer. Consequently, riprap may be required for the channel downstream of the rock-lined basin.

5. It should be recognized that there is a potential for limited degradation to the floor of the dissipator pool for rare event discharges. With the protection afforded by the $2(d_{50})$ thickness of riprap, the heavy layer of riprap adjacent to the roadway prism, and the apron riprap in the downstream portion of the basin, such damage should be superficial.

6. See Standards in HEC No. 11 (Federal Highway Administration, 1967) for details on riprap materials and use of filter fabric.

7. Stability of the surface at the outlet of a basin should be considered using the methods for open channel flow as outlined in the Open Channel, Chapter 10, of this book.

Example Problems

Following are some example problems to illustrate the design procedures outlined above.

Example No. 1

Given: Box culvert - 8 ft by 6 ft
          Design Discharge $Q = 800$ cfs
          Supercritical flow in culvert
          Normal flow depth = brink depth
          $y_o = 4$ ft
          Tailwater depth TW = 2.8 ft
Find: Riprap basin dimensions for these conditions:
Solution: Definition of terms in steps 1-5 can be found in Figures 12-5 and 12-8.
1. $y_o = y_e$ for rectangular section - therefore with y_o given as 4 ft, $y_e = 4$ ft
2. $V_o = Q/A = 800/(4 \times 8) = 25$ ft/s
3. Froude Number = $Fr = V/(g \times y_e)^{.5}$ $(g = 32.3$ ft/s$^2)$
    $Fr = 25/(32.2 \times 4)^{.5} = 2.20 < 3.0$ O.K.
4. $TW/y_e = 2.8/4.0 = 0.7$ Therefore $TW/y_e < 0.75$ O.K.
5. Try $d_{50}/y_e = 0.45$, $d_{50} = 0.45 \times 4 = 1.80$ ft
    From Figure 12-8, $h_s/y_e = 1.6$ $h_s = 4 \times 1.6 = 6.4$ ft
    $h_s/d_{50} = 6.4/1.8 = 3.6$ ft $2 < h_s/d_{50} < 4$ O.K.
6. $L_s = 10 \times h_s$ (L_s = length of energy dissipator pool)
    $L_s = 10 \times 6.4 = 64$ ft L_s min = $3 \times W_o = 3 \times 8 = 24$ ft,
    Therefore use $L_s = 64$ ft
    $L_B = 15 \times h_s$ (L_B = overall length of riprap basin)
    $L_B = 15 \times 6.4 = 96$ ft L_B min = $4 \times W_o = 4 \times 8 = 32$ ft,
    Therefore use $L_B = 96$ ft
7. Thickness of riprap:
    On the approach = $3 \times d_{50} = 3 \times 1.8 = 5.4$ ft
    Remainder = $2 \times d_{50} = 2 \times 1.8 = 3.6$ ft
    Other basin dimensions designed according to details shown in Figure 12-5.

Example No. 2

Given: Same design data as example problem number 1 except: TW = 4.2 ft
          Downstream channel can tolerate only 7 ft/s discharge
Find: Riprap basin dimensions for these conditions
Solutions: Note -- High tailwater depth, $TW/y_o = 4.2/4 = 1.05 > 0.75$
1. Use the riprap basin designed in example 1 with: $d_{50} = 1.8$ ft, $h_s = 6.4$ ft, $L_s = 64$ ft, $L_B = 96$ ft.
2. Design riprap for downstream channel. Utilize Figure 12-9 for estimating average velocity along the channel. Compute equivalent circular diameter D_e for brink area from:
    $A = 3.14D_e^2/4 = y_o \times W_o = 4 \times 8 = 32$ ft^2
    $D_e = ((32 \times 4)/3.14)^{.5} = 6.4$ ft
    $V_o = 25$ ft/s (From Example 1)

3. Set up the following table.

L/D$_e$ (Assume) (D$_e$ = W$_o$)	L (Compute) ft	V$_L$/V$_o$ (Fig. 12-9)	V$_1$ ft/s	Rock Size d$_{50}$ (ft) (Fig. 12-12)
10	64	0.59	14.7	1.4
15*	96	0.37	9.0	0.6
20	128	0.30	7.5	0.4
21	135	0.28	7.0	0.4

*L/W$_o$ is on a logarithmic scale so interpolations must be logarithmically.
Riprap should be at least the size shown but can be larger. As a practical consideration, the channel can be lined with the same size rock used for the basin. Protection must extend at least 135 ft downstream from the culvert brink. Channel should be shaped and riprap installed in accordance with details in HEC No. 11.

Example No. 3

Given: 6 ft diameter CMC   Design discharge Q = 135 cfs
       Slope channel S$_o$ = 0.004 ft/ft   Manning's n = 0.024
       Flow is subcritical   Tailwater depth TW = 2.0 ft
       Normal depth in pipe for Q = 135 cfs is 4.5 ft
       Normal velocity is 5.9 ft/s

Find: Riprap basin dimensions for these conditions:
Solution: Following are the steps in the solution.

1. Determine y$_o$ and V$_o$ From Figure 12-7, y$_o$/D = 0.45
    Q/D$^{3/2}$ = 135/6$^{3/2}$ = 1.53   TW/D = 2.0/6 = 0.33
    y$_o$ = .45 x 6 = 2.7 ft   TW/y$_o$ = 2.0/2.7 = 0.74
    TW/y$_o$ < 0.75 O.K.
    Brink Area (A) for y$_o$/D = 0.45 (Note: from Uniform Flow In Circular Sections Table given in Design Of Culverts, Chapter 9.)
    For y$_o$/D = d/D = 0.45   A/D^2 = 0.3428
    Therefore A = 0.3428 x 6^2 = 12.3 ft^2
    V$_o$ = Q/A = 135/12.3 = 11.0 ft/s
2. For Froude number calculations at brink conditions,
    y$_e$ = (A/2)$^{1/2}$ = (12.3/2)$^{1/2}$ = 2.48 ft
3. Froude number = Fr = V$_o$/(32.2 x y$_e$)$^{1/2}$ = 11/(32.2x2.48)$^{1/2}$ = 1.23<3 O.K.
4. For most satisfactory results - 0.25 < d$_{50}$/y$_e$ < 0.45
    Try d$_{50}$/y$_e$ = 0.25   d$_{50}$ = 0.25 x 2.48 = 0.62 ft
    From Figure 12-8, h$_s$/y$_e$ = 0.75   Therefore h$_s$ = 0.75 x 2.48 = 1.86 ft
    Check: h$_s$/d$_{50}$ = 1.86/0.62 = 3,   2 < h$_s$/d$_{50}$ < 4 O.K.
5. L$_s$ = 10 x h$_s$ = 10 x 1.86 = 18.6 ft   or L$_s$ = 3 x W$_o$ = 3 x 6 = 18 ft,
    therefore use L$_s$ = 18.6 ft
    L$_B$ = 15 x h$_s$ = 15 x 1.86 = 27.9 ft   or L$_B$ = 4 x W$_o$ = 4 x 6 = 24 ft,
    therefore use L$_B$ = 27.9 ft
    d$_{50}$ = 0.62 ft or use d$_{50}$ = 8 in

Other basin dimensions should be designed in accordance with details shown on Figure 12-5.
     When using the design procedure outlined in this chapter, it is recognized that there is some chance of limited degradation of the floor of the dissipator pool for rare event discharges. With the protection afforded by the 3 x d$_{50}$ thickness of riprap on the approach and the 2 x d$_{50}$ thickness

of riprap on the basin floor and the apron in the downstream portion of the basin, the damage should be superficial.

Figure 12-10 is provided as a convenient form to organize and present the results of riprap basin designs.

	TW	y_e	(1) TW/y_e	d_{50}/y_e	d_{50}	h_s/y_e	h_s	(2) h_s/d_{50}	
LOW TW $TW/y_e \leq 0.75$									
	V ALLOWABLE	L/D_e (3)	L	V_{ave}/V_L	V_L				
HIGH TW $TW/y_e > 0.75$									

	Larger of					
Length of Pool	= 10 h_s or 3W_o	=	_____	ft.	(1) $TW/y_e \leq 0.75$ for Low TW Design
Length of Apron	= 5 h_s or W_o	=	_____	ft.	(2) 2 < hs/d_{50} < 4
Thickness of Approach	= 3d_{50} or 2d_{max}	=	_____	ft.	(3) $D_e = [4A/\pi]^{1/2}$
Thickness of Remainder of Basin	= 2d_{50} or 1.5d_{max}	=	_____	ft.	

Figure 12-10 Riprap Basins

Source: HEC-14, 1983

Section 12.8 Baffled Outlets

The baffled outlet is a boxlike structure with a vertical hanging baffle and an end sill, as shown in Figure 12-11. Energy is dissipated primarily through the impact of the water striking the baffle and, to a lesser extent, through the resulting turbulence. This type of outlet protection has been used with outlet velocities up to 50 ft per second and with Froude numbers from 1 to 9. Tailwater depth is not required for adequate energy dissipation, but a tailwater will help smooth the outlet flow. Denver Urban Drainage District has developed a design modification that allows the basin to drain during low flows (UDFCD, 1991).

Design Procedure

The following design procedure is based on physical modeling studies summarized from the U.S. Department of Interior (1978). The dimensions of a baffled outlet as shown in Figure 12-11 should be calculated as follows.

1. Determine input parameters, including:

   h = energy head to be dissipated, ft (can be approximated as the difference between channel invert elevations at the inlet and outlet)

   Q = design discharge, cfs

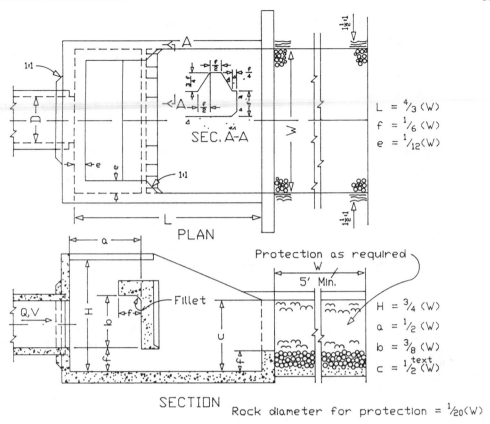

$$L = {}^4\!/_3 \,(W)$$
$$f = {}^1\!/_6 \,(W)$$
$$e = {}^1\!/_{12}(W)$$

SEC. A-A

PLAN

Protection as required

W

5' Min.

$$H = {}^3\!/_4 \,(W)$$
$$a = {}^1\!/_2 \,(W)$$
$$b = {}^3\!/_8 \,(W)$$
$$c = {}^1\!/_2 \,(W)$$

SECTION

Rock diameter for protection = $\frac{1}{20}$(W)

Figure 12-11 Schematic Of Baffled Outlet

Source: U.S. Department Of The Interior, 1978

V = theoretical velocity, ft/s, = 2gh
A = Q/V = flow area, ft^2
d = A$^{0.5}$ = flow depth entering the basin, ft (assumes square jet)
Fr = V/(gd)$^{0.5}$ = Froude number, dimensionless

2. Calculate the minimum basin width, W, in ft, using the following equation:

$$W/d = 2.88Fr^{0.566} \qquad\qquad\qquad (12.3)$$

where: W = minimum basin width, ft
         d = depth of incoming flow, ft
         Fr = V/(gd)$^{0.5}$ = Froude number, dimensionless

The limits of the W/d ratio are from 3 to 10, which corresponds to Froude numbers 1 and 9. If the basin is much wider than W, flow will pass under the baffle and energy dissipation will not be effective.

3. Calculate the other basin dimensions as shown in Figure 12-11, as a function of W. Standard construction drawings for selected widths are available from the U.S. Department of the Interior (1978).

4. Calculate required protection for the transition from the baffled outlet to the natural channel based on the outlet width. A riprap apron should be added of width W, length L (or a 5-foot minimum), and depth f = (W/6). The side slopes should be 1.5:1, and median rock diameter should be at least W/20.

5. Calculate the baffled outlet invert elevation based on expected tailwater. The maximum distance between expected tailwater elevation and the invert should be b + f or some flow will go over the baffle with no energy dissipation. If the tailwater is known and fairly controlled, the baffled outlet invert should be a distance, b/2 + f, below the calculated tailwater elevation. If tailwater is uncontrolled, the baffled outlet invert should be a distance, f, below the downstream channel invert.

6. Calculate the outlet pipe diameter entering the basin assuming a velocity of 12 ft/s flowing full.

7. If the entrance pipe slopes downward, the outlet pipe should be turned horizontal for at least 3 ft before entering the baffled outlet.

8. If the upstream and downstream ends of the pipe will be submerged, provide an air vent of diameter approximately 1/6 the pipe diameter near the upstream end to prevent pressure fluctuations and possible surging flow conditions.

Example Problem

A cross-drainage pipe structure has a design flow rate of 150 cfs, a head, h, of 30 ft, and a tailwater depth, TW, of 3 ft above ground surface. Find the baffled outlet basin dimensions and inlet pipe requirements.

1. Compute the theoretical velocity from:
    $V = (2gh)^{0.5} = [2(32.2 \text{ ft/s}^2)(30 \text{ ft})]^{0.5} = 43.95 \text{ ft/s}$
    This is less than 50 ft/s, so a baffled outlet is suitable.

2. Determine the flow area using the theoretical velocity as follows:
    $A = Q/V = 150 \text{ cfs}/43.95 \text{ ft/s} = 3.41 \text{ ft}^2$

3. Compute the flow depth using the area from Step 2.
    $d = (A)^{0.5} = (3.41 \text{ ft}^2)^{0.5} = 1.85 \text{ ft}$

4. Compute the Froude number using the results from Steps 2 and 3.
    $Fr = V/(gd)^{0.5} = 43.95 \text{ ft/s}/[(32.2 \text{ ft/s}^2)(1.85 \text{ ft})]^{0.5} = 5.7$

5. Determine the basin width using equation 12.3 with the Froude number from Step 4.
    $W = 2.88 \, dFr^{0.566} = 2.88 \, (1.85) \, (5.7)^{0.566}$
    $W = 14.27 \text{ ft (minimum)}$ Use 14 ft, 4 in. as design width.

6. Compute the remaining basin dimensions (as shown in Figure 12-11):
    L = 4/3 (W) = 19.1 ft   Use L = 19 ft, 2 in.
    f = 1/6 (W) = 2.39 ft   Use f = 2 ft, 5 in.
    e = 1/12 (W) = 1.19 ft   Use e = 1 ft, 3 in.
    H = 3/4 (W) = 10.75 ft   Use H = 10 ft, 9 in.
    a = 1/2 (W) = 7.17 ft   Use a = 7 ft, 2 in.
    b = 3/8 (W) = 5.38 ft   Use b = 5 ft, 5 in.
    c = 1/2 (W) = 7.17 ft   Use c = 7 ft, 2 in.
    Baffle opening dimensions would be calculated from f (Figure 12-11).

7. Basin invert should be at b/2 + f below tailwater, or (5 ft, 5 in)/2 + 2 ft, 5 in. = 5.125 ft. Use 5 ft 2 in; therefore, invert should be 2 ft, 2 in. below ground surface.

8. The riprap transition from the baffled outlet to the natural channel should be 14 ft, 4 in. long by 14 ft, 4 in. wide by 2 ft, 5 in. deep (W x W x f). Median rock diameter should be of diameter W/20, or about 9 in.

9. Inlet pipe diameter should be sized for an inlet velocity of about 12 ft/s. $(3.14d)^2/4 = Q/V$; $d = [(4Q)/3.14V)]^{0.5} = [(4(150 \text{ cfs})/3.14(12 \text{ ft/s})]^{0.5} = 3.99$ ft. Use 48-inch pipe. If a vent is required, it should be about 1/6 of the pipe diameter or 8 in.

12.9 Downstream Channel Transitions

Some energy dissipators result in exit velocity and depth, near critical conditions. This flow condition rapidly adjusts to the downstream or natural channel regime; however, critical velocity may be sufficient to cause erosion problems requiring protection adjacent to the basin exit. Figure 12-12 provides an estimate of the riprap size recommended for use downstream of energy dissipators.

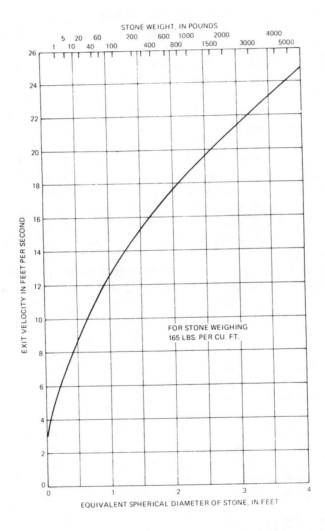

Figure 12-12 Riprap Size For Use Downstream Of Energy Dissipator

Source: Searcy, 1967

The Corps of Engineers (COE) (HDC, 1988) has developed an estimate of riprap size required below stilling basins in moderately and highly turbulent situations. The estimate has proven effective in model studies and in the field. The equation based on the work of Isbash is:

$$D_{50} = \frac{V^2}{2g} \frac{1}{C^2} \left(\frac{\gamma_w}{\gamma_s - \gamma_w} \right)$$ (12.4)

and $W_{50} = (\pi D_{50}^3/6) \, \gamma_s$ (12.5)

Where: W_{50} =   weight of stone for which 50 percent of material by weight contains stone of lesser weight, lbs
   D_{50} =   spherical diameter of stone having weight of W_{50}, ft
   V =   average velocity of flow over the end sill of the stilling basin, ft/s
   g =   acceleration of gravity, ft/s^2
   γ_w =   specific weight of water, 62.4 lbs/ft^3
   γ_s =   specific weight of stone, normally about 165, lbs/ft^3
   C =   coefficient equal to 0.86 for riprap just downstream from stilling basins and 1.2 for riprap dumped as a temporary closure dam in a channel

It should be noted that the USBR performed a series of studies on stone stability downstream from various sites. Their design curve for stability is identical with the high turbulence value (0.86) in the COE equation (Peterka, 1978). The curve provided by the FHWA (Figure 12-12) is identical to the low turbulence value (1.2) in the COE equation. It is therefore recommended that the COE high turbulence values be used for all situations of high velocity flow and for all larger structures while the FHWA curve (or the low turbulence COE values) can be used downstream from smaller structures with low turbulence. The COE low turbulence values are nearly equivalent to the riprap sizing method given in the Open Channel, Chapter 10, but without the adjustment capability.

The gradation and thickness of the stone downstream from stilling basins is recommended to fit the following criteria (LL means lower limit, UL means upper limit) (HDC, 1988):

- UL - W_{50} should satisfy thickness requirements specified below.
- LL - $W_{100} < 2$ LL W_{50}
- UL - $W_{100} < 5$ LL W_{50} and satisfy thickness requirements below.
- LL - $W_{15} > 1/16$ UL W_{100}
- UL - $W_{15} <$ UL W_{50} and satisfy any graded stone filter requirements.
- Bulk volume of stone $< W_{15}$ should not exceed the void volume in the riprap.
- W_0 to W_{25} may be used in place of W_{15} to better meet available stone gradations.
- Thickness of the stone should be the greater of 2 $D_{50 \, max}$ or 1.5 $D_{100 \, max}$.

The riprap should be extended downstream to a point where the velocity has slowed enough for nonerosive conditions in the natural channel. The end of the riprap should be toed into the banks and bed. The extent of the protection downstream has not been settled with either theory or experiment. Rice and Kadavy, based on tests for SAF type stilling basins (baffle blocks with end sill), provide a preliminary direct relation for the downstream length of riprap as:

$$L_s = 4.5 \text{ Fr } D_1$$ (12.6)

Where: L_s =   length of riprap section, ft
   Fr =   Froude number based on velocity and depth at basin entrance

   D_1 = depth at basin entrance, ft

The Denver Urban Drainage and Flood District uses a rule developed by considering an expanding jet as (UDFCD, 1991):

$$L_t = \frac{1}{2\,Tan\theta}\left(\frac{A_t}{Y_t}-W\right)$$ (12.7)

Where: L_t = length of protection, ft
        W = conduit width or diameter, ft
        A_t = Q/V = design discharge/ allowable downstream velocity, ft^2
        Y_t = tailwater depth, ft
        θ = expansion angle of discharging flow, from Figures 12-13 and 12-14 for circular and rectangular conduits.

For multiple conduits see UDFCD, 1991. If unreasonable results are calculated a rule-of-thumb is:

3 times conduit diameter or height < L < 10 times diameter or height

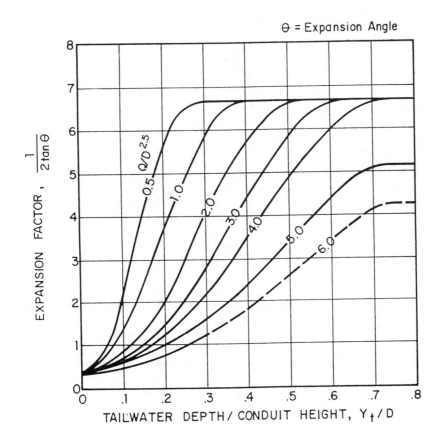

Figure 12-13 Expansion Factor For Circular Conduits (UDFCD, 1991)

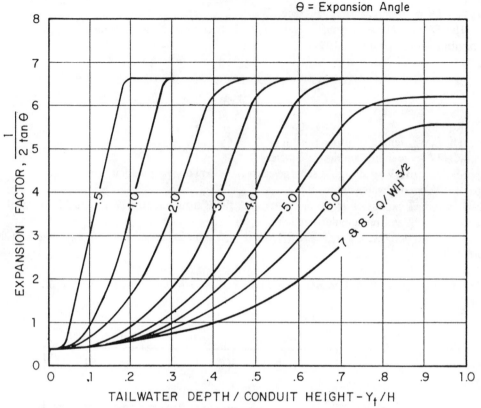

Figure 12-14 **Expansion Factor For Rectangular Conduits** (UDFCD, 1991)

References

Blaisdell, F. W. and C. L. Anderson, "A Comprehensive Generalized Study of Scour at Cantilevered Pipe
    Outlets - Background", J. Hyd. Res., Vol. 26, No. 4, 1988.

Blaisdell, F. W. and C. L. Anderson, "A Comprehensive Generalized Study of Scour at Cantilevered Pipe
    Outlets - Results", Vol. 26, No. 5, 1988.

Blaisdell, F. W. and C. L. Anderson, "Plunge Pool Energy Dissipator", ASCE J. Hy. Engrg., Vol. 117,
    No. 3, 1991.

Denver Urban Drainage and Flood Control District (UDFCD), "Urban Storm Drainage Criteria Manual",
    1969 with changes 1991.

Federal Highway Administration, Debris-Control Structures, Hydraulic Engineering Circular No. 9, 1964.

Federal Highway Administration, Hydraulic Design of Energy Dissipators for Culverts and Channels,
    Hydraulic Engineering Circular No. 14, 1983.

Federal Highway Administration, Use of Riprap for Bank Protection, Hydraulic Engineering Circular No.
    11, 1967.

Peterka, A. J., "Hydraulic Design of Stilling Basins and Energy Dissipators", Engrg. Mono. No. 25, USBR,
    1978.

Rice, C. E. and K. C. Kadavy, "Protection Against Scour at SAF Stilling Basins", ASCE J. of Hy. Engrg.,
    Vol. 119, No. 1, Jan., 1993.

Searcy, James K., Use of Riprap for Bank Protection, Federal Highway Admin., Washington, D. C., 1967.

U.S. Department of Agriculture, Soil Conservation Service, Standards And Specifications For Soil Erosion
    And Sedimentation In Developing Areas, Washington, D.C., 1975.

U.S. Army Corps of Engineers, "Hydraulic Design Criteria (HDC)", USAEWES 1988.

Appendix A - HY8 Culvert Analysis - Energy Dissipators

Introduction

The HYDRAIN Culvert Analysis microcomputer program HY8 consists of three main options:
(1) Culvert Analysis,
(2) Hydrograph Generation, and
(3) Routing.
The culvert analysis portion of the HYDRAIN model includes a sub-model for the design and analysis of energy dissipators. The purpose of this Appendix is to provide the user with analysis approaches to use with the Energy Dissipator portion of the model. Although the model is specifically designed for energy dissipators associated with culverts, by modifying the culvert discharge data, energy dissipators for other storm water management facilities (i.e., weirs from detention facilities, open channels) can also be designed using this model.

The model has several user-friendly features which permit easy data entry, changing input data to determine the effects on the energy dissipator design, and output of summary design information. Very little input data are required which makes the model very easy to use. The design procedures included in the model follow the design procedures used in the Federal Highway Publication - HEC-14. This publication was also used to develop most of the concepts and design procedures included in this chapter.

The Energy Dissipator Model includes the design and analysis of many dissipators that will probably not be used within municipal storm water management applications. Some of the dissipators were designed for large flood control structures and not the usual urban applications. The scour hole estimation procedure and the design of riprap basins are two of the more useful options and thus these portions of the model are included in this appendix.

Example Application Data

The following example application problem will provide instruction in data entry, file modification and energy dissipator design. The user should work through the problem on the computer while following the text so as to become familiar with the program and some of its applications.

The following data are typical data which could be obtained from a culvert design, either using the nomographs or the culvert portion of the HY8 model. The Energy Dissipator Model uses the data from the file established when the culvert analysis portion of the model is used.

Design Discharge = 60 cfs
Velocity at Culvert Outlet = 19.6 ft/s
Depth of Water at Culvert Outlet = 1.34 ft

Scour Hole Estimation

Using the Energy Dissipator Model estimate the scour hole at the exit to the culvert. From the HY8 Program select <4> ENERGY DISSIPATOR DESIGN

FHWA HY8 PROGRAM MENU

TYPE NUMBER OF DESIRED PROGRAM
<1> CULVERT DESIGN
<2> HYDROGRAPH GENERATION
<3> ROUTING
<4> ENERGY DISSIPATOR DESIGN
<5> RETURN TO DOS

Note that this program will only design a dissipator for one culvert at a time. Thus if more than one culvert exists at a site, the user must specify the culvert for which the dissipator is to be designed.

The user must then input the culvert data file name. If you want to design a dissipator for something other than a culvert, input a culvert file name and then edit the file to include the data needed for the design you desire.

INPUT THE CULVERT DATA FILE NAME - TEST
DESIGN MENU

(1) Edit Culvert Characteristics (Within Current Data File)
(2) Scour Hole Geometry Estimate
(3) Design Internal Dissipator (Box or Circular Culvert Only)
(4) Design External Dissipator
(5) Choose A Different Culvert (Within Current Data File)
(6) Exit Program

Select Appropriate #   2

From the Design Menu select (2) Scour Hole Geometry Estimate

Input the following data for scour hole estimation:

Noncohesive or Cohesive Soil
D16, D50, and D84 particle sizes (this indicates the percent - 16, 50, or 84 - particles that are finer than the size input). Thus in the following example, 16 percent of the particles are finer than 1 mm.

CULVERT OUTLET SOIL PARAMETERS

ENTER TIME TO PEAK OF STORM HYDROGRAPH (MIN)	30
*** IF UNKNOWN ENTER 30 MIN ***	
INDICATE SOIL TYPE (1) NONCOHESIVE (2) COHESIVE	1
ENTER A VALUE FOR D16 (mm)	1
ENTER A VALUE FOR D50 (mm)	2.1
ENTER A VALUE FOR D84 (mm)	5

The model will then determine the dimensions of the scour hole for the design conditions.

SCOUR HOLE GEOMETRY

LENGTH = 33.0 FT   WIDTH = 24.7 FT
DEPTH = 3.8 FT   VOLUME = 2674.7 FT
MAXIMUM SCOUR OCCURS 13.2 FT DOWNSTREAM OF CULVERT
DOWNSTREAM CHANNEL BOTTOM WIDTH = 5 FT
NORMAL DEPTH IN DOWNSTREAM CHANNEL IS 1.8 FT FOR A FLOW OF 60 CFS

<PRESS ANY KEY TO CONTINUE>

A summary of culvert data, channel data, scour hole geometry, and soil data can then be printed.

CULVERT AND CHANNEL DATA

CULVERT NO. 1   DOWNSTREAM CHANNEL

CULVERT TYPE: 3.0 FT CIRCULAR   CHANNEL TYPE: RECTANGULAR
CULVERT LENGTH = 75.1 FT   BOTTOM WIDTH = 5.0 FT
NO. OF BARRELS = 1.0   TAILWATER DEPTH = 1.8 FT
FLOW PER BARREL = 60 CFS   TOTAL DESIGN FLOW = 60 CFS
INVERT ELEVATION = 947.0 FT   BOTTOM ELEVATION = 947.0 FT
OUTLET VELOCITY = 19.6 FPS   NORMAL VELOCITY = 6.8 FPS

SCOUR HOLE GEOMETRY AND SOIL DATA

LENGTH = 33.0 FT   WIDTH = 24.7 FT
DEPTH = 3.8 FT   VOLUME = 2674.7 CU FT
MAXIMUM SCOUR OCCURS 13.2 FT DOWNSTREAM OF CULVERT
SOIL TYPE: NONCOHESIVE
SAND SIZES:
    D16 = 1 mm   D50 = 2.1 mm D84 = 5 mm

Energy Dissipator Design

To design an energy dissipator to control the exit velocity from the culvert, from the Design Menu select (4) Design External Dissipator.

DESIGN MENU

(1) Edit Culvert Characteristics (Within Current Data File)
(2) Scour Hole Geometry Estimate
(3) Design Internal Dissipator (Box or Circular Culvert Only)
(4) Design External Dissipator
(5) Choose A Different Culvert (Within Current Data File)
(6) Exit Program

Select Appropriate # 4

From the External Dissipator Categories select (C) At-Stream-Level Structures.

EXTERNAL DISSIPATOR CATEGORIES

(A)	DROP STRUCTURES
(B)	STILLING BASINS
(C)	AT-STREAMBED-LEVEL STRUCTURES

SELECT LETTER OF DESIRED CATEGORY C

From the table of Dissipator Types select number 11 - Riprap Basin (Consult HEC 14 for the description of the other dissipator types).

AT-STREAM-LEVEL STRUCTURES

DISSIPATOR TYPE	FROUDE # FR	TAILWATER TW	SPECIAL LIMITATIONS	FEASIBILITY
10 CSU BASIN	<3	----	----	OK
11 RIPRAP BASIN	<3	----	----	OK
13 HOOK	1.8 TO 3	----	----	OK
14 USBR-6	----	DESIRABLE	Q<400	OK

VO < 50

12 CONTRA			
COSTA	< 3	< .5D	----

NO GOOD (TAIL WATER TOO HIGH)

ENTER # OF APPLICABLE DISSIPATOR -- IF MORE THAN 1, ENTER 1 AT A TIME UNTIL ALL DESIRED #'S ARE ENTERED, PRESS <ENTER> TO COMMENCE DESIGN   11

Choose the condition to be used to compute basin outlet velocity - it is recommended that <C> Critical Depth At Outlet be used except when the culvert is discharging into a lake or ponded area downstream.   D50 and DMAX represent the mean and maximum riprap size respectively.

YOU HAVE CHOSEN TO DESIGN A RIPRAP STILLING BASIN

CHOOSE CONDITION TO BE USED TO COMPUTE BASIN OUTLET VELOCITY   <N> NORMAL DEPTH AT OUTLET <C> CRITICAL DEPTH AT OUTLET   C

ENTER D50 (ft) OF THE RIPRAP MIXTURE   .8
ENTER DMAX (ft) OF THE RIPRAP MIXTURE   1.5

The model will then calculate the design dimensions of the riprap basin, the outlet velocity from the basin and depth of flow at the basin outlet.

RIPRAP STILLING BASIN -- FINAL DESIGN

THE LENGTH OF THE BASIN = 39.4 FT
THE LENGTH OF THE POOL = 26.3 FT
THE LENGTH OF THE APRON = 13.1 FT

THE WIDTH OF THE BASIN AT THE OUTLET = 5.0 FT
THE DEPTH OF THE POOL BELOW CULVERT INVERT = 2.6 FT

THE THICKNESS OF THE RIPRAP ON THE APRON = 3.0 FT
THE THICKNESS OF THE RIPRAP ON THE REST OF THE BASIN = 2.3 FT

THE BASIN OUTLET VELOCITY = 8.8 FPS
THE DEPTH OF FLOW AT BASIN OUTLET = 1.4 FT

IS DESIGN SATISFACTORY? <Y> DESIGN IS OK   <N> REDO DESIGN

The user can then print a summary of the dissipator design.

CULVERT AND CHANNEL DATA

CULVERT NO. 1   DOWNSTREAM CHANNEL

CULVERT TYPE: 3.0 FT CIRCULAR   CHANNEL TYPE: RECTANGULAR
CULVERT LENGTH = 75.1 FT   BOTTOM WIDTH = 5.0 FT
NO. OF BARRELS = 1.0   TAILWATER DEPTH = 1.8 FT
FLOW PER BARREL = 60 CFS   TOTAL DESIGN FLOW = 60.0 CFS
INVERT ELEVATION = 947.0 FT   BOTTOM ELEVATION = 947.0 FT
OUTLET VELOCITY = 19.6 FPS   NORMAL VELOCITY = 6.8 FPS

RIPRAP STILLING BASIN -- FINAL DESIGN

THE LENGTH OF THE BASIN	= 39.4 FT
THE LENGTH OF THE POOL	= 26.3 FT
THE LENGTH OF THE APRON	= 13.1 FT
THE WIDTH OF THE BASIN AT THE OUTLET	= 5.0 FT
THE DEPTH OF POOL BELOW CULVERT INVERT	= 2.6 FT
THE THICKNESS OF THE RIPRAP ON THE APRON	= 3.0 FT
THE THICKNESS OF RIPRAP ON REST OF THE BASIN	= 2.3 FT
THE BASIN OUTLET VELOCITY	= 8.8 FPS
THE DEPTH OF THE FLOW AT BASIN OUTLET	= 1.4 FT

Chapter 13 Structural Best Management Practices

13.1 Introduction

Environmental engineering involving urban storm water is a rapidly evolving science. In part this is due to the limited scientific findings and different perceptions as to what constitutes acceptable practice where sensitive surface waters are concerned; and in part due to new regulatory pressures brought about by Federal, state and local regulations. In almost every municipal engineering project, the designer is involved in surface water environmental issues. Issues may involve such things as:

- adverse pollution increases due to general development;
- specific permitting for industry;
- wetland encroachment;
- endangered species or flora;
- ecological balance issues;
- greenway, park or buffer requirements;
- stream filling or dredging;
- non-point source pollution or maximum stream load issues;
- erosion control;
- structural and non-structural controls;
- land use regulations and limitations;
- mitigation efforts or environmental design criteria; and
- aesthetics.

In dealing with these and similar situations it is good practice to use a multidisciplined team approach because of the complex and inter-related problems involved. The team may involve biologists and ecologists as well as landscape architects and engineers. In general, the municipal engineer's role is to determine the hydrology, hydraulic and pollution related effects and significance on sensitive surface waters and recommend "Best Management Practices" (BMPs) to be used to mitigate many of the detrimental effects.

There are many BMP variations and differing (and often conflicting) design criteria. Many are yet untested over long periods and diverse applications. Data that does exist are often not able to be compared to each other. Due to the complex nature of surface water environmental engineering, this chapter will address only the more reliable and accepted structural BMPs. Procedures for other more complex and experimental practices may be gleaned from the relevant literature. It is believed that the structural BMPs provided in this chapter will address a high percent of the structurally treated surface water environmental problems to be encountered in municipal storm water management applications.

13.2 Surface Waters And Mitigation Measures

The surface waters addressed in this chapter include streams or rivers, ponds or lakes, and wetlands. The characteristics of sensitive surface waters are related to such things as: geographic location, hydrology, climatology, size, chemical and physical quality, ecology, species present, etc. Normally it is a combination of factors which make the specific waters a unique resource. The interrelationship of these factors for sensitive surface waters determines the water's functional values. These values include such things as flood control, wildlife and aquatic habitat, aesthetics, recreation, groundwater recharge, scenic and wild rivers designation, etc. However, all surface waters have unique or valuable characteristics and should be considered for restoration or preservation during the development or redevelopment process. An exception to this is where the municipality allows some level of temporary deterioration of surface waters principally during the construction process or where functional values are redefined to enhance some while losing others.

Pollution Accumulation - Conventional urban pollution studies assert that the surface load of pollutants for pervious and impervious areas accumulates between storms. This load eventually approaches an upper limit, or "steady state" condition. As the surface load accumulates to a particular level for a given land use, the rate of removal by the described processes approximates the deposition rate, and the surface load reaches its upper limit or steady state. This upper limit is a function of the rate at which the pollutants are deposited on the basin's surface and of the dry weather processes that tend to remove the load. These dry weather processes include wind removal, air currents generated by automotive and pedestrian traffic, adsorption, and decay or oxidation of chemicals from one state to another.

Various studies have indicated that certain land uses accumulate and export much greater amounts of petroleum based hydrocarbons, including more exotic priority pollutants such as highly toxic polycyclic aromatic hydrocarbons (PAH) (Schueler and Shepp, 1992, cited in Schueler, 1994). The levels of such pollutants are from several times, to an order of magnitude, higher than the levels found in runoff from typical residential streets and parking areas, and can often be considered to be toxic. The land uses most associated with elevated levels of these pollutants include service stations, vehicle maintenance and washing facilities, areas where many cars are parked for short periods and areas where a large number of cars are parked for long periods. Therefore targeted use of both structural and non-structural controls for these areas make sense.

Pollution Movement Forces - When rainfall occurs, the surface load is subject to a series of physical processes that dictate which particles and how much of the particulate load can be entrained into the surface flow at a given time and how many constituents are dissolved. For a given depth and velocity, the surface flow has a certain energy for moving both dissolved and suspended material. This potential is a function of the velocity of the moving water, the specific gravity of the water, its acidity and temperature and the amount of potential being used to overcome the friction of the overland flow surface.

In order for a given particle to be entrained into the flow, the flow must have sufficient potential to overcome the effects of gravity on the particle and the friction, adhesion, cohesion and other forces that resist its lateral movement. As particles are entrained into the flow, the water begins to use up its excess energy. When the water reaches the point where its suspended and dissolved load are using all of the potential, an equilibrium is established where particles begin to settle out of the flow at a rate equal to that at which they are being entrained.

If the velocity or volume of flow becomes greater by adding more water, changes in surface friction, or increases in surface slope, then more potential is created and thus more surface load can be entrained. Similarly, if the velocity becomes lower, then the potential decreases and part of the load must settle out. This later case is what is normally used in BMP design. By slowing the velocity through the use of ponds, filter strips, or check dams, the potential of the water is decreased and the particulates begin to settle out of the water. This lowering of the water velocity also provides protection of downstream channels against streambank degradation, bridge scour, and flooding. Filtration can also physically block pollutants from movement between the interstices of the granular material. Chemical or biological changes may also reduce (or increase) the amount of pollutants in a flow.

Pollution Availability and Fate - The total storm washoff load is dependent on the amount of pollution available and the magnitude of the force available to carry it away. The factors then include: total depth of rainfall, the duration of the rainfall, the period of time that has elapsed since the last significant precipitation, the pollution accumulation rate, the residual load on the catchment surface at the end of that previous rainfall event and any intervening chemical or biological changes.

As a storm event begins, and velocities of surface runoff are relatively low, the portion of the surface load that is first susceptible to being washed off will be that portion that requires the least energy to entrain. This load is made up of highly soluble and very fine sediments. The term "first flush" refers to this pollutant load that is readily washed off of a catchment's surface. Several nationwide studies, such as NURP (1983), have documented this occurrence. Estimates are that as much as forty to ninety percent of the total pollutant load may be removed during the first half inch of rainfall runoff. Snowmelt pollution can also be important in Northern climates. See Oberts (1994) for a detailed discussion.

One almost never knows enough about the chemical-biological process to be totally comfortable with a diagnosis of a pollution problem. But great good can be accomplished with even a modicum of knowledge of pollution sources and transport mechanisms. Table 13-1 gives an overview of pollution removal mechanisms.

The ability of a pollutant to migrate to receiving waters from its source is a function of its:
- chemical nature,
- physical-chemical properties (such as water solubility and vapor pressure),
- tendency to adsorb to organic matter or sediment, and
- interception by any mitigation or control measures.

Of the major migration processes, the key mechanism for transporting a pollutant originating from urbanization is a combination of sorption and settling. This is due to many of the runoff constituents being in a settled particulate form. This is further enhanced as soluble organic chemicals and heavy metals tend to adsorb to suspended sediments, which in turn will settle given sufficient time and a suitable environment. Once settled, these pollutants will be most commonly transformed through the biological action of degradation, assimilation by microbes or by rooted vegetation. Resuspension can renew the process making some pollutants biologically available once again, and moving pollutants through the system along with the transported bed and suspended sediment load.

Basic Mitigation Measures - Keeping pollutants from surface and ground waters is the primary goal of storm water quality pollution prevention programs. Keeping pollution from the drainage system is done through a combination of "non-structural BMPs" or "source control" measures. Non-structural BMPs include a multitude of programs and practices designed to target certain types and sources of pollution and reduce or eliminate them. Thus, for example, reduction of dumping of used motor oil can be targeted through a combination of: a storm drain stenciling program, convenient used oil turn-in locations, public education, a reporting hotline, point-of-sale pamphlets, comprehensive recycling programs, financial rewards for recycling, strong enforcement, and other activities. For each type of pollutant and point of entry there can be developed a non-structural program attacking it at its source. With each program there logically should also be an attempt to assess the severity and impacts of the problem, the effectiveness of the program and the value of the investment in pollution reduction. The Storm Water Quality Management Plans, Chapter 15, provides a long list of non-structural and structural BMPs.

Non-structural programs will not eliminate all pollution and thus structural measures may be necessary. Once in the system pollution is much more difficult to remove or abate. Pollutants are normally removed from surface waters through some combination of structural BMPs through the basic mechanisms of:

- physical removal and settling,
- filtering,
- chemical reaction, and
- biochemical transformation.

All structural measures to remove pollutants from the drainage system use a combination of these basic mechanisms.

Table 13-1 Fate Of Pollutants By Management Measures

Pollutant	Vegetative Controls	Detention Basin	Infiltration System	Wetlands
Heavy metals	Filtering	Adsorption Settling	Adsorption Filtration	Adsorption Settling
Toxic organics	Adsorption	Adsorption Settling Biodegradation Volatilization	Adsorption Biodegradation	Adsorption Settling Biodegradation Volatilization
Nutrients Solids	Bioassimilation Filtering	Bioassimilation Settling	Adsorption Adsorption	Bioassimilation Adsorption Settling
Oil and Grease	Adsorption	Adsorption Settling	Adsorption	Adsorption Settling
BOD	Biodegradation	Biodegradation	Biodegradation	Biodegradation
Pathogens	Not applicable	Settling	Filtration	Not applicable

Source: Transportation Research Record, 1988.

Conventional BMPs will commonly suffice for most urban development projects. For most pollutants, nonpoint pollution discharges from frequent minor storms are more critical than discharges during infrequent major storms. Measures designed to control storms producing less than one inch of rainfall will control nonpoint pollution discharges for over 90 percent of the storms each year. First-flush conditions result in relatively high pollutant concentrations (analogous to raw sewage) during the initial, relatively low discharge, stages of storm water runoff. This can induce shock-loading and a short term degradation of the water quality of receiving waters. First-flush effects are attributed primarily to the washoff of particles from paved areas.

Pollution management should commonly rely on controls for minor storms having a recurrence interval of one year or less. As such, management techniques that isolate first-flush discharges take advantage of the smaller required storage capacities for these discharges. Placing several BMPs in line ("treatment train") can produce dramatic increases in pollutant removal efficiency.

13.3 Estimation Of Pollutant Concentration And Loads

There are many methods to estimate the concentration and loading of pollutants to surface waters. Physically based models attempt to mimic the accumulation and removal of pollutants as well as the chemical reactions within the receiving streams. More empirical models rely on general data and information on pollution concentrations in surface runoff and then predict pollution through an estimation of surface runoff volumes. Regression equations use significant variables to predict loadings of various constituents based on data sets. This type of model can be used with little or no data but are very rough in their estimates. They are less effective for "what if" analysis which may extend the situation beyond the limits of data bases, nor are they very accurate in prediction of acute or shock loadings. Physically based models require substantial data for calibration over the range of expected conditions but can be very effective when data exists and can simulate the most important physical, biological and/or chemical aspects of the problem.

Pollutant Loads

USEPA (1983) determined that, based on the sampling done during the National Urban Runoff Program (NURP), there are certain pollutants that may be typically found in urban storm water. More detail is contained in the Data Sources And Collection, Chapter 6 of this book. Some of the conventional pollutants show up in significant concentrations in most samples, notably the metals, but most others were present in measurable quantities in less than 15 percent of the samples. Many of these constituents are related to automotive traffic or industrial activities, while others are characteristic of fertilizing and insect control practices. Automotive sources and street locations are generally the two key factors for other than illicit connections and dumping pollutant sources. Common pollutants addressed in studies include: coliform bacteria; total suspended sediment (TSS), total phosphorus (TP), total nitrogen, (TN), 5-day biochemical oxygen demand (BOD5), chemical oxygen demand (COD), total copper (TCu), total lead (TPb), total zinc (TZn), and oil and grease.

Event Mean Concentrations - NURP was designed and executed under the auspices of USEPA in the late 1970's and early 1980's. Its main goal was to provide reliable data and information characterizing runoff from urban sites (USEPA, 1983). Twenty eight sites were monitored from across the United States. While there were some differences in the objectives and procedures of the sites, a common base of information emerged. Later sampling data from municipalities with NPDES permits confirm NURP's findings.

Because of the variability of measurements within storms, among different storms at one site and among sites it was desirable to use a measure which tended to reduce this variability somewhat. The measure of the magnitude of urban runoff pollution chosen is termed the event mean concentration (EMC). EMC is defined as the total constituent mass discharge divided by the total runoff volume for a given storm event. With few exceptions EMC's were found to not vary significantly for similar land uses from site to site for the same constituent and were found to be distributed lognormally. Therefore measures of central tendency (median and mean) and scatter (standard deviation, coefficient of variation), as well as expected values at any frequency of occurrence, could be calculated by using the logarithmic transformation of the raw data. Standard statistical tests and sampling theory can also be used on the lognormally distributed data. See the Data Sources And Collection, Chapter 6, for more information on EMC values.

In selecting a method for estimation of potential washoff loads for a particular site, it is often decided to use methods that estimate washoff loads by land use type. Total loadings are then determined based on event mean concentrations of pollutants and runoff volumes. Table 6-17 in the Data Sources And Collection, Chapter 6, gives typical EMC's for various land uses and percent imperviousness based on the NURP data and other sources. This information should be compared to local information, when available. Initial data from a number of municipalities throughout the east and midwest indicate that, other than lead, most constituents did not vary significantly from the NURP information. The reduction in lead is thought to be based on the use of lead-free gasoline since the NURP data were collected.

Nationwide Regression Equations Method

Reconnaissance studies of urban storm-runoff loads commonly require preliminary estimates of mean seasonal or mean annual loads of chemical constituents at sites where little or no storm runoff or concentration data are available (Colyer and Yen, 1983). To make preliminary estimates, a regional regression analysis can be used to relate observed mean seasonal or mean annual loads at sites where data are available for physical, land use, or climate characteristics. The result of a major study by the U.S. Geological Survey and the U.S. Environmental Protection Agency resulted in the development of regression equations that can be used to estimate mean loads for chemical oxygen demand, suspended solids, dissolved solids, total nitrogen, total ammonia plus nitrogen, total phosphorous, dissolved phosphorous, total copper, total lead, and total zinc (Tasker and Driver, 1988).

The Nationwide Regression Equations Method (NRE) is based on the assumption that explanatory variables, such as drainage area, basin imperviousness, mean annual rainfall, mean minimum January temperature, and general land use categories, can explain regional variation in annual or seasonal storm loads. Coefficients for the regression equations were estimated by a generalized-least-squares (GLS) regression method that accounts for cross correlation and differences in reliability of sample estimates between sites. Table 13-2 presents the 10 regression equations that can be used to predict mean loads for different constituents at unmonitored sites with drainage areas ranging from 0.015 to 1 square mile. The following characteristics are used in the equations.
1. Total contributing drainage area (DA), in square miles.
2. Impervious area (IA), as a percent of total contributing drainage area.
3. An indicator variable (X1) that is 1, if residential land use (LUR) plus non-urban land use (LUN) exceeds 75 percent of the total contributing area, and is zero otherwise.
4. An indicator variable (X2) that is 1, if commercial land use (LUC) plus industrial land use (LUI) exceeds 75 percent of the total contributing drainage area, and is zero otherwise.
5. Mean annual rainfall (MAR), in inches.
6. Mean minimum January temperatures (MJT), in degrees Fahrenheit.

In general, the equations given in Table 13-2 should not be used to estimate mean storm loads at sites whose characteristics are much beyond the range of values shown in Table 13-2. There has been some work directed at extending these equations for larger drainage areas (Tasker, Gilroy, and Jennings, 1990). This work uses a scheme of summing predictions for small areas to make load predictions for large urban basins. The summing scheme assumes that the mean annual load for a large basin can be computed as the sum of mean annual loads for subbasins. Although this method involves many tedious calculations, a computer model is available to make most of the necessary calculations. This method has not at this time been verified by observed data.

Example Application (Tasker & Diver, 1988)

To compute the mean annual load of total nitrogen (TN), in pounds, at a 0.5 mi^2 basin which is 90 percent residential with impervious area of 30 percent and in a region where the mean number

Table 13-2 Water Quality Nationwide Regression Equations
Model is: $W = 10^{[a+bSQRT(DA)+cIA+dMAR+eMJT+fX2]}$ BCF
W, is the mean load, in pounds, associated with a runoff event.

Regression Coefficients For Indicated Explanatory Variables

Dependent Variable W	Regression Constant a	SORT(DA) b	IA c	MAR d e	MJT f	X2 (BCF)	Bias Correction Factor
COD	1.1174	2.0069	0.0051				1.298
SS	1.5430	1.5906		0.0264	-0.0297		1.521
DS	1.8449	2.5468			-0.0232		1.251
TN	-0.2433	1.6383	0.0061			-0.4442	1.345
AN	-0.7282	1.6123	0.0064	0.0226	-0.0210	-0.4345	1.277
TP	-1.3884	2.0825		0.0234	-0.0213		1.314
DP	-1.3661	1.3955					1.469
CU	-1.4824	1.8281			-0.0141		1.403
PB	-1.9679	1.9037	0.0070	0.0128			1.365
ZN	-1.6302	2.0392	0.0072				1.322

Table 13-2 Water Quality Nationwide Regression Equations (continued)

Dependent Variable W	N	Degrees of Freedom	R^2 (note 1)	SE in Logs - γ (note 2)	Average Prediction Error (ASEP) (note 3) Logs	-%	+%
COD	59	56	0.53	0.302	0.311	-51	105
SS	47	43	0.43	0.412	0.433	-63	171
DS	13	10	0.61	0.310	0.349	-55	123
TN	41	37	0.49	0.345	0.363	-57	131
AN	51	45	0.49	0.316	0.337	-54	119
TP	51	47	0.65	0.303	0.316	-52	107
DP	28	26	0.20	0.372	0.388	-59	144
CU	30	27	0.41	0.361	0.381	-58	140
PB	56	52	0.46	0.353	0.368	-57	133
ZN	34	31	0.59	0.310	0.326	-53	128

[1] R^2 is the proportion of variance in W explained by the sample regression.
[2] SE is the standard error of the model.
[3] ASEP is given in log units and percent.   The conversion from logs to percent is: $ASEP(\text{-percent}) = 100[10^{-ASEP(logs)}-1]$
Source: Tasker and Diver, 1988

Table 13-3 Range Of Variables Used In Regression Equations

| Dependent | DA | | IA | | MAR | | MJT | |
Variable	Min.	Max.	Min.	Max.	Min.	Max.	Min.	Max.
COC	0.019	0.070	4	100	8.38	62.00	3.2	58.7
SS	0.019	0.707	4	100	8.38	49.38	3.2	50.1
DS	0.020	0.450	19	99	10.24	37.61	11.4	35.8
TN	0.019	0.830	4	100	11.83	62.00	3.2	58.7
AN	0.019	0.707	4	100	8.38	62.00	3.2	58.7
TP	0.190	0.830	4	100	8.38	62.00	3.2	58.7
DP	0.020	0.707	5	99	8.38	46.18	10.8	35.8
CU	0.014	0.830	6	99	8.38	62.00	15.3	58.7
PB	0.019	0.830	4	100	8.38	62.00	3.2	58.7
ZN	0.019	0.830	13	100	8.38	62.00	11.4	58.7

of storms per year is 79, first compute the mean load for a storm, W, using the appropriate equation from Table 13-2.

$$W = 10^{(X)} * 1.345$$

Where: $X = [a + b(DA^{1/2}) + cIA + fX2]$
$X = [-0.2433 + 1.6383(0.5)^{1/2} + 0.0061(30) - 0.4442(0)]$
$W = 16.9$ pounds

The mean annual load can be calculated by multiplying W by 79, the average number of storms per year, to yield a mean annual load of TN $= 79(16.9) = 1335$ pounds. The 90 percent confidence interval for this example is calculated as follows:

Given: n-p = 41-4 = 37   (degrees of freedom - Table 13-2)
$\alpha = 0.1$ (90-percent confidence/0.1 level of significance)
$t\alpha/2$, n-p = 1.64   (from statistical table - value of z score for two-tailed tests at 0.10 level of significance)
$x_i = (X1\ DA^{1/2}\ X2\ IA)$ (vector of basin characteristics)
$x_i = (1\quad .707\quad 0\quad 30)$

From Table 13-4   U = the following:

0.0346	-0.0449	0.00472	-0.00033
-0.0449	0.1031	0.0130	0.00012
0.00472	0.0130	0.00421	-0.00035
-0.0003	0.00012	-0.00035	0.0000073

$\gamma^2 = (0.345)^2 = 0.119$ (From Table 13-2)
BCF = 1.345   (From Table 13-2)

Formulas: The fit of the regression models may be measured by the R^2 value (Table 13-2), which
            is the fraction of variance in W explained by the model. A measure of how good the
            models are for prediction at unmonitored sites is the average variance of prediction.
            This statistic is computed by averaging over the sites the estimated variance of predic-
            tion at a site used in the regression. The variance of prediction at site V_{pi} is estimated
            by (Tasker & Diver, 1988):

$$V_{pi} = \gamma^2 + x_i U x'_i \tag{13.1}$$

A 100 $(1-\alpha)$ confidence interval for the true mean load for a storm, W_i, at a particular
unmonitored site i can be computed by:

$$\frac{1}{T} \quad \frac{W_i}{1.345} < W < \frac{W_i}{1.345} \quad T \tag{13.2}$$

where W_i is the regression estimate for one of the models in Table 13-2 and

$$T = 10^X \text{ where } X = (t(\alpha/2, n-p)V_{pi})^{1/2} \tag{13.3}$$

in which $t(\alpha/2, n-p)$ is the critical value of the t-distribution for n-p degrees of freedom
and is tabulated in many statistical texts. The variance-covariance matrices, U, needed
to calculate V_{pi} in T for each of the 10 regression equations are given in Table 13-4.

Although the calculations needed to use these formulas can be tedious, a computer
program is available to make most of the necessary computations (Tasker, Gilroy &
Jennings, 1990). This interactive computer program calculates the expected value of
mean load for a storm, for the 10 constituents, given the appropriate basin characteristics along with the 90 percent confidence interval and standard deviation of the
load.

Calculate: $V_{pi} = \gamma^2 + x_i U x'_i = 0.119 + x_i U x'_i = 0.119 + 0.014 = 0.133$

$T = 10^X \text{ where } X = (t(\alpha/2, n-p)V_{pi})^{1/2}$
$T = 10^X \text{ where } X = [1.64(0.133)^{1/2}] = 3.96$

$$\frac{1}{T} \quad \frac{W_i}{1.345} < W < \frac{W_i}{1.345} \quad T$$

$$\frac{1}{3.96} \quad \frac{16.9}{1.345} < W < \frac{16.9}{1.345} \quad 3.96$$

$$3.2 < W < 49.8$$

Therefore a 90-percent confidence interval for W = 16.9 is (3.2, 49.8)

The Simple Method

The following method of computing pollutant loadings is referred to as the "Simple Method"
and is adapted from earlier EPA reports as modified by Schueler (MWCOG, 1987). The simple

Table 13-4 Variance-Covariance Matrices (U)
(Values are in scientific notation - i.e., -1.9363E-02 = 0.019363)

COD	Constant	SQRT(DA)	IA
Constant	1.9363E-02	-2.716E-02	-1.682E-04
SQRT(DA)	-2.716E-02	6.4332E-02	9.8363E-05
IA	-1.682E-04	9.8363E-05	2.7996E-06

SS	Constant	SQRT(DA)	MAR	MJT
Constant	1.2799E-01	-4.440E-02	-3.779E-03	1.0304E-03
SQRT(DA)	-4.440E-02	1.2989E-01	2.8962E-05	-2.991E-04
MAR	-3.779E-05	2.8962E-05	1.4849E-04	-6.608E-05
MJT	1.0304E-03	-2.991E-04	-6.608E-05	7.3585E-05

DS	Constant	SQRT(DA)	MJT
Constant	5.6262E-02	-5.360E-02	-1.248E-03
SQRT(DA)	-5.360E-02	3.9703E-01	-3.337E-03
MJT	-1.248E-03	-3.337E-03	9.9534E-05

TN	Constant	SQRT(DA)	X2	IA
Constant	3.4590E-02	-4.493E-02	4.7196E-03	-3.324E-04
SQRT(DA)	-4.493E-02	1.0309E-01	1.3013E-02	1.1757E-04
X2	4.7196E-03	1.3013E-02	4.2067E-02	-3.484E-04
IA	-3.324E-04	1.1757E-04	-3.484E-04	7.2720E-06

AN	Constant	SQRT(DA)	X2
Constant	1.0696E-01	-5.283E-02	1.8916E-02
SQRT(DA)	-5.283E-02	1.0073E-01	7.7172E-03
X2	1.8916E-02	7.7172E-03	3.8246E-02
IA	-4.887E-04	1.2527E-04	-3.384E-04
MAR	-2.605E-03	2.7698E-04	-5.631E-04
MJT	1.2411E-03	3.6202E-05	3.8766E-04

AN	IA	MAR	MJT
Constant	-4.887E-04	-2.605E-03	1.2411E-03
SQRT(DA)	1.2527E-04	2.7698E-04	3.6202E-05
X2	-3.8246E-02	-5.631E-04	3.8766E-04
IA	6.5176E-06	9.2175E-06	-6.013E-06
MAR	9.2175E-06	9.3696E-05	-5.564E-05
MJT	-6.013E-06	-5.564E-05	4.4696E-05

TP	Constant	SQRT(DA)	MAR	MJT
Constant	5.5400E-02	-2.589E-02	-1.660E-03	6.7787E-04
SQRT(DA)	-2.589E-02	6.1221E-02	5.5267E-05	8.9042E-05
MAR	-1.660E-03	5.5267E-05	7.0813E-05	-4.022E-05
MJT	6.7787E-04	8.9042E-05	-4.022E-05	3.3482E-05

Table 13-4 Continued

DP	Constant	SQRT(DA)		
Constant	3.7200E-02	-9.503E-02		
SQRT(DA)	-9.503E-02	2.8924E-01		

CU	Constant	SQRT(DA)	MJT	
Constant	6.5351E-02	-6.876E-02	-1.122E-03	
SQRT(DA)	-6.876E-02	1.3005E-01	5.8628E-04	
MJT	-1.122E-03	5.8628E-04	3.0108E-05	

PB	Constant	SQRT(DA)	IA	Mar
Constant	7.1623E-02	-4.938E-02	-3.567E-04	-9.865E-04
SQRT(DA)	-4.938E-02	9.1551E-02	2.3984E-04	1.3275E-04
IA	-3.567E-04	2.3984E-04	4.2164E-06	1.7693E-06
MAR	-9.865E-04	1.3275E-04	1.7693E-06	2.5135E-05

ZN	Constant	SQRT(DA)	IA	
Constant	0.4020E-01	-0.4700E-01	-0.3564E-03	
SQRT(DA)	-0.4700E-01	0.9232E-01	0.2412E-03	
IA	-0.3564E-03	0.2412E-03	0.4873E-05	

method provides a quick and reasonable estimate of pollutant loadings with a minimal amount of required data. The basic loading equation is:

$$L = RO * C * A * 2.72 \tag{13.4}$$

Where: $L =$ pollutant washoff load, lbs
         $RO =$ runoff depth, ft
         $C =$ event mean concentration of pollutant, mgl
         $A =$ drainage area, acres
         $2.72 =$ units conversion constant

The runoff depth, RO, is computed from the following formula:

$$RO = (P * Pj * Rv)/12 \tag{13.5}$$

Where: $P =$ rainfall depth, in.
         $Pj =$ fraction of events producing runoff
         $Rv =$ volume runoff coefficient
         $12 =$ units conversion constant

     To use this method the following steps should be taken:

Step 1:   Estimate rainfall depth (P) for the time interval or period of interest. Annual values often range from about 30 to 60 inches in the United States. Seasonal and even individual storm values can be input. Storm durations should be appropriate to the type of BMP being tested.

Step 2:   Rv is similar to the Rational Method C factor but it relates volumes of runoff not rates. Determination of a volume-of-runoff coefficient (Rv) is based on an equation developed by the Metropolitan Washington Council of Governments (1987) after Driscoll (1983).

Regression of computed runoff coefficients and measured imperviousness at 44 small urban catchments resulted in the following equation:

$$Rv = 0.05 + 0.009 * I \qquad\qquad (13.6)$$

where I is the percent imperviousness of the watershed (use 50 for 50%) obtained from site plans or aerial photographs.

Rv does not have a component for baseflow. For smaller or more densely developed watersheds or for pollutants which are not found in appreciable quantities in baseflow this is not important. If baseflow is important a separate pollutant loading for baseflow can be computed given a knowledge of baseflow amounts and mean pollutant concentrations sampled from baseflow. An approximate method is given in COG (1987). Percent impervious values can be estimated from Table 13-5 for different land use categories (Ogden, 1990).

Step 3: Rv is adjusted by the fraction of rainfall from the many smaller storms which produce no runoff at all (Pj) and were under represented in NURP studies. For an individual storm use a value of 1 to avoid double counting. Washington, DC rainfall analysis suggests that about 90 percent of the storms produce runoff (COG, 1987) and a value of 0.9 can be used for Pj.

Step 4: Calculate the total pollutant load from equation 13.1 for a given event mean concentration for the pollutant of interest.

Individual storm pollution estimates are very approximate.

Table 13-5 Land Use Definitions

Land Use	% Imp	Description
Wooded	1	Wooded land
Pasture	1	Meadow, pasture, or open land
Crops	1	Agriculture, row crops
Rural Res	6	Sgl Fam, 1 dwl / 2+ ac
Low Res	12	Sgl Fam, 1 dwl / ac
Med Res	30	Sgl Fam, 2+ dwl / ac; Duplex
High Res	60	Multi-Family, Apartment / townhouse
Commerce	75	Light commercial with landscaping
Industry	90	Light manufacturing, warehousing

Source: Ogden, 1990

Acute Or "Shock" Pollutant Loading Estimates

The U.S. Department of Transportation, based on earlier work, has developed detailed procedures and information dealing with the characterization of individual storm runoff pollutant loads from a statistical perspective from streets and highways, and the prediction of water quality impacts they cause. The procedures developed are based on monitoring data from 993 individual storm events at 31 highway runoff sites in 11 states. The Federal document provides a step-by-step procedure for computing the estimated impacts on water quality of a stream or lake that receives highway runoff (FHWA, 1990). The basic approach is to determine statistics on storms and runoff for a particular area and compute exceedance frequencies. Guidance is provided for evaluating whether or not a water quality problem will result, and the degree of pollution control required to mitigate impacts to acceptable levels. The Federal document provides many worksheets, maps and detailed examples to assist the user in the application of the concepts. This statistical methodology is applicable to many types of acute loading problems, though EMC and other parameters may need to be modified for application to other than roadway drainage. In addition, recently highways have been shown to be a source of polynuclear aromatic hydrocarbons (PAH) a highly toxic pollutant which is released from asphalt on hot sunny days.

Simulation Models

In instances where a municipality has a significant amount of historical data for the drainage areas serviced by storm drain outfalls, including historical precipitation data and receiving water concentration and flow data, the municipality may elect to use more complex dynamic models to derive pollutant loads and to analyze the effects of discharges on receiving waters. Normally these more complex models are used because of the significance of the resource being protected. Details on the theory and types of such models can be found in a number of sources including a summary by Huber (1986).

Dynamic models have an additional benefit over steady-state models in that dynamic models determine the entire discharge concentration-frequency distribution over time. Therefore, dynamic models take into consideration the inherent variability of data associated with urban storm water discharges, including concentration, flow rate, and runoff volume. This would enable the modeler to examine the effects of storm water discharges on receiving water quality in terms of the runoff pollutograph.

Another benefit of using a dynamic model is that it allows the modeler to consider a multitude of "what-if" scenarios. For example, when sufficient historical data are available, the modeler could consider the benefits and risks associated with alternative BMP strategies.

For purposes of computing pollutant loadings, a number of models are available including EPA's Stormwater Management Model (SWMM) and QUAL2E, the Hydrologic Simulation Program (HSPF); U.S. Army Corps of Engineers' Storage, Treatment, Overflow, Runoff Model (STORM); and Illinois State Water Survey's Model QILLUDAS (or Auto-QI).

13.4 General Effectiveness Of Structural Management Measures

Structures designed without specific consideration of storm water quality improvement are generally ineffective for pollutant removal, except for removal of the largest particles. Following are some examples of such measures.

*   Sumped Catch Basins - Catch basins should be considered only as temporary storage locations for pollutant loads. Sediment and adhered pollution may settle out during the recession limb of a storm hydrograph, but the next storm will mobilize and flush out such sediments leaving a new deposit. Cleaning is seldom routine nor sufficiently frequent. The first flush from catch basins is often more septic than raw sewage.
*   Dry Detention Basins - While a municipality may consider dry detention basins as effective for storm water quantity management, a detention time of even a few hours is insufficient to permit settling of the smaller fractions of the suspended materials associated with pollutants.
*   Temporary Filtration Systems - Filtration systems (straw bales, sand bags, filter cloth fences, gravel filters) are used extensively as temporary sediment control measures during construction

and when establishing a new vegetative cover. However, the finer solids are not effectively trapped, and runoff pollution control levels are low.

General BMP Pollutant Removal Effectiveness

Not all urban BMPs can reliably provide high levels of removal for both particulate and soluble pollutants. Effective BMPs include wet ponds, storm water wetlands, multiple pond systems and sand filters. Infiltration BMPs are presumed to be effective in removing pollutants, but are often not reliable given their poor longevity. Other BMPs, such as grassed swales, filter strips and water quality inlets, cannot provide reliable levels of pollutant removal unless their basic design is significantly enhanced. The general effectiveness of a number of BMPs is assessed in Table 13-6.

13.5 Selection Of Structural Management Measures

Introduction

There are many detailed and readily available discussions of structural best management practices. New data and variations on design criteria and structural variations are often being published. In addition, new devices and modifications on older devices now in experimental stages become available on a regular basis. The reader interested in general and detailed design discussions of BMPs should consult any of the following: EPA (1983), MWCOG, (1987,1992), California Stormwater Task Force (1992a), Puget Sound (1992), Austin Texas (1988), Northern Virginia (1992), Southwest Florida Watershed Management District (1986, 1990), Denver Urban Drainage and Flood Control District (1993), Transportation Research Board (1988), Minnesota PCA (1989), local or state guidance, or specific BMP studies from the literature. It should be noted that there are four other major sources of pollution where structural BMPs are applicable (as well as non-structural BMPs) which are adequately covered elsewhere and will not be treated in this book.

* Soil erosion during construction is an important source of pollution of receiving waters. Structural erosion control measures are well covered in State erosion control manuals such as North Carolina (1991), Virginia (1992) or Georgia (undated) and recent EPA publications (EPA 1992a).
* Pollution from non-urban non-point sources (e.g., agriculture, silvaculture) may greatly affect both the incoming quality of surface waters to the urban environment and the ability of the municipality to meet its own targets and goals. Such non-point measures are covered in detail in USEPA, 1993 and numerous other references.
* Pollution from industrial facilities can also be a source of both standard and non-standard urban pollution. There are also certain municipal activities which are considered equivalent to industrial activities requiring NPDES permits (such as vehicle maintenance facilities, landfills, airports, etc.). Development of industrial BMP plans is covered in detail in EPA publications (EPA 1992b) and in state and local guidance (such as California Stormwater Task Force, 1993b).
* Illicit connections and illegal dumping can also be important sources of pollution. The permanent connection to the drainage system is normally illegal without an NPDES permit. Structural controls have been used to treat the storm water prior to discharge or in lieu of discharge, though the controls are similar to those for other industrial and commercial uses (EPA, 1993).

Any BMP has both unique capabilities and persistent limitations. Additionally each site imposes limitations and each overall watershed location requires consideration of the storm water

Table 13-6 Effectiveness And Applicability Of Management Measures

Management Measure	Type	Pollutant Removal Effectiveness			
		Particulates	Heavy Metals	Pesticides	Organics
Curb elimination	Deposition	H	H	N/A	H
Litter control	Source	L to H	L to H	L to H	L to H
Controlled use of					
deicing chemicals	Source	N/A	H	H	H
pesticides/herbicides	Source	N/A	H	H	H
Grassed channels	Runoff	H	H	M	H
Overland flow	Runoff	H	H	M	H
Dry detention basins	Runoff	L to H	L to H	L to M	L to M
Wet retention basins	Runoff	H	H	H	H
Infiltration systems	Runoff	H	H	H	H
Wetlands	Runoff	H	H	M to H	M to H
Street cleaning	Deposition	L to H	L	L	L
Conv. Catch basins	Runoff	L	L	L	L
Porous pavements	Runoff	H	H	N/A	H
Filtration	Runoff	L to M	L	L	L

****

Management Measure	Type	Relative Capital Costs per Acre[1]	Additional Land Requirements	O & M Costs	
				Routine	Nonroutine
Curb elimination	Deposition	L	M to H	O	O
Litter control	Source	L	O	O	O
Controlled use of					
deicing chemicals	Source	L	O	O	O
pesticides/herbicides	Source	L	O	O	O
Grassed channels	Runoff	L	L	L	L
Overland flow	Runoff	L	M to HL	L	L
Dry detention basins	Runoff	M	M	L	L
Wet retention basins	Runoff	H	H	L	L
Infiltration systems	Runoff	M to H	L to M	H	H
Wetlands	Runoff	M to H	M to H	L	L
Street cleaning	Deposition	L	O	H	O
Conv. Catch basins	Runoff	M to H	L to M	H	H
Porous pavements	Runoff	L to H	O	M	M
Filtration	Runoff	L	O	M	M

Ratings: H = high, M = medium, L = low, O = none, N/A = not applicable

[1]Based on additional capital costs required for nonpoint pollution management per acre.
Source: Transportation Research Board, 1988.

and other management objectives for that watershed. Selection of BMPs is always a balance between competing needs, physical limitations, sociological or institutional constraints, financial constraints, and pollution removal needs.

The longevity of some BMPs is limited to such a degree that their widespread use is encouraged only under certain circumstances. Of particular concern are the infiltration practices, such as basins, trenches and porous pavement. The poor longevity of these BMPs is attributable to a number of factors: lack of pretreatment, poor construction practices, application to infeasible sites, lack of regular maintenance, and in some cases, fundamental difficulties in basic design. Very often the life-spans of BMPs can be increased to acceptable lengths if local communities adopt enhanced designs and commit to strong maintenance and inspection programs.

BMP options are adaptable to most regions of the country though there may be extra efforts required for extremely arid regions of the West and colder climates of the North. In these regions, conventional BMP designs may need to be refined to account for high evaporation rates during the growing season or subfreezing and snow melt conditions.

No single BMP option can be applied to all development situations and all BMPs options require careful site assessment prior to design. Pond options are applicable to the widest range of development situations, but typically require a minimum drainage area. On the other hand, infiltration practices often have more limited applications, and require field verification of soils, water tables, slope and other factors.

Several BMPs can have significant secondary environmental impacts, although the extent and nature of these impacts is uncertain and site-specific. Pond systems, which offer reliable pollutant removal and longevity, tend to be associated with the greatest number and strongest degree of secondary environmental impacts. Careful site assessment and design are often required to prevent stream warming, natural wetland destruction and riparian habitat modification.

Relatively limited cost data exist to aid in the assessment of the comparative cost-effectiveness of urban BMP options. Presently, the selection of BMPs is based more on longevity, feasibility, and local design factors than on comparative cost. It is expected that construction costs for all BMPs will increase in the future due to the enhanced designs needed for more reliable pollutant removal and longevity. Costs may also increase in response to increasingly complex permitting requirements. Maintenance costs may rival construction costs over the design life of many BMPs; however, many jurisdictions currently do not have very active BMP maintenance programs.

Many of the conventional urban BMPs need to be enhanced to provide more reliable pollutant removal and greater longevity. In many cases, a systems approach to BMP design is warranted whereby multiple techniques for runoff attenuation, conveyance, pretreatment, and treatment are utilized. A priority should be placed on demonstrating the performance of the enhanced BMPs in the field.

Several fundamental uncertainties still exist with respect to urban BMPs and need to be resolved through basic research. These uncertainties include the toxicity of residuals trapped within BMPs; the interaction of groundwater and BMPs (both ponds and infiltration); and the long-term performance of urban BMPs.

Basic Structural BMP Types

Structural BMPs can be classified by the predominant removal mechanism into the four major categories of: detention basins, filtration devices, vegetative filtration, and special devices.

- Detention basins, as the term implies, hold runoff for a period of time allowing for settling of the solid pollutants. Other removal mechanisms, which are generally of lesser importance, include filtration through vegetation, infiltration, evapotranspiration, and biological and chemical transformation. Basic types include extended detention, retention ponds and constructed wetlands.

- Filtration devices remove pollutants through the natural filtering process of soil and reduction in runoff volume. Infiltration devices generally transport runoff to groundwater. Filtration (often called exfiltration) devices filter water through an engineered layer of soil and then discharge it to the drainage system. Specific devices include infiltration trenches or basins, sand filters, peat filters, exfiltrating storm drain systems and trenches, porous pavement, Austin first flush basins, and variations.
- Vegetative filtration devices operate through the contact of runoff with vegetation. They tend to be used for on-site controls. Pollutant removal mechanisms include sedimentation and vegetal filtration, infiltration, evapotranspiration and biological uptake. Devices include buffers or filter strips with flow spreaders, grassed swales (with or without check dams), and other devices.
- Special devices tend to be manufactured devices or more highly engineered devices, which have a variety of removal mechanisms and are used on smaller sites, industrial type sites or for special applications. They can include such structures as swirl concentrators, oil and grit separators, chemical treatment devices, water quality inlets, coalescing filters, sumped catch basins, various catch basin inserts, floating curtain devices, and others.

Combined Measures

Although the management measures mentioned here can be used individually to remove pollutants, it may be desirable to consider combining two or more of these measures. Examples might include the following.

- Use of infiltration wells in a detention basin to increase pollutant removal while concurrently decreasing the long-term storage requirements.
- Use of overland flow through vegetative filters (strips and channels) and/or wetlands so as to filter suspended sediments from runoff upstream from an infiltration basin or trench.
- Use of wetland marshes adjacent to retention ponds.
- Use of oil and grit separators to remove larger sizes of grit and oil and grease on particular sites upstream from regional basins.

Combining two or more management measures can often have advantages. Combinations of measures may:

- increase the operational life of a given BMP,
- increase pollutant removal effectiveness, or
- overcome any site limiting factors.

Wet detention ponds should be used in combination with vegetative filters to provide sediment removal before the runoff can enter an infiltration system or wetland. Additionally, infiltration basins can be used to provide for the temporary or permanent storage of runoff.

Wetlands should only be used in combination with vegetative filters and/or detention, not with infiltration. Typically, wetlands should receive inflow from vegetated conveyance facilities and/or a wet detention pond. Any planned discharge from a wetland should be into a vegetated conveyance facility. Wetlands should not precede an infiltration system as accumulated sediment and/or decaying matter may clog the infiltration mechanism. It must be recognized that conditions favorable to wetlands (high water table, impervious soils, etc.) are unfavorable for infiltration measures.

Many types of structural BMPs also provide, or can be designed to provide, flood control capabilities. Retention ponds, extended detention, infiltration basins and porous pavement are of particular benefit for flood control at least for the more frequent nuisance floods. Additionally, some older detention facilities or interior drainage pumping sump areas can be retrofitted to maximize pollution removal without severely limiting flood control capabilities. In some municipalities flood control design criteria had been overly restrictive providing excess basin volume. Through a modification of such criteria retrofitting for water quality could be easily accomplished.

Structural BMP Screening

In assessing a watershed or site for the use of structural BMPs there are a number of "screens" which can be used to avoid undue consideration of ineffective or unsatisfactory BMPs. Target pollutant removal abilities are one such screen discussed above. Some authors have developed screening procedures with point scoring systems for the selection of priority sites (Woodward-Clyde, 1990), and BMP limitations (Schueler, 1987, California Stormwater Task Force, 1993). These methods are helpful in trying to compare relative importance of various factors (or at least the method's author's perception of that importance), and in insuring completeness in consideration of all pertinent design factors. They are not a substitute for applied overall engineering judgment and consideration of the myriad of details and factors not included in any methodology.

There are two types of screens used: (1) watershed or jurisdictional screens used to determine necessary, suitable and desirable locations for some type of structural BMPs, and (2) screens to zero-in on suitable BMPs for the selected sites. Each site designer must simultaneously consider how the BMPs will (Schueler, et al., 1991):

* contribute to the specific watershed plan and quality targets,
* mitigate the expected impact to the stream from the site development,
* control the spectrum of future storm water events generated from the site and any future development impacts to be addressed by the BMPs, and
* integrate with other aspects of the proposed site plan.

Realistically it may not be possible to bring about stream restoration or to protect a stream system from degradation. High levels of impervious areas within a watershed will result in degradation regardless of the intensity of BMP design and application. The best protection for a stream is development density restrictions (Burby, et al., 1986). This is clearly not feasible except in the more important watersheds; those which are used for drinking water supply or feed very sensitive waters. In other cases it is vitally important to understand the stream and watershed character and to seek to obtain a realistic balance between degradation and preservation. Often a natural stream can be transformed into an attractive urban stream, trading some natural amenities for well designed urban amenities. Urban greenway systems seek to do this. This approach is to be preferred to the alternative of abandoning the stream to "fend for itself" amid the impacts of urban development.

Non-Structural BMPs - It should be remembered that structural BMPs are only one part of an overall BMP program. The most effective way to maintain clean surface waters is to eliminate the sources of pollution, not to remove pollution once it has gotten into the system. Thus each municipality (and many sites) should also be implementing a "baseline" BMP program which includes nonstructural educational, regulatory and programmatic BMPs. The Storm Water Quality Management Plans, Chapter 15, gives general criteria for the assessment of any type of BMP as well as a procedure for the development of an overall storm water quality management program.

Close attention to the nature of site design and layout can go a long way toward reducing pollution. For example, landscaping and terracing provide natural buffers, rainfall holding and infiltration areas. Use of low maintenance locally derived vegetation reduces the need for pesticides and fertilizer. Directing roof and driveway runoff across grassy areas provides a natural filtration and infiltration. Clustering designs, zoning density limitations, critical area identification and protection, and joint regional design planning can all work toward an inherent lowering of pollution loading levels.

Sensitive Area Identification - Structural BMPs are targeted toward specific critical areas as an overlay on the general baseline BMPs. Critical areas may be defined using any number of criteria. Local public perception and desires always play an important role. However there are generally three criteria used in assessing surface waters for targeting of BMPs (Woodward-Clyde, 1990):
- the nature of the designated beneficial uses of the stream (e.g. body contact or contiguous recreation, drinking water supply, fishing, commercial uses, etc.) and potential of denial of these uses;
- the physical and chemical quality of the water (including numerical standards) and its preservation or enhancement; and
- the ecology of the stream and expected or current impacts of surface water runoff.

Pollution Sources - Specific sources and types of pollution may point toward the use of certain kinds of structural BMPs. Oil and grease problems would normally be addressed close to the site through oil and grit separators rather than in regional BMPs further downstream in the system. Nutrient removal, especially in dissolved fractions, can be accomplished through biological activity inherent in wetlands, retention basins, wet extended detention basins and some types of infiltration if the soils are not overly porous.

Site Goals - The goals for each site can play a large role in the BMP selection process. Schueler (MWCOG, 1987) gives six generic goals which should be considered in BMP selection. BMPs should:
- reproduce the predevelopment hydrologic conditions;
- provide moderate pollution removal;
- be appropriate for the site and its constraints;
- be reasonably cost effective;
- have an acceptable future maintenance burden; and
- have a neutral impact on the natural or human environment.

Table 15-2 in the Storm Water Quality Management Plans, Chapter 15, provides a self explanatory list that has been used in various forms in several locations for BMP program development. Some localities total an actual score and use the form for a numeric ranking of BMPs. Others simply use the pluses and minuses as flags to alert planners to particular BMP strengths or potential problems.

Each of the basic and specific structural BMP types has strengths and weaknesses in any application. Each has limitations based on: physical site characteristics (slope, area, soils, depth to bedrock, erosion loading, etc.), pollutant targeted, cost, institutional support, climate and hydrology and more.

Soils - Structural BMPs which rely on filtration or infiltration are restricted in the types of soils they can use. Infiltration devices can control only the upper layers through construction modifications. The lower soil layers which receive the infiltrated water will eventually control the rate of infiltration. This excludes most SCS hydrologic group C and D type soils from consideration for these BMP types. Very sandy soils make retention of water in wet ponds infeasible in most cases. Table 13-7 gives a general indication of soil permeability feasibility ranges for BMP categories.

Table 13-7 General Restrictions On BMP Use Based On Soil Permeability

	SAND	LOAMY SAND	SANDY LOAM	LOAM	SILT LOAM	SANDY CLAY-LOAM	CLAY LOAM	SILTY CLAY-LOAM	SANDY CLAY	SILTY CLAY	CLAY
Extended Detention Pond									XXXX	XXXX	XXXX
Retention Pond	XXXX	XXXX	XXXX								
Filtration and Infiltration					XXXX						
Vegetative Filtration								XXXX	XXXX	XXXX	XXXX
Min. Infil. Rate (in/hr)	8.27	2.41	1.02	0.52	0.27	0.17	0.09	0.06	0.05	0.04	0.02
Soil Type	SAND	LOAMY SAND	SANDY LOAM	LOAM	SILT LOAM	SANDY CLAY-LOAM	CLAY LOAM	SILTY CLAY-LOAM	SANDY CLAY	SILTY CLAY	CLAY

Shaded - Feasible   XXXX - Marginally Feasible   Blank - Generally Not Feasible
Source: Schueler, 1987

Drainage Area - Drainage area minimums for wet ponds and extended detention ponds are determined by the ability to retain a permanent pool and the minimum feasible orifice size respectively. Infiltration device maximums are set by space and flow velocity limits as well as economics. For larger areas it generally becomes more feasible to use a wet pond or extended detention pond. Drainage area can be modified through imaginative grading and flow routing. For example, to minimize drainage area for an infiltration device sometimes only directly connected impervious areas are routed through the BMP while other areas are routed around it. Table 13-8 gives general considerations for drainage area.

Table 13-8 General Restrictions On BMP Use Based On Contributing Drainage Area

		5	10	15	20	25	30	35	40	100
Detention Basins					XXXXXX	XXXXXX				
Retention Basins					XXXXXX	XXXXXX				
Infiltration Basins		XXXXXX				XXXXXX	XXXXXX	XXXXXX		
Infiltration Trenches			XXXXXX							
Porous Pavement					XXXXXX	XXXXX				
Vegetative Filtration			XXXXXX							
Water Quality Inlet										
Drainage Area Acres		5	10	15	20	25	30	35	40	100

Shaded - Feasible   XXXXXX - Marginally Feasible   Blank - Generally Not Feasible
Source: Schueler, 1987

Other Restrictions and Limitations - Other key restrictions or limitations must be considered when planning BMP placement and use.
• Slope may restrict the use of filtration and infiltration type BMPs due to the high velocities, erosion potential or shortness of contact time. Infiltration devices should not be used on slopes

greater than about 20 to 25 percent. Porous pavement and grassed swales should be limited to slopes of about 5 percent or less. Wet ponds and constructed wetlands cannot be placed on steep or unstable slopes.

- Water availability may restrict the use of vegetative BMPs in dry climates and the use of wet ponds in smaller areas and in dry climates.

- Water table location can restrict the use of both ponds and infiltration devices. If the water table is within 4 feet of an infiltration BMP it will limit the ability of the basin to function (Schueler, 1987).

- Bedrock close to the surface will effectively provide a barrier to infiltration if it is located within 2 to 4 feet (Schueler, 1987). Cost considerations of bedrock excavation may be important for retention and detention ponds and filtration devices such as the Austin filtration basin.

- Sediment load into the site will severely limit the life expectance of most types of BMPs. Areas facing continued construction, and the associated high erosion loads, are not suitable for BMP placement without either pretreatment of the flows or an acceptance of high maintenance costs. Infiltration and filtration devices are especially susceptible to sediment loading.

Basis For Design Criteria For Structural BMPs

The design of structural BMPs is essentially a matter of mass balance of pollutants, in many ways similar to a typical pond level-pool routing equation with one additional term: mass inflow - mass outflow $\pm$ mass creation/destruction = accumulated mass. The mass inflow for a time period is a function of the flow volume times the mean pollutant concentration. The mass created or destroyed is a function of the chemical, biological or physical changes which take place within the BMP. For any given storm there can be an actual net export of pollutants from the BMP due to release of pollutants in some way bound within the BMP (often in the sediments or biomass). The goal then is to design a BMP to, on average, provide sufficient pollution reduction to meet a mass outflow target, standard or goal.

The difficulties in the design become apparent when the complexities of urban rainfall-runoff sequences are compounded with the great number of pollutant types and concentration variability (even within a single storm) and with the myriad of chemical, biological and physical reactions that can take place among the various constituents. In addition the pollutant removal rates of BMPs vary considerably even among carefully constructed, well maintained and similar BMPs at contiguous locations. The result in application is often derived rules-of-thumb or simplifications for certain localities based on statistical or continuous simulation analysis of rainfall and runoff, assumptions about pollutant concentrations and estimated pollutant removal rates (Nix, et al., 1988). Adjustment of theory is left to monitoring and performance observation.

The rules-of-thumb for design purposes are often based on "capture" and treatment of a certain depth of rainfall over the watershed draining to the BMP or based on runoff from all or only the impervious areas of a watershed. Rainfall and runoff in urban storm water pollution calculations are normally related using a Rational Method type C factor (or its volume equivalent) because of the simplicity, the fact that much of the NURP analysis was done in a similar way, the many unknowns in the analysis, and the wide variability of rainfall-runoff produced when using the initial loss estimates of SCS TR-55 methods for small storms (Harrington, 1987). Therefore, for example, capture of the first inch of rainfall can equate to capture of the first half-inch of runoff given an appropriate C factor for the watershed.

The size of event or depth of runoff to be captured and treated depends on several factors including the pollution target, loading rate, type of BMP, its location (on-line or off line), etc. The total pollution removed is a function of both the portion captured and the part of the captured portion that is actually removed prior to discharge from the BMP back to receiving waters. For example treating 1/2 inch of runoff totally through infiltration may be equivalent to treating one inch of runoff partially through wet pond removal mechanisms. Or, diverting some small volume of flow through an off-line filtration system may be more effective than capturing the total flow for a lesser period through an on-line extended detention system.

For practical purposes it can be assumed that the percent of pollution mass captured is proportional to the percent of flow volume captured. Several authors have shown, using various methods, that capture of about 0.25 to 0.5 inch of runoff will capture in the range of 80 to 90 percent of the annual runoff volume and thus that percentage of the pollution (Wanielista and Yousef, 1992, Urbonas, et al., 1990, Roesner et al., 1989). The reason is, of course, that the vast majority of the storms in any given year only produce a small amount of runoff. If the area experiences a pronounced first flush phenomena than the pollution capture will be even greater than the runoff volume capture.

The first flush phenomena is much less pronounced for areas of lower imperviousness. Studies have shown that less than 50 percent impervious will show little first flush. They also show that the higher the imperviousness the more depth of capture is necessary to achieve the same result. A sliding scale should be specified based on storm runoff from a certain rainfall depth not a constant depth (Chang, et al., 1990).

Specific analyses to determine these capture volumes for any location are done using continuous simulation of a long rainfall record or statistical analysis of the rainfall record. A detailed generic method for planning level statistical analysis in areas without developed rules-of-thumb is given by Driscoll (Woodward-Clyde, 1986) and illustrated by Urbonas and Stahre (1993). Nix, et al., (1988) illustrate application of continuous simulation to derive removal efficiencies for trial BMP sizes.

Quite often capture is based on mean annual or mean storm statistics. Driscoll, et al., (1989) provide such statistics for the United States in 15 rainfall zones. Information is for all storms with depths greater than 0.1 inch. Depths less than that were considered to produce no runoff. Figure 13-1 and Table 13-9 present a summary of this information for use in design of both extended detention basins and retention basins where mean storm volume criteria are called for. This local rainfall analysis has been done for many places in the United States resulting in tables of surface area and drainage area relationships for extended detention ponds with permanent pools (for example, Professional Engineers of North Carolina, 1988 or Puget Sound, 1992).

Cv is the coefficient of variation, and an indication of the scatter of the data, it is defined as the standard deviation divided by the mean. The volume runoff coefficient can be found in the discussion on the Simple Method. The runoff volume, given the rainfall volume from the table above, is simply the rainfall volume times the volume runoff coefficient.

For designs using settling velocities, typical settling velocity values for urban particulate pollutants are given in Table 13-10 (Woodward-Clyde, 1986).

Specific Structural Measures

Following are design considerations for representative designs for some of the more effective management measures that can be used for water quality control in urban areas. Each section gives typical recommended specifications, operation and maintenance requirements, and performance standards. Again, it should be noted, that climatic and site specific conditions may lead to major modifications of each of these sets of criteria. The criteria are given to illustrate the considerations necessary for BMP design and are not meant to be an exhaustive or restrictive treatment. There are many variations on the basic device types to accommodate specific site restrictions or take advantage of site opportunities. The following BMPs are included in this section.

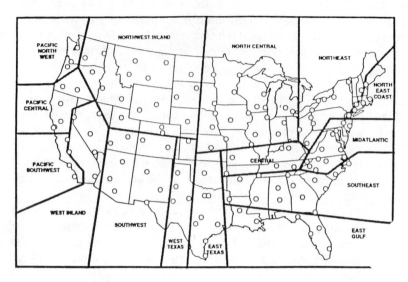

Figure 13-1 Fifteen Rainfall Zones In The United States (Driscoll, et al., 1989)

Table 13-9 Annual And Individual Storm Statistic Typical Values For Fifteen Rainfall Zones In The United States

Rainfall Zone	Annual Storm Depth (in)		Annual No. of Storms		Storm Separation (hours)		Duration (hrs)		Intensity (in/hr)		Volume (inches)	
	Avg.	CV	Avg.	CV	Avg.	CV	Avg.	CV	Avg.	CV	Avg.	CV
Northeast	34.6	0.18	70	0.13	126	0.94	11.2	0.81	0.067	1.23	0.50	0.95
Northeast, coastal	41.4	0.21	63	0.12	140	0.87	11.7	0.77	0.071	1.05	0.66	1.03
Mid-Atlantic	39.5	0.18	62	0.13	143	0.97	10.1	0.84	0.092	1.20	0.64	1.01
Central	41.9	0.19	68	0.14	133	0.99	9.2	0.85	0.097	1.09	0.62	1.00
North Central	29.8	0.22	55	0.16	167	1.17	9.5	0.83	0.087	1.20	0.55	1.01
Southeast	49.0	0.20	65	0.15	136	1.03	8.7	0.92	0.122	1.09	0.75	1.10
East Gulf	53.7	0.23	68	0.17	130	1.25	6.4	1.05	0.178	1.03	0.80	1.19
East Texas	31.2	0.29	41	0.22	213	1.28	8.0	0.97	0.137	1.08	0.76	1.18
West Texas	17.3	0.33	30	0.27	302	1.53	7.4	0.98	0.121	1.13	0.57	1.07
Southwest	7.4	0.37	20	0.30	473	1.46	7.8	0.88	0.079	1.16	0.37	0.88
West, inland	4.9	0.43	14	0.38	786	1.54	9.4	0.75	0.055	1.06	0.36	0.87
Pacific Southwest	10.2	0.42	19	0.36	476	2.09	11.6	0.78	0.054	0.76	0.54	0.98
Northwest, inland	11.5	0.29	31	0.23	304	1.43	10.4	0.82	0.057	1.20	0.37	0.93
Pacific Central	18.4	0.33	32	0.25	265	2.00	13.7	0.80	0.048	0.85	0.58	1.05
Pacific Northwest	35.7	0.19	71	0.15	123	1.50	15.9	0.80	0.035	0.73	0.50	1.09

Source: Driscoll, et al., 1989

Table 13-10 Settling Velocities For Typical Urban Runoff Pollutants

Size Fraction	Percent of Particle Mass in Urban Runoff	Average Settling Velocity (ft/hr)
1	0 - 20 %	0.03
2	20 - 40 %	0.3
3	40 - 60 %	1.5
4	60 - 80 %	7.0
5	80 - 100 %	65.0

Source: Woodward-Clyde, 1986

Detention Ponds
- Extended Dry Detention Basins
- Retention Ponds
- Constructed Wetlands

Vegetative Filtration
- Filter Strips and Flow Spreaders
- Grassed Swales

Infiltration and Filtration
- Austin First-Flush Filtration Basin
- Sand Filters
- Infiltration Trenches
- Porous Pavement

Special Devices
- Water Quality Inlets
- Oil/Grit Separators
- Catch Basin Inserts

Extended Dry Detention Basin

An extended dry detention basin is a detention basin with increased runoff residence time sufficient to remove settleable pollutants to acceptable levels of concentration. There are two basic types of extended detention ponds. The first is normally dry and drains completely in a specified period of time. The second maintains a permanent pool with the extended detention volume above the permanent pool and really is a combination of a wet pond and extended detention (Schueler and Helfrich, 1988). The second type will be covered under wet ponds.

Extended dry ponds are used where lack of water or other multi-use considerations preclude the use of wet ponds or constructed wetlands. Extended ponds have been used for: recreational fields, tennis courts, parking lots, roof tops, nature areas, etc. They also handle flood control objectives more economically than do wet ponds because they do not have extra storage volume set aside for a permanent pool. The flood storage volume is set aside above the pollution control volume, and a multi-stage riser arrangement is used to handle the outflow. They are generally targeted where particulate removal is the focus. Extended ponds can be readily converted from sediment traps used during construction (UDFCD, 1992). They are also applicable where soil with poor infiltration characteristics exists since they do not depend on infiltration for proper operation.

Disadvantages of extended dry ponds include difficulty in keeping some outlet types unclogged, difficulty to mow if wet, rapid accumulation of sediments, swampy eyesore possibility if not maintained, and modifies flow regime if in-line. Maintenance is estimated to be 3 to 5 percent of construction cost annually (MWCOG, 1992).

Figure 13-2 shows a schematic of a typical pond design. Other designs include vaults and tanks for underground applications and a myriad of aesthetic and pollutant removal enhancing features such as micro-pools and marshes.

Erosion of the banks and the pond bottom have been a problem for this type of design and special care should be taken to provide both vegetation which can stand frequent inundation and a non-eroding low flow channel. Structural bank protection could be provided in lieu of vegetation though aesthetic requirements may make it costly.

Pond Sizing - It has been shown by several authors that detention of all storms more frequent than about the 6 month storm for a period of at least 24 hours can affect a long-term suspended solid (TSS) removal rate by about 80%. However, if ponds are designed with a single orifice to detain the larger design storm, the 24-hour period smaller storms will not be detained adequately. Also, the smaller particles attract pollutants more readily than the larger but are settled less easily. Flocculation of the smaller particles may provide better results than expected. To provide at least 24 hours of detention over the complete spectrum of small storms one of two approaches is used (California Stormwater Task Force, 1993, Schueler, 1987, Grizzard, et al., 1986):

- the pond is designed to detain the larger storms for a period approaching 40 hours (in the eastern part of the United States), resulting in about a 24 hour time for the smaller storms; or
- more than one outlet orifice is used with one placed at the pool bottom for the smaller storms and a second placed at a mid-height level for the larger storms, resulting in a 24-hour draw-down time for all storm volumes.

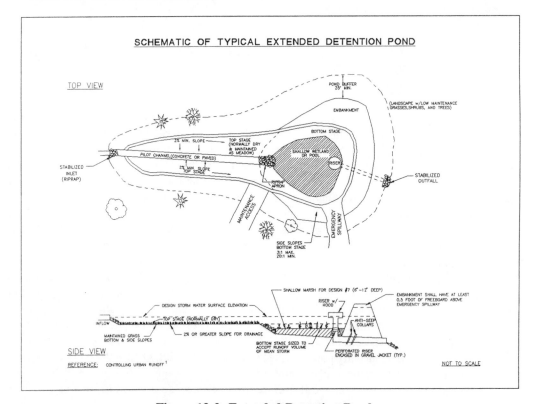

Figure 13-2 Extended Detention Pond

For wet ponds there are analytical methods to estimate pollution reduction. For ponds without a permanent pool below the extended detention volume there is no comparable generally accepted analytic method (Urbonas and Stahre, 1993). Most localities have developed rules of thumb for the pond sizing volume for dry extended detention ponds or ponds with small wet ponds. These rules are designed to remove a high percentage of suspended solids and often provide a volume of basin to volume of mean annual runoff event ratio (VB/VR) of 1.0 to 2.5. Common design volume criteria include the following.

- "First flush" (half-inch) runoff volume should be detained for 24 hours.
- The runoff volume produced by a 1.0 inch storm should be detained 24 hours.
- Detain the runoff from the 6 month 24 hour storm for 24 hours.
- Some multiple of mean runoff volume (see retention pond design for runoff information).
- Design storm - 1 to 2 year storm.

Northern Virginia (1992) provides rules for pond size for a 48 hour drawdown time as:

$$S_a = 4375\,C - 875 \quad \text{or} \quad 32.25\,I_\% \qquad (13.7)$$

Where: S_a = storage per controlled acre, ft^3

C = Rational Method C factor

$I_\%$ = upstream watershed percent impervious, %

Sizing of the pond dimensions is a trial and error procedure fitting the required volume to the site. Generally wider and shallower ponds are more efficient than deeper ponds for the same volume (California Stormwater Task Force, 1993). Distance between inlet and outlet should be maximized, though this is not critical since most of the settling takes place during quiescent periods (Driscoll, 1986).

Outlets - Physically, the extended detention time is provided by an outlet structure with a small opening(s). The Storage Facilities, Chapter 11, gives routing details. There are several ways to accomplish the outlet control arrangement including: gravel packet perforated risers, perforated risers with an internal orifice for control, negatively sloped orifice pipes, v-notched weirs, concrete block structures with multiple openings, concrete manhole riser with submerged internal orifice plate/pipe, etc. Southwest Florida Management District recommends clogging protection for all outlets less than 6 inches in cross sectional area, 2 inches in dimension or 20 degrees for "V" notch weirs (Southwest Florida, 1990).

GKY (1989) provides an equation for single-orifice outlet design for ponds where surface area does not vary greatly with depth (larger ponds with steeper side-slopes or any size with vertical side-slopes).

$$a = \frac{24A(H-Ho)^{0.5}}{3600\,CT(2g)^{0.5}} \qquad (13.8)$$

Where: a = orifice area, in.^{2}

A = average surface area of pond, ft^2

C = discharge coefficient

T = drawdown time, hrs

g = gravitational acceleration, ft/s^2

H = full pond elevation, ft

Ho = elevation when pond is empty, ft

If a perforated riser is used for drawdown control without an internal orifice (i.e. the perforations themselves provide the control) use Figure 13-3 for perforation opening requirements (C value of 0.65 assumed) per row. The rows are spaced vertically four inches on center. Use a gravel pack around the riser and keep perforations larger than about 3/4" to 1". Use of filter fabric against the riser has been shown to clog (California Stormwater Task Force, 1993). This may limit this type of outlet structure to larger pond areas.

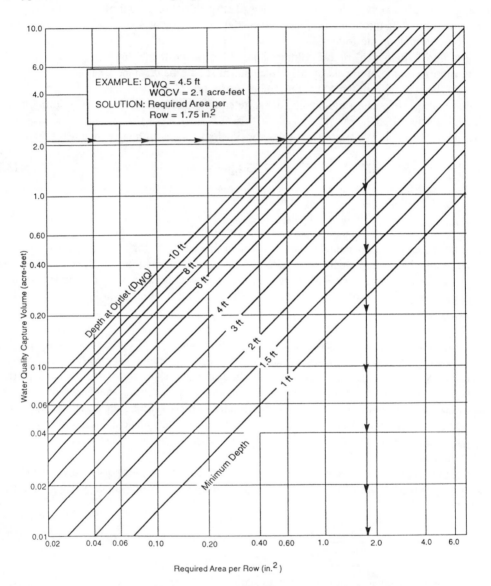

**Figure 13-3 Water Quality Outlet Sizing - Extended Detention
Basin With 40 Hour Drawdown Time Of The Capture Volume**
After UDFCD, 1992, 1994

A much smaller orifice can be used if it is placed inside the riser pipe downstream from the perforated section or inside a riser manhole in a sump-like condition. Figure 13-4 gives a schematic of this arrangement.

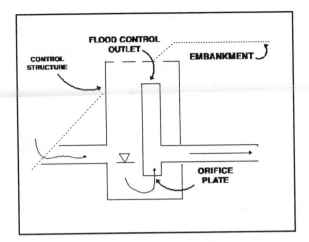

Figure 13-4 Extended Detention Basin With Internal Orifice

Pollutant Removal Efficiency - Soluble pollutant removal rates are low for extended dry detention ponds but can be enhanced either with greatly increased detention time, through the use of shallow marshes to increase biological uptake, or through using an infiltration device downstream from the outlet orifice (Schueler, 1987, Urbonas and Ruzzo, 1986). Approximated removal rates can be estimated from Table 13-11. The lower end of the ranges be used unless experience indicates otherwise. Freezing conditions can be expected to reduce removal rates by 50 percent.

Because of the great cost of sediment removal (perhaps 20 to 40% of first cost) (Schueler and Helfrich, 1988), even at intervals of ten years or more, there is often a paved sediment forebay placed at the entrance of either an extended detention pond or a wet pond. Sediment accumulation in the pond proper can be estimated from the Simple Method for TSS and a suitable trap efficiency estimated from Table 13-11 or from local data and criteria (Schueler, 1987). The size of the forebay is developed from settling theory, and is typically 5 to 10 percent of the water quality control volume (UDFCD, 1992). Urbonas and Stahre (1993) recommend forebays with depths of at least 3.5 feet to avoid resuspension, detention times for the average annual runoff rate to be 5 minutes, and loading rates to be approximately 50 feet per hour.

Table 13-11 Average Annual Pollutant Removal Capability Of Extended Detention Ponds

	Extended Detention Design Type		
Pollutant	6-12 Hr First Flush Detention	1 Inch Rain Detained 24 Hours	Same as Previous w/ Shallow Marsh
TSS	40-60%	60-80%	60-80%
TP	20-40%	40-60%	60-80%
TN	20-40%	20-40%	40-60%
BOD	20-40%	40-60%	40-60%
Metals	40-60%	60-80%	60-80%

Source: Schueler, 1987, 1993

Sediments collected in detention and retention ponds have not been found to be classified as hazardous waste (Carr, et al., 1982, Zanoni, 1986, and Nightingale, 1987). But accumulated sediment must be handled and stored in a way that will not effect surface or ground water. In general it should be disposed of in a location where it will be stable and not in contact with humans (Minnesota PCA, 1989). For on-site disposal it should be covered with at least four inches of topsoil and vegetated. In all cases sediment must be disposed of in accordance with any applicable solid waste regulations.

Sediments in the bottoms of wet ponds also accumulate metals in the first couple inches of soil. Nightingale (1987) recommends monitoring for lead content and removing sediments if the lead concentration exceeds 1000 mg/kg of soil as suggested in a California standard.

General Design Criteria
- Hydrograph centroid should be delayed through control from a small orifice in the release structure.
- Do not locate on fill if seepage considered to be great.
- Extended detention design storm should be limited to one-year runoff events.
- Drawdown time should be at least 24 hours average to 40 hours for maximum design storm.
- Runoff in excess of a two year event should normally be passed through the basin with peak discharge controls only.

Typical Required Specifications
- Pilot channel of paved or concrete material for erosion control (alternately use turf if little low flow). Size such that any event runoff will overflow the low flow channel onto the pond floor.
- Side slopes no greater than 3:1 if mowed.
- Inlet and outlet located to maximize flow length.
- Design for full development upstream of control.
- Riprap protection (or other suitable erosion control means) for the outlet and all inlet structures into the pond.
- One-half foot minimum freeboard above peak stage for top of embankment for design storm.
- Emergency spillway designed to pass the 50-year storm event (must be paved in fill areas).
- Maintenance access (< 15% slope - 10 feet wide).
- Trash racks, filters or other debris protection on control.
- Anti-vortex plates.
- Ensure no outlet leakage and use anti-seep collars.
- Benchmark for sediment removal.

Typical Recommended Specifications
- Two stage design (top stage - dry during the mean storm, bottom stage - inundated during storms less than the mean storm event.)
- Top stage slopes between 2% and 5% and a depth of 2 to 5 feet.
- Bottom stage maintained as shallow wetland or pool (6 to 12 in.).
- Manage buffer and pond as meadow.
- Minimum 25-foot wide buffer around pool.
- On-site disposal areas for two sediment removal cycles.
- Anti-seep collars on barrel of principal spillway.
- Impervious soil boundary.
- Design as off-line pond to bypass larger flows.
- Design a sediment settling basin for pretreatment of the larger particles.

Sizing Example - An extended detention pond for an area of 50 acres is to be sized for a VB/VR ratio of 2.0. The rational coefficient is 0.6 and the percent imperviousness is 35%. The location is Nashville, TN. Since no local information is available Driscoll, et al., will be used for rainfall statistics. The mean annual storm volume is (central area): 0.62 inches (V), and the intensity is 0.097 in/hr.

- The volume runoff coefficient is calculated from Shelley's equation (13.6) as:
  $Rv = 0.05 + 0.009 * I = 0.05 + 0.009 * 35 = 0.365$
- VR is then calculated as:
  $VR = V * Rv * A * 43,560/12 = 0.62*0.365*50*43,560/12 = 41,073$ ft^3
- Then VB is found from:
  $VB = 2.0 \; VR = 82,000$ ft^3

The sediment forebay, if desired, is calculated using Urbonas and Stahre's criteria:

- Mean annual discharge is:
  $Q = C * i * A = 0.6 * 0.097 * 50 = 2.91$ cfs
- Surface Area = Mean Annual Discharge/Loading Rate
  $A = Q * 43,560/12 * 1/50 = 2.91 * 43,560/12 * 1/50 = 211$ ft^2
- Volume for 5 min detention time:
  $V = Q * 5 * 60 = 2.91 * 5 * 60 = 873$ ft^3
- Depth is found from
  $D = V/A = 873/211 = 4.13$ ft
- For a 2:1 length to width ratio the horizontal dimensions are:
  $2 * W^2 = A$ $W = 10.27$ ft $L = 20.54$ ft

Retention Pond

A retention (or wet) pond is designed as a permanent pool of water, often with additional flood control and/or extended detention storage volume available above the permanent pool. Retention ponds generally hold water for release only through evapotranspiration and infiltration, though there are designs which maintain an extended detention volume above the wet pond. Retention ponds have proven to be the most effective common structural BMP type in the Eastern United States when considering all aspects of cost, performance and maintenance.

Pond volumes are designed to control storm related pond overflows to a design standard, and so that the contributing drainage area and/or groundwater is capable of supporting a permanent pool. This pond provides pollutant removal through settling of particulates, infiltration, filtration and biological uptake of soluble contaminants. To be effective in removing pollutants there must be sufficient runoff detention time. Expected performance can be very good, with proper design, depending upon the basin's size relative to the following five characteristics:

- watershed area,
- vegetative cover of watershed and pond,
- seasonal effects and variation,
- soil erodibility and infiltration rate, and
- storm characteristics.

There is some difference of opinion on the requirement for an extended detention pool above the retention pond. There is little evidence to suggest there will be a substantial increase in pollutant removal to justify the additional cost (California Stormwater Task Force, 1993). The Denver Urban Drainage and Flood Control District recommends a draw-down (detention) time of 12 hours for the design storm before reaching the permanent pool level post storm (UDFCD, 1992). Other states or localities have different criteria for the extended detention pool. Figure 13-5 presents a typical wet pond design.

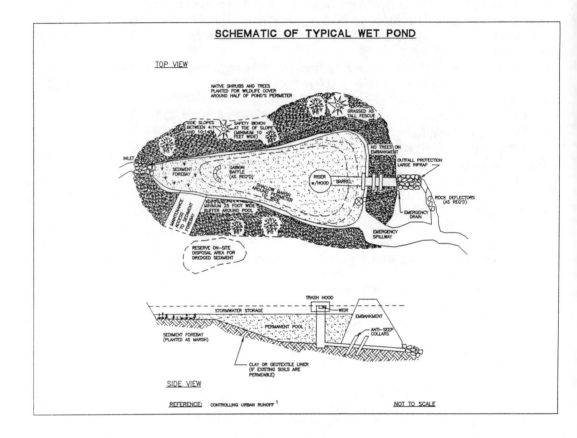

Figure 13-5 Wet Pond

Wet ponds provide opportunities for multi-objective applications involving recreation and nature areas. They are ideal for large tributary areas. "Pondscaping" has become almost an art form in some communities (Karouna, 1992). Effective arrangements have reflected both natural and human interaction themes. Recent applications feature pretreatment for oil and grease, flow spreader berms, and lengthened flow paths. It is important to obtain native plants which can both propagate and stand frequent inundation without significant maintenance. Whitlow and Harris (1979), among others, provide a guide for flood tolerance of plants in each area of the United States.

Emergent vegetation is also used. Approximately 10 foot wide shelves with depths of 1 to 2 feet provide an ideal setting for emergent vegetation and increased pollution removal. If mosquitoes are a problem this shelf could be reduced, pond level fluctuated, steeper slopes added or Gambosia (mosquito fish) added to the pond for control (California Stormwater Task Force, 1993). Non-rooted vegetation or rock filters near the outlet can also assist in pollution removal. Consider designing and managing the pond as a fishing lake. Wet pond costs are generally 25 to 40 percent higher than conventional detention ponds (MWCOG, 1992). Maintenance costs range from 3 to 5 percent of first costs annually. Wet ponds accumulate sediment under water and thus do not need to be cleaned out for aesthetic reasons as often as dry ponds (Hartigan, 1988). The pond muck itself can have pollutant removal capabilities. See Techniques (1994b) for a complete discussion.

Some other possible disadvantages of wet ponds include: safety considerations of a pond, floating debris and scum, waterfowl increasing the nutrient loading, and odors. Resuspension and remobilization of pollutants can be a problem if a large storm passes through the pond during unfavorable chemical conditions (UDFCD, 1992). Pond water quality can become quite poor. If a storm displaces this water the outflow can be more polluted than the inflow, and warmer. Larger ponds also inhibit fish passage, change stream flow regimes and can destroy wetlands adjacent to the stream. Pond "finger printing" is used to reduce environmental impacts through flow diversion, pond shaping ("doughnut ponds"), pond series, and other techniques. Additional environmental factors are discussed by Schueler and Galli (1992). See the extended detention pond discussion for brief consideration of bottom sediment contamination.

Pollution Removal Efficiency - Wet ponds can be very effective in removal of both the soluble and particulate fractions of pollution. Estimates of removal rates vary widely in the literature. Approximated removal rates for wet ponds of different designs can be estimated from Table 13-12. Since a particular detention basin may exhibit variable storm to storm performance characteristics, the long-term average performance should be considered rather than through analyzing individual events. Freezing conditions can be expected to reduce removal rates by 50 percent.

Results from the NURP study (EPA, 1983) provide approximate removal rates for TSS for a design with an average depth of 3.5 feet and an average impervious area upstream of 20 percent (low density residential). Figure 13-6 gives this removal information. These curves can be converted to VB/VR curves through the use of the mean depth and a 0.2 runoff coefficient (Hartigan, 1988). It is a reasonable estimate for TSS and for the particulate fractions of other pollutants as well (Driscoll, 1986).

Table 13-12 Average Annual Pollutant Removal Capability Of Retention (Wet) Ponds

Pollutant	Wet Pond Design Type		
	0.5 Inch Per Impervious Acre	VB/VR = 2.5	VB/VR = 4 Two Week Ret.
TSS	60-80%	60-80%	80-100%
TP	40-60%	50-70%	60-80%
TN	20-40%	20-40%	40-60%
BOD	20-40%	20-40%	40-60%
Metals	20-40%	60-80%	60-80%

Source: Schueler, 1987, 1993

Pond Sizing - There are two basic and differing design methodologies for retention ponds which reflect: (1) consideration of settling as the major removal mechanism or, (2) a further consideration of biological uptake as an additional important mechanism (Hartigan, 1988, Nix, et al., 1988, Wanielista and Yousef, 1993). Since many pollutants have an affinity for suspended solids in runoff, design of a wet pond for pollution removal by sedimentation is a logical approach (Randall, 1982). Basins with average retention times on the order of two weeks during the biologically active growing season will exhibit better removal rates due to the soluble portion of nutrients becoming available for biological uptake (Hartigan, 1988). The most important factors in this nutrient cycling are phosphorus loading and decay rates, hydraulic residence time and mean depth. The two leading models used in the eutrophication modeling of retention ponds are those of Walker (1985, 1987) and Reckhow (1988).

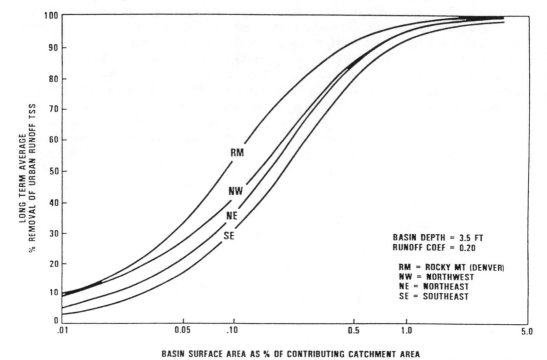

BASIN SURFACE AREA AS % OF CONTRIBUTING CATCHMENT AREA

Figure 13-6 Wet Detention Basins Size Performance Relationship

Source: USEPA, 1983

The detailed statistical method of Driscoll (Woodward-Clyde, 1986, Driscoll, 1988) can be used for pond sizing with permanent pools where there are no other local or state criteria. The necessary rainfall information for this method can be obtained from Driscoll, et al., (1989) if local rainfall statistical estimates are not available, and which was provided earlier in this chapter. Local rainfall analysis has been done for many places in the United States resulting in tables of surface area and drainage area relationships for extended detention ponds with permanent pools (for example, Professional Engineers of North Carolina, 1988 or Puget Sound, 1992).

An alternative method has been proposed by Hvitved-Jacobsen, et al., (1988). In this method the rainfall record is analyzed to determine rainfall statistics for a selected inter-event dry period. For example, to ensure at least a 72 hour recommended retention period the rainfall would be analyzed to determine the mean rainfall (and runoff) volume assuming storms must be separated by dry intervals of at least 72 hours. If there is not at least a 72 hour separation then all the precipitation is considered to be one runoff event. By selecting a storm return frequency (such as the 4 month storm) and an inter-event dry period (such as 72 hours) the designer controls the number of overflows per year (3 on average in this example) and the average time of retention within the pond. While this is equated to 1.06 inches of rainfall in a location in Denmark, the pond must be designed to contain the runoff from 3.25 inches of rainfall for a location in Orlando, Florida, USA (Hvitved-Jacobsen, et al., 1988).

Common pond volume criteria include:

- 0.5 inch per impervious acre;
- 2.5 times the mean runoff; or
- 4.0 times the mean runoff or approximately two weeks retention whichever is greater.

In arid and semi-arid climates these predicted volumes should be considered after decreases due to evaporation and groundwater losses have been accounted for. If there is some question on the pond's ability to hold water, a water balance should be performed taking into account all losses. Infiltration through the pond bottom can be reduced through the use of a clay or plastic liner at and below the permanent water line. These facilities often provide release temperatures significantly higher than the receiving waters. As such, care should be exercised when discharging into cold water fisheries. Some authors recommend adding volume to account for sediment deposition, up to 25 percent. This is especially important when construction may occur within the watershed upstream. In one case a pond was designed for pump back to a treatment plant (Segarra-Garcia and Loganathan, 1993). A sediment forebay sized to ten percent of the total pond volume will extend the pond life from a 10 to 20 year cycle to twice that. The forebay should be 4 to 6 feet deep and have exit velocities less than one foot per second (Techniques, 1994b).

Wet ponds typically take up 0.5 to 2 percent of a watershed area. Calculations of pond volume are similar to those for extended detention when a VB/VR ratio is the design criteria. Two variations on this method are:

$$VR = V * Rv * A * 43,560/12 \qquad\qquad (13.9)$$
$$VR = V * A_i * 43,560/12 \qquad\qquad (13.10)$$

Where: VR = volume of mean annual runoff, in.
V = volume of mean annual rainfall, in.
Rv = volume runoff coefficient
A = total drainage area, acres
A_i = total impervious drainage area, acres

If an extended detention volume is provided above the permanent pool its outlet calculation is handled similar to a dry extended detention pond. One advantage of having a permanent pool is that a negatively sloped pipe or inverted siphon can be used to draw water from below the surface to bleed off the extended detention portion of the pond, reducing clogging. Alternately, the Denver Flood Control District recommends a shorter 12 hour drawdown time for such a design and has provided a figure (Figure 13-7) to determine the required perforated riser area per row similar to the 40 hour drawdown figure presented earlier. Table 11-4 in the Storage Facilities, Chapter 11, gives the arrangement for perforations in the rows four inches apart vertically on center.

It should be noted that performance surveys of detention ponds designed using "state-of-the-art" methods still achieved a wide range of removal rates of TSS (Driscoll and Strecker, 1993). Fifty percent of the ponds surveyed in Maryland had achieved removal rates between 60 and 80 percent, with 25 percent on either side. Assuming a target rate of 80 percent TSS removal, 50 percent of the ponds failed to achieve the target, though 75 percent were above 60 percent TSS removal.

Typical Required Specifications

- Minimum length to width ratio of 3:1 (preferably expanding outward toward the outlet).
- Irregular shorelines for larger ponds provide visual variety.
- Design for full development upstream of control.

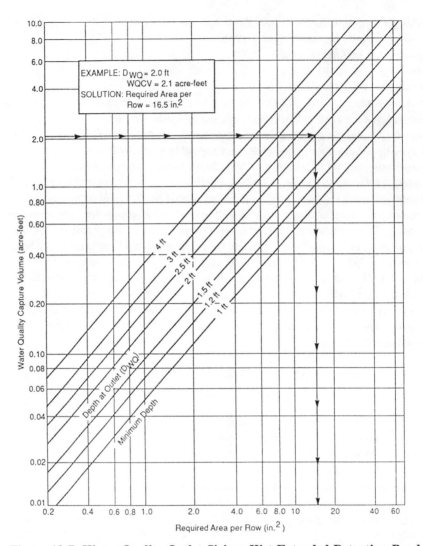

**Figure 13-7 Water Quality Outlet Sizing: Wet Extended Detention Pond
With A 12-Hour Drain Time** (UDFCD, 1992, 1994)

- Inlet and outlet located to maximize flow length. Use baffles if short circuiting cannot be prevented with inlet-outlet placement.
- Permanent pool depth: minimum - 2 to 3 feet, maximum - 9 to 10 feet, average - 3 to 6 feet.
- Side slopes no greater than 3:1 if mowed.
- Riprap protection (or other suitable erosion control means) for the outlet and all inlet structures into the pond. Individual boulders or baffle plates can work for this.
- Minimum drainage area of 10 acres.
- Anti-seep collars on barrel of principal spillway.
- One-half foot minimum freeboard above peak stage for top of embankment.
- Emergency drain; i.e. sluice gate, drawdown pipe; capable of draining within 24 hours.
- Emergency spillway designed to pass the 50-year storm event.
- Bypass greater than 2-year storms.

- Trash racks, filters, hoods or other debris control on riser.
- Maintenance access (< 15% slope and 10 feet wide).
- Benchmark for sediment removal.
- Paved or concrete channel.

Typical Recommended Specifications
- Multi-objective use such as amenities or flood control.
- Landscaping management of buffer as meadow.
- Design for multi-function as flood control and extended detention.
- Minimum length to width ratio of 3:1 to 4:1 (preferably wedge shaped).
- Use reinforced concrete instead of corrugated metal.
- Sediment forebay for larger ponds (often designed for 5 to 15 percent of total volume). Forebay should have separate drain for de-watering. Grass biofilters for smaller ponds.
- Consider artificial mixing for small sheltered ponds.
- Provision should be made for vehicle access at a 4:1 slope.
- Impervious soil boundary to prevent drawdown.
- Shallow marsh area around fringe 25 to 50 percent of area (including aquatic vegetation).
- Safety bench at toe of slope (minimum 10 feet wide).
- Minimum 25 foot wide buffer around pool.
- Mow embankment and side slopes at least twice a year.
- Emergency drain to allow draw-down within 24 hours.
- On-site disposal areas, for two sediment removal cycles, protected from runoff.
- Consider chemical treatment by alum if algal bloom are a problem.
- An oil and grease skimmer for sites with high production of such pollutants.

Typical Maintenance Standards for Extended Detention and Wet Ponds
- Sediment to be removed when 20% of storage volume of the facility is filled (design storage volume must account for volume lost to sediment storage).
- Sediment traps should be cleaned out when filled.
- No woody vegetation should be allowed on the embankment without special design provisions.
- Other vegetation over 18 inches high should be cut unless it is part of planned landscaping.
- Debris should be removed from blocking inlet and outlet structures and from areas of potential clogging.
- The control should be kept structurally sound, free from erosion, and functioning as designed.
- Periodic removal of dead vegetation should be accomplished.
- No standing water is allowed within extended detention pond unless specifically designed for.
- An annual inspection is required, reports to be kept by owner.
- The site should be inspected and debris removed after every major storm.

Constructed Storm Water Wetlands

Wetlands can provide a highly effective management measure for mitigation of pollution from runoff because they have the ability to assimilate large quantities of suspended and dissolved materials from inflow. A basic understanding of wetlands types and function is a necessary prerequisite to the use of wetlands for storm water pollution control. Mitsch and Gosselink (1986) and Hammer (1989, 1992) provide basic information on wetlands, constructed wetlands for wastewater treatment and freshwater wetland creation. Schueler (1992) presents a comprehensive consideration of wetland creation with emphasis on the Eastern United States. Later literature amplifies and extends his findings.

Wetlands functions and values include (EPA Region 7, 1992):
- flood reduction and conveyance,
- sediment control and reduction,
- habitat for fish, waterfowl, rare and endangered species and other wildlife,
- enhancement of water quality,
- wastewater treatment,
- groundwater recharge,
- recreation, education and research,
- food and timber production, and
- open space and aesthetic values.

The design of wetland management measures is very complex being, generally, a function of "nearly everything". But the three most important components of wetland creation and function are: water, soil and vegetation. The "hydrology" of the wetland essentially drives the physical and chemical processes within the wetland itself. These, in turn, drive the ecosystem character and growth. Water carries the nutrients into the wetland system. It also effects such critical primary variables as pH, temperature, degree of substrate anoxia, and soil salinity. Species composition and richness, primary productivity of the vegetation, organic accumulation and nutrient cycling are all primarily controlled by the hydrology of the wetland. Thus the hydrology of the wetland controls its ability to remove pollutants through changes in velocity and flow rate, water depth and fluctuation, detention time, circulation and distribution patterns, seasonal hydrological influences, and groundwater conditions (Livingston, 1988).

Wetlands are an intermediate form between uplands and aquatic systems. Not that they are always transitional in the sense of succession from one to the other, but they share vital features of each. Wetland areas are subject to the random ebb and flow of surface water. As such, they are very sensitive to changes in hydrology, and may respond to minor changes in flow or "hydroperiod" with major changes in species richness and ecosystem productivity (Mitsch and Gosselink, 1986). Wetlands that are intermittently flooded are very different from those which are only intermittently exposed to the drying atmosphere. Saturated soils maintain a different ecosystem than partially saturated systems. The major parameter for the measurement of gross wetland hydrology is the residence time, which is the measure of the average time that water stays in the wetland. The reciprocal of residence time is termed the "turnover rate".

Wetland soils are termed "hydric". This means that the soil is saturated, flooded or ponded long enough during the growing season to develop anaerobic conditions that favor the growth and regeneration of hydrophytic (adopted to wetland conditions) vegetation (SCS, 1985). If the soils contain less than 12 to 20 percent organic materials they are considered mineral soils. More organic material than that and they are considered organic, or peat, soils (Hammer, 1992). The soil is the basic medium in which many of the wetland chemical transformations take place and in which are stored the primary chemicals for wetland plants (Mitsch and Gosselink, 1986). These two basic types of soils support vegetation in differing ways. Organic soils hold water better while mineral soils have more minerals biochemically available to plants. Most urban wetlands will be based on mineral soils.

Contrary to thinking that a wetland is a lush vegetated paradise for plants, a wetland environment can be harsh and demanding on vegetation requiring many adaptations for survival and, often, a narrow range of conditions under which a particular species can survive at all. Nonetheless there are over 500 species of wetland plants in the United States (Hammer, 1992). A wetland plant is defined as a species which can either "tolerate" or "regulate" sufficiently to survive up to 5 days inundation during the growing season. These plants are often classified as vascular (having internal structural transport mechanisms) or nonvascular, and as free floating or rooted. The rooted forms are subdivided into submergent, emergent and floating-leaved types. Woody species are sometimes classified separately but actually fall under the rooted emergent category, though their roots are rarely under water.

Wetlands plants tend to insulate themselves from the changes the environment brings about. They do this through both individual methods and through peat production to stabilize the nutrient and moisture source. All wetland plants have elaborate mechanisms for survival which include both structural methods (such as air spaces) and physiological methods (such as modified respiration and photosynthesis methods). Even with this, the limitations imposed by the wetlands environment, along with plant competition, limit the aerial extent of species. Zones of plant communities tend to be sharply delineated by abiotic factors such as soils, inundation durations, saturation, and water chemistry (EPA Region 7, 1992). "Families" of plant species with fairly similar tolerances group together on these steep ecological gradients (Mitsch and Gosselink, 1986). Horner (1988) found some evidence that urban (versus rural) wetlands tended toward more opportunistic and mon-ocultural vegetative stands, probably due to modified hydrology. Virginia planners have divided these zones into: aquatic, reed and riparian zones. Aquatic is almost always inundated while riparian is normally above the water surface. The reed zone begins slightly above mean low water level and extends into the water to a depth of about 18 inches (Northern Virginia, 1992). Shenot (1993) recorded a number of volunteer species at ponds in Maryland dominated by exotic grasses (Graminea), soft rush (Juncus Effusus), and various sedges. Cattails are also notorious volunteers.

Storm Water Wetlands - Storm water wetlands differ from more natural wetlands in that they are not normally designed with the diversity of species and ecological functions of natural wetlands (MWCOG, 1992). Rather they are designed to maximize pollution removal. The long term viability of constructed wetlands for storm water removal has not yet been tested. Few are older than 10 to 15 years. Horner found that a major difference between urban and rural wetlands was simply the amount of trash in urban wetlands detracting from aesthetic appeal (Horner, 1988). Schueler (1992) gives other differences between natural and urban wetlands in the mid-Atlantic region including: groundwater versus runoff dominated water balance, hydroperiod gradual and buffered versus hydroperiod based on runoff and is rapid, adjusting wetland boundaries versus controlled boundaries, high diversity maintained by seedbank versus low diversity established by planner, self maintaining versus man maintained, complex topography versus simple topography, low sediment supply versus high sediment supply, and high wildlife habitat versus low wildlife habitat. There may be some question about the use of wetlands in arid western climates due to the possible need for irrigation of the vegetation and that the wet season occurs during times of vegetation dormancy (California Stormwater Task Force, 1993).

Schueler (1992) gives four standard wetlands designs used in the mid-Atlantic region: the shallow marsh system, the pond/wetland system, the extended detention wetland system and the pocket storm water wetland. Except for the pocket wetland, as the names imply, the major differences are the incorporation of either wet pond or extended detention pond designs into the wetland marsh system. Table 13-13 gives characteristic information for the four types. Figures 13-8 and 13-9 provide examples of a typical wetland marsh and pocket wetland.

The pond/wetland system consists of two separate cells - a deep pond leading to a shallow wetland. The pond removes pollutants, and reduces the space required for the system.

Pocket wetlands seldom are more than a tenth of acre in size, and serve development sites of ten acres or less. Due to their size and unreliable water supply, pocket wetlands do not possess all of the benefits of other wetland designs. Most pocket wetlands have no sediment forebay. Despite many drawbacks, pocket wetlands may be an attractive BMP alternative for smaller development situations. They have been used successfully at culvert outlets and at parking lot exits in Minnesota.

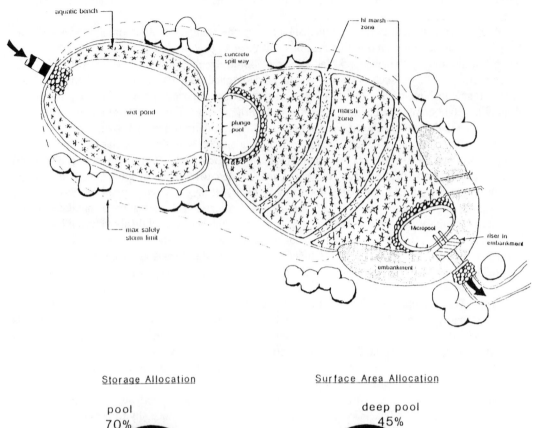

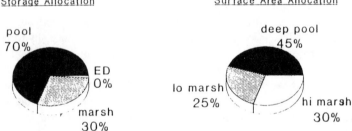

Figure 13-8 Typical Storm Water Wetland (Schueler, 1992)

Another type of wetland feature used most often in the west is a wetland bottom channel. They are used downstream from detention ponds to "polish" the effluent and in their own right wherever there is an ability for wide and gently sloping channels. Wetland channels are designed with a 2-year flow velocity of less than 2.5 ft/s (n = 0.03) and flow depth of 3 to 5 feet, and 8:1 minimum width to depth ratio (UDFCD, 1992). Grade control or drop structures are used to control the longitudinal slope.

Pollution Removal Efficiency - Wetlands can be very effective in removal of both the soluble and particulate fractions of pollution. But attention to details in design is vital. Soils, hydrology and plants must support each other well to achieve high removal rates. Detailed construction supervision is also essential. Estimates of removal rates vary widely in the literature, and many of these are for experimental wetlands. Rough estimates of removal rates for wet lands can

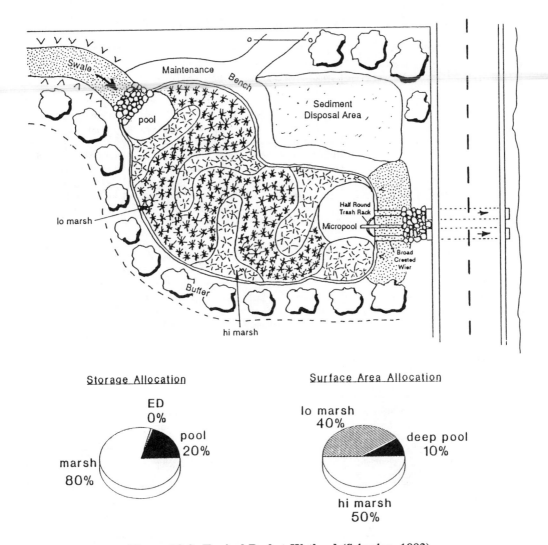

Figure 13-9 Typical Pocket Wetland (Schueler, 1992)

be estimated from Table 13-14, based on studies (Schueler, 1992). The greatest consistency in pollutant removal among the various studies was for BOD, suspended solids and metals (Livingston, 1988). In the non-growing season there may even be nutrient export (especially ammonia and ortho-phosphorus) (MWCOG, 1992). While metals tend to accumulate in wetlands nutrients tend to cycle (Horner, 1988). Pollution removal takes place through biological uptake (microbial, algal, vascular plant), sedimentation, volatilization, adsorption, precipitation, and filtration. Rooted vegetation removes nutrients through the soil while non-rooted vegetation removes it directly from the water.

Pollutant removal efficiency depends physically on aquatic treatment volume, surface area to volume ratio and surface area to watershed ratio. Pollutant removal can be enhanced by maximizing any of these variables and increasing the flow path, providing pretreatment with forebays and using redundant pollutant removal pathways (Schueler, 1992). Large storms can overwhelm smaller

Table 13-13 Attributes Of Four Storm Water Wetland Systems

Attribute	Design 1. Shallow Marsh System	Design 2. Pond/Wetland System	Design 3. Extended Detention Wetland	Design 4. Pocket Wetland
General Description	Flow through a sediment forebay, channels and micropool near outlet. Islands and benches promote circulation. Majority is less than 18 inches deep.	Two celled pond/wetland system with pond upstream from wetland. Hard barrier or other separation between pond and marsh. Micropool at outlet. Wetland fringe on bench around pool. Less area required due to pool pollutant removal.	Forebay with low and high marsh zone. Whole area floods when runoff occurs. May rise as much as three feet with 12 to 24-hour drawdown. Less space requirement due to vertical storage.	Small in size (less than 0.1 acre) with drainage areas less than 10 acres. Normally without forebay and less complex. Small inlet stilling pool and micropool at outlet with trash rack. Fluctuating water levels, may be below groundwater level to support constant moisture.
Pollutant Removal	Moderate reliable removal of sediments and nutrients.	Moderate to high reliable removal of sediments and nutrients.	Moderate less reliable removal of sediments and nutrients.	Moderate with possibility of resuspension and groundwater displacement.
Land Consumption	High	Moderate	Moderate	Moderate but very flexible.
Water Balance	Dry weather baseflow recommended to maintain water elevations. Groundwater not recommended as the primary source of water supply to wetland.			Water supply supplied by excavation to groundwater.
Deep Water Cells	Forebay, channels, micropool	Pond, micropool	Forebay, micropool	Micropool, if possible
Outlet Configuration	Reversed slope pipe extending from riser, withdrawn about one foot below normal pool. Pipe and pond drain equipped with gate valve. Alternate is perforated riser emptying into weir box.			Broad crested weir with half-round trash rack, and pond drain
Sediment Cleanout Cycle	Cleanout of forebay every 2-5 years	Cleanout pond every 10 - 15 years.	Cleanout of forebay every 2-5 years	Cleanout of wetland every 5 to 10 years, on site disposal and stockpile mulch
Native Plant Diversity	High, if complex micro topography is present	High, with sufficient wetland complexity area	Moderate, fluctuating water levels impose physiological constraints	Low to moderate due to small surface area and poor control of water levels
Wildlife Habitat Potential	High with complexity and wide buffer (50 foot)	High, with wide buffer it attracts waterfowl	Moderate, with buffer	Low due to small area and low diversity

Source: Schueler, 1992

Table 13-14 Average Annual Pollutant Removal Capability Of Wetlands

Pollutant	Standard Wetland[1]
TSS	60-85%
TP	40-60%
TN	20-30%
Organic Carbon	10-20%
Lead	70-80%
Zinc	45-55%
Bacteria	2 log reduction

[1] Pocket wetlands will have lower removal rates

[2] Wetlands sized with pool will have higher removal rates on the order of 65% for TP and 40% for TN.

Source: Schueler, 1992, 1993

wetlands, greatly reducing the pollutant removal capabilities. Off-line design practices, or increasing the size, can circumvent this problem somewhat. Plant harvesting prior to winter dieback may not increase the nutrient removal efficiency. Much of the biomass in wetlands is retained in accumulating bottom sediments or roots. Removal of the particulate portions of nutrients is often lower when baseflow is monitored (Techniques, 1994a). Pollutant removal can be enhanced through use of the "BMP treatment train" concept using a pond or sump prior to the wetland.

Wetland Design - Wetland design standards differ somewhat around the country because of the great changes in climate, soils and vegetation types. Therefore only general standards and design guidance can be given here. Table 13-15 gives design criteria summaries for the four types of wetlands.

Wetlands should be sized to capture 90 percent of the total annual storm water volume. In Washington DC this amounts to a 3 month storm and a rainfall depth of 1.25 inches over the whole watershed. This is in excess of any dry weather flow or groundwater seepage. As for wet ponds other rules-of-thumb abound. Calculations are also similar to extended ponds or wet ponds.

A sediment forebay should be provided and should be at least 10 percent of the treatment volume and be 4 to 6 feet deep. It should be separated from the main area by a berm or gabion structure. A cattail forebay can be useful if trapping of oil and grease and trash is important. Lowflow channels can cause short circuiting and should be avoided (California Stormwater Task force, 1993).

The micropool at the outlet creates sufficient depth for the reverse slope pipe. The reverse slope pipe should be about 1 foot below the permanent pond surface. It can discharge into a stilling well riser which can have openings at higher elevations for flood control. The micropool, like the sediment forebay, should be 10 percent of the treatment volume and be 4 to 6 feet deep. A drain for the micropool, as well as the wetland should be placed at the bottom of the micropool.

Table 13-15 Comparative Design Considerations Of Four Storm Water Wetland Systems

Design Criterion	Design 1. Shallow Marsh System	Design 2. Pond/Wetland System	Design 3. Extended Detention Wetland	Design 4. Pocket Wetland
Storage Allocation	40 % pool 60 % marsh	70 % pool 30 % marsh	30 % marsh 20 % pool 50 % extended deten-tion	80 % marsh 20 % pool
Surface Area Alloca-tion	40 % low marsh 20 % deep pool 40 % high marsh	45 % deep pool 25 % low marsh 30 % high marsh	20 % deep pool 35 % low marsh 45 % high marsh	10 % deep pool 40 % low marsh 50 % high marsh
Flow Path length to Width Ratio	1:1	1:1	1:1	NA
Dry Weath-er Path length to Width Ratio	2:1	2:1	2:1	2:1
Wetland Area/ Wa-tershed Area	0.02 minimum	0.01 minimum	0.01 minimum	0.01 minimum
Contributing Watershed Area	> 25 acres with dry weather flow	> 25 acres with dry weather flow	> 10 acres	1 to 10 acres

Source: Schueler, 1992

The maximum depths for most emergent types of vegetation is from 3 to 12 inches. Local experience should dictate. The specifications given below are typical for the eastern United States. The high marsh zone is planted zero to 6 inches deep while the low marsh zone is typically 6 to 18 inches deep. It must be remembered that for the extended detention portion of the design, as well as for fringe areas of the other designs, inundation tolerant species should be selected since they may become wet 10 to 30 times per year. Vegetation is planted using any of four techniques: using container grown plant stock, using wetland mulch, broadcasting seeds (extended detention zone), and allowing volunteer growth (Schueler, 1992, Thunhorst, 1993). Transplanting stock is the most reliable method of establishing preferred vegetation if the plants are well taken care of during transport to the site and pre-planted, and if the peat pots are broken open to facilitate root spreading. The use of donor soils as a mulch can be very effective as both a primary and secondary planting method, should such soils be available. Volunteer growth and broadcast seeding should not normally be relied upon as the primary means of vegetation establishment except for small pocket wetlands and for fill-in areas between plantings. Seeding should be done on the basis of "pure, live seed" to avoid purity and germination problems (Gorbisch, 1994). Careful inspection during the first three to five years will be necessary to reinforce plantings which fail and to control exotic species. Shenot (1993) found four planted species to persist and spread in Maryland; broadleaf arrowhead (Saggitaria latifolia), soft stem bulrush (Scirpus validus), common three square

(Scirpus americanus), and pickerelweed (Pontederia cordata). Wild rice showed high persistence. Other specifications are given below. Figure 13-10 gives a shallow marsh planting strategy.

Adjustment of the water balance can be done by slightly raising or lowering the pool elevation, diverting more flow around or through the pond, reducing infiltration with pond bottom liners installed below rooted vegetation and by changing orifice sizes on outlets.

Permitting - Wetlands permitting has always been a concern. If a wetlands project involves fill into a wetlands, a Section 404 permit will be required. It is not recommended to destroy natural wetlands to create storm water wetlands since their basic functions are different. "Finger-printing" can be used to combine both natural and constructed wetlands in a location (Schueler, 1992). Generally a project will be permitted if: there are no practicable alternatives; toxic effluent or water quality standards are not violated; endangered species or critical habitat will not be jeopardized; waters of the United States will not be degraded; and, appropriate steps are taken to minimize unavoidable adverse impacts. At present permitting for the removal of wetlands vegetation from constructed wetlands does not require a Section 404 permit. The EPA Wetlands Protection Hotline can be called for more information at 1-800-832-7828.

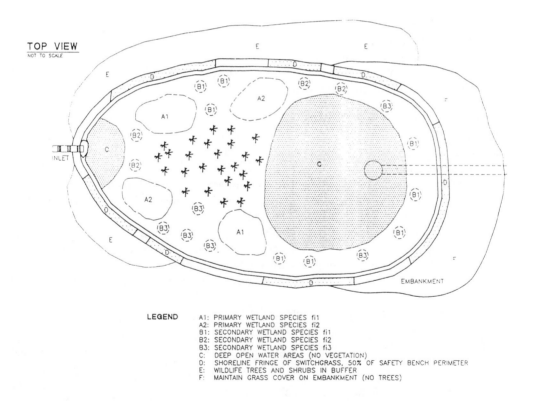

Figure 13-10 Shallow Marsh Planting Strategies

Source: Schueler, 1987

Other laws that may impact activities involving wetlands and 404 permitting include:
- the National Environmental Policy Act (NEPA) for required environmental impact statements or environmental assessments;
- the Endangered Species Act;
- the Fish and Wildlife Coordination Act;
- the National Historic Preservation Act;
- the Food Security Act of 1985;
- the Food, Agriculture, Conservation and Trade Act of 1990
- the Emergency Wetlands Resources Act;
- the North American Wetlands Conservation Act;
- the Water Resources Development Act;
- historic or archeological preservation concerns; and
- state and local laws or zoning overlays.

Design of a constructed storm water wetland proceeds through the development stages of: feasibility evaluation, concept plan development, pond sizing, pond feature development, pondscaping design and planning, construction and grading, wetland bed preparation, pondscape and vegetation establishment, and post construction inspection and maintenance (Schueler, 1992).

Typical Standard Specifications For Constructed Wetlands

Typical Required Specifications
- Inflow of water must be greater than that leaving the basin by infiltration.
- Designed for an extended detention time of 24 hours for the 1-year storm.
- The orifices used for extended detention will be vulnerable to blockage from plant material or other debris that will enter the basin with storm water runoff. Therefore, some form of protection against blockage should be installed (such as some type of non-corrodible wire mesh). Reverse slope pipes are recommended.
- Surface area of the wetland should account for a minimum of 3 percent of the area of the watershed draining into it.
- The length to width ratio should be at least 2 to 1.
- A soil depth of at least 4 inches should be used for shallow wetland basins.
- The deeper area of the wetland should include the outlet structure so that the outflow from the basin is not interfered with by sediment buildup.
- A forebay should be established at the pond inflow points to capture larger sediments and be 4 to 6 feet deep. Direct maintenance access to the forebay should be provided with access 15 feet wide minimum and 5:1 slope maximum. Sediment depth markers should be provided.
- If high water velocity is a potential problem, some type of energy dissipation device should be installed.
- The designer should maximize use of pre- and post-grading pondscaping design to create both horizontal and vertical diversity and habitat.
- A minimum of 3 aggressive emergent wetland species (primary species) of vegetation should be established in quantity on the wetland.
- Three additional emergent wetland species (secondary species) of vegetation should be planted on the wetland, although in far less numbers than the two primary species.
- 30 to 50 percent of the shallow (12 inches or less) area of the basin should be planted with wetland vegetation.
- Approximately 50 individuals of each secondary species should be planted per acre; set out in 10 clumps of approximately 5 individuals and planted within 6 feet of the edge of the pond

in the shallow area leading up to the ponds edge; spaced as far apart as possible, but no need to segregate species to different areas of the wetland.

- Wetland mulch, if used, should be spread over the high marsh area and adjacent wet zones (-6 to +6 inches of depth) to depths of 3 to 6 inches.
- A minimum 25 foot buffer, for all but pocket wetlands, should be established and planted with riparian and upland vegetation (50 foot buffer if wildlife habitat value required in design).
- Surrounding slopes should be stabilized by planting in order to trap sediments and some pollutants and prevent them from entering the wetland.
- A written maintenance plan should be provided and adequate provision made for ongoing inspection and maintenance, with more intense activity for the first three years after construction.
- The wetland should be maintained to prevent loss of area of ponded water available for emergent vegetation due to sedimentation and/or accumulation of plant material.

Typical Recommended Specification

- To minimize maintenance as much as possible, it is recommended that wetland basins be installed on stabilized watersheds and not be used for sediment control.
- Complex topography can be maintained by bioengineering methods (such as fascines) or straw bales and geotextile rolls.
- It is recommended that the frequently flooded zone surrounding the wetland be located within approximately 10 to 20 feet from the edge of the permanent pool.
- Soil types conducive to wetland vegetation should be used during construction.
- The wetland should be designed to allow slow percolation of the runoff through the substrate (add a layer of clay for porous substrates).
- The depth of the forebay should be in excess of 3 feet and contain approximately 10 percent of the total volume of the normal pool.
- As much vegetation as possible and as much distance as possible should separate the basin inlet from the outlet.
- The water should gradually get shallower about 10 feet from the edge of the pond.
- The planted areas should be made as square as possible within the overall design of the wetland, rather than long and narrow.
- The only site preparation that is necessary for the actual planting (besides flooding the basin) is to ensure that the substrate is soft enough to permit relatively easy insertion of the plants.

Operation And Maintenance Requirements

- A storm water management easement and maintenance covenant should be required for each facility. The maintenance covenant should require the owner of the wetland to periodically clean the structure.
- Sediment forebays should be cleaned every 2 to 5 years except for pocket wetlands without forebays which are cleaned after a six inch accumulation of sediment.
- The ponded water area may be maintained by raising the elevation of the water level in the permanent pond, by raising the height of the orifice in the outlet structure, or by removing accumulated solids by excavation.
- Water levels may need to be supplemented or drained periodically until vegetation is fully established.
- It may be desirable to remove contaminated sediment bottoms or to harvest above ground biomass and remove it from the site in order to permanently remove pollutants from the wetland.

Austin First Flush Filtration Basin

The Austin First Flush Filtration Basin has been in use in Austin, Texas for a number of years and, thus, has a proven track record. There are now several hundred in existence there. It is used when there is insufficient water for a retention pond or natural extended detention pond. It consists, as shown in Figures 13-11 and 13-12 of, in turn, a flow energy dissipator, an extended detention sedimentation chamber for settling larger solids, a flow spreader device, a sand filter, and an underdrain outlet. It depends on a hydraulic head to drive the system and thus requires sufficient drop over the structure to operate.

This type of basin can be used for areas from one to 50 acres. Beyond 50 acres the first flush criteria becomes harder to justify. Most of the basins are designed as off-line basins though some have flow through capabilities. Runoff enters the basin by a diversion weir sized to pass the first flush runoff. The height of the weir is such that the water in the basin backs up to the weir height when the capture volume is achieved. Additional runoff then cannot enter the basin due to the backwater and flows past and out to a flood control detention structure. Experience has shown this device should only be used in areas not being developed. Even with a forebay, sediment fines can penetrate the filtration media resulting in a need for replacement (Austin, 1988b).

Pollution Removal Efficiency - Pollution removal takes place through settling and filtration. If peat or possibly leaf compost is used as part of the filtration media the dissolved fraction removal rate will probably improve (California Stormwater Task Force, 1993). Field studies in Austin (1990) indicate the removal estimates shown in Table 13-16. Results varied considerably with one site achieving rates for almost all constituents in the 80% - 90% range, through combined retention with the filter. Negative removal rates can be expected for dissolved solids and nitrate-nitrogen.

Table 13-16 Average Annual Pollutant Removal Capability Of Austin First Flush Basins

Pollutant	Percent Removal
TSS	75-90%
TP	30-60%
TN	30-50%
TKN	40-60%
TOC	40-60%
BOD	30-50%
Metals	40-80% (depending on metal)
Bacteria	40-70%

Source: Austin, 1988, 1990

Design - The storage volume is based on a first flush runoff volume from 1/2 inch of rainfall per impervious acre. Included in the volume is runoff from pervious areas which flow to impervious areas. Austin designs utilize infiltration trenches or basins for treatment. The sedimentation basin is designed for either full or partial sedimentation according to the Austin criteria (Austin, 1988b). This results in two separate but related design criteria (see Figures 13-11 and 13-12).

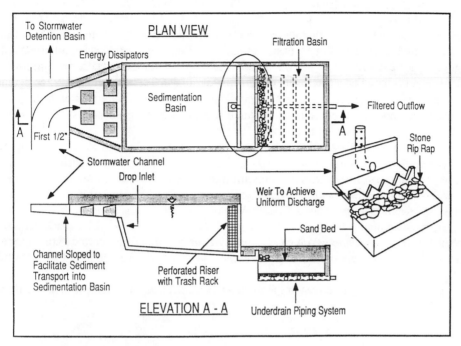

Figure 13-11   Austin First-Flush Basin - Full Sedimentation Design
Source: Austin, 1988

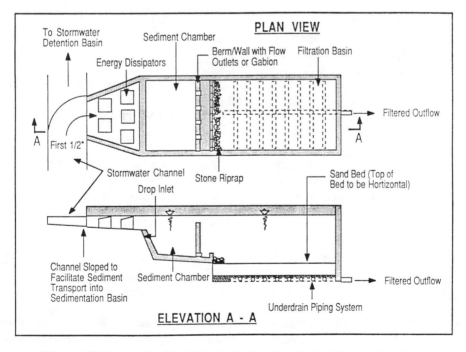

Figure 13-12   Austin First-Flush Basin - Partial Sedimentation Design
Source: Austin, 1988

1. Full Sedimentation Design - For full sedimentation the pretreatment basin is designed to hold the full capture volume and release it over a period of 24 hours and is designed to remove particles with diameters of 20 microns and specific gravity of 2.65. A sediment trap can be provided at 10 percent of the sediment basin volume. The horizontally graded filtration component is designed to achieve drawdown for the design flow. The filtration standard for full sediment capture upstream is based on a coefficient of permeability of 3.5 feet per day, an average hydraulic head on the filter surface of three feet and a sand depth of 18 inches. The equation for the surface area of the sedimentation basin are based on a ten foot maximum depth:

$$A_s = A_d H / D_p \qquad\qquad (13.11)$$

Where: A_s = minimum surface area of the sedimentation basin, acres
A_d = contributing drainage area, acres
H = depth of runoff captured from A_d, ft
D_p = pond depth, ft

The product $A_d H$ is the capture volume in this equation assuming no pervious area additions. Based on the Austin assumptions on size the minimum surface area can be found by substituting 10 for D_p.

The equation for the surface area of the filtration basin depends on the infiltration rate through the sand. The equation is:

$$A_f = \frac{A_d H L}{k(h+L)t_f} \qquad\qquad (13.12)$$

Where: A_f = area of filtration basin, acres
A_d = contributing drainage area, acres
H = depth of runoff captured from A_d, ft
L = sand bed depth, ft (1.5 ft minimum)
k = coefficient of permeability, ft/day (3.5 for full sedimentation and 2.0 for partial sedimentation designs respectively)
h = average water depth over the filter surface, may be taken as one-half the distance from the sand surface to the maximum water surface elevation for 20 percent of the capture volume
t_f = draw down time for filtration basin, days (Austin uses 40 hours, 1.67 days)

Using the typical and minimum values for the Austin design: $L = 1.5$, $k = 3.5$ for full sedimentation and 2.0 for partial sedimentation, $h = 3$, $t_f = 1.67$ the minimum surface area equation becomes for full sedimentation and partial sedimentation respectively:

$$A_f = A_d H / 18 \qquad\qquad (13.13)$$
$$A_f = A_d H / 10 \qquad\qquad (13.14)$$

2. Partial Sedimentation Design - The partial sedimentation basin is designed to settle out only the largest particles. Its volume should be 20 percent of the capture volume minimum and three foot minimum depth. For partial capture the coefficient of permeability for the filtration section is assumed to be 2 feet per day to allow for the faster clogging of the media. The equations for filtration basin sizing given above apply for this case.

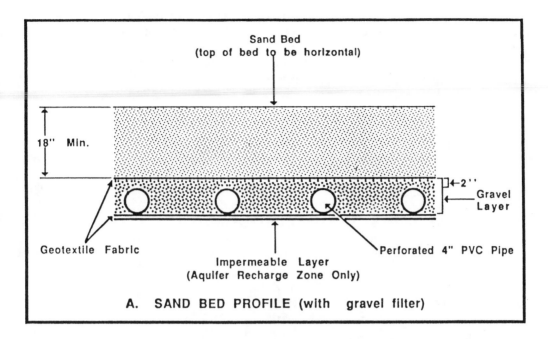

Figure 13-13 Gravel Filter Design (Austin, 1988)

Source: Austin: 1989

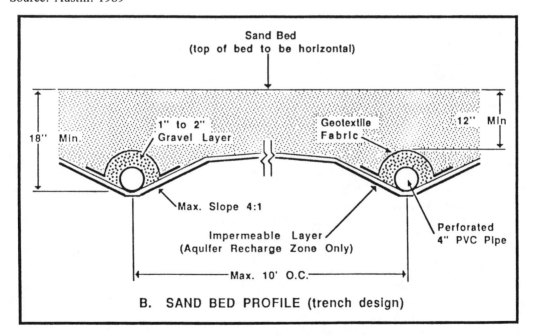

Figure 13-14 Trench Filter Design (Austin, 1988)

Source: Austin: 1989

Filter Design - For both designs the filter is designed either with a sand filter or a trench design. The gravel filter design consists of an 18 inch minimum layer of sand, geotextile fabric, and a gravel filter surrounding 4 inch perforated pipes underlain with another geotextile fabric. The trench design consists of 12 to 18 inches of sand with perforated pipes placed in trenches at the bottom. Figures 13-13 and 13-14 give details.

Typical Design Specifications

Full Sedimentation Basin

- The basin should be designed to capture the 0.5 inch runoff from the 25-year storm and bypass the rest.
- Inlets should discharge water uniformly and at low velocity. Consider an energy dissipator for inlet velocities greater than 3 ft/s.
- Minimum depth of sedimentation basin should be 3 feet.
- Outlets should provide for a drawdown of 24 hours minimum. Recommended outlet is a perforated riser pipe with an internal orifice or throttle plate. They should be protected from blockage.
- An impermeable basin liner should be provided.
- Length to width ratio should be 2:1 or greater.
- Distance from inlet to outlet should be maximized. Baffles may be necessary.
- 0.5 foot freeboard should be provided.

Partial Sedimentation Basin

- The volume should be a minimum of 20 percent of the water quality capture volume.

Filtration Basin

- A flow spreader should be used on the inlet channels or pipe outfalls to spread the incoming runoff more evenly over the surface of the basin to promote better infiltration.
- The filtration basin storage volume above the sand surface should be 20 percent of the capture volume minimum.
- The sand bed may be either a gravel filter design or a trench design.
- Sand should be 0.02 to 0.04 inches in diameter.
- Gravel should be 1/2 to 2 inches in diameter.
- Drain pipe mains and laterals should be 4 inch diameter PVC Schedule 40 pipe with 3/8 inch perforations. The pipe should be no more than 10 feet apart and have a grade of no less than 1/8 inch per foot. The perforation rows should be no more than 6 inches apart. Cleaning access must be provided.
- The surface of the basin should be graded to have a slope close to zero in order to achieve a uniform ponding depth across the entire surface of the basin.
- The side-slopes of the basin should be no steeper than 3:1 (h:v) to allow proper vegetative stabilization.
- Inlet channels leading to the basin should be stabilized to prevent incoming runoff velocities from reaching erosive levels and scouring the basin floor (customarily done by riprapping the inlet channels or pipe outfalls).
- Maintenance access should be provided 10 foot wide and less than 4:1 slope.
- On-site sediment disposal should be provided.

Operation And Maintenance Requirements

- The sedimentation basin should be cleaned prior to putting the site into operation.

- Silt will be removed from the sedimentation basin when it reaches a depth of 6 inches.
- Debris and trash should be removed when necessary.
- Vegetation growing in the basin will be kept mowed less than 18 inches.
- Annual inspection and inspection after major storms, and removal of debris and trash is required.
- When capture volume drawdown exceeds 60 hours or dry drawdown exceeds 96 hours the sedimentation basin needs corrective maintenance.
- Filtration silt removal should be done when it exceeds 1/2 inch on the surface.
- Corrective maintenance is required any time the filtration basin does not draw down the capture volume in 36 hours after the sedimentation basin has emptied.

Design Sizing Example - A commercial site with 20 acres is to be developed. The total impervious area draining to the Austin First Flush Basin is 17 acres. An additional acre of pervious area will drain to the pond. A full filtration design is desired. The capture volume required is: 18 acres * 1/2 inch = 0.75 acre-feet = 32,670 ft^3

Based on site characteristics the sedimentation basin depth is 8 feet. The surface area is: 32,670/8 = 4,084 ft^2

For a 3:1 length to width ratio the horizontal dimensions are: 3 W^2 = 4,084
Therefore: Width = 37 ft   Length = 110.4 ft

If a sediment trap were installed its volume would be: V$_t$ = 0.1 * 32,670 = 3,267 ft^3

For a sand depth of 18 inches, 40 hour draw down time, k = 3.5, and three foot average depth above the filter surface. The surface area of the filtration basin is:

$$A_f = A_d \, H \, L \,/\, k \, (h+L) \, t_f$$
$$= (43,560 * 18 * 0.0417 * 1.5) / (3.5 * (3+1.5) * 1.67) = 1,864 \text{ ft}^2$$

Check of depth: approx. average inflow rate = 32,670/24 = 1,361 ft^3/hr
approx. average outflow rate = 32,670/40 = 817 ft^3/hr
24 hour volume maximum = (1,316 - 817)*24 = 11,976 ft^3 (37% of capture vol)
depth = 11,976/1,864 = 6.4 ft ok

Sand Filters

Sand filters have a number of advantages over both retention devices and infiltration devices such as (Galli, 1992):
- history of successful employment in diverse applications including dry climates and areas with high water table or poor draining soils;
- proven ability for small (less than 10 acres) applications;
- pollutant removal is moderate to high for most pollutants;
- sand filters are more resistant to clogging and easier and cheaper to repair if clogged than infiltration trenches or basins;
- sand filters can be easily coupled with flood control BMPs;
- on a dollar per runoff volume treated basis, sand filters are more economical to construct than infiltration devices or oil and grit chambers;

- removal of contaminated soil and surface scum does not involve costly and elaborate measures; and
- sand filters pose no groundwater contamination threat.

Sand filters are distinguished from infiltration devices by the fact that, normally, an outfall handles the effluent after it has been filtered rather than percolating it to groundwater. Thus they are relatively independent of groundwater depth unless the water table intrudes on a structural feature of the system itself. Nearly all sand filters treat only the first flush and, thus, are constructed off-line. The only general limitations in use are a requirement for 2 or 3 feet of head to ensure gravity flow and a maintenance mechanism to ensure media replacement at intervals. Such devices have not been tested extensively in cold climates where freezing of the media may render them ineffective in winter. They also provide little flood control benefits.

There are a number of configurations and variations of sand filters which involve filtration media on smaller, site scales. These facilities are designed to capture a certain flow volume (normally the first flush), pretreat it for settling of the heaviest particles and filter the flow through an engineered media to an outflow point. The filtration surface is designed using an approach similar to the Austin First Flush Basin filtration surface area equation.

The standard design consists of a sand filter layer over gravel and an underdrain system. Variations include: grassed surfaces, gravel pretreatment, plastic screen pretreatment, compost, peat, stone reservoir storage above the sand layer, grated chambers, and a vertical sand layer with flow moving horizontally. Each design variation has advantages and weaknesses. Table 13-17 shows typical removal rate expectations. Removal can be enhanced with the addition of an organic layer or by combining this BMP with detention or retention ponds.

Linear Filter - The linear filter (Figure 13-15) was tested in Delaware (Shaver, 1991). It basically intercepts sheet flow coming off a parking area or street and filters it through a recommended 18 inches of sand after flowing through a pretreatment chamber, designed like a water quality inlet. The outfall pipe is recommended to be 6 inches in diameter or less. A grate over the outfall opening is wrapped in filter fabric.

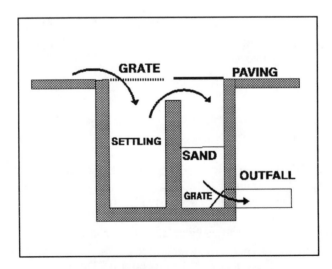

Figure 13-15 Linear Storm Water Filter
(California Stormwater Task Force, 1993)

Peat-Sand Filters - Peat-sand filters are an engineered soil/peat layering used for the filtration of the first flush of runoff. They contain hybrid filtration capability consisting of a nutrient removing cover grass mowed surface, peat layer and sand (or peat/sand) layer underlain by underdrains set in gravel layers.

Figure 13-16 shows a typical peat layering. These layers are placed in filtration beds, similar to the Austin First Flush Basin. Thus there is great flexibility in size and location. Peat-sand Filters have been tested in the Coordinated Anacostia Retrofit Program (CARP) and found to be effective in pollution reduction, especially TP, BOD and bacteria (Galli, 1990, 1992). Peat is an excellent material for promoting microbial growth and assimilation of nutrients. It also has high water retention capabilities and low bulk density. Peat-sand filters were originally used for wastewater treatment, and much of the design information currently comes from those sources (Farnham and Brown, 1972, Farnam and Noonan, 1988, Brown and Farnham, 1976). Sand filtration itself is coming into some prominence for wastewater treatment for small sites.

Pollutant removal for peat-sand filters is presented in Table 13-17. Nutrient removal levels actually increase with increasing pollution concentrations. Negative nutrient removal rates can be expected during startup of the peat-sand filter. Special design modifications including slower infiltration rates and an extra peat layer on the bottom have been proposed to enhance nutrient removal (Galli, 1992).

Peat-sand filters should have a pretreatment pool or detention facility similar to a sediment forebay or extended detention pond to remove the larger sediment particles. A flow spreader should be provided to ensure even flow movement over the bed. The filter is normally an off-line system with a flow splitter to divert larger flows than the capture volume around the basin. Like other filtration systems peat-sand filters are restricted to stabilized watersheds.

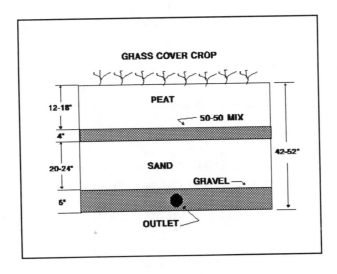

Figure 13-16 Peat-Sand Filter (Galli, 1991)

Peat is classified into three types: fabric, hematic and sapric. Fabric is the most fibrous and light. It has a high hydraulic conductivity (up to 140 cm./hr.) and is brown or yellowish in color. Sapric peat is highly decomposed, non-fibrous, more dense and with a low hydraulic conductivity (down to 0.025 cm./hr.). It is dark grey to black in color. Hematic is intermediate in all respects.

Table 13-17 Average Annual Pollutant Removal Capability Of Sand And Peat-sand Filters

	Pollutant Removal Estimates	
Pollutant	Sand Filters[1]	Peat-sand Filters[2]
TSS	60-80%	80-100%
TP	20-40%	60-80%
TN	20-40%	40-60%
BOD	60-80%	80-100%
Metals	40-60%	40-60%
Bacteria	60-80%	80-100%

[1] 0.5 inch storage per watershed acre with pretreatment for settling.
[2] 0.5 inch storage and exfiltration per impervious acre

Source: Galli, 1992

The peat should be fabric or hematic (not sapric) with less than 30 percent minerals by weight, and should be shredded to a uniform consistency before use (Tomaseck, et al., 1987). It is recommended that a soil scientist test peat supplies prior to bulk purchase.

Peat is placed in 2 to 4 inch lifts, compacted lightly and raked to ensure good contact with the next lift. It should not be allowed to dry out fully during storage or construction. Granular fine ground 200-mesh "Ag-lime" calcitic limestone can be mixed into the top 4 to 6 inches to enhance P removal.

The sand layer should be clean fine to medium grain (0.1 to 0.5 mm) low phosphate. It is placed below the peat layer to draw the water through the peat layer. Between the peat and sand layers there is a 4 inch 50-50 peat-sand mix to ensure a uniform flow and transition from one layer to the next. The bottom layer is a 6 inch thick layer of very fine to medium gravel (2 to 20 mm). Perforated PVC drain pipes are supplied at 10 foot intervals similar to the Austin basin.

A test site in Montgomery County, MD was designed to allow for up to two feet (2/3) of the capture volume storage on top of the peat bed. A 1.0 inch per hour infiltration rate was proposed to avoid saturation of the peat bed and long surface ponding. The drawdown time was then 24 hours.

Peat basins are sized using one of three proposed rules (Galli, 1991). The first is to capture the first half-inch from the watershed drainage area. The necessary surface area is then equal to the capture volume divided by the maximum depth allowed on the surface (two feet in the test site). The second rule is based on a wastewater-like consideration of maximum allowable unit area loading of phosphorus, and involves calculating the total runoff per year and the estimated phosphorus concentration. Hard numbers are difficult to derive by this method. The third rule relies only on not exceeding the N and P removal capabilities of the surface grasses. Values for a Minnesota study of reed, foxtail and fescue filter bed vegetation ranged from 210 to 250 lbs/acre for N and from 33 to 37 lbs/acre P annually (Elling, 1985).

Maintenance requires mowing during the growing season and removal of grass clippings. Sediment and trash removal is important for aesthetic purposes and to keep systems from clogging.

Questions remain concerning use of this system including clogging factors; restoration of the peat layer if clogged; leeway in peat layering and typing; long term stability, repeatability and consistency of pollutant removal; ponding depth; best combinations of ponds and beds; and best cover crops (Galli, 1991).

Catch Basin Inserts

Another, much newer, type of filtration device is an insert that fits inside a standard catch basin. It appears to be an ideal application for retrofitting such areas as parking lots, gas stations, vehicle maintenance areas, "dirty" neighborhoods or industrial areas, etc. It is easy to install (may take as little as 15 minutes), relatively inexpensive, requires no construction or modifications of existing catch basins, easy to maintain by property owners, and is targeted toward the major pollutants from these areas. Several companies produce such inserts, or they can be fabricated from common materials. Early testing in the pacific Northwest indicates that inserts are adequate for sediment, oil and grease control but further testing is needed for verification and measuring effectiveness in controlling other pollutants. Figure 13-17 shows a typical insert.

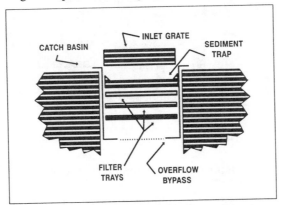

Figure 13-17 Typical Catch Basin Insert

The materials which make up the framework and the trays should be highly resistant to corrosion, easy to install manually and fit standard inlets. Normally the inserts are made of a stainless steel, aluminum or cast iron framework which sits on the lip of the inlet grate frame and hangs down into the catch basin inlet chamber. One or more trays of filtration media are placed into the framework. The flow enters the top of the filtration tray and filters through. Excess flow beyond the capacity of the media bed or due to media clogging is routed over the sides of the tray(s) and out through the bottom or side of the framework. The capacity of the overflow is designed to equal or exceed the capacity of the grate (NCTCOG, 1993, Enviro-Drain, 1994).

Pollutant removal rate information is limited to a few installations, some bench tests and visual inspections, but it appears to be quite high for oil and grease and metals (above 80%-90%) (McPherson, 1992, Enviro-Drain personal communication). The top screen or tray is usually a sediment trap. One or more trays of filtering media, sometimes of different types, are then placed either stacked or in a rack below the sediment trap and screen. Filtering media can be made of activated charcoal (for pesticides, fertilizer and metals removal), reconstituted wood fiber (primarily for oil and grease) or household fiberglass insulation. There is some question concerning the chemical integrity and longevity of the fiberglass in harsher environments. The media can be disposed in a manner similar to oil and grit chamber sediment though it may need to be tested once to see if it is a hazardous waste.

Installation costs range from about $400 to $2000 per catch basin depending on the complexity of the tray arrangement and the types of pollutants desired to be removed. Costs for replacement media are less than $25 per installation and must be replaced periodically depending on the climate and site conditions. Some media can be cleaned and reused. Maintenance requirements include inspecting the flow integrity of the system and replacement of the filtration media. Quarterly replacement is a good starting estimate though the installations should be checked after wet periods and periodically.

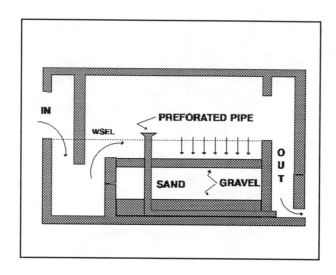

Figure 13-18 Filtration Vault (California Stormwater Task Force, 1993)

Other Filtration Methods - Filtration vaults of different types are also used along the east coast. They appear similar to water quality inlets but with sand beds in the final stage. Figure 13-18 illustrates schematically this approach. In these vaults the depth of the wet pool should be at least three feet and length to width ratio of 3:1 maintained. Pollutant removal for sand filters generally is given in Table 13-17.

Maintenance of sand filter systems is key to long term performance. Surface clogging by fine silts, organics and hydrocarbons diminish the filtration capacity within a year or so. Each year during a storm the system should be inspected to test the infiltration/treatment rate. Austin, among others, have found that it is only the first few inches of media that clog. Removal of this discolored depth and replacement is normally all that is necessary. Filters are especially prone to clogging in areas facing new development as sediment quickly spreads across and through the sand pores. Also, areas of high hydrocarbon generation (such as automotive servicing and fueling areas and high volume parking lots) will quickly clog the sand media. Oil and grit separators are recommended as pre-treatment devices in these areas. The discolored media is not considered a risk for land disposal, though samples should be taken if in a high hydrocarbon generating area (Schueler, 1994).

Techniques to improve the longevity of sand filters include: a surface screen to keep trash and detritus from the media; use of sod grown in sand; limiting use of filter fabrics and using blending layers instead; providing easy access for maintenance (such as a grated inlet); pretreatment for hydrocarbons and sediment; avoiding traffic on the media; and using only for totally impervious areas to reduce organic clogging. See Schueler (1994) for a detailed discussion.

Another filtration type device uses leaf compost as its primary filtration media (Techniques, 1994, Stewart, 1992). The system looks somewhat like the Austin filtration basin with a forebay for sediment removal. The runoff passes into a series of two treatment cells. Each cell has layering as: (1) one-foot depth of composted leaves, (2) non-woven filter fabric, (3) six-inches of small diameter rock with PVC underdrains, (4) two-inches of pea gravel, (5) impermeable polyliner. Runoff filters through the leaf material and is collected in the underdrains. The side-slopes on the compost layer are set at 2V:1H to channel the runoff to the gravel layer below. There is a two percent slope through the system requiring about three feet of head. A horizontally placed notched wooden baffle at the entrance and between the two cells spreads the water. The system was sized to provide 200 square feet of treatment area per cfs of inflow of about 0.10 inches of watershed runoff.

The selection of compost material is important to both physical performance and pollution removal. It must have the following characteristics: (1) mature, (2) humic, (3) low contaminant levels, (4) high permeability, and (5) locally obtainable. Removal rates of TSS were in the 95 percent range TP - 40%, Zinc - 88%, Hydrocarbons - 87% and Copper and Chromium 67% and 61% respectively. Negative removal rates were experienced for dissolved solids and nitrate probably reflecting leaching of the compost medium. Maintenance includes annual or biennial removal and replacement of the compost material and care of the surface to prevent clogging (disking or roughing the surface may be necessary if sediment supply rates are high).

Other filtration methods commonly used in Florida include pond underdrains, bank slope drains, confined filter beds and unconfined filter beds. An experimental rock and vegetation trickle bed is being tested in Orlando. A device called a Vertical Volume Recovery Structure (VVRS) has been used as well (Ghioto, Singhofen and Assoc., Inc., 1988). The VVRS, an experimental structure, consists of two concentric vertically placed corrugated metal pipes with sand contained in the intervening space and flow from the outside inward to the open area in the center. The outer pipe is 72 inches, the inner 21 inches. Perforations allow flow through the outer pipe and into the opening in the center. The sand is placed after both contacting surfaces are covered with filter fabric. The outlet connects to the inner pipe. The structure allows for easy cleaning and back-flushing. It is claimed the device is 1/4 to 1/2 the cost of other underdrain systems.

Infiltration Trenches

Infiltration trenches are one of a number of infiltration devices which accept runoff and exfiltrate it to the groundwater. It is a shallow excavated trench backfilled with stone to create an underground reservoir. The design should include pretreatment of runoff to remove coarse particles and oil and grease, in the form of a sump. The sump should be cleaned out periodically. Unlike other controls infiltration devices can actually reduce the total volume of flow by groundwater recharge. Figure 13-19 shows a typical design.

Trenches are used for on-site type controls and are limited by site characteristics including depth to water table and depth to bedrock. Experience has shown that if sediment is not kept from the trench surface, especially during construction, it will prematurely clog (Galli, 1992b). Maintenance of this BMP is critical to keep the surface permeable. High failure rates have been experienced in Maryland and Florida due to construction related clogging, filtration media compaction, and high groundwater and poor soils. Other primary concerns with infiltration trenches are metals accumulation and groundwater contamination. Groundwater issues should be investigated if that is the source of water supply, the soil is granular and intense non-residential development is present in the drainage area.

Effectiveness can be enhanced through thorough soils investigations prior to design, placing a sand layer at the bottom of the trench, avoiding sites where construction is evident and rototilling the trench bottom to preserve infiltration rates (Schueler, et al., 1992).

Pollutant Removal Efficiency - Pollutant removal is accomplished by a complex series of physical and chemical transformations including straining, sorption, trapping, precipitation and more. Table 13-18 provides estimated removal ranges. For well designed and maintained systems the upper end of the ranges are recommended.

Table 13-18 Average Annual Pollutant Removal Capability Of Infiltration Trenches

Pollutant	Infiltration Trench Design Type		
	0.5 Inch of Runoff Per Impervious Acre	1.0 Inch of Runoff Per Impervious Acre	2-Year Design Storm Treatment
TSS	60-80%	80-100%	80-100%
TP	40-60%	40-60%	60-80%
TN	40-60%	40-60%	60-80%
BOD	60-80%	60-80%	80-100%
Bacteria	60-80%	60-80%	80-100%
Metals	60-80%	60-80%	80-100%

Source: Schueler, 1987

Infiltration Trench Design - Infiltration trenches are very flexible and have been used in a number of ways including: parking lot swales, perimeter strips or median strips, wet wells for down spouts and neighborhood or industrial park swales (Schueler, 1987). Figure 13-19 shows a typical infiltration trench design. Oversized pipes and wet vaults have been used in underground applications to hold water for discharge into trenches through perforations. These are often preceded by an oil-grit chamber. Japanese engineers have retrofitted a whole urban area with infiltrating curb and gutter, wetwells for catch basins and infiltrating (actually exfiltrating) perforated pipe storm drains (Koyama and Fujita, 1989). This type of design is also common in southern Florida. Typical standard trench designs include three basic types (Schueler, 1987). Commercial systems also exist for smaller site applications (e.g. Infiltrator, undated).

- Full exfiltration trench: Runoff can only exit the trench by exfiltrating through the stone into the underlying soils. The storage volume is based on runoff volume of the 2-year storm.
- Water quality trench: The storage volume is based on first flush volume, either 1/2 inch of runoff per impervious acre or 1 inch of runoff on contributing area.
- Partial exfiltration system: The trench is not designed to rely completely on exfiltration to dispose of the captured runoff volume. A perforated pipe is used to drain part of the volume, being placed either beneath or near the top of the trench. The system is not as effective as a full exfiltration system.

Pretreatment minimizes maintenance requirements for trenches. Without such pretreatment life expectancy is less than five years (Schueler, et al., 1992, Maryland, 1987). Suspended sediment loads which will clog the trench can be reduced by requiring that the storm water runoff pass through a sediment forebay, oil and grit chamber or sumped catch basin prior to entering the trench. Grass filter strips have proven ineffective as pretreatment devices. Hydrocarbon loadings (oil and grease) that will clog the filter fabric and sand filter underlaying the trench can be reduced by the use of oil and grit chambers (when receiving large parking lot and roadway runoff).

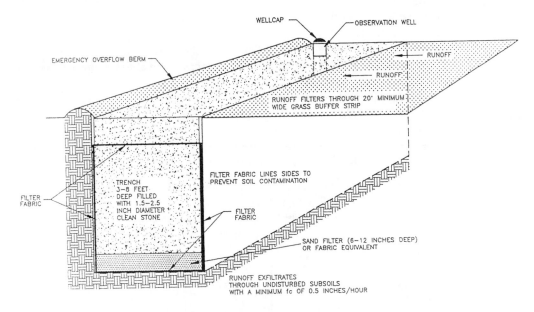

Figure 13-19 Infiltration Trench

Source: Stormwater Management Manual For The Puget Sound Basin, 1992.

There are various volume based design criteria similar to other structural BMPs. The smallest typical volume captured is 0.5 inch of runoff from adjoining impervious areas. There are also a number of design procedures used around the country varying from simple estimates of volumes and infiltration rates to field testing of infiltration and multi-dimensional consideration of infiltration and groundwater movement (Maryland, 1984, Chescheir, et al., 1990). Certainly the level of complexity of analysis should be linked to the level of detail of available site data and information. Often local or regional guidance is available.

Drawdown times should be limited to 48 or 72 hours. Typical times are more commonly 40 hours. The volume to be infiltrated is determined based on methods given previously. Typical infiltration rates for different types of soils are given in Table 13-19.

Knowing the capture volume, the soil porosity and the infiltration rate the trench (designed for sheet flow input through aggregate at the surface) can be sized using equations 13.15 - 13.18.

$$V_w = 3630 \, R_v i A = LWD_w \tag{13.15}$$
$$A_t = V_w/(T_d \, I) \tag{13.16}$$
$$V_t = LWH = LWD_w/n \tag{13.17}$$
$$n = V_w/V_t = D_w/H \tag{13.18}$$

Where: n =   aggregate porosity (can be estimated using effective water capacity) and normally taken as about 0.40 for gravel
       e =   void ratio, equals n/(1-n)
       i =   rainfall depth, in.
       R_v =   volume runoff coefficient
       A =   watershed area, acres
       L =   trench length, ft
       W =   average trench width, ft
       H =   trench depth, ft
       A_t =   minimum bottom surface area of the trench, ft^2
       V_w =   required water storage volume of infiltration basin (equal to capture volume), ft^3
       V_t =   total volume of trench, ft^3
       D_w =   maximum allowable water depth, ft
       T_d =   drawdown time, hrs
       I =   infiltration (exfiltration) rate, ft/hr

Table 13-19 Soil Infiltration Estimates

Soil Texture	Minimum Infiltration Rate (in/hr)	Effective Water Capacity (inches per inch)	SCS Hydrologic Soil Group Classification
Sand	8.27	0.35	A
Loamy Sand	2.41	0.31	A
Sandy Loam	1.02	0.25	B
Loam	0.52	0.19	B
Silt Loam	0.27	0.17	C
Sand Clay Loam	0.17	0.14	C
Clay Loam	0.09	0.14	D
Silty Clay Loam	0.06	0.11	D
Sandy Clay	0.05	0.09	D
Silty Clay	0.04	0.09	D
Clay	0.02	0.08	D

Source: Rawls, et al., 1982

Soils with infiltration rates less than 0.17 inches per hour are rarely suitable for infiltration. Soils with infiltration rates greater than 0.27 inches per hour are recommended. Site materials testing should be done. Infiltration basins with pipes as exfiltration devices would be designed slightly differently in that the volume of storage would include a term for the available pipe volume subtracted from the rock volume.

The depth calculated is the maximum ponding or trench depth. Other site constraints may limit the actual trench depth to something much less. Note that some earlier design methods used the void ratio in place of porosity. The void ratio is a ratio of the volume of voids to the volume of solids while porosity is the ratio of voids to the total volume of voids and solids combined. Also note that the method described does not take into account the infiltrated amount occurring during the storm event, but assumes the time of concentration is small compared to the time of filling of the trench and the infiltration time. If this is not the case more detailed methods (see Maryland, 1984) will need to be used.

Detailed maintenance and inspections criteria can be found in Maryland's Inspector's Manual (Maryland, 1985). Soil investigations details can be found in the Northern Virginia BMP Manual (1992).

Typical Required Specifications
- Used in small drainage areas less than 15 acres.
- Drain the 6-month, 24-hour design storm in 24 hours.
- A minimum of one soils log is required for every 50 feet of trench length, and no less than 2 soils logs for each proposed trench location. Borings should be taken to a depth of at least five feet below the trench depth.
- Each soils log should extend a minimum of 3 feet below the bottom of the trench, describe the SCS series of the soil, the textural class of the soil horizon(s) through the depth of the log, and note any evidence of high ground water level, such as mottling. In addition, the location of impermeable soil layers or dissimilar soil layers should be determined.
- Soil with minimum infiltration rates of 0.17 inches per hour or less are not suitable for infiltration trenches.
- Soils that have a 30 percent or greater clay content are not suitable for infiltration trenches.
- Soils suitable for infiltration systems are silt loam, loam, sandy loam, loamy sand, and sand.
- The use of infiltration systems on fill is not allowed due to the possibility of creating an unstable subgrade.
- A minimum of 2 to 4 feet difference is required between the bottom of the infiltration trench and the groundwater table and 3 feet to bedrock.
- Site slope must be less than 20 percent, trench must be horizontal.
- The proximity of building foundations should be at least 10 feet upgrade.
- A minimum of 100 feet from water supply wells should be maintained when the runoff is from industrial or commercial areas.
- The design infiltration rate should be equal to one-half the infiltration rate found from the soil textural analysis.
- Water quality infiltration trenches must be preceded by a pretreatment BMP.
- If the trench is preceded by a presettling basin, then the combination of both BMPs must be designed to drain the 6-month, 24-hour design storm within 40 hours.
- The aggregate material for the trench should consist of a clean aggregate with a maximum diameter of 3 inches and a minimum diameter of 1.5 inches.
- The aggregate should be graded such that there will be few aggregates smaller than the selected size. For design purposes, void space for these aggregates may be assumed to be in the range of 30 percent to 40 percent.
- The aggregate should be completely surrounded with an engineering filter fabric. If the trench has an aggregate surface, filter fabric should surround all of the aggregate fill material except for the top one foot.
- Runoff must infiltrate through at least 18 inches of soil which has a minimum cation exchange capacity of 5 milliequivalents per 100 grams of dry soil.

- An observation well should be installed for every 50 feet of trench length.
- The observation well should consist of perforated PVC pipe, 4 to 6 inches in diameter, located in the center of the structure, and be constructed flush with the ground elevation of the trench.
- The top of the observation well should be capped to discourage vandalism and tampering.
- Bypass larger flows.

Typical Recommended Specifications
- Can be installed under a swale to increase the storage of the infiltration system.
- Infiltration trenches work well for residential lots, commercial areas, parking lots, and open space areas.
- Infiltration systems should not be constructed until all construction areas draining to them are fully stabilized.
- An analysis should be made to determine any possible adverse effects of seepage zones when there are nearby building foundations, basements, roads, parking lots, or sloping sites.

Typical Operation And Maintenance Requirements
- A storm water management easement and maintenance covenant should be required for each facility. The maintenance covenant should require the owner of the infiltration trench to periodically clean the structure.
- The trench should be monitored after every large storm (>1 inch in 24 hours) for the first year after completion of construction and be monitored quarterly thereafter.
- Sediment buildup in the top foot of stone aggregate or the surface inlet should be monitored on the same schedule as the observation well.

Design Example - A 2 acre site, 100 % impervious is to be treated by an infiltration trench. The soils have an infiltration rate of 1.02 in/hr. The seasonal high groundwater table is at a depth of 9.6 feet. The trench is to be designed to treat the runoff from one inch of rainfall (the 6-hour, 6-month storm). The soil porosity is 0.40. A 40 hour drawdown is the design criteria.

$$R_v = 0.05 + 0.009 \ \%I = 0.05 + 0.009 * 100 = 0.95$$

The runoff volume to be treated is:

$$V_w = 3630 \ R_v iA = 3630 * 0.95 * 1.0 * 2 = 6,897 \ ft^2$$

Then:
$$V_t = V_w/n = 6,897/0.40 = 17,243 \ ft^3$$
$$D_w = T_d \ I/12 = 40 * 1.02/12 = 3.4 \ ft \ of \ water \ maximum$$
$$H = D_w/n = 3.4/0.40 = 8.50 \ ft \ (too \ deep \ for \ groundwater)$$

Select $H = 9.6 - 3 = 6.6 \ ft$

$$A_t = V_t/H = 17,243/6.6 = 2,613 \ ft^2$$

Select a width of 10 feet (W) to fit site and minimize groundwater mounding

$$L = A_t/W = 2,613/10 = 261 \ ft$$

check: $LWHn = V_w = (261)(10)(6.6)(0.40) = 6,890 \ ft^3 \ (ok)$

Note that this example makes no allowance for a safety factor based on unknowns in the site data and information. To complete the design a sensitivity analysis should be done by varying the unknown variables over an expected range and assessing the risk of variable choices. Some designers use a safety factor of from 2 to 10.

Infiltration basins are designed similar to trenches. They have exhibited high failure rates, often becoming de facto wetlands. Use of infiltration basins should be limited to sites with good to ideal conditions and with strict construction control and maintenance.

Porous Pavement

Porous pavement consists variously of: open-graded asphaltic aggregate pavement (gap graded mix or "popcorn" mix), pervious concrete pavement, or concrete or plastic grid paving blocks filled with soil and, normally, vegetated. Further information on pervious concrete can be found in Puget Sound (1992).

The function of the porous pavement is such that water infiltrates through the porous upper layer and into a storage reservoir of stone aggregate below. The runoff eventually either percolates into the ground or runs out of the stone reservoir through an underdrain collection system. Figure 13-20 shows a cross section of a common asphaltic aggregate configuration.

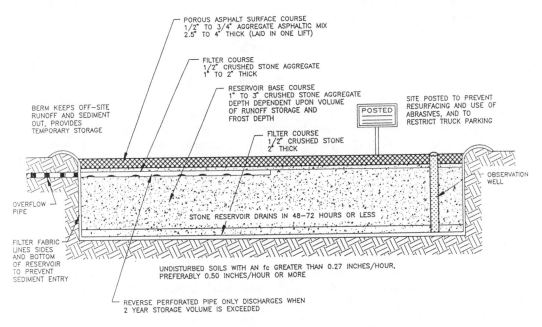

Figure 13-20 Design Schematic For Porous Pavement
Source: Schueler, 1987.

The use of porous pavement is commonly limited to parking lots and light traffic roads. Pretreatment of sediments, oils and greases is required to lengthen the service life of the structure. Porous pavement can be effective in flood peak reductions. One study showed peak reductions by as much as 83 percent through the use of porous pavement (Field, 1985). Freeze and thaw normally does not greatly effect well designed porous pavement.

Porous pavement is typically composed of four layers (Diniz, 1980, Northern Virginia, 1992).
• A minimally compacted sub-base consisting of undisturbed existing soil or, in the worst case,

an imported and prepared base material. Auxiliary drainage structures (French drains, pipe drains) may rest on top of this layer, or, to enhance pollution removal and retention times, placed vertically within the layer to provide some long-term retention below the pipe.

- Reservoir base course consisting of crushed aggregate of 1 to 3 inches in diameter with a thickness depending on both the storage volume required, the bearing strength of the sub-base or the required frost depth.
- Two inches of 1/2-inch crushed stone aggregate to stabilize the reservoir base course surface.
- A thin layer of fine gravel between the two stone layers.
- Porous asphalt bearing course whose thickness is based on bearing strength needed and pavement design requirements. Normal thickness is 2.5 to 4 inches with a void volume of 16 percent. See Thelen and Howe (1978) for detailed design criteria of this layer.

Normally a filter fabric is placed on top of the sub-base to keep fine particles from migrating into the aggregate reservoir. Overlap is recommended. To keep polluted runoff from entering into the ground water, sometimes impervious membranes are placed between the reservoir and the sub-base. Alternate drain lines are provided to convey the water to an outlet point. Often stone overflows are provided as a backup to clogging.

Porous pavement has been used for highway and airport paving since about 1947 and 1967 respectively. It has advantages over normal pavement in that it: reduces vehicle hydroplaning, retains water, enhances groundwater recharge, and can be cost effective if curbs and other drainage infrastructure can be avoided (Thelen, et al., 1972). They have been effective along roadways as filtration strips (Niemczynowicz, 1990).

While porous pavement can be highly effective in both flood control and pollution reduction, it comes at the cost of a high maintenance responsibility. Studies have shown up to a 76 percent failure rate in Maryland when porous pavement was not protected from off-site sediment, not constructed properly or protected during construction or was resurfaced with non-porous material (Maryland, 1991). Without proper maintenance and construction the pores in the upper level can become clogged either during construction (if unprotected) or within three to five years. Clogging of the first 1/2 inch is sufficient to prevent percolation. European and Japanese experience with such devices has been more positive indicating the controls may have more potential than early results indicate.

Care during construction must be taken to avoid spills and sediment export to the site. Construction vehicles may not track across the surface. Vehicle washing must not occur on the surface. Premature rolling of the hot mix may also collapse the pores (Diniz, 1980). Routine maintenance measures include sweeping with a vacuum assist sweeper followed by pressure washing several times a year.

Spills must be cleaned up immediately to restore function to within 95 percent. If not cleaned up for a longer period full restoration is impossible (Diniz, 1980). For minor clogging, drilling holes through the surface can be effective. For major system clogging nothing short of replacement of the surface is required. Alternately grated openings can be cut in the surface and the water drained directly to the aggregate reservoir and finger drains used, though this is not as effective. Avoid placing in high hydrocarbon generating areas. Also, do not drain pervious areas to porous pavement.

Pollutant Removal Efficiency - Porous pavement is designed to remove pollutants from atmospheric deposition onto the surface and runoff from areas immediately adjacent to the site. Porous pavement removes pollution by the same methods and to the same levels as infiltration trenches and other infiltration devices: adsorption, straining, trapping, and microbial decomposition in the sub-soil (Schueler, et al., 1992). Studies have shown they can be very effective in removing pollutants if the majority of the pollution is infiltrated to the groundwater (Schueler, 1987, Thelen and Howe, 1978, Urban and Gburek, 1980, Gburek and Urban, 1980, Day, et al., 1981, Field, et al., 1982, Thelen, et al., 1972, and Diniz, 1976, 1980).

Table 13-20 presents pollutant removal ranges for three alternate designs of asphaltic aggregate: half-inch capture from impervious areas, one-inch capture from impervious areas and 2-year storm exfiltration (erosion control volume too) (Schueler, 1987).

Pollutant removal can be enhanced by increasing the degree of exfiltration, maximizing the drainage time, and performing the routine maintenance mentioned above.

Porous Pavement Design - Typical specifications are provided below. More details can be found from Thelen, et al., (1972), Schueler (1987), Puget Sound (1993), Maryland (1984) or Diniz (1980). A design modelling approach which performs a step-wise routing/flow balance for porous pavement design is presented in Diniz (1976). Table 13-21 gives a typical gradation for the asphaltic aggregate though variations can be found (Smith, et al., 1974, Diniz, 1980, Maryland, 1984). Diniz (1980) or Thelen, et al., (1978) recommend the asphalt mix gradation given in Table 13-22.

Table 13-20 Average Annual Pollutant Removal Capability Of Porous Pavement

Pollutant	Porous Pavement Design Type		
	0.5 Inch of Runoff Per Impervious Acre	1.0 Inch of Runoff Per Impervious Acre	2-Year Design Storm Treatment
TSS	60-80%	80-100%	80-100%
TP	40-60%	40-60%	60-80%
TN	40-60%	40-60%	60-80%
BOD	60-80%	60-80%	80-100%
Bacteria	60-80%	60-80%	80-100%
Metals	40-60%	60-80%	80-100%

Source: Schueler, 1987, Day, et al., 1981

Design volumes include capture of the first half-inch to one-inch from impervious areas or, for volume control, capture of the 2-year storm. Other flows are diverted around or across the area.

The thickness of the reservoir layer depends on the greater of: storage volume required, frost penetration depth and bearing load thickness. The void space is calculated from the method of Smith, et al., 1974. Soils should have infiltration capabilities similar to infiltration trenches. For uniform sized gravel, a voids volume will be in the range of 38 to 46 percent. For a 40 percent typical void volume (porosity) a 12 inch thick reservoir will hold 4.8 inches of water (Thelen, et al., 1972). The frost penetration thickness can be found from standard tables available from asphalt manufacturers or associations and varies from about a foot in Tennessee to about 60 inches in Maine. The bearing thickness can be determined from Table 13-23 for a known traffic level and California Bearing Ratio (CBR).

Table 13-21 Open-graded Asphalt Concrete Formulation

| | | | | Probable Particle Data | |
Material	Screen	Weight (%)	Volume (%)	Width (mm)	Weight (g)
Aggregate	Through 1/2	2.8	2.2	10.7	1.667
	Through 3/8	59.6	46.3	8.0	0.697
	Through #4	17.0	13.3	4.0	0.087
Subtotal Coarse Aggregate		79.4	61.8		
	Through #8	2.8	2.2	2.0	0.0109
	Through #16	10.4	8.0	1.0	0.00136
	Through 200	1.9	1.5	0.06	0.000294
Asphalt		5.5	10.5		
Air		0	16.0		
Totals		100.0	100.0		

Source: Maryland, 1984.

Table 13-22 Aggregate Gradation Limits For Porous Asphalt Mixes

Sieve Size	Gradation (% by Weight) Puget Sound	Gradation (% by weight) Franklin Institute	Gradation (% by weight) Diniz
1/2"	100	100	100
3/8"	95 - 100	90 - 100	90 - 100
#4	30 - 50	35 - 50	35 - 50
#8	5 - 15	15 - 32	15 - 32
#16	-	0 - 15	2 - 15
#200	2 - 5	0 - 3	2 - 15

Source: Diniz, 1980, Thelen and Howe (Franklin Institute), 1978, Puget Sound, 1992

Puget Sound recommends using half the estimated infiltration rate to account for a safety factor for infiltration unknowns (Puget Sound, 1992). Other site limitations are also similar to trenches including depth to bedrock (2 feet), depth to seasonally high water table (2 to 4 feet), and drainage area (1/4 to 10 acres). An additional consideration, most applicable to the arid west, is to limit applications if wind erosion is expected to blow sediments from adjacent barren areas (Schueler, 1987). If the ground is sloped the reservoir size must be increased to account for the level water surface within a non-level reservoir. Alternately terraces can be constructed. Drainage times are limited to 24 hours in the Puget Sound (Puget Sound, 1992 for the 6 month 24-hour storm) and

72 hours in the Washington DC area (Schueler, 1987) to avoid anaerobic or septic conditions in the reservoir or freezing of the water.

Table 13-23 Minimum Thickness Of Porous Paving Top To Sub-base

Traffic Group	General Character	California Bearing Ratio			EAL[1]
		15 plus	10 - 14	6 - 9	
1	Light Traffic	5"	7"	9"	5 or less
2	Med. Light Traffic (Max. 1,000 VPD[2])	6"	8"	11"	6 - 20
3	Medium Traffic (Max. 3,000 VPD)	7"	9"	12"	21 - 75

[1]   EAL = equivalent axle load, equals 18,000 lbs average daily
[2]   VPD = vehicles per day

Note that CBR values less than 5 were improved using gravel mix to CBR = 6.
Source: Thelen and Howe, 1978

Typical Required Specifications
* Geotechnical investigation should be required prior to design including a minimum of one soil log (to a depth of 4 feet below the anticipated bottom of the stone reservoir) for each 5,000 square feet of infiltration surface area with minimum of three logs per BMP.
* Soil infiltration rate should be greater than 0.27 in./hr and clay content less than 30 percent for partial exfiltration system and greater than 0.52 in./hr for full exfiltration system.
* Porous pavement should be designed to exfiltrate a minimum of runoff volume equal to the first one-half inch of runoff from the contributing impervious area.
* Only feasible on sites with gentle slopes (less than 5%).
* Design infiltration rate should be equal to 1/2 of the infiltration rate determined from soil textural analysis.
* Minimum of 2 feet of clearance between stone reservoir level and bedrock.
* Minimum of 2 to 4 feet between stone reservoir level and seasonally high water table.
* Should not be constructed over fill soils.
* Vegetative strip or diversion berm required to protect pavement area from off-site runoff before and during construction.
* If porous pavement areas receive oily runoff from off-site areas, a pretreatment facility should be constructed to remove oil, grit, and sediments before entering the porous pavement.
* Dry subgrade should be covered with engineering filter fabric such as Mirafi #14N or equal on bottom and sides.
* Pavement section consisting of 4 layers as shown on Figure 13-20.
* Stone should be clean, washed, stone meeting municipal roadway standards.
* Reservoir base course should consist of 1 to 3 inches crushed stone aggregate compacted lightly at the depth required to achieve design storage.
* Filter courses to be 1/2 in. crushed stone aggregate at a 1 to 2 inches depth.

- Surface course should be laid in one lift at the design depth with compaction done while the surface is cool enough to resist a 10-ton roller. Only 1 or 2 passes are required.
- After final rolling, no vehicular traffic should be permitted on pavement until cooling and hardening, a minimum of 1 day.
- Stone reservoir should be designed to completely drain within a maximum of 2 to 3 days after design storm event.
- Stone reservoir should be designed for a minimum residence time of 12 hours.
- The porous pavement site should be posted with signs indicating the nature of the surface and warning against resurfacing, using abrasive, and parking heavy equipment.
- An observation well should be installed on the downslope end of the porous pavement area to monitor runoff clearance rates.
- The observation well should consist of perforated PVC pipe, 4 to 6 inches in diameter, constructed flush with the ground, with the top of the well capped.

Typical Recommended Specifications
- Limited in application to parking lots, service roads, emergency and utility access lanes, and other low traffic areas.
- Limited to sites between 1/4 acre and 10 acres.
- Should not be constructed near groundwater drinking supplies.
- Heavy equipment should be prevented from compacting the underlying soils before and during construction.
- Other porous pavement types such as concrete lattice blocks, perforated concrete grid slabs, reinforced plastic pavers, etc., placed over a porous subgrade and filled with soil may be used in lieu of the asphalt surface course.
- Signs should be posted identifying the area and specifying maintenance requirements.

Operation And Maintenance Requirements
- A storm water management easement and maintenance covenant should be required for each facility. The maintenance covenant should require the owner of the porous pavement to periodically clean the structure.
- Sediment should be kept off of the pavement before, during, and after construction to prevent premature clogging.
- Surface of the porous pavement should be vacuum swept at least 3 times a year, followed by high pressure jet hosing, to keep the pores free from clogging.
- Sand or ash should never be applied to porous pavement.
- Spot clogging of the porous pavement layer can be relieved by drilling 1/4 inch holes through the porous asphalt layer every square foot.
- The observation well should be monitored several times during the first few months after construction and be monitored quarterly thereafter. Water depth in the well should be measured at 0-, 24-, and 48-hour intervals after a storm to determine the clearance rate.

Modular Paving Blocks - Modular block paving originated from the use of slotted, flexible steel railways in post-war Europe as a means of temporarily repairing damaged airfields. Later they were used for temporary parking lots. Eventually concrete and then plastic modular blocks evolved (Day, et al., 1981).

Paving blocks come in three types: castellated, poured in place, and lattice. Castellated (such as Monoslabs and Checker Blocks) are roughly cast blocks with "castles" cast into the surface for roughness. Lattice blocks (such as Turfblocks and Grasstone) are normally pre-cast flat concrete slabs with different opening patterns in them. Poured in place types (such as Grasscrete) are poured over plastic forms with reinforcing (Field, et al., 1982). Plastic grids or cells are placed over a porous subgrade and then filled with soil. They have some reported advantages over the concrete modular blocks including an 82 percent surface permeability (Presto Products, 1990).

Paving blocks have the advantage in that they tend to seal less easily than the asphaltic paving. Also, replacing individual blocks is easier than patching porous pavement. However, they tend to be more expensive and less able to resist loadings without misalignment. Consequently they are often used in overflow parking areas, driveways, sidewalks, trails or shoulder areas. Horizontal infiltration cutoff walls keep flow from moving just under the block layer downslope and exiting without infiltrating (Urbonas and Stahre, 1993). Pollution removal rates are generally comparable to those of porous pavement being dependent on the infiltrated amount and the type and depth of soil materials (Day, et al., 1981).

The design volumes, site limitations and soil layers are the same as those for conventional porous pavement with the exception that a Terzaggi filtering layer is placed under the modular block to serve as a bedding and filter layer. Other design criteria are given below.

Typical Design Specifications For Modular Blocks
- Limit impervious tributary area to less than twice the modular block area.
- Six inch thick cutoff walls should be spaced at a distance of 0.54/S (where S is the slope in ft/ft).
- Size downstream structures as if the surface were 30 percent impermeable if infiltration is possible and 65 percent impermeable if underdrains are necessary due to the poor infiltration qualities of the sub-base (UDFCD, 1992).
- Use perforated pipe underdrain in each cell for poorly draining soils.
- Fill modular block with coarse sandy soil or sandy sod.
- Filter course layer should be a minimum of 4 inches of clean, well graded stone 1/8 to 3/4 inches in diameter.
- The stone reservoir should be a minimum of 12 inches thick of 1.5 to 3 inch stone.
- Install a filter fabric on top of the sub-base.

Grassed Swales

Grassed swales are one of a number of landscaping BMPs which primarily use biofiltration and limited infiltration to remove pollutants. The second major type is a filter strip discussed subsequently. A number of other related practices can be used including terracing, clustering, ground cover, runoff diversion and redirection, disconnecting of impervious areas, etc. (Bellevue, 1991). In some flat areas swales are designed essentially as retention basins with sluggish flow (Yousef, et al., 1985, Wanielista, et al., 1981).

A grassed swale looks similar to a ditch but is designed to be wider and to maintain flow below the height of the vegetation up to a certain design flow. They must be designed and constructed carefully to preserve even distribution of flow across the channel, avoid channelization and erosion, and avoid high velocities which will limit water-vegetation contact time and, perhaps, erode the channel and push the vegetation over. There have been problems reported with swales in Florida including standing water and erosion problems. Figure 13-21 is a schematic of a grassed swale.

Swales can be used in place of curb and gutter in low density residential neighborhoods and for drainage on back property lines or through industrial parks with pretreatment if high levels of oil and grease exist. However, the design of the true swale BMP is not an especially efficient channel design, taking up more room and allowing for higher vegetation growth. Use of an efficient hydraulic design as a swale BMP will lower the pollutant removal capability, though data on effects are not available.

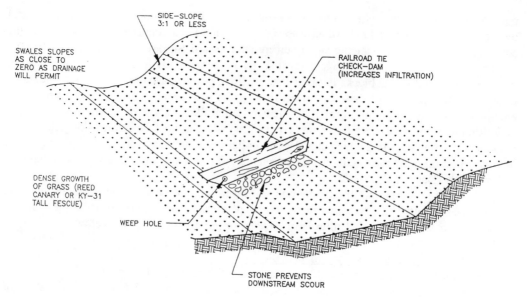

Figure 13-21 Schematic Of A Grass Swale

Source: Controlling Urban Runoff, 1987

Pollutant Removal Efficiency - Pollutant removal efficiency for grassed swales varies widely (Schueler, et al., 1992). Some researchers show moderate to high removal rates while others show little pollutant removal. The reasons for this great variation reflect major differences in design conditions.

Swales appear to be quite effective in the removal of metals and suspended solids, with less long-term effectiveness for nutrients. Runoff from highways indicates a high concentration of metals within the first fifty to hundred feet from the inlet to the swale and within the top 5 cm of soil (Wigington, et al., 1986, Yousef, et al., 1985). A 200 foot swale in Washington State removed suspended solids and lead by 80 percent (Wang, et al., 1982).

The improvement of pollutant removal with check dams or swale blocks included in the design is not known but some improvement can be surmised since they tend to hold the water longer in a particular area, bleeding down through a small hole on the lower side of the check dam (Yousef, et al., 1985). This allows the sedimentation and infiltration mechanisms to work longer and more effectively. Other factors to improve removal include appropriate length, small slope or a series of terraces, more pervious soil, and any other practice which lengthens the contact time with the vegetation.

The effectiveness of swales can also be enhanced by integrating them into the general design of the area through limiting the amount of directly connected impervious area and through educating the public on low fertilizer maintenance procedures and to catch grass clippings.

Grassed Swale Design - Low velocity and shallow depth are the key design criteria. Thus achievement of a low slope and insurance that the flow spreads evenly across the grassed channel are of great importance.

Special care must be taken to establish a healthy stand of vegetation tolerant to frequent inundation. See discussions under the wetlands BMP for more information. Irrigation may be a problem in the dryer climates. Since the system must only work during the wet season it may be acceptable to begin to irrigate in time to establish a grass stand just prior to the onset of the

Table 13-24 Estimates Of Average Annual Pollutant Removal Capability Of Grassed Swales

Pollutant	Low Gradient Swales with Check Dams
TSS	60-80%
TP	20-40%
TN	20-40%
BOD	20-40%
Metals	60-80%

Source: Schueler, 1987, California Stormwater Task Force, 1993, Schueler, et al., 1992

wet season (California Stormwater Task Force, 1993). Local information is available from agricultural extension services on types of vegetation. Information on eastern vegetation can be found in Schueler (1987). Puget Sound recommends grass heights of 6 inches and flow depth for design discharges of 5 inches (1992). Hydraulic design of grassed swales can be found in the Open Channel, Chapter 10, or any of a number of state sediment control manuals.

Flows larger than the swale design flow need to be checked and accounted for. It may be possible to design a compound channel with the bottom designed as a swale and cutout sides designed to handle the higher flows. It may be extremely difficult to keep the flow from concentrating, channelizing and meandering. Bottom hardpoints or grade control, flow spreaders and effective maintenance of the channel and its vegetation are required. Puget Sound recommends grade control every 50 feet and a 200 foot minimal length channel. Designing meandering channels may be necessary to achieve the 200 foot criteria.

Actual, pollutant removal based, design criteria for the design of a swale is highly dependent on empirical evidence and is more applicable regionally than for all parts of the country with differing hydrology, soils and vegetation types. Horner (1988) has applied the method of vegetative channel design presented in the Open Channel, Chapter 10. This method has been used on the West Coast (Puget Sound, 1992, California Stormwater Task force, 1993). The design criteria for the Puget Sound area is the 6-month, 24-hour storm, a length of 200 feet and a maximum velocity of 1.5 ft/s. Design steps then are:

- Determine the runoff rate for the design storm using suitable methods.
- Layout the channel to determine preliminary slope requirements. Make sure it is 200 feet long minimum or wide enough to achieve an equivalent residence time.
- Choose vegetation based on soils, climate, wetness, etc.
- Design the channel using the vegetative method given in the Open Channel, Chapter 10. The stability check in that method should be for the largest design flow required to be carried within the banks. Ensure the velocity for the pollution removal design storm is less than the maximum criteria (1.5 ft/s for Puget Sound).

Alternate design methods are given by Maryland (1984) and Wanielista and Yousef (1993).

Typical Required Specifications
- Grassed swales should only convey standing or flowing water following a storm.
- As a BMP, grass swales should be designed for the 6-month, 24-hour design storm.
- Limited to peak discharges generally less than 5 to 10 cfs.
- Limited to runoff velocities less than 1.5 to 2.5 ft/s.
- Maximum design flow depth to be 1 foot.
- Swale slopes should be graded as close to zero as drainage will permit.
- Swale slope should not exceed 2%.
- Swale cross-section should have side slopes of 3:1 (h:v) or flatter.
- Underlying soils should have a high permeability (fc > 0.5 inches per hour).
- Swale area should be tilled before grass cover is established.
- Dense cover of a water tolerant, erosion resistant grass should be established over swale area.

Typical Recommended Specifications
- As a BMP, grassed swales to be limited to residential or institutional areas where percentage of impervious area is relatively small.
- Seasonally high water table to be greater than 2 feet below the bottom of the swale.
- Check dams can be installed in swales to promote additional infiltration. Recommended method is to sink a railroad tie halfway into the swale. Riprap stone should be placed on the downstream side to prevent erosion.
- Maximum ponding time behind check dam to be less than 24 hours.

Typical Operation And Maintenance Procedures
- A storm water management easement and maintenance covenant should be required for each facility. The maintenance covenant should require the owner of the grassed swale to periodically clean the structure.
- Grass swales should be maintained to keep grass cover dense and vigorous.
- Maintenance should include periodic mowing, occasional spot reseeding, and weed control.
- Swale grasses should never be mowed close to the ground.
- Fertilization of grass swales should be done when needed to maintain the health of the grass, with care not to over-apply the fertilizer.

Filter Strips And Flow Spreaders

Filter strips are vegetated sections of land located between sources of on-point pollution and receiving waterbodies, designed to accept runoff as overland sheet flow from an upstream development or small site. Filter strips may adopt any vegetated form from grassy meadows to forests. The density of the vegetation, the flatness of the surface, the permeability of the soil and the evenness of the flow spread all help determine the pollution removal effectiveness (Schueler, et al., 1992, Northern Virginia, 1992). Buffers, when engineered, can function as filter strips.

Filter strips are commonly used along stream banks (riparian corridor buffers), downstream from agricultural runoff, around area inlets, as pretreatment for other BMPs, and as areas for sheet flow from paved areas to run through. They are often associated with the disconnection of directly connected impervious areas. Landscaping can often take this type of BMP into account almost invisibly through the appropriate use of multi-function flow spreaders and proper structure siting, drainage and grading. Filter strips and buffers can have multiple uses. However not all uses are compatible with pollution removal. By their very invisibility they can fall victim to unintended use by property owners, contractors and even local governments (Heraty, 1993, Cooke, 1991). A typical filter strip is depicted in Figure 13-22.

Filter strips must accept storm water runoff as overland sheet flow in order to effectively filter suspended materials out of the overland flow. In order to function properly, flow entering a filter strip must be spread relatively uniformly over the width of the strip. Filter strip applications should be limited to drainage areas of 5 acres. Even filter strips of 10 to 20 feet in width can have a great effect in removing the coarser particles. In arid climates they may need to be irrigated to maintain a healthy appearance.

Pollutant Removal Efficiency - Pollution removal mechanisms are the same as those for grassy swales. Filter strips can reliably remove sediment and other settleable solids but are less reliable for soluble nutrients and metals. Pollutant removal is dependent on length, slope, soil permeability, flow volume and velocity. Two factors work in opposition to each other when compared to swales. The filter strip is normally not as long as a grassy swale limiting the time of contact between the vegetation and the runoff. But, if designed and maintained correctly, the flow across the buffer strip is in the form of sheet flow moving very slowly across the strip. Table 13-25 gives estimates of pollution removal efficiencies for filter strips and buffers.

Various studies conducted primarily for agricultural sites have shown high removal rates for sediments, pesticides, bacteria and nutrients (Readling, 1992, NCDEHNR, 1991). However, variability is very great from site to site, and few of the sites experienced urban storm water runoff. The values given in Table 13-25 are therefore conservative expressing the thought that many strips are not well constructed nor maintained. Somewhat higher removal rates can be expected for more ideal conditions.

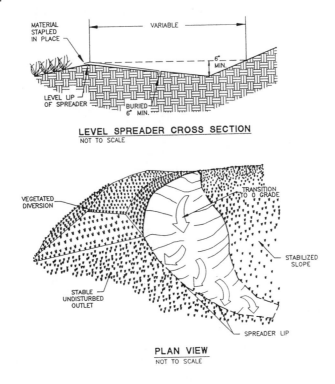

LEVEL SPREADER CROSS SECTION
NOT TO SCALE

PLAN VIEW
NOT TO SCALE

Figure 13-22 Flow Spreader

Source: North Carolina Erosion And Sediment Control Planning And Design Manual, 1988

Table 13-25 Average Annual Pollutant Removal Capability Of Filter Strips

	Filter Strip Design Type	
Pollutant	20 Foot Wide Turf	100 Foot Wide Forested with Level Spreader
TSS	20-40%	80-100%
TP	0-20%	40-60%
TN	0-20%	40-60%
BOD	0-20%	60-80%
Metals	0-20%	80-100%

Source: Schueler, 1987

Because the removal rate is directly related to the contact time it is imperative that the flow be kept from concentrating and short circuiting through the filter strip (Dillaha, et al., 1989). If this happens most of the width of the strip becomes ineffective, simply lengthening the small feeder stream in its course to the receiving waters. Most strips which have failed to remove target pollutants have failed due to poor design, high slopes, narrow widths, erosion, and short circuiting (Schueler, et al., 1992).

There may be site specific factors which greatly affect the ability of the strip to remove nitrogen and phosphorus. Gilliam and Skaggs (1987) found denitrification to greatly reduce nitrate-nitrogen all within the first few yards of the riparian wetland buffer. High water table may influence the ability for vegetation to uptake soluble nutrients as does the time of year and phase of the vegetation growth (Schueler, 1987).

Often the first few yards of the strip accumulates sediment in the shape of fans or ridges if high loads exist (Gilliam and Skaggs, 1987). Eventually the flow is diverted to another crossing area, concentrated or diverted around the strip itself. Also the turf builds up above the pavement section. At this point the turf must be stripped, sediment removed and lowered to reestablish the pollution removal capability.

Design of Filter Strips - As for other types of BMPs there are several design approaches for filter strips. Often a simple width rule-of-thumb is used matched to a specific design criteria and empirical study results, and politics. Filter strip minimums have been established across the country from as little as 25 feet to as much as 300 feet (Palfrey and Bradley, 1982). The use of minimum buffer widths make it simple to apply the criteria uniformly, but may not be as effective as a variable buffer width based on important variables and factors. Minimum buffer widths along streams are given by North Carolina and shown in Table 13-26 (NCDEHNR, 1991).

The Denver Urban Drainage and Flood Control District recommends sizing the buffer such that the runoff from the 2-year storm will load the strip at a rate of 0.05 cfs/linear foot, for a flow of less than one inch of depth for grass that is at least two inches high. The width is the greater of 8 feet or 0.2 times the flow path from the impervious area upstream from the strip (UDFCD, 1992). The Puget Sound manual recommends designing filter strips similar to swales using the vegetated channel design method given in the Open Channel, Chapter 10 (Puget Sound, 1993).

**Table 13-26 Recommended Minimum Widths Of Streamside
Buffer Zones In North Carolina**

Type of Waterbody	Percent Slope of Adjacent Lands				
	0-5	6-10	11-20	21-45	46+
	Streamside Buffer Zone Width				
Intermittent and Perennial	50	50	50	50	50
Perennial Trout Waters	50	66	75	100	125
Public Water Supplies	50	100	150	150	200

Source: NCDEHNR, 1991

Wong and McCuen provide a model for filter strip design, though rather large values can result (Wong and McCuen ,1982). Figure 13-23 gives a nomograph for calculation of buffer width given slope, roughness and target trap efficiency. Yu, et al., (1992) recommend 80 to 100 feet widths.

Swift (1986) gives the following equations for buffer width perpendicular to flow for various conditions. His equations are based on earlier work and studies in southern Appalachian sandy loam soils. His equations are designed to filter out particles greater than 0.05 mm.

- Forested buffer with brush barriers 10 feet wide:

$$W = 32 + 0.4 * \% \text{ Slope} \tag{13.19}$$

- Forested buffer without brush barrier:

$$W = 43 + 1.39 * \% \text{ Slope} \tag{13.20}$$

- Grassed buffer width:

$$W = 74 + 1.39 * \% \text{ Slope} + 6.4 * \% \text{ Slope}/10 \tag{13.21}$$

Where: W = width, ft
% Slope = percent slope of buffer area

The actual length perpendicular to the flow depends on the target pollutants to be removed. Phillips (1989) developed two models to predict buffer length based on sedimentation and removal of small sediments and dissolved solids respectively. The sediment model predicted much narrower strips than did the dissolved nutrient removal model. For the sediment model slope was the determining factor. For nutrient removal length of the flowpath across the strip was the factor as it reflected contact time.

The use of level spreaders is encouraged wherever there is some question on the ability of sheet flow to exist, for runoff from parking lots and for all larger areas. Level spreader design criteria can be found in state sediment handbooks (Virginia, 1992, North Carolina, 1991). The Virginia and North Carolina spreader consists of a diversion berm graded to disperse the flow from the end of the berm into a small Vee shaped ditch graded to zero slope at the lip. Other types of spreaders include small troughs with Vee-shaped "saw-teeth" cut into the downstream side, curbs with holes cut into them, gravel filled trenches graded to zero slope, and curb blocks with openings. A typical level spreader is depicted in Figure 13-24.

The use of buffer strips along roadways is encouraged in Maine (Maine, 1993). Turnouts along the road direct flow off the pavement and across a buffer. They recommend flows less than 1 cfs be released at any one point, and recommend a formula for calculating flow:

$$I_a * 0.000139 + P_a * 0.00069 < 1.0 \text{ cfs} \tag{13.22}$$

Where: I_a = impervious area in ft^2
       P_a = pervious area in ft^2

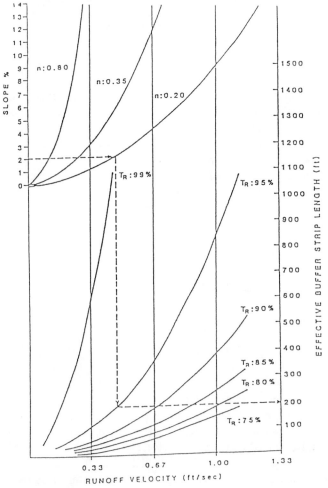

Figure 13-23 Effective Buffer Length Determination For Selected Trap Efficiency
Source: Wong and McCuen, 1982

Aesthetics are important in the design of buffer strips. However, there is some disagreement on the possibility for multi-objective use of buffer areas. Certainly use of the areas as habitat is seen as permissible, and some authors argue that an undisturbed forest area makes the best buffer due to a greater ability to retain nutrients (Maine, 1992, Schueler, 1987). Filter strips tend to attract urban dwelling "edge species" such as songbirds and squirrels (Northern Virginia, 1992). But some authors recommend against use as pedestrian walkways or recreational areas. It appears that site specific factors will dictate whether other uses are permissible. Human interaction with such areas tends to ensure the site is better maintained in terms of trash and erosion control, though there may be some compaction of the soil and wearing away of vegetation if not well designed for human traffic control. Greenways have been effectively used as buffer areas along streams. It may be advisable to plant different vegetation in specific zones depending on inundation, wetness and other factors (New Jersey, 1989, Karouna, 1992). The use of experimental vegetation such as vetiver grass may improve erosion control and pollutant removal (BSTID, 1993). Shade trees and the allowance for some debris along stream edges will enhance biodiversity.

Maintenance is minimal consisting of lawn care and mowing for grassed areas, litter removal, erosion control and inspections. Periodic turf or vegetation replacement may be necessary in areas of die-off or excess sediment build-up.

Schueler (1994) and local experience provide guidance to maintain the integrity of buffers during and after construction.

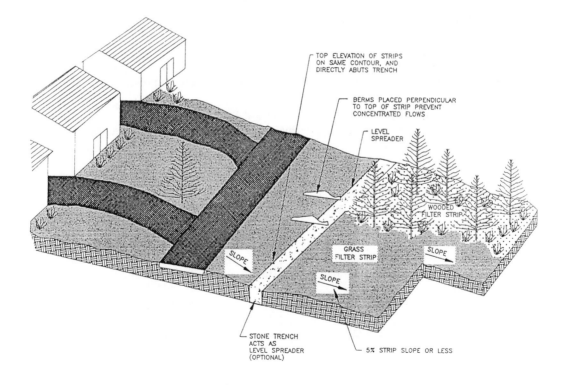

Figure 13-24 Schematic Of A Filter Strip

Source: Schueler, 1987

Planning Stage
- Require buffer limits to be present on all clearing and grading and erosion control plans.
- Record all buffer boundaries on official maps and plats.
- Clearly establish acceptable and unacceptable uses for the buffer by both ordinance and through preparing public education and awareness campaign literature.
- Establish clear vegetation targets and management rules for different lateral zones of the buffer.
- Design plantings to encourage proper use and discourage traffic in sensitive areas. Anticipate human traffic and provide walkways in combination with flow spreaders or with multiple openings to allow flow passage.
- Carefully plan multiple use opportunities and locations which are compatible with the buffers purpose.
- Provide incentives for owners to protect buffers through perpetual conservation easements rather than deed restrictions.

Construction Stage
- Do a pre-construction stake-out of buffers to establish the limits of disturbance.
- Set the limit of disturbance based on the drip line of any forested areas of the buffer to avoid root damage.
- Conduct pre-construction meetings and education to familiarize contractors with the limit of disturbance locations and buffer limits.
- Mark the limit of disturbance with a silt fence barrier, signs or other methods to exclude construction equipment.

Post-Development Stage
- Mark buffer boundaries with permanent signs or attractive barriers in spots.
- Educate property owners and homeowner associations on the purpose, limits and allowable uses of the buffer.
- Conduct periodic walks of the buffer to inspect the condition.
- Encourage the use of volunteer monitoring groups and enforcement reporting.
- Quickly reforest or replant bare or eroded areas.

Typical Required Specifications
- Flow spreaders and filter strips should be limited to drainage areas of 5 acres or less.
- Capacity of the spreader should be determined by 10-year peak flow storm.
- The filter strip length (perpendicular to flow) should be calculated based on the 6-month frequency, 24-hour duration storm.
- Grade of level spreader should be 0%.
- Released runoff to outlet onto undisturbed stabilized areas in sheet flow and not allowed to reconcentrate below the structure.
- Slope of filter strip from level spreader should not exceed 10-15 percent.
- All disturbed areas should be vegetated immediately after construction.
- Filter strip width to be a minimum of 20 feet.

Typical Recommended Specifications
- Top edge of filter strip should directly abut the contributing impervious area.
- Runoff water containing high sediment loads to be treated in a sediment trapping device before release in a flow spreader.

- The top edge of the filter strip should follow the same elevational contour line.
- Spreader lip to be protected with an erosion resistant material, such as fiberglass matting or a rigid non-erodible material for higher flows, to prevent erosion and allow vegetation to become established.
- Wooded filter strips are preferred to grassed strips.

Typical Operation And Maintenance Requirements
- A storm water management easement and maintenance covenant should be required for each facility. The maintenance covenant should require the owner of the filter strip/flow spreader to periodically clean the structure.
- Flow spreader should be inspected after every rainfall until vegetation is established, and needed repairs made promptly.
- After area is stabilized, inspections should be made quarterly.
- Vegetation should be kept in a healthy, vigorous condition.
- Filter strip and flow spreader should be maintained in a manner to achieve sheet flow.

Oil/Grit Separators (Water Quality Inlet)

A water quality inlet is a three-stage underground retention system designed to remove heavy particles and hydrocarbons from storm water runoff. This type of BMP has been used for many years in industrial applications, rather than in urban storm water applications, where a known and steady flow can be assured. In industrial applications great strides have been made in design of grit chambers and in coalescing plate type devices. When translated to storm water applications two basic problems are realized:
- an expectation of removal of pollutants other than grit and oil has been created; and
- storm water runoff can overwhelm devices designed for steady flow measured in gallons per minute, not widely varied flow measured in cubic feet per second.

When adopted for storm water flows a basic mismatch occurs. The conveyance portion of the system is designed for the same flow as the rest of the storm water system (for example, the 10-year flow) while the storage and treatment portion of the chamber is designed for very low flows. Typical design criteria of capture of 400 cubic feet per impervious acre is equivalent to capture of only the first tenth of an inch of runoff. Runoff residence times average less than 30 minutes (Schueler, 1994). Ongoing studies have shown that when the larger flows enter the chamber they resuspend sediments resulting in little actual capture and treatment. Average sediment depth accumulates to only about 2 inches, and varies from storm to storm.

The first problem can be solved by combining this type of application with others designed to remove other target pollution or modifying the structure itself to provide other types of removal mechanisms. The second problem is more difficult to handle since many of these devices are in-line type structures unable cheaply to shunt excess flows past the device and to an outlet. In all new design cases this type of device should be designed or retrofit as an off-line system.

Innovative modifications to these devices which improve pollution removal are in the experimental stages. Some are experimenting with sand filtration and even infiltration type inlets. Some locations have combined oil/grit separators with exfiltrating storm drain systems in high density areas with sandy soils. Internal geometry changes are being tried including flow regulators, tangential inlets and buffers. Chattanooga, TN has developed a hooded baffle arrangement for the outlet pipe which they feel both reduces cost and improves performance.

To be cost effective water quality inlets should be limited to capturing runoff from small high density sites such as maintenance shops, gas stations, and certain industrial areas where high concentrations of oils from small contributing areas are expected. These facilities have a fairly high initial cost and must be frequently cleaned. Cleanout costs can be as high as $1,000 to $2,000 per site annually, though some municipalities use partial cleanout techniques.

Pollution Removal Efficiency - True pollution removal for this type of structure only occurs when the structure is cleaned out. There is a high likelihood that the grit and oil from this type of structure is toxic and requires special disposal actions. Thus many municipalities and industries find themselves in an expensive "catch-22". The pollution removal efficiency has not been reported in the literature and must be inferred. Schueler, et al., (1992) and recent MWCOG study results report that sediment removal of coarse materials appears to be moderate but that resuspension occurs for every moderate storm. However, the types of pollutants found adsorbed to the sediments match those found in receiving water bodies where there are no such devices, indicating that the right target pollutants are being addressed. Different off-line type devices or other experimental modifications should go a long way in improving the pollutant removal capabilities of the system.

In industrial applications, specification sheets give typical oil removal rates (relying on coalescing of oil droplets) of 95 percent for flows of 2 to 3 cfs but infer initial concentrations of 2000 mg/l, much higher than the 2 to 10 mg/l commonly found in urban runoff. But for highly oily runoff, coalescing type structures have shown an ability to reduce concentrations of oil to the 10 mg/l range (Lettenmaier and Richey, 1985). Studies on sumped catch basins in Switzerland showed that only about 10 percent of the solids were removed by the basins (Conradin, 1989).

These devices have limited effectiveness as stand alone treatment of storm water runoff quality, unless hydrocarbons and grit are the only pollutant of concern. When used with infiltration in the bottom, as suggested in Schueler (1987), there is an expectation of clogging. Hydrocarbons in urban runoff can effectively clog the infiltration capacity of underlying soils because they tend to attach themselves to particles in the water column and settle to the bottom of the BMP. Three chamber oil and grit devices may remove from 40 to 60 percent of the hydrocarbons found in parking lot and street runoff if they are cleaned quarterly during wet seasons. Three chamber oil and grit devices may also remove a moderate portion (less than 40 percent) of the suspended sediment and associated adsorbed pollutants. When combined with the high cost of installation (from $6,000 to $10,000 per site) they are not, at present, considered very effective on a unit cost basis. However, there are many types of sites where they are necessary, being one of the only ways to remove their target pollutants. Pollution removal can be enhanced by:

- maximizing volume in first two chambers,
- use of coalescing material type chamber for oil,
- protecting orifice between first two chambers with a trash rack,
- extending inverted elbow at least three feet into the pool to improve oil removal, and
- performing regular maintenance.

Oil/Water Separator Design - There are several types of oil/water separators. The most simple is a minor spill control separator which consists of a simple sumped catch basin with a Tee outlet to remove water from below the surface (Puget Sound, 1992).

Most other water quality inlets are designed to remove sediment and hydrocarbon loadings before they are conveyed through a storm drain, or into an infiltration facility. This is done through settling of the heavier grit and trapping (settling upward) of floatables and oil. Figure 13-25 displays the basic layout of the three chambered type used in the Washington, DC area. A modified design which provides another chamber and a submerged orifice for improved oil removal similar to industrial applications has been suggested by Washington DC engineers (1992). The type of oil/grit separator shown in Figure 13-25 has the following features:

- Chamber one traps particulates and has two, six inch orifices (with trash rack) at mid water depth, through a partition between chambers one and two.

- Chamber two traps floating oil and has an inverted elbow pipe to regulate water levels in both chambers one and two.
- Chamber three receives discharge from chamber two and has an overflow, outfall pipe set with the crown below the horizontal invert of the inverted elbow pipe from chamber two.
- An access manhole should be provided into each chamber.

Inlets must be cleaned at least twice each year. The total wet storage should be, as a minimum, 400 cubic feet per impervious acre. These devices apply to areas less than two acres.

Another type of separator has been described for use in California (Figure 13-26) (California Stormwater Task Force, 1993). Sizing of this conventional separator has been approximated based on industrial sizing criteria. Large droplets of oil rise faster just like large particles fall faster, according to Stokes law. Just as large particles need time to settle, large oil droplets need time to float. The equation for rise rate is:

$$V_p = 1.79 \ (d_p - 1) \ d^2 \ x \ 10^{-8} \ /n \qquad (13.23)$$

Where: V_p = rise rate, ft/s
 d_p = density of oil (between 0.85 and 0.95), gm/cc
 d = diameter of minimum droplet to be removed, microns
 n = absolute viscosity of the water, poises

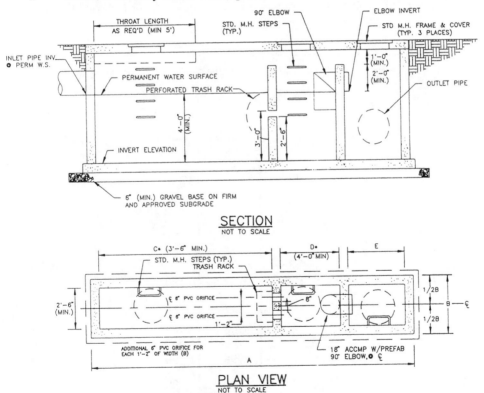

Figure 13-25 Oil/Grit Separator

Source: Maryland Department Of Environment Sediment And Stormwater Administration (1991)

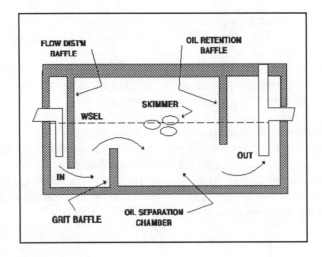

Figure 13-26 Conventional Oil Grit Separator (California Stormwater Task Force, 1993)

The absolute viscosity of water can be reasonably linearly interpolated between values of 0.0179 poises at 32 degrees F and 0.010 poises at 68 degrees F. The distribution of droplet sizes can be estimated from Table 13-27 assuming a water temperature for removal in the winter or wet season.

Table 13-27 Size And Volume Distribution Of Oil Droplets

Drop Size (microns)	Percent by Volume Smaller	Percent by Size Smaller
20	1	4
40	1	6
60	1	12
80	11	47
100	26	67
120	32	80
140	57	88
160	78	95
180	98	98
200	99	99

Source: Branion in California Stormwater Task Force, 1993

Sizing equations based on the American Petroleum Institute procedure (API, 1990) and modified by California Stormwater Task Force (1993) are:

$$D = (Q/RV)^{0.5} \tag{13.24}$$
$$L = VD/V_p \tag{13.25}$$
$$W = Q/(VD) \tag{13.26}$$

Where: D = depth (3 to 8 feet), ft
           Q = design flow, cfs
           R = length to width ratio (e.g. 2:1 equals 2)
           V = allowable horizontal velocity equal to 15 times V_p but less than 0.05 ft/s
           L = length, ft
           W = width (2 to 3 times the depth but less than 20 feet), ft
           V_p = rise velocity of oil droplet - 0.00055 ft/s recommended by Puget Sound, ft/s

If the depth exceeds eight feet, design parallel units and split the flow. The baffle height to depth ratio for top baffles is 0.85 and 0.15 for bottom baffles. The flow distribution baffle is located at 0.1 L from the entrance. Add freeboard. A design flow less than the 6 month discharge should capture sufficient annual volume to achieve a high efficiency. Install a bypass for flows in excess of design. The depth to width ratio is normally 0.3 to 0.5 with a width of 6 to 16 feet (Puget Sound, 1992).

The design procedure then is:

• Select a removal efficiency and then a target diameter from Table 13-27 based either on estimated influent concentration and effluent target ratio (e.g. 10 mg/l out divided by 60 mg/l in = 83 percent efficiency) or on design criteria target (e.g. 80 percent).
• Estimate oil viscosity and water viscosity and calculate rise rate (V_p).
• Estimate a design flow rate using impervious area and design rainfall rate.
• Calculate V and D to place each within allowable limits. Consider parallel units if depth or velocity limits are exceeded.
• Calculate L and W and set the baffle heights and locations.
• Complete the design.

There is some question whether a conventional separator can remove particles less than 150 microns efficiently without coalescing the smaller droplets into larger ones, though there are little data on oil droplet size in storm water. An alternate design criteria has been developed for sizing water quality inlets using the design in Figure 13-27. The coalescing process whereby small droplets are merged with others forming large droplets on the surfaces of parallel plates made of fiberglass or polypropylene has been used and is being modified (using among other things, corrugated filters) for more efficient use in storm drain inlets. The use of coalescing filter media is more commonplace in the west for sites where the oil droplets can be expected to be less than 150 microns (California Stormwater Task Force, 1993).

Packaged coalescing units can be provided for flows up to several cfs. Larger units must be sized by the designer. A number of manufacturers offer coalescing filter materials (or coalescing tubes) so individual specifications and efficiencies must be assessed. The basic equation for coalescing filter design is:

$$A_f = Q/(E_f V_p \text{ Cosine H}) \tag{13.27}$$

Where: A_f = total surface area of filter, ft^2
           H = plate angle measured from horizontal, deg

V_p = rise velocity of oil droplet, ft/s
E_f = efficiency of the specific filter (0.35 to 0.95)

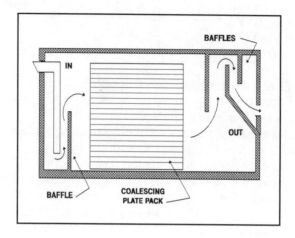

Figure 13-27 Typical Coalescing Plate Separator (City of Bellevue, Washington, 1988)

Spacing between plates is normally 0.75 to 1.5 inch. Add 12 inches below the plates for sediment accumulation and 6 to 12 inches above for oil accumulation. Add one foot of freeboard. Design the forebay for flow distribution and floatable collection. Placement of plates on angles of 45 to 60 degrees allow the solids to slide off the plates to the bottom. Even so, periodic maintenance will be necessary and will be based on local experience, though monthly or post-storm checks are recommended initially. A trash rack with openings smaller than the plates will keep debris out of the plate area. Following are the design steps.
- Collect basic information on available filter types, angles and efficiencies.
- Calculate V_p and Q as in the procedure for the conventional separator.
- Calculate A_f.
- Select spacing (S) and reasonable width (W) and length (L).
- Calculate number of plates as N = A_f/LW.
- Calculate total volume from spacing, number of plates and plate thickness.
- Complete design and baffle placement.

Typical Standard Specifications For Oil/Grit Separators

<u>Typical Required Specifications</u>
- Separators should be sized for the 6-month, 24-hour design storm. Larger storms should not be allowed to enter the separator.
- Separator should be structurally sound and designed for acceptable municipal traffic loadings where subject to traffic loadings.
- Separator should be designed to be water tight.
- Volume of separator should be at least 400 cubic feet per acre tributary to the facility (first two chambers).
- Forebay or first chamber should be designed to collect floatables and larger settleable solids. Its surface area should not be less than 20 square feet per 10,000 square feet of drainage area.
- Separator pool should be at least 4 feet deep.

- Weirs, openings, and pipes should be sized to pass, as a minimum, the storm drain system design storm.
- Manholes should be provided to each chamber to provide access for cleaning.

Recommended Specifications

- Oil absorbent pads, oil skimmers, or other approved methods for removing accumulated oil should be provided.
- Separator to be located close to the source before pollutants are conveyed to storm drains or other BMPs.
- Use only on sites of less than one to two acres.
- Provide perforated covers as trash racks on orifices leading from first to second chamber.
- Center chamber may contain a coalescing medium to enhance the oil flotation separating process.
- The storm drain inlet in the third chamber should be located above the floor to permit additional settling.
- Storm water from rooftops and other impervious areas not likely to be polluted with oil should not discharge to the separator.
- Design to bypass flows above 400 cubic feet per acre.

Operation And Maintenance Requirements

- A storm water management easement and maintenance covenant should be required for each facility. The maintenance covenant should require the owner of the separator to periodically clean the structure.
- Cleaning quarterly should be a minimum schedule with more intense land uses such as gas stations requiring cleaning as often as monthly.
- Cleaning should include pumping out waste water and grit and having the water processed to remove oils and metals.
- For coalescing plate separators the facility should be inspected weekly. Pads are to be replaced when they become clogged and gritty but no less than annually.

References

American Association Of State Highway And Transportation Officials, Model Drainage Manual, 1991.

American Petroleum Institute (API), "Design and Operation of Oil-Water Separators", Publication 421, 1990.

Austin Texas, "Inventory of Urban Non-point Source Control Practices", Env. Resources Div., City of Austin, June, 1988a.

Austin Texas, "Environmental Quality Criteria Manual", 1988b.

Austin Texas, "Removal Efficiencies of Stormwater Control Structures", May, 1990.

Board on Science and Technology for International Development (BSTID), "Vetiver Grass", Nat. Academy Press, Wash. DC, 1993.

Bellevue, City of, Washington, "Standard Drawings", 1988.

Bellevue, City of, Washington, "Water Quality Protection for Landscaping Businesses", 1991.

Brown, J. L. and R. S. Farnham, "Use of Peat for Wastewater Filtration - Principles and Methods", Proc. 5th Int. Peat Congress, Poznan, Poland, pp. 349-357, 1976.

Burby, R. J., E.J. Kaiser, T. L. Miller and D. H. Moreau, "Drinking Water Supplies", Ann Arbor Science, 1983, reprinted 1986.

California Stormwater Task Force, "Best Management Practice Handbook - Vol. 1 Municipal", March, 1993a.

California Stormwater Task Force, "Best Management Practice Handbook - Vol. 2 Industrial/Commercial", March, 1993b.

Carr, D. J., Geinopolos, A., and Zanoni, A. E., "Characteristics and Treatability of Urban Runoff Residuals", EPA-600/2-82-094, Nov. 1982.

Chang, G. J., and P. C. Souer, "The First Flush of Runoff and its Effect on Control Structure Design", Dept. of Environment and Conservation Services, Austin, Tx., 1990.

Chescheir, G. M., G. Fipps and R. W. Skaggs, "Analysis of Stormwater Infiltration Ponds on the North Carolina Outer Banks", NC State U., Rpt. No. 254, Sept., 1990.

Conradin, F., "Study of Catch Basins in Switzerland", *Urban Stormwater Quality Enhancement*, ASCE, 1989.

Cooke, S. S., "Wetland Buffers", Washington Dept. of Ecology, 1991.

Day, G. E., D. R. Smith and J. Bowers, "Runoff and Pollution Abatement Characteristics of Concrete Grid Pavements", Virginia Wat. Resources Research Ctr., Vir. Polytechnic Inst. and State University, Rpt. PB82-168469, Bull. 135, Oct., 1981.

Denver Urban Drainage and Flood Control District (UDFCD), "Drainage Criteria Manual Vol. 3", Denver, CO, 1992.

Dillaha, T. A., R. B. Reneau, S. Mostaghimi, and D. Lee, "Vegetative Filter Strips for Agricultural Non-Point Source Pollution Control", Trans. ASAE, 1989.

Diniz, E. V., "Quantifying the Effects of Porous Pavements on Urban Runoff", Nat. Symp. on Urb. Hydrology, Hydrau., and Sed. Control, Lexington, KY, 1976.

Diniz, E. V., "Porous Pavement - Phase I, Design and Operational Criteria", USEPA 600/2-80-135, August, 1980.

Driscoll, E. D., "Detention and Retention Controls for Urban Runoff", *Urban Runoff Quality*, ASCE, 1986.

Driscoll, E. D., "Long Term Performance of Water Quality Ponds", *Urban Runoff Quality Controls*, ASCE, 1988.

Driscoll, E. D., G. E. Palhegyi, E. W. Strecker and P. E. Shelley, "Analysis of Storm Event Characteristics for Selected Rainfall Gages Throughout the United States", Woodward-Clyde Consultants, prepared for USEPA, Washington, DC, 1989.

Driscoll, E. D. and E. W. Strecker, "Assessment of BMPs Being Used in the US and Canada", Proc. 6th Int. Conf. on Urban Storm Drainage, Niagara Falls, CA, Sept., 1993.

Elling, A. E., "Managing Vegetation on Peat-Sand Filter Beds for Wastewater Disposal", USDA Forest Service, North Central Forest Experiment Station, Research Note NC-333, 3 pp., 1985.

Enviro-Drain, Informational Brochure, Kirkland, WA, 1994.

Farnham, R. S. and J. L. Brown, "Advanced Wastewater Treatment Using Organic and Inorganic Materials, Part 1 - Use of Peat and Peat-Sand Filtration Media", Proc. 4th Int. Peat Congress, Helsinki, Vol. 4, pp. 272-286, 1972.

Farnham, R. S., and T. Noonan, "An Evaluation of Secondary Treatment of Stormwater Inflows in Como Lake, MN Using a Peat-Sand Filter", Como Lake Restoration, EPA Project. S-005660-02, 29 pp., 1988.

Field, R., H. Masters, and M. Singer, "An Overview of Porous Pavement Research", Wat. Res. Bull., AWRA, Vol. 18, No. 2, April, 1982.

Field, R., "Urban Runoff: Pollution Sources, Control, Treatment", Wat. Res. Bull., AWRA, Vol. 21, No. 2, April, 1985.

Galli, J., "Peat Sand Filters", Coordinated Anacostia Retrofit Program, Dec., 1990.

Galli, J., "Peat Sand Filters", *Watershed Restoration Sourcebook*, Metropolitan Washington Council of Governments, 777 North Capital Street N.W., Suite 300, Washington, D.C. 20002-4201, Telephone 202-962-3265, 1992a.

Galli, J., "Analysis of Urban BMP Performance and Longevity in Prince George's County Maryland", MWCOG, 777 North Capital Street N.W., Suite 300, Washington, D.C. 20002-4201, Telephone 202-962-3265, 1992b.

Gburek, W. J. and J. B. Urban, "Storm Water Detention and Groundwater Recharge Using Porous Asphalt - Experimental Site", Proc. Int. Symp. on Urban Storm Runoff, U. of Kentucky, Lexington, 1980.

Georgia State Soil and Water Conservation Committee, "Manual for Erosion and Sediment Control in Georgia", undated.

Ghioto, Singhofen and Assoc., Inc., "Stormwater Rule Design Practices and Methods of Calculation", in *Design of Stormwater Facilities*, Workshop sponsored by NC Div. of Env. Mgmt. and Prof. Engrs. of North Carolina, May 4, 1988.

Gilliam, J. W. and R. W. Skaggs, "Nutrient and Sediment Removal in Wetland Buffers", in <u>Wetland Hydrology</u>, Proc. Nat. Wetland Symp., Assoc. of State Wetland Managers, Chicago, Sept. 1987

GKY, "Outlet Hydraulics of Extended Detention Facilities", report for the Northern Virginia Planning Commission, 1989.

Gorbisch, E., "the Do's and Don'ts of Wetland Planning", Wetland Journal, Vol. 6, No. 1, 1994.

Grizzard, T. L., C. W. Randall, B. L. Weand, and K. L. Ellis, "Effectiveness of Extended Detention Ponds", *Urban Runoff Quality*, ASCE, 1986.

Hammer, D. A., "Creating Freshwater Wetlands", Lewis Publishers, 1992.

Hammer, D. A. ed., "Constructed Wetlands for Wastewater Treatment", Lewis Publishers, 1989.

Harrington, B., "Feasibility and Design of Wet Ponds to Achieve Water Quality Control", Sed. and Stormwater Div., Maryland Dept. of the Environment, 1987.

Hartigan, J. P., "Basis for Design of Wet Detention Basin BMPs", *Urban Runoff Quality Controls*, ASCE, 1988.

Heraty, M., "Riparian Buffer Programs", USEPA, 1993.

Horner, R. R., "Biofiltration Systems for Storm Runoff Water Quality Control", Rpt. to Wash. State Dept. of Ecology, 1988.

Horner, R. R., "Long-term Effects of Urban Stormwater on Wetlands", *Urban Runoff Quality Controls*, ASCE, 1988.

Huber, W. C., "Deterministic Modeling of Urban Runoff Quality", in *Urban Runoff Pollution*, NATO ASI Series, Vol. G10, Springer-Verlag, Berlin, 1986.

Hvitved-Jacobsen, T., Y. A. Yousef and M. P. Wanielista, "Rainfall Analysis for Efficient Detention Ponds", *Urban Runoff Quality Controls*, ASCE, 1988.

Infiltrator Systems, Inc., "The Infiltrator", Old Saybrook, CT, undated.

Karouna, N., "Native Plant Pondscaping Guide", *Watershed Restoration Sourcebook*, Metropolitan Washington Council of Governments, April, 1992.

Kayata, T. and S. Fujita, "Pollution Abatement in Tokyo 'ESS'", *Urban Stormwater Quality Enhancement*, ASCE, 1989.

Lettenmaier, D. and J. Richey, "Operational Assessment of a Coalescing Plate Oil/Water Separator", City of Seattle, 1985.

Livingston, E., "The Use of Wetlands for Urban Stormwater Management", *Urban Runoff Quality Controls*, ASCE, 1988.

Maestri, B. and others, "Managing Pollution From Highway Stormwater Runoff", Transportation Research Board, National Academy of Science, Transportation Research Record Number 1166, 1988.

Maine Dept. of Env. Protection, "Phosphorus Control in Lake Watersheds", Sept. 1992.

Maryland Dept. of the Environment, "Standards and Specifications for Infiltration Practices", Sediment and Stormwater Administration, Feb., 1984.

Maryland Dept. of the Environment, "Standards and Specifications for Porous Pavement", Sediment and Stormwater Administration, Feb., 1984.

Maryland Dept. of the Environment, "Inspector's Guidelines Manual for Stormwater Management Infiltration Practices", Sediment and Stormwater Administration, Dec., 1985.

Maryland Dept. of the Environment, "Minimum Water Quality Objectives and Planning Guidelines for Infiltration Practices", Sediment and Stormwater Administration, April, 1986.

Maryland Dept. of the Environment, "Results of the State of Maryland Infiltration Practices Survey", Sediment and Stormwater Administration, August, 1987.

Maryland Dept. of the Environment, "Water Quality Inlets", Vincent Berg, Sediment and Stormwater Administration, Jan., 1991.

Maryland Dept. of the Environment, "Stormwater Infiltration Practices in Maryland: A Second Survey", Sediment and Stormwater Administration, 1991.

McPherson, J., "Water Quality BMPs: Catch Basin Infiltration", Presented to the APWA Stormwater Managers Committee, Tacoma, WA, 1992.

Metropolitan Washington Council of Governments, <u>Controlling Urban Runoff</u>, T. Schueler, 777 North Capital Street, Suite 300, Washington, D.C., Telephone 202-962-3265, 1987.

Metropolitan Washington Council of Governments, <u>A Current Assessment Of Urban Best Management Practices - Techniques for Reducing Non-Point Source Pollution in the Coastal Zone</u>, 777 North Capital Street, Suite 300, Washington, D.C., Telephone 202-962-3265, 1992.

Minnesota Pollution Control Agency, "Protecting Water Quality in Urban Areas", Div. of Water Quality, 1989.

Mitsch, W. J. and J. G. Gosselink, "Wetlands", Van Nostrand Reinhold, 1986.

New Jersey Dept. of Env. Protection, "Evaluation and Recommendations Concerning Buffer Zones Around Public Water Supply Reservoirs", Rpt. to Governor Kean, 1989.

Niemczynowicz, S., "Swedish Way to Stormwater Enhancement", <u>Urban SW Quality Enhancement</u>, H. C. Torno, ed., ASCE, pp. 156-168, 1990.

Nightingale, H. I., "Accumulation of AS, NI, CU and PB in Retention and Recharge Basins Soils for Urban Runoff", Water Resources Bulletin, AWRA, Vol. 23, No. 4, 1987.

Nix, S. J., J. P. Heaney and W. C. Huber, "Suspended Solids Removal in Detention Basins", ASCE J. of Env. Engrg., Vol. 114, No. 6, 1988.

North Carolina Sediment Control Commission, "Erosion and Sediment Control Planning and Design", NC Dept. of Natural Resources and Community Development, Div. of Land Resources, 1991.

North Carolina Dept. of Env., Health and Nat. Res. (NCDEHNR), "An Evaluation of Vegetative Buffer Areas for Water Quality Protection", Raleigh, NC, Feb., 1991.

North Central Texas Council of Governments, "Storm Water Quality Best Management Practices for Residential and Commercial Land Uses", First Edition, Arlington, TX, July 1993.

Northern Virginia Planning District, "Northern Virginia BMP Handbook", Northern Virginia Planning District Commission and Engineers and Surveyors Inst. of Northern Virginia, Nov. 1992.

Oberts, G. L., "Influence of Snowmelt Dynamics on Stormwater Runoff Quality", in Techniques for Watershed Protection, Vol. 1, No. 2, Summer, 1994.

Palfrey, R. and E. Bradley, "Buffer Area Study", Maryland Dept. of Nat. Res., Coastal Res. Div., Tidewater Admin., 1982.

Phillips, J. D., "An Evaluation of Factors Determining the Effectiveness of Water Quality Buffer Zones", J. of Hydrology, 107:133-145, 1989.

Presto Products Company, "Geoblock Porous Pavement System: Product Overview", 1990.

Professional Engineers of North Carolina, "Design of Stormwater Control Facilities", Workshop Proceedings, May 4, 1988.

Randall, C. W., "Stormwater Detention Ponds for Water Quality Control", ASCE Conf. on Stormwater Detention Facilities, Henniker, NH, 1982.

Rawls, W. J., D. L. Brakensiek and K. E. Saxon, "Estimation of Soil Properties", Trans. ASAE, Vol. 25, No. 5, 1982.

Readling, K., "Flow Spreaders and Buffer Strips", Ogden Environmental BMP Fact Sheet, 1992.

Reckhow, K. H., "Empirical Models for Trophic State in Southeastern U.S. Lakes and Reservoirs", Water Res. Bull., Vol. 24, No. 4, 1988.

Roesner, L. A., E. H. Burgess and J. A. Aldrich, "The Hydrology of Urban Runoff Quality Management", Water Resources Planning and Management, ASCE, 1989.

Schueler, T., "Developments in Sand Filter Technology to Improve Stormwater Runoff Quality," in Techniques for Watershed Protection, Vol. 1, No. 2, Summer 1994.

Schueler, T., "Hydrocarbon Hotspots in the Urban Landscape: Can They Be Controlled", in Techniques, Vol. 1, No. 1, Feb., 1994.

Schueler, T., "Performance of Stormwater Pond and Wetland Systems", in <u>Engineering Hydrology</u>, C. Kuo ed., ASCE, 1993.

Schueler, T. and D. Shepp, "The Quality of Trapped Sediments and Pool Water Within Oil Grit Separators in Suurban Maryland," Washington Council of Governments, 1993.

Schueler, T. R. and Helfrich, M., "Design of Extended Detention Wet Pond Systems", *Urban Runoff Quality Controls*, ASCE, 1988.

Schueler, T., J. Galli, "The Environmental Impacts of Stormwater Ponds", *Watershed Restoration Sourcebook*, Metropolitan Washington Council of Governments, 777 North Capital Street N.W., Suite 300, Washington, D.C. 20002-4201, Telephone 202-962-3265, April, 1992.

Schueler, T., J. Galli, L. Herson, P. Kumble and D. Shepp, "Developing Effective BMP Systems for Urban Watersheds", *Watershed Restoration Sourcebook*, Metropolitan Washington Council of Governments, 777 North Capital Street N.W., Suite 300, Washington, D.C. 20002-4201, Telephone 202-962-3265, April, 1992.

Schueler, T. J., "Design of Stormwater Wetland Systems", Metro. Wash. Council of Govts., 777 North Capital Street N.W., Suite 300, Washington, D.C. 20002-4201, Telephone 202-962-3265, October, 1992.

Segarra-Garcia, R., and V. G. Loganathan, "Design of Stormwater Detention Ponds Subject to Random Pollution Loading", Proc. 6th Int. Conf. on Urban Storm Drainage, Niagara Falls, CA, Sept., 1993.

Shaver, E., "Sand Filter Design for Water Quality Treatment", Delaware Dept. of Nat. Resources, 1991.

Shenot, S., "An Analysis of Wetland Planting Success at Three SW Management Ponds in Montgomery, MD", M.S. Thesis, University of Maryland, 114 pp., 1993.

Smith, R. W., J. M. Rice and S. R. Spelman, "Design of Open-Graded Asphalt Friction Courses", Federal Highway Admin., FHWA-RD-74-2, January, 1974.

Southwest Florida Watershed Management District, "Permit Information Manual", Vol. I, Revised, March 1990 and Vol. IV, 1986.

Stewart, W., "Compost Stormwater Treatment System," W&H Pacific Consultants, Draft Report, Portland Oregon, 1992.

Swift, L. W., "Filter Strip Widths for Forestry Roads in the Southern Appalachians", Southern J. of Applied Forestry, Vol. 10, No. 1, 1986.

Tasker, D.T. and N.E. Driver, "Nationwide Regression Models For Predicting Urban Runoff Water Quality At Unmonitored Sites", Water Res. Bulletin, Vol. 24, No. 5, October 1988.

Techniques, "Innovative Leaf Compost System used to Filter Runoff at Small Sites in Northwest", Vol. 1 No. 1, Feb., 1994.

Thelen, E., W. C. Grover, A. J. Hoiberg and T. I. Haigh, "Investigation of Porous Pavements for Urban Runoff Control", The Franklin Inst., Philadelphia, PA, USEPA 11034 DUY 03/72, March 1972.

Thelen, E. and L. F. Howe, "Porous Pavement", Franklin Inst. Press, Philadelphia, PA, 1978.

Thunhorst, G., "Wetland Planting Guide for the Northeastern United States", Environmental Concerns, Inc., St. Michaels, MD, 179 pp., 1993.

Tomaseck, M. D., G. E. Johnson and P. J. Mulloy, "Operational Problems with a Soil Filtration System for Treating Stormwater", Lake and Reservoir Management, Vol. 3, pp. 306-313, 1987.

United States Environmental Protection Agency, "Storm Water Management for Construction Activities", EPA 832-R-92-005, Sept. 1992a.

United States Environmental Protection Agency, "Storm Water Management for Industrial Activities", USEPA 832-R-92-006, Sept., 1992b.

United States Environmental Protection Agency, "Investigation of Inappropriate Pollutant Entries into Storm Drainage Systems", USEPA 600/R-92/238, Jan., 1993.

United States Environmental Protection Agency - Region 7, "Restoring and Creating Wetlands", USEPA, Sept., 1992.

U. S. Soil Conservation Service, "Soils - Hydric Soils of the United States", Natural Bull. No. 430-5-9, Washington, DC, 1985.

J. B. Urban and W. J. Gburek, "Storm Water Detention and Groundwater Recharge Using Porous Asphalt - Initial Results", Proc. Int. Symp. on Urban Storm Runoff, U. of Kentucky, Lexington, 1980.

Urbonas, B. and W. P. Ruzzo, "Standardization of Detention Pond Design for Phosphorus Removal", *Urban Runoff Pollution*, NATO ASI Series Vol. G10, Springer-Verlag, Berlin, 1986.

Urbonas, B., C. Y. Guo and L. S. Tucker, "Optimization of Stormwater Quality Capture Volume", *Urban Stormwater Quality Enhancement*, Proc. Engrg. Foundation Conf., ASCE, New York, 1990.

Urbonas, B. and P. Stahre, "Stormwater Best Management Practices and Detention", Prentice-Hall, Englewood Cliffs, NJ, 1993.

Virginia Dept. of Conservation and Recreation, "Virginia Erosion and Sediment Control Handbook", Div. of Soil and Water Conservation, Third Ed. 1992.

Walker, W. W., "Empirical Methods for Predicting Eutrophication in Impoundments - Report 3: Model Refinements", USAWES Tech. Rpt. E-81-9, US Army Corps of Engineers, Vicksburg, MS, 1985.

Walker, W. W., "Phosphorus Removal by Urban Runoff Detention Basins", Lake and Reservoir Management: Vol. III, North American Lake Management Society, Washington, DC, 1987.

Wang, T. S., D. E. Spyridakis, B. W. Mar and R. R. Horner, "Transportation, Deposition and Control of Heavy Metals in Highway Runoff", FHWA-WA-RD-39.10, Wash. State University, Dept. of Civil Engineering, Seattle, WA, 1982.

Wanielista, M. P., Y. A. Yousef, B. L. Golding and C. L. Cassagnol, "Stormwater Management Manual", University of Central Florida, Oct., 1981.

Wanielista, M. P., and Y. A. Yousef, "Stormwater Management", Wiley Interscience, 1993.

Washington DC, "Stormwater Design Handout", City of Washington Dept. of Public Works, 1992.

Washington State Dept. of Ecology, "Puget Sound Stormwater Management Manual", Feb. 1992.

Watershed Protection Techniques, "Pollutant Dynamics of Pond Muck", Vol. 1, No. 2, Summer, 1994b.

Whitlow, T. H., and R. W. Harris, "Flood Tolerance in Plants: a State-of-the-Art Review", USAEWES Tech. Rpt. E-79-2, Vicksburg, MS, 1979.

Wigington, P. J., C. W. Randall and T. J. Grizzard, "Accumulation of Selected Trace Metals in Soils of Urban Runoff Drain Swales", Wat. Res. Bull., Vol. 22, No. 1, Feb., 1986.

Wong, S. L. and R. H. McCuen, "The Design of Vegetative Buffer Strips for Sediment and Runoff Control", Stormwater Management in Coastal Areas, Tidewater Admin., Maryland Dept. of Nat. Res., Annapolis, MD, 1982.

Woodward-Clyde Consultants, "Methodology for Analysis of Detention Basins for Control of Urban Runoff Quality", prepared for EPA Office of Water, Washington, DC, 1986.

Woodward-Clyde Consultants, "Urban Targeting and BMP Selection", The Terrene Institute, Washington, DC, 1990.

Yousef, Y. A., M. P. Wanielista and H. H. Harper, "Removal of Highway Contaminants by Roadside Swales", Trans. Res. Record 1017, TRB, Washington DC, 1985.

Zanoni, A. E., "Characteristics and Treatability of Urban Runoff Residuals", Water Research, Vol. 20, No. 5, pp. 651-659, 1986.

Chapter 14 Storm Water Master Planning

14.1 The Role Of Storm Water Master Planning

"Storm water master planning" is a term which has come to mean a wide variety of things. In some municipalities a storm water master plan (SWMP) is nothing more than a collection of general specification drawings with some explanatory paragraphs as an introduction. In others it is a complex analysis of the total major and minor storm water system including storm water quality. In some locations master plans address existing problems while in some they seek to anticipate, and avoid, future problems. Following are some examples.

- In Nashville, Tennessee a SWMP was developed to correct known and future major creek flooding problems.
- In Atlanta, Georgia a SWMP was conceived as a very detailed inventory of drainage structures to assist in the identification of flooding and water quality problems, including combined sewer overflows.
- In a Virginia SWMP planners and engineers defined in broad qualitative terms environmental quality activities necessary to bring a certain stream basin to a pre-determined level of water quality and aesthetic goals.
- In Mecklenburg County, North Carolina a multi-criterion master plan was developed to address, through a series of layers, storm water quantity, quality, erosion and opportunities for multi-objective use of major drainage systems.
- In Raleigh, North Carolina a plan envisioned corrections to storm water quality problems to save a lake resource.
- In Santa Clara County, California a comprehensive plan for the reduction of both point and non-point source pollution was developed.

Thus, any short chapter on SWMP could not possibly cover all the possibilities and combinations of approaches to develop a storm water master plan. However, all good master planning does follow some very well defined steps. It is these steps and application of them that this chapter addresses.

Storm water master planning plays a key role in identifying existing and future problems, opportunities for enhancement of storm water use, cost estimates, impact assessment and assigning priorities for change and improvement. Even master planning cannot address the core issues unless it explores, however summarily, factors other than strictly the technical ones.

14.2 The Master Planning Process

The discussion below defines a generic process which should be followed by any master planner. The process moves through a thorough needs and goals analysis which assesses all physical problems to be addressed by the master plan and uncovers root causes. The master planning process then moves, in turn, through identifying constraints to possible solutions and formation of an appropriate technical approach to find useful solutions.

Step 1: Make A Needs Analysis

The first step in any master plan is to gain a clear understanding of the actual needs and goals of the municipality. There is an old axiom of sales that states that the customer never tells you what he really needs but what he thinks he needs. That is never more true than in storm water master planning. To gain a clear understanding of the real needs a logical and ordered approach must be followed.

The term "storm water planning" means different things to different people. To the municipal administrator it often means a drainage study to "solve Mrs. Smith's backyard flooding". It might mean "floodplain management" to the community FEMA representative; or "water quality enhancement" to the environmentalist. Whatever its meaning or eventual activities, an ordered understanding of its various components is essential to proper control of storm water.

A typical program begins when a particular problem becomes apparent. Usually it is flooding, erosion, water quality and/or drainage infrastructure problems. The solution normally is to fix the individual problem. However, a "quick fix" solution to the apparent problem is rarely a permanent solution because it rarely addresses the underlying issues and root causes.

Chapter 1 this book laid out an ordered approach to understand the various causes of problems. A master plan should seek to understand the problems at many levels. Even if it cannot address solutions on all the levels it should seek to call attention to them.

Goals and objectives of the master plan should be carefully determined or disappointment will invariably result. A master plan is not a magic box out of which instant solutions are pulled. Most problems took a long time to develop and will not be fixed overnight. This becomes even more true as the planner deals with erosion and non-point source pollution problems.

The deliverable product should also be defined. Often master plans gather dust on the shelf as implementation funds are sought. When the time comes to actually build the projects the information may be out of date as zoning changes and new development have changed the situation. A master plan should be a dynamic document tied to a mathematical model where appropriate and that model kept up- to-date periodically. This can be done quite easily through modern technology which ties mathematical models to geographic information systems (GIS) and CADD systems. As changes occur in the field they are registered in the database. The model is automatically updated and run when appropriate. In this way master planning is done "on the fly" as development occurs, changes from assumed conditions are made (there will always be changes) or budgets are modified.

One warning is in order though. If the technology delivered to a municipality is more than the municipality can keep up, either due to lack of training in its use or through burdensome data requirements, it will be less than useless. Several municipalities have spent large sums on complex models tied to GIS systems only to find they were not useful due to data collection procedural problems or levels of sophistication beyond the day-to-day user. This leads directly into step 2.

Step 2 : Determine Constraints To Possible Solutions

There will always be constraints to solutions to problems. Master plans must be realistic. Expensive structural solutions to flooding problems cannot often be afforded. Key pieces of property may not be available and the "political will" may not be there to condemn land even if it is vital to flooding abatement. It is thus important to define the "universe" in which possible solutions can be found.

Constraints to any possible solutions fall naturally into major categories.

- Technical constraints - to any solution might include lack of necessary data. This is often the case for stream flow and almost always the case for water quality studies. Another technical constraint may be the ability of the municipal staff to use the tools developed. It makes no sense to recommend a complex watershed model if no one on the staff will be able to keep the model up-to-date after the departure of some consultant.

- Political constraints - are often the most important to the would be master planner. Some solutions have failed in the past with embarrassing consequences to a political supporter of that approach. The planner should think twice before recommending this approach again even if it is superior. For example, impact fees may have a very bad reputation in some parts of the country where they have been successfully challenged in court. Detention dams (or "dry dams" as the Corps of Engineers calls them) have a very bad reputation in some municipalities where maintenance has proven to be costly. A sense of the political will power is necessary to configure solutions which will be acceptable. This must be tempered with the fact that political climate statements are perceptual factors and often depend on who is making the statement of "fact". Also, political constraints will change often: with each election, after a crisis, when a new regulation impacts storm water, during budget discussions, etc.

- Social constraints - reflect the values of a given community. For example, channel modification may be seen by a certain municipality as anti-environmental and not allowed. Wetlands may be seen as either a very positive environmental amenity or as a waste of valuable development land depending often on the economic condition of the community. Some parts of some municipalities may be "off limits" for location of detention facilities or for investing scarce resources.

- Physical constraints - can be such things as soil type, topography, moisture, bedrock, etc. Physical constraints might also include the availability of materials such as sand or riprap. Urban development may constrain certain solutions due to the presence of structures or roads.

- Legal constraints - reflect local, state and federal restrictions of master planning options. Legal constraints are often environmental in nature, such as the necessity of obtaining permits for activities in a stream. Some municipalities are constrained by state law from exercising very many powers while "home rule" states have given strong powers to local municipalities. General case law and perceived liability may restrain municipalities from certain options for fear of a lawsuit or a ruling of a "taking" of property. Legal boundaries may also constrain some types of solutions that cross outside municipal limits.

- Financial constraints - almost always limit, to some extent, the types of solutions that are possible. While not all studies are assessed on a strict "benefit-cost" basis, most projects have to give the perception that they are worth the expense, both initial and ongoing. Budgets for the master plan also usually limit the type of analysis and level of detail. Sometimes suitable options exist, though a "wild guess" at the front end of a study does not justify rigorous analysis of data based on that guess, later in the study. This is often the case with detailed reliance on hydrology developed from little or no data.

Step 3: Formulate The Technical Approach

As stated in the beginning, the term "master plan" can mean almost anything. When considered herein the term will deal with the technical approach to storm water solutions over a certain area or in a certain drainage system. This will eliminate the primarily institutional approaches to master planning where ordinances or financing studies predominate.

A master plan can concentrate on:
- the major drainage system and/or the minor system,
- quantity and/or quality and erosion control,
- existing and/or future conditions,
- urban and/or rural systems,

- correcting problems and/or planning to avoid problems,
- risk-based solutions or more deterministic approaches,
- "soft" solutions and/or "hard" solutions, or
- individual solutions and/or systems analysis.

The goals of the master plan study determine the level of effort and detail involved, as well as the degree of conservatism built into flow estimates. Flood hazard studies may require only an estimate of peak flows. Other studies may require the development of design storm water runoff hydrographs and the timing of these hydrographs with respect to each other. There are varying complexity of studies and often studies are done in phases (COE, 1992, 1993).

For example, studies involving flooding generally fall into three categories. The categories build on one another with the final phase being the design and construction phase.

1. Flood Hazard Potential Studies - These studies provide flood hazard information for certain frequency floods (for example, identification of the 100-year return period flood elevations for definition of FEMA required floodplains and floodways). Studies vary significantly in scope and complexity. The end product of a flood hazard potential study is the ability to determine, within an acceptable level of accuracy, both the flooding elevation (stage) and the probability of occurrence of a flood (frequency) at points along a stream. This can be done for both existing and future conditions. Measured or transferred data are extremely valuable for accurate flood hazard potential studies. This includes coincident rainfall-runoff records, high water marks, and long term flow records. The stage-frequency relationship is developed from a combination of a stage-discharge curve (rating curve) at a reach developed through hydraulic analysis and a discharge-frequency curve, developed from a stream gage or synthetically through regression analysis.

2. Planning Studies - Information from flood hazard potential studies is used to develop economic evaluations and/or environmental impacts. Examples would be feasibility evaluations of structural or non-structural flood control alternatives or impact evaluations of alternate land use patterns. A stage-damage curve is usually developed for points of interest through the use of property surveys. The stage-damage curve shows the dollar amount of damages which would be expected from a flood of a certain elevation. If the frequency of certain flood levels are also known from the flood hazard potential study, an average annual damage estimate can be made and used as a tool to evaluate and compare various proposed alternatives.

3. Design Studies - These studies use flood flow frequency information in the design of flood control structures, culverts, bridges, levees, detention facilities, etc. Non-structural designs seek to mitigate the effects of a flood, while structural designs seek to alter flood events. Designs move through sketch plans, preliminary planning, design, and plans and specifications.

14.3 Holistic Watershed Management

In the past five years there has arisen an approach to watershed planning and management which increasingly considers the whole watershed as one integrated unit, each part effecting the others (ENR, 1993). This has grown out of a realization that, despite significant strides in reducing point-source pollution impacts to the nation's aquatic resources, nonpoint pollution and development pressures have placed more and more watersheds at ecological risk. Efforts such as the Chesapeake Bay multi-state effort, Santa Clara County's multi-agency task force, North Carolina's 17 watershed coordinated permitting efforts, Oregon's governmental reorganization and the Great Lakes initiative are only the beginning.

The holistic watershed management approach attempts to provide a framework within which stakeholders from many sectors and viewpoints can work together toward defining and bringing about a long term environment which allows both reasonable growth and freedom of land use with ecological protection and conservation. Basically the approach seeks to understand the complex interaction between humans and the ecosystems involved given the background of climate, topography, and the myriad other details of life in the watershed. The idea is to seek to work with nature and find a balance between ecological preservation and conservation and human development pressures.

Amendments to the Clean Water Act, Endangered Species Act and other major legislative efforts will reflect this changing emphasis throughout the nineties and on into the next millennium. The concept is practiced and understood a little differently by each of the major Federal agencies which have a role in some part of watershed stewardship, and by each state active in watershed management. The Soil Conservation Service has developed a nine step process for the planning of its watersheds.

1.   Identify all the various problems within the watershed.
2.   Develop a set of objectives for the watershed.
3.   Inventory the resources within the watershed including: air, soils, water, vegetation, and wildlife.
4.   Analyze the resources within the watershed.
5.   Develop alternatives to addressing the problems and opportunities within a watershed.
6.   Evaluate and analyze the alternatives.
7.   Make decisions concerning the alternatives.
8.   Implement the decisions.
9.   Reevaluate progress and make adjustments.

The Environmental Protection Agency (1993) has developed three cornerstones to its "watershed protection approach".

1.   Problem Identification - Identify the primary threats to human and ecosystem health within the watershed.
2.   Stakeholder Involvement - Involve the people most likely to be concerned or most able to take action.
3.   Integrated Actions - Take corrective actions in a comprehensive, integrated manner once solutions are determined. Evaluate success and refine actions, as necessary.

It is obvious that such an approach can be very controversial from the standpoint of both land use controls and development restrictions on one hand and ecological risk assessment and compromise on the other. The key to success is balanced local stakeholder involvement and control in development and pursuit of a vision for a certain watershed, along with suitable governmental encouragement, funding and broad limits or controls. Experience has shown that sound ecological preservation practices can translate into economic development benefits and higher quality of life. But it is a relearning process for almost everyone.

The process can work at both the macro-scale covering large multi-state areas and at the minor urban scale through the concept of "nesting". Nesting means planning smaller areas or unique character within a larger overall concept. An urban area can plan its own watershed development and that of areas it may eventually annex in conjunction with adjacent authorities and regional or state planners. In fact, it makes sense for urban areas to become involved in the planning process to avoid the problems which can result from some outside agency doing it for them and imposing terms.

EPA (1992) has defined its strategy to bring about such a change. This strategy can serve to guide both local planning and regional efforts.

1. Try It Out - Develop credible methodologies, tools, models, and case studies wherein both the planning framework and its tools and methods can be tried and proven. Success stories serve to encourage other regions and areas to try it too. A pilot project with high visibility and good chances for success is important at the local level. A project such as a multi-objective greenway or ecologically sensitive park area might be appropriate.

2. Advertise It - Promote a broad understanding of the watershed planning and protection concept. Develop regular cooperators in the process among other Federal and state agencies as well as private stakeholder groups. Local efforts can involve such things as citizens task forces, press conferences, pilot project tours, school outings, etc.

3. Align Programs - Seek to modify existing Federal and state or local programs and organizations in a way which removes barriers to both cooperation and funding. Examples at the municipal level might be to form an advisory group made up of local citizen stakeholders to work with a staff task force, or the reorganization of local government to align all water and/or land use related regulatory agencies under one authority in close cooperation with planning. Another example is for a municipality to seek to assist local agricultural extension and soil conservation services in reducing agricultural nonpoint pollution entering the urban stream system from the undeveloped headwaters areas or identifying and acquiring key ecologically important locations well ahead of development.

4. Develop Tools - Tools can include both new or applied technology such as multi-spectral aerial photography, global positioning systems (GPS) and geographical information systems (GIS) and new ways of building a framework for ease in solution identification and implementation. Measures of stream health and biological and chemical monitoring applications are important tools in seeking to integrate traditional engineering flood control applications with ecological impact assessment. Such concepts as Total Maximum Daily Load (TMDL) and even pollution trading have proven helpful in large scale watershed planning.

5. Measure Success - Measures of success must both be developed and applied. Measures may include specific actions taken (such as local reorganization) or watersheds planned and plans implemented. Eventually measures of success must reflect both real changes in the watershed's ecosystem and at least perceived quality of life enhancements for the watersheds human inhabitants.

14.4 Computer Model Choice For Master Planning

Choice of a computer model depends on a number of factors including: answers needed, application of information, physical situation, special considerations (such as pressure flow in pipes), available data, modeler capabilities and those of the final user. In addition, there may not be a need for a computer model at all. Often models are tied to their ability to represent results graphically through a Geographic Information System (GIS), Automated Mapping/ Facilities Management (AM/FM) system or other means.

A number of models should be considered for use; however if a model the user is intimately familiar with will suffice there must be a very good reason not to use it. Simply because a model is complex does not necessarily mean it produces better answers. All models depend on verification or calibration to real data to produce accurate and reliable results. When those data are missing there is no way to be sure of accuracy. The modeler must be sure that the model used can simulate all the key physical processes (unless it is a "black box" regression type model) encountered in the watershed. For example, if a hydraulic model cannot determine energy losses through a long culvert then a supplemental model must be used and the answer inserted into the hydraulic model in the appropriate place.

Physically based (rather than empirical) models have the advantage in that they can be extended into new situations with some confidence. This is because they are based on physical laws rather than fitting some mathematical relationship to a data base.

Though there are a number of ways to categorize models (deterministic vs. stochastic, lumped vs. distributed, continuous vs. event, analytical vs. empirical, single vs. multiple dimension, closed form vs. numerical method, steady vs. unsteady, interactive menu-driven vs. batch, public vs. private) models fall generally into four categories for storm water master plan type studies. Some models span a number of these categories (such as Texas A&M Model, SWMM, and SEDIMOT).

1. Hydraulic Models - Hydraulic models normally are used to determine flow elevations, velocities, distributions and pressures given flow rates and boundary characteristics as inputs. Backwater models such as HEC-2 and WSPRO fall into this category as well as pipe flow models or special purpose models such as hydraulic structure design models. The Federal Highway Administration's HY-8 model for culvert design would be an example. Two and even three dimensional models exist for more complex problems. Unsteady flow models such as UNET, SWMM, DAMBRK or DWOPER may be necessary for dynamic flow conditions. The leap from simple one dimensional models to 2-dimensional requires a great increase in boundary information input. The more the model attempts to mimic nature through complex mathematical formulations the more the modeler must deal with numerical stability as well as trying to obtain accurate answers. For example both the dynamic routing of SWMM's EXTRAN block and the mass conservation in FESWM's two dimensional finite element network can "blow up" with wrong data even if it is physically reasonable.

2. Hydrologic Models - Hydrologic models normally are used to determine flow rates at various points throughout a watershed or basin given the typical inputs of rainfall, basin characteristics and basin structure. There are many models which fit this category or can be used for this purpose such as HEC-1, HYDROS, TR-20, PSRM, HYMO, TR-55, ILLUDAS and many more. The leap from single event to long term continuous simulation requires much more rainfall data but will most often give more accurate results for such things as flow volumes or interactions of flows in reservoirs or retention ponds.

3. Water Quality Models - There are a number of water quality models each with specific capabilities and drawbacks. Very few water quality models have been broadly tested and are fully supported. These models fall generally into the categories of: statistical models, regression models and physically based models. Inputs beyond basin hydrologic inputs include pollutant loading information or parameters. Several of the most common include SWMM, DR3M, QUAL-2E, STORM and HSPF. This category includes models which perform such activities as pollutant loading, transport, chemical or biological reaction or interaction, deposition or other treatment and receiving water quality impacts. Pollutant loading estimates are generated based on known constants, rating curves, buildup-washoff relationships, regression equations, sediment transport relationships, or statistical estimates based on data analysis (Huber, 1986). Water quality models are reviewed in Huber & Heaney (1982) & Whipple, et al., (1983).

4. <u>Sediment Transport Or Erosion Models</u> - These models are used to determine channel stability both in the short term and long term and to predict sediment yield. Inputs beyond basin hydrology or hydraulic parameters include grain size distribution, transport equation parameters, sediment supply, and a variety of parameters for special functions depending on the model (such as settling rates for a sediment yield model for use in the design of sediment traps). Common models include SEDIMOT for sediment yield and HEC-6 for one-dimensional riverine sediment transport.

14.5 Typical Master Plan Types

Even though there are an infinite number of possible detailed approaches to master planning there some general categories of approach or typical types of plans that can be described. For each type typical steps and concerns specific to that type of plan will be discussed briefly. Because of the great variability in approach and objectives a detailed discussion of sediment transport, channel stability and geomorphic analysis will not be covered in this chapter. The reader is referred to Barfield, et al., (1985), SLA, (1982), ASCE (1977), HEC (1977b), COE (1989) or Simons & Senturk (1977) for more information.

Major Creek Flooding

This type of master plan seeks to correct out-of-bank flooding from major creek systems. Typically, these are the creeks that have been studied under the Federal Flood Insurance Program, though often the study continues upstream to a point where intermittent stream flow begins (blue line on a typical USGS Quad map). Flood elevation is often key so that a backwater profile is necessary, though often an existing FEMA study can be used for lower budget studies. Benefit-cost analysis is often of importance since the magnitude of flooding is usually great and the cost for mitigation equally great. Non-structural solutions often predominate. An overview of non-structural controls is given in Appendix B.

The Water Resource Council (1983) has laid out typical steps for the development of flood reduction benefit calculations as given in Table 14-1. These steps serve as guidelines for all agency planning for floodplain management and design projects. They serve as the basis for a typical major creek master plan for flood control.

Table 14-1 Typical Steps In Flood Control Study

Step 1:   Delineate Affected Area
Step 2:   Determine Floodplain Characteristics
Step 3:   Project Activities in Affected Area
Step 4:   Estimate Potential Land Use
Step 5:   Project Land Use
Step 6:   Determine Existing Flood Damages
Step 7:   Project Future Flood Damages
Step 8:   Determine Other Costs of Using Floodplain
Step 9:   Collect Land Market Value and Related Data
Step 10:   Compute National Economic Development (NED) Benefits

Source: Water Resources Council, 1983

Major creek flooding master plans begin by locating flooding areas and by a determination of information about the flooding such as: depth, frequency, expected annual damages, types of structures, velocities, directly contributing areas, channel characteristics, etc. Usually a combination of alternatives are found to be effective in reducing flooding. Often a frequency such as the 25-year flood is used for alternative development since this frequency normally deals with the most cost effective frequencies for major creek flood problems, though floods of 100-year or greater return period are used for locating floodplain boundaries for safety purposes. Two-dimensional modeling may be necessary for very wide and flat floodplain areas where velocities and more exact flood heights are necessary. Unsteady flow modeling may be necessary for calculation of major flood waves during a dam break situation or in steeply sloping terrain. Risk-based analysis is often used to determine explicitly probabilities of failure of possible solutions.

Structural alternatives such as: levees and floodwalls, reduced roughness, channelization, cutoffs, diversions, detention and retention, etc., are assessed in combination with non-structural measures. Concern for the environment and for stable channel design should be in the forefront.

Erosion control design often is based on the 2- to 5-year flood since this is the approximate frequency of bankfull flows for most channels which have not been altered by man. Environmental enhancement or mitigation measures include such things as: high or low flow channels, clearing one side only, artificial meanders, artificial wetlands, in-stream habitat structures, etc.

Development, land use, and flood control decisions are often made on the basis of benefit-cost analysis, though this requires first floor elevations and detailed elevation-damage curves. There are approximate methods which will yield accurate answers putting solutions in rank order and approximate absolute ratio values (Ogden, 1988). Often decisions depend on hard-to-quantify features such as integration of a solution site with such things as parks, greenways or other recreational areas.

After objectives and constraints of the plan are obtained typical technical analysis steps include:
- data collection,
- preliminary field investigation,
- design storm development,
- hydrologic model development,
- hydraulic model development,
- system alternatives development and analysis, and
- benefit and cost analysis.

1. Data Collection - This phase determines all available information and ascertains field conditions. Sub-steps include the following.
- Review of available documents from U. S. Geological Survey, Corps of Engineers, Soil Conservation Service, FEMA, other Federal and State Offices, local records, miscellaneous studies, etc. Determine degree and location of urban development over time from zoning records, annexation records, plat recordings, or other local records and knowledge.
- Obtain historic flow information including discharge-frequency relationships, historic high water marks, rainfall records, stage-discharge relationships, bridge designs, known cross sections, mapping, etc. Determine whether transfer of information is a viable solution.
- Plan hydrologic study. Delineate preliminary sub-basins and routing reaches and structures for hydrologic study.
- Plan hydraulic study. Locate existing or new cross sections and modelled structures for backwater analysis on maps.

2.   Preliminary Field Investigation - This phase estimates all hydrologic and hydraulic parameters in the field as well as identifying unusual conditions. Consider dictating notes as you go. Typical sub-steps include the following.

•   Make field estimates of all hydrologic and hydraulic parameters such as roughness coefficients, bridge openings, culvert sizes and inlet types, overtopping circumstances, flooded structure locations, etc.

•   Perform a field investigation to ascertain conditions which may impact runoff or hydraulic parameters such as ponding and reservoirs, undersized crossings, debris barriers, recent erosion, new development, hydraulic controls, channelization, location of wetlands or other environmentally sensitive areas, etc.

•   Interview local agencies, residents; review newspaper files for historic high water marks or rainfall information, etc.

•   Take pictures of all unusual situations with measurable scale in picture.

•   Finalize cross section locations, overflow bypass pathways, and sub-basin boundaries. Measure cross sections with hand level and tape or mark survey points if necessary.

3.   Design Storm Development - For event modeling the development of a "design storm" is often more art than science. There has been considerable discussion in the literature about proper storm duration, storm movement, distribution of rainfall within storms, adjustment of storm depths for areal distribution, storm types, urban shadow effects and so on. Each of these factors, and more, may be important in specific instances. For example, storm movement may take on importance if the storm of concern tends to be a fast moving thunderstorm or frontal system; storm movement is predominantly in one direction; and the basin to be modelled is situated with a long axis parallel to storm movement. It may also be important in mountainous regions where orographic effects can predominate for the less frequent events (such as the Big Thompson Colorado flood of the late 1970's). The adjustment of point rainfall depths for areal distribution will be important for larger watersheds (perhaps greater than five to ten square miles).

However, for other than very infrequent event modeling in conjunction with such considerations as dam safety (Probably Maximum Storm Development) the modeler should refrain from conjuring a set of theoretical conditions which may have a certain probability of occurrence but whose joint probability would produce an event far more extraordinary than simple frequency analysis would predict.

The rainfall values given in Intensity-Duration-Frequency (IDF) tables or curves are for point rainfall. If storms are developed over a larger area these values will overpredict the storm depths and intensities. Therefore the point values must be adjusted downward using an areal adjustment factor. Equation 14.1 has been fit to data and a figure first developed by the national Weather Service (NWS, 1961) by the Corps of Engineers for the HEC-1 computer model (COE, 1990). Equation 14.2 is a fit of duration versus B_v information provided by the Corps of Engineers.

$$F_r = 1 - B_v * (1 - e^{(-0.015 * A)}) \qquad (14.1)$$
$$B_v = 0.35488859 * D^{-0.42722864} \qquad (14.2)$$

where: F_r = coefficient to adjust point rainfall
         B_v = maximum reduction in point rainfall
         A = area, mi^2
         D = storm duration, hrs

Any hydrologic model seeks to provide accurate flow information at all points in the watershed. In reality, for example, the 100-year flow will not occur at all points in the watershed at the same time nor will an individual storm generate a 100-year flow event throughout a watershed. More typically the 100-year flow level is achieved as a flood wave moves through a drainage system and then only in part of the watershed. A localized thunderstorm may produce such a storm over a few city blocks, several square miles, or only at a few street intersections. Areas adjacent to these peak flow locations will experience some lesser storm event.

For calibration purposes rainfall may be set to vary from sub-basin to sub-basin. On-line radar information or multiple gage locations may be useful for such purposes. Various methods are available for averaging historical storm event rainfall over areas of a watershed. See the Urban Hydrology, Chapter 7.

For modeling purposes it may be appropriate to use a storm of smaller areal extent (and thus a more intense storm) for upland sub-basins and increase the area's extent as computation points lower in the watershed are reached. HEC-1 can accomplish this adjustment automatically, generating a number of adjusted storm intensities depending on the drainage area to the computation point. Flow estimates between such points are estimated through transfers according to equation 14.3.

Accurate and logical estimates of both storm volume and peak intensities are important depending on the type of analysis to be done. For example, a six-hour duration storm may reproduce a frequency distribution of peak flows accurately. However such frequency distributions of volume are often unknown. In the sizing of retention basins or extended detention basins storms of longer duration will provide more volume. Often the choice is a policy decision. In some cases continuous simulation of such structures is the most accurate method of design.

Typical steps in the development of a synthetic design storm include the following.

- Obtain rainfall information from local gage data, the National Weather Service or other publications. Assess the rainfall to determine intensities, durations and frequencies. Develop or use intensity-duration-frequency curves as appropriate.
- In conjunction with hydrologic model calibration choose a design storm frequency and duration.
- Distribute the storm in time using time increments compatible with the watershed size and hydrologic methodology chosen. The SCS distribution methodology given in the Hydrology chapter can be used for a center peaked storm. A "balanced storm" distribution methodology presented by the Corps of Engineers (COE, 1982) provides a similar distribution. The Urban Hydrology, Chapter 7, provides alternate methods for the development of such distributions as first-, second- or third-quartile storms.
- Consider adjustments for storm movement, orographic effects, areal reduction, or other factors.
- Test the sensitivity of assumptions and choices using the chosen hydrologic model.
- Consider the use of different storms for different purposes.

4. <u>Hydrologic Model Development</u> - Recall that the normal objective of the hydrologic model is to accurately reproduce flood events, not to concoct a series of rare factors which together produce a rare event. For example, some planners may combine high antecedent moisture, an adjusted "third quartile" storm for late peaking, worst case storm movement assumptions, fully dense developed conditions, and structural blockage to "reproduce", for example, the 100-year event. In fact, the event reproduced is a much rarer occurrence than the 100-year event since the combination of all the factors must be the joint probability of each independent or semi-independent occurrence. While this may lead to conservative assumptions it does not reflect reality... it does not reproduce a flood event of known frequency.

A typical hydrologic model, either hand or computer generated, accomplishes four functions: runoff hydrograph generation, stream reach routing, pond routing, and combining hydrographs. Runoff hydrograph generation (Hydrology chapter) includes: storm characteristics, losses, and total convoluted hydrograph development. Stream reach routing (Open Channel chapter) involves: moving a stream flood wave from point to point, moving it in time, and changing its shape due to attenuation. Pond routing (Storage Facilities chapter) includes: development of stage-discharge and stage-storage curves. Hydrologic models may be developed for a number of scenarios. Sometimes a model is developed for specific past conditions to assess its ability to mimic historical storm information. Urbanization adjustments to a frequency record can be made in accordance with information in the Hydrology chapter. Often one, or more, future development scenarios are developed to allow the planner to evaluate "what if" questions. Future development normally involves estimates of future land use for a dated "planning horizon" or for full buildout of a basin given existing or anticipated zoning.

Depending on the purpose of the planning various combinations of structural or non-structural practices (such as land use controls) are considered against a developed set of criteria. Usually the hydrologic and hydraulic models are used in tandem since a hydrologic model provides stage-frequency information while a hydraulic model gives stage-discharge data.

Typical steps and considerations in hydrologic modeling are:

- development of basic data and information,
- development and calibration of runoff parameters,
- delineation of sub-areas,
- analysis of historic events and development of unit hydrograph parameters,
- development of a typical existing and future conditions model with one or more future development alternatives,
- analysis with the model of several scenarios (often in conjunction with the hydraulic model), and
- development and testing of recommendations.

There is a wide variety of calibration and modeling approaches depending on available data, complexity of the watershed, detail required and modeling approach. If peak, discharge frequency or flood hydrograph information is available in the watershed for historic events the planner can develop watershed characteristics above the gage location and seek to calibrate the model to these events. Antecedent moisture is accounted for through suitable curve-number (or other parameter) adjustment. It is usually easier to adjust parameters, within reasonable bounds, for volume matching first. Then adjust the shape of the runoff hydrograph through adjustments in lag time estimates, routing parameters and adjustments for inadvertent ponding behind undersized culverts, bridges, etc.

Approximate adjustments for within watershed ponding in swampy areas, behind undersized structures or through detention can be made through adjustment of the peak flow. This can be accomplished in computer modeling through lengthening of the lag time. There may be different adjustments necessary for different frequency floods. This method avoids breaking the watershed into too many small sub-basins while maintaining suitable accuracy for most applications. However, if sub-basin timing of combinations of peaks close to the sub-basin outlet is of key importance this method will delay an individual peak longer than an actual reservoir routing procedure.

The Soil Conservation Service provided a table for such an adjustment for small sub-basins in an early version of its TR55 (SCS, 1981). While this version has been superseded, the adjustment factors have been found, through testing by detailed pond routing of many small basins, to be fairly accurate for a first approximation of the peak reducing impacts of many small ponds within basins or for model calibration when there is reason to believe flat swampy areas or storage behind structures plays an important role in peak reduction. Tables 14-2 and 14-3 reproduce two of these SCS adjustment tables. The factors given are multiplied by the predicted peak flow to give a lower, adjusted peak flow. This peak can then be matched through trial and error adjustment of the lag time in the sub-basin. Often a fit equation can be easily developed for lag time prediction if a number of minor sub-basins are involved. This method should not be used for relatively large basins, nor just upstream from key design points.

If gage information is available on a similar watershed in the area, or another point in the same watershed, transfer of that information can be accomplished if regression equations are available through a simple ratio:

$$Q_1/(aX_1^b * cY_1^d * ...) = Q_2/(aX_2^b * cY_2^d * ...)$$ (14.3)

where: Q_i = regression predicted discharge or volume
X_i, Y_i = regression independent variables such as watershed area, slope, etc.

Table 14-2 Adjustment Factors Where Ponding And Swampy Areas Are Spread Throughout, Or Occur In Central Parts Of The Watershed

Percentage of Pond or Swampy Area	Storm Frequency (years)					
	2	5	10	25	50	100
0.2	0.94	0.95	0.96	0.97	0.98	0.99
0.5	.88	.89	.90	.91	.92	.94
1.0	.83	.84	.86	.87	.88	.90
2.0	.78	.79	.81	.83	.85	.87
2.5	.73	.74	.76	.78	.81	.84
3.3	.69	.70	.71	.74	.77	.81
5.0	.65	.66	.68	.72	.75	.78
6.7	.62	.63	.65	.69	.72	.75
10.0	.58	.59	.61	.65	.68	.71
20.0	.53	.54	.56	.60	.63	.68
25.0	.50	.51	.53	.57	.61	.66

Source: SCS TR55, 1981

Table 14-3 Adjustment Factors Where Ponding And Swampy Areas Are Located Only In Upper Areas Of The Watershed

Percentage of Pond or Swampy Area	Storm Frequency (years)					
	2	5	10	25	50	100
0.2	0.96	0.97	0.98	0.98	0.99	0.99
0.5	.93	.94	.94	.95	.96	.97
1.0	.90	.91	.92	.93	.94	.95
2.0	.87	.88	.88	.90	.91	.93
2.5	.85	.85	.86	.88	.89	.91
3.3	.82	.83	.84	.86	.88	.89
5.0	.80	.81	.82	.84	.86	.88
6.7	.78	.79	.80	.82	.84	.86
10.0	.77	.77	.78	.80	.82	.84
20.0	.74	.75	.76	.78	.80	.82

Source: SCS TR55, 1981

Each recorded event is assessed to determine the best unit hydrograph and loss rate parameters. Adjustments are made and then tested on other storms, if available, to "verify" the choices. The eventual goal is to develop good average parameters to use for synthetic event modeling of design storms. Separate sets of parameters may need to be developed for divergent storm frequencies.

Sub-areas are delineated to balance considerations for: selection of sub-basins of approximately equal size; the time-step selected for the fastest peaking sub-basin; tributary inflow points; stream gage locations; uniform land use or soil type within a sub-basin; the need for computation points; in damage reaches, below pond routing sites; and for future site location considerations.

For larger basins the development of estimates of baseflow may be important. This is especially true if the soil is quite porous; quick return flow is known to exist; or karst or rock entrances may influence the infiltration, inflow and reappearance of flow.

Channel routing methods and parameters can be determined from gage data, from backwater computations, rating curves or estimated. See Open Channel chapter for more information.

Reservoir routing should be done for all major ponding and undersized road crossings with high fill (i.e. potential for significant storage). See Storage Facilities chapter for a discussion of reservoir routing.

Often the timing of the runoff hydrographs is significant for flood reduction purposes. Given storm stationarity assumptions certain sub-basins will contribute most directly to flooding in certain locations. These sub-basins would be initial targets for flow reduction since decreases in peak flow in these sub-basins would translate most directly to decreases in flooding. Several computer models facilitate such analysis indirectly. The Penn State Runoff Model (Aaron & Laktos) provides tables for direct timing comparison.

5. Hydraulic Model Development - For a large open channel system hydraulic modeling may take several forms. Normally it involves developing one-dimensional, steady, gradually varied backwater computations. Use of multi-dimensional or unsteady models are beyond the scope of this text. Hand computations for backwater profiles are almost never done outside of classroom exercises; though it is important to understand the methodologies employed by the computer software used.

Typical steps in the development of a backwater model include the following.
* Select and prepare input for cross section locations.
* Define Manning's "n" values using a combination of field reconnaissance and analytical calculations.
* Determine flood methods for bridges and define openings, overtopping information and routing locations.
* Develop a series of flow profiles for discharges from hydrologic modeling or other information for input for stage-discharge curves for Modified Puls routing as appropriate.
* Calibrate model to known flow and frequency information, regression equations or high water marks for known storms adjusting cross sections for dead flow areas, Manning's n changes or flow combining and timing of peaks.

The two most popular public domain open channel backwater models are the Corps of Engineers HEC-2 (COE, 1982, 1988) and the Federal Highway Administration WSPRO (GKY, 1992, FHWA, 1986). Each model has its advantages and disadvantages. The HEC-2 model allows for the easy calculation of floodways, channel improvements and removal of "dead flow" areas. WSPRO has more flexible, and arguably more accurate, bridge modeling routines and provides the capability for multi-opening modeling. It uses a pseudo two-dimensional analysis of losses. Presentation of output and ease of use of documentation is a matter of preference. Sometimes modelers use both models to check particularly unique bridge situations for comparison purposes. Often the Federal Highway HY8 model (GKY, 1992) is used for culvert rating curve and loss calculations and results input into one of the backwater models. Both models give similar results for standard backwater computations in open channels.

Spacing, location and alignment of cross sections must be selected to fully represent the flow condition as a one-dimensional problem. Cross section spacing depends on changes in roughness, shape, slope, discharge, and abrupt transitions, and locations where calculated information is required. Cross sections should be spaced such that channel conveyance ratio between successive sections is between 0.7 and 1.4. See Open Channel chapter for more information on open channel flow. Conveyance is defined as:

$$k = 1.49/n \ R^{2/3} \ A \qquad (14.4)$$

where: n = is Manning's n
      R = is the sub-section hydraulic radius, ft
      A = is the flow area, ft^2

Flow lines should cross the center line of the section at right angles. For high flows the thalweg of flow tends to shorten and cut across point bars and bends. Flow in overbank areas may be at a different orientation than the main channel. For multiflow profiles flow lines should be an average of the low and high flow lines or separate sections should be input. Select cross section computation points to represent the average cross section.

Flow tends to contract through constricted openings at about a 1:1 downstream to cross-stream slope. Expanding flow tends follow more of a 4:1 downstream to cross-stream slope. Water outside these constriction and expansion lines does not contribute to the flow area. Flow may actually be upstream in some areas as eddy currents are set up. Elimination of these dead flow areas should be accomplished by redefining the cross section or by taking advantage of the capabilities of HEC-2 for encroachment calculations.

Manning's n or other loss coefficients should be appropriate for the flow stage, time of year and land use. For example, resistance to flow would decrease as flow rises above grass level in the overbank but would increase significantly as it encounters the leaves of the lower tree limbs. Typically conservatively high values are chosen for flood height estimates while conservatively low values are chosen for bank protection design.

The starting point of the backwater profile should be sufficiently far downstream that minor errors in assumed starting elevation would even out by the time the profile reaches the design reach. Several starting elevations or methods should be tried to test for model sensitivity if there is no rated control downstream.

Model output should be carefully scrutinized since results can be generated which are in error but do not stop the program computations (Ogden, 1989, Hoggan, 1989). Check for large changes in top width or conveyance, flow, slope, computed elevation, or velocity from section to section as a start. Channel Froude number should be checked to ensure the flow is in the assumed regime: subcritical or super critical. Flow velocities should be examined to see if they are reasonable. Cross sections where critical depth occurs should be noted and checked. If they are few and sparse look for channel geometry or slope errors. If critical depth occurs at several cross sections in sequence it is likely that supercritical flow occurs through the reach and an additional supercritical backwater run may be warranted. Review all bridge sections to ensure flow is reasonable and modelled correctly.

Regulatory floodway determination can be time consuming. Time spent in following a logical approach is usually less than the running of many haphazard trial backwater profile runs. Composite Manning's n values in overbank areas can often give misleading results and should be avoided. The loss of valley storage can actually increase the flows. If, through hydrologic modeling and routing, the flows increase by more than 10 percent after floodways fringe development is in place different flow values may be warranted to establish the floodway locations. Floodways may be restricted if anticipated development within a flat floodway fringe area raises the elevation of the flow such that a large area is brought into the floodplain area beyond the actual 100-year non-constricted limits. Floodway determination generally follows the following steps (COE, 1978, 1988).

- Check locations for special consideration based on specific land use or zoning. Look for any conditions which might dictate a variation from standard floodway determinations.
- Develop and run natural profiles through the reach.
- Check and eliminate locations for constriction which have excessive velocity or existing erosion.
- Using a model (HEC-2 has automatic floodway methods typically starting with methods 4, 5 or 6) develop floodway locations for maximum one foot surcharge and approximately equal conveyance from both over banks.
- Check and adjust for: excessive velocities, excessive surcharge, smooth alignment (no "balloons"), local width minimums or other local requirements. Use Hec-2 Method 1 for this if HEC-2 is used.

6. System Alternatives Development and Analysis - There are many methods for developing an analysis scheme for the watershed. The actual procedure chosen depends on the problems identified in step 1 of the standard master planning procedure. Brainstorming meetings are particularly effective if solutions are not limited to conventional channel improvements or detention. See Appendix A for an example of a process for multi-option development.

After a preliminary assessment of the situation and possible solutions on the map, make field reconnaissance of known flooding or erosion locations taking note of types of structures, relative worth, out buildings, fences, etc. Look for evidence of geomorphic adjustments of the channel in terms of widening, deepening, large movement of sediment, lateral instability, headcutting in the main channel or tributaries, indications from tree shape and size, etc. Look for evidence of debris problems, structural failure of conveyance or crossing structures.

Assess the physical possibility of different types of structural solutions such as channel widening or deepening, levees, possible reservoir or detention locations, high flow channels, diversions, pumping, etc. Assess non-structural solutions such as home raising or moving, flood proofing, flood warning, etc. Checking the timing of hydrographs may reveal other locations for reservoirs which provide indirect peak flow reduction from a side tributary. See Appendix B for a discussion of non-structural solutions.

After screening likely alternatives modify model for each feasible alternative. Modify outflow-storage relationship for reservoirs and hydraulic geometry for channel adjustments. Various combinations of alternatives may serve to enhance any one used alone. Look for opportunities to integrate solutions into the character of the neighborhood or area. Tennis courts and soccer and baseball fields have been used for detention ponds. Greenways, walkways and linear bike trails work well for floodplain areas. Wetlands can make wonderful strolling areas if boardwalks are provided.

7. Benefit and Cost Analysis - Masterplans that seek to compare alternative flood control schemes based on cost objectives will contain, at minimum, a rudimentary benefit-cost or other economic analysis (such as cost effectiveness). Depending on the type of master plan perform either detailed or approximate cost-benefit analysis through the development of stage-damage curves, or the use of standard curves found in publications such as Johnson (1985), Arnell (1989), COE (1985), COE (1988), or the latest information on depth-damage statistics available from FEMA.

Table 14-4 gives generalized residential depth damage information based on FEMA data through 1986 (FEMA, 1988). Its use is appropriate when no other information exists for non-velocity zone damages. The water depth in this table is relative to the top of the first finished floor excluding the basement. Post-flood survey information or site specific appraisals would provide significantly better information than Table 14-4. Losses for non-residential property should be estimated based on field surveys following techniques given in COE (1988), COE (1985) or other available information.

Table 14-4 Residential Depth Percent Damage - Non-Velocity Zone

Water Depth (ft)	Percent Of Structure Value			
	Two Floors No Basement	Two Floors With Basement	One Floor No Basement	Contents First Floor And Above
-2		4.0		
-1		8.0		
0	5.0	10.8	7.6	7.3
1	9.0	15.2	13.5	10.4
2	13.0	20.8	20.6	18.0
3	18.0	23.4	26.8	22.5
4	20.0	27.8	28.8	28.1
5	22.0	32.7	29.9	33.1
6		37.7	40.7	38.9
7	26.0	43.6	42.8	43.9
8		48.6	44.0	49.8
9		50.8	45.0	
10	38.0	52.9	46.3	57.9
11		54.9	47.1	
12		56.9	48.3	
13		58.9	49.0	
14		60.0	50.0	
15			50.1	

Source: FEMA, 1987

Losses (and therefore benefits) from other than physical flooding damage reduction are more difficult to determine. The magnitude of such losses is not well documented and accurate estimation may involve the compilation of data from numerous sources. Such costs are not normally large compared to flood damage, but can be significant in situations such as a location where flood warning and preparedness occurs more often than actual flood damage. Such "shadow costs" fall into nine categories:

- Income Loss Costs - are losses due to loss of income due the halting of the production or delivery of goods and services. Losses are only suffered if they are not recouped by delaying delivery or production or when additional costs are incurred because of a delay. Losses may occur directly due to shut-down of businesses or indirectly due to the closing of roads or shut-down of utilities.

- Emergency Costs - include efforts taken to monitor or forecast flooding; police, fire or National Guard actions; flood fighting efforts; victim relief and aid costs; and administrative costs.

- Traffic Routing Costs - include the costs for added mileage driven by each car rerouted by flooding or its threat and the time costs for each passenger.

- Floodproofing Costs - are incurred for both a prior floodproofing ("dry floodproofing") and actions taken during a flood ("wet floodproofing").

- Costs Of Flood Insurance - are incurred for the administration of every policy written for property in a floodprone area. Costs are in the order of $100 per policy. Specific information is available from the Corps of Engineers for any given year.

- Temporary Relocation and Reoccupation Costs - include living expenses incurred by flood victims who are forced to find temporary shelter during and after a flood. Reoccupation costs include all opportunity and real costs for arranging the repair of the flooded property.

- Costs for Modified Land Use - can be incurred when property is not utilized to its full potential because of the threat of flooding.

- Land Market Value Reduction Costs - are realized when the land prices are depressed due to the location of the parcel in a floodprone area. These costs are determined through comparison with values of adjacent, non-floodprone properties.

- Erosion Costs - can be incurred when flooding and associated high velocities erode property or such erosion damages structures.

A benefit-cost analysis follows the steps indicated in Table 14-5.

Table 14-5 Steps In Typical Benefit-Cost Analysis

Existing Condition Costs

Step 1	Delineate the Affected Area and Select Calculation Reaches
Step 2	Establish Elevation-Frequency Relationships
Step 3	Develop Depth-Damage Relationships
Step 4	Calculate Non-Physical Depth-Damages Estimates
Step 5	Calculate Damage-Frequency Relationships
Step 6	Calculate Expected Annual Damages

Future Condition Costs

Step 1	Establish Economic and Demographic Base
Step 2	Project Land Use Changes
Step 3	Establish New Floodplain Inventory

Step 4   Establish New Elevation-Frequency Relationships
Step 5   Calculate New Non-Physical Damages
Step 6   Calculate New Equivalent Annual Damages

Project Costs

Step 1   Calculate Project Capital and Financing Costs
Step 2   Calculate Project Operating and Maintenance Costs

Benefit Calculation

Step 1   Calculate Inundation Reduction From Structural Measures
Step 2   Calculate Cost Reductions for Non-Structural Measures
Step 3   Calculate Other Benefits

B/C Ratio

Step 1   Calculate Discounted Benefits Over Project Life
Step 2   Calculate Discounted Costs Over Project Life
Step 3   Ratio Discounted Benefits and Costs

Damage-frequency curves are developed for each calculation or damage reach. That is, all damages for a certain damage reach are related to the depth at a certain point within the reach. Damage-frequency curves are plotted for all the damage reaches and, often, by damage category. An aggregated curve gives the total damage-frequency curve for the project location. The expected annual damages can be calculated from combining a depth-damage and depth-frequency curve to develop a damage-frequency curve. The expected annual damage is the area under this curve.

The expected annual damage calculation can be done in tabular form with columns for frequency, frequency interval, damage, average damage for the frequency interval, expected damage for the interval (multiplying the two interval columns) and finally the accumulated damage total which is the sum of the previous column. The narrower the interval (i.e. the more values which are calculated) the more accurate is the procedure mathematically.

Minor System Flooding

The minor or "convenience" system is often designed to the 10-year or even more frequent flood. This system is often overwhelmed by intense thundershowers. Although any one site may not be overwhelmed on the average more often than the design frequency, there may be many such flood events in a major urban area over a design frequency time span.

The approach to minor system master plans is to identify and attempt to quantify flooding problems. To do so this often requires a concerted effort to involve local residents and political leaders. Often a "problem" is not anything more than an inconvenience. Therefore, it is important to determine the design policy and performance level-of-service of the municipality and prioritize problems based on a professionally driven evaluation system such as: life and health, residential structural damage, non-residential structural damage, utility damage and nuisance (Debo, et al., 1979).

Quantifying these problems involves use of a simple methodology for peak (or sometimes whole hydrograph) flow estimation and a determination of whether the problem is isolated or part of a system-wide deficiency. It is important not to simply pass the problem downstream to the next undersized structure. Solutions are normally straight forward and cost and materials estimates are often included.

Data collection and analysis can be very time consuming and expensive. It is often not worth the expense to analyze the whole of a municipality system. If the system is old it has probably been tested above its performance capabilities. If an effective complaints system is in existence those complaints should be the source of determining which parts of the system are analyzed in any detail. Simply because a system backs water up does not constitute a problem per se. It is actually advantageous to store runoff throughout the system in planned locations both for economic reasons having to do with pipe size and for pollution and erosion reduction reasons.

Often municipalities will use various levels of complexity for analysis of the system. For example, if the Rational Method was used for the design of systems it is often applied to much of the system as a first screen. This can be done very efficiently using a GIS or AM/FM system. For those systems which: (1) exhibit flooding based on this analysis, (2) violate the assumptions in the Rational Method (such as high tailwater conditions) or (3) are identified through complaint files or inspection reports as problem areas, a more rigorous analysis using hydrograph routing and, often, full dynamic routing can be done for these portions of the drainage system. Typical steps for this type of approach are provided in Table 14-6.

Table 14-6 Typical Steps In Minor System Flooding Study

Step 1   Determine Design Criteria and Levels of Service Standards for Analysis
Step 2   Assess Complaints and Inspection Reports
Step 3   Develop System and Runoff Information For the Rational Method or other Simplified Hydrologic Method
Step 4   Perform Rational Method Screening on Some or all of the System
Step 5   Prioritize Study Areas and Identify any Other Areas to be Further Studied Based on Other Factors
Step 6   Field Check Sites, Interview Residents
Step 7   Perform More Rigorous Analysis Including Field Data Collection
Step 8   Develop and Assess Alternative Solutions
Step 9   Develop Cost Information and Estimates (B/C Analysis as Appropriate)
Step 10   Prioritize Locations for Capital or Remedial Maintenance Improvements

---

Data collection, if not available, can be performed using automated methods including hand-held dataloggers and automated mapping and facility management systems. Even the modeling can be tied to these systems to provide a complete system design and management package.

Alternative solutions developed under Step 8 can include such things as: detention facilities, inlet modifications, pipe upsizing or ditch enlarging, parallel piping, rerouting, etc. Solutions which include allowing ponded water to escape should be checked downstream through the system. It can be shown that the system should be checked to a point downstream where the site draining through the structure comprises about ten percent of the total drainage area to that downstream check point (Debo & Reese, 1992). Beyond that point in the system the increase will normally have negligible impact.

Pre-Development Planning

This type of a study involves not only the correction of problems but the search for land development alternatives to avoid problems in the future and to enhance the storm water resources of a region. There are a number of ways to go about such a study but they typically involve an estimation of both existing and future flows. Often, more than one future land use development alternative is used. Hydrologic studies are the norm. Much of the process is similar to the major system flood study except alternative solutions have a great deal of flexibility and multi-objective planning plays a greater role.

An example of such a study will illustrate the approach. Master plans were developed for Mecklenburg County, North Carolina. Appendix A gives details of this approach which will only be summarized here. The master planning effort followed the typical three step approach: (1) identify the needs, goals and opportunities; (2) identify and explore constraints on the possible solutions; and (3) develop and execute a set of objectives and an overall technical approach. Step one variables included: major and minor channel flooding and erosion, water quality degradation due to urban runoff, undersized stream crossings, recreational area locations, floodplain management strategies and enhancement locations. A number of political, social, financial and technical constraints were identified and methods developed to incorporate all concerns.

The technical approach consisted of the development of a system of computer-generated overlays using customized software and existing land use conditions and ultimate development conditions for hydrologic modeling. To assist in the location of areas in each of the watersheds where multi-objective needs, goals and opportunities could be met, a series of overlays were developed depicting: all of the problem location variables listed above, critical flora and fauna, historical and cultural locations, parks and recreation locations, planned greenways, wetlands, planned land use conflicts, transportation arteries existing and planned, planned and existing intense development areas, existing water features and suitable topography.

Based on these variables, a number of possible locations were identified where planned facilities or non-structural efforts should concentrate. Each suitable site was then checked against soil suitability, land ownership, utilities and specific environmental concerns. Locations and site-specific approaches were developed, costs estimated and alternatives placed in a priority list for consideration. The second phase of the plan is the actual design and construction of key facilities.

Storm Water Quality Study

There are some factors which may mitigate against performing detailed storm water quality modeling:

- The state-of-the-art in storm water quality modeling is approximate even with available data.
- The availability and quality of existing data for most areas is poor.
- It is often difficult to ascertain cause and effect in surface runoff pollution cases, especially if chemical interaction is of concern.
- Methods for treating surface water pollution problems even if they are well known and quantified are often only marginally effective and expensive to build and maintain.
- Most water quality requirements do not involve modeling but a simple design standard (such as capture of the first half-inch of runoff).

Despite these drawbacks storm water quality modeling is often done on a site specific or watershed scale. Storm water quality master planning is mandated under the 1987 Water Quality Act and implementing regulations for new developments or significant redevelopments. In the future, as the knowledge of the cause and effect of receiving water pollution impacts grow and the knowledge of its mitigation increase surface quality modeling will grow in prominence. Water quality modeling is most useful and effective when there are specific pollutants of concern and known impacts on receiving waters are to be mitigated.

The objectives of a typical water quality study are (Huber, 1986):

- characterize the urban runoff;

- provide input to receiving water analysis;
- determine effects, sizes and combinations of control options;
- perform frequency analysis on quality parameters;
- provide input to cost-benefit analysis.

A water quality study normally includes or concentrates on the reduction of surface water pollution to some receiving water. Storms of interest are less than the 2-year storm since the vast majority of the pollution comes from the many minor storms rather than the few major ones (Roesner, et al., 1990, Livingston, 1986). Often the study must take into account requirements of the 1987 Water Quality Act.

Pollution impacts are both acute (short term) and chronic (long term). The response time for pollution impacts in lakes and bays is in the order of weeks to years (Hydroscience, 1979). However, for the type of ponds and streams commonly encountered in urban runoff the response time for such constituents as dissolved oxygen, bacteria, sediment or nitrogen is in the order of hours or, at most, days. Statistical or regression based models are simple to apply and require little calibration. Physically based models have more flexibility in application, can generate pollutographs for single storms more effectively, and can be extended for "what if" analysis with more confidence.

Sometimes modeling of concentrations and impacts is not strictly necessary. In many cases the inability to show or quantify direct environmental impairment from storm water runoff makes the development of designs to mitigate some specific impairment difficult to quantify or justify. Typically the solution to the cause-effect dilemma is to take one or more steps back from the specific impairment or degradation issue and determine storm water quality master plan features based on: (1) a numerical chemical water quality standard or designated use basis; (2) a biological indicator or assessment standard; or most commonly (3) on a design criteria basis where Best Management Practice (BMP) designs must meet some simple standard (for example, capture and treatment of the first half-inch of excess rainfall). If these simplifying steps are appropriate then detailed water quality modeling to determine cause and effect is not necessary and simple estimates of annual or single storm pollution loadings are all that is required. In these cases a simple loading model is developed and BMPs placed within the watershed to reduce these loadings to some target values or percentages. In many cases storm water or receiving stream sampling in support of the study or to guide its ultimate application and measure its effectiveness is desirable.

There are then a number of variations of storm water quality master plans depending on the results and problems.

- If protection of a lake, bay or estuary is the main objective long term loading estimates and impact assessment is normally done. An attempt is made to link these loadings to land use. In these cases a lake or estuary model is applied in conjunction with the loading estimates to determine long term impacts of loadings. Base flow loadings are also important in these types of studies with concentration values taken from sampling or standard figures for the region. Alternative land use or development controls are assessed in conjunction with structural or non-structural BMPs. Typical of this type of study is a plan to protect a drinking water supply from the impacts of urban development or agricultural practices.
- If estimations of pollutant thresholds, concentrations and frequencies in either runoff or receiving waters are of concern the most effective modeling will generate pollutographs for individual storms or run a continuous simulation for a period to estimate pollutant removal efficiencies of structural BMPs. The SWMM (Huber, et al., 1981), STORM (HEC, 1977a), DR3M-QUAL (Alley & Smith, 1982) or HSPF (Johanson, et al., 1980) models are especially effective for this purpose. Statistical methods for frequency of pollutant level exceedance can also be used as a pseudo-continuous simulation (Hydroscience, 1979, USEPA, 1983). No storm water quality model should be used without sufficient model validation data if actual

estimates of specific pollutants are to be made (Huber, 1986). Relative changes in pollutant loading can be estimated without site specific data using generalized values. The models are better at predicting relative changes in unknown values than in predicting the initial values to begin with.

- If the design of BMPs to meet simple annual loading reduction targets or design criteria is appropriate then simple event mean concentration based annual loading methods can be effective if estimates of pollution reduction for BMP designs can be made. In these cases annual or seasonal loadings are estimated based on rainfall and runoff volumes. Often these are tied to land use both for runoff volume and event mean concentrations of pollutants. Then locations for effective placement of regional or site BMP controls are found and assessed. Solutions include land use controls, regional BMPs, site development restrictions or design criteria, or other programmatic BMPs such as street sweeping, limitations on herbicide or pesticide usage, maintenance practices, etc. This is probably the most common kind of study performed by the non-specialist in storm water quality modeling.

There are a number of simple methods available to make these estimates. The two most popular are the USGS pollution loading regression equation (Tasker & Driver, 1988) and the "simple method" developed by the Washington Metropolitan Council of Governments (Schueler, 1987) and variations of this method. Both of these methods are based on the data base generated during the national Urban Runoff Program (NURP) study (USEPA, 1983). Often these data are supplemented with local data where available and a weighted loading or event mean concentration estimated. This type of loading calculation lends itself well to GIS and to spreadsheet calculations. The Universal Soil Loss equation (Barfield, et al., 1985) can be used to estimate sediment loss and accumulation though there is a wide range of sediment loss and delivery depending on the sediment source, type and delivery mechanisms. Often sampling is the best way to estimate sediment accumulation, or use of regionalized delivery ratio mechanisms tied to watershed area (Scheuler, 1987).

BMPs are designed or developed to treat pollution "problems" with either target percent reductions or determinations of the "maximum extent practicable" for reduction based on various possible combinations of BMPs. Both structural and non-structural BMPs are used. Non-structural BMPs are sometimes all that is possible in highly urbanized areas with little room left for structural solutions. Regional structural treatment is often most cost effective and provides the greatest assurance for a long term commitment to maintenance. Usually rules of thumb can be developed for first phase screening analysis of possible sites for structures using such things as available volume per acre treated or pond surface area related to drainage area. See Storm Water Quality Management Programs chapter for storm water quality plan development and Best Management Practices chapter for BMP design guidance.

14.6 Limitations Of Master Plans

As important as master plans are to any comprehensive storm water program, by themselves they will not solve problems or prevent flooding, drainage or water quality problems. The master plans represent a blue print for action that must be taken if these problems are to be solved or prevented. Too often people see the master plan as the end product and forget that if the plans are not implemented little good will result from the completed work. The real work begins when the master plan is completed.

Each master plan is undertaken in response to specific problems, goals, policies, etc., that relate to the watershed being planned and the implementation tools available. Detailed plans which analyze hydrologic-hydraulic and water quality parameters to a very small scale, say 10 to 20 acres, will be much more expensive to develop than more general plans which only analyze down to several hundred acres. However as a consequence, the general plans will not address flooding and drainage problems associated with small developments and single lots. Thus many of the limitations of any master plan will be related to the scale used for plan development and engineering analysis. Often large scale planning is done over a whole area with key area small scale plans done for certain areas of interest. These "nested" plans fit into the grid of the overall plan and feed data to it.

Other factors that may influence the limitations of specific master plans include data available for plan development, political acceptance of the final plan, acceptance and enforcement of the plan by local citizens and professional groups, and available funding. In addition, keeping the plan up-to-date is very important if the plan is to remain a viable tool in dealing with storm water management problems. The plan should not become a static document but rather it should be a part of a dynamic storm water management program that changes as the problems and local circumstances change.

References

Aaron, G., and D. R. Laktos, "Penn State Runoff Model - Users Manual", January 1980 version, Pennsylvania State University, January, 1980.

Alley, W.M. and P.E. Smith, "Multi-Event Urban Runoff Quality Model", USGS Open File Rpt. 82-764, Resten, VA, 1982.

American Society of Civil Engineers, "Sedimentation Engineering", ASCE Manuals & Reports No. 54, 1977.

Arnell, N. W., "Expected Annual Damages and Uncertainties in Flood Frequency Estimation", ASCE J. of WR Planning and Mgmt., Vol. 115, No. 1, Jan., 1989.

Barfield, B.J., R.C. Warner and C.T. Haan, "Applied Hydrology and Sedimentology for Disturbed Areas", Oklahoma Technical Press, 1985 (Third Printing).

Debo, T. N. and A. J. Reese, "Determining Downstream Analysis Limits for Detention Facilities", Proc. NOVATECH, Lyon France, Nov. 3-5, 1992.

Debo, T. N., D. Westerich, T. Newsom, "Drainage Problem Categorization Study", The Columbus, Georgia Storm Water Management Program, 1979.

Federal Emergency Management Agency, "Flood Insurance Rate Review", 1987.

Federal Highway Administration, "Bridge Waterways Analysis Model - Research Report", Rpt. No. FHWA-/RD-86/108, 1986.

GKY & Assoc., "HYDRAIN - Integrated Drainage Design Computer System Version 4.0", Rpt. No. FHWA-RD-92-061, 1992.

Hoggan, D. H., "Computer Assisted Floodplain Hydrology and Hydraulics", McGraw Hill, 1989.

Huber, W. C. "Modeling Urban Runoff Quality: State-of-the-art", In: Urban Runoff Quality, ASCE Pub. Proc. Engineering Foundation Conf., Henniker, NH, June 23-27, Ben Urbanos and Larry A. Roesner ed., 1986.

Huber, W.C., J.P. Heaney, S.J. Nix, R.E. Dickenson, & D.J. Polmann, "Stormwater Management Model User's Manual, Ver. III", EPA 600/2-84-109a (NTIS PB84-198432), EPA Nov. 1981.

Huber, W. C. and J. P. Heaney, "Analyzing Residuals Discharge and Generation from Urban and Non-urban Land Surfaces", in Analyzing Natural Systems, Analysis for Regional Residuals, EPA 600/3-83-046 (NTIS PB83-223321), USEPA, June, 1982.

Hydrologic Engineering Center, "Storage, Treatment, Overflow, Runoff Model, STORM, Users Manual", Generalized Computer Program 723-S8-L7520, Corps of Engineers, Davis, CA, 1977a.

Hydrologic Engineering Center, "HEC-6 Scour and Deposition in Rivers and Reservoirs", Users Manual, Hydrologic Engineering Center, Davis, CA., 1977b.

Hydroscience, Inc., "A Statistical Method for Assessment of Urban Stormwater Loads - Impacts - Controls," EPA Rpt. 440/3-79-023, EPA, Wash. D.C., January, 1979.

Johanson, R.C., J.C. Imhoff, H.H. Davis, "User's Manual for Hydrologic Simulation Program - Fortran (HSPF)", EPA-600/9-80-015, EPA Athens GA., 1980.

Johnson, W. K., "Significance of Location in Computing Flood Damage", ASCE J. of WR Planning and Mgmt., Vol. 111, No. 1, Jan. 1985.

Livingston, E. "Stormwater Regulatory Program In Florida", In: Urban Runoff Quality, ASCE Pub. Proc. Engineering Foundation Conf., Henniker, NH, June 23-27, Ben Urbonas and Larry A. Roesner ed., 1986.

National Weather Service, Rainfall Frequency Atlas of the United States, Technical Paper 40, U.S. Dept. of Commerce, Wash. DC, 1961.

Ogden, Inc., "Master Plan for McCrory Creek, Nashville, TN", For Nashville and Davidson County, TN, 1988.

Ogden Environmental, "Guidelines for Analyzing HEC-1 and HEC-2 Input and Output", 1989.

Roesner, L. A., E. H. Burgess, J. A. Aldrich. The Hydrology of Urban Runoff Quality Management. Water Resource Planning and Management. pp. 764-780, 1990.

Schueler, T. R., "Controlling Urban Runoff", Metro. Wash. Council of Governments, July, 1987.

Simons, D.B. and F. Senturk, "Sediment Transport Technology", Water Resources Publications, Ft. Collins, CO., 1977.

Simons, Li & Associates (SLA), "Engineering Analysis of Fluvial Systems", SLA, 1982.

Soil Conservation Service, "Urban Hydrology for Small Watersheds" TR 55, 1981.

Tasker, G. D. and N. E. Driver. Nationwide Regression Models for Predicting Urban Runoff Water Quality at Unmonitored Sites. Water Resource Bulletin, Vol. 24, No. 5., pp. 1091-1101, October 1988.

U. S. Army Corps of Engineers, Floodway Design Considerations, 1978.

U. S. Army Corps of Engineers, HEC-2 Water Surface Profiles, Hydrologic Engineering Center, 1982.

U. S. Army Corps of Engineers, "Business Depth-Damage Analysis Procedures", Engineer Institute for Water Resources, Research Rpt. 85-R-5 Sept., 1985.

U. S. Army Corps of Engineers, "National Economic Development Procedures Manual - Urban Flood Damage", Engineer Institute for Water Resources, IWR Report 88-R-2, March, 1988.

U. S. Army Corps of Engineers, "Floodway Determination Using Computer Program HEC-2", Training Doc. No. 5, January, 1988.

U. S. Army Corps of Engineers, "Computing Water Surface Profiles With HEC-2 on a Personal Computer", Training Doc. No. 26, September, 1988.

U. S. Army Corps of Engineers, "Sedimentation Investigations of Rivers and Reservoirs", EM 1110-2-4000, Dec., 1989.

U. S. Army Corps of Engineers, "HEC-1 - Flood Hydrograph Package", Sept., 1990.

U. S. Army Corps of Engineers, "Hydrologic Analysis of Ungaged Watersheds Using HEC-1", Training Document No. 15, April, 1982.

U. S. Army Corps of Engineers, "Hydrologic Engineering Analysis Concepts for Cost-shared Flood Damage Reduction Studies", EP 1110-2-6005, December, 1992.

U. S. Army Corps of Engineers, "Hydrologic Engineering Studies Design", EP 1110-2-6007, February, 1993.

U. S. Environmental Protection Agency, "Results of the Nationwide Urban Runoff Program", Vol 1., Final Report, Washington, DC, 1983.

Water Resources Council, "Economic and Environmental Principles and Guidelines for Water and Related Land Resource Implementation Studies", Wash. D.C., 1983.

Whipple, W., N. S. Grigg, T. Grizzard, C. W. Randall, R. P. Shubinski, and L. S. Tucker, "Stormwater Management in Urbanizing Areas", Prentice-Hall, Inc., 1983.

Appendix A - Example Multi-objective Master Plan

Introduction

Mecklenburg County, North Carolina has responsibility for storm water management in the populous though unincorporated portions of Mecklenburg County that are not within State easements, and for all drainage systems which drain an area greater than one square mile throughout the county. Because of expected rapid development of the rural areas of Mecklenburg County, the County's desire to prevent and reduce potential flooding and erosion problems and a desire to maintain environmental quality, the County retained an engineering team led by Ogden Environmental and Energy Services, Inc. to develop storm water master plans and develop other storm water management related studies and products. Four Westside watersheds, Long Creek, Paw Creek, Coffey Creek and Steele Creek, were chosen initially for analysis. Other areas have been added since the initial analysis was completed.

Goals and Objectives

Goals and objectives were developed by the consultant, a citizens advisory group and the Mecklenburg County Engineering staff during the initial phases of the study. The goals and objectives were the following.

1. Develop an engineering description of the watershed drainage systems and major physical and institutional factors impacting drainage.
2. Develop a GIS system for depiction and management of land use, basin boundary, and soil overlays for use in management of urban growth and analysis of impacts on the drainage system.
3. Develop and calibrate hydrologic models of each watershed in such a way that they are integrated into the total storm water management program for Mecklenburg County and can be used and kept up-to-date by Mecklenburg County Engineering staff.
4. Develop GIS-based pollution models for prediction of pollution loadings and prediction of relative impacts of development and BMPs.
5. Use developed models and field analysis to:
   - determine flows throughout the watershed;
   - identify and analyze existing flooding, pollution and erosional problems;
   - identify and analyze potential flooding, pollution and erosional problems based on probable future development; and
   - identify alternative solutions to prevent or alleviate flooding, pollution and erosional problems.
6. Integrate possible solutions to flooding or erosional problems into the bigger picture of watershed development, environmental enhancement and possible multi-objective uses.
7. Provide conceptual designs for recommended alternatives.
8. Identify institutional and regulatory factors that impact storm water management and recommend possible changes or modifications.
9. Develop an automated inventory and related database of all surface drainage structures of size equivalent to 36 inches in diameter and greater.
10. Research and review all pertinent documents related to storm water management in Mecklenburg County.
11. Prepare a formal report containing all information and recommendations.

12. Involve citizens, the County Drainage Advisory Board and other county staffs in development of the master plans.

Approach

It was determined that a multi-objective master plan was the best approach for this situation. There was a desire to both handle existing, and avoid future, problems. Both minor feeder streams and major creeks were to be involved as well as sources of point and non-point pollution. There was a desire to integrate any structural solutions into the overall planning process for land use, roads, parks and greenways.

In order to do this and to narrow down possible sites in a cost effective way the three-step master planning process was used and an innovative methodology developed.

The Three-step Process

Step 1: Identified Need - A comprehensive search was executed to identify the real and perceived needs, both physical and institutional, within the County.

- Federal Emergency Management Agency (FEMA) profiles and other US Geological Survey backwater profiles were studied to locate homes within or near the regulatory floodplain. Approximate rating curves were developed in areas above the regulatory floodplain where homes were thought to be in or near the floodplain. These locations were identified through use of aerial photos.
- New developments within the floodplain or that would have an expected impact on flooding were identified by county staff.
- Existing and future condition hydrologic models were developed to predict the impacts of development on flood profiles.
- The Mecklenburg County staff was asked to identify and locate any known flooding complaints or problems within each area.
- Neighborhood leaders were identified by the Planning Commission and contacted by letter to help identify any other flood problem areas.
- A literature search was conducted to identify any other studies done to locate flood problems. A number of studies were uncovered.
- Stream quality data was analyzed to determine pollutants and possible sources.
- Each stream was walked to locate areas of channel erosion and areas of obvious water quality degradation based on a visual screening.
- Known point sources of pollution were investigated.
- An inventory of all structures greater than 36 inches in equivalent diameter was done assessing condition, size, location and other parameters.

Apart from a significant number of institutional problems (Ogden, 1989) a set of physical problems and opportunities was uncovered including:

- scattered major creek flooding;
- overall erosion and downcutting of large creeks due to geomorphic processes operating throughout the Piedmont region;
- minor feeder stream nuisance flooding due to lack of regulatory control and maintenance on these streams and urban development impacts;
- minor feeder stream erosion due to uncontrolled urban development and headcutting at the outfalls to larger streams;
- roadway overtopping due to urban development and inadequate state design standards;
- pollution impacts from agriculture, industrial sites and point sources;

- existing pond and flood control structure deterioration or dam-safety based rehabilitation costs which were too much to justify;
- park and greenway planning active in the area leading to possible opportunities for multi-objective sites;
- roadway planning in the area leading to opportunities for combined use of roadway embankments or, at least, bridge sizing for future conditions;
- oversized floodplains which could experience enhanced use through recovery by peak flow reduction;
- urbanization impacts in erosion and peak flow increase predicted (Martens, 1968, Malcom, et al.,1986).

Objectives were set for each watershed based on its unique problems and character.

Step 2: Constraints - A number of constraints to possible solutions to the problems and opportunities existed in Mecklenburg County. Many of them are common to most locations around the country. Many of the constraints had been eliminated by the County in the years since the study including, most importantly, the establishment of a storm water utility for funding and development of an effective public awareness program.

Legal constraints included a lack of control over much of the drainage system; ineffective subdivision regulations; poor detention ordinances; lack of pollution control capabilities; awkward enforcement capabilities and procedures; etc.

Organizational constraints included a dispersion of responsibility through several organizations and manpower shortages.

Financial constraints included the low priority of drainage when compared to other pressing needs. All storm water related work was funded through taxation either through annual appropriations or bond funding.

Sociological constraints included a strong developer influence mitigating tighter regulatory controls; a history of uneven enforcement; and strong state and Federal regulatory environment which prevented certain structural solutions from being realized due to wetlands concerns.

Several technical constraints, having primarily to do with lack of adequate data, existed which limited the accuracy of some of the solutions.

- First floor elevation data for the residences in the analysis were not available. Since surveying was specifically excluded from this study, map derived foot print elevation data were used as a surrogate to actual first floor elevations.
- Surveyed channel cross section information was not available, thus forcing reliance on map derived FEMA cross section and stage-discharge information. There was often great disagreement between Corps of Engineers, map, and FEMA data.
- Specific studies to determine the speed of bank retreat or channel widening were not available. Attempts were made to measure the speed of channel widening in two locations in the Westside basins by comparing field measured information to old FEMA cross sections. Since the FEMA cross sections were map derived, channel data and actual location of the sections were in question. Only in obvious cases were meaningful data derived.
- Specific studies to determine the susceptibility of various types of bank material to erosion (through the use of vane shear tests, etc.) was not available. Therefore, it was not possible, except qualitatively, to estimate possible bank erosion rates and to size detention/retention release rates to ensure that flow velocities will not cause scouring.
- A contract has been let by the County for wetlands mapping to be conducted by the US Fish and Wildlife Service. This mapping might identify a large number of locations which would be considered wetlands. This impacts much of the creek floodplain and greenway system. Mitigation of any impacts of structural controls in these areas would then be necessary. These

locations were unknown and could not be anticipated at the time of the master plan development.

- Water quality data keyed to storm event runoff were not available downstream from industrial and commercial sites in the watershed. Therefore, it was not possible, except qualitatively, to predict any impacts on water quality of developments and to size regional flood control basins to have beneficial water quality impacts.

Step 3: Technical Approach - Based on the first two steps in the overall master planning process a technical approach was developed to zero in on the solutions in a cost effective manner. A three-step screening process was developed to logically consider all the pertinent parameters, eliminate non-useful areas of the watershed and concentrate analysis efforts on the key areas where something needed to be and could be done.

The basic modeling approach was to develop and verify a hydrologic model (HEC-1) to produce reasonable flows at key locations throughout the watershed under existing conditions. Then a future condition land use scenario, based on the 2005 Plan, was developed. Flows of various frequencies were obtained from the models. These models were applied to pre-selected problems or opportunity locations and solutions sought.

Heuristics based on possible pond size, drainage to surface area ratios and volume were developed to quickly assess certain locations for flood control, pollution removal and possible erosion control, without complex analysis. Significant research into erosion in the Piedmont area was conducted. GIS pollution models were used for "what if" analysis.

After these basic tools were developed the screening process developed GIS overlays of various parameters or variables. It then put each watershed through a rigorous project team screening process involving all pertinent staff, the consultant team and citizens. Public meetings were held to obtain more input.

The initial potential site location variables were:
- downstream from an area of existing industrial or regional commercial use,
- site above known flooding location,
- site above known channel erosion location,
- known or suspected pollution impacts,
- site above reach with wide shallow floodplains,
- site below sub-basin with great expected peak flow increase for future conditions,
- site located in conjunction with existing undersized stream crossing with storage capacity upstream, and
- location of existing lakes or detention which may be rehabilitated or enhanced.

The sites that passed through this first screen were then subjected to analysis based on step-two variables. The step-two variables included:
- topography and relief suitability,
- site where new roadway construction could be used for detention dam,
- site located in proposed greenway or park reach,
- location of critical flora or fauna,
- location of wetlands, and
- location of site with historic or archeological significance.

After the sites were analyzed certain "stopper" variables were assessed prior to more detailed pre-design analysis. Step three variables were:
- site availability and land cost,
- soil suitability,
- conflict with existing or planned utilities,
- conflict with Planning Commission land use schemes, and

- environmental concerns.

Based on this three step process detention/retention and other alternatives were identified and approximate benefits and costs generated. However, due to budget constraints, no attempt was made to ascertain exact benefits in a pure B/C ratio type analysis. It is usually difficult to assign benefits to water quality and some erosion objectives, and, without actual inundation information, assign damages for flooding. Rather, benefits were developed in terms of "Cost Effectiveness" using descriptive impacts of the proposed alternatives and quantifying flood flow impacts.

Results Of The Master Planning

This flexible approach can spawn a wide variety of solutions even within the same watershed. Two examples will be given.

Paw Creek - Figure 14-1 illustrates Paw Creek and its solution locations. The basin is developed in the southeast, developing in the northeast and rural in the northwest. The watershed experienced a number of problems including flooding, spot erosion and pollution downstream from an industrial corridor in the northern headwaters. Solutions include:
- the rehabilitation of a dam at site 8 to handle flooding and pollution problems from upstream development;
- maintenance and enhancement of a sister lake wetlands area at site T2-1 to handle flooding and pollution from new development upstream;
- establishment of a sampling program at site 1 to assess water quality and look into possible joint pond construction with Parks and Recreation and greenway extension;
- alternate locations for a pair of ponds upstream from site 1 (U1B-3 and U1B-4);
- enlargement of a crossing above site 1;
- geologic assessment of a grossly undersized railroad crossing for use as planned detention;
- site surveying to determine actual site undersizing and potential for channel enlargement;
- purchase of six individual sites prior to development encroachment for reserve as possible regional pond locations.

Coffey Creek - Coffey Creek (Figure 14-2) is long and narrow and dominated by airport runoff in the upstream end. The rest of the watershed is characterized by commercial and light industrial development on the narrow east side bluffs and a potential large residential development along the rural wide west side plains.
- A string of detention lakes is suggested along the east side. They would be used as regional erosion control ponds during construction and later converted to be first flush settling ponds.
- Existing lakes along the west side will be rehabilitated and turned into regional parks to serve the residential development and handle expected runoff increases and pollution control. Several new lakes will be added.
- A linear wetland along the creek below the industrial sites will be preserved and enhanced into a wetland greenway complete with wooden walkways and nature trails.
- The airport in the north will be controlled by first flush devices according to expected NPDES permit requirements.

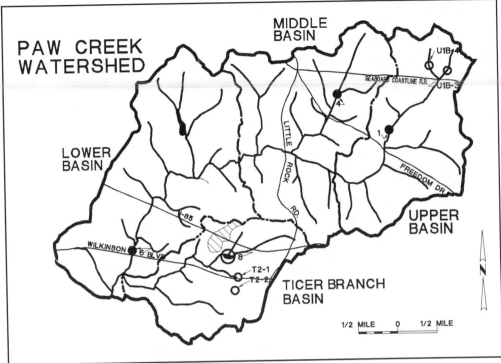

Figure 14-1 Paw Creek Master Plan

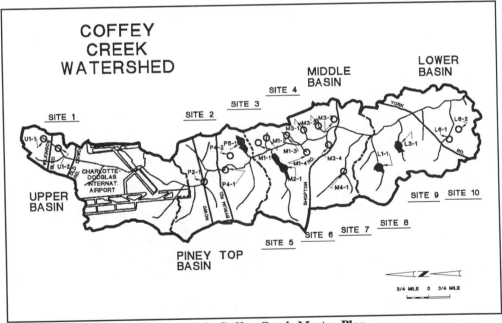

Figure 14-2 Coffey Creek Master Plan

Appendix B - Non-structural Flood Control Overview

Many flood prone areas contain residential and non-residential buildings which are located well within the regulatory floodplain. Many of these can be expected to flood often, some almost yearly. Because of the fully developed nature of the watershed above these locations and the location of structures nearly on the bank of the streams through these flood prone reaches, structural measures such as detention or channel widening are often not feasible or cost effective.

Therefore alternate methods of a non-structural nature are warranted. "Non-structural" measures (sometimes called "flood hazard mitigation") refers to those measures which tend to remove or protect persons and property from floods rather than removing floodwaters from persons and property. Non-structural alternatives are often more cost effective and less environmentally damaging than structural measures.

There is difficulty when investigating non-structural measures in estimating economic feasibility, plan formulation and local acceptance. The greatest difficulty with non-structural alternatives is that they often involve modifying human behavior, which is often more complex and more difficult to predict and control than the natural world. Estimating the effectiveness of such measures is difficult, but procedures exist and are available from the Corps of Engineers Institute for Water Resources.

Most of these measures, when considering a specific site, must be applied on a case-by-case basis. Such level of detail is beyond the scope of some master plans and is, in fact, not possible without development of a municipal non-structural program. The best approach is to institute an arsenal of flood hazard mitigation programs and activities and use them as a workman would use a set of tools. Some are appropriate in a given situation and some are not.

Besides the current floodplain regulation activities under the flood insurance program and the development permitting process, the basic non-structural approaches for these flood prone areas fall into seven general and inter-related categories: (1) floodproofing, (2) structure elevating, (3) flood warning, (4) relocation, (5) flood insurance, (6) public education, and (7) post flood programs.

Floodproofing

Floodproofing is a combination of adjustments and/or additions of features to individual buildings that are designed to eliminate or reduce the potential for flood damage. Floodproofing measures are classified as: permanent, contingent and emergency. Permanent measures include levees, floodwalls, elevation, closures and sealants and watertight cores. Contingent measures include shields, doors, movable walls, etc. Emergency measures include sandbags, earthfill and log barriers used in flood fighting.

Floodproofing can often be economic if the depth of flooding is not too great and, for contingency measures, sufficient warning is available to put the measures in place. Standard masonry walls can often withstand depths of up to six feet while brick veneer can usually withstand depths of about two feet. There are several flood proofing systems available commercially. Even low level flood proofing can provide protection for numerous smaller floods and buy time for evacuation when the larger floods come.

The Federal Emergency Management Agency has numerous publications available to assist communities and individuals to assess their options for flood proofing. Congress has made monies available to fund flood proofing and other mitigation measures for the nations 30,000 repetitive loss structures.

Structure Elevating

The raising of homes is often a cost effective measure and is a common recommendation for flood damage reduction. There are certain National Flood Insurance program regulations which must be complied with when considering home raising when the raising can be considered a substantial improvement to the structure.

The type of structures where raising is most effective are one-story frame houses built over a crawl space of at least 18 inches. Houses over basements, brick or other construction materials or slab-on-grade construction are more expensive to move or raise. This type of activity is also much more difficult for non-residential structures. Detailed publications are available from FEMA covering all technical and regulatory concerns.

Flood Warning

Flood warning has no intrinsic value of its own. It is the response to the warning that saves lives and reduces property damage. A well conceived flood warning system consists of three elements: (1) a flood threat recognition methodology, (2) a warning dissemination mechanism, and (3) an emergency response plan. Without proper attention to each of these three elements a warning system may not be effective and certainly will not achieve its maximum potential.

Whether or not a local flood warning system is appropriate depends on: (1) hydrologic characteristics of the basin and ability to predict a flood and its severity, (2) flooding frequency, (3) flood loss potential, (4) warning time in relation to benefits, and (5) the need for other capabilities which can be incorporated into the system such as road freezing information or tornado warning. With the short warning times on most urban streams a cost-benefit analysis will usually not produce sufficient tangible benefits to justify the expense. There is little time to remove property from the path of the flood, though there may be time to place contingent flood proofing measures. However, loss of life and well being is hard to quantify and may warrant a cost effectiveness consideration rather than a strict cost-benefit analysis.

Relocation

Home relocation through property acquisition is often the only method to prevent flood damages. There are opportunities to use it most effectively to remove properties that have sustained substantial damage, on a voluntary basis immediately after a flood. Some municipalities have set up a fund to purchase flood prone properties at the market rate whenever they become available for sale. In other cases a municipality must use its power of eminent domain to condemn and purchase properties. There are funds available through Federal Insurance payments to flood damaged properties as well as under Section 1362 of the National Flood Insurance Program (NFIP).

A number of cities (For example: Birmingham, Rapid City and Tulsa) have successfully executed massive relocation programs, though most were in direct response to devastating floods. Many of these were combined with urban redevelopment or greenway or river conserva-tion/restoration programs.

Most plans for floodplain evacuation (relocation) have envisioned future uses of the land for open space, interruptible recreational purposes and wildlife. These plans have particular appeal and benefits for those concerned with environmental issues. Greenways have been shown to be cost effective in the broader sense. There is also a need to seek to justify these activities in the broader context of floodplain redevelopment and public health and safety.

Flood Insurance

The National Flood Insurance program (NFIP) was created by Congress in 1968 as a non-structural approach in the prevention of flood damage. It has two main objectives: (1) to enable property owners in flood prone areas to purchase reasonably priced flood insurance; and (2) to discourage future development in floodplains. Community participation in NFIP is voluntary. Any community which chooses to participate must meet and maintain certain minimum regulatory requirements relating to the development process and floodplain regulation.

Many homeowners who qualify for federal insurance either do not know they are in a flood prone area or are not aware of the insurance program or its details. Programs for home buyers alerting them of their home location and floodplain surveys of residents and property owners, forewarn and inform owners to possible hazards and may induce some to purchase flood insurance.

Public Education/Involvement

Public education can be accomplished through slide shows, pamphlets, use of the press, TV and radio programs, or floodplain surveys for home buyers alerting them to their home location and to possible flood hazards (such as is done in Tulsa, OK). Self help warning systems also aid in public awareness. A program of offering technical aid for those owners wishing to flood proof their homes or businesses also provides positive results. The constitution of an interest group may serve a vital role in both public or staff acceptance of flood control activities or requirements and as sources of input in decision making.

Post Flood Programs

The short time period just after a flood has occurred is the most opportune time to implement flood hazard mitigation programs. At this time rebuilding (requiring permits) begins, clean-up activities are ongoing, funds from various sources may be available and flood victims and political leaders have their attention focused on the problem of flooding. A well conceived post-flood mitigation strategy can achieve great progress in the 30 to 60 day period following a major flood event. For example, studies have shown that nearly 70% of all floodproofing by homeowners takes place in this short time period. After this time period the memory of the flood has faded and property owners have turned their attention to other things.

A post-flood program should include many of the non-structural topics covered previously. Property owners are interested in information on clean up and financial assistance. This information can be used as a lead which can be followed by information communicating the mitigation message. Self-help manuals, slide or video shows available from FEMA, door-to-door flyers or visits, public meetings, Disaster Application Centers (DAC) and news releases are all applicable. The State of Illinois and the Association of State Floodplain Managers (ASFPM) have been leaders in the use of DAC Tables for post-flood assistance and have many useful publications on the subject.

Funding sources for disaster assistance are numerous and often confusing. A FEMA Public Assistance Officer is available to assist community preparation for a disaster and in locating funding for post-disaster assistance. Funds from the FEMA Individual and Family Grant and Small Business Administration Disaster Loan programs both provide assistance to private property owners.

Chapter 15 Storm Water Quality Management Plans

15.1 Urban Runoff Quality

It has been known for years that urban runoff contains a wide variety of pollutants, often in concentrations and volumes to cause both acute and chronic environmental impairment of receiving streams and lakes. Numerous studies have been done over the last twenty years to attempt to quantify both the impacts of such runoff and to attempt to formulate tools to analyze the impacts and seek to mitigate them.

A profile of the impacts of storm water discharges was developed by EPA in 1992 (EPA, 1992a). It was found that between one-third and two-thirds of all designated use impairment for our nation's streams was a result of agricultural and urban runoff. While agricultural nonpoint sources comprise the major source of pollution, urban runoff is also very significant. The pollutants found in urban runoff include: sediment (TSS in the range of 35 to 400 mg/l), oxygen demanding substances (BOD in the range of 6 to 20 mg/l), heavy metals (typical range of 30 to 450 micro grams per liter), nutrients (TKN in the range of 1 to 4.5 mg/l), bacteria, oil and grease and other more rare toxic chemicals, pesticides, etc. (APWA, 1991, EPA, 1990).

Impacts on receiving waters are varied and often difficult to measure. Because of the transient and intermittent nature of the pollution loading wastewater measurements and standards do not apply well. However, impairment is often measured in terms of:
- inability to attain a designated beneficial use (drinking water, fishable, body contact, etc.);
- a violation of some chemical or biological numerical standard;
- an ecological impact such as a fish kill, lake eutrophication, reproductive disorders, etc.; or
- a public perception impact such as offensive odor, litter, color, etc.

The water quality impacts of storm water runoff to the receiving water depend on a number of factors including: the magnitude and duration of rainfall events, soil types, time between storms, land use type and specific activities, illicit connections or illegal dumping, and the ratio of the runoff flow volume to the receiving water flow volumes. While actual impacts of storm water runoff are often difficult to ascertain, there is a general understanding that urban runoff contains significant levels of pollution comparable to primarily treated effluent from wastewater treatment plants and is a major source of receiving water pollution. Yet hard data were often scarce, not able to be compared, or contradictory in analysis and interpretation.

The NURP Program

This lack of data eventually led to the development of the National Urban Runoff Program (NURP) (USEPA, 1983). Various municipalities and agencies under NURP collected data from many municipal residential, commercial and light industrial sources in an effort to characterize urban storm water pollution runoff. One of the major findings from NURP is that urban polluters to streams in the urban environment can be divided into four major categories: (1) illicit connections and illegal dumping, (2) industrial dischargers, (3) construction site runoff, and (4) commercial and residential sites (everybody else).

Another finding is that total loadings of wash-off type pollutants are related primarily to the volume of storm water and thus are relatively dependent on percent imperviousness of the land surface. Additionally the event mean concentrations of pollutants (a flow weighted mean concentration) are generally independent of storm duration, intensity, site location, etc. Thus, similar land uses will yield similar concentrations of pollutants regardless of runoff characteristics. Figure 15-1 illustrates typical annual loadings of pollutants (total nitrogen, biological oxygen demand, total suspended solids and chemical oxygen demand) from various land use types for an area of the Piedmont Region in North Carolina. Thus long term (chronic) pollutant loadings (and to a lesser extent short term pollution) can be linked directly to land use and rainfall.

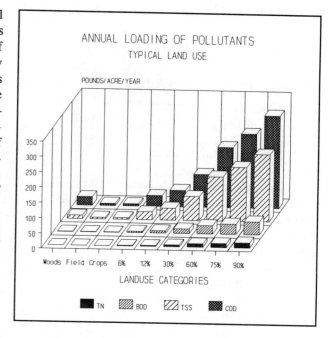

Figure 15-1 Total Loading Of Pollutants

It has been recognized that the best way to control this pollution is through the use of "source controls". There are a wide variety of such source controls both of a structural nature (such as retention ponds) and of a non-structural nature (such as zoning ordinances). These controls are called urban "Best Management Practices" (BMPs). These practices help form the basis of any control program. For example, Figure 15-2 shows the approximate pollutant removal capability of several alternate designs of retention ponds and extended detention ponds.

Another key aspect of urban storm water pollution is that for the more impervious areas, higher concentrations of pollutants are washed in the first flush of the rainfall. Thus, the many smaller storms which occur each year contribute the great majority of pollution to the system. Therefore control systems designed to treat the first flush and the smaller volumes may be potentially useful.

Armed with this knowledge and legal and congressional mandates to proceed, EPA promulgated regulations to require municipalities to develop storm water quality management programs.

15.2 Water Quality Act Overview

The Water Quality Act of 1987, Section 405 amends Section 402 of the Clean Water Act of 1972 to require EPA to provide rules and regulations that establish a permit program for storm water discharges associated with industrial activities and from municipal separate storm drain systems serving populations of 100,000 or more (Federal Register, 1988). This Act and the promulgated regulations have had a major impact on those municipalities affected.

The Act requires that dischargers to "Waters of the State" (WOS) be subject to the National Pollution Discharge Elimination System (NPDES) permit program. However, the requirements, and the philosophy behind these requirements, outlined in the draft regulation are far reaching.

Basic Requirements

The Act, as it applies to municipalities, has three main thrusts.

1. It requires municipalities to effectively prohibit non-storm water discharges into their publicly owned or operated separate storm drain system.
2. It requires municipalities to control discharge of pollution into their systems to the "maximum extent practicable" (MEP).
3. It defines one system-wide permit rather than permits for each individual discharge point.

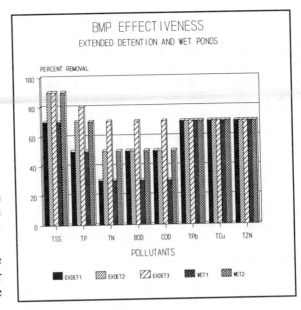

Figure 15-2 **Effectiveness Of Ponds And Detention Used As BMPs**

These "MEP plans" are envisioned to consist of comprehensive storm water quality management programs (SWQMP) which include: management practices, control techniques and systems, design and engineering methods and other such provisions as the Director of the program determines appropriate for control of such pollutants (EPA, 1992b). Senator Stafford is quoted in the congressional record as stating:

> "These are not permits in the normal sense we expect them to be. These are actual programs. These are permits that go far beyond the normal permits we would issue for an industry because they in effect are programs for storm water management that we would be writing into permits.[1]"

The NPDES permit is in two parts. Part 1 concentrates on identification of and description of the system and dischargers. Part 2 concentrates on a program to reduce pollution discharges.

Part 1 of the permit requires identification of known data, structures, outfalls, land uses, water quality problems, etc. It also requires a description of existing program and physical elements such as legal authority, financing, physical characteristics of the area and controls. There is a dry weather field screening requirement wherein any major outfall discharging to waters of the state that shows turbidity, oily sheen, odor or any obvious degradation or has dry weather flow is identified and sampled.

Part 2 of the permit has four major aspects:

1. _Program Description_ entails demonstrating sufficient legal authority, financial capacity and administrative capability to control pollution, prohibit illicit dumping and to require compliance.
2. _Source Characterization_ entails identifying all sources of urban storm water pollution such as landfills, industrial sites that have a pollution problem, etc. It also requires estimation of pollution loadings from all areas of the municipality.
3. _Discharge Characterization_ entails verification sampling of those low flows under Part 1 which demonstrated problems as well as representative sampling of from 5 to 10 outfalls representing

---

[1] Federal Register, Dec 7th, 1988, page 49443

different land use types. This sampling is done during runoff events and is designed to measure both first flush (acute) and long term (chronic) pollution loadings.

4.   Proposed Storm Water Management Program entails the complete description of a comprehensive storm water management program which can deal effectively with pollutant runoff from the four major categories of dischargers. The program will contain such elements as sampling, inspection, enforcement, BMPs, planning, etc. It will also contain an element to regulate storm water discharge from industrial sites that discharge into the publicly owned or operated separate storm water system. This element may have some similarity to the existing industrial pretreatment program.

The costs to develop an application for the permit range from approximately $300,000 to over $2,500,000 depending on what costs are included in the total and what effort was put into the plan (NAFSMA, 1992). It is expected that the costs for those municipalities doing the minimum necessary will even out over the permit period as the types of programs and efforts become more similar.

Municipal Reaction

Unlike most NPDES permits the one for municipal storm water has few or no numeric water quality criteria to meet, few set procedures, forms or activities... and no true precedent. It is a negotiated permit where, within the confines of broad USEPA guidance and sometimes vague regulatory language, a municipality has the responsibility and the leeway to develop an approach to control the discharge of pollutants to public waters to the "Maximum Extent Practicable" (MEP) and to prohibit non-storm water from entering into and being discharged from the storm drainage system.

Many municipalities have asked themselves how to get from a sometimes poorly developed "urban drainage" program to comprehensive storm water quality management. A review of the literature is not very helpful. While there are reams of information and mounds of case studies on how various BMPs may work or have worked, there is little guidance available for helping a municipality through the process of actually planning and implementing a SWQMP which employs these BMPs in a cost effective and prioritized basis. This chapter, gleaned from some of the best storm water programs across the nation, is meant to give some helpful advice. It cannot provide specific detail (each program is unique) but can give a framework to arrive at an effective program and hit the highlights of the approach.

It should be noted that the 1987 Water Quality Act and implementing regulations are only a part of the overall Federal and State web of regulatory programs designed to reduce the amount of pollution to both waters-of-the-United States and coastal waters. For example the Coastal Nonpoint Pollution Control Program under Section 6217(g) of the Coastal Zone Act Reauthorization Amendments of 1990 (CZARA) is intended to foster the development and implementation of management measures for nonpoint source pollution control to restore and protect coastal waters. The North Carolina Watershed Protection Act is designed to control the discharge of pollutants into drinking water supplies through a combination of land use restriction and BMPs.

Concern for adverse impacts of polluted runoff is good stewardship and makes good sense regardless of legal mandates. Each municipality should, to some appropriate extend, develop and carry out sound policies and programs which meet the adjectives of MEP and seek to restore, preserve and conserve environmental resources. With that in mind a typical planning approach is described.

15.3 Basic Approach Philosophy

Two priorities must be balanced in development of a SWQMP. It is desirable to identify and solve real problems, not simply spend money on paperwork or meaningless activities. If it is not known what those problems are (as is the case in most of the municipalities in the United States) there is a need to have a focus on identifying and characterizing the problems and developing strategies to solve them.

However, based on years of data and information, USEPA feels that there are certain storm water runoff pollution problems which exist in almost any municipality. For example, it is expected that sediment from eroding banks, nutrients from fertilizer use and pet excrement, natural and man made litter, chemical runoff from pesticides and herbicides, oil and grease, trace metals and other toxics, bacteria, thermal increases and other pollutants find their sources in activities occurring in commercial and residential neighborhoods and reduce the general water quality of the receiving streams.

There is a USEPA mandated requirement for a minimal type of program even in areas where specific pollution problems are not known to exist but are generally suspected to be present in any urban area. The four general areas included in a program are: illicit discharges and improper disposal, certain categories of industries and waste related businesses, construction site runoff, and commercial and residential sites. Therefore it will be necessary to compare any proposed SWQMP to the baseline minimums for these four areas to ensure overall program completeness. Thus there is a two-pronged attack: (1) identify and solve real problems, and (2) address USEPA program minimums.

Defining "Maximum Extent Practicable"

Figure 15-3 depicts a simplified matrix schematic for MEP definition. For clarity of understanding and consideration, it is convenient to consider four major areas of polluters as identified in the regulations: (1) illicit connections or illegal dumping, (2) industrial, hazardous waste sites and landfills, (3) erosion from construction sites and (4) commercial and residential pollution. These areas are defined in this manner because the approach to deal with pollution from each of these areas is slightly different. Under each of the four major program areas it will be necessary to define a program which contains four major consider-ations: (1) technical approach adequacy and effective-ness, (2) legal authority, (3) financial sufficiency, and (4) administrative and organizational support. In actual analysis the matrix is greatly expanded to con-sider every facet of the proposed SWQMP to assure it is funded, staffed, legal and will actually accomplish environmental or watershed objectives.

THE 'MEP' GRID

	TECH	LEGAL	MONEY	ADMIN
ILLICIT & ILLEGAL				
INDUSTRY				
CONST. SITES				
COM. & RES.				

Figure 15-3 Defining MEP

Technical adequacy and effectiveness of a planned program can be analyzed under each of the four program areas. Under each area it will be necessary to specify both a priority system and a schedule for program development. There are a vast number of possible BMPs that are applicable under each of the categories. It is important to choose only those BMPs (structural and non-structural) which are in some sense "cost effective" for a municipality.

Legal authority must be sufficient to be able to prohibit the discharge of non-storm water to the municipal system and to require compliance with other measures. Legal authority is based on both Federal authority and State delegation. Some states allow strong home-rule capabilities while others may require state legislation to allow municipalities to fully carry out the program mandates. Some states have developed their own storm water management quality programs and regulations which must be implemented by municipalities (see, for example, Virginia, 1990). Other states have simply supplied guidance to municipal officials (see, for example, Maine DEP, 1990, 1992).

Local authority in most municipalities rests in the Municipal Charter, subdivision and zoning ordinances and other miscellaneous provisions. However, some of the legal references apply only indirectly to storm water. It may be desirable to revise these portions to more specifically address storm water quality control management. This will be a dynamic process as the program grows and changes.

Financial sufficiency can be judged on the basis of whether the funding source for the storm water program is: adequate, stable and equitable. Many municipalities are finding that general fund tax based financing is "none-of-the-above" and are turning to user fee based systems (often termed "storm water utilities"). Costs to add a storm water quality program to a well established storm water quantity program can be in the range of a twenty-five to forty percent increase.

Guiding Principles

There are some sources which gives steps for SWQMP development but few give guidance on "how" to think about development of a SWQMP. Some common sense strategic principles applicable to the development of a SWQMP generally and individually to all the program elements can be derived from successful programs now in existence and from discussions with USEPA and other expert personnel. These might be called the "ten commandments of SWQMP development". They have been expressed in "popular" language to help in understanding and application among even non-technical decision makers (Reese, 1993, 1994).

1. If It's Not Broke Don't Fix It.
Seek reasonably (see commandment 2) to understand the problems before spending large amounts of money on the solutions... sample, monitor and test first. Concentrate on identifying pollution sources, levels, trends and impacts through a strong and well targeted monitoring program. Remember the goal is to identify and solve real problems whether they are measured in terms of designated use impairment, violations of chemical standards, biological measures or impairment, aesthetics or actual impairment of organisms or human health and safety.

2. Go Ahead, It's Not Rocket Science.
There are certain things which make sense to correct without resorting to prolonged study and monitoring programs. For example, a western city cleaned up pollution problems on a small island for less than one-third the cost of a proposed comprehensive monitoring program to assess the problem. Do what you can do that makes sense now. For example, develop a strong illicit connections program and approach if illicit connections are a known problem.

3. If You Can't Fix It - Don't.
Focus on those situations which you have a reasonable opportunity to improve first. Employ measures which are cost effective (by a consistent definition), sustainable and reliable. Taking

advantage of excess capacity in wastewater treatment plants may make sense. Constructing storm water treatment plants probably does not.

4. Fix The Worst - First.

Get the most out of scarce resources by experiencing early successes. Remember that public perception is often key. "The worst" may be defined as the ones which threaten health, safety, wildlife, or property or are the most visible or well known or based on some other criteria.

5. Don't Build New Problems.

While figuring out how to fix existing problems take the necessary steps to avoid building new problems into new development...control and guide development. This may mean pushing for subdivision or zoning ordinance changes early. It may also mean master planning for quality and, as a minimum, purchasing key locations for regional BMPs ahead of development even if the actual size and configuration is not known.

6. Lay The Financial, Organizational And Legal Foundations - First.

No SWQMP can flourish and grow without the key underpinnings of a sound organizational structure with sufficient legal authority and financing. Actual physical urban pollution problems are usually the result of misguided institutional policies or the lack thereof.

7. Emphasize Things You Can Modify Easily - First.

Try cheaper and more flexible non-structural and programmatic BMPs and institutional changes prior to physical construction. Public education, awareness, and reporting and other non-structural programs can be easily modified or abandoned if they prove to be ineffective or to address a non-problem. Employ measures which have high public, political and "stakeholder" acceptance such as the "Stream Teams" of Bellevue, WA or the "Rain Rangers" of Charlotte, NC.

8. Make The Most Of What You Have.

Employ measures which are consistent with or extensions of other programs and regulations. Maximize what you have to get what you need. This may make "strange bedfellows" of departments which normally do not interface. Fire Departments may perform storm water inspections along with Hazardous Materials Team (HAZMAT) inspections. All field related municipal employees could serve as the "eyes and ears" of the SWQMP reporting system.

9. Get The Timing Right.

Aim to reduce both total pollution (chronic - long term) and pollution concentrations (acute - short term) in sensitive areas as appropriate. Use BMPs and measuring activities which are appropriate to the types and time frame of pollutants. Annual pollutant models, while easy to employ, are normally inappropriate for transient dissolved oxygen sag problems in a stream.

10. Wade In Don't Dive In.

Define programs to be developed in well defined phases and steps with checks of effectiveness at every major milestone. Use pilot studies and demonstration projects. Proceed slowly and cautiously with in-course corrections. Look for methods with few environmental side effects, and which do not introduce unacceptable risks to health or safety.

15.4 The SWQMP Development Process

Each regulatory program touching on storm water management has developed a planning process which can roughly be expressed as: (1) determine existing conditions, (2) quantify sources and assess impacts, (3) assess alternatives, (4) develop and implement the recommended plan, and (5) assess results (USEPA, 1993d). Inherent in any approach is the setting of goals and objectives that are quantitative, measurable and flexible on a site, watershed or municipal-wide basis.

One approach to the development of a SWQMP used in several municipalities, is a nine-step process which proceeds rationally through narrowing down possible approaches to the few which have the most promise. It is very important to have a systematic way to go about planning the storm water quality management program. This will not only provide a logical, building block, program development but will provide ample support should litigation result from program implementation measures. The approach seeks to define goals, set objectives, take actions, and assess program effectiveness. Many factors working together can be thought of as a "chain" linked together. If one link is weak the whole program suffers (Shaver, 1988, Ogden, 1992a). Four key characteristics of a successful SWQMP are that it is: comprehensive, integrated, balanced and dynamic. Many facets of municipal government work together, some of which may not have been thought of as part of a storm water management team before. However, they can create synergy as they work together. The result is a program that is balanced between studying and doing, and effort is expended commensurate to the problems generated. The program also changes as new information is gleaned.

The 9-Step Process

The nine-step process is illustrated in Figure 15-4. It is specific to the municipal NPDES permit requirements for storm water in that it shows an initial two pronged approach to the problem identification phase. But the process also serves those municipalities not involved in the NPDES process in two ways: (1) it positions them to more easily meet any new NPDES regulations for smaller municipalities or cooperate with adjacent municipalities already in the program, and (2) it takes advantage of limited prior knowledge of the pollution problems in the system through both problem identification and doing those minimal things which are deemed necessary in almost every municipality. The process uses as a guiding philosophy the approach of solving real problems and meeting program minimum requirements. Several steps will go on simultaneously for various portions of the program development. There may be a need to cycle back through several steps one or more times to arrive at an acceptable MEP strategy.

EPA has added a mandatory description for each proposed management program. These items should be kept in mind and documented:
- the nature of the controls,
- the sources addressed and the area served by the system that will be affected,
- inter-governmental coordination necessary,
- the proposed nature of the implementation and the implementation schedule,
- staff and equipment necessary to implement the program, and
- procedures for monitoring implementation and evaluating effectiveness of the management program component.

Step 1. Identify and quantify (if possible) known "problems". Problems can be defined as anything from aesthetic appearance to actual known environmental impairment. Often the inability to attain a designated stream use is considered a problem. If urban runoff contributes to that non-attainment then it will be addressed in the development of the SWQMP.

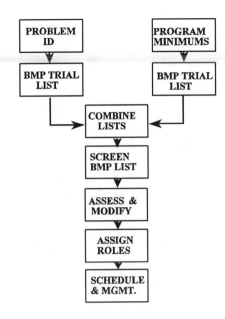

Figure 15-4 A Nine-step Planning Process for SWQMP Development

If little information exists the initial NPDES permit period may be used to develop more information and put off commitment to expensive BMP programs until a minimum of information is available to guide the development of these programs. Known problems can be gleaned from Federal, state and local reports and records; maintenance records or the knowledge of field personnel; field screening; citizen environmental groups; etc.

Seek, as possible, to determine cause and effect or at least possible sources. If possible seek to quantify problems. One method of this is to develop a stream segment rating system for various types of pollution or impairment. One city (Ogden, 1992b) developed rating systems for key indicator pollutants which would place stream segments in "action", "watch", "no action", or "unknown" categories based on sample information from 47 receiving stream sites. The system was placed into a Graphic Information System (GIS) for both display and long term management. Stream segments could be "switched on" to display the category of each segment for each pollutant of concern. When overlaid with land use, industrial sites, outfall locations and neighborhood age a clear picture of cause and effect began to emerge, EPA (1993d) gives more information.

At this stage an initial development of goals and objectives is done. However the details and final goals and objectives are reserved for later steps when more information is available to realistically set limits and determine tradeoffs.

Step 2. Develop a preliminary brainstorm list of potential solutions. There are hundreds of BMPs and variations. A long laundry list helps stimulate discussion and imagination. The solution direction may, at first, appear obvious but, because of the integrated nature of such solutions, there may be other aspects (such as citizen monitoring or a new technology) which may add significantly to mitigation. A good source of information on education products is published by EPA (1993b). There are many other sources of BMP information including: APWA (1992), City of Austin (1988, 1991), EPA (1990, 1992c, 1992d, 1992e, 1992f, 1993c), Storm Water Quality Task Force (1993),

Ogden (1992a, 1992b), MWCOG (1987, 1992), UDFCA (1992), Maine DEP (1990, 1992) and State of Minnesota (1989).

   Step 3. Identify program minimums not adequately covered. After all known or suspected problems are dealt with there may still be program areas thought to be important by EPA and generally common to all municipalities which have not been adequately addressed. Table 15-1 gives the program areas under each of the four major categories of polluters.

Table 15-1 EPA Program Minimum List

Residential And Commercial Activities

- Maintenance activities and a maintenance schedule for drainage facility normally including: ditches, structural BMPs, pipes, streams and creeks, inlets and curb and gutter.
- Public or private BMP master planning for areas of new development or significant redevelopment.
- Environmentally sensitive roadway operations and maintenance programs.
- A program to require new flood control projects to assess the potential environmental impacts from their construction and operation and a requirement to look into retrofitting existing flood control projects to increase their ability to reduce pollution or remove pollutants.
- Municipal waste handling and disposal facilities.
- Controls for the use of pesticides, herbicides and fertilizers for the categories of: commercial use, public right-of-way use and private use.

Illicit Connections And Improper Disposal

- A program to prevent illicit connections to the storm drain system.
- A program to screen outfalls for the presence of improper or illicit connections.
- A program to perform follow-up investigations and removal of illicit connections.
- A spill prevention, containment and response program.
- A program to facilitate public reporting of illicit connections and improper disposal.
- A program to promote proper use and disposal of toxic materials and substances.
- A program to reduce pollution from overflowing and leaking sanitary sewers.

Industrial Activities

- Development of inspections and control prioritization procedures.
- A program to monitor significant industrial dischargers.

Construction Activity

- Water quality and BMP assessment during site planning.
- Site inspection and enforcement programs and procedures.
- Erosion control training for developers and contractors.

Source: EPA, 1990

Step 4. Develop a preliminary brainstorm list of potential solutions for these program minimums. A later section gives long brainstorm lists for each program area.

Step 5. Combine lists for problem solutions and program minimums, blend programs and modify. This shortened list is organized and analyzed to determine how they will function singularly and in conjunction with other program elements and how and by whom they will be implemented. Conflicts and overlaps are eliminated.

Another part of this analysis is to determine ranges of BMP application to allow for development of alternative programs and to get a feel for cost sensitivity. Normally a minimal, moderate and maximum program are assessed to some extent, keeping in mind realistic upper limitations to application (legal, financial, political, physical, social, etc.) and lower limits where the programs have no effectiveness. This is particularly applicable to the program minimums where no specific known problems are targeted.

Step 6. Finalize goals and priorities and develop storm water quality programs. Goals and objectives were set in a preliminary way during the problem and program minimums identification stages. In this step they are analyzed and finalized based on the problems and program minimums identified in steps 1 and 3 and the combined list of solution possibilities developed in step 4. It is important to involve the public in each step of the goal and priority setting and refinement process to ensure community support. See the chapter on public involvement for further information on this.

Goals, objectives and priorities may be set on a municipal-wide basis, watershed basis or to protect a specific resource or environmental amenity. For example, Austin, Texas set a goal to maintain the quality of the water in Town Lake in the center of the city. Interim objectives and priorities were then set to attain this goal. Another city determined that illicit connections were a major problem for a local river and set goals to eliminate them by a certain date. Actions, objectives and priorities were then set up around this goal. A Northwestern city set as a goal the maintenance of salmon runs within its jurisdiction. A southern county wanted to maintain a particular wetland area for both passive recreation and habitat. A Midwestern city focused on reduction of erosion from its stream banks and the elimination of illicit connections and dumping from industrial hotspots.

Inevitably there will be fewer financial or manpower resources than are necessary to address each of the programs or needs fully. Because, at this point, there are so many options and subprograms being considered in a "big picture" way, it is not advisable to get too detailed in cost estimation. There will be development of priorities and trade-offs based more on a "feel" for costs and other resources than a detailed estimate. Programs are adjusted and shifted with an emphasis on the overall goal of a sound and comprehensive program than to maximize cost effectiveness, that is the next step.

There are several methods to attempt to assign priorities and goals to environmental programs (USEPA, 1987, 1993d) which include:

- assessing the resources impacted by pollution in terms of types of resource (water, biological, aesthetic, etc.), values of resources, desired use of resource and extent of impairment;
- assessing institutional concerns related to the resource impairment including: legal authorities available, organizational capabilities, technological capabilities, financial resources available, legal penalties and enforcement issues of non-compliance, and public and political perceptions;
- assessing the pollutant source in terms of type, magnitude, impact, transport mechanisms, wet/dry weather trends;
- assessing how the problem or program fits overall goals and objectives including: land use goals, water resource goals, site or watershed-specific goals, and planning and development goals.

Screening of the combined BMP list to meet the goals can be done in a number of ways. Table 15-2 provides a list of screening criteria used in some form by several municipalities to provide a consistent rating methodology. The next section gives more information on Table 15-2 and screening methods. Some programs are scaled back, delayed, or planned to be accomplished in phases to make way for higher priority programs and activities.

The end result is a preliminary comprehensive SWQMP which attempts to meet the goals and objectives set by the municipality in light of EPA requirements using the best combination of BMP strategies in a cost effective and responsible manner. The next step refines this process using cost determinations.

Step 7. Estimate overall program costs and pollution reduction effectiveness, modify program as necessary. Some measure of effectiveness is better than none. From a municipality's point of view the best measure of permit condition attainment is one which simply looks at accomplishment of prescribed tasks rather than actual pollution reduction or expenditure of funds. Actually measuring pollution reduction should be accomplished but not used as a standard for permits; at least until the level of cause and effect is improved in storm water quality management.

Cost estimates have been made for each of the municipal permits already applied for. They are, at best, first order accuracy. Experience has shown that the addition of a medium level of storm water quality to an advanced storm water quantity program will increase costs by from thirty to fifty percent. Costs for the initial programs have ranged from quite conservatively small programs (less than $400,000 annually) to very large programs (over $3,000,000 annually). However, it is difficult to make a comparison since municipalities have lumped different things into their costs which may inflate them (such as GIS system development, mapping, etc.). APWA (1992) provides a comprehensive table of cost estimates for various types of BMPs. Experience with municipalities having similar sub-programs indicates that the costs can also be quite variable ("a litter program is not a litter program is not a litter program"). Probably a good average is that the program costs for medium sized municipalities (between 100,000 and 250,000 population) is in the order of $1M to $2M annually for the initial program.

It should be remembered that the MEP standard is not normally numerical. Specific objectives, performance specifications, or other means of measurement of performance for pollution reduction have not been specified by regulatory agencies. In many cases, particularly for non-structural BMPs, it is very difficult to assign pollution reduction numbers without better data and information. One is often compelled to solve a problem which has not been fully defined, and use poorly documented means which have not been well measured in terms of effectiveness. Great care and engineering judgment must then be exercised.

There is a need in any comprehensive program development to go back and look at the whole assembled puzzle after suitable examination of each of the pieces. There will almost always be a need for adjustment, compromise, and combining of programs at this stage.

Step 8. Describe roles and responsibilities to implement the program. All key tasks must be identified and clear direction given as to responsibility and actions expected. The goal is to get all parties to attain a sense of ownership of their portion of the program and a sense of teamwork in execution of the overall program.

Step 9. Develop a schedule for implementing the program including management and feedback loops. Feedback information may allow for changes in direction or emphasis in any program element. These changes can be negotiated during preparation of the annual report. Decision points can be built right in as permit conditions. The annual reporting requirement is a good time to make adjustments. Such freedom to modify proposed programs should be built into the schedule, if possible.

The end result of this step is a schedule and budget for implementation of the program. It is important to assess the success of the programs at every step and build into each program ways to measure that success. This may be through specially designed feedback from the program implementors, through data collection and monitoring, public awareness polls, or other means.

15.5 Evaluation Of Best Management Practices

Best management practices (BMPs) are used to control and reduce the discharge of pollutants. There are many ways to categorize BMPs including: structural or non-structural; source control or treatment control; structural or programmatic; etc. However BMPs are categorized, there must be a consistent way to assess their effectiveness. The evaluation of best management practices is more an art than a science. EPA (1993d) and others describe several methods for the evaluation of BMPs. These methods are most often used together to assess various parts of a SWQMP.

The holistic approach is a combination of intuition, engineering judgment, basic analysis and cost assessment which can lead to a set of BMPs and a strategy. This method can be very effective but requires the involvement of experienced personnel from a number of technical disciplines and from senior staff or political leadership. Often an ad hoc committee is selected which develops a set of objective criteria and then goes through a process, like the nine-step process above, to arrive at a final program.

Cost-benefit analysis is also an approach which can be used to develop SWQMPs. Such measures as cost per pound of pollutant removed or cost per day of activity can be used. This method is good for making decisions on which approach to use first and for how long. For example, it may be possible to achieve a reduction in a certain pollutant (such as petroleum hydrocarbons) using public education and regulatory processes. At some point though, structural controls may be necessary to complete the program. Cost-benefit analysis is typically used for more site or project specific activities rather than municipal-wide or more subjective non-structural type applications.

Optimization has been used in other sciences for decision making. Optimization consists of finding the maxima or minima of an objective function subject to a series of constraints each expressed in terms of decision variables. The objective function can be expressed in terms of cost or some complex combination of cost, benefit and judgment. Nonlinear relationships among the variables and the fact that the numerical complexity can hide the subjectivity of many of the inputs often limits this method to narrow applications targeted toward a single site or pollution concern.

Matrix Comparison is a more subjective method (unless numeric ratings are used) whereby a table is constructed with various BMP or BMP programs along the horizontal axis and evaluation criteria along the vertical axis. Each alternative program is then rated in terms of each of the criteria. A numeric rating (say 1 through 5) or a simple + or - rating can be used. Table 15-2 provides a list that has been used in various forms in several locations for BMP program development. Some localities total an actual score and use the form for a numeric ranking of BMPs. Others simply use the pluses and minuses as flags to alert planners to particular BMP strengths or potential problems. Not all criteria apply to each BMP. Another list may be developed specifically for subcategories of BMPs.

Following is a discussion of the items included in Table 15-2.

1. Human Risk, Public Safety And Potential Liability. Some BMPs carry with them some inherent risk (such as the risk of drowning in a wet pond) as well as value. In many cases this risk can be minimized through sound design. Risk must be assessed even informally to limit potential human risk and liability.

2. Environmental Risk And Implications. Some BMPs carry little human risk but may carry a high environmental risk. For example, infiltration basins in a groundwater recharge zone may

Table 15-2 Best Management Practice Screening Criteria

BMP:				
Criteria Description	**+**	**0**	**-**	**CMTS**
1. Human Risk, Public Safety And Potential Liability				
2. Environmental Risk And Implications				
3. Ability To Control Key Targeted Pollutants				
4. Costs To Implement And Continuing Costs				
5. Acceptability To Public, Stakeholders, Staff, Political Leadership				
6. Equitability To Impacted Persons				
7. Reliability And Consistency Over Time				
8. Sustainability Of Maintenance Or Program Management				
9. Ability To Be Applied Universally Throughout The Municipality, Or, On A Specific Watershed Basis				
10. "Fit" With Other Operations And Programs				
11. Relationship To Federal, State Or Local Requirements				
12. Amenity or Multi-use Value				
Totals				

be unacceptable if the pollutants infiltrated would enter the drinking water supply. If the risk of BMP failure is high the use of the BMP may need to be avoided or modified.

3. Ability To Control Key Targeted Pollutants. Each BMP should be targeted toward certain pollutants and certain problems. The BMP should be able to reduce to an acceptable extent the targeted pollutant. This may require consideration of chemical forms the pollutant may take, long term ability, extreme environmental conditions, etc.

4. Costs To Implement And Continuing Costs. Every BMP must be scrutinized for its "cost effectiveness". First costs may be low but if life cycle costs are high the BMP may not prove effective or economically sustainable.

5. Acceptability To The Public, Stakeholders, Staff And Political Leadership. Some BMPs may not have a high acceptance in certain areas or for certain uses. For example, some types of zoning control may not be acceptable if it is seen as a "taking" of land. Use of some BMPs can be made more palatable through the use of incentives such as user fee credits for the use of detention ponds.

6. Equitability To Impacted Persons. All BMPs inconvenience someone. The level of this inconvenience must be weighed against the public good and ways found to reduce the impact

on those most affected by the BMP. The application of the BMP must be fair and reasonable and have considered ways which would reduce the impact.

7. <u>Reliability And Consistency Over Time</u> . Reliability refers to the ability of the BMP to continue over a relatively long period of time without failing. Consistency refers to its ability to reproduce its mitigation ability without changing or varying appreciably in its ability to mitigate the pollution problem.

8. <u>Sustainability In Terms Of Maintenance Or Program Management.</u> All BMPs require program or structural upkeep. If the program cannot be sustained modifications in its application should be made. If maintenance cannot be assured reasons for non-assurance should be eliminated or other BMP alternatives should be sought.

9. <u>Ability To Be Applied Universally Throughout The Municipality, Or, On A Specific Watershed Basis.</u> BMPs have an advantage if they can be applied widely. This supports fairness, flexibility and possible economies-of-scale. While the use of certain BMPs may be appropriate in only some areas, it is desirable to be able to use them wherever necessary.

10. <u>"Fit" With Other Operations And Programs.</u> Storm water quality management must be integrated with, not simply added on to, the existing storm water management program. BMPs have a better prospect for success if existing programs can be simply modified to incorporate storm water quality aspects.

11. <u>Relationship To Other Federal, State Or Local Regulatory Requirements.</u> There are many overlapping program requirements. Watershed protection, zoning, wetlands permitting and other programs may have bearing on how and where the BMP is used. It is advantageous if a BMP fits within and does not violate the requirements of these other programs.

12. <u>Amenity Or Multi-use Value.</u> Structural BMPs should be both functional and beautiful. Program BMPs should mutually support each other and other community objectives. Parks and greenways can be easily integrated with structural BMPs. Education programs and public schools or citizen volunteer organizations often gain synergy from each other and can assist in gaining acceptance of multi-use projects.

15.6 BMP Lists For The Four Program Areas

The pollutants from each of the four major program areas may be similar. However, intensities, delivery mechanisms and, often, control mechanisms differ sufficiently to warrant four separate but somewhat parallel approaches to dealing with the types of pollutant discharges from each of the four major program areas.

Illicit Connections And Illegal Dumping

This is part of the program authorized by Congress wherein municipalities must be able to prohibit non-storm water discharges to the municipal separate storm drain or drainage system. While some discharges of non-storm water to the system are not illegal, per se, they must be covered by a separate NPDES permit and subject to NPDES conditions (technology or water quality based standards) similar to any other discharge covered under sections 402 and 301 of the Clean Water Act.

Illicit connections and illegal dumping were not specifically sampled under the NURP program. However, it was concluded that illicit connections can result in both high bacterial counts and the introduction of priority pollutants into the urban drainage system. Studies have shown that illicit connections to storm drains can cause severe and wide-spread contamination problems. The problem is seen as more severe in older neighborhoods and older industrial and commercial areas where approval processes allowed for such connections. These areas should be targeted as a priority because there is the possibility of dramatic improvement in water quality in these areas.

The regulations require a dry weather "first phase" screen of outfalls for the presence of illegal connections. The purpose of the screening is to allow municipalities to develop a program and priorities to detect and remove illicit connections and to control improper disposal to the storm drain system. While the first phase screen required under the regulations has the potential to locate, in a more general way, a number of such connections, there is no substitute for physical or remote inspection, smoke testing, etc. Not all illicit connections discharge all the time. Part 2 of the permit requires a plan for such program development including schedule and priorities.

Development of a coherent illicit connections and illegal dumping investigation program includes the following steps (USEPA, 1993a):
- mapping and preliminary watershed evaluation,
- selection of tracer parameters,
- initial field screening activities,
- data analysis to identify problem outfalls and flow components,
- watershed surveys to locate inappropriate pollutant entries, and
- corrective techniques.

Low cost techniques have been proven to be cost effective in Ft. Worth, Texas (Falkenbury, 1987). Detailed manuals for field and laboratory procedures have been developed for a number of municipalities.

The types of pollutants to look for in an illicit connection program include: sanitary wastewater sources, septic tank systems, automobile washing and maintenance centers, irrigation sources, laundry waste waters, dewatering operations, improper disposal of household hazardous waste, spills, sump disposal, gas station materials and substances, industrial rinse water, cooling water, swimming pools, etc. Industrial sources can be related to raw materials, waste materials and the final product (EPA, 1993a).

Illicit Discharges And Improper Disposal Program BMPs -

Legal Authority
- Describe the process for development of the storm water only portion of the municipal code from develop to implement and enforcement.
- Develop the basic outline of an ordinance for illicit connections and illegal dumping.
- Do a basic staff and resource estimate.
- Strengthen dumping ordinances and enforcement.
- Strengthen illegal connection ordinances and enforcement.
- Strengthen spill ordinances and enforcement.
- Coordinate with state and Federal authorities for stronger enforcement.
- Streamline enforcement and injunction process.
- Train and empower inspectors.

Screening
- Develop a screening process for the rest of the outfalls:
  - train inventory crews for field screening,
  - inventory crews note problems and annotate record for that structure,
  - sampling crews return for colormetric sampling of suspected problem outfalls, and
  - train stream walk personnel for field screening.
- Review parameters list and other operations.
- Low cost screening techniques.

- Low cost toxicity testing.
- Citizen and school groups.

Follow-Up
- Describe and assess current procedures for calling in state authorities for enforcement actions.
- Develop procedure for follow-up if the screen finds something.
- Begin follow-up of priority sites after procedures developed.
- Make basic estimate of manpower needs.
- Describe the use of complaint records for follow-up.
- Alert inventory crews and stream walk crews to presence of high priority areas.
- Low cost screening techniques.
- Low cost toxicity testing.
- Citizen and school groups.
- Smoke testing.
- Camera testing.
- Dye testing.
- Public reporting.
- Surveys and education.
- Ordinances and enforcement.
- Publication of enforcement activities.
- Use fire inspectors.
- Use pre-treatment inspectors.
- Monitoring programs.

Spills
- Describe in detail the current response activities.
- Describe spill prevention plans.
- Plan to use the current program with few changes as the final program for permit submittal.
- Improve coordination.
- Improve reporting.
- Hotlines.
- Education.
- Improved enforcement and public reporting.
- Pre-positioned equipment.
- Chemical spill prevention and control plans at certain locations.
- Storage and handling requirements.
- Diversion or containment site designs.
- Use fire inspectors.
- Use pre-treatment inspectors.

Public Reporting
- Rain hotline.
- Public education programs.
- Public awareness programs.
- Adopt-a-stream.
- School children teams.
- "Rain Rangers".

Used Oil And Toxics
- Stenciling.
- Promote use of existing commercial used oil disposal sites.
- Investigate small quantity generators.
- Develop description of pretreatment and fire inspection programs as well as existing used oil and household hazardous wastes planning or operations.
- Coordinate the private collection locations.
- Fill in locations where none exist.
- Household hazardous wastes turn-in days.
- Permanent household hazardous wastes locations.
- Education on proper disposal.
- Curb-side pickup of household hazardous wastes.
- Proper disposal of some household hazardous wastes in landfill education.
- Paint recycling.
- Paint solidification.
- Household hazardous wastes recycling.

Sanitary Waste
- Describe the current infiltration and inflow program as it pertains to storm water.
- Develop a monitoring program to partner with wastewater and water utility.
- Describe the current septic tank program.
- Infiltration and inflow program looks at exfiltration too.
- Septic system inspection, certification and enforcement program.
- Septic system education program during and after permitting process.
- Coordination with county health department.
- Monitoring.
- Physical separation and disconnection.

Industrial

Pollution from industrial sites includes both illicit connections and illegal dumping and general storm water runoff from rainwater coming in contact with potential pollutants through inadequate material management practices.

Those industries which discharge directly into waters of the state must obtain a storm water permit from the state or EPA. Under the proposed programs of most municipalities, the municipality will either assume, or share with EPA, responsibility for enforcing such a program within their jurisdiction for dischargers to their separate storm drainage system and, often, to waters-of-the state directly. What this amounts to is, in effect, a storm water program in some ways similar to the industrial pre-treatment program for dischargers to the wastewater treatment system.

Permit Standards - While municipalities have to meet the MEP standard, industrial dischargers are expected to meet applicable standards found in sections 301 and 402 of the Clean Water Act which require the use of Best Available Technology (BAT) and water quality standards. EPA expects that proper application of the MEP standard to industries discharging to the municipal system will result in similar levels of control on these dischargers as conditions placed in individual NPDES permits for storm water discharges associated with industrial activity. Therefore municipalities must become familiar with current NPDES standards for industry and typical NPDES permit conditions.

Not all industries are covered, but only those which fall under the term "associated with industrial activity". This term is generally defined as "directly relating to manufacturing, processing or raw materials storage at an industrial plant". This term does not include such areas as employee parking lots but does include plant areas which were once but are no longer used for industrial activities. Examples of the types of categories included are:

- facilities subject to effluent limitations guidelines, new source performance standards or toxic pollutant effluent standards;
- SIC Codes 10 through 14 (mineral industry), 20 through 39 (manufacturing industry), 40 through 45 and 47 (transportation facilities with vehicle maintenance shops) and 15 and 16 (Construction sites though this will be covered under another portion of the program);
- hazardous waste treatment sites subject to RCRA;
- landfill and land application sites subject to RCRA;
- significant recycling facilities; and
- public owned treatment works (POTW) lands used for land application, sludge disposal or handling, or chemical storage and handling.

EPA has established effluent guidelines for many subcategories of industrial dischargers including: cement manufacturing, petroleum refining, phosphate manufacturing, steam electric, coal mining, and ore mining and dressing. These, and some others, may already have NPDES permits which include storm water and they will generally be required to maintain these permits.

Industrial Assistance To The Municipality - Industries which discharge to the municipal system are required to assist the municipality by supplying certain information to them. In reality, the municipality must take the lead in informing the industries of this requirement and providing forms to allow for consistent reporting and to facilitate database input. The information the industries must provide includes:

- name, location and type of activity;
- location of discharges;
- a certification that the discharge has been tested for the presence of non-storm water discharges and has been found to consist entirely of storm water; and/or
- a certification that the discharge is in compliance with any conditions of an NPDES permit issued to the municipality for the portion of the sewer system into which the industry discharges, providing the industry has been notified of such conditions; and
- a certification that the discharge does not contain a hazardous substance in excess of reportable quantities established by 40 CFR 117.3 or 40 CFR 302.4.

There are two ways an industry, which would otherwise fall under the municipality's program, can be placed under the direct jurisdiction of the state or federal permit writer:

1. the Administrator or NPDES state determines that the discharge contributes to a violation of a water quality standard or is a significant contributor to pollutants to the waters of the state; or
2. a municipality can show that it cannot adequately control a certain industrial discharger. The discharger will then, at the discretion of the Administrator, be required to obtain an individual permit or become a co-permittee to the municipal system permit by establishing conditions specific to that discharger.

Industrial Program - The municipality must propose a program and priorities for control of discharges associated with industrial activity to their systems. The program, at a minimum, must contain provisions for identification of dischargers, monitoring and appropriate control measures. Identification of applicable industries has often proven to be a difficult and a time consuming job. Some municipalities have included only those industries required to obtain a separate NPDES storm water permit and relied on the state or EPA to identify the required list. Various databases are probably available to assist this effort including: current NPDES permit holders, Title III (SARA) listed facilities, and the pre-treatment listing.

Monitoring programs for industry can be time consuming and possibly expensive. While pollutant discharge can be identified to some extent at the outfall to waters of the state, there will be a need to trace pollutant discharges to individual industrial locations. Many municipalities are taking an approach of monitoring only if a pollution problem is suspected through random inspection of waters or complaints. At a minimum monitoring should include those categories of effluent considered in the individual industrial permit section, namely:

- any pollutant limited in an effluent guideline for the industrial subcategory where applicable;
- any pollutant listed in the facilities NPDES permit for its process wastewater if it has one;
- oil and grease, COD, pH, BOD5, TOC, TSS, total phosphorous, total nitrogen; and
- any information required under 40 CFR 122.21(g)(7)(iii) and (iv).

Storm Water Pollution Prevention Plans - Industries covered in the NPDES program must develop a storm water pollution prevention plan (SWPPP). Each plan will be unique both to the type and location of industry and also to the specific site. Group plans and guidance have been developed for similar industries which have formed groups under the regulations (see, for example, Ogden, 1993). General permits also cover many categories of industrial discharges which lay out the approach for SWPPP development and monitoring. EPA has issued multi-sector permits as guidance. All SWPPPs are to be developed using the same basic procedure (EPA, 1992e).

- Planning and Organization - develop team and review other plans.
- Assessment Phase - map and inventory materials; list spills and leaks; test for non-storm water discharges; assess data; and summarize pollutant sources and risks.
- BMP Identification Phase - develop baseline BMPs; select activity and site-specific BMPs.
- Implementation Phase - train employees and implement BMPs.
- Evaluation/Monitoring - annual site inspection; keep records; review and revise plan.

Industrial And Related Facilities Program BMPs -

Inspections
- Develop comprehensive list based on state permits and other industries the municipality finds it must regulate and maintain as a database.
- The initial program will include only those categories of industries included in the regulations and those sites the municipality finds are a problem.
- Investigate the current inspection programs of the fire department, water utility and wastewater utility, seek possible changes and develop implementation schedule.
- Perform inspections on a random basis primarily through these cross-trained inspectors.
- Pre-treatment program expansion.
- Illicit connection methods.
- Fire HAZMAT program expansion.
- Posted materials plans.
- Spill response plans.
- On-site instruction and training.

Control And Training Implementation
- Obtain industry management plan if problems are found or suspected, perform investigation and assist in enforcement or modification of permit conditions and plan to better facilitate pollution reduction.

- Obtain and employ any available State BMP manual for industries.
- Technical guidance.
- Wet detention ponds.
- Extended dry detention ponds.
- Off-line first flush detention.
- Infiltration/exfiltration trenches.
- Dry wells (rooftop runoff control for office parks, apartments, etc.).
- Oil/water/grit separators.
- Sand filters.
- Pretreatment.
- Sumped catchbasins.
- Exfiltrating curb and gutter (quality and quantity control).
- Velocity control structures (to avoid in-stream scouring velocities).
- Porous pavement (roadway).
- Vegetated pavement (parking).
- Storage and pump back to sanitary system.
- Spill plans and cleanup.
- Controlled wash down.
- Covering loading areas.
- Material handling.
- Diversion of flows.
- Parking lot sweeping.
- At site instruction and visits.
- Video tapes.
- Instruction hotline.

Monitoring
- Develop a monitoring program for industry which includes inspections and sampling.
- Grab sampling.
- Illicit connections testing methods.
- On-site inspections.
- Use of pre-treatment system.
- Fire HAZMAT program expansion.
- Self policing.

Construction Sites

Runoff from construction sites can involve sediment runoff several orders of magnitude greater than that of the pre-developed site. Sediment surges can have a number of adverse impacts on the environment including: bonding of other pollutants to the sediment, clogging of fish gills, turbidity, burying of benthic organisms important in the food web, increasing the cost of drinking water treatment, filling and clogging storm water conveyance and retention structures, aesthetic degradation and changing downstream flow lines through bed and overbank deposition.

Construction sites can also generate high loadings of nitrogen and phosphorus from fertilizer, pesticide and herbicide runoff, petroleum products, chemicals solid wastes and other toxics.

The regulations include activities involving SIC Codes 15 and 16 (general building contractors and heavy construction contractors) and pre-construction activities with certain exceptions.

While a number of municipalities have some sort of erosion control ordinance or program, many of them are not effective due to lack of inspection and enforcement capability. Also, there are often problems in the ordinances themselves which allow, for example, residential subdivisions to sell off individual lots which are often then re-graded without erosion control provisions.

Therefore, any erosion control program must emphasize inspection and enforcement, meeting the requirements and limits of the regulations, and closing holes in existing ordinances. Under the regulations a municipality must propose a plan for a program to implement and maintain structural and non-structural BMPs for controlling storm water runoff at construction sites meeting the requirements stated above.

There are a large number of structural and non-structural BMPs for use in erosion control. More complete lists can be found in the better state or county erosion control manuals such as those published by Virginia, North Carolina or Georgia.

Storm Water Pollution Prevention Plans - Like industrial sites, storm water pollution prevention plans (SWPPPs) must be developed and implemented for construction sites. Development of such SWPPPs under the Federal general permit for construction sites is required. Other state general permits or individual permits will also require a form of the SWPPP. Development and implementation of the SWPPP is a six phase process (EPA 1992d, 1992f).
* Site Evaluation and Design Development - collection of site information; site plan design; and pollution prevention site map.
* Assessment - measurement of site area; determination of drainage; runoff coefficient.
* Control Selection/ Plan Design - review state and local requirements; select BMP controls; locate controls; prepare inspection and maintenance plan; develop control timing and coordination; prepare construction sequence.
* Certification and Notification - certify the plan; submit notice of intent to comply with general permit; post plan.
* Implementation/Construction - implement controls; inspect and maintain controls; update plan; report releases as appropriate.
* Final Stabilization/Termination - perform final stabilization; post notice of termination; retain records.

Construction Site Program BMPs -

Planning And Operations
* Seek to facilitate the general permit application throughout the municipality.
* Modifications to site plan submittals.
* Scheduled grading.
* Temporary measures.
* Technical guidance.
* Videos.
* Training courses.
* Field trips.
* Pamphlets.

BMPs
* Make use of available erosion and sediment control manuals for your area.
* Stabilization.
* Mulching.
* Straw bales.
* Silt fences.
* Brush barriers.

- Storm drain inlet protection.
- Diversions.
- Dikes.
- Sediment traps and basins.
- Filtering media.
- Filter strips.
- Slope drains and flumes.
- Outlet protection.
- Riprap.
- Check dams.
- Drop structures.
- Level spreaders.
- Streambank stabilization - structural.
- Streambank stabilization - vegetative.
- Temporary stream crossings.
- Subsurface drains.
- Surface roughening.
- Topsoiling.
- Soil tacking.
- Seeding and sodding.

Inspections
- Train inspectors in general permit requirements.
- Employ a public relations program in making developers and the general public aware of erosion problems.
- Certifications.
- Inspector checklists.
- Single lot requirements.
- Building inspector training for erosion inspection.
- Joint permitting for use and occupancy (U&O) permits and erosion control.

Commercial And Residential Areas

This area requires appropriate controls, planning activities, public roadway activities, integration of flood control programs with SWQM and chemical application programs (fertilizer, pesticides, herbicides). Structural BMPs are most applicable in areas facing development where advance planning can achieve a drainage design with integrated structural features, regional approaches and water quality enhancing individual design features.

Commercial And Residential BMPs -

Maintenance And Inspections
- Develop levels of service for maintenance.
- Perform inventory of system.
- Develop master plan of system.
- Identify and begin to work off backlog of maintenance needs.
- Develop maintenance requirements and procedures and analysis for water quality benefit.
- Develop monitoring and sampling program for maintenance in inlets for total solids and to see if maintenance improves water quality.

- Do a feasibility study of maintenance procedures.
- Establish an inspection schedule for key portions of the drainage system.
- Plan for maintenance program expansion for both quantity and quality.
- Begin public participation and education in system maintenance.
- Develop and implement complaint response hotline.
- Investigate yard waste recycling program and look for improvements.
- Investigate existing pet ordinance for appropriate modifications.
- Program for maintenance of structural controls.
- Program for maintenance of catch basins.
- Program for maintenance of channels and ditches.
- Inspection program.
- Housekeeping litter program downtown.
- Trash racks for detention.
- Litter/housekeeping ordinances.
- Leaf removal - yard waste recycling.
- Pet ordinances.
- Mowing/clipping disposal.
- Disposal ordinances.
- Street sweeping for leaves and litter.
- Adopt-a-stream, stream walking.

New Development And Regional Master Planning
- Develop a process for change in the development procedure including:
    - consensus building;
    - changes in various ordinances, regulations and policies;
    - changes in technical manual;
    - changes in review process;
    - changes in permits; and
    - develop education and public awareness on development procedure changes and the use of BMPs in new developments.
- Designate sensitive areas based on requirements of any watershed protection regulations and develop storm water management changes for these select areas.
- Proceed with the development of master plans that take into account storm water quality and the possibility of regional treatment where appropriate.
- Technical manual changes.
- Education on techniques.
- Credits.
- Examples on how to modify existing plans for storm water quality.
- Landscaping guidance.
- Modify zoning regulations for density or development type controls.
- Changed open space requirements.
- Pollution capture or reduction requirements (e.g. 1/2" capture and treat).
- New policies on development submittals (e.g. site pollution reduction plan).
- Educate inspectors and engineers.
- Sensitive area mapping or overlays in zoning regulations.
- New design specifications.
- Regional master planning for structural and non-structural controls.
- Regional master planning partnership with developers.
- Sensitive area planning and mapping.

- Coordination with sewer extensions.
- Key area acquisition.
- Intergovernmental coordination.
- Coordination with parks and highway construction.

Streets
- Obtain and analyze air quality information developed under the clean air act.
- Analyze existing deicing program, plan changes and develop implementation schedule.
- Analyze existing vegetation management (herbicide and fertilize use) program, plan changes and develop implementation schedule.
- Analyze existing street maintenance program, plan changes and develop implementation schedule.
- Analyze existing litter control program, plan changes and develop implementation schedule.
- Analyze existing street sweeping program, plan changes and develop implementation schedule.
- Analyze existing catch basin cleaning program, plan changes and develop implementation schedule.
- Perform research into structural BMPs applicable to streets.
- Develop a feasibility study and monitoring program to assess the use of structural BMPs.
- New street designs for sumped catch basins.
- Coalescing filters for streets.
- First flush devices.
- Wet or filtration ponds.
- Streets without curbs.
- Grassed swales.
- Street retrofitting.
- Infiltration/exfiltration trenches.
- Screened catch basins.
- Exfiltrating curb and gutter (quality and quantity control).
- Porous pavement (roadway).
- Vegetated pavement (parking).
- Oil/grit separators.
- Velocity controls.
- Direct discharge reduction.
- Salt control practices.
- Salt storage management.
- Use of other than salt for deicing.
- Environmentally conscious fertilizer programs.
- Herbicide plans.
- Integrated pest management (IPM) programs.
- Xerigraphic vegetation and use of natural low maintenance vegetation.
- Vegetation maintenance on embankments.
- Catch basin cleaning.
- Environmentally conscious street maintenance and construction practices.
- Require the municipality to employ as strict a practice as others.
- Modified street sweeping practices.
- Litter programs.
- Traffic reduction.
- Clean air act compliance changes.

Flood Control Structures And Retrofitting
- Perform drainage structure size and condition inventory.
- Develop procedures for the evaluation of all new projects for environmental impacts or possible modification for environmental purposes.
- Consider the possibility of cost effectively retrofitting structures during the master planning process.
- Evaluate the existing detention design criteria and several example sites designs using this criteria for the possibility of retrofitting on-site detention for water quality purposes.
- Analyze existing greenway program, plan changes and develop implementation schedule.
- Employ macro invertebrate richness sampling in the larger streams to assess the impacts of channel erosion on the benthic community, where appropriate.
- Analyze existing channel maintenance program, plan changes and develop implementation schedule.
- Analyze existing floodplain designation program, plan changes and develop implementation schedule.
- Analyze existing drainage easement requirement, plan changes and develop implementation schedule.
- Retrofitting flood control structures.
- Flood control structure operation changes.
- Detention abandonment.
- Environmental impact statements.
- In-stream mitigation measures.
- Multi-objective planning of structures.
- Natural or minimal bank protection methods.
- Grade control.
- Regulatory requirements for bank protection during development.
- Buffer strips and flow spreaders.
- Greenways and corridors.
- Vegetated bank protection.
- Habitat restoration practices.
- Artificial overbank wetlands.
- Erosion control detention and retention.
- Monitoring and repair practices.
- Geomorphic studies and alignment.
- Extension of floodplain regulation to smaller streams.
- Modified floodplain regulation.

Municipal Waste
- Coordinate with and facilitate application of the general permit to these sites.
- Assess current monitoring program at open and closed landfills, plan changes and develop implementation schedule.
- Monitoring.
- Inspections.
- Modified practices.
- Structural controls (see Industry).

Pesticides, Fertilizer, Herbicide
- Analyze existing park and recreation and street maintenance public programs, plan changes and develop implementation schedule.

- Develop a pesticide, fertilizer and herbicide public education program in coordination with the ongoing storm water public relations program.
- Investigate current requirements on commercial applicators before coming up with high priority BMPs in this area.
- Public education measures.
- Commercial applicator training and certification checks.
- Permits for applicators.
- Applicator inspection and enforcement program.
- Integrated pest management (IPM) concepts.
- Collection points for excess materials.
- Right-of-way and public lands application controls.
- Golf course and cemetery controls.
- Use of more natural or less damaging alternate fertilizers, pesticides and herbicides.
- Use of natural vegetation.
- Environmentally conscious fertilizer programs.
- Herbicide plans.
- Xerigraphic vegetation and use of natural low maintenance vegetation.
- Buffer zones for applications.
- Ordinances for use and application.

15.7 Non-Structural Best Management Practices

Best management practices (BMPs) are the basic mitigation measures used in the storm water quality management plans to control pollutants within the municipality. Chapter 13 presented the details of structural best management practices and their use within the municipal drainage system. The other major category of BMPs include the many non-structural or source control practices that can be used for pollution prevention and control of pollutants. In most cases it is much easier and less costly to prevent the pollutants from entering the drainage system than trying to control pollutants with structural BMPs. Thus within the "treatment train" concept, the non-structural BMPs should be the first line of defence in protecting the receiving stream within the municipality. If used properly, the non-structural BMPs can be very effective in controlling pollutants and greatly reduce the need for structural BMPs. In addition, non-structural BMPs tend to be less costly, easier to design and implement and easier to maintain than structural BMPs. Following is a brief discussion of some non-structural BMPs that can be used within a storm water quality management plan (North Central Texas, 1993)

Public Education/Participation

Public education/participation, like an ordinance or a piece of equipment, is not so much a best management practice as it is a method by which to implement BMPs. Public education/participation are vital components of many of the individual source control BMPs. A public education and participation plan provides the municipality with a strategy for educating its employees, the public, and businesses about the importance of protecting storm water from improper use, storage, and disposal of pollutants. Municipal employees must be trained, especially those that work in departments not directly related to storm water but whose actions affect storm water. Residents must become aware that a variety of hazardous products are used in the home and that their improper use and disposal can pollute storm water and groundwater supplies. Businesses, particularly smaller ones that may not be regulated by Federal, State, or local regulations, must be informed of ways to reduce their potential to pollute storm water.

The public education and participation plan should be based on four objectives:
- promote a clear identification and understanding of the problem and the solutions,
- identify responsible parties and efforts to date,
- promote community ownership of the problems and the solutions, and
- integrate public feedback into program implementation.

Each public education/participation program is specifically designed and developed for the needs of the particular municipality. Typical target audiences include:
- Political - elected officials, chambers of commerce, and heads of departments, agencies, and commissions;
- Technical - municipal department and agency staffs, State agencies;
- Business - commercial and industrial, including trade associations;
- Community Groups - fraternal, ethnic, hobby, horticulture, senior citizen, and service;
- Environmental;
- General Public/Residential;
- Schools/Youth Groups;
- Media - print and electronic, and
- Pollutant-defined - groups of individuals defined by the specific pollutant(s) they discharge (e.g., used motor oil, pesticides)

For these target audiences the activities within the public education/participation plan can include surveys, presentations, school activities, development of working committees, development of literature and media campaigns, workshops, etc. All of these activities can be an important part of controlling local storm water management problems. For more details on public education and participation see Chapter 3, Public Awareness and Involvement.

Land Use Planning/Management

This BMP presents an important opportunity to reduce the pollutants in storm water runoff by using a comprehensive planning process to control or prevent certain land use activities in areas where water quality is sensitive to development. It is applicable to all types of land use and represents one of the most effective pollution prevention practices. Subdivision regulations, zoning ordinances, preliminary plan reviews and detailed plan reviews, are tools that may be used to mitigate storm water contamination in newly developing areas. Also, master planning, cluster development, terracing and buffers are ways to use land use planning as a BMP in the normal design for subdivisions and other urban developments. These are planning tools that municipal agencies can use to require conditions of approval or establish improvement/construction standards to meet the water quality objectives within specific watersheds.

An impervious cover limitation is one of the more effective land use management tools, since nationwide research has consistently documented increases in pollution loads with increases in impervious cover. For example, in compliance with State water supply watershed regulations, municipalities in North Carolina have adopted impervious area limitations. These limitations are quite complex since there are five different classifications of watersheds, depending on the amount of existing development. However, following are the general requirements:
- Critical Areas (near water supply points) -
  12% impervious area without structural BMPs
  12-30% impervious area with structural BMPs
- Entire Watershed -
  24% impervious area without structural BMPs

24-50% impervious area with structural BMPs

In addition, BMPs for agriculture, forestry and transportation are required. Other features include provisions for cluster developments, no landfills, sewer lines encouraged, control of one inch storm for structural BMPs.

In addition to controlling impervious area cover, directly connected impervious areas are kept to a minimum. This is especially important for large impervious areas such as parking lots and highways and it can also be effective for small impervious areas such as roof drainage.

Land use planning/management frequently addresses sensitive public issues and will generally develop policies that will not be universally accepted within the municipality. As such, implementation of the necessary restrictions on certain land uses required to mitigate storm water pollution may not be politically feasible.

Material Use Controls

There are three major BMPs included in this category:
1. Housekeeping Practices
2. Safer Alternative Products
3. Pesticide/Fertilizer Use

In housekeeping practices, the goal is to promote efficient and safe practices such as storage, use, cleanup, and disposal, when handling potentially harmful materials such as fertilizers, pesticides, cleaning solutions, paint products, automotive products, and swimming pool chemicals. In addition, the municipality can promote the use of less harmful products. Alternatives exist for most product classes including fertilizers, pesticides, cleaning solutions, and automotive and paint products.

Pesticides and fertilizers have become an important component of land use and maintenance for municipalities, commercial land uses and residential land owners. Any usage of pesticides and fertilizers increases the potential for storm water pollution. BMPs for pesticides and fertilizers include education in their use, control runoff from affected areas, control times when they are used, provide proper disposal areas, etc.

For the general public, municipalities should establish a public education program that provides information on such items as storm water pollution and the beneficial effects of proper disposal on water quality; reading product labels; safer alternative products; safe storage, handling, and disposal of hazardous products; list of local agencies; and emergency phone numbers. This information can be provided through brochures or booklets that can be made available at a variety of places including municipal offices, public information fairs, and places where such products are sold. Education should also be developed for municipal employees and commercial and industrial establishments.

Material Exposure Controls

There are two major BMPs included in this category:
1. Material Storage Control
2. Vehicle Use Reduction

Material storage control is used to prevent or reduce the discharge of pollutants to storm water from material delivery and storage by minimizing the storage of hazardous materials onsite, storing materials in a designated area, installing secondary containment, conducting regular inspections, and training employees and subcontractors.

Vehicle use reduction is used to reduce the discharge of pollutants to storm water from vehicle use by high-lighting the storm water impacts, promoting the benefits to storm water of alternative transportation, and integrating initiatives with existing or emerging regulations and programs.

The major problem with storage controls is providing storage areas and containers and developing programs for their inspection and maintenance.   Education programs will be needed to implement storage controls and for vehicle use reduction. Areas for storage of materials and reduction of vehicle use should be considered in all land use plans.

Material Disposal And Recycling

There are three major BMPs included in this category:
1.   Storm Drain System Signs
2.   Household Hazardous Waste Collection
3.   Used Oil Collection

Stenciling of the storm drain system (inlets, catch basins, channels, and creeks) with prohibitive language/graphic icons discourages the illegal dumping of unwanted materials.

Household hazardous wastes are defined as waste materials which are typically found in homes or similar sources, which exhibit characteristics such as: corrosivity, ignitability, reactivity, and/or toxicity, or are listed as hazardous materials by the EPA.

Used oil recycling is a responsible alternative to improper disposal practices such as dumping oil in the sanitary sewer or storm drain system, applying oil to roads for dust control, placing used oil and filters in the trash for disposal to landfill, or simply pouring used oil on the ground.

Storm drain system signs act as highly visible source controls that are typically stenciled directly adjacent to storm drain inlets.   The signs contain brief statements that discourage the dumping of improper materials into the storm drain system.   Graphical icons, either illustrating anti-dumping symbols or images of receiving water fauna, are effective supplements to the anti-dumping message. The intent of such a storm drain system stenciling program is to enhance public awareness of the pollutant effect on local receiving waters from storm water runoff and also to discourage individual's habitual waste disposal actions (e.e., automotive fluids and landscaping wastes).   An important aspect of a stenciling program is the distribution of informational flyers that educate the neighborhood (business or residential) about storm water pollution, the storm drain system, and the watershed, and that provides information on alternatives such as recycling, household hazardous waste disposal, and safer products.

While it is generally recognized that the potential exists for hazardous household materials to come in contact with storm water runoff, it is unclear at present how significant this source of contamination is. As such, it is difficult to quantify the benefits to water quality from household hazardous waste collection programs. However, such programs are a preventative, rather than curative measure, and may reduce the need for more elaborate treatment controls. Programs can be a combination of permanent collection centers, mobile collection centers, curbside collection, recycling, reuse, and source reduction. Public education is extremely important in implementing this BMP. The municipality should also assist small commercial establishments in the proper disposal of their wastes.

The two basic factors to be addressed during the implementation of a used oil recycling program are: 1) used oil collection, and 2) hauling/disposal. One commonly used oil collection alternative is a temporary "drop off" site on designated collection days. The public is notified that used oil can be brought to these specific locations and if successful, permanent site can be developed. Some communities rely solely on private collectors such a automobile service stations, quick oil change centers and auto parts stores to serve as collection sites. Well before oil collection commences, reliable local used oil haulers and recyclers should be contacted.   In addition, information flyers should be distributed at the point of purchase informing users about the proper disposal of wastes.

Spill Prevention And Cleanup

There are two major BMPs included in this category:
1. Vehicle Spill Control
2. Aboveground Tank Spill Control

The purpose of a vehicle spill control program is to prevent or reduce the discharge of pollutants to storm water from vehicle leaks and spills by reducing the chance for spills by preventive maintenance, stopping the source of spills, containing and cleaning up spills, properly disposing of spill materials, and training employees. It is also very important to respond to spills quickly and effectively.

Major spills on roadways and other public areas are generally handled by highly trained Hazmat teams from local fire departments or environmental health departments. Major involvement of the municipalities is in the area of education and prevention.

Aboveground tank spill control programs prevent or reduce the discharge of pollutants to storm water by installing safeguards against accidental releases, installing secondary containment, conducting regular inspections, and training employees in standard operating procedures and spill cleanup techniques.

Accidental releases of materials from aboveground liquid storage tanks present the potential for contaminating storm water with many different pollutants. Materials spilled, leaked, or lost from tanks may accumulate in soils or on impervious surfaces and be carried away by storm water runoff.

Proper handling and storage of materials is very important and should include proper labeling; development of storage and handling procedures, secondary containment procedures, spill response procedures; and adequate training and education for those involved with this BMP.

Dumping Controls

This BMP addresses the implementation of measures to detect, correct, and enforce against illegal dumping of pollutants on streets and into the storm drain system, streams, and creeks. Substances illegally dumped on streets and into the storm drain system and creeks include paints, used oil and other automotive fluids, construction debris, chemicals, fresh concrete, leaves, grass clippings, and pet wastes. All of these wastes can cause storm water and receiving water quality problems as well as clog the storm sewer system itself. See section 15.6, Illicit Connections and Illegal Dumping, for more details on dumping controls.

Connection Controls

There are three major BMPs included in this category:
1. Illicit Connection Prevention
2. Illicit Connection Detection and Removal
3. Leaking Sanitary Sewer Control

Illicit connection protection tries to prevent unwarranted physical connections to the storm drain system from sanitary sewers, floor drains, etc., through regulation, regular inspection, testing, and education. In addition, programs include implementation control procedures for detection and removal of illegal connections from the storm drain conveyance system. Procedures include field screening, follow-up testing, and complaint investigation.

Leaking sanitary sewer control includes implementing control procedures for identifying, repairing, and remediating infiltration, inflow, and wet weather overflows from sanitary sewers into the storm drain conveyance system. Procedures include field screening, testing, and complaint investigation. See section 15.6, Illicit Connections and Illegal Dumping, for more details on illicit connections and leaking sanitary sewer controls.

Illegal connections can occur in new as well as existing developments. Improper connections in areas of new development can be prevented through inspection and other verification techniques. The first measure to prevention is to make sure that existing municipal building and plumbing codes prohibit any unwarranted, non-permitted physical connections to the storm drain system. Building and plumbing code inspectors, in addition to new land development project inspectors, must visually inspect to ensure that illegal connections are not being physically tied to the storm conveyance system. Proper documentation and record keeping is essential to the function of such inspections. Documentation helps catalog the storm drain system and is required by Federal regulations. Visual inspection, however, is not a very reliable means of verifying the status of new physical connections and their final destination. Continued monitoring throughout the entire development phase would be necessary to guarantee the new physical connections between the sanitary sewers and storm drains had been prevented through the inspection process.

Public education programs will also aid in the monitoring of illegal connections and leaking sanitary sewers by making individuals aware of evidence of unwarranted discharges to the storm drain system. A community hotline for reporting such evidence can greatly supplement the storm water department's field screening efforts.

Street/Storm Drain Maintenance

There are seven major BMPs included in this category:
1. Roadway Cleaning
2. Catch Basin Cleaning
3. Vegetation Controls
4. Storm Drain Flushing
5. Roadway/Bridge Maintenance
6. Detention/Infiltration Device Maintenance
7. Drainage Channel/Creek Maintenance

Roadway cleaning may help reduce the discharge of pollutants to storm water from street surfaces by conducting cleaning on a regular basis. However, cleaning often removes the larger sizes of pollutants but not the smaller sizes. Most pollutants are deposited within three feet of the curb which is where the roadway cleaning should be concentrated. Catch basin cleaning on a regular basis also helps reduce pollutants in the storm drain system, reduces high pollutant concentrations during the first flush of storms, prevents clogging of the downstream conveyance system and restores the catch basins' sediment trapping capacity.

Vegetation control typically involves a combination of chemical (herbicide) application and mechanical methods. Mechanical vegetation control includes leaving existing vegetation, cutting less frequently, handcutting, planting low maintenance vegetation, mulching, collecting and properly disposing of clippings and cuttings, and educating employees.

Storm drains can be "flushed" with water to suspend and remove deposited materials. Flushing is particularly beneficial for storm drain pipes with grades too flat to be self-cleansing. Flushing helps ensure pipes convey design flow and removes pollutants from the storm drain system. However, flushing will only push the pollutants into downstream receiving waters unless the discharge from the flushing is captured and removed from the drainage system.

Roadway/bridge maintenance is used to prevent or reduce the discharge of pollutants to storm water from roadway and bridge maintenance by paving as little are as possible, designing bridges to collect and convey storm water to proper locations, using measure to prevent runoff from entering the drainage system, properly disposing of maintenance wastes, and training employees.

Proper maintenance and siltation removal is required on both a routine and corrective basis to promote effective storm water pollutant removal efficiency for wet and dry detention ponds and infiltration devices. Also, regularly removing illegally dumped items and material from storm drainage channels and creeks will reduce pollutant levels.

Given the costs in involved with these BMPs, public education is needed to alert the public to the need of maintenance and the consequences of not developing maintenance programs. Regular maintenance of all portions of the drainage system is necessary to ensure the proper functioning of the system. Studies have indicated that detention basins may have a life as short as five year without maintenance. Lack of maintenance can also result in drainage and flooding problems which could result in significant damages and legal problems for the municipality. For the drainage system to operate as designed for both water quality protection and storm conveyance capacity, maintenance is essential. The public must become aware that everyone uses and benefits from the drainage system and all must share in the cost of maintenance.

Permanent Erosion Control

There are three major BMPs included in this category:
1. Erosion Control - Permanent Vegetation
2. Erosion Control - Flow Control
3. Erosion Control - Channel Stabilization

Vegetation is a highly effective method for providing long term, cost effective erosion protection for a wide variety of conditions. It is primarily used to protect the soil surface from the impact of rain and the energy of the wind. Vegetation is also effective in reducing the velocity and sediment load in runoff sheet flow.

Channel stabilization addresses the problem of erosion due to concentrated flows. Concentrated flows occur in channels, swales, creeks, rivers and other water courses in which a substantial drainage area drains into a central point. Overland sheet flow begins to collect and concentrate in the form of rills and gullies after overland flow of as little as 100 feet. Erosion due to concentrated flow is typically extensive, causing large soil loss, undermining foundations and decreasing the flow capacity of watercourses.

Proper selection of ground cover is dependent on the type of soil, the time of year of planting, and the anticipated conditions that the ground cover will be subjected. In addition, mulching is a form of erosion protection which is commonly used in conjunction with establishment of vegetation. It typically improves infiltration of water, reduces, runoff, holds seed, fertilizer and lime in place, retains soil moisture, helps maintain temperatures, aids in germination, retards erosion and helps establish plants in disturbed areas.

Once flow is allowed to concentrate, it is more difficult to control erosion problems. Thus every effort should be made to maintain sheet flow conditions for runoff. Where concentrated flows are unavoidable, the following techniques can be used to control erosion and resulting water quality problems.

- Rip Rap
- Level Spreaders
- Gabions
- Armor Protection
- Check Dams
- Diversions

For more information on erosion control see Section 15.6, Construction Sites, given earlier in this chapter.

References

American Public Works Association, "Water Quality: Urban Runoff Solutions", APWA Rpt. #61, May, 1991.

American Public Works Association, "Study of Nationwide Costs to Implement Municipal Stormwater Best Management Practices", May, 1992.

City of Austin, "Inventory of Urban Nonpoint Source Pollution Practices", June, 1988.

Denver Urban Drainage and Flood Control District, "Urban Storm Drainage Criteria Manual", Vol. 3 Best Management Practices, Denver, CO., Sept., 1992.

Falkenbury, J., "Water Quality Standard Operating Procedures", City of Fort Worth Public Health Dept., Ft. Worth, TX., 76107, 1987.

Federal Register, December 7th, 1988 pp. 49416-49487.

Maine DEP, "Watershed: An Action Guide to Improving Maine Waters", Dept. of Env. Prot., April, 1990.

Maine DEP, "Environmental Management: A Guide for Town Officials", Dept. of Env. Prot., April, 1992.

Metropolitan Washington Council of Governments, "Controlling Urban Runoff", T.R. Schueler, July, 1987.

Metropolitan Washington Council of Governments, "A Current Assessment of Urban Best Management Practices", March, 1992.

NAFSMA, "Municipal Separate Storm Sewer (MS4) Permit Application Costs", June, 1992.

North Central Texas, "Residential/Commercial BMP Manual", July 1993.

Ogden Environmental, "Stormwater Quality Management Plan Development", Nashville, TN., Memo, 1992a.

Ogden Environmental, "Stormwater Quality Permit Application for the City of Charlotte, NC", May, 1992b.

Reese, A. J., "Developing Municipal Storm Water Quality Management Programs", Proc. 6th International Conference on Urban Storm Drainage, Sept. 12-17, 1993, Ontario, Canada.

Reese, A. J., "The 10 Commandments of Municipal Stormwater Quality Management Programs", APWA Reporter, March 1994.

State of Minnesota, "Protecting Water Quality in Urban Areas", Minnesota Poln. Control Agency, Oct., 1989.

Stormwater Quality Task Force, "California Storm Water Best Management Practice Handbooks - Vol. 1, Municipal", State of California, Water Control Board, March, 1993.

Shaver, H. E., "Stormwater Management Issues", in Design of Urban Runoff Quality Controls, ASCE, L.A. Roesner, B. Urbonas and M.B. Sonnen, eds., 1988.

USEPA, "Environmental Impacts of Stormwater Discharges", EPA 841-R-92-001, June, 1992a.

USEPA, "Guidance Manual for the Preparation of Part 2 of the NPDES Permit Applications for Discharges From Municipal Separate Storm Sewer Systems", The Cadmus Group, August, 1992c.

USEPA, "Guidance Specifying Management Measures for Sources of Nonpoint Pollution in Coastal Waters", EPA 840-B-92-002, Jan., 1993.

USEPA, "Handbook - Urban Runoff Pollution Prevention and Control Planning", EPA/625/R-93/004, Sept. 1993d.

USEPA, "Investigation of Inappropriate Entries into Storm Drainage Systems", EPA/600/R-92/238, Jan., 1993a.

USEPA, "NPDES Permit Application Workshop For Storm Water Discharges From Municipal Separate Storm Sewer Systems and Industrial Sources - Speaker's Notebook", USEPA Region IV, July, 1992b.

USEPA, "Results of the Nationwide Urban Runoff Program", Vol. 1 -Final Rpt., PB84-185552, Dec., 1983.

USEPA, "Setting Priorities: The Key to Nonpoint Source Control," Office of Water Reg. and Stds., 1987.

USEPA, "Storm Water Management for Construction Activities - Development of Pollution Prevention Plans and Best Management Practices", EPA 833-R4-92-005, Sept., 1992d.

USEPA, "Stormwater Management for Industrial Activities", Stormwater Permit Manual Reprint, Thompson Publishing Group, Sept., 1992e.

USEPA, "Storm Water Management for Construction Activities - Development of Pollution Prevention Plans and Best Management Practices", Summary Guidance EPA 833-R4-92-001, October, 1992f.

USEPA, "Urban Runoff Management Information/Education Products", Office of Wastewater Enforcement and Compliance, Permits Division, Feb., 1993b.

USEPA, "Urban Targeting and BMP Selection", Region V Water Division, Nov. 1990c.

Virginia DCR, "Stormwater Management Regulations and Act", State of Virginia Dept. of Cons. and Rec., Richmond, VA., 1990.

Chapter 16 Documentation

16.1 Overview

An important part of the design or analysis of any storm water management facility is the documentation. Appropriate documentation of the design of any storm water management facility is essential because of:

- the importance of public safety,
- justification of expenditure of public funds,
- future reference by engineers (when improvements, changes, or rehabilitations are made to the facilities),
- information leading to the development of defense in matters of litigation, and
- public information.

Frequently, it is necessary to refer to plans, specifications and analysis long after the actual construction has been completed. Documentation permits evaluation of the performance of structures after flood events to determine if the structures performed as anticipated or to establish the cause of unexpected behavior, if such is the case. In the event of a failure, it is essential that contributing factors be identified in order that recurring damage can be avoided.

The definition of hydrologic and hydraulic documentation as used in this chapter is the compilation and preservation of the design and related details as well as all pertinent information on which the design and decisions were based. This may include drainage area and other maps, field survey information, source references, photographs, engineering calculations and analyses, measured and other data, and flood history including narratives from newspapers and individuals such as highway maintenance personnel and local residents who witnessed or had knowledge of an unusual event.

16.2 Purpose Of Documentation

Although the amount and detail of documentation should be commensurate with the size, scope, and importance of the project, this chapter presents the documentation which should be included in the design files and on the construction plans. While the documentation requirements for existing and proposed storm water management facilities are similar, the data retained for existing facilities are often slightly different than that for proposed facilities, and these differences are discussed. This chapter focuses on the documentation of the findings obtained in using the other chapters of this book, and thus readers should be familiar with all the hydrologic and hydraulic design procedures associated with this chapter. This chapter identifies the system for organizing the documentation of storm water management facility designs and reviews so as to provide as complete a history of the design process as is practical.

The major purpose of providing good documentation is to define the design procedure that was used and to show how the final design and decisions were arrived at. Often there is expressed the myth that avoiding documentation will prevent or limit litigation losses as it supposedly precludes providing the plaintiff with incriminating evidence. This is seldom if ever the case and documentation should be viewed as the record of reasonable and prudent design analysis based on the best available technology. Thus, good documentation can provide the following.

- Protecting the design engineer and the local agency by proving that reasonable and prudent actions were, in fact, taken (such proof should certainly not increase the potential court award and may decrease it by disproving any claims of negligence by the plaintiff).
- Identifying the situation at the time of design which might be very important if legal action occurs in the future.
- Documenting that rationally accepted procedures and analysis were used at the time of the design which were commensurate with the perceived site importance and the flood hazard, (this should further disprove any negligence claims).
- Providing a continuous site history to facilitate future reconstruction.
- Providing the data necessary to quickly evaluate any future site problems that might occur during the facilities service life.
- Expediting plan development by clearly providing the reasons and rationale for specific design decisions.

16.3 Types Of Documentation

There are three basic types of documentation which should be considered. The types are preconstruction, design, and construction or operation.

Preconstruction

Preconstruction documentation should include the following if available or within the budgetary restraints of the project.
- aerial photographs,
- contour mapping,
- watershed map or plan including:
    flow directions,
    watershed boundaries,
    watershed areas,
    natural storage areas,
- surveyed data reduced to include:
    existing hydraulic facilities,
    existing controls,
    profiles - roadway, channel, driveways,
    cross sections - roadway, channels, faces of structures,
- flood insurance studies and maps by Federal Emergency Management Agency (FEMA),
- Soil Conservation Service soil maps,
- field trip report(s) which may include:
    video cassette recordings,
    audio tape recordings,
    still camera photographs,
    movie camera films,
    written analysis of findings with sketches, and
- reports from agencies (local, State or Federal), newspapers, and abutting property owners.

Design

Design documentation should include all the information used to justify the design, including:
- reports from other agencies,
- hydrological report,
- hydraulic report,
- computer analysis output, and
- any approvals received.

Construction Or Operation

Construction or operation documentation should include:
- plans,
- revisions,
- as-built plans and subsurface borings,
- photographs, and
- record of operation: during flooding events, complaints and resolutions.

It is very important to prepare and maintain in a permanent file the as-built plans for every storm water management structure to document subsurface foundation elements such as footing types and elevations, pile types and (driven) tip elevations, etc. There may be other information which should be included or may become evident as the design or investigation develops. This additional information should be incorporated at the discretion of the designer.

16.4 Documenting The Plan Development Process

Documentation should not be considered as occurring at specific times during the design or as the final step in the process which could be long after the final design is completed. Documentation should rather be an ongoing process and part of each step in the hydrologic and hydraulic analysis and design process. This will increase the accuracy of the documentation, provide data for future steps in the plan development process, and provide consistency in the design even when different designers are involved at different times of the plan development process. Accurate documentation should be provided during the following steps or phases of the plan development process:
1. surface water environmental (environmental impact study) phase,
2. reconnaissance phase,
3. location phase,
4. survey phase (drainage surveys),
5. design phase,
6. revised design phase,
7. construction phase to include "As Built" plans, and
8. operational phase - documentation should be continuous over the structure's life cycle.

The designer should be responsible for determining what hydrologic analyses, hydraulic design, and related information should be documented during the plan development process. The designer should make a determination that complete documentation has been achieved during the plan development process which will include the final design. To assist in this determination refer to the Documentation Project Check List contained in Appendix A at the end of this chapter.

16.5 Documentation Procedures

A complete hydrologic and hydraulic design and analysis documentation file for each project or design should be maintained by the designer. Where practicable this file should include such items as:

- identification and location of the facility,
- photographs (ground and aerial),
- engineering cost estimates and actual construction costs,
- hydrology investigations,
- drainage area maps,
- vicinity maps, topographic and contour maps,
- interviews (local residents, adjacent property owners, and maintenance forces),
- newspaper clippings,
- design notes, and correspondence relating to design decisions,
- history of performance of existing structure(s), and
- assumptions.

The documentation file should contain design/analysis data and information which influenced the facility design and which may not appear in other project documentation.

Following are some recommended practices that the designer should follow related to documentation of hydrologic and hydraulic designs and analyses.

1. Hydrologic and hydraulic data, preliminary calculations and analyses and all related information used in developing conclusions and recommendations related to storm water management facility requirements, including estimates of structure size and location, should be compiled in a documentation file.
2. The designer should document all design assumptions and selected criteria including the decisions related thereto.
3. The amount of detail of documentation for each design or analysis should be commensurate with the risk and the importance of the facility.
4. Documentation should be organized to be as concise and complete as practicable so that knowledgeable designers can understand years hence what was done by predecessors.
5. Circumvent incriminating statements wherever possible by stating uncertainties in less than specific terms - (e.g., the culvert may back water rather than the culvert will back water).
6. Provide all related references in the documentation file to include such things as published data and reports, memos and letters, and interviews. Include dates and signatures where appropriate.
7. Documentation should include data and information from the conceptual stage of project development through service life so as to provide successors with all information.
8. Documentation should be organized to logically lead the reader from past history through the problem background, into the findings, and through the performance.
9. An executive summary at the beginning of the documentation should provide an outline of the documentation file to assist users in finding detailed information.
10. Include all completed forms used by local agencies, copies or references of standards used, and any items required for submittal, related to facility design.

16.6 Storage Requirements

Where and how to store and preserve records is an important consideration. Ease of access, durability, legibility, storage room required, and cost are the prime factors to consider when evaluating alternative methods of storage and preservation. For instance, microfilm and microfiche systems require relatively small storage spaces. However, the support systems are expensive, subject to "down time", and may produce poor copies, and both microfilm and microfiche systems require controlled humidity and temperature. Conversely, the storage of actual plans and documents requires much more space, and the records are not as durable. However, better reproductions are usually obtained, and the temperature and humidity requirements are not as critical. Combinations of the two are sometimes used.

The designer should maintain the documentation files including: microfilm, microfiche, magnetic media, etc. in a location where it will be readily available for use during construction, for defense of litigation and for future replacement or extension. Only that documentation need be retained which is not retained elsewhere. Original plans, project correspondence files, construction modifications, and inspection reports are the types of documentation which usually do not need to be duplicated in the documentation file.

Hydrologic/Hydraulic documentation should be retained with the project plans or in some other permanent location, at least until the storm water management facility is totally replaced or modified as a result of a new storm water management facility study. Procedures should be established to determine when documentation can be destroyed.

16.7 Specific Content Of Documentation Files

The following items should be included in the documentation file. The intent is not to limit the data to only those items listed, but rather establish a minimum requirement consistent with the storm water management design procedures as outlined in the different chapters of this book. If circumstances are such that the storm water management facility is sized by other than normal procedures or if the size of the facility is governed by factors other than hydrologic or hydraulic factors, a narrative summary detailing the design basis should appear in the documentation file. Additionally, the designer should include in the documentation file items not listed below but which are useful in understanding the analysis, design, findings, and final recommendations.

Hydrology

The following hydrologic related items used in the design or analysis should be included in the documentation file:
- contributing watershed area size and identification of source (map name, etc.);
- design frequency and decision for selection;
- hydrologic discharge and hydrograph estimating method and findings;
- copies of all computer analyses;
- flood frequency curves to include design, 100-year flood, any other flood frequencies used for design or analysis, discharge hydrograph, and any historical floods; and
- expected level of development in upstream watershed areas over the anticipated life of the facility (include sources of and basis for these development projections).

Bridges

The following items should be included in the documentation file related to the design and analysis of bridges:
- design and 100-year highwater for undisturbed, existing and proposed conditions;
- stage-discharge curve for undisturbed, existing and proposed conditions;

- cross-section(s) used in the design highwater determination;
- roughness coefficient ("n" value) assignments;
- information on the method used for design highwater determination;
- observed highwater, dates, and discharges;
- velocity measurements or estimates and locations (include both the through-bridge and channel velocity) for design and 100-year floods;
- performance curve to include calculated backwater, velocity and scour for design, 100-year floods, and 500-year flood for scour evaluation;
- magnitude and frequency of overtopping flood;
- copies of all computer analyses;
- complete hydraulic study report;
- economic analysis of design and alternatives;
- risk assessment;
- bridge scour results;
- roadway geometry (plan and profile); and
- potential flood hazards to adjacent properties.

Culverts

The following items should be included in the documentation file related to the design and analysis of culverts:
- culvert performance curves;
- allowable headwater elevation and basis for its selection;
- cross-section(s) used in the design highwater determinations;
- roughness coefficient assignments ("n" values);
- observed highwater, dates, and discharges;
- stage discharge curve for undisturbed, existing and proposed conditions to include the depth and velocity measurements or estimates and locations for the design, 100-year and other check floods;
- performance curves showing the calculated backwater elevations, outlet velocities and scour for the design, 100-year and any historical floods;
- type of culvert entrance condition;
- culvert outlet appurtenances and energy dissipation calculations and designs;
- copies of all computer analyses and standard computation sheets given in the Design of Culverts, Chapter 9, of this book;
- roadway geometry (plan and profile); and
- potential flood hazard to adjacent properties.

Open Channels

The following items should be included in the documentation file related to the design and analysis of open channels:
- stage discharge curves for the design, 100-year and any historical water surface elevation(s);
- cross-section(s) used in the design water surface determinations and their locations;
- roughness coefficient assignments ("n" values);
- information on the method used for design water surface determinations;
- observed highwater, dates, and discharges;
- channel velocity measurements or estimates and locations;
- water surface profiles through the reach for the design, 100-year and any historical floods;

- design or analysis of materials proposed for the channel bed and banks;
- energy dissipation calculations and designs; and
- copies of all computer analyses.

Storm Drains

The following items should be included in the documentation file related to the design and analysis of storm drain systems:
- computations for inlets and pipes, including hydraulic grade lines;
- copies of the standard computation sheets given in Storm Drainage Systems, Chapter 8, of this book;
- complete drainage area map;
- design frequency;
- information concerning outfalls, existing storm drains, and other design considerations; and
- a schematic indicating storm drain system layout.

Storage Facilities

The following items should be included in the documentation file related to the design and analysis of storage facilities:
- routing calculations including inflow and outflow hydrographs and volumes for the design and 100-year floods;
- design calculations of any riser pipes and outflow devices;
- computations for dam design including typical cross-section;
- storage volume estimates;
- design or analysis of materials proposed for the banks and bottom of the storage facility;
- emergency spillway design;
- outlet velocity estimates;
- energy dissipation calculations and designs;
- maintenance plan; and
- copies of all computer analyses.

Pump Stations

The following items should be included in the documentation file related to the design and analysis of pump stations:
- inflow design hydrograph from drainage area to pump;
- flood frequency curve for the attenuated peak discharge;
- maximum allowable headwater elevations and related probable damage;
- starting sequence and elevations;
- sump dimensions;
- available storage amounts;
- pump sizes and operations;
- pump calculations and design report; and
- line storage and pit storage capacity.

In addition whenever computer programs are used for design or analysis, input data listings and output results of selected alternatives should be included in the documentation file. The file should also document what computer model was used, the version, and any other information which would enable future designers to duplicate the computer analysis that was used. If computer models are changed or new versions purchased, copies of the model or versions used for analysis or design should be kept.

References

American Association Of State Highway And Transportation Officials, Highway Drainage Guidelines, 1982.

American Association Of State Highway And Transportation Officials, <u>Model Drainage Manual</u>, 1991.

Appendix A

Documentation Project Check List
(Check Appropriate Items)

Engineer   _____

Project   _____

City/County   _____

Description   _____
                  _____
                  _____
                  _____

REFERENCE DATA   HYDRAULIC DESIGN

Maps:

USGS Quad   Scale   Date
USGS 1:250 00
Other Maps:
Local Zoning Maps
Flood Hazard Delineation (Quad.)
Floodplain Delineation (HUD)
Local Land Use
Soils Maps
Geologic Maps
Aerial Photos   Scale   Date

Studies By External Agencies:

SCS Watershed Studies
Local Watershed Management
USGS Gages & Studies
USACE Flood Plain Infor. Report
Interim Flood Plain Studies
Water Resource Data
Regional Planning Data
Forestry Service
Utility Company Plans

Calibration Of High Water Data

Discharge and Frequency of H.W. el.
Influences Responsible for H.W. el.
Analyze Hydraulic Performance of
        Existing Facility For Min.
        Flow Through (100-year)
Analyze Hydraulic Performance of
        Proposed Facility For Min.
        Flow Through (100-year)

Design Appurtenances:

Dissipators
Rip Rap
Erosion & Sediment Control
Fish & Wildlife Protection

Technical Aids:

Storm water Management Design Manual
FHWA Policies & Directives
AASHTO Model Drainage Manual
Technical Library

Studies By Internal Sources:

Quarterly Reports
Planning Reports
Public Works Reports
Consultant Reports
Flood Records (High Water, Newspaper)

Computer Programs:

Culvert Inlets
Direct Step Water Surface Profile
Standard Step Water Surface Profile
USACE HEC-2 Water Surface Profile
FHWA Bridge Backwater
Energy Dissipation
Storm Drain Design
Hydrology
Other

HYDROLOGY   HYDROLOGIC-HYDRAULIC

Technical Resources:

Storm water Management Design Manual
Directives
Technical Library

Discharge Calculations

Drainage Areas
Rational Formula
U.S.G.S. Equations
HEC-1
SCS
Gaging Data and Reports
Computer Programs

High Water Elevations:

Internal and External Surveys
Personal Reconnaissance

Flood History:

External Sources
Personal Reconnaissance
Maintenance Records
Photographs

Data Reports:

Municipal Data
Environmental Reports
Surface Water Environmental Study
Reconnaissance Reports
Drainage Survey Inspection Reports
Hydraulic Design Reports
Construction Reports
Operation Reports

Chapter 17 Construction

17.1 Introduction

Lack of constructability and lack of construction supervision are two of the major reasons storm water systems do not function properly. In most municipalities different personnel perform the design and construction functions. Adequate communication between the two is essential. Design personnel should be aware that there are construction related design considerations and construction personnel should be aware that there are design related construction considerations.

When problems arise, they should be discussed and a workable solution decided upon and documented. A forum for design-construction related discussion sessions should be provided on a regular basis. Redundancy and adherence to archaic methodology could be overcome by using this forum to develop solution(s) and utilize expertise from both functions. Primarily these sessions would be at the pre-bid stage. Open communication between design and construction personnel should lead to good working relationships. These relationships will help improve designs and ensure that projects are constructed as envisioned without causing major problems or field revisions.

Construction related hydraulic considerations are a necessary part of the planning and design phases. Factors which will affect construction timing and methods need to be kept in mind as project development proceeds. Those responsible for contract administration and actual construction may need to coordinate their scheduling and construction procedures with the designer in order to achieve results intended. Any special or unique construction requirements should be communicated to the designer prior to the final design phase of the project.

The designer should be present at the preconstruction conference to explain special features and planned construction phasing where these considerations are necessary to the proper functioning of the design. It may be advisable and necessary to specify certain time limits and special instructions as to how the work will be accomplished. Phased construction to accommodate seasonal variations, floods, fish passage, irrigation, etc., may be needed. In addition, the need for special considerations related to needed temporary work, detours, and public safety issues can be outlined and discussed. It should be emphasized by the designer that any and all revisions of storm water management designs as contained on construction plans should be discussed with the designer prior to execution.

17.2 Construction Costs

Cost is an important consideration in any design. The primary components of first costs are related to materials, land and labor. Future maintenance costs are an important related design consideration. Often life cycle costing and value engineering are used to make design and materials decisions. The designer must achieve the proper balance of material and construction costs. Ordinarily, material costs are optimized by: using available materials in a consistent manner;

recycling materials; research programs to identify potential construction materials, and how they may be utilized efficiently; using reasonable safety factors in design; and encouraging and allowing alternatives where possible. In some cases, the least expensive material may not be the proper choice because construction costs are greater than for a more expensive material or future maintenance/replacement costs override the material cost advantage (ASCE, 1992).

Construction costs are affected by the:
- relative difficulty of construction,
- laws, rules, and/or regulations governing construction procedures,
- degree of competition among contractors,
- construction latitude allowed by the specifications,
- quality of the construction plans, and
- degree of supervision and inspection provided by the municipality.

The choice of a more complicated and expensive construction procedure may be proper if it allows the use of more economical materials, decreases maintenance costs, and eliminates or reduces the need for replacements.

Cost estimating and benefit-cost analysis should both be performed for a reasonable set of project alternatives after constructability and suitability screening is accomplished. Initial cost estimates are normally done on a unit or rule-of-thumb basis. These are refined as the project moves through a sketch plan phase, through preliminary design and to final design and specifications. Actual costs based on bidding still vary considerably (often more than 20%) based on individual construction company circumstances and equipment. Construction cost estimates must look beyond first costs and should include realistic consideration of such factors as (AASHTO, 1992, ASCE, 1992):
- engineering, surveying, mapping and design,
- land and easements,
- site preparation,
- materials,
- installation requirements such as trenching or tunneling,
- sizes of structures of all types,
- excavation requirements and special excavation needs such as blasting,
- water quality enhancement and permitting costs,
- de-watering or flow rerouting costs,
- replacement costs over the life of the project for different material options,
- annual costs such as operation and maintenance,
- annual flood or other losses for various options,
- public involvement and input costs,
- costs related to liability and legal costs, and
- construction related costs such as traffic delays, service disruption, etc.

"Life-cycle costing" has some advantages in this type of consideration in that it allows a comparison basis for all the various cost factors mentioned above, each reduced to some base year taking into account the impacts of estimated interest rates and inflation. The actual interest rate and inflation factors are not often as important as the fact that they are consistent across all options.

17.3 Construction Plans

The designer must be aware of the above relationships and how they affect costs, and consider them in design. Additionally, construction plans should reflect these considerations by containing:

- suggested construction sequences that consider construction costs and environmental considerations as well as public convenience,
- subsurface soil borings,
- complete descriptions of utilities, and
- consistent plan format which enhances the contractor's ability to assimilate and understand the municipalities' plans.

Despite the best of efforts, construction changes occasionally occur. The designer should be consulted when these changes may affect the proper functioning of the storm water management facilities.

The post-construction inspection following completion of the project should document any deviations from the original plans as well as an initial assessment of the hydraulic performance. Construction personnel should be encouraged to inform the designer of any design-related difficulties which are encountered and suggestions to improve future designs. Changes should be incorporated into "as-built" plans for future reference.

Plans should be checked to verify that site conditions have not changed between location surveys and construction, and also between location survey and preliminary plan completion. Meander migration, bank caving, aggradation, headcutting or other natural or man-induced changes in the channel may have occurred which would require that the designer reconsider decisions made on the basis of conditions which were different from those which exist at the beginning of construction. This is best accomplished with a joint field inspection by design, construction, and maintenance personnel. Additional objectives of the inspection are to assure location survey accuracy and to ascertain if the designer has properly visualized existing situations and designed accordingly.

The changed conditions may require river control works, revisions to pier locations and orientation, rearrangement of spans, or other modifications of the design to accommodate the changes that have occurred. Plan changes required because of differences between location surveys and construction field inspections should be made in consultation with the designer. Some changes could significantly affect either the hydrology or the hydraulic performance of the storm water management facility designed for the site.

Land use changes in the watershed can modify the hydrology and debris considerations used in the design. New development located near the project could change damage risk considerations for the design. Dependent upon the time that has elapsed between completion of the design plans and the beginning of construction, changes in land use could significantly affect the validity of design considerations. Commercial mining of materials for construction is a rather common practice that can change flow velocities, volume and character of bedload, and flow direction and distribution at the crossing site. Land clearing for agricultural purposes may create a need to reconsider the location and size of waterway openings and the need for spur dikes. Land development near the site could change damage risk considerations for the crossing. The designer should be consulted regarding the need to modify the design at any storm water management facility site if the site conditions have changed significantly from the conditions which existed during design.

Changes in stream alignment and profile can result in different flow conditions than those used in the outfall or cross drain design. Drainage area changes due to diversions or site grading can affect inlet and outlet locations and type, as well as storm drain or roadside ditch designs.

Utilities added after the survey may require extensive redesign of storm drain systems to avoid conflicts; this reinforces the need for good utility surveys prior to design in order to forestall costly redesigns and delays. See Table 17-1 for additional discussion of some typical changes which may affect storm water management facility design.

In some cases, a considerable amount of time may elapse between design and construction. In other cases, designs may change before construction is begun. Any changes in the plans, specifications, and estimates should be reflected in the final plans. If questions arise, the construction personnel should check with the designers to determine if changes have been made and how construction should proceed.

Table 17-1 Changes Which May Affect Storm Water Management Facility Designs And Are Detectable During Field Inspection

Possible Errors Or Omissions
- Incorrect existing structure and/or invert elevations
- Incorrect drainage area size
- Channel alignment, profile
- Unreported utility
- Existing structure condition as related to service life, outlet scour, siltation, etc.
- Local flooding not documented
- Existing slope erosion not reported
- Sensitive receiving waters not reported
- Unreported debris problems
- Attractive nuisance problems

Possible Changes
- Increased development
- Channel improvements
- Diversion or site grading changes
- New utility
- Loss of outfall due to development

Miscellaneous
- Incorrect typical section choice, and/or incorrect grade

17.4 Preconstruction Conference

It is important for the designer to be present at the preconstruction conference to explain special features of the designs and planned construction phasing, where these considerations are necessary for proper functioning of the design. It may be advisable and necessary to specify certain time limits and special instructions as to how the work will be accomplished. The designer should answer any construction-design questions that the contractor's construction personnel have or endeavor to obtain answers as soon as possible. The purpose of the meeting is to discuss the design and construction aspects of the project, thus affording all parties a common understanding of the proposed work and the problems and possible solutions which may be expected.

The designer should go over the job with several key personnel including the resident engineer, contractor's agent, right-of-way agent, traffic engineers, materials engineer, maintenance superintendent, surveyors, and others who may have a direct interest in the project. Such a review at this time will aid materially in clearing up reasons for certain design features such as: right-of-way obligations; signing and traffic handling difficulties; materials sites; selected material; foundation treatment; potential slides; environmental commitments; and potential drainage and maintenance problems, including erosion control and water pollution.

Several other concerns should be discussed at the preconstruction conference including storm water management facility maintenance during construction, water pollution, and erosion control.

Drainage work on some projects may be completed several months before total project completion. During this period, vegetative erosion control measures are not well established and maintenance to correct erosion and sediment deposition in the newly constructed channels is important to achieving the results intended. The municipality should provide for maintenance by the contractor during the term of the contract, require interim protective measures, and/or advance its own maintenance schedule to assure that minor damage will not develop into major damage which will require costly repairs or replacement when it assumes the permanent maintenance responsibility.

During the preconstruction conference, provisions of the contract relating to pollution control should be reviewed. The contractor should submit a program to control water pollution prior to beginning work or waive the specified requirement if such action is in order.

17.5 Construction Considerations

Problems may be avoided during construction when important drainage or other water-related factors are considered during the location and planning phases of the project. If at all possible, problem locations should be avoided. A site may be considered a problem location because of geological aspects, environmental concerns, other existing facilities, or other reasons which might conflict with the proposed project.

The concerns of erosion and sediment, where they might occur, and how to control them, must be considered, at least in broad terms, during the early phases of location. As an example, the designer may be involved in the geological investigations because of underground water so that proper measures can be taken to prevent problems before they occur.

The time of the year and total construction time should be taken into consideration. Certain elements, such as embankments along a stream, should be completed before the anticipated flood season. In some areas, work cannot be performed in the streams during fish spawning runs. In other areas, the stream may be an irrigation supply and flows cannot be interrupted nor can the pumping and distribution system be contaminated with sediment.

The use of temporary structures must also be planned. Often a temporary crossing can be smaller than normal if it is only going to be utilized during the dry months. If it will be used for more than one year, perhaps it needs to be sized for a lower frequency flood. This consideration may change the concept of the project or at least the type of structure designed.

Many construction related hydraulic problems are related to scheduling. Although these problems will be studied in more detail during the design phase, they should be initially considered, at least in a preliminary manner, as early as possible. Commitments regarding water resource related items made in the Environmental Impact Statement (EIS) must be made known to the personnel who will be involved in the actual construction. Some commitments that "sound nice" may not be feasible to build. In other cases, construction occurs so long after the EIS has been prepared that those commitments are forgotten or not included in the plans or contract documents. A "commitment list" which follows the project through the various stages of development should be prepared to ensure these items are, in fact, incorporated into the project.

The designer can work with other disciplines to devise and construct mitigation measures which reduce adverse effects. The designer can recommend locations and sizes for storm water management facilities such as for culverts, bridges and channels; also spoil disposal areas and geometry, and various construction alternatives can be identified. The designer may also assist in developing programs for protecting surface waters during construction. These programs include such things as levees and ponds to collect various types and quantities of pollutants including those from construction equipment or which are accidentally spilled, methods to reduce erosion and sedimentation, and replacement of surface waters that assimilate the hydrologic and hydraulic regime of those that are affected.

17.6 Hydrology Considerations

Construction and maintenance of storm water management facilities may require knowledge of low flow discharge properties such as discharges, flow stages, flow durations and related flow variables. For example, the construction of a culvert or a bridge may require knowledge of the time frame at which flows are below certain levels, or below certain magnitudes. This knowledge might be useful in scheduling construction, or designing temporary construction facilities. With some facilities it is often necessary to avoid long periods where the facilities are unavailable to the user due to prolonged occupation of a portion of the facility by frequent low flows.

Annually the USGS publishes "Water Resources Data" for gaged streams listing mean daily discharge. Based on these daily records, a low flow analysis may determine an acceptable discharge for the hydraulic design of temporary construction facilities. A rigorous flood frequency analysis is not generally required for these low flow studies. Flow discharges may be cursorily determined based on a visual examination of monthly mean discharge data as determined from the mean daily discharge values for all years of record and with consideration given to construction timing and degree of risk. Data for the monthly mean discharge may be obtained from the local USGS field office.

If hydrographs or other hydrologic data or temporary structure sizes are furnished to the contractor or included in the plans for the contractor's use in planning and scheduling operations, the contract documents should indicate that the plots or data are for information only and that the municipality assumes no responsibility for conclusions or interpretations made from the records.

Water quality of streams and lakes has become a very sensitive issue. Storm water management facility construction may deliver such things as sediment and chemicals to streams, rivers and lakes unless precautions are taken. Annual runoff hydrographs may indicate that very low stream flows will occur during the late fall and winter months. During these periods even small amounts of additional sediment or chemicals entering the stream from construction areas could be detrimental because of the low dilution effect provided by the receiving waters. The effects of sediment or chemicals due to construction during the low flow periods should be investigated for those sensitive areas such as where stream flow is used for municipal water supply. This investigation may include periods of water quality monitoring and testing. If the investigation concludes that the amount of sediment or chemicals will exceed an acceptable threshold value, the construction periods may have to be rescheduled or mitigation measures taken.

17.7 Erosion And Sediment Control Considerations

By the time construction begins, all erosion and sediment control considerations made during the planning, location, and plan development phases should be contained in the plans, specifications, and special provisions provided to the contractor and municipal personnel for accomplishment of the project construction. The resident engineer should thoroughly familiarize himself and the municipal inspection staff with the erosion and sediment sensitive areas of the project and the control measures contained in the plans. This information should be shared with the contractor for formulation of a work plan.

The contractor should utilize an erosion and sediment control schedule which sets forth the proposed construction sequences and the erosion control measures that will be employed. This schedule allows the contractor and engineering personnel to plan ahead and control erosion and sediment before it becomes a problem rather than adding measures after damages have occurred.

Adequate inspection during construction is essential for erosion and sediment control. If deficiencies in the design or performance of control measures are discovered, the supervising engineer should take immediate steps for correction, including notification of the designer to avoid a reoccurrence of the problem. Periodic field reviews and inspections by the design and construction personnel to correct deficiencies and improve control procedures are highly recommended.

An important consideration in the decision to utilize any erosion or sediment control measure is its effectiveness in the particular circumstances of planned use. There is no better way to answer this question than through experience. For this reason it is very critical to the development of a good erosion and sediment control program that communication exists between design and construction personnel. One method of establishing communication is to have regularly scheduled project field reviews or meetings involving those responsible for design and construction. During these meetings, problems and successes with particular items can be evaluated. Different ideas and procedures that have been successfully employed by a contractor can be studied to determine if they warrant consideration for widespread use. Also of importance for discussion is possible modification to standard design items that would facilitate their construction and/or perhaps reduce their cost.

This feedback procedure extends beyond construction into the long term maintenance of erosion related items. Maintenance personnel must check and correct any deficiencies in the permanent erosion control measures. Design personnel should be apprised of any persistent problems so that an analysis can be made to determine if any alteration of design or construction practices is warranted to reduce maintenance problems.

17.8 Culvert Considerations

The plans, specifications, and other construction documents should be reviewed to ensure that the design fits current site conditions. Design personnel should be informed and involved in all changes to the plans and specifications.

As soon as final locations are determined, the contractor should be furnished a revised culvert list, including those culverts which have been added or altered by change order. Assembly or construction, bedding and backfill are as important to culvert service as the hydraulic and structural design.

Culverts should be protected from damage during construction operations and should be periodically inspected. A particularly critical time for inspection is upon completion of grading operations and prior to the start of surfacing operations. It is as important to inspect culverts which are not under the roadway as it is for those structures that are under the roadway. Prior to the acceptance of the installation, all culverts should be inspected and cleaned as necessary.

Records should be kept of the construction of each culvert installation. The final location and slope of the culvert should be recorded on the "as-built" plans. This information is useful for evaluating overall performance of the installation. The following records should be kept for each installation:
*   inspection tags;
*   location and layout including:
        station,

skew(s),
location of inlets, outlets, junctions,
flowline elevations in inlets, outlets, junctions,
camber,
alignment, and
grade;
- daily reports;
- structure summary sheet containing:
measurements,
calculations,
pay quantities.

17.9 Bridge Considerations

The responsibility for construction-related hydraulic considerations of bridge construction ordinarily rests with the contractor, but, in some cases, the municipality may include construction-related details in the plans and specifications, in order to mitigate potential environmental effects or to assume or reduce the risk of failure during construction. In addition, other special provisions related to the construction phase of the bridge construction may be specified in the plans. Whether the municipality or the contractor assumes the risk and responsibility, hydraulic considerations during construction usually differ from the design considerations for the completed facility.

A hydrograph of superimposed mean daily flows and a plot of the rating table for a stream gaging station near the crossing site are useful for the design of cofferdams, falsework and temporary crossings, in the municipality scheduling of the work, and in selecting the location of work and material storage areas. If the hydrograph is furnished to the contractor or included in the plans for the contractor's use in planning and scheduling operations, the contract documents should include that the plot is for information only and that the municipality assumes no responsibility for conclusions or interpretations made from the records. In the event a gaging station is not located near the stream crossing site, records from upstream or downstream gages may be useful as an indication of the usual magnitude, duration and time of flood events.

Any site conditions that might impact the foundation design or create unusual scour problems should be discussed with design personnel to determine if design changes are indicated. If specific elements of the bridge design are dependent on special foundation or scour considerations, construction personnel should be informed so they can identify possible problems.

Cofferdams, falsework, and occasionally contractor's equipment, such as barges, constrict the stream channel more than the completed substructure and consequently have greater potential for causing scour and bank caving, and for collecting debris. Scheduling of work to avoid flood seasons is especially important if these types of operations will be involved.

Temporary stream crossings necessary for the construction of bridges are usually the responsibility of the contractor. It may be desirable in some instances, however, for the municipality to design such crossings in order to minimize or mitigate the adverse effects on the stream environment, to facilitate securing permits, or to reduce the risk assumed by the contractor and thereby reduce construction costs. In addition, stream crossings for detours are built to much lesser standards than permanent crossings. The criteria used for the hydraulic design of detour stream crossings should be based on risk factors which should be evaluated considering the probability of flood exceedance during the anticipated service life of the detour, the construction period for the crossing, the risk to life and property, and traffic service requirements. Figure 17-1 can be used to assist in designing temporary stream crossings and detours.

As in the case of the design of highway stream crossings, detour designs should accommodate floods larger than the event for which they are designed in order to avoid undue liability for damages from excessive backwater and to reduce the probability of losing the detour stream crossing structure during a larger flood. In most instances, the conveyance of floods larger than the detour design flood is provided for by a low roadway profile which allows overflow without creating excessive velocities or backwater.

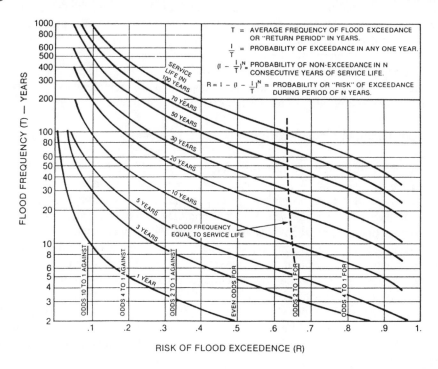

Figure 17-1 Risk Of Exceedance - Flood Event vs. Service Life Of A Highway Encroachment

Source: AASHTO Model Drainage Manual, 1992.

Minimum disturbance of the banks and bed of a stream during the construction period will reduce erosion damage to the banks, sedimentation and harm to fish and wildlife. Embankments in or along streams should be constructed of erosion resistant material and/or protected against erosion to avoid adverse sediment concentrations which contribute to the turbidity of the stream.

Consideration should be given to precluding in-stream operations that would cause turbidity during the spawning season of certain types of fish. Detours and construction roads are other sources of turbidity and should either be constructed at a time that fishery activities will not be disturbed, or provisions should be made to control any harmful effects of erosion. Silts and clays will generally flush out of the substrate over a period of time but sands tend to become embedded. Gravel and rock similar to the gradations found in the existing substrate will do the least damage to the aquatic habitat. Municipalities may want to identify local agencies that can be contacted to obtain information about the presence of fish and the seasons during which protection is necessary.

Pumping of cofferdams and other dewatering operations may have a discharge of unacceptable quality to the receiving stream. Mitigation measures such as settling basins may be necessary if the ecosystem of the stream would be upset by the temporary degradation of water quality.

Most designers do not have an opportunity to participate in the construction of the works that they have designed. For this reason, designs that could be improved upon for construction purposes tend to be perpetuated simply because the designer is not informed of the deficiencies. Designers are encouraged to visit construction sites to discuss problems with designs and possible improvements in future designs. This is especially important for major projects like bridge construction. Upon completion of a project, a design critique conducted jointly by designers and field personnel can be a very useful learning experience for both. This critique should include difficulties encountered in the construction and possible design changes to prevent such difficulties in the future. This will also give the designer an opportunity to present why some difficulties in construction are necessary because of specific design considerations.

17.10 Open Channel Considerations

Many of the construction considerations for open channels are the same as for culverts and bridges (i.e., plans, specifications, and special provisions; hydrologic information, timing and scheduling; environmental and ecological aspects; feedback); thus sections 17.8 and 17.9 should be reviewed as they relate to open channel construction. The following discussions will concentrate on those construction considerations that are more unique to open channels.

Bank stabilization is an important aspect of open channel construction. Since in some cases a considerable length of a stream or channel system may be disturbed by construction, great care should be exercised in scheduling and implementing stabilization measures. Immediately prior to the commencement of construction of bank stabilization measures, the designer should inspect the site to ensure that measures proposed are not inappropriate because of bank movement subsequent to completion of design surveys. Recognizing that an entire reach of stream may require stabilization, municipal responsibility may well be much more limited in scope and a total solution not possible.

Channel excavation work on some projects may be completed several months before total project completion. The time between completion of channel excavation and total project completion is usually longer when grading and structure projects are separated from the contract for paving or stabilization. During this period, vegetative erosion and control measures are not well established and maintenance to correct erosion and sediment deposition in the newly constructed channels is important to achieving the results intended. The municipality should provide for maintenance by the contractor during the term of his contract, require interim protective measures, and/or advance its own maintenance schedule to assure that minor damage will not develop into major damage which will require costly repairs or replacement when it assumes the permanent maintenance responsibility.

Damaged channels can be both expensive to repair and hazardous to the public. To facilitate repair and maintenance, channels should be designed recognizing that periodic maintenance, inspection and repair will be required. Where possible access should be incorporated for personnel and equipment during the construction period and afterward. Consideration should be given to the size and type of equipment which will ordinarily be required in assessing the need for access easements, entrance ramps and gates through right-of-way fences, and purchase of right-of-way.

17.11 Storm Drain Considerations

Many storm drain construction considerations are similar to those encountered in culvert and open channel construction. Thus, Sections 17.8 and 17.10 should be reviewed as they relate to storm drainage construction considerations. The following discussions will concentrate on those considerations which are more unique to storm drainage systems.

Plans for subsurface drains are seldom as complete as those for culverts. The discovery of damaging amounts of groundwater during preliminary materials investigation is difficult. During dry seasons, or following a long dry cycle, indications of groundwater problems may be missing entirely. However, with the return of a wet season serious problems may occur if needed subsurface drains are not installed.

Installations should be carefully reviewed and plans revised as necessary to fit field conditions, in consultation with the designer. It is seldom necessary to decrease the number of planned subsurface drains; the contrary is usually the case. Also the location of subsurface and other drains may need to be changed to locate these facilities in stable areas and at low points or other locations where the drainage of surface water can be intercepted and allowed to efficiently enter the storm drain system.

During the clearing and grading operations, groundwater problems may become evident. Swamps, bogs, springs, and areas of lush growth are possible indicators of excess groundwater. Fill foundation areas should be inspected minutely before starting embankments. Ravines and draws are especially suspect. As excavation progresses, perched water or various aquifers may be encountered in the area of slopes or at grade.

17.12 Temporary Storm Water Management Facilities

Temporary storm water management facilities include all channels, culverts or bridges which are required for haul roads, channel relocations, culvert installations, bridge construction, temporary roads, or detours. They are to be designed with the same care which is used for the primary facility. These designs are to be included in the plans for the project. Approval is required from the municipality having jurisdiction for those designs which they regulate.

It is recommended that drainage systems for these facilities are to be designed for a two-year frequency if the roadway is required for a year or less and a five-year frequency if required for longer than a year. All other temporary storm water management facilities connected with these roads are to be designed for frequencies as determined by using municipal guidance.

Storm water management facilities for haul roads which cross or encroach into a watercourse are to be designed for a frequency as determined by using a Design Risk of 50 percent. As a general rule, to avoid excess upstream flooding, the profile of the road should connect the tops of the channel embankments and the road designed to be overtopped by those events which exceed the design discharge. Sufficient cover must be provided over the temporary conduit to ensure structural integrity. The structural analysis of the conduit is to be included with the design. The plan is to include a warning to the contractor that this road is expected to be under water during certain rainfall events for undetermined lengths of time.

Selection Factors

The selection of a design flood frequency for the remaining temporary storm water management facilities involves consideration of several factors. Following is one procedure that can be used to quantify these factors. In this procedure, the factors are rated considering their severity as 1, 2, or 3 for low, medium or high conditions. Following are the selection factors.

Potential Loss Of Life - If inhabited structures, permanent or temporary, can be inundated or are in the path of a flood wave caused by an embankment failure, then this item will have a multiple of fifteen applied. If no possibility of the above exists, then loss of life will be the same as the severity used for the average daily traffic.

Property Damages - Private and public structures (houses, commercial, or manufacturing); appurtenances such as sewerage treatment and water supply; utility structures either above or below ground are to have a multiple of ten applied. Active cropland, parking lots, recreational areas are to have a multiple of five applied. All other areas should use the severity determined by site conditions.

Traffic Interruption - Includes consideration for emergency supplies and rescue; delays; alternate routes; busses; etc. Short duration flooding of a low volume roadway might be acceptable. If the duration of flooding is long (more than a day), and there is a nearby good quality alternate route, then the flooding of a higher volume highway might also be acceptable. The severity of this component is determined by the detour length multiplied by the average daily traffic projected for bi-directional travel.

Detour Length - The length in miles of an emergency detour by other roads should the temporary facility fail.

Height Above Streambed - The difference in elevation in feet between the traveled roadway and the bed of the waterway.

Drainage Area - The total area contributing runoff to the temporary facility, in square miles.

Average Daily Traffic (ADT) - The average amount of vehicles traveling through the area both ways in a twenty-four hour period.

Example Format

Table 17-2 gives an example format which illustrates a method of determining the design discharge. The severity and rating of each component is determined and entered in the Impact Rating Table. The total impact rating determines the Percent Design Risk, and the construction time is then considered to find the design frequency. A ratio corresponding to the frequency is used with the 50-year and 100-year storm to determine the design discharge. The municipality may wish to change the rating selection criteria to fit local conditions.

Note: If sufficient discharges have been developed either by the designer or a Flood Insurance Study then a frequency curve should be plotted to determine the Design Discharge instead of the final formula using the ratio.

17.13 As Built Plans

"As Built" plans serve many functions related to the design and construction process including documentation of:
* the final location of all elements of the drainage system and storm water management facilities,
* any changes that were made in the design during the construction process (i.e., size of facilities, materials used, addition or elimination of facilities), and
* any variation between the original plans and specifications and the final installed facilities.

The completion of accurate and complete "As Built" plans can be invaluable in documenting changes that can be incorporated in future designs and to future investigations of the project if problems are encountered or there is some need to analyze the facilities performance. Possible future legal action makes the documentation of "As Built" plans very important.

Table 17-2 Impact Rating Table

RATING SELECTION

Factor	Rating		
	1	2	3
Loss of Life	See Instructions		
Property Damage	See Instructions		
Traffic Interruptions	0-2000	2001-4000	4001-6000
Detour Length (mi.)	< 5	5-10	> 10
Height Above Streambed	< 10	10-20	> 20
Drainage Area (sq. mi.)	< 1	1-10	> 10
Rural ADT	0-400	401-1500	> 1500
Suburban ADT	0-750	751-1500	> 1500
Urban ADT	0-1500	1501-3000	> 3000

IMPACT RATING

Loss of Life x 15 =   _____
Property Damage x 10 or x 5 =   _____
Traffic Interruption =   _____
Detour Length =   _____
Height Above Streambed =   _____
Drainage Area =   _____
Average Daily Traffic =   _____

Total Impact Rating = (sum of the above) =   _____

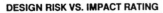

DESIGN RISK VS. IMPACT RATING

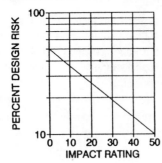

DESIGN FREQUENCY (YEAR)

Design Frequency = _____ years

Year	Ratio	Year	Ratio
2.0	.8	10.0	1.9
2.33	1.0	25.0	2.7
3.0	1.2	50.0	3.7
4.0	1.3	100.0	5.0
5.0	1.4		

Ratio = _____ x 0.27 (Q_{50} _____) = ____ cfs

Ratio = _____ x 0.20 (Q_{100} _____) = ____ cfs

References

American Association Of State Highway And Transportation Officials, Model Drainage Manual, 1991.
American Society of Civil Engineers, "Design and Construction of Urban Stormwater Management Systems",
    Manual and Rpt. of Engr. Practice No. 77, 1992.

Chapter 18 Maintenance

18.1 Introduction

Storm water management facilities perform the function of removal of water from street, highway sections, parking areas, and other drainage areas and the protection of the facilities from the effects of the water. These storm water management facilities include drop inlets, storm drains, culverts, underdrains, ditches, slope protection, detention facilities, and erosion control devices.

In order for these facilities to function as designed and constructed, they must be properly maintained. In a recent survey of North Carolina Cities, 20 percent of the cities stated that local flooding problems within their jurisdiction were attributed to maintenance problems (Roenigk, et al., 1992). Detention pond surveys conducted in Maryland found a 70 percent failure rate for one set of detention ponds and about 50 percent failure rate for ponds from all three surveys conducted (Maryland, 1986). Another Maryland survey found high failure rates for infiltration basins (Galli, 1992). A major reason was poor or nonexistent maintenance.

Full consideration must therefore be given to maintenance during the design process. Designing storm water management facilities that are as maintenance free as practical will often result in cost savings that, over the service life of the facility, equal or exceed initial construction cost. Good storm water management facility design practices recognize that all structures require periodic maintenance inspections and repairs. Reasonable access for maintenance personnel and equipment must be provided for this necessary function. For example, proper access spacing in a storm drain system provides a relatively easy way to clear sediment and debris blockages or to isolate a portion of the system for repairs. Provision of a sediment trap and machinery access area for detention design greatly reduces annual maintenance cost for larger detention and retention ponds.

Communications between designers and maintenance personnel are essential. Design personnel are encouraged to contact maintenance personnel for their input on difficulties they identify in maintaining storm water management facilities. Conditions which appear to require extensive repair or which incur frequent recurring maintenance should be investigated. Investigation may reveal that a complete redesign is more cost effective than repetitive repair. Reports by the maintenance forces of both effective and non-effective installations aid designers in future work.

Improper maintenance of a storm water system does not just affect the system but has spill over effects on surrounding property and other infrastructure. For example, typical storm water management facility maintenance problems that affect streets are ponding of water that softens the subgrade, secondary ditches along the permanent edge that erode the material that supports the pavement edge, and breaks in storm water management facilities that lead to erosion of pavement supporting material.

Maintenance Categories

Maintenance generally falls into three categories: routine, remedial and capital improvements.

- Routine maintenance includes those activities which happen on a periodic basis, which may be driven by the passage of time not the specific deterioration of the system.
- Remedial maintenance corrects specific deficiencies in the existing system without upgrading its capacity. Remedial maintenance makes the best use of a system even if it is deficient.
- Capital improvements replace deficient systems with larger or improved designs. They become, in effect, new systems.

Routine maintenance of storm water management structures and facilities include the following.

- Keeping water courses free from accumulations of debris and vegetation and storm drains free of silt, sand, and debris.
- Correcting malfunctioning parts of the systems. Settlements and breaks are the most common types of failure.
- Anticipating problems and making minor modifications.
- Detention facilities should receive periodic inspection to ensure that siltation and debris have not decreased the storage volume and that the outlet structure is not clogged. They should also be checked after major storms.

The routine cleaning and minor repairs of drainage features often require that labor intensive hand methods be used. Adequate access for maintenance personnel and equipment to get to the site and do work on storm water management facilities should be provided for in the designs. In addition, most costly maintenance work might easily be avoided, or more efficiently accomplished, if designers were to give more attention and thought to the shape and location of drainage features. For example, a "V" shaped roadside gutter that is contiguous to the shoulder can be efficiently reshaped and cleaned with a motor grader. Small trapezoidal and other shaped roadside ditches may require hand cleaning or special equipment.

Inspection

Inspection of storm water facilities serves as a method for both scheduling and controlling work crews and for responding or anticipating specific problems within the system. It is often cheaper to send a single inspector to a complaint site or suspected trouble spot than a whole crew. Also, in some municipalities inspectors serve as "scouts" determining when, for example, inlets need cleaning or ditches need mowing or erosion control.

Maintenance of storm water management facilities is very important both during the development and construction of a project and afterwards. In some areas maintenance of natural drainage systems presents minimal problems while in other areas major resources will need to be allocated for maintenance related tasks. Storm water system inspections should be made quarterly in other than arid climates, and/or during and after each major storm to confirm that satisfactory conditions exist, or to evaluate the need for cleanup and repair. Inspection schedules should include mandatory inspection of known trouble areas, and inspection of other areas as time and resources permit.

The best time to look at storm water management facilities is often during a storm. It is easy then to see where water ponds and where storm water management facilities are over flowing. Often there is no gainful work to be performed at this time, so personnel are available for this inspection. It is felt in some municipalities that the same individual should always inspect the same drainage area. In this way, the inspector can spot any changes that might have occurred. The inspector should be alert to any pavement cracks or ground settlements that appear after a severe storm even if these defects are small as they may be evidence of an erosion caused by a break in the pipes. Areas that generate large amounts of sediment and debris and locations within the drainage system where debris and sediment accumulate should be identified and included in any preventive maintenance schedule.

A record of the inspection should be kept with any deficiencies referenced by street name and house number.

Specific inspection items are contained under a description of each type of storm water management component in later sections and in the chapter covering best management practices for storm water quality. From these sections inspector checklists can be developed according to the preferences and needs of a specific municipality. For example the Corps of Engineers has developed an inspection guide for local flood control projects which has specific questions and a form for inspectors to fill out (COE, 1973). It has also published a four volume inspector's guide which includes all types of storm water components (COE, 1960).

18.2 Developing Municipal Maintenance Programs

Introduction

The development of an effective maintenance program begins with the development of clear goals and objectives along with a number of supporting policies. While there are a number of ways to consider grouping of these policy issues, the grouping used by several of the leading storm water management programs in the United States has been combined and serves as a convenient and intuitive structure.

The remainder of the program is the myriad of details required to carry out the objectives and policies and to make appropriate adjustments as the program continues to grow. The resources and policies to carry out the policies can be organized into the categories of: organization and staffing, technical approach and activities, and financial resources and procedures.

Goals

Typical goals related to the maintenance of a storm water system can be developed along the lines of the major duties of local governments and stated as follows (MSD, 1988).
- Provide, operate and maintain a system of storm water management facilities, controls, criteria, and standards which will mitigate the damaging effects of uncontrolled and unplanned storm water runoff, to improve public health, safety, and welfare, protect property and lives, and maintain and enhance the environment.
- Establish consistent levels of protection for local and regional collection and transmission systems against flooding for existing and future land development conditions, and to correct existing physical drainage problems.

Objectives

Comprehensive objectives developed to meet these goals can include such typical considerations as the fourteen objectives listed below.
- Adopt, review and update regulatory and other legal measures which control storm water rates, volumes and pollution loadings into streams and ground water.
- Review all plans and permit applications to ensure compliance with design criteria, master plans and sound engineering judgment in design.
- Inspect effectively all construction of storm water management facilities to ensure compliance with all design criteria, conditions and plans.

- Develop and implement specific levels of service for design and operation of all storm water systems on public and private property to provide satisfactory protection of individuals, property and the environment.
- Enforce all standards, ordinances and policies in an effective manner to deter, reduce or eliminate activities which would harm life, property or the environment.
- Conduct public education and awareness programs that emphasize protection of the natural and structural storm water management facilities.
- Train maintenance crews to be able to respond effectively to the full range of drainage and flooding maintenance complaints and activities.
- Develop and implement an operation and maintenance program designed to achieve the lowest life-cycle costs for storm water management facilities consistent with adopted standards and criteria without violation of environmental standards and regulations.
- Develop and maintain an up-to-date inventory of the complete storm water system including identified deficiencies.
- Seek to automate and integrate system maintenance operations with planning, engineering and environmental management through the use of joint databases and geographical information systems.
- Develop and implement a prioritized annual capital construction and remedial maintenance program based on documented needs.
- Develop and implement design and maintenance standards and practices which are effective, reduce the need for maintenance and take into account environmental considerations and opportunities for multi-objective and integrated uses of storm water systems and facilities.
- Provide adequate resources to ensure the storm water system is operated and maintained according to standards, criteria and policies.
- Develop and implement operations and maintenance financing mechanisms which maximize equitability, stability of funding, and the targeting of special charges and fees to those requiring or causing the need for special services.

Policies

To carry out the maintenance objectives a series of basic and more specific policies need to be developed. These policies cover a large number of both generic and municipal-specific topics under the general categories of organization and staffing, technical approach and activities, and financial resources and procedures. The more generic policies include the following.

- Organizational Responsibility - which personnel will be responsible for which activities and how will they be organized, staffed and controlled.
- Privatization of Functions - which activities will be conducted with in-house resources and which will be contracted out.
- Intergovernmental Coordination - how will the municipality coordinate with surrounding entities or regional authorities, state government and federal agencies.
- Physical Extent of Service - the geographic limits and categories of drainage facilities for which the municipality will have some responsibility.
- Levels of Service - which services will be provided where or which standards will apply where.
- Setting Maintenance Priorities - how will priorities for use of scarce resources be set and implemented.
- Financing Operations, Maintenance and Capital Projects - which mix of funding methods (including cost sharing) will be used for maintenance and capital improvements for which types of projects.

- Use of Geographic Information Systems (GIS) and Infrastructure Management Techniques -
  how will modern automation be utilized to manage the program and how will it integrate with
  other programs.

There are also a large number of procedural policy decisions on inspections, enforcement
procedures, maintenance procedures, activity types and frequencies, etc.

Policy Resolution

Resolving some policy issues can be a complicated exercise, especially when it must be done
in the midst of major changes like those lying ahead for many municipalities facing growing water
quality demands. However, adequate maintenance is essential in order to successfully upgrade
the program, and can only be established on the basis of clear and consistent policy decisions.

Although policy-making in the highest sense is reserved to the Chief Executive and Council,
day-to-day policy decisions are in fact made at several levels in municipalities. The municipal
Council formally adopts many of the major policy decisions which guide the municipality. The
Chief Executive makes policy decisions based on Council positions. However, other decisions
are made by the Municipal Manager and staff administrators pursuant to the general directives
spelled out by the Chief Executive and Council. Recognizing this dispersed policy- making environment, a simple hierarchy is recommended for the level of review of the issues listed in this
section.

An initial screening of possible issues must consider:

- impacts of policy decision alternatives on costs and manpower;
- the appropriate level(s) of municipal government at which the issue should be addressed and
  resolved;
- the relationship of each specific issue to other policy issues; and
- the priority and timing associated with the issue given the municipality's objectives.

Extent And Level Of Service

Extent of service refers to how much of the system the municipality will take responsibility,
in some way, for. Level of service refers to what type of service or service standards will be
incorporated in its extent of service. Service levels can be design based (e.g. design, construct
and maintain for the 10-year storm), condition based (e.g. culverts will be cleaned if they become
more than 1/3 blocked), or performance based (e.g. clean culverts every five years).

Extent Of Service - Extent of service includes a definition of how much of the system in some
way "belongs" to the municipality and which it will provide services for. It is important to
recognize that the storm water network and flow conveyance system is made up of many components which can be characterized in different ways, including the following.

- Some of the system is located on private property while some is in the public right-of-way
  or on other public lands.
- Some of the system carries significant runoff which comes from public property such as streets
  while some carries only runoff from private property.
- Some of the system is an important part of the roadway system (such as curb, gutter, storm
  drains, side ditches, cross ditches, etc.) while some is remote from the road system.
- Some of the system can be considered "major" drainageways which carry large amounts of
  water while some is relatively "minor", draining small areas.
- Some of the system is underground and out of sight while some is on the surface, such as
  ditches, swales and streams.

- Some of the system is considered by EPA to be "waters-of-the-state" while some is considered to be part of the public or private system.
- Some of the system may be considered by the courts to be the legal responsibility of the municipality while some is not.

Each of these ways of characterizing the storm water infrastructure system may have implications for the municipality in deciding how much of the system should be maintained, regulated or planned for and at what level.

Level Of Service - The basic tenant of level of service is that similarly situated properties are treated in a similar manner (Reese and Suggs, 1992). Level of service can be used for all sorts of storm water management related purposes including such things as: regulatory functions or development plans review, engineering standards and design criteria (termed "performance" levels of service), capital project programs, overall condition descriptors, and others. One practical use of level of service, is in assisting day-to-day routine maintenance management operations.

Conditions Standards And Performance Standards

When considering routine maintenance operations two types of level of service standards are used: condition standards and performance standards.

- Condition standards refer to the physical condition and function goals or objectives of the various parts of the drainage system.
- Performance standards refer to rate of work or production goals or objectives of maintenance crews.

Condition standards have the advantages in that: they are physically based; are directly related to the ability of the system to fulfill its design purpose; and tend to reflect the way customers would look at or lodge complaints about a system. Performance standards have the advantage in that they are easy to manage and monitor, although there is not always a good correspondence between meeting some performance objective (such as cleaning 50 catch basin inlets per crew per day) and the actual condition of the municipality's catch basin inlets.

Condition measures which can be addressed through the routine maintenance process (as distinguished from the remedial maintenance and capital improvement processes) can be identified for each type of storm water management facility. For example, conditions which can be both assessed and addressed for channels and ditches might include: erosion, channel obstruction, and vegetation growth factors. The single standard for storm drains may be obstruction, or the municipality can also include minor structural damage as a standard (if this can be addressed through the routine maintenance process).

Each of these condition factors are given a standard or objective which is to be achieved and maintained individually or for a certain percent of the structures in an area. For example, a conditional level of service standard might state that "95 percent of all grate inlets must be less than 20 percent obstructed at any one time". Cities such as Tampa, Florida and Bellevue, Washington have used such condition based levels of service with some success.

To be useful, condition levels of service must have a strong basis in the actual adverse impacts of exceeding the condition. Engineering, economic and policy judgments must be made on, for example, how much obstruction can be allowed in a culvert opening before it must be cleaned out.

Furthermore, the adverse impacts of even minor obstruction of some culverts may be deemed to be intolerable due to location near flood prone structures and/or lack of adequate size. In other cases a system could be almost totally obstructed without significant consequences. Thus there may be differing conditional levels of service for the same type of structural elements but with differing situations and locations. The basic philosophy, that similar situations are treated in a similar manner, requires judgment that must be carefully applied. Initially a uniform level of service is applied to the entire service area. In practice, as experience is gained in system capacity and function some portions of the system may be cleaned more often or to a more exacting standard.

When a program is developing these conditional levels of service, initial estimates must be made on such things as: numbers of structures, crew production rates, and activity interval to maintain desired standard. From these estimates program costs can be developed. Developing these estimates relies on experience in other municipalities, local experience, pilot studies and other factors.

In practice, after sufficient experience and knowledge of the system is gained, leading municipalities usually shift from operating on a conditional basis to operating on a more easily managed performance basis supplemented with inspections (Reese and Suggs, 1992). These performance standards are derived empirically to meet the conditional levels of service. Thus, for example, it might be found that in a certain neighborhood or watershed annual cleaning of catch basins or bi-weekly mowing of ditch banks is adequate to maintain desired conditional standards. In a recent survey of North Carolina cities less than half the 86 cities stated they did any routine maintenance except mowing and catch basin cleaning (Roenigk, et al., 1992).

The operation of such a program must be integrated with an effective inspection, complaint reception and work order processing program. Obviously, some culverts would clog faster than others. Inspections trigger work orders for some parts of the system (such as obstruction of larger culverts) while some parts are serviced on a routine basis without the need for independent precursor inspection (such as changing filtration media in first flush devices). Some maintenance activities also are triggered by specific need (such as clearing an obstruction from a detention basin) while some are routine (mowing).

Infrastructure Management

The use of computerized databases and GIS or facilities management (FM) systems is becoming more commonplace in storm water management. These computerized systems have grown out of water, wastewater and pavement management practices. Various firms have developed commercial applications of complex work order and tracking systems, with some tied to sophisticated GIS software. Others, for smaller towns or simpler applications, can be operated as a shell over simple and popular CADD software. All these systems will become even more popular as probabilistic and reliability concepts in maintenance management are brought into general practice (Lapay, 1990, Mays, 1991, Wonderlich, 1990, 1991, Quimpo & Shamsi, 1991) and data becomes available for more reliable estimates of replacement and maintenance scheduling (Lund, 1990).

Maintenance management is a subset of general infrastructure or facilities management. It is a systematic approach, complex or simple in function, to: develop work programs, budget and allocate resources, schedule work, and report and evaluate performance and cost (NACE, 1992, Karaa, 1989).

- Work programs specify the kinds and amounts of maintenance work required to provide a desired level of service. Work programs can be used to measure results by expectations and to make suitable adjustments in activities, schedules, equipment or other aspects of the program.
- Budgeting is a result of a well developed work program. Labor, equipment and material needs are derived from the work plan. The budget allocates these resources among field operating units. Unit costs are derived from the work program and experience, and used to feed the next year's budget process. In this way the efficiency of the program is able to be reviewed

and realistic estimates of new or lost labor and equipment can be made in the face of annexation or budget reductions.

- Work program objectives must be communicated effectively to field supervisors who develop schedules. Through good communication the well developed work program will be used as a tool by the field supervisors for scheduling. The plan is translated by them into weekly and daily work schedules or work "calendars" which consider program objectives and goals balanced with unexpected needs and cost demands.
- Timely reporting of pertinent and accurate data is of vital importance to any maintenance program. This allows management decisions to be made on the basis of facts and in-course corrections to be made prior to problems getting out of hand. It also allows for more accurate input into the budgeting process. Accounting for the "cost" of work by activity can lead to significant cost savings. This type of reporting greatly assists in convincing public officials of the need for infrastructure rehabilitation (Martin, 1984).

Experience has shown that the benefits of a well designed and implemented maintenance management system include (NACE, 1992):

- improved resource utilization;
- equitable resource allocation;
- timely and accurate budget evaluation; and
- increased employee morale.

Common problems in the implementation of maintenance management systems include:

- insufficient management support to maintain the system or use it as a management tool;
- unrealistic refinement and paperwork to meet too detailed expectations;
- lack of hands-on orientation and training of supervisory and worker staff; and
- labor union opposition to new and "unproven" work habit changes.

Components of a typical facilities management system may include: a database of the system's elements, database management capabilities, a complaint processing and management system, a work order processing system, mapping capability, report generating, financial and cost accounting and reporting, customer service systems, remote field input capabilities, GIS, internal graphics or CADD linkages, spreadsheet capabilities, multi-user access using LAN, and system security. Future systems will increasingly involve remote data collection and real time operations. They will continue to merge all these overlapping operations into one seamless package with a user-friendly interface for the day-to-day user. Table 18-1 shows the relationships between mapping and facilities management systems.

Table 18-1 Relationship Between Mapping GIS And FM Systems

Topic	Automated Mapping	Facilities Management
Technology	Picture Processing	Data Processing
Strengths	Graphics Management	Text Management
Weaknesses	Text Management	Graphics Management
Transaction Volume	Initial Drafting & Design	Daily Work Orders
User Focus	Building Facilities	Managing Facilities
Database	Hierarchical & Flat File	Relational & Extended
Product Focus	Work Print	Work Order
Management Focus	"What If"	"What Now"

Source: Network, 1989

There must also be a detailed and maintained system inventory and database. Collection of this information is both faster and more error free with the use of automated data collection systems. Consisting of programmed hand held data collectors these systems allow for plain English inputs, logical question and answer systems and complex database schemes. They can then dump data directly into other databases accessible by both GIS and database management software.

Infrastructure Inventories

One of the first activities necessary to take better control of the storm water infrastructure is to perform an inventory of the system. But prior to expending the resources to accomplish the inventory there are several questions that must be answered and preparations made.

If the purpose of the inventory is simply to keep track of approximate locations of the system then a far different level of effort and accuracy is needed than if the purpose is maintenance management or master planning. The key uses for any storm water inventory include the following.

* Maintenance Management - the key information to collect is type, size, material, location (both physical and legal), age, condition (structural and performance), ownership if necessary, and a priority rating if appropriate.

* Water Quality - the key information includes much of the size and location information from the previous list plus pollutant monitoring information, screening information, land use, and whether the pipe or ditch is a major outfall.

* Master Planning - in addition to the above information there is a necessity to determine roughness, slope, drainage area, land use upstream, connectivity, location of flood-prone structures, etc.

* Cost Estimates - the inventory must be a statistically valid sample of the total system based on both age and land use type if an accurate extrapolation of the system condition and cost to upgrade is important.

Depending on the eventual purpose of the data possibly only the largest pipes and ditches would be inventoried. If more detail is needed, driveway culverts and backyard swales might be included. The eventual use of the data and method of use will also determine the collection method. Automatic hand-held data collectors, global positioning, and automatic data dumping to an electronic database are now the state-of-the-art. In one municipality actual voice files were filed digitally along with scanned images of major outfalls for field screening. But if automated mapping does not exist then plotting should be done on paper or mylar sheets.

Obviously it is important to maintain the database as activities occur in the watershed. Such things as maintenance, storms, new construction, annexation, subdivision, and simple inspections could trigger a change to the database. Some of the change can be done on a periodic basis. Some must be done continually if the user needs daily updates for, for example, maintenance scheduling.

Quality control during the conduct of the inventory is of extreme importance. Hand-held data collectors can have built in program logic checks to, for example, catch a negatively sloped pipe or a larger pipe dumping into a smaller one. A pilot study is recommended to work the kinks out and determine efficient and accurate methods for collection.

Things which drive the cost of an inventory include: surveying the elevations, obtaining underground data, and collecting information on smaller system components.

Privatization Of Services

Because storm water maintenance for most municipalities is on an as-needed basis, when the system fails, contracting for storm water maintenance services has been mainly equated to contracting for remedial and capital construction (APWA, 1990). The privatization of storm water routine maintenance services has been limited in the past to activities such as mowing, channel cleaning, litter removal, and other seasonal unskilled labor jobs. However, all aspects of privatization are growing in importance in municipalities across the country. In some municipalities there has been serious consideration given to a totally privatized storm water management function similar to a wastewater treatment function. Some municipalities now contract for activities such as pressurized pipe cleaning and vacuum trucks, infiltration studies and repairs, channel and ditch cleaning using advanced and high cost equipment, etc. Of over 1000 survey responses, APWA (1990) categorized the types of services contracted out for storm water management. Table 18-2 gives a summary of their findings.

Statements of work are normally based on standard specifications and unit pricing is the norm. Charlotte, NC has developed a unit pricing and contracting scheme which has proven effective for all types of system remedial maintenance (Charlotte, 1993). Denver has used a privatization scheme for routine and remedial maintenance on major flood control channels for years (UDFCD, 1983). It solicits bids and selects contractors based on rates, experience and past performance. A weighting scheme has been used with unit costs counting about 43 percent and experience and past performance splitting the other percentage points evenly.

Table 18-2 Summary Of Storm Water Maintenance Contracts

Topic	Percent of Respondents	Topic	Percent of Respondents
No work contracted	33	Mainline Replacement	28
Pump Repairs	27	Emergency Repairs	22
Channel Cleaning	19	Dredging	18
Vegetation Control	18	Concrete Repair	15
Catch Basin Repair	12	Litter/Debris Removal	12
Pipe Cleaning	12	Catch Basin Cleaning	12
Grass Maintenance	10	Sewer Repairs	10
Erosion Control	10	Inlet Repairs	10
Leak Detection/Sealing	9	Culvert Cleaning	9
Headwall Repair	7		

Source: APWA, 1990

18.3 Detention Facilities

Detention facilities are often used in urban drainage systems to temporarily store or detain excess storm water runoff and then release it at a regulated rate to downstream areas. Using this approach, runoff is stored in constructed or natural basins from which it is released continually until the water elevation in the facility reaches its design dry-weather stage. To function properly the storage volume must be maintained at its design level and outlet facilities must be kept open and free from obstructions or clogging. A North Carolina survey found that detention problems were the most pronounced of all problems for storm water systems (Roenigk, et al., 1992).

Detention facilities are susceptible to many maintenance problems that can affect how the facility will function including:

- weed growth,
- grass maintenance,
- sedimentation buildup,
- detention basin and streambed deterioration,
- mosquito, rodent, and snake control,
- outlet stoppages,
- soggy surfaces,
- inflow water pollution,
- algal growth,
- fence maintenance,
- inadequate emergency spillway capacity, and
- dam leakage and failures.

Often these facilities will require more maintenance than other elements within the drainage system. Maryland found that detention basins fail most often for five major reasons: vegetation overgrowth, debris clogging, erosion, sedimentation, and outlet or riprap failure (Maryland, 1986).

Inspection of major detention facilities should be made as frequently as experience shows necessary, perhaps quarterly as a minimum and more often in wet seasons. Where debris is a problem, inspections must be spaced according to debris generation. In any event, it is important to conduct inspections and cleanup work following major individual runoff events. It is sometimes necessary to make inspections during rainstorms when intense rainfall occurs.

During inspections, besides identifying facilities where debris blockages should be removed, mechanical equipment such as generators, float valves, pumps, discharge controls, and other electrical and mechanical equipment should be checked and adjusted as necessary.

Maintenance Tasks

Maintenance tasks related to detention facilities can be grouped in three general categories:
1. aesthetic maintenance,
2. nuisance maintenance, and
3. operation maintenance.

Of these, the most important, from the standpoint of health and safety, is operation maintenance.

Operation maintenance can be characterized as that level of maintenance required to ensure against failure of major structural components and/or flow controls, and to ensure that the facility continues to function as designed. Neglecting this level of maintenance could cause dam failure and subsequent property damage as well as possible loss of life. In addition, neglect often causes a facility to cease functioning as it was originally designed to do. A program of scheduled, periodic inspections of the facility is essential to recognize potential structural maintenance needs. The following is a partial list of items that should be checked periodically and corrective action taken as required:

- settling of dam,
- woody growth on the dam (roots can create channels for dam leakage and eventual failure),
- signs of piping (leakage),
- signs of seepage or wet spots on the downstream face of a dam (may require toe drains or chimney drains to solve problems),
- riprap failures,
- deterioration of principal and emergency spillways,
- various stage/outlet controls,
- effectiveness of debris racks,
- outlet channel conditions,

- safety features (access controls to hazardous areas),
- mechanical and electrical equipment (pumps, generators, automatic controls, etc.),
- access for maintenance equipment,
- availability of manufacturer's mechanical and electrical information manuals, and
- availability of design information such as rating curves and tables for spillway flow, bypass flow, total flow, and storage and pumpout calculations.

In addition, the following actions should be taken on each dam as required to ensure that the dam will not present operation or safety problems.

- Replace soil removed by rodent burrows.
- Inspect drainage systems and relief wells annually for proper functioning and clean out or replace as necessary.
- Maintain riprap or other wave-protective measures and replace as needed.
- Remove and/or stabilize slide material as soon as practical. It may be necessary to construct a berm or flatten the slope.
- Replace eroded material and establish vegetation in eroded areas in emergency spillways, swales and other areas.
- Repair any unusual seepages, boils or settlements in fill areas, or sinkholes in pool areas.

Observations should be made of any changes in topography, downstream drainage systems or upstream land uses that may have a bearing on the operational effectiveness and safety of the detention facility.

Storage Volume Control

One of the most important variables in the design of a detention facility is the volume available for storage of runoff. If a detention facility is allowed to accumulate sediment and debris, decreasing the storage volume, the ability of the facility to function as designed will be reduced. To facilitate the inspection of the facilities for volume control, it is recommended that a sediment accumulation gage marker be installed in the detention facility to indicate the maximum level for silt buildup before the facility must be dredged or cleaned. This marker could be a small pipe with a stripe or suitable indicator at the cleanout level. A suitable indicator could also be placed on the inlet or outlet device or in some location which can be easily identified during the inspection process.

Often detention facilities are used as temporary sediment basins. To control the maintenance of these facilities some criteria must be established to determine when these facilities should be cleaned and how much of the available storage can be used for sediment storage. Following is an example of possible criteria and procedures that can be used for sediment basin maintenance.

Figure 18-1 can be used to estimate sediment trap efficiency for sediment basins with different volumes. Following is the procedure for using this figure.

1. Establish sediment generation criteria (e.g., 0.5 acre-inches per disturbed area draining to the sediment basin).
2. Estimate total volume available for sediment storage from the geometric shape of the basin (e.g., .446 acre-feet).
3. Calculate minimum silt storage needed given the silt generation criteria (e.g., 0.5 acre-inches per disturbed area x 10 acres of disturbed area x 12 inches/foot = 0.417 acre-feet).
4. Trap efficiency can be estimated from Figure 20-1 as follows
   - 0.446 available storage x 12 in./ft = 5.35 acre-in.
   - 5.35 acre-in./10 acres = 0.535 in. - storage/drainage area
   - From Figure 18-1 at 0.535 in. - Trap Efficiency = 72%
5. If you were required to clean out the facility when the efficiency reached 50%, the cleanout elevation could be determined as follows.

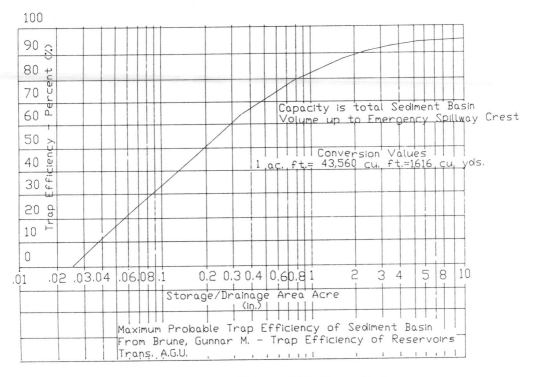

Figure 18-1 Efficiency Of Sediment Basins

Source: AASHTO, 1992

- From Figure 18-1 at 50% trap efficiency, the storage/drainage area acre = 0.19 in.
- Storage = 0.19 in. x 10 acres x 12 in./ft = 0.158 acre-ft.
- Given the basin geometry, try different depths until you have 0.158 acre-ft of storage still available for sediment storage. This is then the depth where the basin should be cleaned to ensure that the trap efficiency does not go below 50 percent.

This gives a procedure that can be used to estimate trap efficiency and also establish cleanout levels for a given efficiency. Different efficiencies might be established according to the damage potential downstream. Another procedure is found in Schueler, 1987.

18.4 Storm Drains

Storm drains move the water collected from catch basins, drop inlets, and sometimes detention facilities to the natural water courses. The maintenance involved in storm drain systems is the removal of any sand, silt, or debris and the maintenance of a tight seal at each pipe joint. There are occasions where abrasive material is present in the water (or some chemical that has a deleterious effect on the pipe) that causes the pipe material to be worn away. This necessitates relining the pipe to preserve its integrity. It is recommended that the entire storm drain system be inspected every 10 years and the catch basins twice yearly. In a North Carolina survey of 86 cities, 50 percent of the cities indicated that pipe failure accounted for a significant portion of system failures and complaints.

When storm drains do not adequately accommodate the storm water, they may be of inadequate size; but just as probably, they are partially clogged. Water flushing and heavy duty vacuum equipment can remove some clogs. Storm drains can be cleaned by inserting a rodding machine (heavy duty sewer snake) in one manhole and running it through to the next manhole. A line is attached and the tool is pulled back through to the first manhole.

A cable machine can also be used by placing it at each manhole. The cable from the drum of the first machine is attached to a line, and then it is pulled through to the opposite end. A torpedo shaped cleaning bucket is hung from the cable on the second machine. The cable is then attached to the bottom of the cleaning bucket making it the connecting link between the cables of the two machines. A sheave is suspended from each machine and braced directly over the center of the pipe to facilitate changing the direction of the cable. As the first machine pulls the bucket through the storm drain, the jaws of the bucket automatically open as it meets resistance. The first machine continues pulling the bucket through the material until it is full. When the direction of the cable pull is reversed to retrieve the bucket, the jaws close to retain the material and form the bottom of the bucket. A laborer in the manhole empties the bucket when it reaches the manhole. The traditional cleaning machines can be used with various bucket sizes depending upon the diameter of the pipe and the amount of material that must be removed.

18.5 Culverts

Culverts are openings under a roadway or embankment which permit the natural flow of water from one side of the roadway to the other. They may be constructed of corrugated metal, reinforced concrete or other materials. Culverts must be kept free of obstructions. Sand or sediment deposits should be removed as soon as possible. During storms, critical areas should be patrolled and the inlets kept free of debris. Inlet and outlet channels should be kept in alignment and vegetation should be controlled in order to prevent any significant restriction of flow. In some areas the culverts are normally self cleansing when large storms occur. Inspection in this case identifies permanence of the clogging material indicated by vegetation or a compacted material.

Culverts may become clogged if the flow-line grade prevents self-cleaning. A permanent correction is to relay the pipe on a steeper grade, but this is not always possible and is often very expensive. The alternative is to clean the pipe frequently.

Reinforced concrete box culverts require little maintenance, but they should be inspected annually for cracks, bottom erosion and undermining. Undermining is the result of high outlet velocities or poor control of seepage. Correction of tailwater undermining usually requires adding an energy dissipator.

Small culverts may be cleaned by flushing away debris with water pressure. A water truck or other vehicle equipped with a pump and hose attachment is used to direct the stream from the hose into the outlet end of the culvert. Thus, the water dislodges and washes away debris and sand.

Culverts over 3 feet in diameter must usually be cleaned by hand. A small sled or wagon is useful for transporting material from inside the barrel to the culvert ends. A small rubber-tired tractor, equipped with a push blade, may be used to remove sand and silt deposits from the larger reinforced concrete box culverts.

If the invert of a metal or concrete culvert becomes worn or eroded, it can be repaired by relining with concrete grout, gunite, or asphalt cement. If the hydraulic capacity of the culvert is not critical, it can be lined or a smaller pipe can be placed inside, and the space between the pipes filled with pressure pumped portland cement grout, or gunite.

High velocity flows, containing large quantities of stone and rock, scour the culvert bottom. Scour may be reduced by securing steel plates longitudinally along the bottom. Scour around footings, cutoff walls, and headwalls is repaired by replacing the eroded material in kind or by filling the void with riprap or sacked concrete. In an emergency a bituminous mix may be used.

When concrete pipe culverts settle, joints pull apart. Joints are repaired by tamping or rodding grout into the cracks. Grout consists of 1 part portland cement, 2 parts sand, and 1/5 part hydrated lime, with sufficient water to produce a plastic mix. Sand must be well graded and of such size that all will pass a No. 8 sieve.

In order to prevent erosion, energy dissipators are sometimes placed at outlets of culverts and drains. It is important that these be inspected periodically, particularly after major flows, to ensure that they are in place and functional.

18.6 Bridges

Bridges must be kept free of obstructions. Debris and vegetative growth under a bridge may contribute to scour, create a potential fire hazard and reduce freeboard for ice and debris during high-water flows, resulting in a serious threat to the bridge. A reduced effective flow area under the bridge may also result in excessive bridge backwater damage, more frequent roadway overtopping and a hazard to the traveling public.

Maintaining a channel profile record and revising it as significant changes occur provides an invaluable record of the tendency toward scour, channel shifting, and degradation or aggradation. A study of these characteristics can help predict when protection of pier and abutment footings may be required. Being able to anticipate problems and taking adequate protective steps will avoid or minimize the possibility of future serious difficulties.

Maintenance inspection must be commensurate with the risk involved. Where probing and or diving are necessary, the inspection should be scheduled at the season of lowest water elevation. High-water, high-ice, and debris marks with the date of occurrence should be recorded for future reference.

Following are some of the maintenance problems that can be encountered related to bridges.
- Clogging of bridge deck drains and scuppers, which may create a hazard to traffic and contribute to deck deterioration.
- Discharges of bridge deck drains that are detrimental to other members of the bridge, and those spilling onto a traveled way below. In addition, discharges that may cause fill and bank erosion.
- Clogging of air vents in the superstructure or deck of bridges subject to overtopping which may increase buoyancy forces and the possibility of bridge washouts.
- Accumulation of debris in the open space between the handrails of bridges subject to overtopping which may induce additional lateral forces on the bridges and increase the risk of washouts.
- Channel aggradation or degradation.
- Scour at piers and abutments caused by accumulation of debris and or excessive velocities.
- Damage to bridge approach embankment caused by channel encroachment.
- Loss of riprap due to erosion, scour, and wave action.
- Damage to bridge elements due to debris, ice jams, and excessive velocities.
- Missing navigational signs and lights over navigable channels.

Maintenance measures that should be undertaken to ensure bridge integrity and design standards include the following.
- Repair of damaged bridge elements.

- A schedule for removal of debris after major floods.
- Removal of sand and gravel bars in the channel that may direct stream flow in such a manner as to cause harmful scour at piers and abutments.
- Cleaning bridge deck drains and keeping their outlets away from traffic underneath. Also providing riprap or other means of protection at outlets to avoid fill and bank erosion.
- Removal of debris caught between bridge handrails and opening vent holes designed to reduce buoyancy.
- Making a channel change when necessary to redirect the flow away from bridge approaches and in line with the bridge skew.
- Dredging of channels that are subjected to a high degree of aggradation in order to maintain waterway adequacy.
- Constructing cutoff walls to reduce or stop progressive channel degradation.
- Replacing lost dirt in scour holes and constructing riprap mats or other means of protection for undermined piers and abutments.
- Replacing missing riprap on embankment slopes, channel banks, spur dikes, etc.
- Replacing missing or damaged navigational signs and lights.
- Constructing additional openings to accommodate increased urbanization in the drainage area upstream from the bridge.
- Modifying or increasing existing protective measures when needed.

18.7 Ditches

Ditches convey water away from roadways and other areas to locations where the water can flow without causing erosion or ponding. Ditches may be unlined or lined with portland cement concrete, gunite, masonry, riprap, or bituminous concrete. Ditches must be kept free of silt, debris, large amounts of vegetation, or any other material that restricts the flow of water.

The flow lines of unlined roadside ditches can be maintained by motorized equipment supplemented with hand work. A pass is made with a motor grader having the blade positioned about 120 degrees to the direction of travel and with the blade set approximately to the slope between the outside edge of the shoulder and the ditch flow line. This removes unwanted material from the ditch and deposits it in a windrow near the edge of the shoulder. Then this material is loaded into a dump truck with a rubber-tired, front-end loader or by hand shovels, and it is hauled to a disposal site. Hand work will be required to remove unwanted material at locations inaccessible to the motor grader, e.g., near pipe culverts.

Large roadside ditches are sometimes located at an elevation well below the roadway and not accessible to a motor grader. These may be reached with a truck mounted hydraulic excavator operated from the shoulder. In this situation unwanted material is placed directly into a dump truck and hauled away. The equipment operator should exercise care to prevent undercutting the flow line grade. Such undercutting would result in undesirable ponding.

Interceptor ditches on slopes, and along excavation or embankment benches, and outlet ditches from culverts may require hand cleaning by using shovels and wheelbarrows.

Joint separation is a common problem associated with concrete lined ditches. Once water gets under the concrete or asphalt, the underlying soil is removed and the deterioration may be rapid, so frequent inspection is vital and fast repair a necessity if the investment is to be protected. Joints are sealed with hot rubber asphalt. Heating kettles, for hot rubber asphalt compound, should be the double-jacketed, oil bath type to avoid damage to the compound by overheating. Before any compound is used, joints and cracks should be cleaned. Enough sealing material should be placed to fill the crack. When filling deep cracks, the cooled sealer may shrink and additional sealer must be added to fill the joint so the sealer is flush with the surface.

Ditch erosion is the loss of soil caused by rapid flow of water. It is controlled by paving the ditch with bituminous asphalt aggregate mix, placement of masonry, grouting rock, establishing erosion resistant vegetation, or by constructing wash checks. Ditches lined with bituminous material oxidize or weather rapidly and should be sprayed with asphalt emulsion.

The growth of vegetation is often encouraged. The vegetation may be maintained by adjoining property owners in residential areas but in other areas it often must be maintained by the municipality. One of the major problems when vegetation is used to control erosion in ditches is the control of weeds. Weeds become a major problem in turf when the grass loses its vigor and density and cannot compete with them. Clover and knotweed may take possession in areas where nitrogen levels are low. Crabgrass is a serious pest in many areas where high summer temperatures check the growth of grass. Weed encroachment is often the result and not the primary cause of poor turf. Weed eradication often will not result in permanent improvement unless conditions which weakened the turf are corrected.

The best weed control chemicals available are often nitrogen, phosphorus and potassium (i.e. fertilizers) applied in the correct amounts, at the proper times, and in the correct ratio. A healthy turf will compete with and drive out most weeds. Turf specialists agree that the best weed control is a dense vigorous growth of grass. Herbicide chemicals, however, do have a place in any turf maintenance program. Weeds in the right-of-way are unsightly, and most of them can be eliminated by a good program of spraying and mowing. A good program of spraying the entire right-of-way for three consecutive years will eliminate most of the weeds, except possibly for small areas of weeds that may require spot spraying. A good mowing program goes hand-in-hand with a good spraying program in elimination of weeds before they go to seed.

Where weeds have been destroyed and short grasses cover the unsurfaced areas of the roadway, the mowing expense can be reduced and the municipality will still have a neat, well kept right-of-way. Certain steep slopes are not to be mowed, so these must be sprayed to control weed growth. Weed spraying should not be done on new seeding, except to kill noxious weeds and then only by spot spraying. The spraying of new seeding will kill out the desirable legumes and young grass. New seeding should be at least three years old before an overall spray is given, so if the area is weedy, the area should be mowed instead of sprayed.

There are many different formulations of herbicides on the market under different trade names. Many of these are special purpose herbicides which may or may not have application in ditch maintenance. Also, concern for stream protection is producing a rapidly changing picture of the use of pesticides and fertilizers. The municipality should keep up with new products, equipment and the findings of studies, and should at various times recommend chemicals and equipment for test or general usage.

18.8 Storm Water Inlets

Storm water inlet structures are designed to intercept water in gutters and drainage courses. They also act as settling basins to collect heavy solids, and they prevent debris from entering culvert systems. Mobile heavy duty industrial vacuum equipment is used to clean sediments from catch basins.

Grates on catch basins are used to prevent large objects and debris from entering the system. Frequent inspections during run-off periods are required because debris such as pieces of cardboard, newspapers, or flat metal can prevent water from entering the catch basin. Many times the time of rainfall is too short to allow for self-cleaning of the system and debris tends to buildup over time. Grates are usually designed to be placed with bars parallel to the flow. However, they can be turned perpendicular to the curb and sized so that they do not allow bicycle tires to drop in the opening, although this will affect the efficiency of the grate.

Large catch basins constructed without a grate may collect large quantities of rock. This rock may be removed by lowering a clam or backhoe bucket into the catch basin. Hand work will be required to load the bucket. The loaded bucket is lifted from the catch basin and the rock is dumped into a truck and hauled to a disposal site. Muck may be removed by an orange peel bucket.

18.9 Slope Drains

Slope drains are paved, or metal troughs or pipes used to carry water from a collector drain, gutter, or ditch, into a roadside channel or natural water course. They should have firm contact with the supporting surface. If connected to pavement or dike, a tight seal shall be maintained. Cracks should be sealed with hot rubber asphalt compound. The outlet end of slope drains should be inspected regularly for erosion. Eroded areas should be repaired with broken concrete, sacked concrete, riprap, etc. Sometimes the repair may include extending the drain.

Metal pipes used for slope drains on high embankments or benched excavations should be rigidly attached to the surface with pipe anchors. Anchors are designed to prevent the pipe from separating at joints, but in spite of this, separation occasionally will occur. Repairs consist of removing, re-installing and anchoring all pipe below the separation.

Once water gets under a paved drain, it quickly erodes away the bedding material and dislodges the gutter. This erosion can be reduced or eliminated by installing a cutoff wall at the beginning of the gutter and at intervals throughout the length. Those cutoffs should be 12 to 18 inches deep and extend the full width at the gutter, and should be made of portland cement or asphalt concrete.

18.10 Wash Checks And Energy Dissipators

Wash checks are used to collect water, redirect its flow, provide sediment basins for siltation control, and control rate of flow. They are constructed of reinforced concrete or grouted stone. Wash checks are inspected for undermining after each period of heavy water runoff. Undermining should be repaired by filling the void with sacked concrete.

Energy dissipators are used at the outlets to culverts and pipes flowing with inlet control and therefore supercritical velocities. All energy dissipators lessen outlet velocities so that downstream channels can more nearly sustain their integrity. The energy dissipator may be a level spreader, discharge apron, or hydraulic jump. They should discharge into an area stabilized by vegetation as a minimum. They should be inspected for undermining after periods of heavy runoff, and any undermining should be repaired.

Undermining is the major cause of failure of this type of bank protection. Undermining leaves a void under the grouted riprap which ruptures and may cause the riprap to collapse into the void. Repair consists of filling the voids and collapsed surfaces with rock and grouting.

18.11 Bank Protection

Bank protection is required when stream flow or wave action endangers highway embankments or structures. Erosion may be controlled by rock slope protection, grouted riprap, sacked concrete riprap, concrete slope paving, gunite slope paving, retards, permeable jetties, retaining walls, and cribs. The protection should make optimum use of local materials. Velocity of flow and direction of currents are critical factors in selecting materials. With any bank protection, care should be exercised to prevent the protection from being undercut by erosion. The leading edge of the protection should be protected against erosion.

Undermining causes failure of many bank protection materials. Undermining leaves a void under the bank protection which may lead to the failure of the protection to stabilize the banks. Repair consists of filling the voids and repairing the bank protection. Cracks and other damage to the bank protection should be repaired to prevent seepage and further damage.

18.12 Underdrains

Underdrains consist of a trench lined with filter cloth and backfilled with filter material. Perforated galvanized corrugated metal pipe, semicircular pipe, and tile pipe may be placed near the bottom of the trench. Underdrains are used to remove groundwater and to lower the water table. Local conditions determine whether the installation should be along the shoulder line, or the toe of slope, or herringbone system under the traveled way.

Underdrains should be checked in the early part of the wet season to ensure that they have not become clogged with sand or roots and that outlets are free to drain. Presence of silt or dirty water coming out of an underdrain indicates a possible break in the pipe. This should be reported to the municipality at once so that materials investigation and remedial measures can be initiated.

Suction pumps, high pressure pumps, or powered rotary sewer cleaners are required to clean out long sections of clogged pipe. When the pipe becomes clogged, the filter material probably has become silted and its effectiveness has been reduced to a level that make it necessary to consider replacement.

Filter cloth is available which reduces silting when used in underdrain construction. Filter cloth made of polyproplene, polyester, or nylon is highly permeable, lightweight, strong, and flexible. It can be used with or without pipe. The cloth is filled with crushed stone or open graded gravel. There are many types of filter cloths; choose a type suitable for the purpose. Some types cannot be used in sun light because they will disintegrate.

18.13 Pavement Edge Drains

Water enters the pavement structure in several ways: cracks in the pavement, infiltration through the shoulders, melting of ice layers, water through the pavement base, and a high water table. This free water under the pavement is of particular concern because it can decrease the strength of the pavement by reducing the cohesion, reducing the friction between particles, and by increasing more pressure. As a result, the bearing capacity is reduced, and the pavement then must support loads without the proper support from below resulting in cracking and deflection.

New pavements are designed with subdrainage systems for infiltrated water. The infiltration of free water into the pavement structure, its effect on material strength, and its normal removal by vertical flow or a lateral subsurface drainage system should be an integral part of the pavement structural design process.

Where infiltration of water has become a problem on an existing pavement some action should be taken to prevent this infiltration or remove it. The prevention includes sealing of cracks in the pavement, positive slope on shoulders, and deeper ditches. Where this prevention does not result in the elimination of the subsurface water, longitudinal drains installed as edge drains may be successful. A 9-12 inch slotted plastic tubing is laid in the trench and the trench is then backfilled with permeable material. The tubing must have a positive outlet, and rodent screens should be provided. The edge support for the pavement must not be damaged when the drain is installed, and the material adjacent to the drain which needs to be drained must be sufficiently permeable to allow the free water to reach the longitudinal drain.

References

American Association of State Highway and Transportation Officials, Manual For Maintenance Inspection Of Bridges, 1983.

American Association of State Highway and Transportation Officials, Model Drainage Manual, 1991.

American Public Works Association, Urban Stormwater Management, Special Report No. 49, American Public Works Association, 1313 East Sixtieth Street, Chicago, Illinois, 1981.

American Public Works Association, "Managing Public Equipment", Special Report 55, 1989.

American Public Works Association, Contracting Maintenance Services, Special Report 58, American Public Works Association, 1313 East Sixtieth Street, Chicago, Illinois, 1990.

Charlotte, "Storm Water Maintenance Contracts", personnel communication memo., 1993.

Denver Urban Drainage and Flood Control District (UDFCD), "Contracting for Restoration Maintenance", Flood Hazard News, 1983.

Galli, J., "Analysis of BMP Performance and Longevity", Metro Wash. Council of Govts., 1992.

Karaa, F. A., "Infrastructure Maintenance Management System Development", ASCE J. of Prof. Issues in Engineering, Vol. 115, No. 4, Oct., 1989.

Lapay, W. S., "Probabilistic Basis for Managing Maintenance", Proc. ASCE Symp. on Water Resources Infrastructure, Ft. Worth, TX., April, 1990.

Louisville MSD, "Storm Water Drainage Master Plan", 1988.

Lund, J. R., "Cost-effective Maintenance and Replacement Scheduling", Proc. ASCE Symp. on Water Resources Infrastructure, Ft. Worth, TX, April, 1990.

Martin, J. L. "Selling Elected Officials on Infrastructure Needs", ASCE J. of Prof. Issues in Engineering, Vol. 110, No. 2, April, 1984.

Maryland, Sediment and Storm Water Division, "Maintenance of Stormwater Management Structures, a Departmental Summary", July, 1986.

Mays, L. W., "Models for Optimal Maintenance of Hydraulic Structures", Proc. ASCE Symp. on Water Resources Infrastructure, Ft. Worth, TX., April, 1990.

National Assoc. of County Engineers, "Maintenance Management", Action Guide Series, 1992.

Network Magazine, Hansen Software, Sacramento California, Fall 1989.

Quimpo, R. G., and U. M. Shamsi, "Reliability-based Distribution System Maintenance", ASCE J. of Irr. & Drainage, Vol. 117, No. 3, 1991.

Reese, A.J. and Suggs, J.B., Discussion of "Levels of Service Applied to Urban Streams", ASCE J. of Water Resources Planning and Management, Sept-Oct 1992, Vol 118, No.5 pp. 582-584.

Roenigk, D. J., R. G. Paterson, M. A. Heraty, E. J. Kaiser and R. J. Burby, "Evaluation of Urban Storm Water Maintenance in North Carolina", Rpt. No. UNC-WRRI-92-267, Dept. of City & Regional Planning, UNC, Chapel Hill, June, 1992.

Schueler, T.R., "Controlling Urban Runoff", Metro. Wash. Council of Governments, July, 1987.

U.S. Army Corps of Engineers, "Inspection of Local Flood Protection Projects", ER 1130-2-339, Oct. 1973.

U.S. Army Corps of Engineers, "Construction Inspector's Guide", 4 Volumes, 1960.

U.S. Department of Transportation, Federal Highway Administration, Culvert Inspection Manual, FHWA-IP-86, July 1986.

U.S. Department of Transportation, Federal Highway Administration, Integration of Maintenance Needs into Preconstruction Procedures, FHWA-TS-78-216, 1978.

Wonderlich, W. O., "Probabilistic Concepts in Hydroproject Maintenance", Proc. ASCE Spec. Conf. on Hydraulics, Nashville, TN, 1991.

Wonderlich, W. O., "Probabilistic Analysis of Maintenance Operations", Proc. ASCE Symp. on Water Resources Infrastructure, Ft. Worth, TX., April, 1990.

Index